113

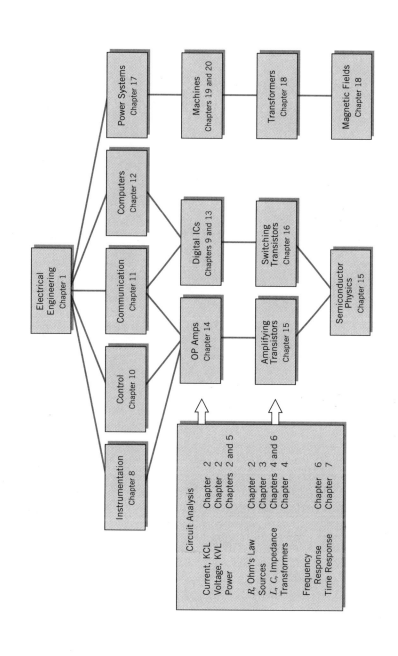

ELECTRICAL ENGINEERING

SECOND EDITION

ELECTRICAL ENGINEERING
For All Engineers

WILLIAM H. ROADSTRUM
Worcester Polytechnic Institute

DAN H. WOLAVER
Microwave Logic, Inc.

JOHN WILEY & SONS, INC.

New York • Chichester • Brisbane • Toronto • Singapore

Acquisitions Editor   Steven Elliot
Marketing Manager   Susan Elbe
Senior Production Editor   Savoula Amanatidis
Text Designer   Kevin Murphy/David Levy
Cover Designer   David Levy
Illustration Coordinator   Edward Starr
Manufacturing Manager   Andrea Price

Coventry University

Cover Art: Mondrian, Piet. *Broadway Boogie Woogie,* 1942/43. Oil on Canvas, 50″ × 50″, The Museum of Modern Art, NY. Given Anonymously. Photograph © 1994, The Museum of Modern Art.

This book was typeset in Baskerville by CRWaldman Graphic Communications and printed and bound by R.R. Donnelley in Crawfordsville. The cover was printed by Lehigh Press, Inc.

Recognizing the importance of preserving what has been written, it is a policy of John Wiley & Sons, Inc. to have books of enduring value published in the United States printed on acid-free paper, and we exert our best efforts to that end.

Library of Congress Cataloging-in-Publication Data
Roadstrum, William H. (William Henry), 1915–
    Electrical engineering : for all engineers / William H. Roadstrum,
Dan H. Wolaver.—2nd ed.
        p.   cm.
    Includes index.

    1. Electric engineering.   I. Wolaver, Dan H.   II. Title.
TK146.R48   1994
621.3—dc20   93-23818
                                                  CIP

Printed in the United States of America

10 9 8 7 6 5 4 3 2 1

Preface

This text is a broad and immediately useful introduction to electrical engineering for students and practicing engineers in other disciplines. Their special needs are met by the following features of the text:

1. Thorough quantitative introductions to the subareas of electrical engineering: instrumentation, control, communications, computers, power (see diagram inside front cover, left to right).

2. A structure that in general allows topics to be taken up in any order. The particularly suggested order is from broad application to increasing detail—for example, from control system, to compensation filter, to operational amplifier, to transistor (see diagram inside front cover, top to bottom).

3. The presentation of a new concept, first by a specific example (such as a motor speed regulator), and then proceeding to the generalization (such as control systems).

4. Frequent use of analogies to build on the reader's rich background in other areas. The reader's understanding of levers is used to help give him or her a feel for the operation of op amp circuits.

5. Three levels of practice problems—drill, application, and extension. This separates the tasks of learning the basic concepts, coordinating them, and expanding on them.

6. A book short enough and so written as to reduce its formidableness for nonelectricals.

The authors believe this text is unique in bringing these features together. The intent has been to design a text to meet the problems facing engineers today.

CHANGES IN SECOND EDITION

The second edition retains the features described above while expanding and re-arranging the material to meet current needs.

The chapter on ac power has been moved forward to follow Chapter 4 on ac circuit analysis. This may better accommodate those users who wish to go next to Part III on power and machinery after developing the basic language and analysis of circuits in Chapters 1–5. In response to requests from users in introductory courses with major EE students, we have incorporated at the end of Chapter 7 elementary material on the Laplace transform, with some additional work in Appendix D.

Each of the first five chapters in Part II on electronics now includes a case study. This draws together the tools learned in the chapter and shows a realistic, practical application of the material. Chapter 13 on digital integrated circuits now covers programmable logic devices (PLDs), and Chapter 16 on switching transistors has a new section of MOSFET's and CMOS logic. Chapter 10 on control has been revised extensively.

All chapters now have study questions following the problems at the end. New problems have been added, especially easy drill problems. In all, we have tried to respond to users' requests and to keep the book current.

THE CURRENT PROBLEMS OF ENGINEERING

Electrical engineering today serves as a home for many amazingly diverse fields—satellite communication, helicopter control, speech recognition, digital stereo recordings, megawatt power generators. By contrast, electrical engineering in its infancy dealt only with power applications—motors, lighting, heating, and power transmission. At that time the whole field could be studied in both breadth and depth in a reasonable length of time. Over the years it has become less and less feasible for an electrical engineer, let alone a nonelectrical, to study the whole of electrical engineering in both breadth and depth. How then is study time best spent?

A common (but less than satisfactory) answer has been for nonelectricals to concentrate primarily on circuit theory and devices. The assumption was that these tools would allow the students to grasp any of the many applications of electrical engineering that would be encountered later. In fact, such students are in a poor position to appreciate the higher-level concepts (signal conversion and processing; information coding, multiplexing, transmission, processing, and storage; system response, stability, and compensation; power conversion). These concepts, which are of increasing importance to all fields of engineering, require at least as much attention as basic circuit theory and devices.

Fortunately, the advance of technology has made it possible to relegate many of the details of circuit design to the designers of integrated circuits, leaving more time to study the systems concepts. Most engineers never get involved at a device level. Nonelectricals especially are interested in electrical engineering on a systems level, where it interfaces with their own discipline through instrumentation, control, communication, computation, and power systems.

Another challenge facing all engineers is that the already huge bulk of electrical engineering is continuing to grow at an ever-increasing pace. If an engineer is to keep up with the field after his or her formal education, he or she must be in a position to absorb new concepts as they develop—artificial intelligence and integrated optics, for example. Lifelong learning is a necessary part of an engineer's career.

PURPOSE OF THE TEXT

This text is designed to help readers (whether in a class or in industry) to meet these challenges, that is, to think and learn on their own, to maintain a broad perspective, and to understand the elements of technology in the context of their application. This is done in large part through the unique structure of the book.

After carefully establishing the language—the fundamentals of circuits—the book deals immediately with electronic systems. This emphasizes the higher-level concepts so important to engineers today, and it provides a context for the component-level concepts that follow. Instructors have the great advantage of knowing where they are going, being able by virtue of prior knowledge to relate to the subject as a whole each detail they present. The book attempts, by top-down organization and style, to extend the same advantage to the reader. Almost never is a topic built up from details to final structure, but rather the reverse. This organization is just as useful to an industrial reader practicing his or her lifelong learning.

A topic is often introduced by a simple example to establish nomenclature and provide a concrete focus. This is followed by a more rigorous and comprehensive generalization. The authors believe that this is the way engineers properly think in practice: from the specific need to a generalization of the problem, only then to solution alternatives and details. It would appear good pedagogy to have the student adopt professional thought practices as soon as possible.

End-of-chapter problems are graded into ''Drill,'' ''Application,'' and ''Extension.'' The easy ''Drill'' problems (with answers provided) are intended to get the reader off to a fast, running start. The ''Application'' problems are a mix of typical problems that might be encountered in engineering practice. ''Extension'' problems include some more challenging examples, but are designed mostly to aid an instructor in delving into other areas of the chapter topic not covered in detail in the text. All three categories include problems that take advantage of programmable calculators or computers to gain a wider outlook on various phenomena.

USING THE BOOK

The book is divided into three main parts, I—Circuit Analysis (six chapters), II—Electronics (nine chapters), III—Machines and Power (five chapters). A *Solutions Manual* (which also describes use of chapters) is available to instructors from the publisher. Enough material is provided to form the base of two 14-week terms. However, courses can be configured easily for a single term or into other formats. One-or two-term courses can cover all three major topics, or just circuits and electronics, or circuits and power. In using this material the authors have found that many students majoring in other engineering fields can go directly from this text into a second-level electronics course or machine courses for EE majors.

The top-down structure of the text makes it particularly easy for instructors to add or omit, without confusion, chapters or topics in accordance with the students' and instructor's interests. Indeed, after the basics of circuit theory are covered, the chapters can be taken up in almost any order desired. For example, Chapter 7 (Transients) can be omitted with little effect on later chapters

If the instructor desires, it is entirely possible to present electronics with the more conventional "bottom-up" approach. Chapters 15 and 16 (on transistors) could be taken up first, followed by Chapters 14 (Operational Amplifiers), 8 (Analog Signals and Instrumentation), 10 (Controls), and 11 (Communication). After this "analog" package, digital electronics could be covered similarly in the sequence of chapters: 9 (Digital Signals and Logic), 13 (Digital Integrated Circuits), and 12 (Microcomputers). Chapter 13 can be omitted easily if the sequence needs to be shortened.

Because Chapter 17 (Plant Power Systems) stands alone, Chapters 18 through 20 can be omitted where less emphasis on power is desired. Note also that Chapter 19 (Motors) precedes 20 (Rotating Machinery Basics) in accordance with the idea of showing the context before the details. This need not, however, prevent an instructor who desires the more conventional order from taking up Chapter 20 first.

It is recommended strongly that Chapter 21 (Electrical Safety) be included in some way in every course.

Minor electrical engineering courses are more meaningful when it is possible to provide some type of laboratory experience. Equipment requirements can be modest.

ACKNOWLEDGMENTS

The authors acknowledge with gratitude help from many sources.

From WPI: Professor Archie McCurdy for contributing much over the years to courses on which many of these ideas are based; the late Professor Harit Majmudar for insightful reviewing of the power sections of the manuscript and important ideas for improvement; Professor David Cyganski for valuable review of the chapter on communication; Professor John Mayer for help with power and energy data; Professor Donald Howe for information on cables and radio transmission; and Professor Alex Emanual for information on measurement transducers.

Contents

List of Abbreviations

ABBREVIATIONS

ac	alternating current
ADC	analog-to-digital converter
AM	amplitude modulation
ALU	arithmetic and logic unit
ANSI	American National Standards Institute
B&S	Brown & Sharpe wire gage system
BCD	binary-coded decimal
CMOS	complementary metal oxide semiconductor
dc	direct current
DAC	digital-to-analog converter
DIP	dual-in-line package
EE	electrical engineering
EEI	Edison Electric Institute
ECL	emitter-coupled logic
EMI	electromagnetic interference
emf	electromotive force
FM	frequency modulation
FET	field-effect transistor
FCC	Federal Communications Commission

FSK	frequency shift keying
GBP	gain-bandwidth product
IC	integrated circuit
IEEE	Institute of Electrical and Electronic Engineers
I/O	input-output
KCL	Kirchhoff's current law
KVL	Kirchhoff's voltage law
LED	light-emitting diode
LSB	least significant bit
LSI	large scale integration
MMF	magnetomotive force
MSB	most significant bit
MSI	medium scale integration
NEC	National Electric code
NEMA	National Electrical Manufacturers Assn.
NFPA	National Fire Protection Association
NRZ	nonreturn to zero (PCM)
OSHA	Occupational Safety & Health Administr.
pc	personal computer
PM	phase modulation
PAM	pulse amplitude modulation
PCM	pulse-code modulation
PPM	pulse-position modulation
PSK	phase-shift keying
PWM	pulse-width modulation
QED	which was to be proved
RAM	random access memory
ROM	read-only memory
rms	root mean square
SAE	Society of Automotive Engineers
SCR	silicon controlled rectifier
SNR	signal-to-noise ratio
SSB	single-sideband
SSI	small scale integration
THD	total harmonic distortion
TTL	transistor-transistor logic
UL	Underwriters' Laboratory
VCO	voltage-controlled oscillator
vhv	very high voltage
VLSI	very large scale integration

CHAPTER 1

Introduction

Electrical engineering is the application of electrical and electronic technology to the daily needs of people. It touches almost every aspect of their lives and occupations.

The many uses for electricity group naturally into two major areas: electrical *power engineering* and electrical *signal engineering*.

It is suggested that readers think carefully about this brief introduction with the following objectives:

1. Understand the main power and signal applications of electrical engineering.
2. Develop a general idea of how to take what they need from the book.
3. Make a preliminary selection of chapters that will interest them.

1.1 POWER APPLICATIONS

Power applications include heating, lighting, and motors of various kinds. Here the engineer is interested in generating and transporting the energy needed, and providing it in a convenient and controllable form.

In heating, sufficient heat must be provided at exactly the right place. Electrical heating is particularly applicable to precise tailoring of the heat to the material or area to be heated. Electrical heaters of various kinds are familiar to everyone—

electric stoves, microwave ovens, water heaters, heating tapes, radiant heaters, for example. Electrical heating in industry includes all these and may even extend to firing boilers in critical applications. Industrial uses include particularly induction and dielectric heating with electronic generators.

In the case of lighting the right amount and kind of light must again be applied to the correct area. For lighting there is hardly any practical source except electricity. A wide variety of lamps are in use today, from the small light-emitting diodes used extensively for computer displays to powerful and intense floodlamps for highway lighting. More kinds of lighting systems and devices are being developed continually.

In motor applications the engineer is interested in providing mechanical motion—force or torque exerted over some appropriate distance or angle or at an appropriate speed—again in the desired amount and in such a way that the whole process can be readily controlled. Electric motors and other power applications are ubiquitous. The average household has several dozen that run refrigerators, rotisseries, furnaces, fans, record players, recorders, door bells, door openers, can openers, laundry equipment, computers, and so on. Industrial uses include also relays, positioners, robots, mill drives, mixers, printing presses, and many others. The simpler aspects of these power applications will be found in the chapters of Part I. Part III goes further into them.

Power electrical engineering usually involves relatively large amounts of energy compared to signal engineering.

1.2   SIGNAL APPLICATIONS

The other broad classification of electrical applications—signal engineering—is concerned with transmitting, processing, and using information. Power is not a primary consideration here, and the power involved in this work is usually relatively small. But it must be large enough that the signal information is not lost. However, very large powers are used in some signal applications such as large radio transmitters and radars.

It is convenient to further subdivide signal or information applications into instrumentation, communication, computers, and control.

Instrumentation is concerned with measurement applications—sensing and amplifying parameters such as speed, temperature, and sound. There are many biomedical applications such as monitoring a heartbeat or measuring blood viscosity. Instrumentation has found increasing use in cars for monitoring engine temperature, manifold pressure, etc. Industrial processes also require the monitoring of parameters such as temperature and pressure.

Communications includes such applications as telephone, data transmission and handling, radio, and television. Modern telephone systems illustrate the basics of communication in a way known to almost everyone. They make extensive use of electronic devices of all kinds. There are several important concepts included in communication systems—the idea of information transmission, message, encoding, a carrier modulated with the message information, a medium through which the carrier moves; and at the receiving end demodulation, decoding, and information use. Every communication system has all or most of these elements, whether it is designed for communication between people or machines.

Computer systems are primarily processors of (often large) quantities of information, which they handle very rapidly. The large "main frames" are usually not dedicated in the sense of being limited to specific tasks but provide large computing capabilities which can be applied to various needs, often serving a group of many typewriter-and-television-like terminals through which users can have access to them. They perform such tasks as accounting operations (for example, payroll calculation and check writing), analytical services (as in engineering design), the provision of graphic design capabilities (again extensively used in engineering design), and general list maintenance and manipulation (as in commercial work and periodical circulation, and in the insurance industry).

Small computers, the "micros" and "PCs" (personal computers) are more usually dedicated to some specific task or to the work of some specific group or individual, have smaller memories, operate more slowly, and have a more limited set of instructions to which they will respond.

Computers are made up of "hardware," the physical devices and connections of the machine itself, and "software," the programs of sequences of instructions to which the hardware responds. The software is entered into and "stored" in computer "memory."

Rapid recent computer development has come about in both hardware and software areas. "ICs" (integrated circuits or "chips") provide extreme miniaturization, low cost, low power consumption, higher reliability, and hardware standardization. They are used repetitively in a way which simplifies both the circuits themselves and their conception and design. Software has been developed that applies computers to tasks of word processing, structural analysis, music synthesis, graphics production, electrical circuit simulation, and game playing, to mention only a few areas.

Control systems are electronics to provide precise and quick adjustment in tasks such as positioning a robot's arm to pick up a part from a storage bin, automatically applying brakes to a skidding automobile wheel, or running the air-conditioning system of a building. In most of these applications the concept of "feedback control" enters. Feedback means that the result of the controlling action is being observed to decide how much more change is needed. For instance, in controlling a robot's arm its position is continuously observed in order to decide in what direction and how far it must be farther moved.

The division of signal engineering into four areas—instrumentation, control, communication, and computers—often has fuzzy boundaries. Computers or computerlike functions enter into control systems and more and more into communications. Much control information must be communicated from one point to another. Computers now talk to each other and exchange information. For a number of years more information has been carried on our telephone systems as digital data than as voice transmissions. Further, power systems often utilize complex control subsystems. Thus all phases of information engineering and power engineering have at least some interest for any engineer.

Information systems can also be classified into *digital* and *analog.* As we will see in subsequent chapters, much information (music and the human voice are examples) can be represented by electrical voltage waves whose amplitude is proportional to (analogous to) the quantity being represented (air pressure, in the sound examples).

In a quite different way, such information can be represented by digital quantities—numbers. Using numbers instead of analog representation greatly reduces design and production quality problems. Such digital devices are well suited to computer systems and similar data processing systems.

Signal or information engineering is often taken as synonymous with the term *electronics*. This is because electronic devices are most commonly used in these applications. But strictly speaking we should reserve the term electronics for that technology dealing with electron devices—devices such as transistors and diodes and their development in integrated circuits. Electronics is increasingly being applied to electric power uses also. The term *electrical* includes *electronic* as a subset.

1.3   ELECTRICAL ENGINEERING AND ENGINEERS

From the above brief listing of applications it is plain that every engineer of whatever discipline will be faced with using and operating electrical equipment and systems in his or her own practice. Mechanical engineers will need motors to drive their machines. Chemical engineers apply heat and drive pumps. Civil engineers operate construction sites and apply electronic surveying devices. Further, all these activities need instrumentation and control equipment that is largely electrical. Electrical engineers will be particularly concerned with developing devices and advanced circuit and system techniques.

On first encounter, analysis and operation of electrical systems seems abstruse and difficult to most people. This is because electricity can't be seen or heard or felt except in a few unusual manifestations such as lightning or electric shock. (And we must admit that neither of these is likely to inspire a desire for closer acquaintance.) Contrast this with mechanical motion, for instance, which we learn intimately on our tricycles. Or consider how directly intuitive a structure becomes to the person who grows up in houses and buildings of various kinds and makes sand castles at the beach. The beginning chemical engineering student builds on lifelong experience in the family kitchen.

But this lack of intuitive feeling for electric circuits is misleading. Electrical phenomena are in fact less complicated and more predictable than the phenomena in other engineering fields. The reader can be confident of this fact. The battle in learning about electrical engineering is not with any intrinsic or esoteric difficulty but with lack of intuitive appreciation for the material being dealt with. Therefore we recommend that the reader make a deliberate effort in the early chapters of this book to drill an intuitive feeling for simple circuits into his own thinking. After that has been accomplished he or she will find the remaining material readily understandable. This point will be addressed further in the next chapter.

It is the goal of this text to assist its readers of whatever engineering discipline to intelligently

1. operate and maintain plant power systems, select and apply equipment such as motors and heating devices to routine applications;

2. operate and maintain measurement and control systems, select equipment for and configure routine systems both digital and analog;

3. work with

   • EE consultants,

   • specialized electrical managers, contractors, and technicians,

   • vendors of electrical equipment;

4. establish a practical and theoretical basis on which an engineer can build further electrical engineering ideas and information as his or her interests and work may require.

1.4   CIRCUITS AND CIRCUIT ANALYSIS

Electrical engineering, as is the case with other engineering disciplines, is concerned with several levels of thinking. These might be listed (with examples) as

• *systems* (plant power system; temperature control system for a certain chemical process)

• *apparatus* or *equipment* (motors, amplifiers, oscilloscopes)

• *devices* (transistors, inductors, integrated circuits, light-emitting diodes)

Generally speaking the devices are assembled into equipment and the equipment into systems.

Engineers, being primarily concerned with the application of these things to some immediate practical end, are first interested in the system. They ask, "For this application, what system would most effectively (from both technical and economic considerations) meet the need?" Going down the above list they next ask, "What equipment will be needed and available to configure such a system?" And finally they ask, "What devices are needed and available to construct such equipment?" Taking up the questions in this order is called a *top-down* approach; it starts with the broad picture and proceeds with increasing detail. The ordering of the topics in this book follows a top-down structure, patterning the usual thinking of engineers and providing context and motivation for the reader.

Of course the answers to these questions in this order are often tentative and will depend on the possibilities developed at lower levels. The nonelectrical engineer in particular, who is simply applying electrical means to accomplish something in a mechanical, chemical, or structure-building system, is not much interested in the lower levels for their own sake. But there is always the danger that in neglecting to consider the device problems involved, he or she will overlook some limitation on the application of the selected equipment and systems. For example, a mechanical engineer not aware of the temperature limitations of transistors and integrated circuits may decide to mount electronic equipment in a closed box without ventilation.

Happily there is a unifying, integrating discipline in electrical engineering practice—circuit analysis—which is completely common to equipment and system thinking, and to both power and information engineering. Devices are joined together with conductors (wires) into equipment. Equipment is wired together into systems. These resulting *physical circuits* (the aggregation of conductors and devices or equipment) are the focus of attention in 90% of the work of any engineer dealing with electrical matters.

Circuit analysis is actually performed on circuit models, which are simplifications of physical circuits, equipment, and devices. Engineers of every type are *modelers*. We find reality too complex to deal with exactly and are forced to make simplifying models of it. The mechanical engineer may assume perfect stiffness in structural members and think in terms of a free-body diagram. The civil engineer assumes uniform floor loading in a structure. The chemical engineer neglects turbulence in some system and models it as uniform flow. Similarly the electrical engineer may assume distributed systems are lumped, wires are lossless, and resistors are linear. Carefully chosen, these models serve amazingly well in almost every situation.

Dealing with electrical engineering situations means dealing with circuits. Thus the reader's first task will be to learn basic circuit analysis, the "language" of EE. Chapter 2 introduces this useful and not difficult skill.

The driving force in electrical circuits is some kind of voltage generator. These generators or sources can produce voltage waves (varying electromotive force patterns) at various frequencies. The special case of zero frequency—direct current—makes for such easy analysis and simple mathematics that learning the circuit concepts and circuit analysis is quite simple. And there are many direct-current (dc) situations encountered in practice—particularly in electronics. Hence Chapters 2 and 3 of this text begin with dc analysis.

1.5   USING THIS BOOK

The reader may find it helpful to approach the remaining chapters in the following way:

Chapters 2 and 3—dc circuit analysis. These ideas and techniques should be thoroughly mastered and drilled on before going on to the rest of the book. With a practical grasp of this material the reader will find the remaining chapters much more understandable. When difficulties arise later, returning to these chapters can often help.

Chapters 4, 5, and 6—the circuit analysis material of the first two chapters extended to alternating current (ac) circuits. An easy and natural transition.

Chapter 7—simple electrical transients. This material is not vital to the rest of the book and can be omitted on first reading if desired.

Chapters 8 through 16 constitute the electronics and signals portion of the text, divided as follows:

Chapters 8 and 9 introduce analog and digital signal concepts and general practices. They are intended as an introduction, but may be all that many nonelectricals will want from this portion.

Chapters 10, 11, and 12 are systems chapters. Chapter 10—control systems and feedback—while included in the signal section of the book, is of broad interest. It is in this important technical area that electrical systems commonly interface with the systems structures of other disciplines. The chapter should help to tie electrical thinking to systems books and courses in the reader's own discipline.

Chapter 11 on communications presents the important basics of this area. In this "information age" there is an ever-increasing need for engineers to understand the nature of information and the various means of handling and conveying it. For example, plant control is taking advantage of high-capacity communication through use of digital transmission. Managers must choose among alternative communication systems to link people and machines.

Chapter 12 on microprocessors develops in a simple way the nature of computers and data processing systems, and will expand the horizons of those readers who know only the keyboard operation of computers, as well as enabling some readers to use and specify small data processing applications.

Chapters 13 through 16—digital and analog electronics—take the reader into more detail in electronics and practical electronic devices. Nonelectrical engineers do very little design at this level. However, knowledge of these basics allows them to act more effectively at higher levels.

Chapters 17, 18, 19, and 20—power systems, transformers, motors, and electrical machinery—provide some basic orientation in plant power systems operation for nonelectricals with these responsibilities.

These last two groupings, power (machinery) and signals (electronics), can be taken up in any order; either can be omitted in accordance with the interests of the reader.

Chapter 21—electrical safety is an important area for all engineers. This chapter also includes material on the National Electrical Code, the legal basis for much of electrical practice.

FOR FURTHER STUDY

Vincent Del Toro, *Electrical Engineering Fundamentals*, 2nd ed., 1986, Prentice-Hall, Englewood Cliffs, NJ. Chapter 1 of this text provides a stimulating description of electrical engineering as based on only six fundamental laws.

STUDY QUESTIONS FOR CHAPTER 1

1.1. What are the two principal divisions of electrical engineering applications?

1.2. What are the three main power uses of electricity?

1.3. For what kinds of applications is electric heating particularly advantageous?

1.4. List the purposes for which motors are used in your home.

1.5. List the many signal applications of electricity in your home.

1.6. What are the four main classes of signal applications?

1.7. Is *electronics* synonymous with electrical information or signal engineering?

1.8. Why does electrical engineering seem difficult to most people?

1.9. What can you do to overcome this?

1.10. What are the four goals of this textbook?

1.11. What are the three "levels" of engineering thought discussed?

1.12. What is the integrating factor that ties almost all electrical engineering work together and really characterizes the subject?

1.13. Why is it advantageous to study dc circuits before ac?

1.14. Why is circuit analysis drill important?

PART ONE

CIRCUIT ANALYSIS

CHAPTER 2

Basic Circuits

Since circuit analysis is the principal activity of engineers when dealing with electrical matters, we begin this study with circuits.

A *circuit* is simply a collection of electrical devices *and* the wires (conductors) connecting them. Figure 2–1 shows a *schematic diagram* of a typical circuit from Chapter 15—in this case an amplifier. Diagrams like this are intended to make it easy to understand the operation of the circuit. These contrast with *wiring diagrams*, which are intended to facilitate physical circuit construction.

The symbols for devices (for example, the transistors, which are marked Q in Figure 2–1, or the battery and bulb of Figure 2–2) are typically quite simplified in a circuit diagram. The engineer analyzes his or her circuit in terms of this simplified circuit representation.

Engineers talk about *direct-current circuits* (dc) and *alternating-current circuits* (ac), meaning by this circuits in which the currents and voltages are essentially constant (dc) or those in which the driving voltages and resulting currents vary with time as shown in Figure 2–14. Alternating-current circuits for power applications are usually sinusoidal, their currents and voltages varying with time as sine waves. Signal circuits have time variations according to the signals—in pulse form, or in voice waves, and so on. Electronic equipment often has dc and ac currents and voltages mixed together. We will study these varying voltages and currents as ''ac'' in Chapters 4, 5, and 6.

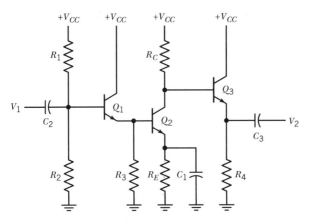

FIGURE 2–1 Typical circuit diagram. This circuit combines transistors (marked Q), resistors, and capacitors to make up an amplifier.

Circuit analysis means determining the voltages across and currents through each of the devices in the circuit. We define and explain these terms in this chapter.

The best way to learn both dc and ac circuit analysis is through starting with dc alone, as this chapter and the next do. This is because ac circuits follow exactly the same laws and are approached and analyzed in exactly the same way as dc

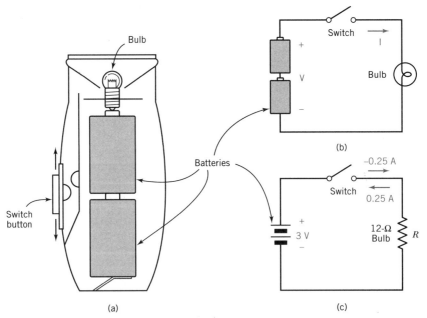

FIGURE 2–2 Circuit modeling for a two-cell flashlight. The complex physical device is reduced to an electric circuit using standard symbols and simplifying assumptions about how they work.

circuits. They are a little harder to think about and the numerical calculations, which involve complex algebra, are more challenging. A thorough mastery of simple dc circuit analysis can be gained quickly without the distracting and more or less peripheral difficulties of ac. After this it is relatively easy to transfer dc understanding and skills to ac circuits.

This chapter considers the basic circuit ideas of voltage and current and Kirchhoff's two important circuit laws. Extensive drill is provided by application of these ideas to many simple circuits. Chapter 3 will bring more formal and powerful analysis methods and more drill.

The reader's objective in this chapter could be:

1. become familiar with the circuit concept;
2. after first reading, review each of the italicized terms for comprehension;
3. become confidently proficient in solving the Easy Drill problems at the end of the chapter with emphasis on Kirchhoff's two laws.

2.1   MODELING, CURRENT, AND CHARGE

Engineers model electrical equipment and systems in terms of circuits. In analyzing an electrical circuit, they use simple models of the devices in the circuit. A simplified representation of the real situation if the model is properly conceived, will help predict quantitatively how the actual circuit behaves under various circumstances.

The two-cell flashlight treatment in Figure 2–2 is a simple example of this circuit modeling idea. Figure 2–2a is a cutaway sketch of a typical flashlight with drawn metal case. Figures 2–2b and c are circuit representations of the flashlight. Figure 2–2b lays out carefully the path through which current flows from the battery (comprising two D cells) to the bulb via a switch and back to the cells. Part of the current flow is, of course, through the metallic case, but in this circuit model the conductors connecting these parts are shown as wires. Electric current (flowing around the circuit) is represented here by the symbol I and the arrow.

The three *elements* in the circuit (battery, switch, lamp) could be laid out physically as in Figure 2–2b using some pieces of wire to make the closed current path (instead of the metal case) and would work just as well as they do in the case, but be a lot less handy. The circuit diagram* of Figure 2–2c uses the typical stylized symbols for the three elements, an easier representation.

Note that this model ignores everything except the electrical aspects of the flashlight. Case color, reflector size, total weight, and so on may be important but cannot be found from the circuit diagram. Even the electrical circuit model used is idealized in the sense that not only the symbols but also the elements themselves of Figure 2–2c are simplified.

The term *wiring diagram* refers to a circuit diagram of sorts but one that is drawn to show builders how to connect the parts of equipment. It is not very easy

---

*Often an engineer will loosely refer to a circuit diagram as a "circuit."

to use for understanding the operation of a circuit. An engineer faced with a strange piece of equipment and only a wiring diagram to go by, usually finds it helpful to redraw it into a more conventional circuit diagram.

What goes around the circuit? *Current* is thought of as the flow of *electric charges* through a conductor. At the top of the circuit diagram, Figure 2–2c, current flows out of the positive terminal of the battery (marked +), through the closed switch (the open switch shown here must be closed), on through the resistance R representing the lamp, and around the circuit back into the negative terminal of the battery. Positive charges are thought of as flowing out of the battery around the circuit—in this case clockwise. The direction that current flows is the direction in which the positive charges move.

Electric current, usually represented by the symbol *I*, is measured in the unit of *amperes* (abbreviated A). An ordinary 115-volt, 100-watt household lamp takes slightly less than 1 ampere of current to bring it to full brilliance. (Lamps work by the heating effect of the current raising lamp filament temperature to incandescence.) Power machinery may take hundreds or thousands of amperes, while transistor radio circuits may operate with thousandths or millionths of an ampere flowing through them. The current in the flashlight might be typically 250 mA.*

We have already noted that current flow is actually the flow of electric charges. The unit of electric charge is the *coulomb* (C). A flow of one coulomb per second is an ampere. Current is the time derivative of charge passing a given point in the circuit, the coulombs per second.

An arrow indicating current flow (as for example in Figure 2–2b) is called a *current reference direction*. In this case it is to the right on top of the circuit. It shows the direction in which the current labeled ''0.25 A'' flows. Note that this is equivalent to having a current reference direction arrow pointing left with the notation on it '' −0.25 A'' as shown in c. When the current is known the reference is usually chosen in the positive current direction. When starting to analyze a circuit with unknown currents, it is often convenient to select an arbitrary reference direction with a label such as ''*I*.'' Then algebraic manipulation during circuit analysis may result in an *I* either negative or positive.

Current flow through a conductor is quite analogous to the flow of water through a pipe, and electric charge is analogous to the drops of water. Voltage is analogous to the force of water pressure pushing water through the pipes. Friction on the pipe walls drops the pressure (head is lost) and corresponds roughly to the electrical resistance of the circuit.

Persons learning to think carefully about electric current for the first time are sometimes confused about the direction of current flow. They may have previously learned that in copper or aluminum (and in all metallic conductors) the moving charges that make up the current flow are negative. Charge carriers in metallic conductors are the free electrons, and the charge on each tiny electron is -1.6×10^{-19} C. So it is true (for metallic conduction) that current flow to the right in

---

*SI electrical units are commonly modified by these prefixes to change their values decimally: pico 10^{-12}, nano 10^{-9}, micro 10^{-6}, milli 10^{-3}, kilo 10^{3}, mega 10^{6}, and giga 10^{9}. Their abbreviations as applied, for example, to amperes are pA, nA, μA, mA, kA, MA, and GA.

the top conductor of Figure 2–2b or c is actually carried out physically by electrons moving to the left. But negative charges traveling to the left have *exactly* the same current effect as positive charges traveling to the right. It is helpful to always think of *current* (sometimes called *conventional current*) as the flow of positive charges regardless of how this process is being carried out. There are nonmetallic conducting situations (some chemical baths, for instance) where positive charges do the moving. This current flow convention will cover both situations interchangeably. The reader is advised to simply forget the matter of electrons for the present.

2.2   VOLTAGE AND ELEMENTARY METERS

What pushes the current around the circuit of Figure 2–2? *Electromotive force* (emf). Its unit is the *volt* (V) and the term *voltage* is commonly used instead of emf. But it is helpful to think often of the words "electromotive force" when first getting acquainted with circuits. These words drill in the idea of a voltage force pushing or pulling the charges around a circuit to make current flow. The battery of Figure 2–2 produces about 3 V of emf to operate the flashlight.

Household voltage in the United States varies from place to place but is commonly about 118 V (often referred to as "110").

This electrical force, voltage, appears *between two points*. For example, in Figure 2–4 there are 3 V appearing between points a and b, the terminals of the battery. (The term *battery* means an assemblage of one or more cells. Manufacturers designate the most common flashlight cell as size D.) Each of the two D cells in the flashlight has a voltage of approximately 1.5 V, and the voltages of two cells in series add to about 3 V. The battery (by chemical and physical means) tends to push positive charges out of its positive terminal (indicated by a long thin line) and pull them into its negative terminal (indicated by the short fat line), thus pushing them around the circuit. The circuit is said to be *closed* if a complete closed path is provided for current flow, or *open* if there is some interruption as, for example, having the switch open.*

In specifying a voltage between two points it is usually necessary to specify the *polarity* (which point is positive with respect to the other) in addition to the numerical value. As with current, this is handled by a *voltage reference direction* and labeling. The reference indicates the polarity by + and − signs on the battery, as in Figure 2–2c. The voltage label here is "3 V," and the polarity markings mean that the top of the battery is three volts plus with respect to the bottom. Voltage always occurs *between two points*. Again, when the actual voltage is unknown, an arbitrary reference direction is chosen and labeled with an algebraic symbol "*V*" to simplify analysis.

---

*Note that we have used the word *circuit* up to this point in several ways: as a physical group of electrical devices or equipment connected together in some manner, as a model mental conception of such an interconnected system, and as a diagram of such interconnections. It is further used in precise thinking about circuits to mean a "closed" circuit—one in which current is flowing. In this last sense, an "open" circuit has not yet become a circuit.

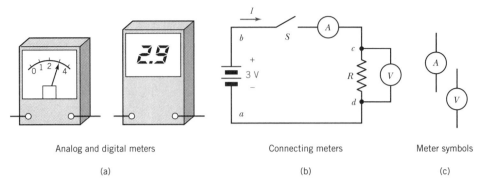

Analog and digital meters   Connecting meters   Meter symbols

(a)   (b)   (c)

FIGURE 2–3 Voltmeters are connected *across* (or *between*) two points whose voltage difference is to be measured. Ammeters must be inserted *into* the circuit so the current to be measured is forced to go through them.

Current or voltage can be measured by means of ammeters and voltmeters. Figure 2–3a shows two common forms of meter. One, an analog meter, has a needle that moves up a calibrated scale. The other is the digital meter in which numbers appear in a window. Each type has two terminals. Figure 2–3c shows the circuit symbols for voltmeters and ammeters.

In order to measure current at some given place in a circuit, for example in the top conductor of Figure 2–2, the circuit must be broken open at that point and the ammeter *inserted into* the conductor, as shown in Figure 2–3b. Ammeters connected across (between) two points with significant voltage difference will probably be damaged. For example, if the voltmeter in Figure 2–3b were carelessly replaced with an ammeter, the ammeter would probably be damaged.

Two elements are said to be *in series* if one end of one is joined to one end of the other and there is no other conductor connected to that junction. The battery current to be measured is forced to go through the meter in Figure 2–3b. The meter and battery are in series. Note, however, that in that same circuit the resistor R is not in series with the battery (unless the voltmeter is neglected). Although terminals a and d are connected together, there is a third connection there going to the voltmeter. The essence of a series connection is that the same current must go through each element.

To measure the voltage between two points, say across the flashlight lamp (between c and d in Figure 2–3b), the voltmeter is connected across them, in parallel with the two-terminal device whose voltage is to be measured. Note that this circuit arrangement agrees with the previous statement that voltage appears between two points.

To speak of "voltage at a point" is meaningless unless some other reference point is implied against which it is to be measured. Voltage exists or occurs between *two* points. It is also helpful to understand that voltage doesn't move or go through any element or conductor. Current and charge are the things that move.*

---

*The voltage between two points can be more carefully defined as the work done in moving a unit positive charge from one point to the other.

Two two-terminal elements are *in parallel,* if the terminals of one are connected to the terminals of the other, for example, the voltmeter and resistor of
Figure 2–3b. It doesn't matter whether there is another connection at these junctions or not. The essence of a parallel connection is that all elements in parallel
have the same voltage across them.

2.3   RESISTANCE AND OHM'S LAW

In Figures 2–2c, 2–3b, and 2–4, the *resistance R* represents the flashlight lamp.
Resistances are important two-terminal circuit elements in electrical engineering
and have the property that the voltage across them is proportional to the current
through them. That is, the ratio between voltage and current is constant for a given
resistance. Thus,

$$R = V/I. \tag{2.1}$$

The letter R stands for this ratio, the resistance. Resistance is measured in units of
ohms. The ohm is abbreviated or symbolized by the capital Greek letter omega (Ω).

Thus 17 V placed across a 10-Ω resistor would produce a current of 1.7 A
through it. The term *resistor* refers to a practical, physical device that can be bought
in an electronics store. A resistor has a certain value of resistance which can be
expressed, for instance, in ohms or megohms. Thus resistance is a property of a
resistor, or a resistance is an idealized element in a circuit diagram.

Experienced engineers carelessly mix the two terms resistor and resistance.
The beginner will want to think and speak carefully to avoid confusion.

Equation (2.1), in addition to being the definition of resistance, is also called
Ohm's law and is the most common circuit relation in EE. The end of a resistor into
which a current is flowing is always positive with respect to the other end.

It is helpful in thinking about circuits to visualize a *V-I characteristic* for two-
terminal devices. The shape of this graph of V vs I varies from one kind of two-
terminal device to another. For a resistance it is, of course, a straight line [Equation
(2.1)] as shown in Figure 2–4b.

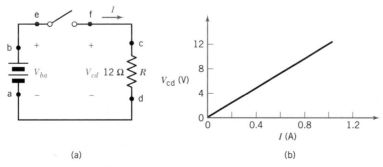

(a)   (b)

FIGURE 2–4 Each of the three elements has two terminals. These terminals
are connected by conductors. (b) A *V-I characteristic* for R on terminals c, d.

2.4   VOLTAGE SOURCES AND CURRENT SOURCES AND IDEAL ELEMENTS

So far batteries, switches, lamps, resistors, voltmeters, and ammeters have been encountered. Two additional elements of common use in circuit modeling are voltage and current sources as depicted in Figure 2–5a. A voltage source could have been used in place of the battery symbol of Figures 2–3 and 2–4. It is defined simply as a two-terminal element which produces a specific voltage between its terminals—in this case exactly 3 V. Note that an ideal voltage source will maintain its nominal voltage regardless of the current drawn from it.

Figure 2–5b shows a voltage source of V volts connected to a variable resistance R. As R varies, I will change also. Suppose R is 100 Ω and V is 9 V. Then according to Ohm's law [Equation (2.1)] the current will be 90 mA (0.09 A). But if the resistance R is 100 $\mu\Omega$, this simple circuit model indicates the current will be 90 kA! Suppose the voltage source in Figure 2–5b is being used to represent one of the popular small 9-V batteries used in portable radios. Surely it is unreasonable to suppose that so small a device could produce 90,000 A. As a matter of fact, taking more than a few hundred milliamperes for long would probably ruin the battery.

Practical engineers would conclude that the model was inadequate to represent such a high-current situation. They speak of the voltage source, resistance, and the conductors that connect them in Figure 2–5b as *ideal*, meaning that they are artificial representations of practical batteries, resistors, and conductors. This language implies that the real elements will not act exactly (as simply) as the ideal models are defined.

A 9-V battery, for example, will have a "dropping" *V-I* characteristic as shown in Figure 2–5c. That is, as the current drawn is increased, the terminal voltage of the battery will go down. The ideal voltage source used to represent the actual battery would have the "flat" *V-I* characteristic shown by the dotted line. The reader might consider what combination of the elements considered so far could better represent this drooping *V-I* characteristic than a simple voltage source does. (See end of chapter problems P2.3 and P2.6.)

Thus a real battery will not act exactly as an ideal voltage source. Similarly a practical resistor will not act exactly as a constant resistance. Actual conductors (wires) will not have zero resistance (as assumed so far). Despite these cautions,

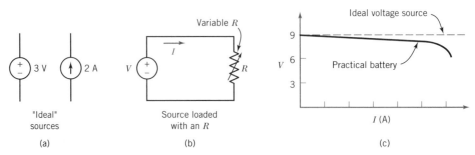

FIGURE 2–5 Source models (voltage or current) are used to represent batteries or other practical sources. The *V-I* characteristic (c) shows the difference between a real battery and an ideal voltage source.

the simple models usually serve quite well when equipment such as batteries or resistors are used within their intended ratings.

The current source of Figure 2–5a is similarly defined as a two-terminal element through which 2 A (or whatever specific current the source is designed to maintain) is always flowing. It is a little harder to relate this ideal source to any real element such as a battery. Perhaps the transistor comes closest, as Chapter 15 will show.

It is helpful to note that the voltage source and the current source each produce both voltage and current. The difference between them is that one maintains a constant voltage (providing whatever current is needed to do this). The other maintains constant current (generating whatever voltage is needed).

2.5   POWER AND ENERGY

Electric power is measured in units of watts (W) or often kilowatts (kW). In the flashlight example and Figure 2–4, when the switch is closed, the battery supplies power to the circuit. The lamp (represented as an R) is taking power from the circuit, and dissipating it in the form of light and heat. Energy flows from left to right in the circuit. The whole energy process is this: The battery converts its potential chemical energy into electrical energy. The lamp converts the electrical energy it absorbs from the circuit into heat and light energy.

Energy is work or heat or the potential to produce them. One can speak of the potential energy in a gallon of gasoline or the kinetic energy in a heavy, speeding car. Power is the rate of producing or expending energy, as the power of a generator or the power dissipated in a lamp. Energy is commonly measured in units of *joules* (J). The joule is a watt-second.

Thus using W as the symbol for energy and P for power,

$$ P = \frac{dW}{dt} \quad \text{or} \quad W = \int P \, dt, \tag{2.2} $$

where the units are watts, joules, and seconds. (Note that W is the usual symbol for energy. It is also an abbreviation for watts—the unit of power. In this latter use it always follows a number—20 W, for example.)

The basic power concept in two-terminal electrical equipment is*

$$ P = VI. \tag{2.3} $$

Going back to the flashlight, if the current is 250 mA, the 3-V battery is supplying 3 V × 0.250 A = 750 mW of power. The 3 V of the battery is directly across the lamp so the lamp power is also 3 V × 0.250 A or 750 mW. This result should not

---

*This fact is a consequence of the definition of voltage (see footnote to page 16).

be surprising—the power taken from the battery is the same as the power dissipated by the lamp.

EXAMPLE 2.1 Gasoline contains about 125,000 Btu or 131,850 kJ of potential energy per gallon. A gasoline engine driving an electric generator produces electric power which feeds a lamp. Assume the conversion processes from gasoline burning to electricity are in total 8% efficient.

Find
1. How long a 20-W lamp can be lit with a gas supply of one gallon.
2. If the system operates at 12 V, what is the lamp current?
3. How much electric charge passes through the lamp in 30 s?

Solution: 131.9 million × 8% = 10.552 million J (W-s). Dividing this figure by 20 W gives a little over <u>146 h and 30 min</u>.

$$I = P/V = 20/12 = \underline{1.67 \text{ A}}.$$

1.67 A is 1.67 C/s or <u>50 C</u> in 30 s.

Note that in the flashlight example, current is flowing out of the plus terminal of the battery and into the plus terminal of the lamp. Figure 2–6 makes an important generalization on this point. Consider a two-terminal element. If current flows into the positive voltage terminal the element is absorbing power. The figure illustrates this point and its converse. Note that in Figure 2–6b the battery is actually absorbing power, being charged.

Figures 2–6d,e generalize this for any two-terminal circuit. A make-believe fence* encloses one or more circuit elements. Either power is going into the fence (circuit inside is *absorbing power*) or power is coming out of the fence (circuit inside is *generating power*). Whether that part of the circuit is absorbing or generating is determined just as for the two-terminal elements of Figure 2–6a,b,c. We will refer to these facts as the *Power Convention* when there is need for them later. Figure 2–6 sets forth the facts about power and two-terminal circuits or elements.

A resistance *dissipates* power whenever current is flowing through it. Dissipating is a special case of absorbing in which the power is turned into heat and cannot be recovered. Hence for resistors the Power Convention requires that current always flow into the positive terminal, or, putting it another way, the end into which current flows is always positive. By substituting Ohm's law [Equation (2.1)] into (2.3), we obtain two other useful relations:

$$P = VI, \qquad P = V^2/R, \qquad P = I^2R. \tag{2.4}$$

---

*This device is frequently used in analyzing circuits. Its three-dimensional equivalent is a "black box." The black box or fence helps distinguish between what happens in that part of the circuit inside from that part outside. In this case it also emphasizes that the details of the inside part of the circuit are immaterial to the discussion. In some other disciplines this idea is known as a "controlled volume" or CV.

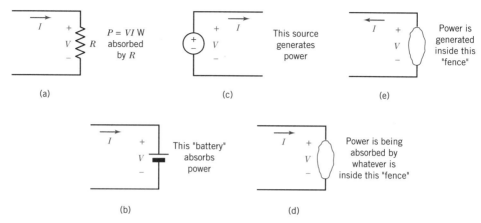

FIGURE 2–6 The power convention. Does the equipment or device between two terminals absorb or supply power?

It is important to understand that the V's and I's in these equations are the voltage across and current through the two-terminal element whose power is being investigated.

In the circuit we have been dealing with (e.g., Figure 2–4a) we can speak of the 12-ohm resistor as *loading* the 3-V battery. In engineers' language, by Ohm's law it *draws* 250 mA from the battery when the switch is closed; a smaller resistor, say 8 ohms, would draw more current (375 mA) and thus have a greater loading effect.

This is analogous to a mechanical engineer loading the motor of a table saw by connecting a mechanical load, a saw blade, to the motor shaft (probably with pulleys and a belt). In one case, power flows down a pair of wires electrically. $P = VI$ watts. In the saw, power flows down a rotating, torqued shaft; or down two sides of a speeding, differentially pulling belt. Here $P = 2\pi nT$ again in watts where n is in rps. Pushing a piece of oak into the blade "torques" the motor, draws more power down the shaft and the belt. Just as the battery falls off in terminal voltage as more current is drawn (see Figure 2–5c), similarly the motor falls off in speed if torqued too heavily.

Current through a current source or voltage source can flow into either the plus or the minus terminal. Thus these so-called sources can be acting either as sources or as sinks—generating or absorbing power. A "source" that is absorbing (really a *sink*) can be compared to a charging battery or, as will appear in Chapter 20, to the back voltage that a motor generates while turning. Looking at the circuit as a whole there will always be conservation of power. Power generated equals power absorbed. The same is true for energy over a given period of time.

2.6   KIRCHHOFF'S CURRENT AND VOLTAGE LAWS

Before proceeding to circuit drill we add one more topic—Kirchhoff's laws—to the list of analytic tools developed so far.

Consider Figure 2–7a. By Ohm's law currents I_b and I_c are 5 and 2 A, respec-

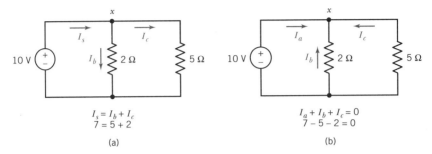

FIGURE 2–7 Kirchhoff's current law (KCL). At any node the current into the node must equal the current out.

tively, since the 10 V appears across each of the two resistors. At node x, current I_s must equal the sum of I_b and I_c. A *node* is the junction of two or more conductors. The current law (KCL) says simply that at a node the current in is equal to the current out. If the reference directions of the currents are all chosen as going into the node, as shown in Figure 2–7b, this law can be stated algebraically:

$$\Sigma\ I_{\text{in}} = 0. \tag{2.5}$$

The numbers for this particular case are shown under both circuits in the figure.

KCL is sometimes stated: "The sum of the currents out of a node is zero." By this is meant that if all currents are considered as going out of the node, as indicated by their reference directions, their sum must be zero. For example, in Figure 2–7b all three reference current directions could be selected going out of node x, which would simply reverse each of the three signs in the arithmetic. The sum of these three currents at the node must still equal zero. Note that a current of 7 A going into a node is equivalent to a current of -7 A coming out of it. The a and b parts of Figure 2–7 are actually the same. What would the arithmetic equation be for this node if all the I's were considered to be coming out?*

The current law is certainly intuitively true. It's like water flowing through a piping system. Water out equals water in. Gallons per minute (gpm) out must equal gpm in. This law is especially useful in parallel circuits, as in Figure 2–7.

Similarly, Kirchhoff's voltage law (KVL, Figure 2–8) is particularly good for series circuits. It says that the sum of the voltage drops around any closed path in a circuit is equal to zero. Being specific mathematically,

$$\Sigma\ V_{\text{drops}} = 0. \tag{2.6}$$

A *loop* is any closed path in a circuit. *Mesh* is the term for a loop that has no branch intersecting it. Thus in the circuit of Figure 2–7 there are three loops, two of which are meshes.

---

*$-7 + 5 + 2 = 0.$

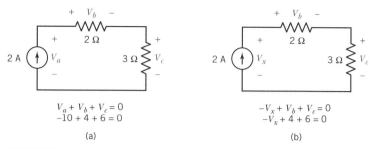

$$V_a + V_b + V_c = 0$$
$$-10 + 4 + 6 = 0$$

(a)

$$-V_x + V_b + V_c = 0$$
$$-V_x + 4 + 6 = 0$$

(b)

FIGURE 2–8 Kirchhoff's voltage law (KVL). Around any closed path the sum of the voltage drops must be zero.

Consider the series circuit of Figure 2–8a: since the voltage drops V_b and V_c are known from Ohm's law, then V_a, the voltage of the current source, can be calculated using KVL to be -10 V as shown. The reference polarities for voltage on each of the three elements in Figure 2–8a have been selected so that in going clockwise around the loop they are all drops. The minus sign in this result simply means that the source reference was taken "wrong" and that the true polarity of the source is plus on top. But as with the KCL currents, voltage reference polarities can be selected in any direction. Figure 2–8b is an example of this. The resulting KVL equation has a negative drop (a positive rise) for V_x. Solving again for V_x yields 10 V plus. But since the reference direction is the opposite of that chosen in Figure 2–8a, the result is the same, plus on top. An interesting exercise is to redo this problem moving counterclockwise around the loop.

Walking around a closed path on the ground is a helpful analogy for the KVL. Voltage drops correspond to decreases in elevation. Going uphill and down, a walker finally arrives back at the starting point. Since there can be no net gain in elevation, the sum of the drops (a rise being counted as a negative drop) must be zero.

The word *potential* is often used for voltage at a node with respect to a reference node, just as a location on the ground has *elevation* with respect to sea level.

2.7   RESISTANCES IN SERIES OR PARALLEL

Note that a further consequence of Kirchhoff's voltage law is: *resistances in series should be added to get total resistance.* Thus the 2-A source in Figure 2–8 "sees" $2 + 3$ or $5\ \Omega$ resistance.*

Similarly, considering the work done with the two parallel R's in Figure 2–7, the source sees the current-voltage ratio at its terminals or 10 V/7 A $= 1.43\ \Omega$. A little algebra (see problem P2.2) will generalize these resistor combinations as

---

*This odd but useful language implies that the 2-A source "looks out" of its two terminals and sees the voltage-current ratio there.

follows: for series R's,

$$R_s = R_1 + R_2 + R_3 + \cdots; \tag{2.7}$$

for parallel R's,

$$\frac{1}{R_p} = \frac{1}{R_1} + \frac{1}{R_2} + \frac{1}{R_3} + \cdots. \tag{2.8}$$

For the special case of only two R's in parallel Equation (2.8) reduces to

$$R_p = R_1 R_2 / (R_1 + R_2). \tag{2.9}$$

Equation (2.9) (simply the product over the sum) is particularly convenient for mental calculations; use it for more than two R's by combining them two at a time.

EXAMPLE 2.2 Consider the circuits a and d of Figure 2–9.

Find
1. the equivalent R each source sees;
2. V_x and I_x.

Solution: Circuit 2–9a: reduce to b by product over sum [$3 \times 6/(3 + 6)$]; reduce to c by adding series resistors in b. Then $V_x = RI = 4 \times 7 = \underline{28\ V.}$
For circuit 2–9d: proceed as indicated in sketches 2–9e, 9f, 9g. $I_x = V/R = 12/5.6 = \underline{2.14\ A.}$

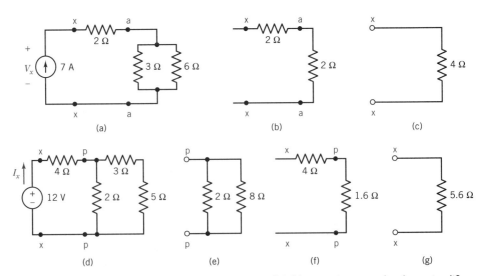

FIGURE 2–9 Combining resistances in series or parallel. Many engineers make these simplifications in their heads as they look at a circuit.

2.8   INTUITIVE CIRCUIT ANALYSIS

Circuit analysis—solving for the currents and voltages present in any part of a circuit—is an interesting and satisfying skill. Like other skills (tennis, for example) it is learned by thoughtful drill and practice. The following material on circuits is designed to supply such drill at a very simple level. It is also useful because the great majority of electrical and electronic applications the reader will encounter in practice are at about this level of circuit simplicity.

EXAMPLE 2.3 Consider the circuit of Figure 2–10a.

Find
1. The power involved in each of the three elements.
2. Indicate whether it is being generated (G) or absorbed (A) in that element.
3. Fill the answers in on the form shown in Figure 2–10b.

Solution: Such a problem can be solved by direct mental application of Ohm's law, KVL, and KCL without writing complicated equations. *There is no routine procedure to follow.* As in any worthwhile problem, engineers look at the situation carefully to see what it suggests.

Note that all three elements are in series. So the current I will be the same throughout the circuit—the current source will force two amperes counterclockwise. (Note that KCL has been applied three times in the last three statements.) Thus the power involved in the 4-V source is 8 W ($P = VI$) and is absorbed since current is entering the plus terminal of that source (this "source" is actually charging or absorbing). It is most helpful to pencil

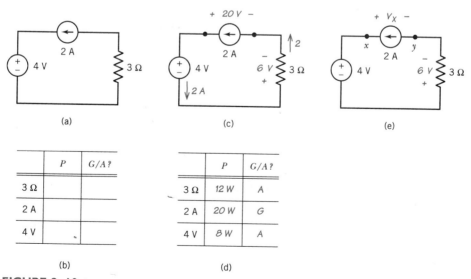

	P	$G/A?$
3 Ω		
2 A		
4 V		

(b)

	P	$G/A?$
3 Ω	12 W	A
2 A	20 W	G
4 V	8 W	A

(d)

FIGURE 2–10 Intuitive circuit analysis problem. Find the power involved in each element. Is it generated or absorbed?

all currents and voltages right on the circuit diagram as they are found (Figure 2–10c).

Much effective analytical thinking is done backwards. The analyst here wishes to get the power in each element from one of the expressions $P = VI$, $P = V^2/R$, $P = I^2R$ (the latter two are of course applicable only to resistors). Looking next for another element to which to apply one of the three, the 3-Ω resistor suggests itself. $P = I^2R$ gives $P = 4 \times 3 = 12$ W. Enter the powers on the chart as each is worked out (Figure 2–10d). Since resistors always absorb power (never supply or generate it) a plus sign should be put on the bottom of the 3-Ω resistor. Note also that there will be 6 V dropped across it (a result of Ohm's law).

Entering this power information in the table (as in 10d) and marking all currents and voltages and their directions and polarities right on the diagram (as in 10c) will serve to suggest further steps. (Ignore for the present the "10 V" marked on the current source.)

It remains to find the power involved in the 2-A source.* Only $P = VI$ will give this, so the V across that source is needed. V_{xy} (the voltage of x with respect to y) is the voltage drop clockwise from x to y. (This is also the negative of the voltage drop counterclockwise from y to x.) So the voltage across the 2-A source must be 10 V (Figure 2–10c) and plus must be on the left. This result comes from intuitively applying KVL. In order for the sum of the voltage drops around this loop to equal zero, the voltage across the 2-A source must be 10 V *and* the polarity must be plus on the left. This finding is then marked on the diagram as in Figure 2–10c. The resulting power is 2 A $\times$ 10 V or 20 W. It is generated: current is emerging from the plus end of the source.

The result of the example problem then is the form Figure 2–10d. A necessary (but not sufficient) check is that the sum of the powers generated is equal to the sum of the powers absorbed.

The technique just illustrated of applying KVL by summing voltages around a closed path in the circuit may take a little practice. If the reader finds it difficult, the method of Figure 2–10e will help while learning. Assign an algebraic value V_x to the unknown voltage across the 2-A source. Assign an arbitrary polarity—say plus on the right. Write KVL clockwise starting in the lower left-hand corner as "the sum of the drops around a closed loop is 0." So: $-4 - V_x - 6 = 0$. Hence $V_x = -10$ V. The minus sign simply says that the wrong polarity was chosen for V_x; the plus should be on the left.

Although it isn't necessary since V_x, polarity and magnitude, is now known, it is instructive to rewrite KVL by choosing the plus on the left. ▪

In the above example we have manipulated polarities and jotted values on paper to help in the application of KVL. With surprisingly little practice the new circuit analyst can do problems like this mentally, as most experienced engineers do.

The next example is from the circuit in Figure 2–11a.

---

*Engineers commonly read this kind of expression as "two-amp source," although amp is not an officially recognized term.

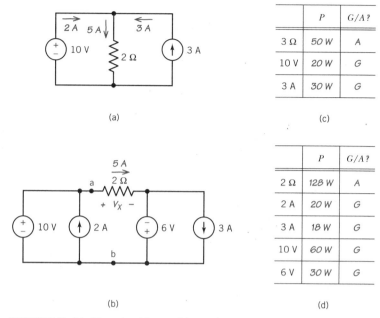

	P	G/A?
3 Ω	50 W	A
10 V	20 W	G
3 A	30 W	G

(a)

(c)

	P	G/A?
2 Ω	128 W	A
2 A	20 W	G
3 A	18 W	G
10 V	60 W	G
6 V	30 W	G

(b)

(d)

FIGURE 2–11 More intuitive problems. As engineers work out current and voltage information they pencil results right onto the circuit sketch.

EXAMPLE 2.4 Consider the circuit of Figure 2–11a. (Neglect the current values shown.)

Find: the power involved in each element and whether it is generated or absorbed.

Solution: Note that with all the elements in parallel each must have 10 V across it. The plus is on top as the voltage source requires. Searching for further ideas, Ohm's law requires that there be 5 A down through the 2-Ω R. Because of the 3-A source there will be 3 A coming into the top node from the right. Considering the top node and applying KCL, there must be 2 A coming in from the top of the 10-V source. This information is again penciled right on the circuit as shown.

The relations of Equation (2.4) give the powers (as entered into the form 11c). Again it can be noted that power is conserved by the answer. This technique of summing currents at a junction is similar to the previous problem where voltages were summed around a loop.

As a last worked-out example:

EXAMPLE 2.5 Consider the circuit of Figure 2–11b.

Find: powers involved in each of the five elements and whether absorbed or generated.

Solution: After some thoughtful inspection of the circuit, an analyst will conclude that there must be 16 V across the 2-Ω resistor with the plus on the left
(KVL around the loop containing the two voltage sources and the *R*). Hence
8 A go from left to right (Ohm's law applied to the 2-Ω *R*). Now examining
the two top nodes (pencil in the 8 A leaving or coming into them) it must be
concluded that 6 A comes up out of the 10-V source and 5 A goes down into
the 6-V source (KCL at each of these nodes). And again power is properly
conserved (Figure 2–11d).

For further help in applying KVL intuitively we can offer two procedures: first,
looking at the circuit Figure 2–11b, suppose the voltage V_x across the 2-Ω resistor
is wanted. Starting at the terminal under the minus sign the analyst asks, "If I go
to the plus terminal how many volts have I gone up?" There are several ways to get
there besides through the resistor itself. One might move clockwise through rises
of 6 and 10 V for a total of + 16 V. It will be instructive to ask the question in other
ways and move from the left terminal to the right one. Is the result always the same?

As a second procedure consider Figures 2–12a and b. If one desired to know
the voltage across the 4-Ω *R* of Figure 2–12a we could simply redraw the circuit as
in Figure 2–12b. Now it is quite easy to see the electrical parallelism. Thus the 4-Ω
R has 10 V across it with the plus on the right. Circuits can be pulled and bent
around in any helpful way so long as the connections are not changed or wires
made to touch each other. Sometimes it is helpful to position the nodes so that
their height on the drawing corresponds roughly to their potential. Potential is
taken relative to *ground* or some other reference that is common to the entire circuit
(the reference is considered the zero-potential node). The reference node is then
placed usually at the bottom of the circuit sketch.

Using the technique of twisting circuits about so they are more easily understood, consider Figure 2–13 as a redrawing of 2–11b. The circuit is unchanged,
but how easy it is to see now that there must be 16 V across the 2-Ω resistor with
the plus terminal on top (to the left in Figure 2–11). Experienced analysts twist
circuits or parts of circuits around in their heads while examining them. While
readers are learning the technique, they will want to redraw the circuits on paper
frequently. Circuit drawing and redrawing is a key to good, productive thought for
circuit analysts.

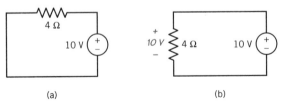

(a)   (b)

FIGURE 2–12 Redrawing circuits (on paper or mentally)
will often help in solving.

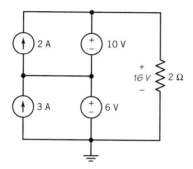

FIGURE 2–13 A redrawing of the circuit in Figure 2–11b.

2.9   VARYING VOLTAGES AND CURRENTS

It is useful to extend this circuit thinking by asking what effect would be produced if one were to drive a resistive circuit with a varying voltage source. Figure 2–14a reproduces the circuit of Figure 2–8 except that the dc current source is replaced with a sinusoidal voltage source which produces a *voltage wave* (a voltage varying in time) as shown in 2–14b. This particular wave varies between plus 14.1 V and −14.1 V.

The answer is that at any instant there is some particular voltage supplied by the source, 10 V for instance, and a 2-A current will flow at that instant. When the source is producing 5 V, a 1-A current will flow, and so on. When the source goes

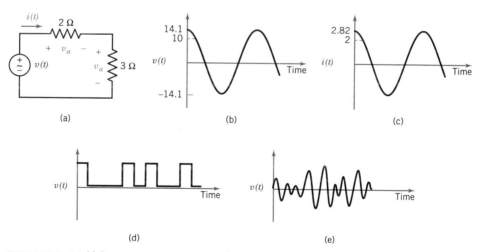

FIGURE 2–14 Voltage sources may vary with time. For a resistive circuit, solutions can be made by dc methods at any instant. The sinusoidal driving voltage (b) produces the resulting $i(t)$ form current sketched in (c). Other more complex driving voltage sources may be pulses (d) or waveforms such as (e).

negative the current will reverse. Thus over a period of time a current wave will flow in the circuit as shown in 2–14c, peaking at plus and minus 2.82 A and at exactly the same time the voltage wave peaks.

The *sinusoidal* wave shown is typical of ac power. But varying signal voltages could have a variety of different forms, for example as indicated in Figures 2–14d and e. Even so, one could find the instantaneous currents (and any other effects) in the same way used throughout this chapter and the next. The current would change in proportion to voltage and maintain the same relation to the voltage supplied, in this case by a factor of 1 to 5. The important point is that Ohm's law and Kirchhoff's laws apply at each instant of time.

Readers should be warned, however, that there are two other passive ac circuit elements besides resistors, called inductors and capacitors, with symbols as shown in Figure 4-1. In circuits containing either of these two elements the effect of time-varying driving voltages is not so simple and cannot be calculated on an instant-to-instant basis. For them we resort to the methods of Chapters 4 and 5.

In summary then, for purely resistive circuits (no inductors or capacitors) varying voltages and currents can be handled on a simple instantaneous basis. Their variations produce no unexpected effect on the circuit.

2.10   CIRCUIT ANALYSIS DRILL

Chapter 1 noted the need to establish an intuitive familiarity with circuits—equivalent, for example, to the direct mechanical understanding developed from years of driving and parking cars. One learns such basics most quickly and thoroughly by doing *many easy problems*. Hard problems are poor vehicles for initial learning. They are important for other reasons but come later.

In addition to problem solving, other ingredients in good, easy, rapid learning are drawing pictures (in this case drawing circuits) and making up circuits (problems) of the reader's own choosing. The reader is strongly urged to drill himself or herself in these ways. In the problems of Figures 2–7 through 2–11, for example, change a value or a polarity in the circuit and re-solve. By this procedure one can have as many drill problems as desired. (For those persons who are using this book as part of a formal course, it is suggested that learning will be faster and more pleasant if some study and problem-solving time is regularly shared with another student.)

But in making up circuits for problems it is easy to go beyond what is possible by intuitive analysis. For example, as simple a situation as Figure 2–15a cannot be solved by intuitive methods. The next chapter's methods are needed for this. But even getting into this kind of trouble is instructive. You should be able to easily solve Figure 2–15a (find all the currents and voltages) by trial and error methods. (*Hint:* Guess and reguess the current through the 3-Ω R. Each time find the voltages on the other two R's and their currents using KVL and Ohm's law. Continue until KCL is satisfied at the top node.)

It should be apparent that the circuits in Figures 2–15b,c cannot be allowed. Herr Kirchhoff would turn over in his grave. They are impossible—actually an anomaly of modeling. We address this paradox in Chapter 3.

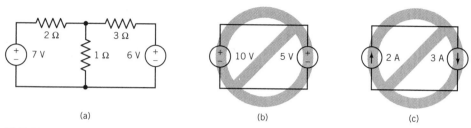

FIGURE 2–15 Some circuits, for example (a), cannot be solved by intuitive methods. Some model circuits such as (b) or (c), are not allowed—impossible.

2.11   SUMMARY

Circuits are the electrical models for equipment and systems and consist of all electrical parts plus the wires connecting them. The term also refers to the stylized diagrams that represent circuits.

Electrical current (in amperes) is forced through two-terminal devices and parts of circuits by the voltage (electrical force—in volts) across them. Where the ratio of voltage to current between the two terminals is about constant, it is described as the resistance between those terminals, and Ohm's law $V = RI$ may be applied.

Ideal voltage and current sources model practical batteries and generators. The voltage source maintains a specified voltage across its terminals regardless of any current drawn from it. The current source maintains a specified current through itself regardless of the voltage across it. Voltmeters and ammeters measure voltage between two terminals, or current passing through the meter, respectively. Voltmeters draw very little current (zero for the ideal) and ammeters have very small resistance (zero for the ideal).

Power going into a two-terminal circuit branch is VI in watts, where the current is going into the positive voltage terminal. That fact is known as the power convention.

Kirchhoff's current and voltage laws (KCL and KVL) state that the sum of currents flowing into a junction must be zero (charge cannot be created or destroyed), and the sum of the voltage drops around a closed path (in a circuit) must be zero. Resistances add in series and combine as the reciprocal of the sum of reciprocals in parallel.

For many practical circuits, analysis can be done "intuitively" without extended calculations, by using KVL, KCL, and Ohm's law.

FOR FURTHER STUDY

Ralph J. Smith and Richard C. Dorf, *Circuits, Devices and Systems*, 5th ed., John Wiley & Sons, New York, 1992. Chapter 1.

PROBLEMS

Easy Drill Problems (answers at end of chapter)

D2.1. Sketch (make circuit diagrams for) the following circuits:
   (a) three resistors of 2, 3, and 4 Ω are placed in series across the terminals of a 6-V battery;
   (b) the same resistors are placed in parallel across the battery;
   (c) the parallel combination of 2 and 3 is placed in series with the 4 and the battery;
   (d) find the current through the 2-Ω resistor in each of the three cases above.

D2.2. Seventeen 50-Ω resistors are placed in parallel and six 1.5-Ω resistors in series.
   (a) Find the total resistance if the two groups are placed in series.
   (b) The six 1.5-Ω R's are now placed in parallel and the seventeen 50-Ω R's in series. These two groups are placed in parallel. What is the total resistance?

D2.3. Find I_x and V_x in the two circuits sketched.

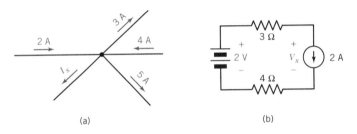

(a)   (b)

D2.4. Sketch the simple circuits involved when solving the following:
   (a) A current of 1 mA flows for 7 ks. Find the amount of charge transferred in units of MC.
   (b) In part **(a)** above, suppose the current is passing through a 1-kΩ resistor. How much energy is involved in GJ?
   (c) Long-distance power transmission is sometimes carried out with high-voltage dc lines. Suppose that one such line is operating at 1.2 MV and a current of 43 A flows. What is the power transferred in kilowatts?

D2.5. Solve by intuitive methods for the currents, voltages, and powers in the circuit sketched. (Be sure to pencil out the circuit again and enter voltages and currents on it as they are found. Make an answer chart to fill in as in Figure 2–11c or d.)

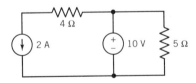

D2.6. Repeat for this circuit.

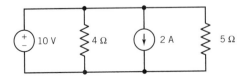

D2.7. Repeat for this circuit.

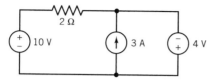

D2.8. Repeat for this circuit.

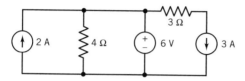

D2.9. Two generators, A and B, supply two loads on the same bus as sketched. Generator A produces 10 kW. Load 2 absorbs 8 kW. Generator B produces half as much power as load 1 absorbs. How much power does load 1 absorb?

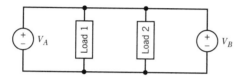

D2.10. Find I_x in this circuit by at least two methods. (*Hint:* Don't forget the possibility of trial and error solution. What is the relation between the voltages across the two resistors? How much resistance does the 24-A source see?)

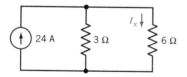

D2.11. (a) Find I_x, I_y, V_x, and V_y in the sketch.
(b) Find the power absorbed by each of the three elements.
(c) Find the power generated by each of the three elements. (*Hint:* Use signs very carefully.)

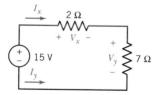

D2.12. A 2-A current source, a 5-V voltage source, and a 10-Ω resistor are placed in a series loop. Make a tabulation of the voltage, current, and power (*G* or *A*) involved with each of the three elements.
(a) With the current source feeding into the positive end of the voltage source and;
(b) with the current source reversed.

D2.13. The three elements of the previous problem are all connected in parallel. Repeat your tabulation for the **(a)** and **(b)** cases.

D2.14. A 20-Ω resistor is connected across an adjustable voltage source.
 (a) Tabulate the current that will flow for voltages from 0 to 100 V in steps of 10.
 (b) Add power dissipated by the resistor to your tabulation.

   If its rated power handling capability is 750 W, what is the maximum safe voltage that can be used on it?

D2.15. A 120-V battery is connected to a 20-Ω resistive load. How many coulombs of charge flow through the resistor each minute? How much electrical energy in joules does the resistor convert into heat each minute?

D2.16. Redraw Figure D2.8 to provide instruments (voltmeters and ammeters) to show: the current down through the 4-Ω resistor, the current to the right in the top, center branch, the voltage across the 3-Ω resistor, the voltage across the 3-A source, and the current through the 6-V source.

D2.17. It is found that a certain worn-out automobile battery can be approximately modeled (represented) by a 12-V source in series with a 1-Ω resistance.
 (a) For currents of 0–10 A in steps of 2, tabulate the battery voltage versus current drawn.
 (b) What would this model predict as the short circuit current? (That is, what current would flow if the terminals were connected together with a piece of heavy copper wire?)

D2.18. In Figure 2–9(d) relabel the four resistors with their conductance values in siemens and combine them into a single conductance connected across the 12-V source. Check your result against the answer in (g).

D2.19. A house has the following intermittent electrical loads: water heater (3 kW), dryer (2 kW), kitchen range (4 kW), and miscellaneous appliances (1.5 kW).
 (a) What is the maximum kW load that can occur?
 (b) If the supply voltage is 235 V, what maximum current might be drawn from the mains?

D2.20. In Figure 2–11(a) reverse the polarity of the 10-V voltage source so that the + is on the bottom. Mark the three new currents on your diagram and fill out the power chart as in Figure 2–11(c).

D2.21. In the circuit of Figure D2.7 reverse the 10-V source, pencil in the currents and voltage, and fill out a power chart as in Figure 2–10.

D2.22. In the circuit of Figure D2.7 reverse the 4-V source, pencil in the currents and voltage, and fill out a power chart for each element as in Figure 2–10.

Application Problems

P2.1. Make up two good intuitive circuit problems of your own and solve them.

P2.2. Prove with Ohm's law, KCL, and KVL that series resistors add, and parallel resistors combine as the reciprocal of the sum of their reciprocals.

P2.3. A two-terminal circuit is formed by connecting a 5-V source to a series R of 10 Ω. Make a "V-I" characteristic plot for the combination. (*Hint:* Connect a load resistor to this combination and vary it from zero to infinity, observing current drawn and voltage at the two terminals. See Figure 2–5c.)

P2.4. **(a)** You find a resistor and wish to determine its value with a voltmeter (drawing negligible current) and an ammeter (with negligible resistance) and a battery.

(see sketch a). You find that when the voltmeter reads 5.92 V, a current of 56 mA flows in the ammeter. What is the ohmic value of the resistor?

(b) Now for the same readings suppose that the ammeter has resistance of 10 Ω and the voltmeter draws a current of 6 mA. What is the R being measured?

(c) Repeat **(b)** with the connections shown in sketch b.

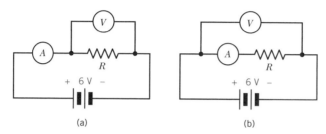

(a)   (b)

P2.5. A 12-V automobile battery is connected via switches and horn button to two head-lamps, a horn, and a radio. Each of these devices requires the full 12 V. Sketch a circuit showing how they are connected.

P2.6. A D cell is found to have a nearly linear "drooping" *V-I* characteristic (see Figure 2–5). At 0 A current it provides 1.48 V. At 400 mA it provides 1.39 V. Sketch a circuit model for this battery made up of a voltage source in series with a resistance. Find the values of the source in volts and resistance in ohms. (*Hint:* Do problem P2.3 first to get a feel for the relationship involved.)

P2.7. A two-wire transmission line connects a generating station to a factory 5000 ft away. It is made with #00 standard annealed copper wire. (*Note:* #00 is a designation from the "American" or "B&S" wire gauge for a conductor 0.3648 inches in diameter and with a resistance of 0.07793 Ω per 1000 ft at 20°C.) The factory will use 50 kW of power, and receives power at 500 V dc. There is more on the wire gauge system in Chapter 8.

(a) Find the power lost in the line due to $I^2 R$ losses, and find the efficiency of transmission (power delivered at the receiving end of the line divided by power put into the sending end) for receiving-end operation at 500 V. (*Caution:* Don't forget that there are two wires.)

(b) Repeat for receiving-end operation at 1000 V dc (50 kW is still the power required at the load end).

(c) If the average cost of power is 8 cents per kWh, what is the lost energy cost per year of operating the line at each voltage? Assume the plant operates an average of 40 h per week for 50 weeks per year. [*Note:* A *kilowatt-hour* (kWh) is the energy involved in using a power of one kilowatt for one hour.]

P2.8. A current with the waveform shown flows through a 1000-Ω resistor.

(a) Sketch the circuit of the current source and resistor, and copy the current wave as a function of time. Sketch the power dissipated in the resistor as a function of time on the same axes.

(b) Find the energy dissipated per second. What is the average power?

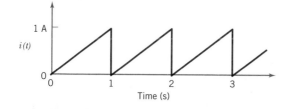

P2.9. Two engineers have a portable gasoline-powered electric generator at their Maine fishing camp. They wish to measure its effectiveness in converting gasoline into electric energy. They put in a measured 2 gallons, turn on three 100-W lamps, and find that the fuel lasts for 5 h and 23 min.
(a) What is the output of the generator in kilowatt-hours per gallon of gasoline?
(b) What is the percent efficiency of the total conversion process if gasoline contains 125,000 Btu per gallon and 1 Btu is equivalent to 1054.8 J?

P2.10. Replace Figure 2–11(a) with two circuits: one with the voltage source replaced with a short circuit (a piece of wire); the other with the current source replaced with an open circuit (two terminals with no connection).
(a) Find the current through the 2-Ω resistor in each circuit.
(b) What is the relation between these two answers and the original answer?

P2.11. It is important to realize that ammeters must be connected in series with the current to be measured (breaking into the circuit to accomplish this). Suppose a careless beginner, wishing to measure the current through the 6-volt source in Figure D2.8, connects his ammeter across the source (in parallel with it). (No voltmeter is yet connected to interrupt the 4-Ω resistor's path.) Suppose further that the ammeter's resistance is 0.1 Ω.
(a) What current tends to flow through the 4-Ω resistor?
(b) What current tends to flow through the ammeter (and probably destroy it)?
(c) Sketch this circuit with a properly connected ammeter.

P2.12. It is important to realize that voltmeters must be connected between (across) the two terminals whose voltage difference is to be measured. Suppose a careless beginner, wishing to measure the voltage across the 4-Ω resistor in Figure D2.8, connects the voltmeter in series with it. (The errant ammeter of P2.11 has been removed (–smoking?). Suppose further that the voltmeter has a resistance of 100 ohms.
(a) What current will flow through the 4-Ω resistor?
(b) What voltage will exist across the resistor?
(c) How much will the voltmeter read?
(d) Sketch a circuit with properly connected voltmeter.

P2.13. Practical analog meters usually consist of a *movement* and a large series resistor for a voltmeter, or a movement and a small parallel *shunt* resistor for an ammeter. The movement is typically a 1-ma full scale meter with a resistance of say 10 ohms.
(a) Design (sketch circuit with values) a 250-V (full scale) voltmeter using this movement.
(b) Design (sketch circuit with values) a 1-A ammeter constructed with this movement.

P2.14. A laboratory technician needs a 38-Ω resistor (accurate to within +/– 1 Ω). He has a box of 10-Ω resistors. Help him design the 38-Ω resistor using as few 10-Ω resistors as possible. Show your design and give its resistance to 2 decimal places.

Extension Problems

E2.1. A transistor is a three-terminal device as sketched. Considering two terminals at a time, the "input" circuit is B and E (base and emitter), and the "output" circuit is C and E (collector and emitter). A small input current makes a proportional but much larger (100 times is typical) output current. In the transistor circuit sketched, a current of 1.5 mA flows from the load resistor R (5000 Ω) into the collector (C). Emitter current is this 1.5 mA plus the base current. The voltage supply is 10 V.

(a) What is V_{CE} (the collector voltage with respect to the emitter)?
(b) What is the power dissipated in R?
(c) What is the power dissipated by the transistor?
(d) What is the power taken from the battery supply?
(e) Is power conserved?

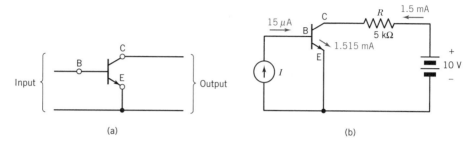

(a)   (b)

E2.2. A motor drives a mechanical load requiring 100 horsepower. [*Note:* One horsepower
(hp) is equivalent to 746 W.] It operates at 82% efficiency.
   (a) How many joules of energy does it take from the line in 20 min? How many does
   it deliver to the load?
   (b) If the load cycle is changed so that for the second ten minutes the load is 50 hp
   (motor efficiency 73%), recalculate the energy the motor receives and delivers in
   20 min.

E2.3. *Ratings* are assigned to devices by their manufacturers so that they can be used properly.
For instance a resistor is rated in terms of its resistance and power dissipation. I^2R
losses appear as heat, and the temperature of the resistor rises above ambient (the
surrounding temperature) until its thermal radiation out is just equal to power in.
Then the resistor gets no hotter.
   (a) For an R rated at 100 Ω, $\frac{1}{2}$ W, find the maximum current allowed.
   (b) Assume that the R's temperature rise above ambient is proportional to the power
   dissipated, and that at full rating, temperature rise is 40°C. How many degrees
   temperature rise will there be at half the current of (a)?
   (c) If the heat radiated by a resistor at some given temperature above ambient is
   proportional to surface area, how much larger in diameter will a 2-W resistor be
   than a 1-W? Assume they are cylindrical with the same length to diameter ratio,
   and are both allowed a 40°C temperature rise. (*Note:* The physical size of a practical
   resistor has nothing to do with its resistance.)

E2.4. A silver cyanide electroplating bath (see sketch) is used to plate silver uniformly onto
objects with a total area of one square meter. A current of 600 A flows for 20 min.
Assuming each coulomb transports 1.1179 mg of silver to the work, what is the weight
of silver deposited in grams per square centimeter?

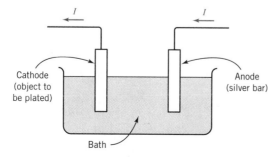

E2.5. Resistivity ρ can be defined as the resistance of a one-centimeter cube of material measured between opposite faces. The resistivity of copper is often taken as 1.7241×10^{-6} Ω-cm.

(a) Derive a general expression for the resistance of a conductor of length l and cross-sectional area A and resistivity ρ. (*Hint:* The expressions for series and parallel resistors will be useful here.)

(b) Show that the resistance per 1000 feet of #00 wire is as indicated in problem P2.7.

E2.6. The resistance of metals increases with temperature in an approximately linear fashion so that $R_{T_2} = R_{T_1} [1 + \alpha_1 (T_2 - T_1)]$. Here α_1 is called the *temperature coefficient of resistance.* For annealed copper and T_1 of 20°C, $\alpha_1 = 0.00393/°C$.

(a) How hot can the weather get in Fahrenheit degrees without increasing the resistance of a transmission line by more than 10%?

(b) A precision resistor wound with iron wire is rated at 100 Ω at 20°C. The temperature coefficient for this wire is 0.006/°C. The resistor's heat loss is 3.2 W/°C above ambient. At an ambient temperature of 20°C how much current can be allowed in the resistor without changing its 100-Ω value by more than 1%?

E2.7. In the circuit of Figure 2–15(b) recognize that a real voltage source, no matter how excellent, will have at least some small series resistance. In most circuits this is negligible compared to the resistances it is connected to.

(a) Assume source resistances of 0.1 Ω each. Find the bus voltage (between top and bottom wires) and the current in the circuit. What are the powers involved in the four elements?

(b) Make a similar study of Figure 2–15(c) using shunt resistances of 100 Ω to represent the current sources as lossy.

E2.8. A cube is constructed by using 10-Ω resistors for its 12 edges.

(a) What is the resistance between opposite corners? (*Hint:* This is a classical and somewhat difficult problem. It may be helpful to assume a specific current to start. The concept of *symmetry* may help you decide how this current divides. Note that there are three paths leading from each corner and that each must have the same resistive network connected to it as the others and therefore have the same current. Ohm's law will help at that point.)

(b) What is the resistance between opposite corners taken on the same face? (*Hint:* a 10-Ω R is equivalent to two 20-Ω R's in parallel.)

Study Questions

2.1. What is a circuit?

2.2. How are electric charge and electric current related?

2.3. What are the resistances of perfect voltmeters and ammeters?

2.4. Make a careful comparison of voltage pushing current around a circuit and an engine pulling three cars down a track.

2.5. What are the electrical units of power and energy? How are they related to V and I? How are they related to each other?

2.6. State the *power convention* and illustrate it with a sketch.

2.7. How are KCL and KVL similar? How do they differ?

2.8. What is the routine by which simple circuit problems are solved "intuitively"?

2.9. Define carefully the circuit concepts of *parallel* and *series.*

2.10. What characteristic is common to all elements in parallel? To all elements in series?

2.11. Why is circuit analysis drill indispensable to the engineer?

ANSWERS TO DRILL PROBLEMS

D2.1*d.* 667 mA, 3.00 A, 692 mA

D2.2. 11.941 Ω, 0.2499 Ω

D2.3. -2 A, -12 V

D2.4. 7×10^{-6} MC, 7×10^{-9} GJ, 51.6×10^{3} kW

D2.5. (4) 16 W A, (5) 20 W A, (2) 4 W A, (10) 40 W G

D2.6. (4) 25 W A, (5) 20 W A, (2) 20 W A, (10) 65 W G

D2.7. (2) 98 W A, (3) 12 W A, (4) 40 W G, (10) 70 W G

D2.8. (3) 27 W A, (4) 9 W A, (2) 12 W G, (3) 9 W G, (6) 15 W G

D2.9. 4 kW

D2.10. 8 A

D2.11. 1.67 A, -1.67 A, 3.33 V, -11.67 V; 5.6 W, 19.4 W, -25 W; -5.6 W, -19.4 W, 25 W

D2.12. (VS, R, CS) 10 W A, 40 W A, 50 W G, CS is feeding into VS+

D2.13. (R, VS, CS) 2.5 W A, 7.5 W A, 10 W G

D2.14. $\cdots$ [30 V 45 W] $\cdots$ [100 V 405 W] $\cdots$ [122.5 V 750 W MAX]

D2.15. 360 C; 43.2 kJ

D2.16. See Figure 2–3

D2.17. $V = 12 - I$; Isc $= 12$ A

D2.18. 0.179 S; equivalent to 5.60 Ω

D2.19. 10.5 kW MAX; 44.7 A MAX

D2.20. (VS, R, CS) 80, 50, 30 W

D2.21. (VS, R, CS, VS) 30, 18, 12, 0 W, GAA-

D2.22. (VS, R, CS, VS) 30, 18, 12, 24 W, GAGA

CHAPTER 3

Circuit Analysis

This chapter builds on the basic circuit concepts introduced in the previous chapter, providing more general and more powerful circuit analysis tools.

We continue the development of circuit analysis in the context of dc voltages and currents. As explained in Chapter 2, this is not at all limiting; all these tools will apply to ac circuits also.

Thus the reader's objective in this chapter is:

1. Continue to develop his or her circuit analysis skills with many drill problems.
2. Become proficient with Thevenin's and Norton's theorems, mesh and nodal analysis, source substitution, dividers, superposition.

3.1   MESH ANALYSIS

At the end of the last chapter the five-element circuit in Figure 3–1a was offered as an example that could not be solved intuitively. Was it possible to find I_x by trial and error? The answer is 3 A. It's easy to see that this is correct: 3 A will produce 3 V across the 1-Ω resistance, plus on top. By KVL there will be 4 V across the 2-Ω resistance and so 2 A (left to right) through it. By KVL there will be 3 V (plus on right) across the 3-Ω resistance and so 1 A current through it. KCL is satisfied at the top node. QED! A redrawing of the circuit with the values of V and I labeled

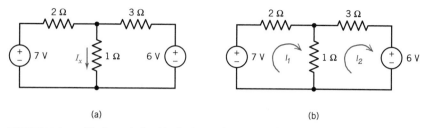

(a)   (b)

FIGURE 3–1 Mesh analysis. Algebraic *mesh currents,* I_1 and I_2 are assigned to each mesh. A KVL equation is written around each mesh.

will help make the solution clear. Suppose the 1-Ω resistance were changed to 4 Ω. What would the value of I_x be?*

Trial and error methods lead eventually to a solution, and they often provide insight. But it would be satisfying to find a more systematic procedure. *Mesh analysis* is such a method. Figure 3–1b illustrates it. Mesh currents are chosen clockwise for each mesh and labeled—in this case I_1 and I_2. A mesh current is considered to travel all the way around the mesh. Thus the current in the 2-Ω resistance is I_1 moving from left to right clockwise; the current in the 3-Ω resistance is I_2 left to right clockwise. The current through the 1-Ω resistance is made up of both I_1 and I_2, one going down and the other up. So the total current going down in the 1-Ω is $I_1 - I_2$. Similarly the total current going up is $I_2 - I_1$.

This procedure will always work, as can be easily shown with KCL and some algebra. Mesh† currents are chosen only for the open-faced loops—those with no elements cutting through them. There is also a loop all the way around the outside, but it is not an open-faced mesh because the 1-Ω resistance cuts through it.

KVL is written around each mesh, yielding two equations in two unknowns— I_1 and I_2. These simultaneous equations are solved for unknowns I_1 and I_2, and from them any current or voltage in the circuit can be quickly determined. A systematic way to do this is to write $\Sigma\ V_{\text{drops}} = 0$. Starting at the lower left-hand corner of each mesh,

$$\Sigma\ V_{\text{drops}} = 0;$$

for mesh 1:

$$-7 + 2I_1 + 1(I_1 - I_2) = 0, \tag{3.1}$$

---

*1.27 A.

†It is useful to distinguish precisely between the terms *mesh* and *loop.* A loop is defined as a closed path through all or part of a circuit. KVL can be written around any loop. A mesh is a loop which is open faced; that is, has no closed current paths inside the loop. The circuit of Figure 3–4b is a single loop which is also a mesh. The circuit of Figure 3–1a contains three loops, two of which are meshes.

The term *circuit* itself is used ambiguously by engineers. It was defined in the previous two chapters. But it also often implies a path along which current can flow in a continuous closed path. Figure 3–4b is also a circuit in that sense.

for mesh 2:

$$1(I_2 - I_1) + 3I_2 + 6 = 0.$$

Simultaneous solution* yields $I_1 = 2$ A and $I_2 = -1$ A. Thus $I_x = I_1 - I_2 = 2 - (-1) = 3$ A. This powerful method will work for practically any circuit. (To use this method, the circuit must be linear—which can be taken for the present to mean that ohmic values of resistors are fixed, independent of the current through them, or that the resulting equations are themselves linear. Almost all circuits encountered in elementary electrical engineering practice are effectively linear.) It will work for any number of meshes.

EXAMPLE 3.1 Consider the circuit of Figure 3–2a.

Find: the three mesh currents.

Solution: Write the KVL equation for each of the three meshes, collect terms into the standard form,† and have

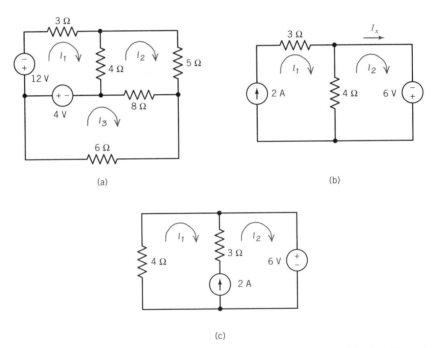

(a)

(b)

(c)

FIGURE 3–2 Mesh analysis will solve practically any linear circuit. The handling of current sources is especially interesting.

*It is assumed that the reader is familiar with these algebraic methods. Any may be used. Determinants are a particularly orderly and easy way. Appendix A gives a short refresher course in determinants.

†The standard form is as illustrated by Equation (3.2)—terms with unknowns on left, in order; terms without unknowns on right.

$$7I_1 - 4I_2 + 0I_3 = -8 \qquad (3.2\text{a})$$

$$-4I_1 + 17I_2 - 8I_3 = 0 \qquad (3.2\text{b})$$

$$0I_1 - 8I_2 + 14I_3 = -4 \qquad (3.2\text{c})$$

These three equations in three unknowns are solved simultaneously by determinants to yield the results

$$I_1 = -1.529 \text{ A}, \qquad I_2 = -0.676 \text{ A}, \qquad I_3 = -0.672 \text{ A}$$

These results should be checked with KCL at the center node, and KVL around the outside loop. The check works.

Other methods of simultaneous solution are perfectly satisfactory, particularly for two-mesh systems, which are easier numerically. For three equations or for complex number problems, the systematic method of determinants is desirable. For more than three, most computer languages (such as FORTRAN) have simultaneous equation solution routines entered from the standard form.

Experienced analysts write mesh equations in standard form directly from the circuit. Looking at Equations (3.2) above, note that the self-resistance of each mesh (R's added up around the mesh) are the coefficients of the terms on the principal diagonal of the matrix (7, 17, 14). They are all positive.

The mutual resistances between each pair of meshes (the resistance common to each of two meshes) are the coefficients of the other terms on the left ($-4, 0, -8$). A minus sign is used with each of these. (Of course with the zero the sign makes no difference.) Thus the coefficient of the I_2 term in Equation (3.2a) and of the I_1 term in Equation (3.2b) is the mutual resistance between meshes one and two, 4 Ω. Similarly, the mutual resistance between meshes one and three (0 Ω) is the coefficient of I_3 in Equation (3.2a) and I_1 in Equation (3.2c). Terms on the principal diagonal are positive. Other terms on the left are negative.

The driving terms on the right are the voltages of batteries or sources in the circuit. They are positive if the polarity of the source is such as to tend to drive the mesh current in the clockwise direction. A further assumption of this analysis is that voltage sources have zero resistance. All this can be easily demonstrated by writing KVL equations for the meshes and putting them in standard form. (Do this experimentation on a two-mesh circuit like Figure 3–1.)

Writing these equations in standard form directly from circuits is not only time saving but mistake saving.

What happens if there are current sources in the circuit? Let us try the mesh method on the circuit of Figure 3–2b. This circuit would be simple enough to solve intuitively. Can you see immediately that I_x is 0.5 A from Chapter 2 methods?

But to pursue the mesh solution: unfortunately the voltage drop across the current source is unknown, and so the usual KVL equation for mesh one cannot be written. But we can write mesh two:

$$4I_1 + 4I_2 = -6.$$

Thus we have one equation in two unknowns. Where can another be found? The object of the mesh method is to find the mesh currents. From them anything else in the circuit can be easily determined. But I_1 is known if there is a current generator on the left. I_1 must be 2 A. That's the other equation. KVL can't be written for a mesh with a current generator in it. But there is no need to write it in this case.

What if the current generator is mutual to two meshes? In Figure 3–2c neither mesh KVL equation can be written. But searching for two equations, we find around the outside loop:*

$$4I_1 + 6 = 0 \qquad \text{(KVL)}$$

and
$$I_2 - I_1 = 2 \qquad \text{(KCL)}.$$

So such a system is solvable. Regardless of where the elements are placed, mesh analysis plus common sense will always yield enough equations for a solution.

Chapter 2 showed that it can be advantageous to bend or stretch circuits into different shapes to analyze them. Figure 3–2b is really just Figure 3–2c bent around a little. The circuit is the same; no connections have been changed. Some thought along these lines before rushing into math will often make circuit problems easier to analyze. As in other engineering problems, solutions are usually undertaken tentatively (or iteratively) until the best possibilities are seen.

3.2   NODAL ANALYSIS

Nodal analysis, another general form of circuit solving, will solve almost any linear circuit. It also provides certain insights and may yield fewer equations than mesh analysis.

In a way completely analogous to mesh, nodal analysis sets up unknown voltages (instead of currents), writes KCL equations (instead of KVL), and again solves them simultaneously.

In Figure 3–3a, for example, there are three nodes that have three or more conductors coming into them, two in the top of the circuit and one at the bottom. (It might be supposed that there were two such nodes at the bottom. But there is no element between them so they count as one. Figure 3–3c illustrates this.)

Node pairs are of interest here since any voltage must occur between two points. Select any one node (the bottom in this case) as a reference node (marking it with a ground symbol, although it need not be connected to an actual ground point). Name the other two node voltages V_1 and V_2 with respect to the reference, and solve for V_1 and V_2.†

Writing KCL equations at each of these named nodes,

$$\Sigma \, I_{\text{out}} = 0;$$

---

*Often called a *supermesh.*

†Voltages at the labeled nodes are *potentials* with respect to the reference node, a concept mentioned in Chapter 2.

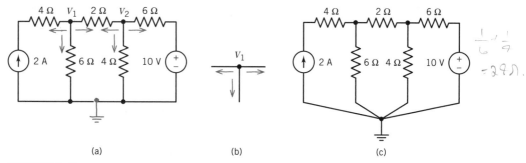

(a)   (b)   (c)

FIGURE 3–3 Nodal analysis is based on node pair voltages and KCL equations. V_1 is the voltage between node 1 and the reference node (bottom line).

at node 1:

$$-2 + \tfrac{1}{6}V_1 + \tfrac{1}{2}(V_1 - V_2) = 0,$$

at node 2:

$$\tfrac{1}{2}(V_2 - V_1) + \tfrac{1}{4}V_2 + \tfrac{1}{6}(V_2 - 10) = 0.$$

The three terms on the left of these equations will be seen to be (in order) the current to the left, the current down, and the current to the right, for each node as shown in Figure 3–3b. Solving these simultaneous equations yields $V_1 = 7.38$ V and $V_2 = 5.85$ V. Remember node voltages are with respect to the reference node. From these node voltage pairs any other current or voltage can be quickly found.

As the presence of current sources tended to complicate mesh analysis a little, so voltage sources will modify nodal analysis in a commonsense way. For example, in Figure 3–2c the presence of the 6-V source right across the node makes KCL impossible since there is no immediate way of telling the current through the source. But again, the purpose of nodal procedures is to determine node pair voltages. The voltage across the node pair in that circuit is known to be 6 V. There is no need to write an equation.*

3.3   THEVENIN'S THEOREM

More than a hundred years ago a French telephone engineer discovered the theorem which bears his name—a theorem of great usefulness to every circuit analyst. It is encountered often, and is helpful in calculation, in working with equipment, and in developing a practical understanding of many electrical situations.

Figure 3–4 illustrates Thevenin's theorem. A network (circuit) (a) is con-

---

*Sometimes a fence and its contents are called a *supernode* since regardless of the number of lines going through it, and regardless of the linear circuit within it, KCL can be applied to the sum of these penetrating currents.

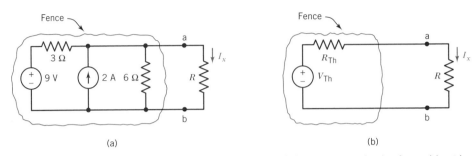

FIGURE 3–4 Thevenin's theorem replaces that part of the circuit inside the fence (a) with a Thevenin's generator—shown inside the fence of (b).

tained inside an imaginary fence or black box with two terminals, a and b, sticking out. Resistor R is connected to them. Thevenin said that the simple circuit inside the fence of Figure 3–4b (a *Thevenin generator* or *Thevenin circuit*) will produce the same current in the resistor R (or the same voltage across R) as the original complicated circuit, and it will do so for any value of R. Of course the voltage of the source, V_{Th}, and the value of the Thevenin resistance, R_{Th}, must be correctly chosen.

Stated more generally, the theorem says: Any two-terminal linear circuit (no matter how complicated) can be replaced with a single voltage source in series with a single resistance which will produce the same effects at the terminals. These terminal effects can be described as the "*V-I* characteristic" at the terminals.

In circuits like Figure 3–4a engineers often speak of the resistance R as a load being driven by the network inside the fence. The V_{Th} to use in the Thevenin equivalent circuit is the *open-circuit* voltage on terminals a, b of Figure 3–4a. "Open circuit" means with nothing connected, as in Figure 3–5a. To find this voltage V_{OC}

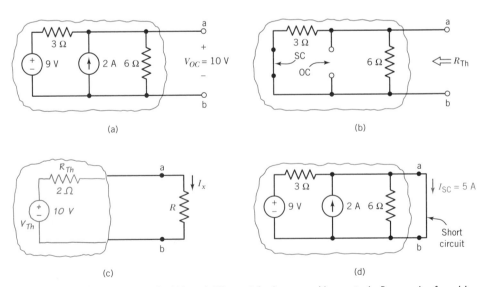

FIGURE 3–5 Solving circuit 3–4(a) with Thevenin's theorem. Alternatively R_{Th} can be found by dividing I_{sc} into V_{oc}.

will take some circuit analysis in itself. (In laboratory work it can be measured instead of calculated.) In this simple case it can be seen as 10 V by summing voltage rises from the bottom terminal to the top.

The Thevenin resistance R_{Th} is the resistance seen looking back into the terminals a, b with sources killed. "Killing" the two sources in Figure 3–4a means replacing the voltage source with a *short circuit* (zero volts) and any current sources with an *open circuit* (zero amperes). This replacement is shown in Figure 3–5b. A little mental (or pencil) twisting of 3–5b shows that the 3- and 6-Ω resistances are in parallel and R_{Th} is 2 Ω.

Another interesting method of finding R_{Th} (and an essential with the dependent sources of Section 3.10) is from the relation $R_{Th} = V_{OC}/I_{SC}$. In Figure 3–5a replace the open circuit with a short circuit as in sketch 3–5d. Solve for I_{SC}. It will be seen here by inspection that $I_{SC} = 5$ A; (V must equal 0 across a short circuit; so by Ohm's law no current flows through the 6-Ω resistor; again by Ohm's law and KVL the I flowing out of the voltage source is 9 V/3 Ω or 3 A; by KCL $2 + 3 = 5$); and $V_{OC}/I_{SC} = R_{Th} = 10/5 = 2$ Ω, agreeing with the previous solution of sketch (c). Thus circuit Figure 3–5c is the result of this particular Thevenin transformation looking into terminals a $-$ b.

EXAMPLE 3.2 Consider the circuit of Figure 3–6a.

Find
1. Find the current I_x through resistor R by using Thevenin's theorem. Assume first that R is 1 Ω.
2. Plot this current for different values of R.
3. Sketch the *V-I* characteristic for the network that drives the center resistor R.

Solution: Remove R from the circuit. Identify terminals as a, b. Enclose circuit in a fence with a, b sticking out. The result is as shown in 3–6b. But to get this circuit into a more standard form (power or signals moving from left to right) redraw it as in 3–6c. (For those who may still be having trouble twisting circuits about like this, perhaps the step-by-step procedure of Figure 3–7 will help. Nothing in the circuit has been changed in progressing from any of these circuits to the next.)

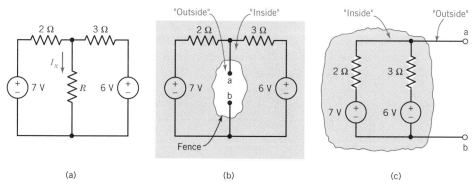

(a) (b) (c)

FIGURE 3–6 Starting a Thevenin's analysis for I_x in (a).

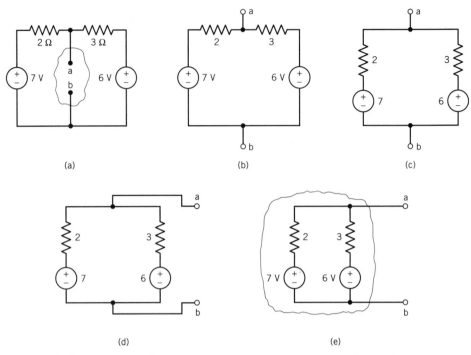

FIGURE 3–7 Example of redrawing a circuit to make it easier to work with, based on Figure 3–6(a).

V_{Th} is the open-circuit voltage on a, b. Finding V_{Th} requires analysis of 3–6c. Note that with open circuit at a, b there will be a clockwise current of V/R A or $(7-6)$ V/$(2 + 3)$ Ω = 0.2 A. Thus the drop on the 2-Ω R is 0.4 V and V_{ab} is $7 - 0.4$ or $\underline{6.6\ V}$, the open-circuit voltage V_{Th}.

Killing the two voltage sources by replacing them with short circuits as in Figure 3–8a gives the result $R_{Th} = 1.2$ Ω. The final resulting equivalent circuit with load R is 3–8b. If R is 1 Ω, $I_x = V/R$ (total) = $6.6/(1.2 + 1)$ = $6.6/2.2 = 3$ A. This is the same problem (Figure 3–1a) solved earlier by mesh methods. It has of course the same result.

If resistor R is to take on different values, how much easier it is to solve

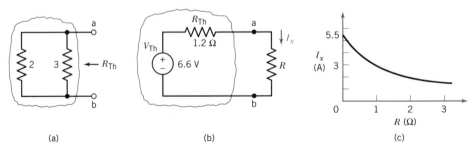

FIGURE 3–8 Thevenin's method continued.

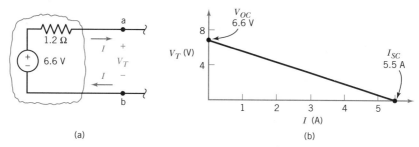

FIGURE 3-9 The *V-I* characteristic of a two-terminal linear circuit developed with Thevenin's theorem.

this problem from the Thevenin equivalent circuit of 3–8b than to start all over again with 3–6a. The plot for the second answer is in 3–8c. (It would be desirable to check the values shown.)

The third answer, the *V-I* characteristic, is particularly interesting. Again it is clearly easier to work from the Thevenin circuit. It would be possible to utilize the results of the I_x vs R plot, but instead, for tutorial purposes, reason directly from the circuit 3–9a. If no current is drawn from the Thevenin generator (combination of V_{Th} and R_{Th}) all the voltage will appear on the terminals. This produces a point on the curve (3–9b) at 6.6 V and 0 A.

If a *short circuit* (a wire) is placed across the output terminals a and b, the voltage will be zero and the current 6.6 V/1.2 Ω = 5.5 A, another point on the graph. Since the circuit is linear it is only necessary to connect these two points for the whole curve. Thus 3–9b is a straight line. (Again checks at a few intermediate points would be instructive.) The slope of this curve (with a negative sign) gives the internal resistance R_{Th} of the source in the fence.

In electronics and electric power analyses these ideas will occur over and over. They apply just as well to ac circuits as to dc. The Thevenin resistance becomes an impedance, and the voltage source must be an alternating source.

3.4　NORTON'S THEOREM

In the 1920s Norton, an American telephone engineer, announced a dual of Thevenin's theorem.*

The Norton equivalent generator is shown in Figure 3–10a, and in 10b specifically for the circuit of 3–6a. It consists of a current source in parallel with a resistor; R_n is the same as R_{Th} and found in the same way, and I_n is the short-circuit current through a short across terminals a, b.

---

*There is a systematic duality possible in circuit analysis which replaces current sources with voltage sources, voltage with current, mesh equations with nodal equations, etc.

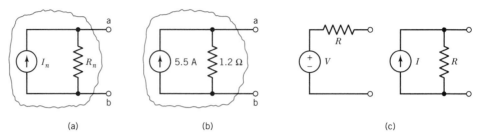

(a)   (b)   (c)

FIGURE 3–10 Norton's equivalent circuit. Sketch (c) defines *lossy* voltage and current sources.

Since both Figures 3–9a and 3–10b offer themselves as equivalents of the same circuit, 3–6a, they must be equivalent to each other; they must have the same *V-I* characteristic. Readers are asked to satisfy themselves that when a load resistor *R* is connected to terminals a, b of 3–10b, the graphs of 3–8c and 3–9b will again result. The next section will generalize this substitution.

It is useful to notice that these equivalent circuits are equivalent *outside* the fences (or black boxes) in which they are drawn (or contained). They are not at all the same inside. Inside the fence many things are different about them, including power relations. It is in their terminal characteristics *outside* that they are equivalent.

3.5   LOSSY VOLTAGE AND CURRENT SOURCES; SOURCE TRANSFORMATION

The above example (Figures 3–6, 3–8, 3–9, 3–10) can be usefully generalized as follows.

Lossy sources, both voltage and current, are defined by the circuits of Figure 3–10c. Each is made up of an ideal source and a resistor. These are of course nothing but the equivalent circuits of M. Thevenin and Mr. Norton.

An ideal voltage source has zero resistance (zero R_{Th}). An ideal current source has infinite resistance (infinite R_n). But in reality there is no such thing as an ideal current or voltage source. All batteries, for example, have some internal resistance. So do transistors. So there will always be some resistance connected with any source—a resistance which is neither zero nor infinite.

In circuits where the series resistance is much greater than R_{Th}, it is generally safe to model a source without that resistance. Similarly, where parallel resistance is much smaller than R_n, a circuit can be modeled without an R_n in the Norton equivalent. Now it is clear why the circuits of Figures 2-15b and c are not allowed. And, happily, they will never come up in practice. When an engineer runs into this problem in modeling she need only put the value of resistance R in, turning the model into one using lossy sources. Then everything will work out.

Source transformation is useful in solving circuit problems. For instance in Figure 3–5a the 9-V source and 3-Ω resistance can be considered a lossy voltage source and be replaced by a lossy current source consisting of a 3-A source paralleled with a 3-Ω resistor. This is done in Figure 3–11a through f to obtain another form of solution. In source transformation the following relations hold:

$$R_{\text{Th}} = R_n = R,$$
$$V_{\text{Th}} = RI_n, \tag{3.3}$$
$$I_n = V_{\text{Th}}/R.$$

Fences should always be used to separate the imaginary world of circuit analysis from the real-world characteristics of the circuit. They remind us to be careful in drawing conclusions about the circuit (before transformation) from anything inside the fence or black box (after transformation). Within the fence, information is lost about the real circuit.

In Figure 3–11a the voltage source to be transformed is carefully fenced in and its terminals just outside identified as a, b. Figure 3–11b shows the transformed circuit inside the fence and its values according to Equation (3.3). Next redraw the

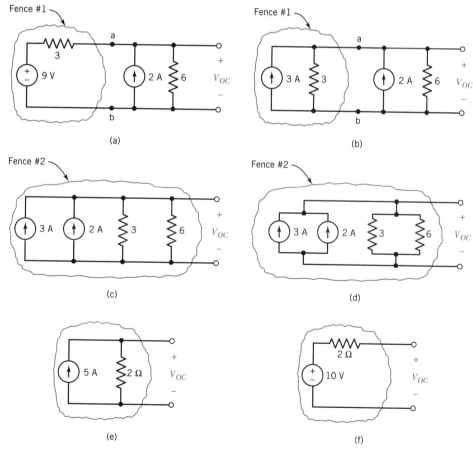

FIGURE 3–11 Reducing a two-terminal circuit (Fig. 3–5a) to Thevenin's or Norton's equivalent by successive transformation.

circuit (Figures 3–11c and d) to see the parallelism of sources and resistors. Current sources in parallel add (KCL) and resistances in parallel combine as product over sum yielding the final result of Figure 3–11e. Note that two different fences were used—one for the source transformation and one to bound the final Thevenin equivalent.

In these powerful transformations one begins to see some of the possibilities of circuit analysis.

3.6 SUPERPOSITION

The *superposition theorem* sheds considerable light on the operation of circuits. It tells us that for any linear circuit the effect of each source (whether current or voltage) can be calculated independently and the total effects simply added.

Suppose it is desired, for example, to find I_x in Figure 3–2b. Replace all sources with their equivalent resistances (zero for voltage sources and infinite for current sources). One at a time return each source to the circuit and see what component of I_x it alone would produce. The actual I_x is the sum of these components. It will be helpful to sketch these circuits as they are used below.

First replace the voltage source with a short circuit (zero resistance). Then I_x due to the current source will be 2 A. The two amperes coming out of the right end of the 3-Ω resistance will all choose to go through the short circuit replacing the 6-V source. With no voltage across the short circuit there can be no voltage across the 4-Ω resistor and hence no current through it. Alternatively, this result comes most easily from the current divider theorem discussed in Section 3.7 below.

Next replace the current source with an open circuit (infinite resistance). The voltage source now produces an I_x of -1.5 A (Ohm's law). Thus the total I_x is $2 - 1.5$ or 0.5 A, agreeing with the previous solutions for this circuit.

Until the beginning analyst is quite familiar with circuits these last two paragraphs will be meaningless unless circuit sketches are drawn for the circuit of Figure 3–2b, first with the voltage source replaced with a short circuit, then with the current source simply removed and an open circuit left in its place. More examples of superposition will be found in the end of chapter problems.

In signal circuits it is quite common to deal with one part of a signal at a time, using this theorem as justification. Many electronic circuits have both dc supply voltages and currents and ac signal voltages and currents mixed together. Superposition thinking is most helpful in those cases. (Chapter 15 has examples.)

3.7 VOLTAGE DIVIDERS AND CURRENT DIVIDERS

Two simple but useful ideas, voltage and current dividers, are encountered again and again in electrical work. In Figure 3–12a suppose the source voltage is 100 V. By the circuit scheme shown one can divide off any voltage desired (V_{out}), less than 100 V, by properly adjusting R_1 and R_2. Similarly the current divider of Figure

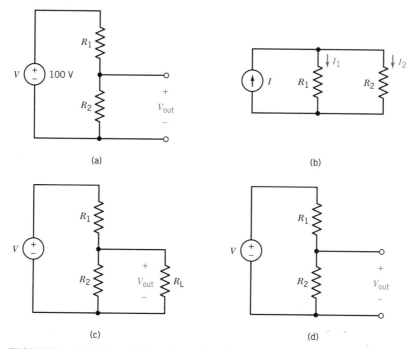

FIGURE 3–12 *Voltage Dividers* (a, c, d) and a *current divider* (b). These standard circuits appear as parts of many larger circuits.

3–12b divides the source current I between the two resistors. A little analysis shows the relations for the voltage divider circuit:

$$V_{out} = VR_2/(R_1 + R_2), \qquad (3.4)$$

and for the current divider:

$$I_2 = IR_1/(R_1 + R_2),$$
$$I_1 = IR_2/(R_1 + R_2). \qquad (3.5)$$

The resistance fractions $R_2/(R_1 + R_2)$, etc., are called the voltage divider or current divider *ratios*. If there is a load resistor R_L connected to the voltage divider, as in Figure 3–12c, formula (3.4) will not work. But by simply combining R_L and R_2 in parallel to find a new R_2' the difficulty is overcome.

EXAMPLE 3.3 In the voltage divider of Figure 3–12a, R_1 is 30 Ω and R_2 is 20 Ω.

Find

1. the voltage out;
2. a 60-Ω load is now placed across the output; what is the new V_{out}?

Solution: 1. $V_{\text{out}} = VR_2/(R_1 + R_2) = 100(20/50) = \underline{40 \text{ V}}$. For the loaded divider in part 2, R_2' (the modified R_2) is the parallel combination of 60 and 20 or 15 Ω. Thus $V_{\text{out}} = V$ times the voltage divider ratio, here $100(15/45)$ = $\underline{33.3 \text{ V}}$.

Alternatively a Thevenin's solution is especially informative here. V_{OC} is already found to be 40 V. R_{Th} is seen to be R_1 and R_2 in parallel or 12 Ω. Thus a 60-Ω load would draw $V/(12 + 60) = 556$ mA and the load voltage would be $IR = 0.556 \times 60 = 33.3$ V.   ▨

3.8   INVERSE RESISTANCES—CONDUCTANCES

If *conductance G* is defined as the reciprocal of resistance $(1/R)$, Ohm's law becomes

$$I = VG. \tag{3.6}$$

Just as the unit of resistance is the ohm, the S.I. Unit (System International) of conductance is the siemens. The word siemens was selected to honor the British Engineer William Siemens and is thus a singular noun.

Now resistors in parallel can be combined by simply adding G's, an easier procedure by far than the reciprocal of the sum of the reciprocals or product over the sum. (This can be made clear by sketching a circuit with two or three resistors in parallel across a voltage source and finding all the currents involved.) A mho or siemens is the conductance of a resistor across which one volt will produce one ampere—I/V, the reciprocal of the ohm definition. Now either resistances or conductances can be thought of as the same circuit element. Analysts will find this kind of inverse unit especially useful in parallel circuits and nodal analysis. But while convenient, the conductance concept is not essential to circuit analysis.

3.9   NONLINEAR RESISTORS AND DIODES

The *V-I* characteristic of a model (linear) resistance is a straight line through the origin as in Figure 2-4b. While real resistors do not function (in their *V-I* characteristic) exactly as model resistances, most are so close that any deviation can be neglected. But if large enough this discrepancy must be dealt with.

Temperature effects cause much nonlinearity. Most materials change resistance with temperature, that of metals increasing as temperature goes up, but that of most nonmetals decreasing. Figure 3–13 shows both types. The typical tungsten lamp rises to a high temperature when incandescent. As voltage increases, more power is dissipated, raising the filament temperature. Further increments of voltage produce smaller increments of current—the resistance is increasing.

Figure 3–13 shows the opposite effect with the thermistor, which is made of a nonmetal, silicon carbide. Additional increments of voltage produce larger increments of current. Thermistors are small resistors useful in control and signal

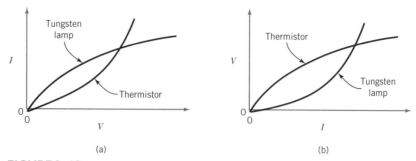

FIGURE 3–13 Temperature effects produce two kinds of circuit nonlinearities. They may be displayed in either of two ways.

applications because of their marked and predictable nonlinearity. They make good thermometers.

Diodes are another common nonlinear element, but do not depend on thermal effects. They are two-terminal, electrical, switchlike devices, symbolized in Figure 3–14a. The diode conducts current in one direction (the direction of the arrow part of the symbol) but not in the other. An ideal diode is a short circuit in the forward or conducting direction. It is an open circuit for current attempting to go the other way. The sketch 3–14b illustrates this.

More information on the diode is presented in Chapter 16. But as a tool to sharpen circuit analysis skills, consider the actual (nonideal) V-I relationship of the typical diode as shown in Figure 3–14c. For a germanium diode it can be expressed as

$$I = I_0(e^{V/a} - 1) \tag{3.7}$$

where a is 0.026 V. For silicon, another common diode material, a is about 0.052 V.

In circuit 3–14c, I_0 has been selected to be one microampere. For larger diodes I_0 will be correspondingly greater, simply changing the scale on the y axis,

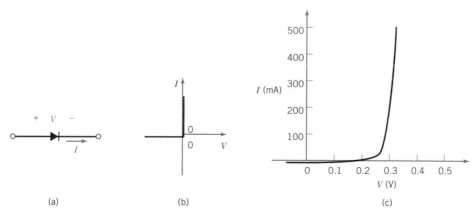

FIGURE 3–14 Diode symbol, and V-I characteristics for ideal (b) and practical (c) diodes.

with the curve remaining the same. Thus for voltages of more than a few volts the ideal on-off characteristic (Figure 3–14b) is a good approximation. The other common diode material is silicon with the on voltage being taken as about 0.6 V instead of 0.3.

A common analytical procedure when there is one nonlinear device in a circuit is the *load-line* technique. The following example will illustrate it.

EXAMPLE 3.4 Given the circuit of Figure 3–15a,

Find: current I_x; the diode's characteristics are those of Equation (3.7) or Figure 3–14c.

Solution: Two equations for I in terms of V are available: Equation (3.7), relating V_d to I, and another relating V_R to I. These two V's are then related by KVL to 0.7 V as

$$0.7 = RI + V_d,$$

and

$$I = I_0(e^{V_d/0.026} - 1),$$

or for this case,

$$I = 10^{-6}(e^{V_d/0.026} - 1).$$

KVL is always applicable, even in a nonlinear circuit. Unfortunately there is no mathematical solution to these equations since one is transcendental. However, a graphical solution can be had by plotting V/I for both resistor and diode, and finding their intersection. As shown in 3–15b, the plot for R is a straight line drawn between the supply voltage, 0.7 V, and the current that would flow through the 2-Ω resistor if all the supply voltage were across it, 350 mA. This is called a *load line*. The voltage across the resistor, V_R, for any current can be found on the graph by renumbering the abscissa as shown.

As one moves back and forth on the x axis, the sum of V_R and V_d is always equal to the supply voltage, thus fulfilling KVL. KCL requires that in this series

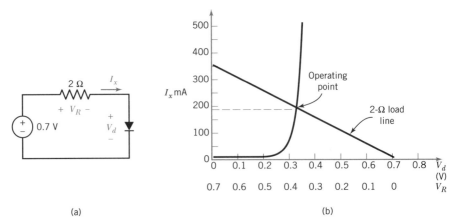

(a)   (b)

FIGURE 3–15 Series diode circuit and *load-line* solution to find current.

circuit the current be the same for both elements. This can only happen at the intersection of the two curves. Hence I_x (from the graph) is about 180 mA. Trimming this figure up with trial and error in Equation (3.7) finds I_x close to 190 mA and V_d about 0.316 V. ▬

The nonlinear curve used in the load-line method can be developed from a mathematical expression as in this example. Or it can be determined experimentally. The circuit of Figure 3–15a can be considered a Thevenin equivalent supplying energy to the nonlinear device. Thus the load-line technique will be easily applicable to any circuit with a single nonlinear device.

In circuit analysis, as in any other engineering work, it is essential to observe carefully the physical system being analyzed. Look carefully at the circuit. The amateur tends to think of formulas or math. The engineer thinks physically of what is going on in the circuit. In the example just completed, the novice's attention is focused on Equation (3.7) and the graphical procedures. The engineer's attention is on the circuit and the way the supply voltage is shared between the resistor and the diode.

3.10   DEPENDENT SOURCES

There is a special form of the current and voltage sources, which were first encountered in Figure 2–5a, called *dependent sources* and illustrated in the circuit of Figure 3–16. Here the two voltage sources are simply ordinary sources because their values, 2 V and 20 V, are specified as constants. But the current source is a dependent source because its value depends on another current or voltage value in the circuit. In this case it depends on the current I_E as shown. So the current through that source, which is I_C, will always be 0.98 I_E. It is possible for either voltage or current sources to be dependent.

To solve circuit 3–16 implicitly for the three currents, first observe that I_E must be 20 mA since there are 2 V across a 100-Ω resistor. Then the dependent source makes $I_C = 0.98I_E = 0.98 \times 20 = 19.6$ mA. Then using KCL at the junction, $I_B = I_E - I_C = 20 - 19.6 = 0.4$ mA.

As in this illustration, any analysis method can be used in a circuit with dependent sources. They simply add a little algebraic complexity to most solutions. But particular caution must be observed with Thevenin's theorem (or Norton's).

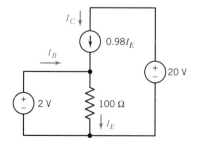

FIGURE 3–16 *Dependent sources such as the current source have values that depend on some other circuit current or voltage. Analysis is like that of other current or voltage sources except that their internal resistance is no longer zero for voltage sources nor infinite for current sources.*

The method of looking into the open-circuited terminals with sources killed to find the Thevenin's resistance will no longer work. A dependent voltage source does not necessarily have a zero resistance, nor a dependent current source an infinite resistance. While interesting theoretically, it is often not very productive to apply Thevenin's theorem to these circuits. If it is to be attempted, R_{Th} is found as the ratio of the OC voltage to the SC current on the terminals involved.

Dependent sources are important in the analysis of transistor circuits, and more will be seen of them in Chapter 15.

3.11   CIRCUIT ANALYSIS DRILL

Heavy and thoughtful drill on the methods of this chapter with dc circuits will prepare the reader for ready handling of the ac circuits to follow. There are many drill problems provided at the end of the chapter.

A useful and interesting exercise is to solve a particular problem by as many of this chapter's methods as possible. The following example carries out this procedure.

EXAMPLE 3.5 In the circuit of Figure 3–17a,

   Find: V_x by a number of different analysis methods.

   Solution: *First* by nodal analysis:

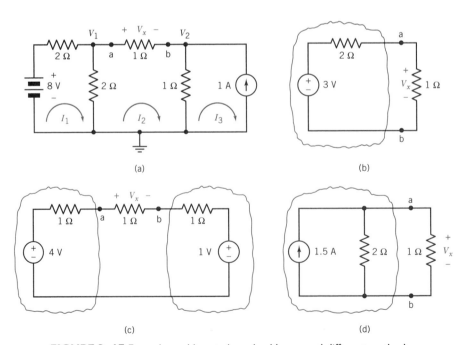

FIGURE 3–17 Example problem to be solved by several different methods.

Two unknown voltages, V_1 and V_2, have already been designated in Figure 3–17a along with a reference node.

At node 1:

$$\tfrac{1}{2}(V_1 - 8) + \tfrac{1}{2}V_1 + \tfrac{1}{1}(V_1 - V_2) = 0.$$

At node 2:

$$\tfrac{1}{1}(V_2 - V_1) + \tfrac{1}{1}V_2 = 1 = 0$$

or
$$2V_1 - V_2 = 4,$$

$$-V_1 + 2V_2 = 1,$$

so
$$3V_2 = 6, \qquad V_2 = 2,$$

$$V_1 = 2V_2 - 1 = 3$$

and
$$V_x = V_1 - V_2 = \underline{1\ \text{V}}.$$

Second by mesh analysis:

Three unknown mesh currents have been set up already in Figure 3–17a. Thus

$$4I_1 - 2I_2 - 0I_3 = 8,$$

$$-2I_1 + 4I_2 - I_3 = 0,$$

$$I_3 = -1,$$

so
$$6I_2 = 6, \qquad I_2 = 1$$

and
$$V_x = 1I_2 = \underline{1\ \text{V}}.$$

Third by Thevenin's theorem:

V_x appears across a 1-Ω resistor. Remove that resistor from the circuit leaving terminals a, b as indicated in Figure 3–17a. The open-circuit voltage there will be 3 V with plus at terminal a [4 V contributed by circuit on left (use voltage division) and -1 V contributed by portion of circuit on right]. Looking back into the circuit from terminals a, b, R_{Th} will be 2 and 2 in parallel plus 1 in series or 2 Ω. These results are as indicated in Figure 3–17b. By voltage division $V_x = 3 \times \tfrac{1}{3} = \underline{1\ \text{V}}$. Beginning analysts may need to make several intermediate sketches here.

Fourth by source transformation:

Figure 3–17c illustrates this method. The portion of the circuit to the left can be transformed twice, and the portion on the right once for the result shown. The result, $V_x = \underline{1\ \text{V}}$, is then obtained directly by adding the sources and using voltage division.

Fifth by Norton's theorem:

The short-circuit current between terminals a and b can be solved to be 1.5 A from a to b. (Perhaps this is most easily seen in Figure 3–17c, but any analysis method can be used to find it.) The R_n is the same as R_{Th} or 2 Ω. Thus Figure 3–17d represents the Norton generator connected to the 1-Ω load in question and V_x = 1.5 A × 0.67 Ω (the parallel combination) or 1 V.

Sixth by successive transformations:

Note that either the Thevenin or Norton final circuit could have been derived from successive transformations (source transformations, series and parallel combinations of R's, series combination of voltage sources, parallel combinations of current sources). This method obtains Figure 3–17c and could then carry it on to either 3–17b or d.

Which was the best method?

3.12   SUMMARY

Circuit analysis consists of determining currents and voltages everywhere in a circuit. There are many tools applicable. Mesh analysis (based on KVL) and nodal analysis (on KCL) are nearly universal methods.

Thevenin's and Norton's theorems are powerful methods for reducing any linear two-terminal network to a source and one resistor. They help the analysis outside the two terminals involved. They also assist in intuitive appreciation of the effect of loading on power or signal sources.

Practical (lossy) voltage sources can be interchanged by source substitution methods with current sources (or vice versa), which often simplifies a circuit. Voltage and current divider networks of resistors (in ac, of impedances) are encountered everywhere. Inverse resistances (conductances) facilitate understanding and simplify parallel circuit work.

For nonlinear devices, such as lamps and diodes, special analytic care is required, and special techniques such as load-line analysis are available.

FOR FURTHER STUDY

Ralph J. Smith and Richard C. Dorf, *Circuits, Devices and Systems*, 5th ed., John Wiley & Sons, New York, 1992. Chapter 2.

Easy Drill Problems (answers at end of chapter)

D3.1. For the circuit of Figure D3.1, find I_x by mesh methods.

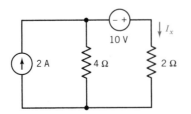

D3.2. Repeat using nodal methods.

D3.3. Repeat using Thevenin's theorem.

D3.4. Repeat using Norton's theorem.

D3.5. Repeat using superposition.

D3.6. Repeat using circuit reduction and simplification (apply source substitution, for instance).

D3.7. For the circuit of Figure D3.7 find V_x using mesh methods.

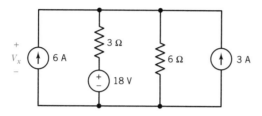

D3.8. Repeat using nodal methods.

D3.9. Repeat using Thevenin's theorem.

D3.10. Repeat using Norton's theorem.

D3.11. Repeat using superposition.

D3.12. Repeat using circuit reduction and simplification.

D3.13. A voltage divider is used to reduce the voltage from a 50-V source. R_1 and R_2 each equal 500 Ω. Find the voltage supplied to loads of 1 million, 100,000, 10,000, 1000, 100, 10, and 1 Ω. Is Thevenin's theorem of any practical use here?

D3.14. Four resistors have G's of 1000, 2000, 3000, and 4000 μS. Find the resistance and conductance of
  (a) the four in series;
  (b) the four in parallel;
  (c) the series combination of 1000 and 2000 in parallel, and 3000 and 4000 in parallel.

D3.15. In Figure D3.7 reverse the polarity of the 18-V source. Find V_x by mesh methods.

D3.16. Repeat using nodal methods.

D3.17. Repeat using Thevenin's theorem.

D3.18. Repeat using Norton's theorem.

D3.19. Repeat using superposition.

D3.20. Repeat using circuit reduction and simplification.

D3.21. A two-terminal circuit has an open-circuit voltage of 17 V, and when a 12-Ω resistor is connected to the terminals a voltage of 8.5 V. Sketch a lossy voltage source with values that will properly represent this circuit. Don't forget the "fence."

D3.22. Repeat problem D3.21 if the voltage across the 12-Ω load is 3.5 V.

D3.23. Assume that in the circuit of Figure 3–12b R_1 is 12 Ω and R_2 is 8 Ω. Let the source current I vary from 0 to 100 A in steps of 20 A. Make a tabulation of I, I_1, I_2.

D3.24. Assume that in the circuit of Figure 3–12a R_1 is 50 Ω and R_2 75 Ω. The ideal source voltage V varies from 0 to 100 V in steps of 20 V. Make a tabulation of V_{out} vs V.

D3.25. Repeat D3.24 if a 50-Ω load R_L is connected as in Figure 3–12c.

Application Problems

P3.1. You have a large box of 1000-Ω resistors and a 50-V source. You desire to provide a dc voltage of 38 V accurate to within 3 V for an experiment. Your application will draw negligible current. Devise a circuit to accomplish this using as few resistors as possible.

P3.2. As shown, a Norton lossy source drives a 1000-Ω load by means of a voltage divider circuit. It is desired to have 5 mA flow through the load. If R_1 is 300 Ω, what value should R_2 have in order for 5 mA to flow through the load? (*Hint:* Consider source substitution for simplification here.)

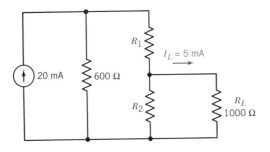

P3.3. Suppose in the preceding problem that R_1 and R_2 can be assigned any values you wish.
 (a) What is the smallest value that R_2 can have?
 (b) What is the largest value that R_1 can be given?
 (c) Allowing R_1 and R_2 to take on any values you wish, sketch a plot of R_1 as a function of R_2 if the load is to have 5 mA. (*Note:* A programmable calculator or computer terminal will be useful here.)

P3.4. For the circuit shown it is desired to have 3 V output with no appreciable load current.
 (a) What is the relation between R_1 and R_2?
 (b) Find values of R_1 and R_2 such that the resistance looking back into terminals x, y is 1000 Ω.

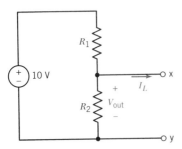

P3.5. Suppose the circuit of problem P3.4 is to be loaded so that I_L can vary from 0 to 1 mA. Select values for R_1 and R_2 so that the load voltage varies from 3.00 to 2.85 V as I_L varies over its range.

P3.6. Develop the simplest possible formula to find the numerical answers to problem D3.13.

P3.7. In the circuit shown the resistance of a nonlinear device is specified as a function of current. Plot this information carefully. Sketch a plot of I_x vs V_g.

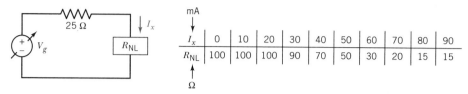

mA										
I_x	0	10	20	30	40	50	60	70	80	90
R_{NL}	100	100	100	90	70	50	30	20	15	15
Ω										

P3.8. Solve this circuit for I_x by **(a)** mesh and **(b)** nodal analysis.

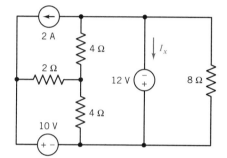

P3.9. Solve the circuit of problem P3.8 by either Thevenin's or Norton's theorems.

P3.10. **(a)** Considering the analysis of the preceding two problems, can you make any generalizations on the effect of a resistor (the 8-Ω R) which is directly across a perfect voltage source (the 12-V source)? How do you recommend this situation be best handled in circuit analysis?

(b) The dual of this situation is a resistor in series with a perfect current source. Devise a relatively simple circuit that will illustrate this situation. Can you generalize on the impact of this configuration on circuit analysis? Is there an easy way to handle it?

 P3.11. One section of an interesting four-terminal "ladder line" circuit can be devised by combining two resistors as sketched. This idea is used in modeling communication

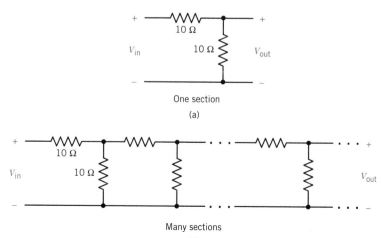

One section

(a)

Many sections

(b)

lines. For long lines as many sections can be connected (output to the next input) as desired. Using all 10-Ω resistors:
(a) Find the resistances looking into the input terminals of such lines composed of 1, 2, 3, or 4 sections.
(b) What resistance is seen for an infinite number of sections? First make a careful guess. Then invent a way to solve. (*Hint:* If you take one section off the front of an infinite line, will that change the resistance seen?)

P3.12. In this sketch find the current I_x:
(a) by Thevenin's, and
(b) intuitively.

[*Hint:* omit the 10-Ω *R* when starting part (b).]

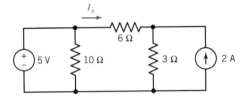

Extension Problems

E3.1. In the *half-wave rectifier* circuit sketch, a sinusoidal source voltage $v_s(t)$ drives a load R_L through an ideal diode. The source voltage peaks at plus and minus 20 V. Sketch the load voltage $v_L(t)$.

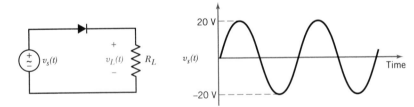

E3.2. The same circuit and source voltage are used as in the previous problem but a 10-V dc source is added in series. (See sketch.) Sketch the load voltage again as a function of time.

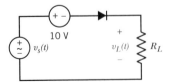

E3.3. Reverse the dc source of the previous problem and repeat.

E3.4. For the *clipper circuit* sketched assume sinusoidal voltage $v_s(t)$ peaks at plus and minus 10 V. Sketch $v_L(t)$ as a function of time.

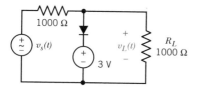

E3.5. (a) Reverse the diode of the previous problem and repeat.

    (b) Sketch the output waveform if the battery is removed and the diode connected directly to ground (the bottom line).

E3.6. Due to the temperature coefficient of resistance and the resulting nonlinearity of tungsten lamps, there is typically, when they are turned on, a surge of current several times as great as rated. This phenomenon lasts for only a few hundredths of a second, after which they have reached operating temperature. But it is large enough to cause trouble in some systems. Investigate this effect numerically. How much larger do you find the surge current to be? The temperature coefficient of resistance for tungsten is 0.0045/°C. The filaments run at more than 1000°C.

E3.7. Circuit *a* shows a three-terminal transistor (previously discussed in problem E2.1). With fixed base current of 30 μA the *V-I* characteristic of the output terminals (V_{CE} and I_C) are nonlinear as shown in *b*. The exact shape of the curve is tabulated as in *c*. You are to operate the transistor with 20-V supply V_{CO} connected to the output circuit through a load resistor of 5000 Ω as shown in *d*. Find the collector current and V_{CE}.

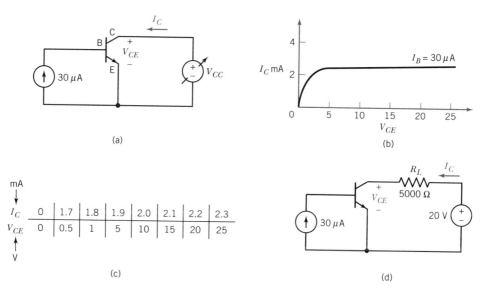

(a)

(b)

mA ↓								
I_C	0	1.7	1.8	1.9	2.0	2.1	2.2	2.3
V_{CE}	0	0.5	1	5	10	15	20	25
↑ V								

(c)

(d)

E3.8. Devise a means and solve problem P3.7 by the load-line method. Assume V_g = 2 V. (*Hint:* Modify the nonlinear table to include voltage.) Do you check your earlier work?

E3.9. Devise a means and solve problem E3.7 by the load line method. Do you check your earlier work?

E3.10. Let the left-hand circuit of Figure 3–10(c) represent a dc generator (mechanical power in and electrical power out). Generated voltage V is 120 V, internal resistance R is 2 Ω. (Nonelectricals are hereby authorized to visualize a mechanical machine shaft coming out of the center of the Thevenin's voltage source of the figure. See the "armature circuit" of Figure 20–6a.) Use this dc machine to drive a 15-Ω lighting load and find:

(a) the load current and
(b) the load voltage.

E3.11. Continuing problem E3.10, look into the power and find:
(a) electrical power delivered to the load,
(b) mechanical power in, (allow 90 W for mechanical losses),
(c) power supplied by the generator, and
(d) efficiency of power conversion from shaft input to lights.

E3.12. In this circuit, find V_x, I_1, I_2.

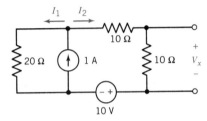

Study Questions

3.1. Mesh and nodal analyses are based on which laws?

3.2. Describe a means of telling how many mesh equations will be required for a given circuit. Repeat for nodal equations.

3.3. Describe how to determine the signs of all terms on the left-hand side of the standard circuit equations in mesh analysis.

3.4. What determines the signs of the driving terms?

3.5. Suggest two means for handling current sources in mesh analysis.

3.6. How are voltage sources handled in nodal analysis?

3.7. What is a "two-terminal V-I characteristic?"

3.8. Why is Thevenin's theorem considered so powerful?

3.9. Show why superposition will not work if there is a nonlinear element in the circuit being analyzed.

3.10. Devise a method for handling a circuit that has two nonlinear devices in series.

ANSWERS TO DRILL PROBLEMS

D3.1.–D3.6. 3 A

D3.7.–D3.12. 30 V

D3.13. 24.99, 24.94, 24.39, 20.00, 7.14, 0.96, 0.10 V

D3.14. 2083 Ω (480 μS), (10,000) 100, 476 (2101)

D3.15.–D3.20. 6 V

D3.21. $V = 17$ V; $R = 12$ Ω

D3.22. $V = 17$ V; $R = 46.3$ Ω

D3.23. $I_1 = 0.4$ I; $I_2 = 0.6$ I

D3.24. $V_{out} = 0.6$ V

D3.25. $V_{out} = 0.375$ V

CHAPTER 4

Alternating-Current Circuit Analysis

Almost all electrical systems, either signal or power, operate with alternating currents and voltages. Fortunately circuit analysis for ac is essentially the same as that presented in Chapters 2 and 3 for dc. It is necessary only to develop some understanding of ac currents and voltages (and of the complex arithmetic by which they are calculated) to be able to apply this same circuit knowledge to them.

Before starting, it might be helpful to look at a typical, simple ac problem to see how much like dc it really is, and to see what new ideas will have to be developed to apply Chapter 3 techniques to ac.

Consider the circuits of Figures 4–1a and 1b. Figure 4–1a is a series circuit with an ac voltage source and three passive elements in series. Besides the resistor R there are two new elements: C is a *capacitor* and L an *inductor*. Figure 4–1b is a similar dc circuit with the same configuration. All three passive elements here are resistors.

From Chapter 2 basics it is possible to come immediately to the solution for 4–1b: $I = V/(R_1 + R_2 + R_3) = 12/11 = 1.09$ A. The 12-V dc driving voltage produces a 1.09-A dc current.

The solution for circuit 1a is almost the same:

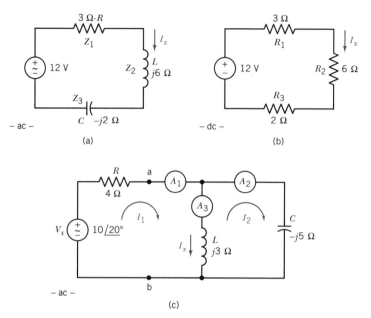

FIGURE 4-1 Steady-state ac analysis (a) is identical to dc (b) except for the use of complex arithmetic—including the j operator.

$$\mathbf{I} = \mathbf{V}/\mathbf{Z}_t,$$

$$\mathbf{Z}_t = \mathbf{Z}_1 + \mathbf{Z}_2 + \mathbf{Z}_3,$$

$$\mathbf{I} = 12/(3 + j6 - j2),$$

$$\mathbf{I} = 12/(3 + j4) = 12\,\underline{/0°}/5\,\underline{/53°},$$

$$\mathbf{I} = 2.4\,\underline{/-53°}\ \text{A}.$$

Here the capital **V** and **I** stand for ac voltages and currents. As might be guessed, the 12-V ac driving voltage produces an ac current—in this case 2.4 A. The angles describe the phase associated with ac currents and voltages as will be explained in Section 4.3.

The ac *impedance* (**Z**) of an element is a more general form of dc resistance (*R*). Or put another way, in addition to *R*, ac circuits have two other forms of impedance. The operator j (which distinguishes between the resistive kind of impedance and that produced with capacitors or inductors) has special mathematical properties. But these are easy to learn.

Figure 4–1c is a standard two-mesh circuit similar to dc circuit Figure 3–1b. What currents will the three ammeters read? This circuit will be solved for ac mesh currents $\mathbf{I}_1$ and $\mathbf{I}_2$ later in the chapter. As in Chapter 3, the ammeter A_3 reads their difference.

The reader's objective then in Chapter 4 is simply to

1. investigate the nature of ac current and voltage waves,
2. look into the impedance properties of L's and C's,
3. learn the use of complex operator j, and
4. drill these ideas into reflexes by applying them to many circuit problems.

Chapter 4 is further simplified by using only the North American power frequency of 60 Hz. Chapter 6 will present some mathematical justification for these methods, consider other frequencies, and add special ac topics such as resonance. Chapter 5 discusses the way in which ac power, both average and instantaneous, is handled in calculations.

4.1   ALTERNATING-CURRENT WAVES—SINUSOIDS

In Figure 4–2a a 10-V ac voltage source drives a 2-Ω resistive load. The resulting ac current (Ohm's law is still used as in Chapter 2) is 5 A. Note that the ac voltage source symbol contains a little sine wave in addition to the polarity marks. (Even in ac a reference polarity is needed.)

Figure 4–2b shows the appearance of the 10-V voltage wave—its instantaneous voltage as a function of time. The wave shown is sinusoidal* and could also be written mathematically as $v(t) = 14.1 \cos(2\,\pi\,60\,t)$,

$$v(t) = 14.1 \cos(377\,t). \tag{4.1}$$

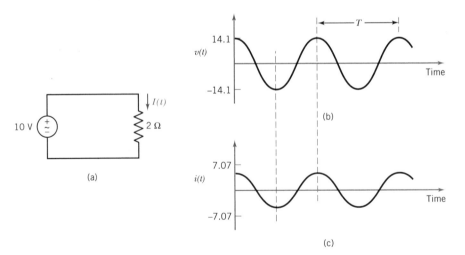

FIGURE 4-2 Sinusoidal voltage and current waves in a simple ac circuit.

---

*Both sine and cosine waves (or any combination of them at the same frequency) are classified as sinusoids—sine-shaped waves.

Thus this same instantaneous ac voltage wave can be described either mathematically or graphically—by Equation (4.1) or by Figure 4–2b. The number 377 will be explained shortly.

Figure 4–2 illustrates how this sinusoidal wave appearing at (or emerging from) the source's two terminals varies with time, up and down in voltage, from −14.1 to +14.1 V at a frequency of 60 times a second, crossing the 0-axis 120 times per second.

The reason a sinusoid, which varies between 14.1 and −14.1 V, is called a 10-V wave will be demonstrated mathematically in Section 4.2 following. But for the moment, consider simply that this wave has the same *effective magnitude* (average heating effect) as a 10-V dc source. And it is more convenient in writing or calculating to be able to use a single number. Or more generally, Section 4.2 will show for sinusoids,

$$V = V_{peak}/\sqrt{2} \quad \text{and} \quad I = I_{peak}/\sqrt{2} \qquad (4.2)$$

Observe that the capital letter symbols V and I in ac, used without subscript, and also the words *voltage* and *current* unmodified, always refer to the effective magnitude of a wave.

The *period* of the wave (symbol T) is the time needed for one complete cycle, for instance, the time between positive peaks in Figure 4–2b. (A *cycle* is a complete set of variations of the wave—for example, a 60-Hz wave goes through a complete cycle in a period of about 16.7 ms—$\frac{1}{60}$s.) *Frequency* (f) is the rate at which the voltage (or current) goes up and down in cycles per second. Its unit is the hertz (*Hz*). One Hz is a variation of one complete cycle per second. In general,

$$T = 1/f. \qquad (4.3)$$

An expression for $v(t)$ for a 10-V wave at any frequency f is

$$v(t) = 14.1 \cos(2\pi f t),$$

or
$$v(t) = 14.1 \cos(\omega t).$$

The quantity $2\pi f$ is called the *radian frequency*, with symbol ω. Most engineers remember the ω for 60 Hz, the standard U.S. and Canadian power frequency, as very closely 377 radians per second ($2\pi \times 60$). For any frequency and any voltage,

$$v(t) = V_{peak} \cos(\omega t)$$

or
$$v(t) = V\sqrt{2} \cos(\omega t).$$

Note again that in ac the symbols V and I, capitalized and without subscript, always mean effective values.

4.2 EFFECTIVE (ROOT-MEAN-SQUARE) VOLTAGES AND CURRENTS

It remains yet to justify the assumed effective value for current and voltage sinusoids, an assumption made in Section 4.1. It was stated that any sinusoidal ac current wave flowing through a resistor R produces the same average heating effect (power) as a dc current equal to the peak ac current divided by $\sqrt{2}$. This is also true for an ac voltage across R or indeed for the product of voltage and current as illustrated by Equation (5.1).

To prove this assertion let us take a practical current time wave of any type—$i(t)$—(not limited to sinusoids) that continuously repeats itself and calculate the average power it puts into a resistor by the concept that $p = i^2 R$. We will employ elementary calculus to compute the average power P:

$$p(t) = i^2(t)R, \tag{4.4}$$

$$P = \frac{1}{T}\int_0^T i^2(t)R \, dt.$$

Defining an *effective* value I, which when squared and multiplied by R gives P, yields

$$I^2 R = P;$$

then

$$I = \sqrt{P/R}$$

or from Equation (4.4),

$$I = \sqrt{\frac{1}{T}\int_0^T i^2 \, dt}. \tag{4.5a}$$

Thus the effective value of an ac wave is its root-mean-square (rms) value. First square the $i(t)$ function, then find its average or mean, then take the root of this mean value.

To apply this idea to a sinusoid substitute $I_p \cos(2\pi f t)$ for $i(t)$. It will be found that $I = I_p/\sqrt{2}$. Other important shapes are symmetrical triangular and square waves, where rms values are peak/$\sqrt{3}$ and peak, respectively.

By a similar development rms voltage can be shown to be

$$V = \sqrt{\frac{1}{T}\int_0^T v^2 \, dt}. \tag{4.5b}$$

EXAMPLE 4.1 A voltage source produces 4 V for 2 s and -2 V for the next 3 s. This 5-s wave is then repeated over and over.

Find: the effective value of the voltage wave.

Solution: It is helpful to sketch the waveforms described in this problem and

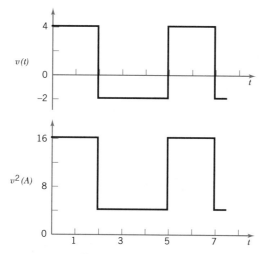

FIGURE 4-3 Example 4.1 The 5-second voltage wave is first squared and then averaged. Then the square root of the average is taken.

its solution (see Figure 4–3). Applying Equation (4.5b), $v^2(t)$ is 16 V^2 for 2 s followed by 4 V^2 for 3 s. The mean of this squared function is $(32 + 12)/5 = 8.8$ V^2. The square root of this is 2.97 V. Thus a dc voltage of 2.97 V would produce the same heating in a resistor connected across it as this time-varying source would produce in the same resistor. ▪

In summary, to find the rms of a waveform: first square it, then find the mean square, then take the square root of the mean square.

4.3   PHASE

The most general expression for a sinusoid includes three dimensions: *magnitude, frequency,* and *phase.*

$$v(t) = V_p \cos(\omega t + \theta) \qquad (4.6)$$

or, using f,

$$v(t) = V_p \cos(2\pi f t + \theta). \qquad (4.7)$$

Here θ is called the *phase angle* and has an effect as shown in the voltage waves of Figure 4–4. A positive θ moves the wave to the left on the axes (4d) and a negative θ to the right (4c). A positive θ makes things happen sooner (farther to the left on the axis). (The subscript p stands for peak.)

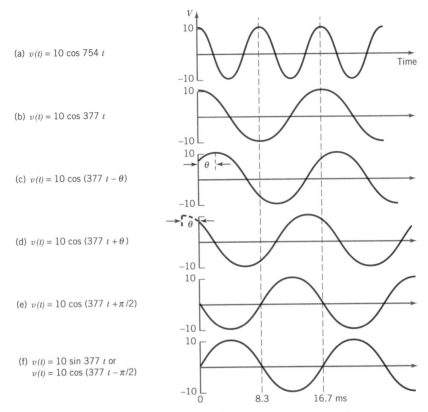

(a) $v(t) = 10 \cos 754\ t$

(b) $v(t) = 10 \cos 377\ t$

(c) $v(t) = 10 \cos (377\ t - \theta)$

(d) $v(t) = 10 \cos (377\ t + \theta)$

(e) $v(t) = 10 \cos (377\ t + \pi/2)$

(f) $v(t) = 10 \sin 377\ t$ or
$v(t) = 10 \cos (377\ t - \pi/2)$

FIGURE 4-4 Mathematical sinusoids and their graphical representation. The reader can use a hand calculator to verify and extend these curves.

The argument of the cosine function is here in radians, but it could be expressed in degrees if desired. (A radian is about 57.3° and there are 2π radians, 6.28 or so, in a full circle.) The term *phase* is often used instead of phase angle if no confusion will be caused. It can be seen that the value of θ used in Figures 4–4c and d is 45°. How many radians is this?*

Figure 4–4a is a 120-Hz wave; the waves in Figure 4–4b-f are 60 Hz.

The cosine function is used typically in electrical mathematics. It would also be entirely possible to describe any sinusoid with a sine function. But it is suggested that to avoid confusion, any sines encountered be converted to cosines before proceeding further. Figure 4–4f illustrates the relation between sine and cosine waves. The conversion is as follows (where the angle is in radians):

$$\sin(\alpha) = \cos(\alpha - \pi/2). \tag{4.8}$$

*$\pi/4$.

▉▉▉▉▉▉▉ **EXAMPLE 4.2** A 1000-V transformer feeds a lighting load of 100 Ω at 60 Hz.

Find

1. the instantaneous voltage and current at exactly 6 P.M. plus 9.5347 s if the voltage wave is zero and going negative at 6 P.M.;
2. how many cycles have occurred since 6 P.M.

Solution: It is necessary to find appropriate numbers for the equation:

$$v(t) = V\sqrt{2} \cos(2\pi f t + \theta),$$

so $v(t) = 1000\sqrt{2} \cos(2\pi 60t + \theta)$ V. There are an integral number of periods in a second for this wave; thus we can equivalently start at 9 s, after 540 periods. That is, $v(0.5347 \text{ s})$ is the same as $v(9.5347 \text{ s})$. The wave (starting down from 0 at time 0) looks like Figure 4–3e. This is a cosine wave moved to the left $\pi/2$ radians. Then from the last equation above:

$$v(t) = 1414 \cos(377 \times 0.5347 + \pi/2),$$

$$v(t) = \underline{-697 \text{ V}}.$$

Similarly and using Ohm's law, $i(t) = v(t)/100 \text{ Ω}$, so

$$i(t) = -697 \text{ V}/100 \text{ Ω} = \underline{-6.97 \text{ A}}.$$

It is most helpful to sketch this wave carefully, showing its intercepts with the v and t axes, while solving this problem.

For the second question, $T = 1/f = 1/60 = 16.67$ ms. Thus 9.5347 s will contain $9.5347/0.01667 =$ about $\underline{572.1}$ cycles or complete alternations. (Or simply multiply 9.5347 by 60.)   ▉▉

It is interesting to note that while 377 is very close to $2\pi \times 60$, it is not quite exact and will cause a slight error over many cycles—one-quarter cycle error for every 10,000 cycles.

4.4   PHASOR REPRESENTATION

The thinking about $v(t)$ and $i(t)$ waves above yielded two ways of describing them: by mathematical expression, and by a time graph. Both these methods deal with instantaneous voltage or current as a function of time.

There are two additional and particularly useful methods of describing ac waves, both called *phasors*. To develop these methods first make a set of axes on a piece of paper (dotted lines of Figure 4–5a). Next imagine an arrow cut out of cardboard. Sticking a pin for an axle through the origin of the axes, set the arrow rotating counterclockwise.

Now that its movement is understood, the whole idea can be represented more simply with the arrow of Figure 4–5b. The arrow is rotating smoothly at a constant speed counterclockwise. Consider the projection of the tip of the arrow on the horizontal axis. This projection goes back and forth sinusoidally. Measuring the angle (or the passing of time) from the right-hand horizontal axis, the function is specifically a cosine wave.*

A time graph (with time positive downward as in Figure 4–5b) illustrates the wave produced by the projection of the arrow's tip on the horizontal axis. If it represented a voltage wave, for example, voltage would be positive to the right and negative to the left of the origin. Or it might, of course, represent an ac current.

The rotating arrow (called a *phasor*) by projection generates the ac $v(t)$ wave. But more than that, the arrow contains all three elements of the ac wave. Consider Figure 4–5c. The peak magnitude of the ac wave is represented by the arrow's length. Its frequency is represented by the rotation speed in radians per second (ω). And its phase angle, θ, is shown by the angle of the arrow at $t = 0$.

The use of j and $\Re$ to designate the vertical and horizontal axes need not concern us at this point; these designations have been chosen to facilitate the more complete and rigorous treatment of this transformation given in Section 6.3.

If the rotating arrow contains all three properties of the ac wave, why not work with the arrow instead of the $v(t)$ and $i(t)$ functions? And that is exactly what phasor

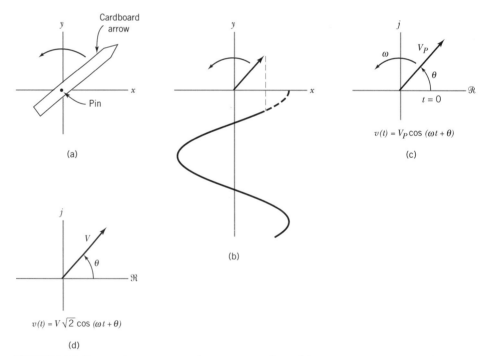

FIGURE 4-5 Phasors (rotating arrows) are an unusually useful way to represent sinusoidal waves.

---

*In physics this to-and-fro motion (sinusoidal wave) is called *simple harmonic motion.*

ac methods are—dealing with the phasors instead of the waves themselves. The result of this kind of transformation (phasor for cosine wave) greatly simplifies ac analysis and calculations. By this means all the dc circuit theorems and methods of Chapters 2 and 3 can be used for ac.

But first, there is another way to represent a phasor besides the illustrations of Figure 4–5. To develop a mathematical notation, write

$$V_p \underline{/\omega t + \theta}$$

where V_p is the peak value of the wave and θ the phase angle. The ωt term (cf. Figure 4–5c) produces the rotation of the arrow. But it is most common to draw the diagram at $t = 0$. (After all, diagrams on paper can't actually spin.) In this case there is no reason to carry the term ωt along. Engineers make a mental note that the speed of rotation is ω radians per second. So a simpler notation is

$$V_p \underline{/\theta}.$$

Taking one further step with this phasor notation, engineers habitually think of ac voltages and currents in terms of their effective or rms value—peak/$\sqrt{2}$. The finished phasor notation looks like this:

$$V \underline{/\theta},$$

where

$$V = V_p/\sqrt{2}.*$$

And we can write

$$\mathbf{V} = V \underline{/\theta}, \tag{4.9}$$

where the bold-faced notation $\mathbf{V}$ includes both magnitude and angle.

Phasor sketches also are almost always made with the length of the rms value, as in Figure 4–5d. It is clear from Figure 4–5b that unless the rotating arrow's length has peak value the proper sinusoid will not be generated. Nevertheless, for convenience, most engineers draw rms-length phasors and keep in mind that they must be multiplied by $\sqrt{2}$ to get peak values.

Thus ac voltage or current can be represented either by a sketch on cartesian axes as in Figure 4–5d, or in the mathematical form of Equation (4.9). Either is called a phasor. In fact they are the same thing. Four ways of representing an ac wave have now been developed:

1. by math $v(t)$ or $i(t)$ expression (as in Figure 4–4)

---

*This rms convention is not always used in other literature involving phasors.

2. by graphical $v(t)$ or $i(t)$ sketch (as in Figure 4–4)
3. by math phasor expression [as in Equation (4.9)]*
4. by phasor sketch on complex axes (as in Figure 4–5d)

EXAMPLE 4.3 Show the four representations for a typical 110-V North American house voltage.

Solution: Since 110 is effective voltage the peak voltage is $110 \times \sqrt{2} =$ 155.6 V. The standard U.S. power frequency is 60 Hz and so $\omega = 2\pi \times 60 =$ 377. Phase is somewhat ambiguous. Since the voltage wave at a house outlet has gone on for a long time, perhaps years (or at least since service was restored after the last power outage), there is a question as to when zero time occurred. So there is a choice to make. This choice is usually made so as to make the problem to be solved as easy as possible. In the case of this problem let us select a phase angle of 0°. This gives the results portrayed in Figure 4–6 as the answers. Answers 4–6a and b are often said to be in the *time domain*, while 4–6c and d (the phasors) are said to be in the *frequency domain*. Note that in 4–6c and d the rotational speed (or frequency) does not explicitly show, but it must be 60 revolutions per second (60 Hz) or 377 radians per second if these are to be correct representations.

Before leaving this example it can be observed that voltage designations like "110 V" or "220 V" are usually considered to be *nominal* voltages; they are approximations. Actual house voltage often varies from perhaps 110 to 120 V from one part of the United States to another. There may also be significant variations at any one place over a 24-h period.

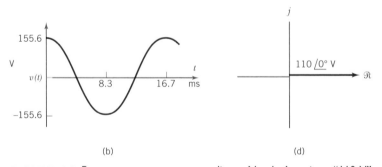

(a) $v(t) = 155.6 \cos 377t$ V   (c) $V = 100 \; \underline{/0°}$ V

FIGURE 4-6 Four ways to represent ordinary North American "110-V" household voltage.

---

*A more formal mathematical expression for the phasor employs a complex exponential. We present this form in Chapter 6.

Besides the implicit frequency, phasor forms have two quantities: magnitude and angle. Thus in dealing with ac voltages or currents one must deal with two numbers at once. Such pairs of numbers are called *complex numbers* and are manipulated mathematically by *complex algebra*. Appendix B provides a simple refresher course in the required ideas of complex algebra.

4.5   VOLTAGES AND CURRENTS AS COMPLEX NUMBERS

The preceding section introduced a mathematical form of phasor notation for voltages and currents as, for example, $V\underline{/\theta}$. This is a very handy and simple notation and widely used. The voltage (or current) so described has two parts—here a magnitude and an angle. Thus ac voltages and currents are complex. Complex numbers are described by the three representations shown in Appendix B, and complex arithmetic must be employed in dealing with them.

Figure 4–7 illustrates these three representations for a specific ac voltage—the graphical representation on complex axes, the polar form, and the rectangular form.

Figure 4–7a shows the graphical representation—here an arrow of length 10 rotated 37° counterclockwise from the positive *x* axis. The numbers 10 and 37 need not be placed on the diagram, although this is often helpful.

Polar form representation (7b) states the same thing in numbers. Both the diagram and this polar form simply designate the position of the point at the tip of the arrow.

Rectangular representation (7c) also locates the tip—in this case by two other voltage numbers. The first is the projection of the 10-V arrow of Figure 4–7a on the horizontal or real axis. The second is the projection of this voltage phasor on the *j* or vertical axis.

As with any complex number, the two voltage numbers are distinguished from each other by the symbol *j*, which always goes with a number on the vertical axis.

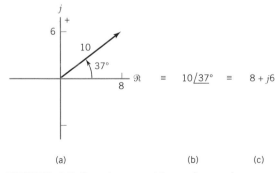

FIGURE 4-7 Complex quantities such as voltage, current, and impedance can be represented in these three equivalent ways.

Consider the equation

$$\mathbf{V} = V\underline{/\theta}.$$

It is essential to be clear on whether one is talking about the entire number (magnitude and angle) $\mathbf{V}$ or just the magnitude V. To make this distinction carefully here, bold face is used for complex numbers like phasors $\mathbf{V}$ and $\mathbf{I}$, and italic for their magnitudes, V and I. Again, these magnitudes are rms unless otherwise marked. Beginners can make this distinction in pencil problem-solving by underlining or overlining. But ac phasor voltages and currents are *always* complex, and magnitudes are always real numbers, so there is little need in engineering practice for a special distinction. Engineers quickly abandon any such notation except when confused.

Modern calculators will handle conversions of complex electrical quantities from rectangular to polar form and back again.

4.6   IMPEDANCE—Z

Chapter 2 showed that in dc analysis Ohm's law, $R = V/I$, was the definition of resistance. That is, knowing the current flowing into any two-terminal, passive, linear dc network as in Figure 4–8a, and the voltage across it, the resistance R can be found with Ohm's law.

Similarly in ac circuits, Figure 4–8b, engineers define *impedance* $\mathbf{Z}$ by Ohm's law:

$$\mathbf{Z} = \mathbf{V}/\mathbf{I} \ \Omega. \tag{4.10}$$

(Just as in the case of resistance in dc circuits it is assumed that impedance is linear or nearly so. That is, the ratio between $\mathbf{V}$ and $\mathbf{I}$ remains constant as $\mathbf{V}$ and $\mathbf{I}$ change in magnitude.) Using the rules of complex arithmetic as set out in Appendix B and

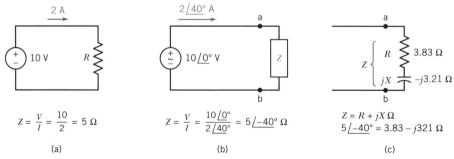

FIGURE 4-8 Impedance $\mathbf{Z}$ is defined by Ohm's law in the same manner as resistance R in dc. $\mathbf{Z}$'s real part is resistance. The J part is reactance (inductive or capacitive).

the data from Figure 4–8b, we have

$$\mathbf{Z} = 10 \, \underline{/0°} / 2 \, \underline{/40°},$$

$$\mathbf{Z} = 5 \, \underline{/-40°} \, \Omega.$$

Impedance (with symbol $\mathbf{Z}$) is the generalized form of ac "resistance." Like ac currents and voltages it is a complex number. Unlike them it is *not a phasor* since it does not rotate on its complex axes but remains stationary. Neither does it have $\omega \, t$ associated with it.

Converting $\mathbf{Z}$ from polar to rectangular form gives the results of Figure 4–8c,

$$\mathbf{Z} = 5 \, \underline{/-40°} = 3.83 \; \Omega \; - \; j3.21 \; \Omega \; = \; R \; + \; jX.$$

The rectangular components R and X are called *resistance* and *reactance*. Resistance is the identical concept used in dc. Current through a resistor is always in phase with the voltage across it. Reactance is the j portion of impedance. Its effect in circuits is to produce a phase angle between voltage and current. The element giving rise to the negative reactance in this circuit is a capacitor, but in some cases there will be a positive reactance corresponding to an inductor.

It is necessary to be precise here: X is itself a real number and is called reactance; jX is the reactive component of the impedance, and taken together is a j number. The units of resistance, reactance, and impedance are all ohms.

Resistors have the quality of resistance. There are two physical circuit elements that have the quality of reactance: inductors (sometimes called *coils*) and capacitors.

4.7   CAPACITORS AND INDUCTORS

Chapter 6 takes these two elements up in detail. For the purpose of the present phasor analysis work of this chapter, let us simply note that inductors and capacitors are passive two-terminal elements, like resistors, and are denoted, respectively, by the symbols L and C (instead of R).

When, as in Figure 4–8b and c, dividing the voltage across a two-terminal element by the current through it results in a negative jX term, that reactance is capacitive, caused by a capacitor. When the jX term is positive, the reactance is inductive, caused by an inductor (or coil).

Figure 4–9 illustrates the voltage and current relations for the three kinds of passive ac elements—R, L, and C (resistance, inductance, and capacitance). Note that for a resistive impedance (9b) the current is *in phase* with the voltage. That is, the current and voltage arrows on the complex axes coincide in angle. Remember that these two phasors are rotating counterclockwise 60 times a second. In the case of an R, neither leads nor lags the other. Note that the three arrow diagrams in this figure have all been stopped (as if by a camera snapshot) with the voltage phasor at the same place, about 40°.

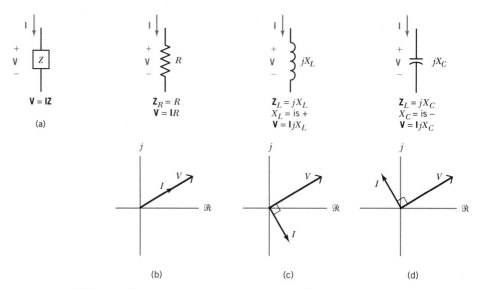

FIGURE 4-9 The three passive elements *R*, *L*, and *C*, and their *V-I* relations.

In Figure 4–9c, for an inductor, the **V** arrow is ahead of the current arrow. Voltage across an inductance leads the current through it by 90°. Or it could be said that the current lags the voltage by 90°.

Similarly in Figure 4–9d for a capacitance, current leads voltage by 90° or the voltage lags the current by 90°. These facts can be taken as experimental. But mathematically (using Ohm's law) the *j operator* in the impedance expression will produce this same 90° current and voltage lead or lag (*j* in polar notation is $1\underline{/90°}$).

Note that the load convention is used for these ac definitions (current going into the positive reference terminal) just as in dc.

4.8   PHASOR ANALYSIS

Up to this point Chapter 4 has established the nature and representation of sinusoids through time functions and time graphs; their corresponding graphical and mathematical representations by means of phasors; the representation of phasor voltages and currents by complex numbers; and a beginning look at the nature of inductance, capacitance, reactance, and impedance. It remains now only to use this erudition in applying to ac circuits the analysis earlier developed for dc.

As a first example, a simple three-element series circuit, Figure 4–1a, was solved in the introductory paragraphs of this chapter. Its dc equivalent was also considered. This technique of making and solving the equivalent dc circuit (Figure 4–1b) is often helpful when getting started with ac analysis. Surely it must be agreed that there is no essential difference in the two solutions, except for the use of complex arithmetic for the ac circuit.

The following mesh example is more challenging but susceptible to the same kind of thinking.

EXAMPLE 4.4 In the circuit of Figure 4–1c

Find
1. $\mathbf{I}_x$ by mesh analysis;
2. $\mathbf{I}_x$ by finding $\mathbf{I}_s$ through A_1 and then using current division.

Solution: Proceeding exactly as with a two mesh dc circuit, except that mesh currents $\mathbf{I}_1$ and $\mathbf{I}_2$ are complex and must be dealt with by complex arithmetic, we have for mesh 1:

$$(4 + j3)\mathbf{I}_1 - j3\mathbf{I}_2 = 10 \underline{/20°},$$

and for mesh 2:

$$-j3\mathbf{I}_1 + j(3 - 5)\mathbf{I}_2 = 0.$$

Solving these complex equations simultaneously:

$$\mathbf{I}_1 = 1.18 \underline{/-41.9°} = 0.88 - j0.79,$$
$$\mathbf{I}_2 = 1.77 \underline{/138.1°} = -1.32 + j1.18.$$

Appendix A presents determinant procedures for solving simultaneous equations. (Other methods than determinants are usable but not so systematic.) Details of solving the above simultaneous equations for the answers indicated are as follows:

$$\Delta = \begin{vmatrix} 4 + j3 & -j3 \\ -j3 & -j2 \end{vmatrix} = -j8 + 6 + 9$$
$$= 15 - j8 = 17 \underline{/-28.1°}.$$

$$\Delta_1 = \begin{vmatrix} 10 \underline{/20°} & -j3 \\ 0 & -j2 \end{vmatrix} = 20 \underline{/-70°},$$

$$\Delta_2 = \begin{vmatrix} (4 + j3) & 10 \underline{/20°} \\ -j3 & 0 \end{vmatrix} = 30 \underline{/110°}.$$

Now
$$\mathbf{I}_1 = \Delta_1/\Delta$$
$$= 1.18 \underline{/-41.9°} = 0.88 - j0.79 \text{ A}.$$

And
$$\mathbf{I}_2 = \Delta_2/\Delta$$
$$= 1.77 \underline{/138.1°}$$
$$= -1.32 + j(1.18) \text{ A}.$$

But
$$\mathbf{I}_x = \mathbf{I}_1 - \mathbf{I}_2 = 2.20 - j(1.97)$$
$$= 2.95 \underline{/-41.9°} \text{ A.}$$

To solve the second part of the example, replace the parallel circuit to the right of points a, b (product over sum) by a single $\mathbf{Z}_p$:

$$\mathbf{Z}_p = j3(-j5)/(j3 - j5) = 15/(-j2)$$
$$= 7.5 \underline{/90°} = j7.5 \ \Omega$$

and
$$\mathbf{I}_s = \mathbf{V}/(4 + j7.5) = 10 \underline{/20°}/8.5 \underline{/61.9°}$$
$$= 1.18 \underline{/-41.9°} \text{ A.}$$

Note that
$$1/j = -j.$$

By current division to the right of point a in the circuit

$$\mathbf{I}_x = 1.18 \underline{/-41.9} \ (-j5/-j2)$$
$$= 1.18 \underline{/-41.9°} \times 2.5$$
$$= 2.95 \underline{/-41.9°}.$$

This result is in good agreement with the mesh solution in spite of rounding off.

This example shows how convenient and easily applied Chapters 2 and 3 ideas like parallel impedances or current division are. All the analytical methods of these chapters can be applied to ac circuits. Simply substitute the general ac $\mathbf{Z}$ for the dc R and recognize $\mathbf{V}$, $\mathbf{I}$, and $\mathbf{Z}$ as complex quantities. But complex arithmetic is tedious enough that some check is needed such as solving by two different methods.

Skill in solving this type of problem depends on practice in both mesh analysis and complex arithmetic. Drill problems provide good practice. As mentioned before, a hand calculator that has polar/rectangular conversion saves time.

Note that a minus sign in front of a phasor quantity can be eliminated simply by applying 180° (plus or minus) to its angle. This is a result of the conversion formulas of Appendix B, but it is more easily understood by drawing the phasor on its axes. A minus sign simply reverses the arrow (or rotates it 180°). Engineers manipulating phasors or any complex quantities find it helpful to think in terms of the phasor diagrams. Beginners getting started with this thinking should draw the phasor diagram and work with it continually while solving problems.

Alternating-current voltmeters and ammeters are nearly always designed to read rms magnitude. The three ammeters in this circuit (Figure 4–1c) would read values of 1.18, 1.77, and 2.95 A. An interesting question is: "Is KCL satisfied at the top node in this circuit? Does $1.18 = 1.77 + 2.95$?" The answer is: "Yes, it does if their phases are considered." (Add the phasors tail to head in a phasor diagram to see this.)

4.9   PHASOR, IMPEDANCE, AND ADMITTANCE DIAGRAMS

There are three principal ac graphical constructions of great help to analysis: the phasor diagram, the impedance diagram (or alternatively the admittance diagram), and the complex power diagram. The last one is considered in Chapter 5.

A *phasor diagram* for some of the voltages and currents involved in the circuit just solved (Figure 4–1c) is illustrated in Figure 4–10a. It is constructed on a complex plane (complex axes: $\mathcal{R}$ and j) and contains only voltage and current phasors—in this case the voltage of the source (10 V at an angle of 20°) and the three currents.

In phasor diagrams all phasors rotate counterclockwise and are usually shown at $t = 0$. It is helpful to distinguish **I** from **V** phasors by closing the arrowhead of the **I**'s. Phasor arrows are drawn not only at the correct angle but also to approximate scale. Separate scales for voltage and current are chosen to make the phasors a convenient length.

The diagram suggests many things to the analyst. For instance $\mathbf{I}_1$ and $\mathbf{I}_2$ are exactly out of phase (180° apart). Is this reasonable? Yes, the L and C are in parallel and so must have the same voltage across them; one has a j impedance while the other is $-j$. The j's represent 90° of rotation. Note also that source current lags voltage by about 60°. This is reasonable since the total impedance the source sees is inductive (the parallel L-C circuit is purely inductive since the L has a smaller Z than the C). In fact, many problems can be solved by precise construction of a phasor diagram. Even approximate sketches are a helpful check on numerical work.

The *impedance diagram* (Z diagram) of Figure 4–10b is also drawn on a complex plane (complex axes R and jX). It includes only R, X, and **Z**. All are stationary. Nothing rotates. Impedances in series can be easily added on this diagram. Figure 4–10b shows the impedance that the source sees in the circuit of Figure 4–1c made up of the 4-Ω resistance in series with the 7.5-Ω inductive **Z** of the parallel circuit. The total $\mathbf{Z}_s$ seen is then $4 + j7.5$ or $8.5\,\underline{/62°}$.

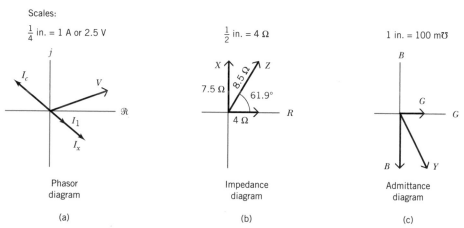

Scales:

$\frac{1}{4}$ in. = 1 A or 2.5 V $\frac{1}{2}$ in. = 4 Ω   1 in. = 100 m℧

(a) Phasor diagram

(b) Impedance diagram

(c) Admittance diagram

FIGURE 4-10 Phasor, impedance, and admittance diagrams clarify circuit concepts and provide checks on calculation.

The reciprocal of the impedance diagram (Figure 4–10c) is also used by electrical engineers—an *admittance diagram* (complex axes G and jB). It was observed earlier that G is the reciprocal of R. Similarly the reciprocal of $\mathbf{Z}$ is $\mathbf{Y}$, the *admittance*. It is, of course, also complex. Whereas $\mathbf{Z}$ is defined as $\mathbf{V}/\mathbf{I}$, $\mathbf{Y}$ is defined as

$$\mathbf{Y} = \mathbf{I}/\mathbf{V} = G + jB. \tag{4.11}$$

As in Chapter 2, G is conductance; B is the j part of $\mathbf{Y}$ and is called *susceptance*. Positive values of B are capacitive, negative are inductive. (This is just the reverse of the signs with X.) $\mathbf{Y}$, G, and B all have the units of siemens. Further ideas on admittance are developed in the problems.

4.10   CIRCUIT ANALYSIS DRILL

At this point the principal need is for ac circuit analysis drill. While the complex calculations are tedious to begin with, they quickly become routine. A systematic checking procedure is essential. Frequent sketchings of phasor diagrams and circuit transformations will aid thought.

The engineer becomes intimately familiar with his or her particular calculator and develops almost unconscious routines with it for this arithmetic. A sophisticated, modern machine with one-button polar/rectangular transformation or even complete freedom in using either polar or rectangular forms is desirable. But it is not difficult to use the square root, arctangent, sine, and cosine transformations (Appendix B) on simpler calculators. One must become so expert and fast with complex calculations that the algebra will not interfere with thought processes about the circuit and its analysis.

The following Thevenin's theorem problem will help make this discussion of circuit analysis drill clearer.

EXAMPLE 4.5 Consider the circuit of Figure 4–1c.

Find: $\mathbf{I}_x$ by Thevenin's theorem.

Solution: Figure 4–11 illustrates the steps involved. In 4–11 L (the element in question) is removed from the circuit, leaving terminals x, y. A Thevenin's generator is substituted (4–11c). (Note that what is inside the fence in 4–11c represents what is outside the fence in 11b.) In 11d the voltage source is replaced with a short circuit. Looking back into terminals, x, y the $\mathbf{Z}$ is 4 Ω real in parallel with 5 Ω capacitive reactance or (product over sum) $20\,\underline{/-90°}/(4 - j5) = 3.12\,\underline{/-38.66°}$ or $2.44 - j1.95\ \Omega$.

Thevenin's voltage, $\mathbf{V}_{\text{Th}}$ (by voltage division) is $10\,\underline{/20°} \times [-j5/(4 - j5)] = 7.81\,\underline{/-18.7°}$ V. This gives us the equivalent circuit of Figure 4–11e, and hence $\mathbf{I}_x = \mathbf{V}/\mathbf{Z} = 7.81\,\underline{/-18.7°}/(2.44 + j1.05) = 2.94\,\underline{/-41.9°}$ A. Within roundoff this agrees with the earlier solution of this problem by mesh analysis.

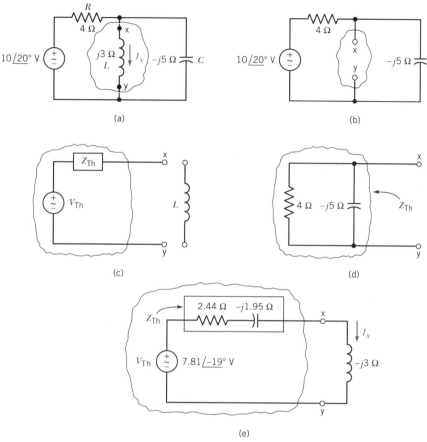

FIGURE 4-11 Using Thevenin's theorem to solve an ac circuit problem. The Thevenin's voltage has an angle. $\mathbf{Z}_{Th}$ is also complex, represented here by an R and a C.

Warning: In the circuit of Figure 4–11(d) two impedances in parallel are correctly combined using the relation $Z_p = Z_1 Z_2 / (Z_1 + Z_2)$. A surprisingly common error made by beginning ac analysts is to erroneously use the series expression $Z_s = Z_1 + Z_2$. If the reader has any doubts about combining impedance in parallel he or she should go back to dc problems like Figure 2–9 for further drill, contrasting particularly sketches (e) and (f).

4.11   ELEMENTARY TRANSFORMERS

Figure 4–12a shows an elementary iron-core transformer symbol. Most transformers are made by winding two windings on a common magnetic core as in Figure 4–12b. There are N_1 turns on the left winding and N_2 turns on the right. In this section we present without justification a simple model for transformers that will serve in most applications. Chapter 18 will consider transformers in more detail, developing the theoretical basis.

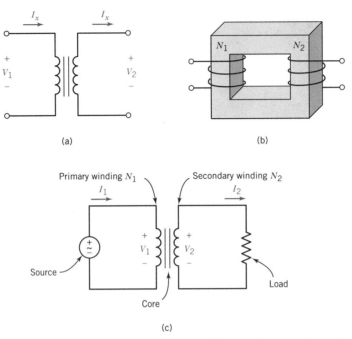

FIGURE 4-12 Iron-core transformers enable ac voltages to be easily stepped up or stepped down. $V_2 = V_1 N_2/N_1$ where N_1 and N_2 are the number of turns of wire of the two coils.

Figure 4–12c shows a power source connected to the terminals on the left (making them the *primary* terminals). A load, represented by resistor Z_L, is connected on the right side to the *secondary* terminals. The principal characteristic of a transformer is that it can change an applied ac voltage from the source to some more advantageous ac voltage for the load. This is a major advantage and has led to the universal use of ac for power systems. (Transformers will not work on dc and will burn up if connected to it.)

Power flows from left to right in Figure 4–12c. Practical transformers will have some losses, but they are very efficient—typically 95% or higher. Neglecting losses in this present simple analysis, assume that all power taken from the source is transferred to the load.

Under these assumptions,

$$V_2 = V_1 N_2/N_1, \qquad\qquad (4.12)$$

where N_2/N_1 is called the *transformer ratio* or *turns ratio*. Suppose, for example, that N_1 is 1000 turns and N_2 500 turns. Then if 220 V were applied from the power source, the load would be driven with 110 V. For the currents on the two sides,

$$I_1 = I_2 N_2/N_1. \qquad\qquad (4.13)$$

Note that a transformer that reduces voltage in the output also reduces current in the input. The V and I phase angles are the same on both sides *or* all of them in the secondary are switched 180° depending on the reference directions chosen.

Because of its voltage and current-transforming properties, the transformer also transforms impedances. The load impedance of Figure 4–12c is represented by V_2/I_2 and is designated R_L or Z_L. The impedance looking into the primary, Z_{in}, is V_1/I_1. Thus,

$$Z_{in} = (N_1/N_2)^2\, Z_L \qquad\qquad (4.14)$$

Impedances are transformed by the inverse square of the turns ratio.

EXAMPLE 4.6 In the circuit of Figure 4–12c the load is an 8-Ω resistor (this might represent an 8-Ω speaker in a music system). A 60-V ac source supplies audio power to the transformer, which has the smaller number of turns on the secondary. The turns ratio, as it has been defined above, is 0.125 (but most engineers would refer to it as "8 to 1"). Transformers are said to be connected either *step-up* or *step-down*, depending on whether the smaller number of turns is connected to the source or the load.

Find: The primary and secondary voltages and currents.

Solution: Placing 60 V on the primary will produce $60/8 = \underline{7.5}$ V from the secondary and on the load. By Ohm's law $I_L = V/Z = 7.5/8 = I_2 = \underline{938}$ mA. (The load impedance is assumed here to be real—all resistive.) Then $I_1 = 938/8 = \underline{117}$ mA.

The cause and effect relationship here is important. Looking at the circuit for this problem (Figure 4–12c), we see that the source voltage from the left placed on the primary winding creates a secondary voltage. This secondary voltage impressed on the load produces a secondary current according to Ohm's law. The secondary current, in turn, produces a primary current drawing power from the source.

4.12   SUMMARY

Sinusoidal ac current or voltage waves have three dimensions—magnitude, frequency, and phase. The period of the wave (the length of time it takes to go through its complete variation) is the reciprocal of frequency in hertz. In their mathematical, function-of-time representation the peak magnitude is used. But engineers usually describe I's and V's by their rms or effective magnitude, which is peak/$\sqrt{2}$. Frequency can be represented in terms of hertz, f; or in terms of radians per second, ω; $\omega = 2\pi f$.

Phasors, rotating arrows, are a convenient transformation for ac sinusoids, and turn the ac function-of-time trigonometry analysis into complex algebra. With this transformation, all the dc analysis techniques and theorems of Chapters 2 and 3 are usable with sinusoidal ac. Complex algebra involves the use of complex ($A + jB$) numbers, manipulated according to the rules set forth in Appendix B.

Sinusoids can thus be represented in four ways: as mathematical or graphical functions of time (the time domain), and as mathematical or graphical phasors (the frequency domain).

The ratio of **V** to **I** (phasor numbers) is now complex and is named impedance **Z**; **Z** = $R + jX$, where R is resistance as used in dc analysis and X, the reactance, is caused by two new passive elements, inductance L and capacitance C. The reciprocal of impedance **Z** is admittance **Y**.

Phasor, admittance, and impedance diagrams (on complex axes) are helpful aids in circuit analysis.

Transformers provide an efficient means of converting ac power from one voltage level to another. Voltage, current, and impedance can be converted in this way.

___

FOR FURTHER STUDY

Ralph J. Smith and Richard C. Dorf, *Circuits, Devices and Systems*, 5th ed., John Wiley & Sons, New York, 1992. Chapter 5.

___

PROBLEMS

Easy Drill Problems (answers at end of chapter)

D4.1. A two-terminal device is driven with an ac voltage source and has a corresponding current through it. Using cross-sectioned paper carefully plot the following waveforms (v or i vs t) on the same axes. (Make your plots large and clear. Select a time scale such that x-axis calculations will be easy on the axes you have drawn. Select v and i scales for easily-read graphs. The frequency is 100 Hz. Values for these plots can be had from your hand calculator.)
   (a) a 7.07-V voltage wave (peak amplitude 10 V);
   (b) a 3.53-A current wave (peak amplitude 5 A) which lags the voltage by 30°.

D4.2. **(a)** Represent the voltage wave of problem D4.1 above by the four representational methods for an ac wave.
   (b) Repeat for the current wave. (Use simple sketches.)

D4.3. Alternatively, solve problem D4.1 by programming and running a computer to make these plots. If possible, put the two waves on the same axis.

D4.4. The European standard domestic electric supply is at 220 V and 50 Hz.
   (a) Write an expression for $v(t)$ assuming the usual cosine wave and zero phase angle. What is ω for this system? What is the period in milliseconds? What instantaneous voltage will appear at 0.347 s? (Don't overloook the sign involved if any; don't forget that the 220-V specification is rms).
   (b) Carefully sketch quantitatively several cycles of this wave centered roughly on 0.347 s (use a hand calculator for data).
   (c) Alternatively, make a computer plot of this region.

D4.5. Assume the voltage of the previous problem is applied to a two-terminal load. Write an expression for $i(t)$. By what angle in degrees does the current lead or lag the voltage? Assume the load is a 100-Ω resistor.

D4.6. Repeat for an inductor of 100-Ω reactance.

D4.7. Repeat for a capacitor of 100-Ω reactance.

D4.8. Repeat for a load made up of the 100-Ω R and 100-Ω L in series.

D4.9. (a) Draw careful phasor diagrams showing voltage and current for problems D4.5–D4.8. (Follow the usual engineering practice of using rms phasors.)

    (b) Express these phasor quantities in mathematical notation.

D4.10. (a) Complex numbers **X** and **Y** are, respectively, $5\underline{/20°}$ and $2 - j6$. Sketch **X** and **Y** each on its own complex axes and give the polar and rectangular form of each.

    (b) Show the rectangular, polar, and axes presentations for $\mathbf{X} + \mathbf{Y}$, $\mathbf{X} - \mathbf{Y}$, $\mathbf{X} \times \mathbf{Y}$, $\mathbf{X}/\mathbf{Y}$, and $\mathbf{X}^{1.8}$.

D4.11. Solve the circuit sketched for $\mathbf{I}_x$ by

    (a) mesh analysis;

    (b) nodal analysis;

    (c) Thevenin's theorem;

    (d) Norton's theorem;

    (e) circuit reduction and simplification.

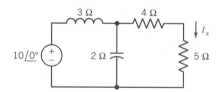

D4.12. Evaluate the following expression: $(1 + j2) + (2 + j3)(4 - j5)$.

D4.13. Three impedances are 100 Ω each. One is an R, one an L, and one a C. Find the total impedance if

    (a) they are in series;

    (b) in parallel;

    (c) R is paralleled with L, and C in series with that combination;

    (d) R is paralleled with C, and L in series.

D4.14. (a) Draw an impedance diagram for D4.13a showing how the three impedances add to the answer.

    (b) Convert the three impedances to admittances and draw an admittance diagram for the parallel case of D4.13b showing how the three admittances add to the answer. Convert the answer to impedance for a check.

D4.15. A transformer converts 4000 V coming from a distribution line to 220 V for domestic use. The equipment on each side of the transformer must be appropriately insulated to withstand the maximum voltage to be encountered. To at least what level of peak voltage must the insulation be designed on each side of the transformer?

D4.16. Complex numbers **X** and **Y** are, respectively, $4\underline{/45°}$ and $5 + j5$. Sketch **X** and **Y** each on its own complex axis and give the polar and rectangular form of each.

D4.17. For the two numbers in problem D4.16 tabulate the rectangular, polar, and axes presentations for $\mathbf{X} + \mathbf{Y}$, $\mathbf{X} - \mathbf{Y}$, $\mathbf{X} \times \mathbf{Y}$, $\mathbf{X}/\mathbf{Y}$, and $\mathbf{X}^3$.

D4.18. Find the polar and the rectangular 2-terminal impedance of four 10-Ω circuit elements, A, B, C, D with A and B in series and C and D in parallel, these two groups then connected in series. A and C are resistors, B and D an inductor and a capacitor, respectively.

D4.19. Solve D4.18 if A and B are an inductor and a capacitor, and if C and D are resistors.

D4.20. Solve D4.18 if A and B are a capacitor and a resistor, and if C and D are an inductor and a resistor.

D4.21. Solve D4.18 if A, B, and C are resistors and D an inductor.

D4.22. Solve D4.18 if A, B, and C are capacitors, and if D is a resistor.

D4.23. Solve for I_x in this circuit by mesh analysis and nodal analysis.

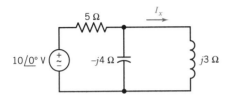

D4.24. Solve D4.23 twice again using Thevenin's and Norton's theorems.

D4.25. Solve D4.23 again by circuit simplification plus current division.

D4.26. A transformer has 100 turns on the primary and 300 on the secondary; a load resistor of 100 ohms is connected. 50 volts are applied from a source. Find I_p, I_{load}, V_{load}.

D4.27. A transformer has a turns ratio of 4 to 1. An ac voltage source is connected to the low (primary) side, and a 100-Ω resistive load is connected to the high side, which supplies it with 440 V. Find the two currents. What impedance does the primary see?

D4.28. The secondary (low side) of a 3/1 transformer is loaded with 50 Ω. The primary is connected in series with a 450-Ω inductor, and a 500-V 60-Hz source. Find the voltage across the 50-Ω resistor in polar form. Take source on real axis.

D4.29. In problem D4.28 find the primary and secondary currents.

Application Problems

P4.1. (a) What is the impedance of the two-terminal device of problem D4.1? Give this result in polar and rectangular form and displayed on complex axes.

(b) Sketch a series circuit of two elements (from R, C, L) that will represent this impedance, and give values of both elements in ohms.

P4.2. Given two two-terminal circuits A and B. Circuit A has a resistor, R_s, in series with an inductor, X_s. Circuit B has an inductor, X_p, in parallel with a resistor, R_p. The two circuits have the same total impedance at their terminals. Sketch the two two-terminal circuits A and B.

(a) Using complex algebra find expressions for R_p and X_p in terms of R_s and X_s.

(b) Find expressions for R_s and X_s in terms of R_p and X_p.

(c) If R_s and X_s are 5 and 3 Ω, respectively, what are R_p and X_p? Check your answers to see that they both have the same impedances at their terminals.

P4.3. Solve for $\mathbf{I}_x$ in the circuit sketched, by mesh methods. [Remember that a formal mesh equation cannot be written (and is not needed) for a mesh containing a current source.]

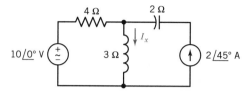

P4.4 Repeat using nodal methods.

P4.5. Repeat using Thevenin's theorem.

P4.6. Repeat using Norton's theorem.

P4.7. Repeat using superposition.

P4.8. For the circuit sketched, write
  (a) the mesh equations;
  (b) the nodal equations.
  (c) Solve for I_2 using whichever set of equations seems better.

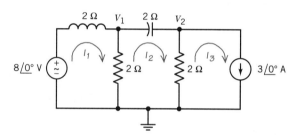

P4.9. (a) Sketch a voltage divider made up of Z_1 a 1000-Ω resistor and Z_2 a 1000-Ω inductive reactance. What is the complex voltage divider ratio?
  (b) Repeat substituting a capacitor for the inductor.
  (c) Does interchanging Z_1 and Z_2 make any difference in the previous parts?

P4.10. Make up a two-mesh circuit of your own containing at least two sources. Select some voltage or current as the unknown. Solve for it by as many methods as you can apply.

P4.11. A series circuit of four elements (50-Ω R, 50-Ω L, 50-Ω R, 50-Ω C) is connected to a 500-V ac source. Find the current in this series circuit. Draw a phasor diagram with I on the real axis. Add the five voltages on the diagram head-to-tail. How does this diagram illustrate KVL?

P4.12. Repeat P4.11 with this string of impedances: 10-Ω R, 20-Ω C, 30-Ω R, $(40 + j20)$-Ω device.

P4.13. A 10-Ω inductor and a 15-Ω resistor are connected in series to a 100-V, 60-Hz source. Draw a phasor diagram of this situation. Find the rectangular and polar forms of the circuit current.

P4.14. The situation in problem P4.13 is unchanged except that a 25-Ω capacitor is also connected across the source. Make a phasor diagram with the 100 V on the real axis and showing the three currents (C, R-L, source). Find the rectangular and polar forms of the three currents.

P4.15. A certain induction motor draws with an impedance of 139 Ω + j123 Ω is connected to 118-V ac voltage source

  (a) Find the current drawn in polar form.

  (b) Draw an accurate, large phasor diagram of V and I with V on the real axis.

  (c) To minimize the mains current an engineer places a capacitor across the motor input to reduce the source current. How much capacitive reactance should he provide for minimum source current? (*Hint:* add the capacitor current to your phasor diagram first to see how it will affect the current drawn.)

  (d) Will the capacitor change the motor current?

P4.16. Solve problem P4.8 by Thevenin's theorem.

P4.17. Solve problem P4.8 by a method you have not used previously.

Extension Problems

E4.1. **(a)** In problem P4.2 consider the parallel elements in their inverse form—G instead of R, B instead of X, **Y** instead of **Z**. Derive again the transformation from a parallel to an equivalent series circuit.

  (b) Repeat for a series to a parallel circuit. Can a capacitor ever replace an inductor by this transformation?

  (c) Check your work again using the values 5 and 3 Ω.

E4.2. Make an investigation to determine how resistive loading affects the voltage divider in problem P4.9(a). Plot on the same axes

  (a) the magnitude of the voltage divider ratio vs R_L;

  (b) the phase angle of the voltage divider ratio vs R_L.

E4.3. A two-terminal circuit is made up of a 50-Ω resistor in series with the parallel combination 50-Ω capacitive X and an inductor with reactance X_L as sketched. Plot the impedance (both magnitude and phase on the same axes) vs X_L as X_L varies from 0 to 1000 Ω.

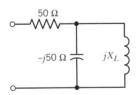

E4.4. The previous problem is a little unrealistic in that practical inductors always have some resistance associated with them. It can be represented in series or parallel as in problem P4.2. Assume that there is such an R_L that can be represented as 2 Ω in series with the L.

  (a) Replot the terminal **Z** vs X_L again with this small resistance added.

E4.5. Replace the series resistor R_L with an appropriate parallel resistor (see P4.2).

  (a) Replot the total circuit impedance again vs X_L.

  (b) Compare the results of **(a)** with the previous plot. Wherein do they differ?

Study Questions

4.1. Compare ac analysis to dc for a simple circuit. What is similar and what is different?

4.2. How does impedance **Z** differ from resistance R?

4.3. Describe with sketches the effect of varying the "phase" of an ac current or voltage wave.

4.4. What is the relation between the *effective value* of a sinusoidal wave and its *peak* value?

4.5. What is the period of the sinusoidal radiation from station WTAG at 580 kHz?

4.6. Describe how the "phasor representation" of an ac wave is related to its "instantaneous representation."

4.7. What are the four forms of representation for ac currents and voltages?

4.8. What are the three complex-number representations of the phasor form for **I** and **V**?

4.9. How do capacitors and inductors enter into impedance?

4.10. Name the three common ac diagrams, list the ac elements shown on each, and state which of the diagrams rotate.

ANSWERS TO DRILL PROBLEMS

D4.2. $10\cos(628t)$, $5\cos(628t - 30°)$

D4.4. $311\cos(314t)$ V, 314, 20, -183

D4.5. $3.11\cos(314t)$ A, 0°

D4.6. $3.11\cos(314t - 90°)$, 90° lag

D4.7. $3.11\cos(314t + 90°)$, 90° lead

D4.8. $2.21\cos(314t - 45°)$, 45° lag

D4.9. $220\,\underline{/0°}$ **V** for each, $2.2\,\underline{/0°}$ **A**, $2.2\,\underline{/-90°}$, $2.2\,\underline{/90°}$, $1.56\,\underline{/-45°}$.

D4.10. $6.70 - j4.29$, $2.70 + j7.71$, $19.7 - j24.8$, $0.02 + j0.79$, $14.6 + j10.6$

D4.11. $1.85\,\underline{/-146°}$ A

D4.12. $24 + j4$ or $24.3\,\underline{/9.5°}$

D4.13. 100, 100, $50 - j50$, $50 + j50$

D4.14. $100 + j100 - j100 = 100$; $.01 - j.01 + j.01 = .01$

D4.15. $5657+$ V; $311+$ V

D4.16.

	Rect	Polar
X	$2.83 + j2.83$	$4\,\underline{/45°}$
Y	$5 + j5$	$7.07\,\underline{/45°}$

D4.17. $11.1\,\underline{/45°}$; $3.1\,\underline{/-135°}$; $28.3\,\underline{/90°}$; $0.57\,\underline{/0°}$; $64\,\underline{/135°}$

D4.18. $15.8\,\underline{/18.4°}$ Ω

D4.19. $5\,\underline{/0°}$ Ω

D4.20. $15.8\,\underline{/-18.4°}$ Ω

D4.21. $25.5 \underline{/11.3°} \; \Omega$

D4.22. $25.5 \underline{/-78.7°} \; \Omega$

D4.23. 24, 25. $3.01 \underline{/-67°} \; A$

D4.26. 4.5 A; 1.5 A; 150 V

D4.27. 17.6 A; 4.4 A; 6.25 Ω resistive

D4.28. $117.9 \underline{/-45°} \; V$

D4.29. $786 \underline{/-45°} \; mA; \; 2.36 \underline{/-45°} \; A$

CHAPTER 5

Alternating-Current Power

Alternating-current power concepts involve both *instantaneous power* and *average power*. As with ac voltages and currents, a small letter p or $p(t)$ is used to designate instantaneous values. Unlike capital V and capital I (which stand for effective or rms values) capital P designates the *average value* of power over an entire cycle of the ac wave. When engineers speak of "power" without any modifying adjective, they always mean capital P, the average power of the ac wave.

In investigating these matters this chapter will show that it is impossible to put average power into inductances or capacitances. The only passive circuit element which can absorb average power is a resistance.

Analysis will show that the frequency at which power pulsates in an ac system is twice the frequency of the voltage or current. The chapter discusses complex power briefly (its application to power factor correction will be considered in Chapter 17) and ends with a presentation of three-phase systems, the almost universal mode in which power is generated and transmitted.

Reader objectives should include a thorough understanding of

1. Average and instantaneous power and, if interested in power use:
2. Complex power
3. 3-wire and polyphase systems

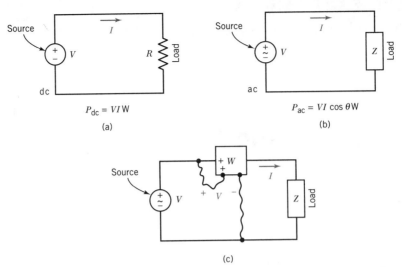

$$P_{dc} = VI \, \text{W}$$

(a)

$$P_{ac} = VI \cos \theta \, \text{W}$$

(b)

(c)

FIGURE 5–1 Alternating-current power calculations require one additional factor, the *power factor*, cos θ. θ is the phase angle between current and voltage. Wattmeter must have both current and voltage connections. From them it also senses θ.

5.1   AVERAGE POWER *P*

Figure 5–1a and b illustrate the similarity and difference in ac and dc power calculation for a two-terminal load. It will be shown that for ac,

$$P = VI \cos \theta. \tag{5.1}$$

It is important to note that the *V* and *I* in this formula are absolute values (simply the rms magnitudes without angles). Compared to dc there is one additional factor beyond the simple formula $P = VI$. This factor, cos θ, is called the *power factor* and varies between 0 and 1. It is usually expressed in percent. θ is the *power factor angle*, and is the angle between current and voltage.

Figure 5–1c is identical to 1b except that it shows a wattmeter connected to read the power transmitted from left to right. In accordance with Equation (5.1) the meter will need both voltage and current inputs and some way of knowing the power factor. As shown, there are four terminals (two for voltage and two for current). One terminal of each pair is marked for polarity, in this case with a plus sign (although sometimes simply with dots). Current flows through the meter from left to right; voltage input is at the bottom of the meter symbol. If current flows into the current plus terminal and the voltage positive lead is connected to the voltage plus terminal, the meter will read upscale for power moving from left to right. This agrees with the power convention established in the second chapter.

The same wattmeter can measure either ac or dc power. Responding to $v(t)i(t)$, it integrates this instantaneous product over the ac cycle—in many meters by mechanical inertia—and thereby senses the difference in phase between *V* and *I*.

5.2   INSTANTANEOUS POWER $p(t)$ AND POWER FACTOR

Let us see how the power factor concept comes about. Consider the instantaneous values of current and voltage for circuit 5–1b. Figure 5–2 illustrates some of the possibilities. If **Z** is a purely resistive load, current will be in phase with voltage and the resulting wave relations for v, i, and p will be as in Figure 5–2a. The current goes through its zeros at the same points as the voltage. The current is positive when the voltage is positive.

Calculating instantaneous power going into the load, at any instant t, we write

$$p(t) = v(t)i(t). \tag{5.2}$$

Figure 5–2a carries out this multiplication, multiplying each value of v by its corresponding value of i to get the curve marked $p(t)$. Voltage and current waves shown have a period of 2 s ($f = 0.5$ Hz). Peak values of the voltage and current waves shown are 100 V and 2 A. For the portion between 0 and 0.5 s both v and i are positive and must have a positive product. For the portion between 0.5 and 1.5 s both are negative and must therefore have a positive product. Thus the peak power is 200 W. The power curve varies between 0 and 200 W (and will be seen below to be sinusoidal).

Note also that the power wave alternates at twice the frequency of the voltage or current. But this should be no surprise to anyone acquainted with elementary trigonometry since

$$(\cos x)(\cos x) = \tfrac{1}{2} + \tfrac{1}{2}\cos(2x),$$

or for the specific case where $p(t) = v(t)i(t)$,

$$p(t) = 100 \cos(2\pi ft)2 \cos(2\pi ft).$$

But $f = 0.5$ Hz,

so $$p(t) = 100 + 100 \cos(2\pi t), \tag{5.3}$$

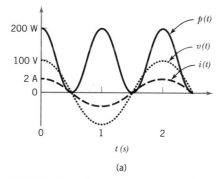

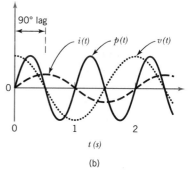

t (s)

(a)

t (s)

(b)

FIGURE 5–2 Relations between voltage, current, and power. At any instant, $p = vi$: (a) for a resistor, (b) for an inductor.

as illustrated in Figure 5–2a. And averaging the power wave over a full cycle, the sinusoidal component of Equation (5.3) contributes, of course, nothing. This leaves 100 W from $(V_p/\sqrt{2})(I_p/\sqrt{2})$ or VI. Here the power factor is $\cos\theta = \cos 0° = 1$, where θ is the phase difference between V and I.

More generally, from trigonometry,

$$\cos x \cos y = \tfrac{1}{2}\cos(x - y) + \tfrac{1}{2}\cos(x + y). \qquad (5.4)$$

Taking the most general forms of voltage and current sinusoids—$V_p\cos(2\pi ft + \theta_V)$ and $I_p\cos(2\pi ft + \theta_I)$—and inserting them in Equation (5.4) will produce

$$p(t) = \tfrac{1}{2}V_pI_p[\cos(\theta_V - \theta_I) + \cos(4\pi ft + \theta_V + \theta_I)]$$

Or using the effective values (peak/$\sqrt{2}$) for V and I,

$$p(t) = VI\cos\theta + VI\cos(2\omega t + \theta_V + \theta_I),$$

where $\theta = \theta_V - \theta_I$, the phase angle between the voltage and current, and the power factor is

$$\text{pf} = \cos(\theta) \qquad (5.5)$$

Thus the power wave is indeed a sinusoid of double the frequency plus a dc term. And the expression for average power P (averaged over a whole cycle) is established as

$$P = VI\cos\theta. \qquad (5.6)$$

Going back to Figure 5–1b, suppose now that the **Z** load is a pure inductive reactance of 50 Ω. As shown in Chapter 4, current will lag voltage by 90°. Multiplying $v(t)$ by $i(t)$ point by point, the $p(t)$ wave of Figure 5–2b will result. Note that this wave will have a zero wherever either v or i has a zero. There will be 90° portions of the waves where v and i have opposite signs, producing negative power. In fact symmetry will show that the resulting sinusoid is centered on the zero axis with an average power of zero.

The interpretation of negative power (see *power convention* in Chapter 2) is that the load is putting power back into the source—their roles are temporarily reversed. Thus for an inductor, power flows from source to load during two parts of the cycle. Power then flows back into the source for two other parts of the cycle. Similar reasoning will show the same thing for a purely capacitive load.

Power is the rate of energy flow. Thus energy flows from source to inductance or capacitance and then flows back from inductance or capacitance to source. The energy is not expended (turned into heat as in a resistor). It is stored temporarily in the magnetic field of the inductor or in the electric field of the capacitor. Chapter 7 shows that these storage phenomena produce short transient effects when switches are operated or other changes made in circuits.

It is instructive to develop the graphical result of Figure 5–2b through the analytical means of Equation (5.5).

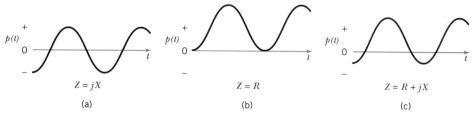

FIGURE 5–3 As the power factor angle of a load is changed, the power versus time curve moves vertically. *Average power, P,* for a purely reactive load is zero.

A yet more general case is developed in situations where the load is a combination of resistance and inductance or resistance and capacitance (or all three). This is of course the normal situation in a practical ac electrical load. There must be a resistive component or no work is being done. However, industrial loads often include large numbers of induction motors which produce a lagging power factor (*I* lags *V*).

To summarize, there are three types of results for the $p(t)$ wave as the nature of the load **Z** in Figure 5–1b is changed. These are illustrated in Figure 5–3. In effect this sinusoidal power wave is simply moved up and down on the axes, as the load's phase angle changes. The X's could be either positive or negative.

EXAMPLE 5.1 In the circuit of Figure 5–1c, V = 100 V at 60 Hz; **Z** = $2 - j5\ \Omega$. Assume that the voltage phasor lies on the real axis at $t = 0$.

Find: $p(t)$ and P. What is the instantaneous power at $t = 0.01$ s? Which way is it flowing?

Solution: This load can be represented as a 2-Ω resistor in series with a capacitor whose reactance is $-5\ \Omega$. Its polar form is 5.39 $\underline{/-68.2°}\ \Omega$. **I** = **V**/**Z** = 18.6 $\underline{/68.2°}$ A. Whence $i(t) = 18.6\sqrt{2}\cos(377t + 68.2°)$ A. $v(t) = 100\sqrt{2}\cos(377t)$. Using the trigonometric expression of Equation (5.4), $p(t) = v(t)i(t) = 1857\cos(-68.2) + 1857\cos(754t + 68.2) = \underline{690 + 1857\cos(754t + 68.2°)}$ W. Then $P = VI\cos\theta = 100 \times 18.6\cos68.2 = \underline{690\ W}$. Remembering that in the hybrid expression for $p(t)$, $754t$ is in radians and 68.2 in degrees, $p(0.01\ s) = \underline{-737\ W}$. At this instant power is flowing out of the capacitive portion of the load and back into the source.

5.3   COMPLEX POWER

Engineers involved in electric power systems find the following development of a *complex power* concept useful. Other people seldom use it.

Consider the 5-Ω complex load Z in Figure 5–4a, driven by a 100-V source. $I = V/Z = 100/5 = 20$ A. $P = VI\cos\theta$. Assume the power factor to be 80% lagging. Then $P = VI\cos\theta = 1.6$ kW. We could draw a *power triangle* or *complex power diagram* as in (b). The hypotenuse is the *apparent power S* in units of volt-amperes (VA) or

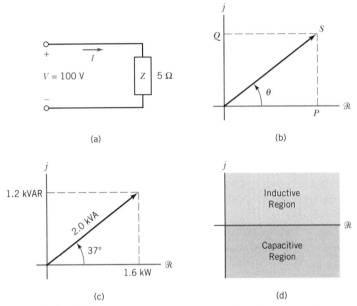

FIGURE 5–4 The *complex power diagram* develops from power P an *apparent power, S,* and *reactive power Q,* where $P = VI \cos \theta$, $S = VI$, $Q = VI \sin \theta$.

more usually kVA; the horizontal component is the power P in watts (W) or kW (with which we are familiar). The vertical component, *reactive power, Q* is in units of VAR or kVAR, volt-amperes reactive. From the diagram:

$$P = VI \cos \theta = 1.6 \text{ kW} \tag{5.7}$$

$$Q = VI \sin \theta = 1.2 \text{ kVAR} \tag{5.8}$$

$$S = VI = 2.0 \text{ kVA.} \tag{5.9}$$

Taking θ as 37°, sketch (c) gives the numerical results.

This can also all be done with complex numbers, like this:

$$\mathbf{S} = S\underline{/\theta} = P + jQ = \mathbf{VI}^*$$

where $\mathbf{I}^*$, the *complex conjugate* of $\mathbf{I}$, forces a lagging power factor situation into the upper right quadrant of the complex axes, by changing the sign of θ. (If $\mathbf{I} = A + jB$, then $\mathbf{I}^*$ *is defined as* $A - jB$.)

But rather than be confused about signs, it is simpler to think in terms of absolute values of θ, V and I, while recognizing that kVAR, when absorbed by a load, is customarily considered positive if the load is inductive (which is the typical case because of the many induction motors in industry). The unit kVARC is sometimes used for kVAR going into a capacitive (leading power factor) load. Power engineers also speak of a capacitive load as generating reactive power and of inductive loads as absorbing it.

It is important to recognize that the unmodified term, power, always means $VI\cos\theta$, and appears on the horizontal (real) axis of the power diagram. Of course, reactive power Q isn't power at all. It simply represents the storage and giving back of energy in capacitors and inductors.

EXAMPLE 5.2 A 220-V source supplies a load, Z, $(5 + j5)$ Ω.

Find: pf, pf angle θ, P, Q, S, and sketch the complex power diagram for the load.

Solution: see Figure 5–5. $\theta = \arctan 5/5 = 45°$; $Z = \sqrt{(5^2 + 5^2)} = 7.07\ \Omega$; $pf = \cos 45 = 70.1\%$; $I = V/Z = 220/7.07 = 31.11$ A; $P = VI\cos\theta = 220 \times 31.11 \times .707 = 4.84$ kW; $S = VI = 220 \times 31.1. = 6.84$ kVA; $Q = S\sin 45 = 6.84 \times .707 = 4.84$ kVAR. The upper right quadrant is selected because the load is inductive. See Figure 5–5.

5.4 SINGLE-PHASE THREE-WIRE POWER SYSTEMS

Power to households, commercial establishments, and other small users is usually distributed single phase but with two available voltages—220 and 110 V. (As observed earlier these two numbers are nominal, and the actual voltages vary by up to ten percent or so.) Figure 5–5 shows how the two voltages are derived from the center-tapped secondary of a small (often only 10 kVA) transformer for ac systems.

The advantage of this supply system is that small loads such as lamps and hand tools can be accommodated with more safety at 110 V, while larger loads—particularly heavy heating devices—can have the advantage of higher voltage and lower current. Furthermore, if the 110-V loads are reasonably balanced between the two sides, some of the advantages of lower current are retained even for them. For safety reasons (and according to National Electrical Code requirements, discussed in Chapter 21) the center wire is grounded as shown in Figure 5–6a. A further advantage is that no part of the circuit, even for 220-V appliances, is more than 110 V above ground.

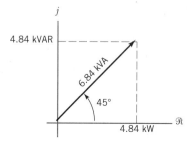

FIGURE 5–5 Solution to example 5.2

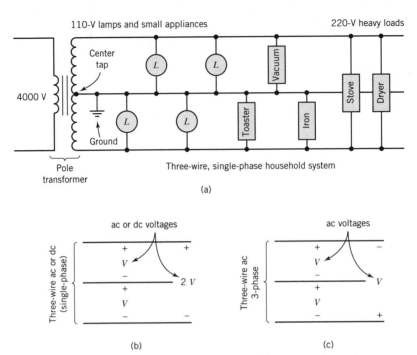

FIGURE 5–6 Three-wire, single-phase systems and three-wire, three-phase systems are distinguished easily by taking voltmeter readings between the three pairs of conductors.

It is easy to determine whether a three-wire system is connected according to this scheme (Figure 5–6b). Of the three voltage readings between wires, two will be equal and the third twice as great. This two-voltage three-wire system is also used infrequently with dc.

5.5   THREE-PHASE DELTA SYSTEMS

Power is universally produced, transmitted, and distributed with *three-phase systems*—systems with three or four conductors. Such arrangements are simply three single-phase (two-wire) systems combined. The three single-phase systems are constrained in a certain phase relationship with one another.

Figure 5–7 illustrates the development of a three-phase delta system. In the three single-phase systems (5–7a), voltages V_1, V_2, V_3 are equal in magnitude and 120° apart in phase as shown in the phasor diagram (5–7b). If each is 100 V and each of their load resistances is 10 Ω, currents I_1, I_2, I_3 will be 10 A. These complex currents will lie along the corresponding voltage phasors and will consequently be 120° apart also. Suppose now, in the interest of saving copper (saving wire) the six wires are combined into three as shown in Figure 5–8a. None of the three load currents will be affected (nor their corresponding currents in the three generators

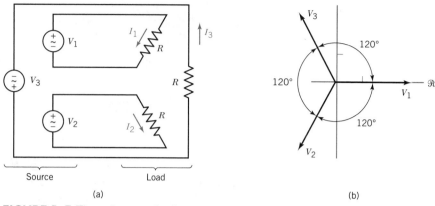

Source   Load

(a)   (b)

FIGURE 5–7 These three single-phase systems can be combined into the three-phase Δ system of Figure 5–8(a). The voltage sources must be related in phase as sketched. V_2 and V_3 may be interchanged.

or sources).* The generators and loads should work as before. It might be objected that the three sources have been short-circuited on each other. But adding three equal and 120°-apart voltages around the source loop on the left side of the circuit makes a zero sum. In practice when these systems are set up, great care must be exercised that one voltage not be connected backwards, destroying the zero-sum effect. In this system it will be noted that the *line currents*, labeled $\mathbf{I}_a$, $\mathbf{I}_b$, $\mathbf{I}_c$, are each the difference of two *phase currents*—for instance $\mathbf{I}_a = \mathbf{I}_1 - \mathbf{I}_3$. But a carefully drawn diagram will show that these phasors ($\mathbf{I}_1$ and $-\mathbf{I}_3$) when combined are 60° apart, so that with three equal loads

$$I_{\text{line}} = \sqrt{3} I_{\text{ph}}. \tag{5.10a}$$

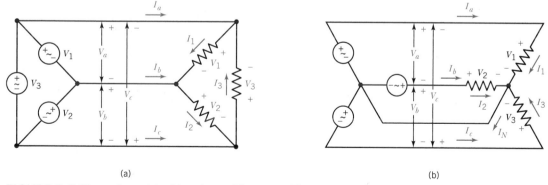

(a)   (b)

FIGURE 5–8 Three-phase delta (a) and wye (b) systems. To maintain voltages uniform under unbalanced load, the Y system requires a fourth (neutral) line (shown dashed). Note the carefully chosen, symmetrical notation (signs and arrows) for V's and I's.

*In plant power systems these sources are usually the secondaries of transformers.

And in any delta,

$$V_{ph} = V_{line}. \qquad\qquad (5.10b)$$

Note that $\sqrt{3} = 1.732 = 2 \cos 30°$.

There are two principal advantages of three-phase systems. Each line carries 1.732 times as much current as any of the six lines of Figure 5–7a. If the same size wire were used, I^2R losses would be three times as great per wire as for any of the six lines. But there are only three lines. Increasing their cross-sectional area by only 50% would bring the losses down to the original case. Thus 25% of the copper is saved. This saving of copper in transmission lines and machines is significant.

The other advantage is even more important: power flow is smoothed. Single-phase 60-Hz ac power pulses 120 times a second. Thus all single-phase electrical machines must be constructed to withstand not only peak torques and stresses but also a continual 120-Hz vibration. It can be easily shown that in three-phase machines power flow is completely uniform. Machines can be smaller for the same ratings; vibration problems are reduced.

Systems such as the above example with equal phase loads are called *balanced*. Phase loads in balanced systems must be the same in angle as well as in magnitude. Equation (5.10a) is true only for balanced delta systems.

5.6   THREE-PHASE "WYE" SYSTEMS

There is a second way to connect three-phase systems called "*wye*" or Y and illustrated in Figure 5–8b. The central node to which all sources are connected (and referred) is called the *neutral* or *N* point. It is often referred to as *ground*, and is frequently actually grounded for safety reasons (Chapter 17). Similarly there is a neutral point in the load connection. Here the phase currents (again marked I_1, I_2, I_3) are equal to the line currents (I_a, I_b, I_c). But the *line voltages* (sometimes called "line-to-line") V_a, V_b, V_c, are $\sqrt{3}$ times the phase voltages V_1, V_2, V_3.

If this system were to have only three wires there would be some interesting problems with unbalanced loads. The voltages on the load phases could adjust themselves differently for each phase (to enable Kirchhoff's current law to be satisfied). If the loads were lamps, some might be very dim and others burn out. Engineers refer to this situation in the load as "the neutral rising above ground." Of course it cannot happen on the source side where the generators (or transformers) establish the voltages and their phasing.

To overcome this voltage problem a fourth wire (the neutral wire) is usually run in wye systems (dashed in Figure 5–8b). This neutral wire carries the unbalance currents and keeps voltage magnitudes equal for all load phases. Since the system is designed to be quite well balanced, neutral wire currents will be small, and there is no need to use a large wire for this purpose. And since the neutral wire is run at ground potential it has no need for expensive and bulky insulation. Single-phase loads on wye systems are generally taken from one hot wire and the neutral. (The term *hot wire* refers to one of the three phase wires as opposed to the ground or neutral line.)

Thus for four-wire wye systems, whether balanced or not, and for balanced three-wire wye systems,

$$V_{\text{line}} = \sqrt{3} V_{\text{ph}} \qquad\qquad (5.11a)$$

and
$$I_{\text{line}} = I_{\text{ph}}. \qquad\qquad (5.11b)$$

While wye systems have the disadvantage of requiring this fourth (usually smaller) wire, they have several advantages over delta. There is a natural ground in the system—often quite helpful for single-phase loads. For a given phase voltage, transmission is at higher voltages with resulting smaller line currents and further copper savings. Older industrial delta systems can be increased in capacity almost 75% by adding a fourth wire, raising distribution voltage, and reconnecting loads in wye.

The wye and delta systems of Figures 5–8 can be generalized as follows: All three-phase systems comprise three ac voltages equal in magnitude but 120° apart. Their currents depend, however, on the nature of the load. If the three load impedances are equal in magnitude *and* phase angle, the three phase currents will be equal in magnitude and the three line currents will be equal in magnitude. This is called a *balanced system.* Only for balanced systems will the $\sqrt{3}$ factor relating line to phase currents in the delta and line to phase voltages in a three-wire wye prevail.

Seriously unbalanced systems sacrifice to some degree the advantages of three-phase operation. Most such systems can usually be analyzed one phase at a time or, at worst, by mesh methods. There are many single-phase loads—lights are a prime example. Plant engineers place any single-phase device across one of the three phases (between two of the three wires or, in a four-wire wye, usually between one phase line and neutral). Such loads are distributed among the three phases in such a way as to balance them. As time goes on and new equipment is added, loads tend to develop an imbalance and must be periodically reconnected for better balance.

In moderate size high-rise buildings the three-phase line is taken up a utility shaft and one phase is picked off for the lighting and office power on each floor. For individual houses the 220/110-V system probably comes from a single-phase transformer mounted on a nearby utility pole. This transformer is supplied from a single-phase distribution circuit at some such voltage as 2300 or 4000. Tracing the distribution circuit back farther, it will be found that it comes from a single hot wire of a four-wire three-phase wye line on a main street and the ground carrier of that line.

There are two possible orders of *phase sequence.* Standing at the point marked *x* in Figure 5–9b, one sees the three phasors (120° apart) rotating counterclockwise (as discussed in Chapter 4). First $\mathbf{V}_a$ passes us, then $\mathbf{V}_b$, then $\mathbf{V}_c$. The phase sequence seen is abc. If $\mathbf{V}_b$ and $\mathbf{V}_c$ were interchanged in position the sequences would be acb. These two sequences are often called *positive* and *negative,* respectively. A little thought will show that these are the only two possibilities. For any other sketch that can be drawn, sooner or later as the phasors rotate counterclockwise the diagram will match one or the other of these. In a well-maintained industrial plant, the line conductors at various points in the distribution system may be actually tagged with labels, "a," "b," "c."

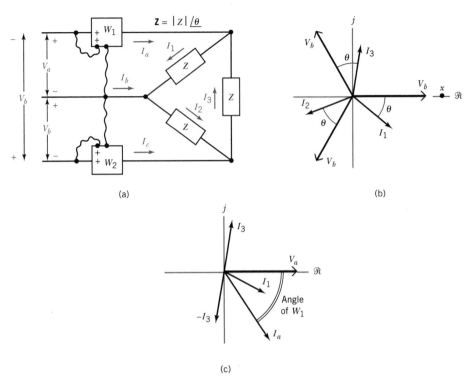

(a)   (b)

(c)

FIGURE 5–9 The "n-1" wattmeter method (two wattmeters for three lines). $W_1 + W_2$ is the total power in the three loads.

There is no reason that delta loads cannot be used on systems supplied by wye-connected generators or transformers and vice versa. But unbalanced three-wire wye loads require a neutral (fourth wire) and there is no convenient place to connect it on a delta system.

5.7   THREE-PHASE POWER

Power is a scalar quantity, and so the three-phase power is simply the addition of the powers in the three phases. Or in a balanced system $P = 3P_{\text{ph}} = 3V_{\text{ph}}I_{\text{ph}} \cos \theta$. But for a delta,

$$V_{\text{line}} = V_{\text{ph}} \quad \text{and} \quad I_{\text{line}} = \sqrt{3}I_{\text{ph}},$$

and for a wye,

$$I_{\text{line}} = I_{\text{ph}} \quad \text{and} \quad V_{\text{line}} = \sqrt{3}V_{\text{ph}}.$$

Hence for either

$$P = \sqrt{3}V_{\text{line}} I_{\text{line}} \cos \theta. \tag{5.12}$$

This expression is the basic power formula for a *balanced* system in either wye or delta. Note that the power factor of a three-phase system, cos θ, is the *power factor of the phase*. Thus the term *power factor* has no meaning for a badly unbalanced system.

In single-phase systems one wattmeter is required to measure the power, connected as in Figure 5–1c. [A wattmeter reads a power which is $VI \cos \theta$ where the V and the I are (rms) magnitudes (not complex phasors) connected to it and θ is the angle between I and V.]

In general it takes $n - 1$ wattmeters to measure power in a transmission system with n wires. The $n - 1$ wattmeters are connected in any of the n lines (Figure 5–9a). The voltage return (minus terminal) for each wattmeter is connected to the line with no wattmeter in it. This method will work whether the system is balanced or not (and will work, in fact, whether the system is three phase or single phase). The readings of the wattmeters are simply added together and the sum is the total power going down the three-wire (or *n*-wire) line.

In balanced three-phase, three-wire systems wattmeters W_1 and W_2 will read the same (half the total power) if the load is purely resistive. As the power factor worsens (becomes smaller), the two readings will differ more and more until with the power factor angle at 60° (pf = 50%) one wattmeter will read zero. For worse power factors than that, one meter reads negative and must be subtracted from the other. But that situation is rare since industrial plants are not designed to have such poor power factors. Reversing the phase sequence will simply interchange the two wattmeter readings.

For a four-wire wye system three wattmeters are needed. Polyphase wattmeters and watt-hour meters are often built with two (or three) sets of coils on the same shaft to mechanically totalize their readings. Such a meter (in this case a watt-hour meter to record *energy* used) is found in the three-wire single-phase system of dwellings and small commercial establishments.

EXAMPLE 5.3 Given the three-wire three-phase 4000-V power system illustrated in Figure 5–9a. The impedance load in each phase is $40 + j30 \ \Omega$. Phase sequence is positive (abc).

Find: the magnitudes of all voltages and currents and the readings of the two wattmeters.

Solution: From the problem information, including phase sequence, construct the phasor diagram shown in Figure 5–9b. The three voltages are equal in magnitude. The angle θ is calculated as $\arctan(30/40)$ or 36.9°. Since the system is balanced (the same magnitude and angle of **Z** is each load) the three line currents will be 120° apart. It is helpful to think of the phase-current phasors as welded together and rotating jointly with respect to the fixed three voltage phasors if the phase load angles should be changed. $\mathbf{Z} = 40 + j30 = 50 \underline{/36.9°}$. And the magnitude of $I_{\text{ph}} = V/Z = 4000/50 = \underline{80 \text{ A}}$. The angles for the various phase currents shown in the diagram. The line voltage is 4000 V. *System voltage* always refers to line (line-to-line) voltage. It is evident from the circuit diagram that in delta systems the three phase voltages are identical to the three line voltages. Thus $V_{\text{ph}} = \underline{4000 \text{ V}}$. Line current is $\sqrt{3} \, I_{\text{ph}}$

or <u>138.6 A</u>. System power factor is 80% (cos 36.9°). $P_{ph} = I^2R = 80^2 \times 40 = 256$ kW. $P_{3-ph} = 3 \times 256 = 768$ kW.

Wattmeters read $VI \cos \theta$ where this θ is not the system power factor angle but the angle between the V and I connected to that particular wattmeter's terminals. W_1 reads $V_a I_a \cos \theta$. But $\mathbf{I}_a = \mathbf{I}_1 - \mathbf{I}_3$ by KCL. This subtraction has been carried out on the phasor diagram of Figure 5–9c. $\mathbf{I}_1$ and $-\mathbf{I}_3$ are 60° apart. The angle between V_a and I_a is seen to be 30° + 36.9°. Then $W_1 = 4000 \times 138.6 \times \cos 66.9° = \underline{217.4 \text{ kW}}$. Similarly W_2 reads $V_b I_c$ cos of angle between $\mathbf{I}_c$ and $-\mathbf{V}_b$. (This minus sign is found from the relative polarity markings on V_b and the wattmeter in the circuit of Figure 5–9a and is *absolutely essential* in the angle calculation.) This calculation should be carried out with a diagram similar to 5–9c. $\mathbf{I}_c = \mathbf{I}_3 - \mathbf{I}_2$. Thus $W_2 = 4000 \times 138.6 \cos 6.9° = \underline{550.4 \text{ kW}}$. (Note that the minus sign on $\mathbf{V}_b$ did not affect the magnitude number entered for the voltage, 4000.) Summing the two wattmeter readings, 217.4 + 550.4 = 767.8 kW which checks well with the original calculation of $3P_{ph}$.

5.8   SUMMARY

For the average power P a third factor, cos θ [the power factor (pf)] enters the equation so that $P_{ac} = VI \cos \theta$, where θ (the power factor angle) is the angle between V and I. V and I are rms magnitudes. For instantaneous ac power $p(t) = v(t)i(t)$, or more commonly, $p = vi$. These equations yield results in watts.

Because of the cos term and 90° phase difference between V and I, neither inductor nor capacitor can receive any average power P. The instantaneous power put into them during parts of the ac cycle comes back out during other parts.

A "complex" ac power $\mathbf{S}$ can be postulated by multiplying phasor voltage and phasor current together so that $\mathbf{S} = \mathbf{V} \times \mathbf{I}^*$, where the term $\mathbf{I}^*$ denotes the complex conjugate of $\mathbf{I}$. To find rectangular components, $\mathbf{S} = P + jQ$, where P is the average power as above, and Q the "reactive" power. $Q = VI \sin \theta$. Q is absorbed by inductors and given out by capacitors. These ideas can be sketched on a complex power diagram with axes P and jQ. $\mathbf{S}$ is placed in the upper right quadrant for inductive (lagging) loads and in the lower right hand quadrant for capacitive (leading) loads.

Essentially all electric power is generated and distributed as three-phase power (using three- or four-wire systems). Machine construction and power transmission are more economical than with two-wire (single-phase) systems. Three-phase systems are designed and operated to run in a nearly "balanced" condition—with the three loads (between the three conductors) equal in magnitude and phase.

There are two common three-phase connections, delta (Δ) and wye (Y). For either of these, operating under balanced conditions, total three-phase power can be calculated as $P = \sqrt{3}\, V_{line} I_{line} \cos \theta$, where cos θ is the power factor of the loads. In the balanced delta system $I_{line} = \sqrt{3}\, I_{ph}$ and $V_{ph} = V_{line}$. In the balanced wye connection, $I_{line} = I_{ph}$ and $V_{line} = \sqrt{3}\, V_{ph}$.

It takes $n - 1$ wattmeters to measure the power on n lines, regardless of whether they are three phase or whether they are balanced.

FOR FURTHER STUDY

Ralph J. Smith and Richard C. Dorf, *Circuits, Devices and Systems*, 5th ed., 1992, John Wiley & Sons, New York. Chapter 7.

PROBLEMS

Easy Drill Problems (answers at end of chapter)

Sketch circuits involved with each problem.

D5.1. A 220-V source drives a resistive load of 100 Ω. Find the power involved.

D5.2. A 220-V source drives a load **Z** of 12 + j5 Ω. Find the power involved in source and load.

D5.3. Sketch a wattmeter in the circuit of D5.2.

D5.4. Sketch a phasor diagram and a Z-diagram for the circuit of D5.2.

D5.5. Sketch a complex power diagram for the circuit of D5.2.

D5.6. In Figure 5–2b a steady-state point-by-point power wave is developed for an inductor. Repeat this graphical analysis for a capacitor which would produce the same current magnitude. *V* is still 100 V peak; *f* is still 0.5 Hz. How much power is delivered to the capacitor at 0.4, 0.8, and 1.2 s?

D5.7. Repeat the graphical analysis of D5.6 for a load of 35.4 + j35.4 Ω.

D5.8. Add a wattmeter to the circuit of D5.7. Show that its reading, $VI \cos \theta$, is the same as $I^2 R$.

D5.9. A 115-V source drives a parallel load comprising a 70-Ω *R* and a 70-Ω *C*. What power is going into the load?

D5.10. Sketch a circuit for D5.9 with two wattmeters, one measuring the source power and the other the power into *R*. How are these powers related?

D5.11. Show that the rms value of a symmetrical triangular wave, as sketched, is $1/\sqrt{3}$.

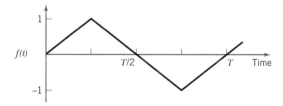

D5.12. A symmetrical trapezoidal wave rises from 0 to 1 in 1 unit of time, remains at 1 for the next unit, falls to 0 during the third unit, and then repeats the pattern in the negative direction. Find its rms value.

D5.13. A voltage wave is made up of symmetrical half circles and has a peak value of 1 V. Find its rms value.

D5.14. A load **Z** is fed by a 2-kV, 60-Hz source and receives a current of 10 A lagging by 45°.
(a) Draw the complex power diagram for this load.
(b) Find *P* and *Q*, give units.
(c) Sketch a series representation of this load with 2 elements. Give values in ohms.
(d) Repeat with paralleled elements.

D5.15. Redo D5.14 the current leading by 45°.

D5.16. In a balanced 3-phase delta situation (see Figure 5–8) the voltage is 440 V and each phase load is 10 Ω resistive. Find the following quantities as meters would read them (magnitudes only): I_{line}, V_{line}, I_{ph}, V_{ph}, pf, power per phase, total 3-ph power.

D5.17. Repeat D5.16 with the 10-Ω load R's in wye. Line-to-line voltage is still 440.

D5.18. A 300-V 60 Hz single phase source drives an R-L series circuit paralleled with an R-C series circuit. $R = R = 50$ Ω and the L and the C are both 50 Ω. By $VI \cos \theta$ what power is supplied to the circuit? Check your result with I^2R.

D5.19. In D5.18, connect one wattmeter at the source and two others across the paralleled branches. Does $W_1 = W_2 + W_3$?

D5.20. For D5.19, sketch a large, accurate phasor diagram showing the 3 currents and 5 voltages.

D5.21. For D5.18, sketch a large, accurate complex power diagram showing VA in from source, and the two P's and two Q's o.

Application Problems

P5.1. The half-wave rectifier, as sketched, supplies a 5-Ω resistive load, providing half-wave current pulses peaking at 2.82 A.

    (a) What is the rms value of the current? (*Hint:* Each pulse is 180° of a sinusoid. The dotted portions of the full wave are missing.)

    (b) What percentage of peak is this?

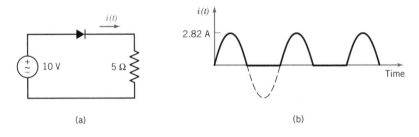

(a)   (b)

P5.2. The sources for this full-wave rectifier circuit are shown as the secondary of a transformer (transformers are covered in Chapters 4 and 18). You may substitute two in-phase sinusoidal ac sources if you wish. As in P5.1, this circuit provides the same 2.82-A pulses to the same load, but twice as many.

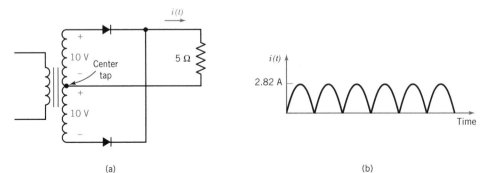

(a)   (b)

(a) What is the rms value of this current?

(b) What percentage of peak is this? This kind of circuit is commonly used to supply dc from an ac source.

(c) Find the value of the dc current (average) in both circuits.

(d) What percentage of peak is this in both cases?

P5.3. (a) What is the rms value of a square wave in terms of peak? Assume the square wave symmetrical with respect to the abscissa—that is, no dc component.

(b) Repeat assuming the square wave is all positive—that is, a dc component equal to peak/2.

P5.4. (a) A symmetrical square wave varies between $+1$ V and -1 V. What is its rms value? (see previous problem).

(b) The same square wave is moved up until it is varying between plus 2 V and zero. What is its rms value?

(c) The same square wave is moved down until it is varying between $+\frac{1}{2}$ and $-\frac{3}{2}$ V. What is its rms value?

(d) Check your results in the previous parts with the formula $I = \sqrt{I_1^2 + I_2^2 + I_3^2 + \cdots}$ where the I's are rms values and I_1, I_2, etc. are components of I. (*Hint:* Components in this case are the dc average and the symmetrical square wave.)

P5.5. In the circuit sketched the wattmeter reads 960 W and the ammeter 6 A. $R = 10\ \Omega$, $R_L = 6\ \Omega$, $X_L = 8\ \Omega$. Find V_S, V_C, I_C, I_L, X_C. (*Hint:* Make a careful phasor diagram.)

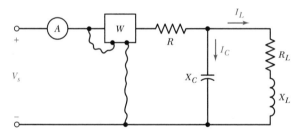

P5.6. In the circuit sketched W_1, W_2, W_3 read 610 W, 350 W, 80 W. R_2, $R_3 = 30$ and $20\ \Omega$. $X_C = -10\ \Omega$.

(a) What is the rms value of the current? (*Hint:* Each pulse is 180° of a sinusoid. The dotted portions of the full wave are missing.)

(b) What percentage of peak is this?

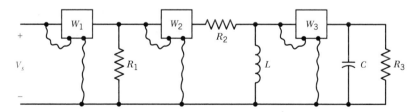

P5.7. A 2400-V, 60-Hz, balanced, three-phase delta circuit (see Figure 5–8a) has phase load impedances of $40 + j60\ \Omega$.

(a) Draw a careful, large-scale phasor diagram of the three voltages and six currents involved.

(b) Find the following quantities as meters would read them (no angles): V_{ph}, V_{line}, I_{ph}, I_{line}, P_{ph}, P total, power factor.

P5.8. Repeat the previous problem with the three loads connected in wye.

P5.9. Put wattmeters in the center and bottom line of the system described in problem P5.7.

  (a) Using a phasor diagram find the quantities (V, I, angle) to which each wattmeter will respond.

  (b) Find the readings of the two wattmeters. Does this check with the results of P5.7? (Remember that the return voltage coil leads go to the line without a wattmeter.)

P5.10. A 100-V three-phase system operates with unbalanced three-wire wye-connected loads of 10, 10, and 20 Ω, all purely resistive.

  (a) Find the currents in and voltages across each of the three single-phase loads.

  (b) If these three loads represent parallel groups of incandescent lamps, rated at $100/\sqrt{3}$ V, describe the operation of each group on this system.

  (c) Solve again **(a)** and **(b)** if a fourth wire is connected as a "neutral."

Extension Problems

E5.1. Show that in a balanced three-phase system the total instantaneous power flow is constant with respect to time.

E5.2. Assume a balanced three-phase 440-V system with delta phase loads of Z at an angle of θ, where the magnitude Z is 10 Ω and the angle θ can vary from 0° to 90°. Investigate the relative readings of two wattmeters connected in two of the lines as a function of θ. In particular show that at an angle 60° one meter reads 0 and beyond that goes negative. (*Hint:* A phasor diagram is, as usual, useful here.) What do the wattmeters read for purely resistive loads?

E5.3. In the two-wattmeter method for measuring the power flowing through a three-wire line, there are three possible ways to install the wattmeters. Investigate the effects of these three methods on the wattmeter readings.

E5.4. Investigate the effect of phase sequence on the two wattmeter readings, installed in accordance with the $n-1$ wattmeter method.

E5.5. In the balanced three-phase system described in problem P5.7 assume that there is line resistance of 10 Ω in each of the three lines, effectively reducing the 2400-V supply before it reaches the delta-connected loads. Find again these quantities at the load as meters would read them, V_{line}, V_{ph}, I_{line}, I_{ph}, pf, P_{ph}, P total. (*Hint:* The only definition of system pf is the pf of the load phase.)

E5.6. Power engineers work with balanced systems by postulating an *equivalent wye* of four wires. The fourth or neutral wire is assumed to be perfectly conducting. This equivalent load circuit produces the same line currents, the same voltage with respect to ground or neutral, and the same power and line power factors, etc. They then make a *one-line diagram*, considering only a single-phase wire of this equivalent system with an assumed neutral. Power results are multiplied by three.

  (a) Derive a transformation system for converting balanced Δ phase loads to the equivalent wye phase load.

  (b) Apply this system to problems P5.7 and E5.5 and compare results to those solutions. (*Hint:* Start your investigation with a resistive load. The key here is to have the same line currents and phase powers.)

E5.7. One advantage of the delta connection in industry is the possibility of using the *open delta* in case one source is lost. In the sketch the three-phase sources are shown as three transformer secondaries—the common industrial situation. If one transformer is burned out or faulty, as shown, it can simply be left out and the system will still be capable of supplying a reduced amount of three-phase power. For $V = 100$ V and $Z = 10\ \Omega$ resistive,

(a) Investigate the load currents and transformer currents that will flow in the two cases. Make a tabulation of them.

(b) Transformers are rated in terms of kilovolt-amperes and voltage—actually current and voltage limitations. What percentage of full system kilovolt-ampere rating can be had with a balanced load on open delta?

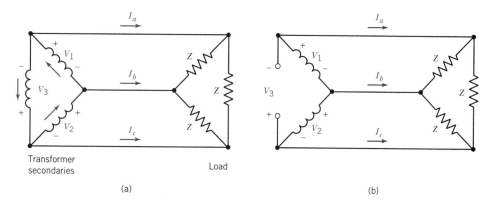

(a)   (b)

Study Questions

5.1. What does the term *average power* (*P*) mean in ac? Over what period of time is it being averaged? How is it differentiated from *instantaneous power* [$p(t)$]? Which do engineers mean when they speak simply about *ac power* and use the concept of $VI \cos \theta$? Are the *V* and *I* in that formula magnitudes or complex quantities?

5.2. Explain with sketches how in a 60-Hz circuit the power can pulse at 120 Hz.

5.3. What do the terms *power factor* and *power factor angle* mean?

5.4. Explain with sketches why the power factor is required in ac power calculations. Does it apply to *P* or $p(t)$? In what way, if any, does the power factor affect $p(t)$?

5.5. Show by algebra and phasor diagrams the relation between the angle of the load impedance and the power factor angle.

5.6. Explain the concept of *effective* value of an ac voltage or current wave.

5.7. List the steps of a calculus procedure to find the effective value of any waveform.

5.8. Sketch a complex power diagram for a load **Z** and explain its component parts.

5.9. Explain what *reactive power* (*Q*) is and relate it to the *V* and *I* across and through capacitors and inductors.

5.10. Sketch delta and three- and four-wire wye systems. Do the loads have to be connected in the same scheme as the sources?

5.11. Why do most wye systems have a fourth (neutral) wire? Why is the neutral wire normally made of smaller-gauge copper?

5.12. What is meant by a *balanced* three-phase system? What is actually supposed to be balanced? How could you tell in a plant whether a system is balanced?

5.13. How many wattmeters are needed to properly measure the power in a four-wire wye system?

5.14. Why is it that the same wattmeter will usually work on either ac or dc without modification?

5.15. Give two reasons why three-phase systems are used so universally.

ANSWERS TO DRILL PROBLEMS

D5.1. 484 W

D5.2. 3.44 kW

D5.3. Figure 5–1(c)

D5.4. Sketch

D5.5. 3.44 kW; 1.43 kVAR; 3.72 kVA

D5.6. −58.8, 95.1, −95.1 W.

D5.7. −55.77, 194.07, −61.95 W

D5.8. 70.6 W; 70.6 W.

D5.9. 189 W

D5.10. 189 = 189 W

D5.11. Answer given with problem

D5.12. 0.745 A

D5.13. 0.817 A

D5.14. 14.14 kW, 14.14 kVAR; 141 Ω, 374 mH; 282 Ω, 748 mH.

D5.15. C = 9.38 μF; parallel capacitor.

D5.16. 76.2 A, 44 A, 440 V, 100%, 19.36 kW, 58.08 kW.

D5.17. 25.4 A, 25.4 A, 254 V, 100%, 6.45 kW, 19.36 kW.

D5.18. 1.80 kW

D5.19. 1.8 kW = 0.9 kW + 0.9 kW

D5.20. 300 $\underline{/0°}$ V; 212 $\underline{/-45°}$ V twice; 212 $\underline{/45°}$ V twice; 6.0 $\underline{/0°}$ A; 4.24 $\underline{/-45°}$ A; 4.24 $\underline{/45°}$ A.

D5.21. 1.8 kVA; 0.9 kW twice; 0.9 kVAR; −0.9 kVAR.

CHAPTER 6

Alternating-Current Topics

Chapter 6 extends ac circuit concepts in several directions, providing a more detailed look at L's and C's, some mathematical justification for the phasor methods previously presented, filters, and the important concept of frequency response.

Frequency response is familiar to most people through high-fidelity music systems. Amplifiers and speakers tend to discriminate against notes whose frequency is either very low or very high. Thus there is a pattern of response versus frequency which should be considered in purchasing them. They are recommended and sold very largely on the basis of the low and high cutoff frequencies. The speakers in particular, driven by the amplifiers, respond differently to different notes.

The reader might select as his or her immediate objectives in this chapter:

1. Understand capacitors and inductors.
2. Develop some understanding of the mathematical basis of the complex voltage, current, and impedance of Chapter 4.
3. Become familiar with the idea of frequency response.

6.1   DRIVING FORCE AND RESPONSE

When a car is pushed, its resulting response is to move forward. This push can be by hand or by its engine. The push force can be thought of as a function of time—for example, it might be a constant force that starts at $t = 0$, or it could even be sinusoidal. The responding car movement is also a function of time.

This concept of *driving force* and *response* is common to all physical systems including electric circuits. Earlier chapters have occasionally used that language without identifying it. The driving forces can be current sources or voltage sources. The response can be any voltage or current, anywhere in the circuit.

A particularly useful form of driving force and response thinking for electrical circuits is illustrated in Figure 6–1a where V_{in} can be thought of as a driving signal into some device or circuit network and V_{out} is the response, the signal coming out. Such circuits are called *two-port* circuits or networks because they have two "doors," one for a signal going in on the left and one for a modified signal coming out on the right. The network between the two ports might be, for example, one channel of a stereo amplifier. Or it might be as simple as the two-element, low-pass filter circuit of Figure 6–1b. Chapter 8 looks at two-port networks in more detail.

In most of these two-port circuits the ratio between V_{out} and V_{in} depends on frequency. This is particularly apparent in 6–1b, where the voltage divider ratio (between Z_C and the total series impedance of the two elements) must decrease as frequency increases. The ratio of V_{out} to V_{in}, as a function of frequency, is called the *transfer function*. This matter is examined further toward the end of this chapter.

As in the automobile example above, these driving force, response, and transfer function concepts occur in many areas of technology. In linear systems, if the driving voltage is sinusoidal, the steady-state or forced response will be sinusoidal, although its phase and magnitude may be changed.

Strictly speaking the response has two parts—a short, decaying transient of some kind, and then the steady state mentioned above. In the automobile pushing example that started this section consider a force exerted on the body of the car standing still. When the constant pushing force is first begun, the car will move forward easily on its springs until a limit is reached where the wheels must begin to follow. At that point some rather jerky transient response movements will occur

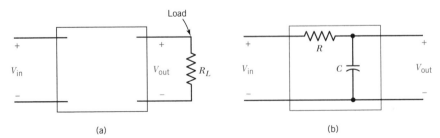

(a)   (b)

FIGURE 6–1 Two-port (four-terminal) networks convert a driving force (V_{in} for example) to a response at the output (V_{out}). (b) is a simple two-port example, a low-pass filter circuit.

until the whole system settles down to an even movement forward, which will be-come the steady-state response.

The *steady-state* part of the response (steady forward movement) is considered to take place from time $t = 0$, the beginning of the push event. The *transient* (temporary) movement, which dies out in a few seconds or so, is considered to be superimposed upon (added to) the steady state. But its effect is seen for only a short time. This double response is characteristic of all physical systems, and Chapter 7 will show that it comes from the solution of the differential equation govern-ing them.

In electrical systems (circuits) the transient current and voltage surges last often for only a small fraction of a second before the circuit settles down to its steady-state operation. These transients are sometimes important, damaging elec-trical equipment or propagating troubles sequentially through systems. There are also some electrical systems which rely on transient phenomena—pulse circuits are an example. But for many practical circuits careful examination will show electrical transients can be safely ignored.

Transients are best studied in the time domain; Chapter 7 treats this topic more fully. Steady-state response is the topic of the present chapter. This is best studied in the frequency domain using the phasors introduced in Chapter 4.

6.2   CAPACITORS AND INDUCTORS

At this point our collection of two-terminal elements includes capacitors and in-ductors as well as resistors. It is useful to look at L's and C's in more detail.

The greatest value of capacitors and inductors is that their impedances de-pend on frequency. It has already been seen in the frequency domain (phasor representation) that these impedances include the imaginary operator j, which introduces a phase shift in circuits. This section examines the point more fully, and will also consider the operation of capacitors and inductors in the time domain (dealing with their voltages and currents as functions of time).

All passive electrical elements are defined by the relation between the voltage across them and the current through them. A resistor has a resistance R,

$$v = Ri, \tag{6.1}$$

and its unit is the ohm. Similarly a capacitor has a capacitance C and an inductor has an inductance L; their defining equations are

$$i = C\frac{dv}{dt}, \tag{6.2}$$

$$v = L\frac{di}{dt}. \tag{6.3}$$

The units of capacitance and inductance are the *farad* (F) and *henry* (H). The phasor analysis of Chapter 4 specified capacitors and inductors in ohms. The simple conversion from farads and henries to ohms will be described below.

These derivative relationships may also be integrated, so that

$$v = \frac{1}{C} \int_0^t i \, dt, \, + \, v(0) \tag{6.4}$$

$$i = \frac{1}{L} \int_0^t v \, dt + i(0) \tag{6.5}$$

Figure 6–2 summarizes these elements and their phasor relations.

It is possible to handle all three elements properly in circuit analysis solely on the basis of these mathematical expressions. However it will help in remembering and applying these relationships to know something of the physical construction of inductors and capacitors and to keep some analogies in mind.

A capacitor (Figure 6–3a) is essentially two conducting plates separated from each other by some dielectric (insulator). The dielectric is often air for variable capacitors. Cylindrical capacitors are common. Two thin sheets of foil separated by two thin sheets of plastic dielectric are rolled up together. Capacitance is proportional to area (A) and inversely proportional to the dielectric thickness (d) separating the two plates. The nature of the dielectric (expressed through its dielectric constant ϵ) also enters into the relation. In suitable units,

$$C = A\epsilon/d. \tag{6.6}$$

For air and in SI units, $\epsilon = 8.84 \times 10^{-12}$ F/m, so two plates 1 mm apart and 10^{-3} m^2 in area form a capacitor of 8.84 pF.

If Equation (6.2) is integrated it will be seen that at any instant

$$q = Cv, \tag{6.7}$$

where q is the charge in coulombs on the capacitor and C its capacitance in farads.

Figure 6–3 shows a capacitor being *charged* by a constant current source (its

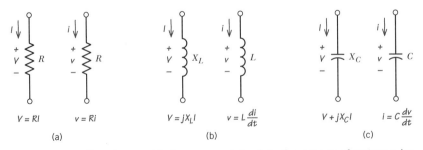

$$V = RI \qquad\qquad v = Ri \qquad\qquad V = jX_L I \qquad\qquad v = L\frac{di}{dt} \qquad\qquad V + jX_C I \qquad\qquad i = C\frac{dv}{dt}$$

(a)   (b)   (c)

FIGURE 6–2 Capacitors and inductors are defined, in the manner of resistors, by the relation of current through them to voltage across them. This can be done in either the time or frequency (phasor) domain.

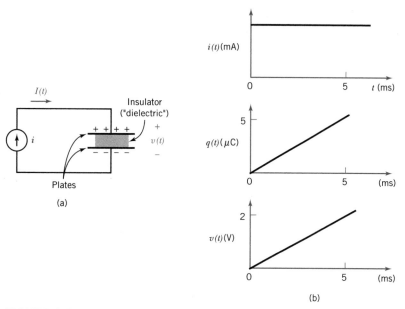

FIGURE 6-3 An initially empty capacitor (shown in cross section) being charged by a constant-current source produces the $q(t)$ time functions shown in (b).

cross section is illustrated). Typical curves of current, charge, and voltage, all versus time, are shown.

EXAMPLE 6.1 A capacitor is charged with a 1-mA current source and is to have a voltage of 2 V after 5 ms. The capacitor is initially discharged (that is, $v = 0$ when $t = 0$). Figure 6–3 illustrates this situation.

Find: the size of the capacitor in microfarads.

Solution

$$v = \frac{1}{C} \int i\ dt;$$

for constant i: $\qquad v = It/C = q/C,$

hence $\qquad C = It/v = (1 \text{ mA} \times 5 \text{ ms})/2 \text{ V}$

$$C = \underline{2.5\ \mu\text{F}}$$

Physical appreciation for the charging process comes from an analogy in fluid dynamics. Compare the beaker being filled with water in Figure 6–4 with the capacitor being charged in Figure 6–3. For a constant flow (current) the beaker (capacitor) fills with water (charge) at a uniform rate. Since the height (voltage) is proportional to the amount of water (charge), it also in-

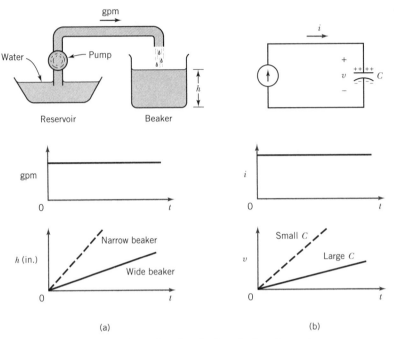

FIGURE 6–4 Hydraulic analogy for the functioning of capacitors.

creases at a uniform rate. If the beaker is narrower (capacitance less) the water height (voltage) goes up faster. Note that a capacitor can't pass a constant (dc) current indefinitely. An unlimitedly large voltage would build up on it and puncture the dielectric. What is the fluid analogy to capacitor voltage rating in this system? What happens when the rating is exceeded?*

EXAMPLE 6.2 Water flows into a basin with vertical walls at the rate of 3 ft³/s.

Find

1. What should the cross-sectional area of the basin be if water is to rise 2 ft in 5 s?
2. A capacitor is charged from zero volts with a current of 3 μA for a 2-ms period. How large should the capacitor be if it has a voltage rating of 100 V and 80% of the rating is not to be exceeded?

Solution

1. In 5 s there will be 15 additional cubic feet of water in the basin. This is to produce an increase in height of 2 ft. 15 ft³/2 ft = 7.5 ft².

---

*Beaker height. Water spills over top of beaker.

2. Charge delivered to capacitor is $\int i \, dt = it$ (for this case of constant current)
$= 3 \times 10^{-6} \times 2 \times 10^{-3} = 6 \times 10^{-9}$ C.

$$C = Q/V = 6 \times 10^{-9}/80 = \underline{75 \text{ pF}}.$$

A consequence of the definition of C is that capacitors in parallel add. Capacitors in series combine as the reciprocal of the sum of the reciprocals, or (for two) the product over the sum. That is,

$$C_p = C_1 + C_2$$

and
$$C_s = C_1 C_2/(C_1 + C_2).$$

This is exactly the reverse of resistance combinations. Does the hydraulic analogy above lead to this same conclusion?

An inductor (a practical form of inductance L) is formed by a coil of wire (ideally a perfect conductor). Sometimes the inductance is increased by winding the coil on a *magnetic core*—usually made of iron or steel. Current through the coil produces a magnetic field in the core and in the air or space around it, which comes out one end and bends around on all sides back into the other end, as shown in Figure 6–5.

When a voltage is applied across the winding, current begins to flow and increase, and a magnetic field spreads outward, cutting through the wires. Magnetic flux cutting the conductors generates a voltage in them which opposes the change of current and change of flux. Thus the current can build up only gradually. Similarly, if the current is to be reduced or stopped, a voltage must be applied across the coil in the other direction. As current reduces, magnetic flux reduces, cutting

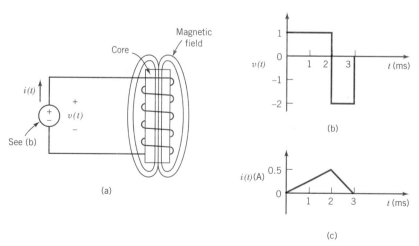

FIGURE 6–5 This inductor is made of wire wound on an iron core. A voltage source (see b waveform) charges it first in one direction and then in the other with resulting current waveform (c).

back through the wires and generating a voltage opposing this reduction. Thus the current can be reduced only gradually.

The inductance L of a single coil is often called *self-inductance* as opposed to the *mutual inductance* between two coils. It is caused by flux (created by coil current) cutting the same coil's conductors to induce a voltage in them. This fact explains why winding wires into a coil form creates an inductor. In that form a stronger flux is generated and more conductors are cut when the flux changes. Mutual inductance (caused by the flux created by one coil cutting the conductors of another coil) is examined later in this chapter.

A useful analogy for a coil comes from a simple mechanical system: a mass slides (with no friction) along a table under the push of a force. Figure 6–6 makes this comparison. The mass (inductance) remains at rest (has zero current) until it is acted on by the force (voltage) for at least some small period of time. Its speed (current) continually increases as long as the force is applied. Similarly if the sliding mass is to be slowed or stopped (current reduced or again made zero), force (voltage) must be applied in the opposite direction.

The definition of inductance Equation (6.3) leads to the conclusion that inductors combine as resistors—*inductors in series add*, and *inductors in parallel combine* as the reciprocal of the sum of the reciprocals or (in the case of two) as the product over the sum. These relations are easily demonstrated with algebra.

EXAMPLE 6.3 A 4-mH inductor has no current through it at $t = 0$.

Find: the current as a function of time if a constant $1\ V$ is applied across it for 2 ms followed by $-2\ V$ for the next 1 ms, as in Figure 6–5.

Solution

$$i = \frac{1}{L} \int_0^t v\ dt.$$

Then for v constant and t in milliseconds ($L = 4$ mH),

$$i = \tfrac{1}{4}vt \text{ A}.$$

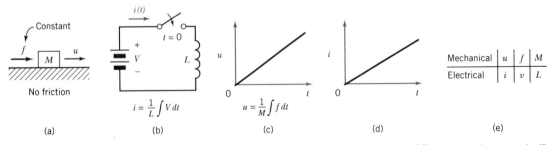

FIGURE 6–6 $F = Ma$ is a good mechanical analogy for inductors. The time integral of f/M (starting from standstill) yields velocity u just as the time integral of v/L yields current i.

Dividing the solution into zones (see Figures 6–5b and c), we have for $0 < t < 2$ ms

$$i = \tfrac{1}{4}t \text{ A},$$

and for $2 < t < 3$ ms

$$i = 0.5 - \tfrac{1}{2}(t - 2) \text{ A}.$$

This result is shown in Figure 6–5.

Inductance and capacitance can be specified in either their own units—henries for inductance, farads for capacitance—or in ohms of reactance at a specific frequency. Reactance or impedance values will change if frequency changes. The relation between reactance X (or impedance $\mathbf{Z}$) and capacitance or inductance is

$$X_L = \omega L \quad \text{and} \quad X_C = -1/\omega C. \tag{6.8}$$

Or
$$\mathbf{Z}_L = j\omega L$$
and
$$\mathbf{Z}_C = -j/\omega C = 1/j\omega C \tag{6.9}$$

These relations are derived in Section 6.3. Thus the impedance of an inductor (coil) increases in proportion to frequency. And the impedance of a capacitor is inverse to frequency. A perfect resistor has a value of resistance independent of frequency.

Let us once more summarize the definitions of resistance, inductance, and capacitance exhibited in Figure 6–2. If the signals applied to them are sinusoidal we can use the phasor form for analysis. These expressions can then be considered in either phasor (capital letter notation) or in instantaneous (small letter) form. They say in words:

- A resistance (symbol R, unit ohm) is a two-terminal element in which the current through it is proportional to the voltage across it. The phasor definition shows current to be always in phase with voltage. $i = v/R$, $\mathbf{I} = \mathbf{V}/R$.
- An inductance (symbol L, unit henry) is a two-terminal element in which the voltage across it is proportional to the rate of change of current through it. $v = L\,di/dt$. In the phasor form: Current lags voltage by 90°, and voltage is proportional to current and proportional to frequency. $\mathbf{V} = j\omega L\mathbf{I}$.
- A capacitance (symbol C, unit farad) is a two-terminal element in which the current through it is proportional to the rate of change of voltage across it. $i = C\,dv/dt$. Current leads voltage by 90°. $\mathbf{I} = \mathbf{V}j\omega C$.

EXAMPLE 6.4 The inductor of Figure 6–2b has inductance 0.2 H and frequency 1000 Hz (or $2 \times \pi \times 1000 = 6283$ rad/s). Its current has a peak value of 5 A and at $t = 0$ is 30° ($\pi/6$ rad) past its positive peak.

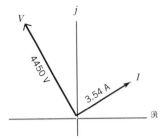

FIGURE 6-7 Phasor sketch for example 6.4.

Find: the voltage across the inductor in phasor notation and as a function of time. Show on the complex axes the current and voltage phasors. Find the instantaneous value of the voltage at $t = 6$ ms.

Solution: Using the impedance concept and phasor notation, $\mathbf{Z}_L = j\omega L = j(6283)(0.2) = j1257\ \Omega$. $i(t) = 5\cos(6283t + \pi/6)$. Then $\mathbf{I} = (5/\sqrt{2})\underline{/30°}$. $\mathbf{V} = \mathbf{ZI} = 1257\underline{/90°} \times 3.54\underline{/30°} = 4450\underline{/120°}$ V. The phasors are shown in Figure 6-7. Expressed as a function of time this answer for voltage would be $v(t) = V_p\cos(\omega t + \theta)$ or $v(t) = 6293\cos(6283t + 120°)$. At $t = 6$ ms, $v(t) = 6293\cos[6283 \times 0.006 + 120\ (\overline{1/57.3})] = \underline{-3140\ \text{V}}$ (the $1/57.3$ factor converts degrees into radians).

Practical resistors are usually quite good examples of resistance. At frequencies much higher than those for which they were designed, they begin to show some inductance (and later capacitance) in addition to their resistance. They are rated in ohms and watts.

Since inductors are made of coils of wire they usually have a significant amount of resistance and are appropriately represented as an inductance in series with a resistance. In addition the iron in cored inductors adds further power loss. Both resistance and core establish definite practical limits on current. More will be said about these points in Chapter 18. Air-core inductors are used principally at radio frequencies—roughly one-half megahertz and higher.

Capacitors also have losses which can be represented by a small series resistance (called *ESR* or *equivalent series resistance*). This is usually evident in very large capacitors, which are chemical (electrolytic) devices. At high frequencies they also exhibit some series inductance. Losses can alternatively be represented by parallel *R*, sometimes called *leakage resistance*. Practical capacitors have a limited voltage capability before they break down or punch through. This rating is specified along with the capacitance.

6.3   THE MATHEMATICS OF IMPEDANCE AND PHASORS

Previous sections have presented the plain facts of phasor representation and the effect of alternating currents on inductors and capacitors with the resulting complex impedance concept. These facts and mathematical expressions are entirely adequate to understand and solve ac circuit problems. But it is enlightening to consider these matters in more depth. This section provides a more extended mathematical treatment of phasors and impedance. It will also facilitate comprehension of control theory as presented in Chapter 10.

Suppose a current $i = I_p \cos(\omega t)$ flows through an inductance L. What will the resulting voltage across the inductance be? Figure 6–2b illustrates the problem. According to the definition of inductance, $v = L \, di/dt$ so if

$$i = I_p \cos(\omega t)$$

then
$$v = -\omega L I_p \sin(\omega t)$$

or
$$v = \omega L I_p \cos(\omega t + \pi/2). \tag{6.10}$$

From this it appears that if the ωt and the $\pi/2$ could somehow be brought out in front of the whole expression, Equation (6.10) would lead toward the phasor impedance expression, $V = j\omega L I$.

A useful identity (due to the mathematician Euler) states that

$$e^{jx} = \cos x + j \sin x \tag{6.11}$$

so that
$$I_p \, e^{jx} = I_p(\cos x + j \sin x).$$

Thus the current $I_p \cos(\omega t + \theta_I)$ is the real part of the above complex exponential expression, or in notation form,

$$I_p \cos(\omega t + \theta_I) = \mathrm{Re}[I_p \, e^{j(\omega t + \theta_I)}] \tag{6.12}$$

This is simply the idea, earlier mentioned, that the rotating phasor has two components—a cosine wave on the horizontal (real) axis and a sine wave on the vertical (j) axis. Only the cosine wave is needed to express ac currents or voltages. To use this idea in analysis it is customary to postulate a complex current and a corresponding complex voltage, $\hat{i}$ and $\hat{v}$, which have both the cosine and j sine parts according to Euler's identity presented above. These complex values are used with the understanding that we are actually interested only in their real parts. This is useful because complex algebra has the property of being simpler to apply here than real algebra just as the phasor representation is simpler than the sinusoid it replaces. With this understanding,

$$i(t) = I_p \cos(\omega t + \theta_I),$$
$$v(t) = V_p \cos(\omega t + \theta_V),$$

whence
$$\begin{cases} \hat{i} = I_p \, e^{j(\omega t + \theta_I)} \\ \hat{v} = V_p \, e^{j(\omega t + \theta_V)}; \end{cases} \tag{6.13}$$

and since
$$\hat{v} = L \frac{d\hat{i}}{dt},$$

then
$$\hat{v} = L\frac{dI_p\, e^{j(\omega t + \theta_I)}}{dt},$$
(6.14)

so,
$$V_p\, e^{j\theta_V} e^{j\omega t} = j\omega L I_p\, e^{j\theta_I} e^{j\omega t}.$$
(6.15)

Note that in this equation the factor $e^{j\omega t}$ is (according to Euler) the mathematical rotator making both current and voltage (right- and left-hand sides of the equation) rotate counterclockwise. It will in fact be present in every voltage or current term in phasor analysis. Engineers therefore simply omit this factor with the understanding that it is everywhere present. Dividing both sides of Equation (6.15) by $e^{j\omega t}$ leaves us with

$$V_p\, e^{j\theta_V} = j\omega L I_p\, e^{j\theta_I}.$$
(6.16a)

But in the phasor notation of Chapter 4 (where $\underline{/\theta}$ represented $e^{j\theta}$),

$$\mathbf{V}_p = V_p\,\underline{/\theta_V}$$

and
$$\mathbf{I}_p = I_p\,\underline{/\theta_I}.$$

Dividing both sides of Equation (6.16a) by $\sqrt{2}$ and rewriting, we have

$$\mathbf{V} = j\omega L\mathbf{I}.$$
(6.16b)

But
$$\mathbf{V} = \mathbf{Z}\mathbf{I},$$

so
$$\mathbf{Z}_L = j\omega L,$$
(6.17a)

nicely establishing the impedance concept for an inductor. Note that the $\mathbf{V}$ used here includes its angle, θ_V. The same sort of thinking will just as easily establish

$$\mathbf{Z}_C = -j/\omega C.$$
(6.17b)

In Equation (6.16b) the factor j has rotated the factor $\omega L I$ by 90° or $\pi/2$ radians counterclockwise, thus positioning $\mathbf{V}$ 90° ahead of $\mathbf{I}$ for an inductor. Equation (6.16) holds for whatever angle is associated with the complex $\mathbf{I}$.

This use of phasors or complex numbers to represent sinusoidal voltages or currents can be more intuitively visualized as follows. Assume that Figure 4–7a shows a peak phasor which is rotating counterclockwise at a rate of 60 times a second (or whatever the frequency associated with this form is). At time $t = 0$ the phasor has an angle of 37° with the real axis. As it sweeps around, its horizontal projection (on the real axis) is producing a sinusoid which has a positive maximum when $\omega t = -37°$.* Its vertical projection (on the j axis) is another sinusoid but one that

---

*Or whatever the radian equivalent of this angle is—engineers seem to speak and think in degrees or radians interchangeably but never mix them up when it is time to calculate.

has little practical interest. The two terms of Euler's identity represent these two sinusoidal components. The phasor mathematical representation includes both these sinusoids.

Now imagine turning Figure 4–7a on edge and looking at it from the bottom of the drawing. The projection on the real axis moves back and forth sinusoidally with all the amplitude, phase, and frequency information. But of course that is actually the voltage at any instant.

Thus the short-hand phasor notation of Section 4.4, for example $V \underline{/\theta°}$, is justified. The complex impedance concepts of Section 4.6 are shown to be based in mathematical fact. Equations (6.8) and (6.9) are justified.

6.4   RESONANCE AND FREQUENCY RESPONSE

To return to the concepts of driving force and response developed in Section 6.1, consider the *resonant* circuit of Figure 6–8a, which diagrams the front end of a radio receiver with antenna and first tuning circuits. The antenna, intercepting electromagnetic (radio) waves, develops a voltage between antenna and ground, that is, between points a and b. This voltage appears as $\mathbf{V}_a$ and is further processed and utilized in the remainder of the receiver circuits to the right (not shown).

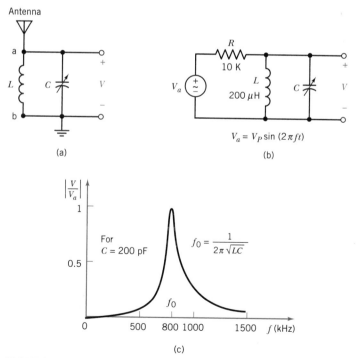

FIGURE 6–8 A radio receiver circuit as an example of frequency response.

Using Thevenin's theorem it is helpful to replace the antenna and ground with a Thevenin source comprising $\mathbf{V}_a$ and R (Figure 6–8b); R represents antenna resistance—not shown in Figure 6–8a.

The point of interest here is this: suppose the frequency of $\mathbf{V}_a$ is varied over the entire standard broadcast band (500–1500 kHz). What will the gain $\mathbf{V}/\mathbf{V}_a$ be as a function of frequency? Take as values for the circuit $R = 10$ kΩ, $\mathbf{V}_a = 100$ μV, $L = 200$ μH, $C = 200$ pF.

Of the various ways to calculate $\mathbf{V}/\mathbf{V}_a$ we choose the voltage divider concept:

$$\mathbf{V}/\mathbf{V}_a = \mathbf{Z}_2/(\mathbf{Z}_1 + \mathbf{Z}_2),$$

where $\mathbf{Z}_2$ is the parallel impedance of L and C, and $\mathbf{Z}_1$ is R. Then

$$\mathbf{Z}_1 = R$$

$$\mathbf{Z}_2 = \frac{j\omega L(-j/\omega C)}{j\omega L - j/\omega C} = \frac{j\omega L}{(j\omega)^2 LC + 1}.$$

Then

$$\frac{\mathbf{V}}{\mathbf{V}_a} = \frac{j\omega/RC}{(j\omega)^2 + j\omega/RC + 1/LC} = \frac{2\alpha(j\omega)}{(j\omega)^2 + 2\alpha(j\omega) + \omega_0^2} \qquad (6.18)$$

where
$$\alpha = \frac{1}{2RC}, \qquad \omega_0 = \frac{1}{\sqrt{LC}}.$$

The magnitude of this complex frequency response (see Appendix B) is:

$$\left|\frac{\mathbf{V}}{\mathbf{V}_a}\right| = \frac{2\alpha\omega}{\sqrt{\omega^4 + (4\alpha^2 - 2\omega_0^2)\omega^2 + \omega_0^4}} \qquad (6.19)$$

$$= \frac{4\pi\alpha f}{\sqrt{(2\pi f)^4 + (4\alpha^2 - 2\omega_0^2)(2\pi f)^2 + \omega_0^4}}.$$

Suppose Equation (6.19) is calculated for, say, every hundred kilohertz from 500 to 1500 kHz. With these data Equation (6.18) can be calculated over the same range and a plot made of V/V_a (the magnitude of $\mathbf{V}/\mathbf{V}_a$) vs ω or vs f. (A programmable hand calculator would be helpful here.) The result should be similar to Figure 6–8c. $\mathbf{V}$ is dependent on the frequency of the source (the radio station being received). The gain $\mathbf{V}/\mathbf{V}_a$—considered as a function of f (Figure 6–8c)—is called the *frequency response* of the circuit.

This particular frequency response curve illustrates a phenomenon called *resonance*. The parallel circuit of L and C (sometimes called a *tank circuit*) exhibits a resonance or peaking effect in impedance as frequency is varied. Its *resonant frequency* (f_0) occurs where the impedance magnitude of the L is equal to that of the C. Thus f_0 can be found by setting ωL equal to $1/\omega C$. It will be found then that

$$f_0 = 1/(2\pi\sqrt{LC}),$$

or $$\omega_0 = 1/\sqrt{LC}. \tag{6.20}$$

By making C variable, f_0 can be *tuned* to lie on the frequency of the signal to be received.

The reasonableness of the Figure 6–8c curve can be seen by examining three points. At $f = 0$ (dc) the coil will be a short circuit and should provide zero response. At $f = \infty$ the capacitor will be a short with similar response. At $f = f_0$ the coil and capacitor resonate to provide an infinite impedance, thus producing a voltage divider ratio of one.

In addition to *parallel resonance* there is a *series resonance* effect, which can be explored by taking the L and C values given above and a resistor of 100 Ω, and using them in the circuit of Figure 4–1a. Calculate $\mathbf{I}_x$ over a range of frequencies which will include f_0. Equation (6.20) will still hold.

These resonance effects are used in tuning radios and TVs and other frequency-selecting equipment to some desired frequency (station). It normally takes a number of circuits (effectively in cascade) to make the tuning characteristic (frequency response) sharp enough to satisfactorily discriminate between two stations fairly close together in frequency. Resonance is useful in some measuring devices and causes unexpected problems in other circuits.

Since any practical inductor or capacitor has losses, better circuit representation for them would be an inductor and resistor in series or a capacitor and resistor in series or parallel. When these resistances are included in the circuit modeling, the infinite impedance of a parallel tank circuit at resonance and the zero impedance of a series inductance and capacitor will be replaced by finite and nonzero values.

The simple parallel and series resonant circuits discussed above are often called *second-order* circuits because a differential equation written to describe their action will be a second-order equation. Chapter 7 will develop this idea further.

Even more basic to frequency response are the *first-order* circuits—circuits which have a single L or C in them, or can be reduced to a single L or C. The next section considers these in some detail.

6.5   FREQUENCY RESPONSE OF FIRST-ORDER CIRCUITS

Consider the two-port circuit (a circuit with two input terminals and two output terminals) of Figure 6–9. What is $\mathbf{V}_{out}$ for a given $\mathbf{V}_{in}$? The ratio of $\mathbf{V}_{out}$ to $\mathbf{V}_{in}$ is usually designated $\mathbf{H}$ or $\mathbf{H}(\omega)$. If there is no significant loading of the output (hardly any current drawn from it) the frequency response will be of the form shown in Figure 6–9b. The mathematical expression from which this curve is taken (it is a simple voltage divider circuit) is

$$\mathbf{H}(\omega) = \mathbf{Z}_C/(R + \mathbf{Z}_C) = 1/(1 + j\omega RC). \tag{6.21}$$

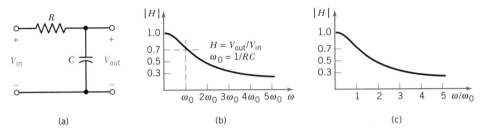

FIGURE 6-9 A low-pass, first-order two-port network with its frequency response curve. $V_{out} = V_{in}H(\Omega)$.

The reasonableness of this result can be seen by noting that the effective $\mathbf{Z}_C$ is large (approaches an open circuit) at low frequencies—making a nearly one-to-one voltage divider ratio. $\mathbf{Z}_C$ becomes progressively smaller (approaches a short circuit) at higher frequencies—rolling off the response as sketched.

This type of curve is encountered frequently in signal circuits. It has a nearly flat portion followed by an S-shaped dropoff in response. After the S-shaped portion the curve continues to fall away to zero in the manner of a hyperbola.

Let us replot the curve using a logarithmic scale on the ordinate and $\log(\omega/\omega_0)$ as the abscissa. The result in Figure 6–10a is striking in that the curve is made up principally of two straight lines. It is asymptotic to these lines at both

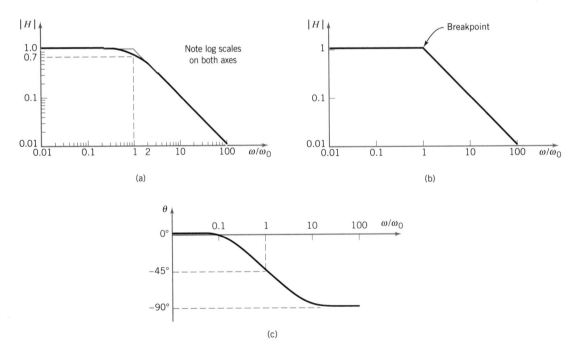

FIGURE 6-10 Transfer function magnitude $H(\omega)$ is usually plotted with log-log scales. In this form it is closely asymptotic to two straight lines. Its *rolloff* above the *breakpoint* is one decade per decade. Sketch (c) shows its phase angle.

ends. These asymptotes are shown in Figure 6–10b. It can be shown that the greatest deviation of the curve from the asymptotes is at ω_0, where it approaches 0.3. The intersection of the asymptotes is called the *breakpoint*. Analysis will show that on the straight-line portion to the right of the breakpoint, H, the magnitude of $\mathbf{H}$, is falling off at a rate of one octave (6 db) per octave or one decade (10 db) per decade. (An octave is a doubling of frequency.)

The flat portion of the curve occurs at low frequencies where the capacitor is nearly an open circuit compared to the resistor. At what frequency does it cease being flat? For measurement purposes engineers select the point at which the response has fallen to $1/\sqrt{2}$ of the flat portion. Some calculation will show that this is where magnitude $X_C = R$. It is usually designated ω_0 but is not to be confused with the resonant frequency of a second-order circuit, which bears the same notation. Thus

$$\omega_0 = 1/RC. \tag{6.22}$$

Substituting (6.22) into (6.21) produces a very useful result:

$$\mathbf{H}(\omega) = \frac{1}{1 + j\omega/\omega_0}. \tag{6.23}$$

This result is plotted in Figure 6–9c, ω_0 occurring at 1 on the horizontal axis. As the ratio of two voltages, $\mathbf{H}$ is dimensionless.

From Equation (6.23), $\mathbf{H}$ is in general complex and must have an angle associated with it. Figure 6–10c shows this angle $\theta(\omega)$, again as a function of log ω. This angle is 45° at the breakpoint frequency. Again the curve is asymptotic—this time at the breakpoint frequency—to a straight line with a slope of about 45° per decade. The error at the 0.1 ω_0 and 10ω_0 frequency ratios is 5.7°.

EXAMPLE 6.5 In the circuit of Figure 6–9a, C is 0.001 μF and $R = 10,000\ \Omega$.

Find

1. the breakpoint frequency ω_0 in radians per second and hertz;
2. H at 90% and 110% of the breakpoint frequency;
3. the percent error in part 2 of accepting the asymptotes as an approximation instead of the curve itself.

Solution

1. The breakpoint occurs where X_C is equal to R, whence $1/\omega_0 C = R$ or $\omega_0 = 1/RC = \underline{100{,}000\ \text{rad/s}}$, or $f_0 = \omega_0/2\pi = \underline{15.92\ \text{kHz}}$.
2. The complex $\mathbf{H}$ (90K rad/s) $= 1/(1 + j0.9) = \underline{0.743}$ at an angle of $\underline{-42.0°}$. Similarly $\mathbf{H}(110\text{K}) = \underline{0.673}$ at an angle of $\underline{-47.7°}$.
3. The asymptote for $\omega/\omega_0 < 1$ has the value 1.0, and for $\omega/\omega_0 > 1$ it has the value ω_0/ω. Then at $\omega = 0.9\omega_0$, the asymptote value is 1, and for $\omega = 1.1\omega_0$ its value is 0.9. Each of these values is about 34% greater than the actual magnitudes calculated in part 2.

The circuit of the examples above (Figure 6–9a) is called a *low-pass filter* since it tends to pass low-frequency signals from left to right but to discriminate against higher frequencies. If the resistor and capacitor are interchanged in that circuit, it becomes a *high-pass filter* as the following example will illustrate.

EXAMPLE 6.6 Given the same two-port network of Example 6.5 but with the C and R interchanged, as shown in Figure 6–11a,

1. Sketch the H and θ vs ω/ω_0 responses as done in Figure 6–10.
2. What is ω_0?
3. Find the error at 5% and 500% of ω_0 caused by accepting the straight-line 45° per decade slope for θ.

Solution

1, 2. Again $\omega_0 = 1/RC = 100,000$ rad/s ($f_0 = \omega_0/2\pi = 15,916$ Hz). Using the voltage divider concept again, $\mathbf{H} = R/(R + jX_C) = 1/(1 - j\omega_0/\omega)$. Thus at frequencies substantially away from ω_0 the decade per decade (or octave per octave) rolloff occurs but this time as frequency goes down. Solution sketches are provided in Figure 6–11.

3. At 5% the straight-line θ vs log ω is already at 90°. At 500% the result is 45° log 5 or a dropoff of 31.5° so that $\theta = 3.5°$. But calculation with the voltage divider formula shows the actual angles to be arctan 20 or 87.1° and arctan 0.2 or 11.3°. Thus the errors in angle are about 2.9° and 7.8°.

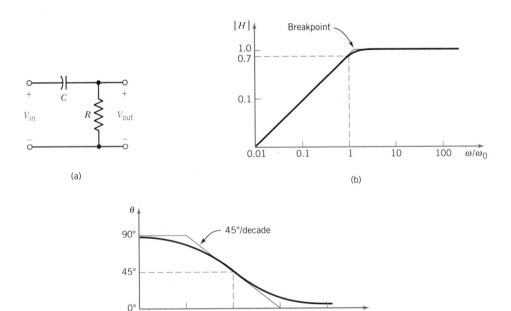

(a)

(b)

(c)

FIGURE 6–11 Transfer function plots for the high-pass filter circuit of Example 6.6.

6.6 MUTUAL INDUCTANCE

In a transformer the coupling effect between two coils comes about from magnetic flux created by the current in one coil linking the turns of the second coil and inducing voltage in both of them. Figure 6–12a illustrates this action. Current in coil N_1 creates the magnetic field. Some of this field links the coil N_2. As the field's linkage with N_2 changes, voltage $v_2(t)$ is induced. Chapter 18 will examine this action in more detail.

In circuit theory this coupling effect between two coils can be accounted for with the concept of *mutual inductance, M*, between the two coils as suggested in Figure 6–12b. (When a distinction is necessary, regular inductance L is called *self-inductance.*) M is defined in a similar manner to L:

$$v_2 = M \frac{di_1}{dt}. \tag{6.24}$$

Compare Equation (6.3). As with L, the units of M are henries. The two dots—one at the end of each coil—are polarity markings. Voltages at dots rise and fall together, and when current flows into one dot, it is flowing out of the other. M can also be used, in phasor calculation with sinusoidal voltages and currents just as L is. Thus where two coils are coupled (have the turns of one linked by magnetic flux created by current in the other) there are three inductances involved: L_1, L_2, and M.

Consider now the circuit of Figure 6–12c. The 30-V source causes current to flow through L_1. This will in turn induce a voltage in the second coil L_2 by mutual

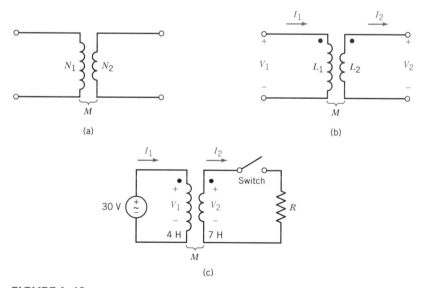

FIGURE 6–12 The magnetic coupling between two coils can be described in henries as a mutual inductance M. Varying the current in either coil induces a voltage in the other as well as in itself.

inductance. If the switch is open, things are perfectly straightforward in this calculation. But if the switch is closed, a secondary current will flow through the load and the L_2 coil, producing an additional voltage in coil 1 and reducing $\mathbf{I}_1$ to some extent.

Writing mesh equations for the two loops (and putting the 30-V source on the real axis), we have

$$j\omega L_1 \mathbf{I}_1 - j\omega M \mathbf{I}_2 = 30,$$

$$-j\omega M \mathbf{I}_1 + (R - j\omega L_2)\mathbf{I}_2 = 0.$$

These equations can be solved simultaneously for the two currents and from them any other circuit aspect worked out.

EXAMPLE 6.7 In Figure 6–12c a 30-V generator is connected to a 10-Ω load through two mutually coupled coils L_1 and L_2. L_1, L_2, and M are 4, 7, and 2 H, respectively; $f = 60$ Hz.

Find: V_2, I_1, and I_2
1. with switch open;
2. with switch closed.

Solution: The reactance for L_1, L_2, and M are simply ωL or ωM (ω is 377 for 60 Hz). So these three reactances are respectively 1508, 2639, and 754 Ω. For part 1 with switch open there is no i_2 and hence no voltage induced in L_1 by i_2. Whence $I_1 = V/Z = 30/1508 = \underline{19.9 \text{ mA.}}$ And $\mathbf{V}_2 = jX_M \mathbf{I}_1 = 754 \times 0.0199 = \underline{15 \text{ V.}}$ The reader is encouraged to solve part 2 by mesh equations.

In using mesh equations for these circuits care must be taken with the dots. In the above example with both dots on top (or on the bottom) care in writing KVL equations will show them to be as indicated above. With alternate dot locations the mutual terms change sign. Physically the dots are determined by the direction in which each coil is wound.

This mutual inductance theory, when first encountered, often appears contrary to the elementary transformer theory of Chapter 4. The difference is that in the previous iron-core transformer theory it is assumed that perfect coupling exists between the two coils—all the flux created by one coil links the other.

The degree of coupling is often expressed by

$$M = k\sqrt{L_1 L_2}, \tag{6.25}$$

where k, the *coefficient of coupling*, is a number that lies between 0 and 1.

6.7   SUMMARY

Capacitances and inductances can be characterized, like resistance, from the relations of voltage across them to current through them, in either the time domain or the frequency domain. In the time domain these equations involve derivatives and integrals. Figure 6–2 summarizes them. The impedance of either has a 90° phase angle, plus for L and minus for C.

The complex phasor representation for either V or I is based formally in a complex number which includes as its real part a term corresponding to the electrical quantity. The remaining portion, the j part, is simply discarded or ignored after calculations have been made. The advantage of dealing with the whole complex phasor is that it replaces a troublesome function of time with an algebraic quantity, making for much easier, although complex, calculations.

Including both L and C in a circuit leads to resonance effects as frequency is varied. In circuits with two input terminals and two output terminals this concept can be expanded into a general circuit response (V_{out}/V_{in}) versus frequency with the usual notation $H(\omega)$ as the transfer function. The frequency response of first-order circuits (containing only one inductor or capacitor with any number of resistors) can be analyzed in terms of breakpoints. Plotting the logarithms of both frequency and response for these circuits leads to plots made up of nearly straight-line segments meeting at easily-calculated breakpoints. Simple R-L and R-C filters provide standard examples of these plots. Subsequent chapters will show that in many more complicated circuits it is still possible to combine these breakpoint analyses.

Two coils whose magnetic fields are mutually interacting can be analyzed in a manner similar to simple inductances, by developing the concept of a mutual inductance between them.

FOR FURTHER STUDY

Ralph J. Smith and Richard C. Dorf, *Circuits, Devices and Systems,* 5th ed., 1992, John Wiley & Sons, New York. Chapters 3, 8.

PROBLEMS

Easy Drill Problems (answers at end of chapter)

D6.1. A capacitor of 100 pF is connected across a voltage source of variable frequency.
  (a) Find the reactance in ohms at 100kHz, 1MHz, and 10 MHz. Sketch a rough quantitative graph of reactance vs frequency from 0 to 100 MHz.
  (b) Repeat more carefully using a log-log plot.

D6.2. Three capacitors of 100, 60, and 40 μF are available. Find their total capacitance:
  (a) if they are connected in series;
  (b) if they are connected in parallel;

(c) if the 40 and 60 are connected in parallel and the 100 in series with that combination.

D6.3. For an inductor of 100 mH:
  (a) Calculate the inductive reactance at 100 kHz, 1 MHz, and 10 MHz and sketch a rough graph of inductive reactance vs frequency.

D6.4. Sketch the series impedance Z_s of a 10-mH inductance and a 100-pF capacitor vs f. Plot magnitude and phase angle, one under the other.

D6.5. Three inductors of 40, 60, and 100 mH are available. Find the total inductance:
  (a) if they are connected in series;
  (b) if they are connected in parallel;
  (c) if the 40 and 60 mH are connected in parallel and the 100 mH in series with that combination;
  (d) if the 40 and 60 mH are connected in series and the 100 mH in parallel with that combination.

D6.6. A 1-kΩ resistor is placed in series with an inductance of 0.1 H.
  (a) Sketch the series impedance of this combination as a function of frequency from 0 to 10 kHz. Sketch both magnitude and phase angle, one under the other.
  (b) Over what range of f can L be neglected and still have Z within 1%? Over what range can R be neglected?
  (c) What is the phase angle where R is 10% of X_L? What is the phase angle where X_L is 10% of R?

D6.7. In problem D6.6, add another 0.1-H inductor in parallel with the first one. Repeat part (a).

D6.8. In problem D6.6, reconnect the R and L in parallel and repeat the problem.

D6.9. A 1-mH inductor has been connected to a 10-V 50-Hz ac cosinusoidal source for a long time. Plot three cycles of the voltage wave and on the same axis three cycles of the corresponding current wave.

D6.10. In the sketched circuit:
  (a) plot the magnitude of V_2/V_1 vs f from 0 to 20 kHz.
  (b) Plot the phase angle of V_2/V_1 vs f.
  (c) Replot part (a) with a log-log plot. Sketch in the two straight-line asymptotes. At what frequency do they intersect?

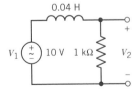

D6.11. In Figure D6.10, add another 1-kΩ resistor in series with the inductor. Make a log-log plot of the transfer function and sketch in the straight lines. What is ω_0?

D6.12. For the two-port of problem D6.11 make a plot of the phase angle of **H** vs frequency. What is the angle at $f = 10$ kHz?

D6.13. Sketch a simple two-port as in Figure D6.10 with the top (horizontal) leg made up of a single 1-kΩ resistor and the right hand (vertical) leg made up of a 0.5 μF capacitor. Make a log-log plot of the transfer function and include the asymptotic lines. What is the breakpoint frequency?

D6.14. Repeat problem D6.10 substituting an 0.5 μF capacitor for the inductor.

D6.15. In the parallel resonant circuit of Figure 6–8 it is found that an inductor of 100 μH resonates with a given capacitor at 3.475 MHz. What is the capacitance of the capacitor in pF?

D6.16. A transformer has 100 turns on the primary and 300 on the secondary; a load resistor of 100 Ω is connected. 50 V are applied from a source. Find I_p, I_{load}, V_{load}.

D6.17. (a) Calculate the power delivered to the load in D6.16. What power is taken from the source?

(b) What impedance does the source see? (*Hint: Z = V/I*).

D6.18. A transformer has a turns ratio of 4 to 1. A 100-Ω resistive load is connected to the high side, which supplies it with 440 V. Find the two currents. What impedance does the primary see?

D6.19. The secondary (low side) of a 3/1 transformer is loaded with 50 Ω. The primary is connected in series with a 2-henry inductor, a 5-μF capacitor, a 500-Ω resistor, and a 500-V 60-Hz source. Find the voltage across the 50-Ω resistor.

D6.20. In problem D6.19, find the primary and secondary currents.

D6.21. In problem D6.19, find the powers dissipated by each of the two resistors and the power supplied by the source.

D6.22. A circuit like Figure 6–8 is designed to tune in WTAG (580 kHz) when the variable capacitor is set at 350 pF. What should the value of the inductance be?

D6.23. Repeat problem D6.22 using a capacitance value of 2.25 pF to tune in an fm station at 89.7 MHz.

Application Problems

P6.1. In the analogy of Figure 6–4 assume that the outlet pipe of the pump extends down into the water in the beaker. The pump flow is sinusoidal. Its peak flow is 1 gpm. The cross-sectional area of the beaker is 23.1 in.^{2}. Its height is 10 in.

(a) What is the effect of frequency on the maximum height of water in the beaker? 1 gal = 231 in.^{3}

(b) What is the lowest frequency in cycles per minute the system can tolerate (without spilling water)?

(c) Make a rough plot of the necessary beaker height versus frequency.

P6.2. Devise a circuit analogous to the situation of problem P6.1 and using an electric capacitor. Select specific component values and answer the same questions as in the preceding problem. What is the electrical equivalent of having the water spill over the top of the beaker? To overcome that difficulty in the water system one could use a taller beaker. What would the equivalent solution be electrically?

P6.3. Show mathematically that capacitances add in parallel and inductances in series. (*Hint:* Use the definitions of capacitance and inductance.)

P6.4. In the ignition systems of automobile engines the high-voltage spark is produced by breaking the current through a coil as sketched. The effect is further enhanced by transformer action. Assume that the primary inductance is 1 H and that the step-up ratio between primary and secondary is 150. The primary current before break is 1.0 A. A sparking voltage of about 30,000 is needed. If we assume for simplicity that the decay of current on break is as shown in Figure P6.4b, over what period of time T must the primary current be reduced to zero?

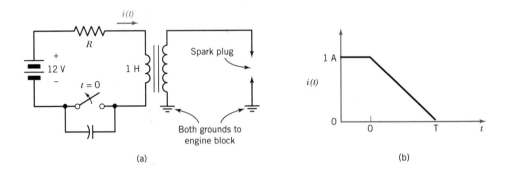

(a)   (b)

P6.5. In the circuit sketched find I_x and I_y. (*Hint:* First find I_x by replacing the transformer and load at points x,x with the transformed load impedance.)

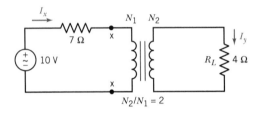

P6.6. **(a)** For the two-port network sketched derive an expression for the transfer function.
   (b) Plot separately the magnitude and phase of this function against frequency. Go a decade above and below the breakpoint frequency. Use a log-log plot for the magnitude and a log$f$ plot for phase.

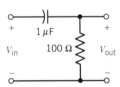

P6.7. Two coils, A and B, of 100 and 200 mH have a mutual inductance of 40 mH. They are connected in a circuit as sketched. The frequency of the source is 200 Hz. Find the primary and secondary currents and the coil voltages.

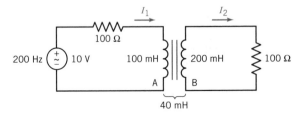

P6.8. In the sketch a load **Z** is connected to the secondary of two coupled coils with L_1 of 25 mH, L_2 of 50 mH, and M of 20 mH. The frequency is 500 Hz. Find the current I_1 with the switch open and again with the switch closed for a load $\mathbf{Z}_L$ of
(a) 100 Ω resistance;
(b) 31.8 mH inductance;
(c) 3.18 μF capacitance.

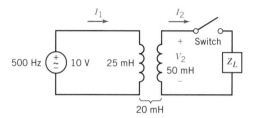

P6.9. (a) Using the results of problem P6.8, determine what impedance the 10-V generator sees in each of the six cases.
(b) For each of the three loads determine the additional impedance caused by closing the switch.
(c) Does the capacitive load add capacitive impedance in the primary? The inductive load? (*Note:* These calculations involve complex numbers.)

P6.10. (a) For the circuit sketched make a rough quantitative plot of the magnitude of $\mathbf{Z}_x$ (the impedance "looking in") as a function of frequency from 0 to 100 kHz.
(b) On the same axes sketch the phase of $\mathbf{Z}_x$.

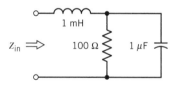

P6.11. A two-port filter is made with a 10-mH inductor in the horizontal arm and a 1000-pF capacitor in the vertical. Plot on the same axes the magnitude and angle of the transfer function. Will log-log plotting give straight lines?

P6.12. Recognizing that the filter of problem P6.11 must have some losses, represent them with series resistors in each arm of 500 Ω and replot the transfer function and its angle. How does it differ from the lossless version?

P6.13. Given 3-elements in series ($L = 1$ H, $C = 1$ μF, R is variable). Plot a series of resonance impedance curves on top of each other for $R = 0$–15 Ω in steps of 3.

P6.14. Two identical 220/110V (N_1/N_2) transformers are each connected to a 220-V source, and their 110-V secondaries connected in series adding. Two 5-Ω resistive loads are connected in series and placed across the combined secondaries. Find the currents and voltages on the loads. Find the load powers and the power taken from the source. Assume that now one load burns out and leaves an open circuit. Re-solve.

P6.15. A 10 to 1 ratio transformer is connected on the high side to a two-terminal load made up of a 10-mH inductor in series with the parallel combination of a 6-Ω resistor and a 700-μF capacitor.

(a) What is the power delivered to the load if a 100-V 60-Hz supply is connected to the low side?

(b) What is the 60-Hz impedance seen looking into the low side?

 P6.16. Plot the answer to P6.15b as a function of frequency (magnitude and angle) from 10 to 500 Hz.

Extension Problems

E6.1. Unlike resistors, which convert electrical energy directly to heat, capacitors and inductors store energy and then return it to the circuit.

(a) For Example 6.1 calculate the energy stored in the capacitor at 5 ms. (*Hint:* For instantaneous power, $p = vi$. Energy, w, is the time integral of this power.)

(b) Derive a general formula for the energy stored in a capacitor as a function of voltage and capacitance.

E6.2. Assume that a 10-V sinusoidal source with a frequency of 100 Hz is placed across a 1-μF capacitor.

(a) Make a sketch of the energy stored in the capacitor as a function of time over a whole voltage cycle.

(b) Replace the capacitor with a 100-mH inductor and replot.

(c) What effect would doubling the frequency have on (a) and (b)?

E6.3. Section 6.3 developed a mathematical justification for the concept of X_L and $\mathbf{Z}_L$. Make a similar analysis to establish the validity of $-j/\omega C$ as $\mathbf{Z}_C$.

E6.4. In Figure 6–12c with the switch closed it was found necessary to resort to mesh equations to solve the circuit. Alternatively, it is possible to postulate a "coupled impedance" in the primary circuit in series with L_1 to account for the effect of secondary current.

(a) Derive an expression for this coupled impedance in terms of M, L_2, R, and ω.

(b) What is the value of this impedance in Figure 6–12c with the 10-Ω load connected?

(c) Resolve circuit 6–12c using this coupled-impedance method to find the two I's and the load voltage. (*Hint:* Sketch the primary circuit with the coupled impedance included. Reason from Ohm's law.)

E6.5. Show that if the voltage induced in a coil can be described as $N\,d\phi/dt$, where N is the number of turns linked by flux ϕ, then the inductance of a coil is $N\,d\phi/di$.

E6.6. For the circuit sketched,

(a) Plot the log-log response (V_2/V_1) vs f from 100 Hz to 1 MHz.

(b) Repeat assuming C is a short circuit.

(c) Repeat assuming C is an open circuit.

(d) Repeat assuming $|Z_C|$ both much less than 10 kΩ and much greater than 100 Ω so that $i = V_1/10\text{k}\Omega$ and the 100-Ω R is negligible.

(e) List the frequency regions over which (b), (c), or (d) is a good approximation to (a).

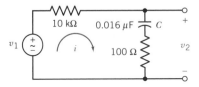

E6.7. A two-port network (Figure 6–1a without load resistor) is fed by a low-pass filter (Figure 6–9a) and followed by a high-pass filter (Figure 6–9a with R and C interchanged). High-pass constants are R-10,000 Ω, C-0.01 μF; low-pass constants R-1000 Ω, C-0.0001 μF. The two-port network is an amplifier with several megohms input impedance, a gain of one ($V_{out} = V_{in}$), and only a few ohms output impedance. Plot for the whole system
  (a) a log $H(\omega)$ vs log ω characteristic, and
  (b) a θ vs log ω characteristic, both analogous to Figure 6–10 except for the abscissas.
  (c) How many breakpoints does the system have and what are their frequencies?

E6.8. A 1-μF discharged capacitor is connected by a switch to a 10-V 50-Hz ac cosinusoidal source. The switch is closed at $t = 0$. Plot the first three cycles of the voltage wave and on the same axis the corresponding current wave. What would you guess happens in the first instant after the switch is closed? Assume a resistance of 0.1 Ω in the source. (*Hint:* write KVL around the capacitor-source loop.) After making a judgment call on that, consider that just as the source must have at least a little resistance, it probably has a little bit of inductance. Qualitatively what would that do to the shape of the first few instants of your current curve?

Study Questions

6.1. How many electrical terminals are needed to provide a "port"? Can you give some reasons why this is? What goes in or comes out of a "port"?

6.2. What are the two parts of a complete "response" to a "driving force"? How long does each last in time?

6.3. What are the defining equations for R's, L's, and C's? State them in both symbols and words.

6.4. What is the relation between charge and voltage on a capacitor? Is this relation independent of time? What are the units of the dielectric constant ϵ?

6.5. What is the relation between an inductor's current and the "volt-seconds" of voltage applied to it? Is this dependent on time? Is it dependent on the initial current when the volt-seconds measurement starts?

6.6. What is the difference between mutual inductance and self-inductance?

6.7. In the analogy between inductance and mass of Figure 6–6, what electrical quantity corresponds to how far the mass has moved (its position)?

6.8. Using Euler's identity [Equation (6.11)] and the rotating-phasor idea of Chapter 4, explain how the phasor representation of a voltage includes more than the actual $v(t)$.

6.9. What is the difference between series and parallel resonance? Explain with circuit sketches and response sketches.

6.10. Numerically how are transformer voltages, currents, and impedances related to the turns ratio?

ANSWERS TO DRILL PROBLEMS

D6.1. 15,920, 1592, 159.2 Ω

D6.2. 19.35, 200, 50 μF

D6.3. 62,832, 628,320, 6,283,200 Ω

D6.4. resonant at 159.2 kHz

D6.5. 200, 19.35, 124, 50 mH

D6.6. 0–226 Hz; 11.23 kHz up; 84.3°; 5.7°

D6.7.

f kHz	$Z\,\Omega$	$\theta°$
0	1000	0
3	1374	43
10	3297	72
inf	inf	90

D6.8. R ok 11.16 kHz up; L ok 227 Hz down. 5.7°; 84.3°

D6.9. $i(t) = 45.0 \cos \omega t - 90°$

D6.10. 3.98 kHz; low pass; H starts at 1 and falls off.

D6.11. 50 krads/s; low pass; falls off from 0.5.

D6.12. $-51.5°$;

D6.13. 2000 rad/s; low pass; falls off from 1

D6.14. 318 Hz; high pass; rises to 1

D6.15. 21 pF

D6.16. 4.5 A; 1.5 A; 150 V

D6.17. 225 W; 225 W; 11.1 Ω

D6.18. 4.4 A; 17.6 A; 6.25 Ω

D6.19. 76.8 V

D6.20. 512 mA; 1.54 A

D6.21. 131 W; 119 W; 250 W

D6.22. 215 μH

D6.23. 1.40 μH

CHAPTER 7

Transients

Suppose an electric switch is turned on or off in some equipment, for example in a lamp circuit. For a short period after this switching event, *transient* currents occur. Then the circuit settles down again to a steady-state situation.

Usually these transient currents are over after only a few micro- or milliseconds. But in some circuits they may last for a few seconds or more. For many applications, designs and simple analyses, transients can be neglected. But in a few cases, they are important and their neglect can cause serious problems in equipment operation.

The discussion of driving force and response at the beginning of Chapter 6 pointed out that electrical circuit transients are simply one example of the general transient-plus-steady-state response of all physical systems.

Consider the spring-platform in Figure 7–1a onto which a mass of clay, M, is dropped. Before the clay is dropped the platform is at an initial steady-state height h_1 above the table. Some time after the clay is dropped the platform will have settled down to height h_2. But for a short while after the clay hits the platform there will be a *transient*—the platform will oscillate down and up several times until wind resistance or spring losses *damp* the transient out and the platform settles down to *steady-state* position h_2. The drawings of Figures 7–1a–e show the *oscillation*. Figure 7–1f graphs it against time.

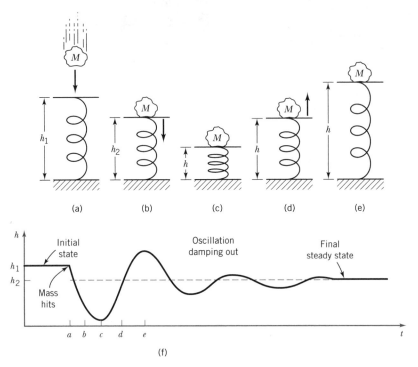

FIGURE 7–1 A typical transient. When the clay mass *M* falls on the spring table there is a period of oscillation before it settles to a new height.

The circuit of Figure 7–2a is analogous in some degree to the spring-platform system. Capacitor voltage v_C goes through a transient response similar to platform height *h*, but in this particular case the voltage rises rather than falls. Losses in electrical resistance *R* play the part of mechanical wind resistance and spring imperfections. The inertial (mass) effect of an inductance was noted in Chapter 6.

When the switch is closed, capacitor voltage v_C will go from *initial* value 0 to *steady-state* value *V*. But in doing so it will oscillate for a short while just like the platform height *h*. The principal difference between electrical and mechanical tran-

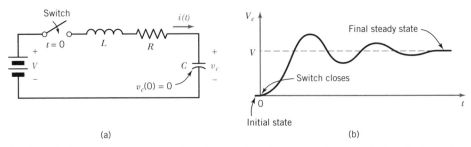

FIGURE 7–2 Electrical transient. When the switch is closed, there is a period of oscillation before the capacitor voltage settles down to *V*.

sients is their duration—fractions of a second versus seconds. Figure 7–2b shows the transient response of the electrical circuit immediately after the switch is closed.

This chapter looks carefully at transient situations in order to predict currents and voltages as functions of time. Such a study will promote an understanding of physical systems in general as well as electric circuits.

Objectives for the reader in this chapter can be:

1. to gain a practical understanding of what electrical transients are and their relation to those in other physical systems.
2. to develop a thorough understanding of how a total system response is made up of a combination of transient and steady state responses.
3. to develop some basic proficiency in handling at least first order systems quantitatively.

7.1   A SIMPLE TRANSIENT

The simpler circuit of Figure 7–3 has only a 5-Ω resistance and a 2.5-H inductance. With the switch open, current must be in initial steady state 0 A. But immediately after the switch closes, current is still 0 since *the current through an inductor cannot be changed instantly.* (Otherwise di/dt would be infinite, and since $v_L = L\,di/dt$, infinite voltage would be required.)

With no current there is no voltage across the resistor. From KVL it is clear that the entire source voltage (10 V) appears across inductor L. Since $v = L\,di/dt$, the rate of current increase at $t = 0^+$ can be calculated as $di/dt = V/L = 10/2.5 = 4$ A/s.* This starting current, beginning at time 0^+, is plotted in Figure 7–3b as a straight line of slope 4 A/s.

Suppose this rate of increase goes on for 0.1 s, until $i = 400$ mA. The drop

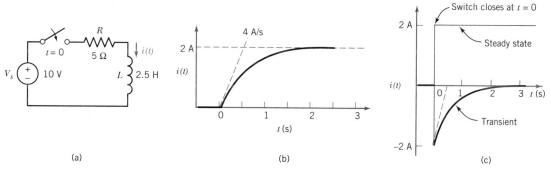

FIGURE 7–3 A first-order RL transient.

---

*The notation 0^+ is often used to designate the moment immediately after a switch is closed or opened at time $t = 0$.

across the 5-Ω resistance will then be 2 V, thus leaving only 8 V across L. So at this point the rate of increase of i will be reduced to 8 V/2.5 H = 3.2 A/s. And the curve of i vs t doesn't go up quite so fast, and begins to bend over. Thus as i builds up, its increase becomes slower and slower. Recognize that sketch (b) is the total solution, steady-state (ss) plus transient. Since the steady state is considered to run from 0^+ time, the transient must be located as shown on the axes in sketch (c).

How large will i get before there is no further increase? When i is large enough that the voltage drop on the resistance is equal to the source voltage (i = 2 A), there can be no more voltage left for the inductance L and no increase in the current. So the curve of Figure 7–3b, which starts out at zero current but with a finite slope, gradually becomes asymptotic to V/R or 2 A with zero slope.

An interesting exercise is to plot short straight-line segments of this current increase to develop the entire curve. Any segment however short will introduce some error since during the segment interval the slope should actually be changing a little. This problem can be overcome by using very short segments and a computer to generate them.

7.2   DIFFERENTIAL EQUATIONS

The transient in the circuit of Figure 7–3 can be found mathematically by writing Kirchhoff's voltage law around the circuit (for the time period *after* the switch is closed), to obtain the *differential equation*

$$V_s = L\frac{di}{dt} + Ri$$

or

$$10 = 2.5\frac{di}{dt} + 5i;$$

whence in standard form

$$\frac{di}{dt} + 2i = 4. \tag{7.1}$$

The solution to a differential equation (an equation containing differentials) is some function of t. To determine whether a given function $i(t)$ is a solution, one inserts $i(t)$ and its derivative in the original equation (7.1). When the answer function has been found, inserting it in Equation (7.1) will provide a check. The answer function is substituted for i, and its derivative for di/dt.

The general way to solve differential equations is to guess reasonable functions that would have a physical shape like the expected answer, and simply try them out by substituting back into the original equation.

Equation (7.1) is the well-known general *first-order* equation for which (in view of Figure 7–3b) it would be reasonable to guess that

$$i(t) = A(1 - 1\,e^{-at}).$$

Then

$$\frac{di}{dt} = aAe^{-at}$$

and substituting in (7.1) we have

$$aAe^{-at} + 2A(1 - e^{-at}) = 4.$$

To evaluate A and a:
Letting $t = 0^+$ (just beyond 0) then yields

$$aA = 4.$$

For t becoming infinite,

$$2A = 4.$$

From these last two equations

$$A = 2, \qquad a = 2,$$

and the solution equation for $i(t)$ is

$$i(t) = 2(1 - e^{-2t})\ \text{A},$$

and from this solution to the differential equation one can find i at any time after the switch closes.

Solving this circuit (Figure 7–3) again, but algebraically using literals V_s, R, and L instead of the numbers the result comes out:

$$i = (V_s/R)(1 - e^{-Rt/L}),$$

so that

$$a = R/L \quad \text{and} \quad A = V_s/R.$$

But it is more customary to write

$$i = A(1 - e^{-t/\tau}), \tag{7.2a}$$

where τ is L/R, and is the reciprocal of a; τ is called the *time constant*. Its units here are seconds.

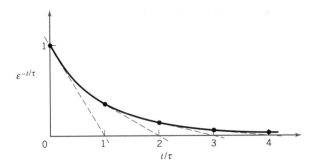

FIGURE 7–4 The basic decaying exponential is easy to sketch, falling to 37% of its previous value with every time constant that passes. The slope at every point is as if it would finish in one more time constant.

By distribution Equation (7.2a) can be written

$$i = A - Ae^{-t/\tau}, \qquad\qquad (7.2b)$$

which makes it very clear that the steady-state solution is simply the first term A, for this particular example, and the transient is the second term $Ae^{-t/\tau}$. As t grows larger and larger the transient, because of its negative exponential factor, diminishes and disappears, leaving only the steady state.

The time constant τ has interesting properties. Since e's exponent must be dimensionless, τ has the dimension of time. At a time of one time constant ($t = \tau$) the exponent of e is -1, at 2 time constants ($t = 2\tau$) it is -2, etc. Figure 7–4 plots the value of the negative exponential versus t/τ. It starts out at 1 (anything to the zero power is 1) and starts downward at a finite slope, eventually becoming asymptotic to 0. Most engineers are familiar with this common curve and remember that its values at 1, 2, 3, 4, and 5 time constants are about 0.37, 0.14, 0.05, 0.02, and nearly 0. Each additional time constant applies the multiplier 0.37 to the value of the previous time constant. Furthermore the slope of the function at $t = 0$ would just bring its value down to zero in one time constant; and this is true for each succeeding time constant. With this information at hand an engineer is in a position to mentally estimate the answers to many problem situations that may come up in practice.

EXAMPLE 7.1 For the circuit in Figure 7–5a

Find: the differential equation using KVL and solve for $v_x(t)$:
1. for $V = 20$ V;
2. for $V = 10$ V.

Solution: $v_x(t) = 0$ for $t < 0$. For $t > 0$, writing KVL: $(1/C) \int i \, dt + V_C(0) + Ri = V$. Differentiating both sides to get rid of integral sign: $R \, di/dt + i/C = 0$. It is necessary to look next at the circuit to find a reasonable function to guess as the transient solution to this differential equation.

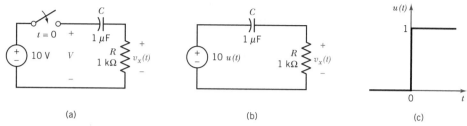

FIGURE 7–5 The *unit step function* u(t) is a convenient notation for a voltage source that switches from 0 to 1 at time $t = 0$. But the source impedance when $V = 0$ is greatly changed from open switch in sketch (a).

Assuming zero charge (and therefore voltage) on the capacitor initially, immediately after the switch is closed, the capacitor voltage is still zero. This is because *the voltage across a capacitor cannot change instantaneously* (since a finite current into the capacitor cannot accumulate any charge—or any voltage—in zero or infinitesimal time).

All the source voltage will appear across the resistor, and current at time $t = 0^+$ is V/R A. As current flows, a charge (and voltage) will build up on the capacitor, reducing both resistor voltage and current. When the capacitor builds up to V, current will have stopped. It is helpful to sketch this expected curve carefully in the manner of Figure 7–4. Thus a decaying exponential could be expected of the form $i = Ae^{-t/\tau}$. Differentiating this form, $di/dt = -A/\tau\, e^{-t/\tau}$.

Substituting these two expressions in the differential equation obtained by differentiating the KVL expression, $\tau = RC$—the time constant in seconds. And recognizing that at $t = 0^+$ the current must be V/R, A then equals V/R. So the solution to the differential equation is $i(t) = (V/R)e^{-t/RC}$. But $v_x(t) = Ri(t)$. So for the resistance and capacitance specified, for part 1, $v(t) = 20e^{-t/0.001}$, and for part 2, $v(t) = 10e^{-t/0.001}$ where t is in seconds. Many engineers would write the time constant as simply 1, using milliseconds, and would then evaluate the result in milliseconds for t. ▨

It might be noted in passing that Example 7.1 has 0 V for a steady-state solution, and with the previous discussion is of somewhat limited generality. Completely general first-order solution forms appear in the next section.

Actually it is far more difficult to write all this down than to get the answer. Once the problem solver sketches the result he or she expects physically, with initial and final values, the problem almost solves itself except for getting the time constant.

In these simple first-order circuits the time constant τ (in seconds) is

$$\tau = L/R \quad \text{or} \quad RC. \tag{7.3}$$

An interesting notational variation is shown in Figure 7–5b where the dc voltage source and switch of the problem circuit have been replaced with a *step-*

function voltage source. The unit step function, $u(t)$, is defined as one whose value is 0 for $t < 0$ and 1 for $t > 0$. It is graphed in Figure 7–5c. Thus it includes the switching function in the source. It is important to note, however, that when a switch is open its impedance is infinite. But the resistance of an ideal voltage source is always zero.

There is, however, a more efficient and intuitive method of solution.

7.3 STANDARD FIRST-ORDER FORMS

First-order circuits contain a single inductance or a single capacitance and any number of resistances and sources (or can be reduced by series and parallel combinations to one of those two forms). They lead to first-order differential equations.

There are only two possible variations of first-order transients as shown in Figure 7–6: a transient can decay either up or down. And in decaying, it can move from any level to any other level (including crossing the x axis). Its general expressions are

$$y = B + (A - B)e^{-t/\tau} \tag{7.4}$$

and

$$y = B - (B - A)e^{-t/\tau} \tag{7.5}$$

for downward and upward decay, respectively. As shown, the transient goes from level A to level B.

In both equations, response y is made up of a steady-state term plus a transient term. The coefficient of the transient term is the difference between beginning and ending values of the variable y (i or v). The fixed term [B in Equations (7.4) and (7.5)] is the steady-state response—a dc response in this case corresponding to a dc driving force. There could also be a sinusoidal steady-state response if the driving force were sinusoidal.

The steady-state value is considered as continuing from the switching instant

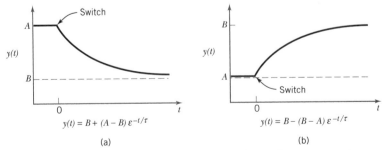

FIGURE 7–6 There are two general cases of the first-order transient—increasing and decreasing.

forever—or until some other switching event takes place. The transient part of the differential equation solution can be thought of as a temporary phenomenon which has just the right value, so that when added to the steady state, it makes the correct value for the total physical response—particularly at time 0^+.

Solving first-order transient problems usually involves simply selecting the right form [(7.4) or (7.5)] and setting the coefficients at whatever values are needed to make the representation correct. The coefficients can usually be evaluated at $t = 0$ and $t = \infty$ for the function (and for higher-order systems, for one or more of the derivatives).

EXAMPLE 7.2 For the circuit of Figure 7–7a
1. Find $v_x(t)$.
2. Sketch a graph of the solution with particular care about the shape of the exponential decay.

Solution: This is clearly a first-order system since it has only one C or L. The switch opens (removing the short on the 10-Ω resistor) at $t = 2$ μs. For t less than 2 μs, $v_x = 6$ V (seen from the voltage divider effect). The capacitor is an open circuit for the dc conditions prior to $t = 2$ μs. Thus v_x starts at 6 V. A long time after the switch is opened, only the steady-state response is left— dc in this case, with the impedance of the capacitor again infinite (an open circuit). Again by voltage divider, $v_x = 3$ V.

Thus under the influence of the transient voltage the system moves from 6 to 3 V. Equation (7.4) fits this situation with $A = 6$ V and $B = 3$ V. To find the time constant, RC, determine the Thevenin's generator (Figure 7–7b) to which the capacitor is attached for $t > 2$ μs. Thus $\tau = RC = 4.2$ $\Omega \times 1\mu$F $= 4.2$ μs. And the final solution is $v_x(t) = \underline{3 + 3e^{-(t-2)/4.2}}$ where t is in microseconds. The factor $(t - 2)$ adjusts the transient for a switching event starting at 2 μs. Substituting $t = 2$ and $t = \infty$ will confirm that this expression gives

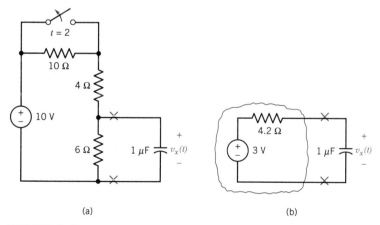

(a)   (b)

FIGURE 7–7 To solve circuit (a) Thevenin's theorem is invoked to produce for $t > 2$ the standard RC circuit (b).

the correct initial and final values. Figure 7–6a is a plot of this solution where A is 6 V and B is 3 V. ▨

These expressions for the time constant ($\tau = L/R$ or RC) have been developed for simple RL and RC circuits. If a circuit has one inductor or capacitor but a number of resistors, it is still first order. But the resistance to use in determining τ is that resistance which the inductor or capacitor sees from its terminals. Thevenin's theorem will help here.

7.4 SECOND-ORDER TRANSIENTS

The circuit of Figure 7–2, with both an L and a C, leads to a second-order differential equation. In accord with procedures covered in any differential equations text, setting the source voltage to zero* gives an equation for the transient itself:

$$\left(\frac{d}{dt}\right)^2 v_C + \frac{R}{L}\frac{d}{dt}v_C + \left(\frac{1}{LC}\right) v_C = 0. \tag{7.6}$$

There are three different forms of the solution, for this second-order transient, determined by the values of R, L, and C. The solution details are not presented here. Figure 7–8 illustrates the three solutions (to include the steady state) the underdamped, critically damped, and overdamped.

In order to simplify the mathematical expressions for these solutions, it is helpful to define some parameters in terms of R, L, and C:

$$\omega_0 = 1/\sqrt{LC}, \qquad \text{series } \alpha = R/2L, \tag{7.7}$$

$$\omega_N = \sqrt{\omega_0^2 - \alpha^2} \tag{7.8}$$

$$p_1 = \alpha + \sqrt{\alpha^2 - \omega_0^2} \tag{7.9a}$$

$$p_2 = \alpha - \sqrt{\alpha^2 - \omega_0^2} \tag{7.9b}$$

Using these shorthand notations, the expressions for the transient for the three cases can be written,

1. For the *underdamped* ($\omega_0 > \alpha$):

$$v_C(t) = Ae^{(-\alpha t)} \cos(\omega_N t + \theta). \tag{7.10}$$

As in the first-order situation, the steady state can be determined by the circuit analysis of Chapters 2, 3, or 4 [depending on the nature of the driving

---

*Mathematics texts use the term *homogeneous equation* to describe an equation derived from the original equation by setting the driving force to 0. The solution of this homogeneous equation produces the transient solution to the original equation. Some mathematics texts use the expressions *natural* or *undriven* response as equivalent to transient response.

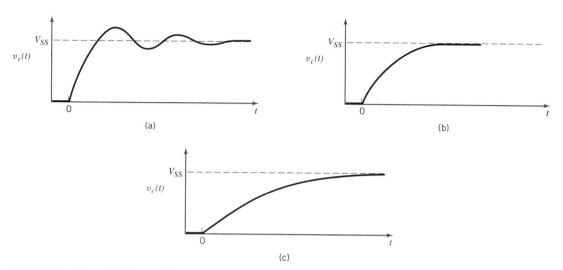

FIGURE 7–8 In an RLC circuit (Figure 7–2) there are three possible second-order transient forms (underdamped, critically damped, and overdamped) depending primarily on the value of R.

force (voltage)—whether it is dc or sinusoidal]. The steady-state solution is then added to the appropriate transient form—Equation (7.10) here. With the steady-state solution added to the transient solution equation, constants θ and A can be evaluated from the known initial conditions $(t = 0^+)$ for $v_C(t)$ and its first derivative.

2. For the *critically damped* $(\omega_0 = \alpha)$:

$$v_C(t) = A_1 e^{-\alpha t} + A_2 t e^{-\alpha t}, \tag{7.11}$$

with constants A_1 and A_2 evaluated in a similar way again with initial conditions applied to the total solution (transient plus steady state).

3. for the *overdamped* case $(\omega_0 < \alpha)$:

$$v_C(t) = A_1 e^{(-p_1 t)} + A_2 e^{(-p_2 t)} \tag{7.12}$$

again with constants A_1 and A_2 evaluated as above. The criteria for which case to use come from the argument of the square root in the expression for the p's (7.9). This argument is called the *discriminant*. If the discriminant is negative, the circuit is underdamped; if zero, the circuit is critically damped; if positive, the circuit is overdamped.

The factor α is called the *damping coefficient*. The larger it is, the faster the oscillating transient will be damped out—so that the response (in this case the capacitor voltage) will approach the steady state more quickly. A large value of α will also tend to reduce the oscillation frequency in the underdamped case.

Figure 7–8a illustrates the *ringing* nature of the underdamped response. The voltage *oscillates* while settling down to its new steady-state value. In attempting to come to the new level it *overshoots*. On the other hand, the overdamped transient does not overshoot but moves slowly and asymptotically to the new level. In the critically damped case the circuit moves to its new value as quickly as possible without overshoot, still approaching it asymptotically.

The analysis for a parallel RLC circuit is identical except that

$$\text{parallel } \alpha = 2\,RC. \tag{7.13}$$

Note again that the steady state must be combined with the transient solution to evaluate the arbitrary constants. At this point, it is still necessary to have the steady state also as part of the total solution. The following example will illustrate this. It is very instructive to solve and graph it also with a sinusoidal driving voltage and consequent sinusoidal steady state.

EXAMPLE 7.3 For the circuit of Figure 7–9a with initial conditions $v_C(0) = 0$ and $i_C(0) = 0$,

Find
1. the steady-state output v_c after the switch is closed,
2. the transient output such that the total response (steady-state plus transient) satisfies the initial values given above.

Solution
Steady State. Using the techniques in Chapter 6, we find the frequency response of the transfer function $\mathbf{H}(\omega) = \mathbf{V}_c/\mathbf{V}_{in}$.

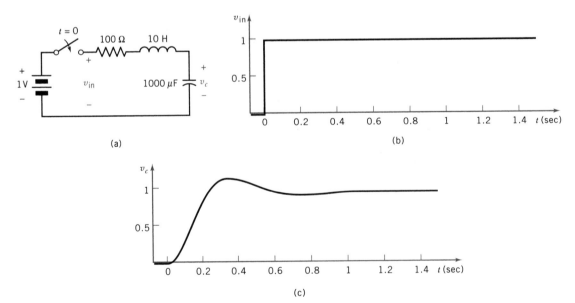

(a)

(b)

(c)

FIGURE 7–9 In sketch (a) an *RLC* circuit is driven by dc 1 V for $t > 0$. Find the voltage on the capacitor. This solution function is made up of two elements: a steady state as sketched in (b) plus one of the standard second-order transients from equations 7.10, 7.11, or 7.12 as sketched in (c).

$$\mathbf{H}(\omega) = \frac{-j/\omega C}{R + j\omega L - j/\omega C} = \frac{1/LC}{(j\omega)^2 + j\omega R/L + 1/LC}$$

$$= \frac{\omega_0^2}{(j\omega)^2 + 2\alpha(j\omega) + \omega_0^2}$$

where, from Equation (7.7),

$$\omega_0 = \frac{1}{\sqrt{LC}} = 10 \text{ rad/s}, \ \alpha = \frac{R}{2L} = 5 \text{ sec}^{-1}.$$

Then

$$\mathbf{H}(\omega) = \frac{100}{(j\omega)^2 + 10(j\omega) + 100}. \tag{7.14}$$

The input signal (after the switch closes) is a dc voltage $v_{in} = 1$V (see Figure 7–9b). Since dc is a frequency of zero, the transfer function for this frequency is $\mathbf{H}(0) = 1$. Then the steady-state component of the output is a dc voltage:

$$v_C(t) = \mathbf{H}(0)v_{in} = 1\text{V}.$$

Transient. Since $\omega_0 > \alpha$, we have an underdamped transient as given in Equations (7.8) and (7.10):

$$\omega_N = \sqrt{\omega_0^2 - \alpha^2} = 8.66 \text{ rad/s}$$

$$v_C(t) = Ae^{-\alpha \cdot t} \cos(\omega_N t + \theta) = Ae^{-5t} \cos(8.66t + \theta).$$

The constants A and θ will be chosen so the total solution satisfies the initial conditions.

Total Solution. The steady-state output added to (superimposed on) the transient output gives the total solution:

$$v_C(t) = 1 + Ae^{-5t} \cos(8.66t + \theta). \tag{7.15}$$

The initial conditions (for $t = 0$) require $v_C(0) = 0$. Then from Equation (7.15)

$$v_C(0) = 1 + A \cos \theta = 0. \tag{7.15a}$$

Since the initial current $i_C(0)$ in the capacitor is specified as zero, the derivative dv_C/dt is initially zero. But from Equation (7.15),

$$dv_C(t)/dt = -5Ae^{-5t} \cos(8.66t + \theta) - 8.66Ae^{-5t} \sin(8.66t + \theta).$$

Therefore,

$$dv_C(0)/dt = -5A \cos \theta - 8.66 \ A \sin \theta = 0. \qquad (7.16)$$

Solving Equations (7.15a) and (7.16) simultaneously, we get

$$\theta = 150°, \ A = 1.155,$$

and the total solution is

$$v_C(t) = 1 + 1.155e^{-5t} \cos (8.66t + 150°).$$

This is plotted in Figure 7–9c. The overshoot of $v_C(t)$ beyond its steady-state value indicates the transient is underdamped. Note that the initial conditions are satisfied. ▄

In Example 7.3 we found the steady-state output by applying the circuit's frequency response to the frequency content of the steady-state input. After the initial step (see Figure 7–9b), the steady-state input is dc, which may be expressed as

$$v_{in} = V_{in} \cos \omega t, \quad t > 0, \qquad (7.17)$$

where $\omega = 0$ and $V_{in} = 1$V. The steady-state output was then

$$v_C = V_C \cos \omega t, \quad t > 0, \qquad (7.18)$$

where $\omega = 0$ and

$$V_C = \mathbf{H}(0) \ V_{in} = 1 \times 1V = 1V. \qquad (7.19)$$

Note that this holds only for $t > 0$ after the step is over. In the next section we will include the step by resolving it into many superimposed frequencies and applying the circuit's frequency response to find the total output including the transient in one operation.

7.5   LAPLACE TRANSFORMS

It can be shown (see Appendix D) that the step waveform in Figure 7–9b can be represented by the sum of an infinite number of sinusoids of the form $e^{\sigma t} \cos (\omega t + \theta)$. The frequencies are so close as to be continuous in ω. Therefore, the summation becomes an integral:

$$v_{in}(t) = \frac{1}{\pi} \int_0^\infty |V_{in}(j\omega + \sigma)| e^{\sigma t} \cos(\omega t + \theta) d\omega$$

where $V_{in}(j\omega + \sigma)$ is a complex function of ω whose magnitude gives the amplitude of each sinusoidal component and whose angle is θ. As in Section 5.3 this can be represented more simply by

$$v_{in}(t) = \frac{1}{\pi}\int_0^\infty \mathrm{Re}[V_{in}(j\omega + \sigma)e^{(j\omega+\sigma)t}]\,d\omega$$

$$= \frac{1}{2\pi}\int_{-\infty}^\infty V_{in}(j\omega + \sigma)e^{(j\omega+\sigma)t}\,d\omega$$

$$= \frac{1}{2\pi j}\int_{-j\infty+\sigma}^{j\infty+\sigma} V_{in}(s)e^{st}\,ds \tag{7.20}$$

where the complex frequency is represented by $s = j\omega + \sigma$. The complex amplitude of each sinusoidal component with frequency s is given by $V_{in}(s)$. The operation described by Equation (7.20) is called the *inverse Laplace transform,* and $v_{in}(t)$ is called the inverse Laplace transform of $V_{in}(s)$. The relationship in Equation (7.20) can also be represented by

$$v_{in}(t) = \mathcal{L}^{-1}[V_{in}(s)].$$

The transform only holds for time functions that are zero for $t < 0$.

The reverse operation is to find $V_{in}(s)$ from $v_{in}(t)$. It can be shown that

$$V_{in}(s) = \int_0^\infty v_{in}(t)e^{-st}dt. \tag{7.21}$$

(See reference by Lathi at the end of this chapter, for instance.) The operation described by Equation (7.21) is called the *Laplace Transform,* and $V_{in}(s)$ is called the Laplace transform of $v_{in}(t)$. This can also be written as

$$V_{in}(s) = \mathcal{L}[v_{in}(t)].$$

Again, $v_{in}(t)$ must be 0 for $t < 0$. For the case here of $v_{in}(t) = 1$ for $t > 0$, Eq. (7.21) gives

$$V_{in}(s) = 1/s. \tag{7.22}$$

Equation (7.20) with its infinitely many frequencies for the total input is analogous to Equation (7.17) with one frequency for the steady-state input. In a similar manner we can express the total output by

$$v_C(t) = \frac{1}{2\pi j}\int_{-\infty}^\infty V_C(s)e^{st}ds. \tag{7.23}$$

This equation is analogous to Equation (7.18).

TABLE 7–1

Laplace Transform	Time Function
$1/s$	$u(t)$
$1/s^2$	$t\,u(t)$
$1/(s + a)$	$e^{-at}\,u(t)$
$1/(s + a)^2$	$t\,e^{-at}\,u(t)$
$1/(s^2 + a^2)$	$(1/a)\,\sin(at)\,u(t)$
$s/(s^2 + a^2)$	$\cos(at)\,u(t)$
$1/(s^2 + 2\alpha s + \omega_0^2)$	$(1/\omega_N)\,e^{-\alpha t}\sin(\omega_N t)\,u(t),\ \omega_0 > \alpha,\ \omega_N = \sqrt{\omega_0^2 - \alpha^2}$
$s/(s^2 + 2\alpha s + \omega_0^2)$	$e^{-\alpha t}\,[\cos(\omega_N t) - (\alpha/\omega_N)\sin(\omega_N t)]\,u(t),\ \omega_0 > \alpha$

$V_C(s)$ represents the complex amplitude of the frequency components of the output signal. This is obtained by weighting the amplitude $V_{in}(s)$ of the frequency components of the input signal by the transfer function $H(s)$ of the circuit:

$$V_C(s) = H(s)\,V_{in}(s). \tag{7.24}$$

This equation is analogous to Equation (7.19). $H(s)$ is the response of the circuit to the complex frequency s. Starting with $\mathbf{H}(\omega)$, we get $H(s)$ by defining $H(j\omega) = \mathbf{H}(\omega)$ and then substituting s for $j\omega$. Then from Equation (7.14),

$$H(s) = \frac{100}{s^2 + 10s + 100}, \tag{7.25}$$

and from Equations (7.22) and (7.24),

$$V_C(s) = \frac{100}{s(s^2 + 10s + 100)}. \tag{7.26}$$

The total output $v_C(t)$ can be found from $V_C(s)$ by Equations (7.23) and (7.26), but this integration is usually avoided by using Laplace transform tables. For example, such a table tells that the Laplace transform of the step function of unit height in Figure 7–9b is $1/s$, as we have found. Conversely, if you know the Laplace transform is $1/s$, the table will tell that the corresponding time function is the unit step [represented by $u(t)$]. A brief table of Laplace transforms is given in Table 7–1; more extensive tables are widely available (see references at the end of the chapter). The function $u(t)$ in the table is defined by

$$u(t) \equiv \begin{cases} 0, & t < 0 \\ 1, & t \geq 0. \end{cases}$$

It is important to note that, while $V_{in}(s)$, $V_C(s)$, and $H(s)$ are all functions of s, $H(s)$ is not a Laplace transform of a time function but rather a circuit transfer function.

EXAMPLE 7.4

Find: the time function $v_C(t)$ corresponding to the Laplace transform

$$V_C(s) = \frac{100}{s(s^2 + 10s + 100)}.$$

Solution: First put the expression for $V_C(s)$ in a form that the Laplace transform table has listings for. $V_C(s)$ can be expanded into three terms:

$$V_C(s) = \frac{1}{s} - \frac{10}{s^2 + 10s + 100} - \frac{s}{s^2 + 10s + 100}.$$

(See Appendix D for examples of how to expand a product into terms.) We already know the first term corresponds to a unit step $u(t)$. The second term has the form

$$-10/(s^2 + 2\alpha s + \omega_0^2),$$

where $\alpha = 5$, $\omega_0 = 10$, and $\omega_N = 8.66$. Then from Table 7-1 the corresponding time function is

$$-(10/8.66)e^{-5t}\sin(8.66t)\ u(t).$$

The third term has the form

$$-s/(s^2 + 2\alpha s + \omega_0^2),$$

where $\alpha = 5$, $\omega_0 = 10$, and $\omega_N = 8.66$. From Table 7-1 the corresponding time function is

$$-e^{-5t}[\cos(8.66t) - (5/8.66)\sin(8.66t)]\ u(t).$$

Summing the time functions corresponding to the three terms, we get

$$\begin{aligned}
v_C(t) &= u(t) - (10/8.66)e^{-5t}\sin(8.66t)\ u(t) \\
&\quad - e^{-5t}[\cos(8.66t) - (5/8.66)\sin(8.66t)]\ u(t) \\
&= u(t) - e^{-5t}[\cos(8.66t) + 0.577\sin(8.66t)]\ u(t) \\
&= u(t) - 1.155e^{-5t}\cos(8.66t - 30°)\ u(t) \\
&= u(t) + 1.155e^{-5t}\cos(8.66t + 150°)\ u(t)
\end{aligned}$$

This time function is plotted in Figure 7-9c. ▬

In Example 7.4 we have solved the same problem as in Example 7.3 but with less effort. Table 7-1 gives the Laplace transform of the input signal. Once the circuit transfer function is known, multiplication gives the Laplace transform of the output signal, and the table gives the corresponding output signal. We have obtained both the steady-state and transient solutions in one step.

7.6   SUMMARY

A linear circuit always responds to any switching event (or step function) in the form *steady state plus transient*, where the transient portion dies out completely (in the manner of a decaying exponential) often within a fraction of a second. The form of the transient portion of the response is determined solely by the elements of the circuit itself—not by the driving function.

The steady-state portion of the response has already been presented in previous chapters. It comes from the analysis in Chapters 3 or 4. The steady-state portion is, in general, similar to the driving function. If the driving function is dc the steady state is dc; if ac there is a corresponding steady state although it may be changed in phase.

The amplitude (and for ac the phase) of the transient (A in some of the above examples, $B - A$ in others) is then adjusted so that when added to the steady state the correct value (in both magnitude and slope) is had for time 0^+.

A fundamental in analyzing transient circuits is that the current through a coil or the voltage across a capacitor cannot change instantly.

While the first- and second-order transients discussed may seem a very limited selection from many possibilities, they cover many practical cases and provide a good intuitive appreciation of the subject. Extension problems at the end of this chapter will go a little further.

In modern engineering analysis, transient circuits are usually handled with an operational mathematics which transforms differential equations into algebraic systems so that they can be solved simultaneously with computers. The mesh methods used previously with dc and phasor analysis can also be used at instants of time as in the beginning of Section 7.2. This procedure will yield enough simultaneous differential equations to handle any linear system. Nonlinear circuits are generally handled incrementally, again with computers.

FOR FURTHER STUDY

Ralph J. Smith and Richard C. Dorf, *Circuits, Devices and Systems*, 5th ed., 1992, John Wiley & Sons, New York. Chapters 4, 6.

B. P. Lathi, *Signals, Systems, and Communication*, Wiley, New York, 1965.

Vincent Del Toro, *Electrical Engineering Fundamentals*, 2nd ed, Prentice Hall, Englewood Cliffs, New Jersey, 1986. Chapter 6.

PROBLEMS

Easy Drill Problems (answers at end of chapter)

D7.1. In a circuit like Figure 7–3 the voltage source is 20 V, $R = 10\ \Omega$, and $L = 20$ H; the switch closes at $t = 0$ s.
    (a) Write and solve the differential equation for $i(t)$ in this circuit.
    (b) Sketch the result using care with respect to the slope and value of the function at each time constant.
    (c) Find the value of the current at the end of each of the first four seconds.

D7.2. Repeat D7.1 using microhenries instead of henries, and microseconds instead of seconds.

D7.3. Repeat D7.1 using a 1-μF capacitor in place of the inductor and using microseconds. (*Hint: i* cannot change instantly through an *L*, and *v* cannot change instantly across a *C*. But the reverse is not true.)

D7.4. Solve D7.3 assuming there is an initial charge of 10 V on *C*.

D7.5. **(a)** In the circuit shown the switch opens at $t = 0$. Find and sketch $i_x(t)$.
    (b) What is the value of i_x at $t = 2$, 4, and 6 s?

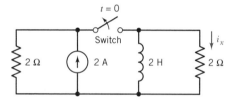

D7.6. Solve D7.5 assuming the switch opens at $t = 3$ s instead of 0 s.

D7.7. In the D7.5 circuit replace the 2-A source with a $2u(t + 2)$-A source and solve for $i_x(t)$. (*Hint:* Source goes on at $t = -2$ s.) Replace switch with a short circuit.

D7.8. In a first-order circuit a switch operates at $t = 0$, and the resulting transient $i(t)$ decays from 0 to 5 A in 10 s (5τ). Write the solution expression for $i(t)$.

D7.9. A switch operates at $t = -2$ s initiating a first-order transient, which starts at -4 A and completes 5τ when $t = 3$ s at 4 A. Write the solution expression for $i(t)$.

D7.10. In a first-order circuit a switch opens at $t = 0$ initiating a voltage transient, which starts at 10 V and appears to be headed for 5 V in one τ. But at $t = 2$ s, another switch opens and the resulting modified transient runs its course of 5τ to 0 V. The time constant after the first switch opens is 2 s; after the second switch opens, it is 4 s. Write the solution expression for v in the interval $-2 < t < 7$ s.

D7.11. Devise a circuit that will produce the results of D7.10.

D7.12. In the circuit sketch, switch 1 opens at $t = 0$ and switch 2 opens at $t = 3$ s. Sketch carefully the voltage $v_x(t)$. What is v_x at $t = 2$ s?, 4 s?, 6 s?

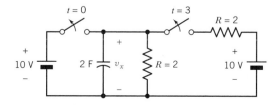

D7.13. Carefully plot, on the same time scale, $v_x(t)$ and the current going up through the capacitor $i_c(t)$ from D7.12. Cover the time interval from -2 to 10 s.

Application Problems

P7.1. Write and solve the differential equation for the circuit of Figure 7–3. Carry this out first with literals for R, L, and V.

P7.2. (a) For the circuit of Figure 7–3, write a program to generate the curve of $i(t)$ vs. t with a computer using short finite increments of time as if slope were constant over that increment. (*Hint:* Have program read an input DELT, the time increment. Start current at 0 for first increment. Using the current and time t from previous increment generate in sequence old i, V_R, V_L, di/dt, δi, new i, and t. Arrange if possible for computer to plot every nth point. You may choose n as a function of DELT.)

(b) From a differential equation solution, compare the current after 2 s with the corresponding result from this program. To what small fraction of a time constant must DELT be taken to satisfactorily reduce the error?

P7.3. In this circuit the switch, as usual, has been open for a long time. It then closes for a millisecond, then opens for a millisecond and repeats this pattern continuously. Is the time constant the same for the two switch positions?

(a) Sketch the first few cycles of vx.

(b) Sketch several cycles of v_x after a long period of time has elapsed.

(c) Between what values of voltage does v_x alternate in **(b)**?

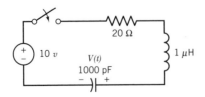

P7.4. (a) For this circuit and using the expressions presented in Equations (7.7)–(7.12), sketch carefully $i(t)$. [*Hint:* To evaluate the amplitude it will be necessary to set the derivative of $i(0)$ at the proper value. Note also that if the factor plus or minus $Ae^{(-\alpha t)}$ is plotted, this plot will constitute an *envelope* for the graph within which the oscillations are damped.]

(b) At what frequency does the decaying exponential oscillate? To what value should R be changed to produce critical damping?

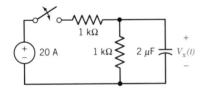

P7.5. In the circuit shown, what value should R have so that the new steady state will be first attained as quickly as possible without overshooting by more than 10%? (*Hint:* A computer will help here.)

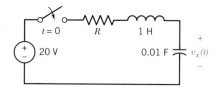

P7.6. In the circuit shown, after being open for a long time, the switch closes at 0 s and opens again at 3 s. Find and sketch on the same axes i_x and v_x.

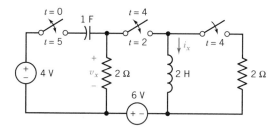

P7.7. For the circuit shown, find and sketch $i_x(t)$ and $v_x(t)$. The left-hand switch (after being closed for a long time) opens at $t = 0$ s and closes again at $t = 5$ s. The center switch closes at $t = 2$ s and opens again at $t = 4$ s. The right-hand switch closes at $t = 4$ s. Is this a first- or second-order circuit?

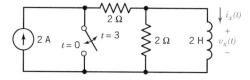

P7.8. In the circuit shown, the left-hand switch opens at $t = 0$ s and closes again at $t = 2$ s. The right-hand switch closes at $t = 1$ s. There is an initial 6-V charge on the capacitor.
  (a) Find and plot on the same axes $v_x(t)$ and $i_x(t)$. (*Hint:* With a firm 12-V source across them, will the right-hand part of the circuit have any effect on the left-hand part or vice versa?)
  (b) Suppose there is a 1-Ω resistor in series with the 12-V source. Write the three simultaneous differential equations for the circuit for the period after 2 s.

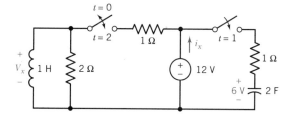

P7.9. In the circuit shown, the notation $u(t - 4)$ designates a unit step function which switches from 0 to 1 at $t = 4$ s. Find $i(t)$. Sketch $i(t)$ for $-2 < t < 10$ s.

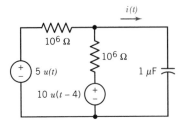

P7.10. In the circuit shown, find $i(t)$. Sketch $i(t)$ for $-2 < t < 10$ s. [See problem P7.9 for explanation of $u(t - 4)$.] (Note that the current-source units are microamperes.)

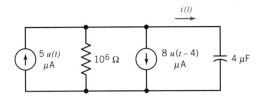

P7.11. Solve problem P7.4 using Laplace transforms. [*Hint:* First find the transfer function $I(s)/V(s)$.]

P7.12. Solve problem P7.6 using Laplace transforms. Closed switch opens at $t = 0$.

Extension Problems

E7.1. For the circuit of Figure 7–2 devise a computer program to find $i(t)$ by incremental steps after the switch is closed. Let $V = 10$ V, $L = 100$ μH, $C = 0.001$ μF, and $R = 200$ Ω. (*Hint:* You will have to make some assumption about the initial charge on C. 0 V is usual. What happens if the initial charge is assumed to be equal to the source voltage?)

E7.2. Using the program at E7.1, investigate the effect of varying R on the response curve.

E7.3. Write and solve the differential equation for the circuit of Figure 7–2. Check against the information in Section 7.4.

E7.4. For the circuit shown, find $v_x(t)$. Select different values of R to illustrate the three second-order cases. [*Hint:* The material in Section 7.4 for the series case will suggest solutions to be experimented with. The α will be different from Equation (7.7) for this parallel case.]

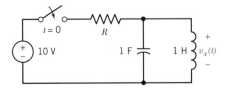

E7.5. In the circuit shown, an ac source drives an *RL* series circuit.
  (a) If the switch closes at $t = 0$, find and sketch $i(t)$.

(b) Is there a time when the switch can be closed that will eliminate the transient part of the response?

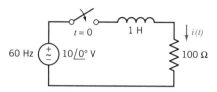

E7.6. In the circuit shown, the switch closes at $t = 0$ s. Find and plot several cycles of the response current (*Hint:* First sketch the transient and steady-state parts separately. But both must be considered together to evaluate A and θ.)

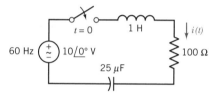

E7.7 A square-wave voltage generator is one that continuously alternates between two values of voltage. In the circuit of problem P7.3 replace the source and switch with a square-wave source alternating between 0 and 20 V, remaining in each state for 1 ms. Find $v_x(t)$. (*Hint:* Don't overlook the fact that a voltage source has zero resistance. Will the time constants be changed in any way from those determined for P7.3?)

E7.8. Repeat the previous problem with a square-wave source varying between plus and minus 10 V.

Study Questions

7.1. In Figure 7–2 the voltage across the capacitor is taken as the response to the forcing step-function voltage of the battery. List several other circuit variables that might be considered a response to this forcing function.

7.2. How does a detailed knowledge of the basic decaying exponential curve help an engineer sketch any first-order response (such as Figure 7–3) quickly and accurately? List the steps in making such a sketch.

7.3. Precisely what percentage of the first-order transient change is completed in 5 time constants? What percentage remains to be completed? How long will it take to make 100% of the change?

7.4. Is there any difference between the battery and switch of Figure 7–3a and the step-function source of Figure 7–5b? What?

7.5. Sketch four first-order transients moving from initial value Y to steady-state value X, for the four cases where both X and Y may be positive or negative. Write the equations for each. (*Hint:* Algebraic signs must be carefully considered.) How many different equations are there?

7.6. Suppose in the circuit of Figure 7–3 the switch had been initially closed for a long period of time. At $t = 0$ it is opened. What is the voltage across the switch at $t = 0^+$ if the current were to instantly stop (an impossibility)? What do you suppose will happen in an actual situation? What safety implications does this have?

7.7. A charged capacitor with 100 V on it is suddenly short-circuited. Describe the current flow as a function of time. (*Hint:* Can the small resistance inherent in the capacitor and in the short circuit be neglected?)

7.8. Make an analogy between the voltage of Figure 7–8a and the swinging of a pendulum in air. Then suspend the pendulum in SAE 10W oil and make an analogy with Figure 7–8b and c. What happens as the oil temperature changes? In a grandfather clock why does the pendulum's oscillation not damp out?

7.9. What are the two important transient rules about L's and C's?

7.10. In question 7.6 above, what difference would putting a large resistor R directly across the inductance make? What is the effect of the value of R used?

ANSWERS TO DRILL PROBLEMS

D7.1. $t < 0$: $i(t) = 0$; $t > 0$: $i(t) = 2(1 - e^{-t/2})$ with t in seconds; 0.79, 1.26, 1.55, 1.73 A

D7.2. The same answers as in D7.1 except t is in μs

D7.3. $t < 0$: $i(t) = 0$; $t > 0$: $i(t) = 2e^{-t/10}$ with t in μs; 1.81, 1.64, 1.48, 1.34 A

D7.4. $t < 0$: $i(t) = 0$; $t > 0$: $i(t) = e^{-t/10}$ with t in μs; 0.90, 0.82, 0.74, 0.67 A

D7.5. $t < 0$: $i(t) = 0$; $t > 0$: $i(t) = -2e^{-t}$; -0.271, -0.037, -0.005 A

D7.6. $t < 3$: $i(t) = 0$; $t > 3$: $i(t) = -e^{-(t-3)}$; 0, -0.74, -0.10 A

D7.7. $e^{-(t+2)/2}$ A

D7.8. $5 - 5e^{-t/2}$ A

D7.9. $4 - 8e^{-(t+2)}$ A

D7.10. $(-2, 0)$ 10 V; $(0, 2)$ $5 + 5e^{-t/2}$ V; $(2, \inf)$ $6.85e^{-(t-2)/4}$ V

D7.11. see circuit D7.12; operate switches at 0 and 2 s

D7.12. 6.84 V; 4.76 V; 2.89 V

D7.13. (<0) 0; (0^+) 5 A; $(0, 3)$ $5e^{-(t/2)}$ A; $(3-)$ 3.06 A; (>3) $1.12e^{(t-3)/4}$ A

PART TWO

ELECTRONICS

CHAPTER **8**

Analog Signals and Instrumentation

We saw in Chapter 1 that electrical engineering applications are divided into two broad categories—signal and power. Chapters 8 through 16 deal with signal or information engineering applications and devices. In this chapter we explore the ways in which instrumentation systems use analog signals to measure physical parameters. Electronic measurement systems originally displayed readings on an instrument panel, hence the name "instrumentation." Today the readings are often stored in computer memory or displayed on a computer screen, but the discipline retains its original name. Mechanical, chemical, and civil engineers most often come in contact with electronics through the need to make measurements. They need to gather data on parameters such as fluid temperature, pressure, viscosity, and velocity; on engine temperature, torque, strain, and rotational velocity; on distance, density, humidity, and radiation. This is done with instrumentation systems that convert the parameters into analog electrical signals, process the signals, and display or store the data.

The reader's objectives in this chapter are to:

1. learn how to specify instrumentation systems to meet one's needs. These specifications include range, sensitivity, resolution, bandwidth, response time, noise, and distortion.

2. be able to model and apply some basic electrical components: transducers, amplifiers, and transmission lines.

3. become familiar with analog signals and how they are characterized in the time and frequency domains.

In electrical engineering an *analog signal* is a voltage or current that varies in the same manner as some other (usually nonelectrical) parameter such as pressure or temperature. An engineer may wish to monitor the pressure in a pipe, where the pressure is a function of time as plotted in Figure 8–1a. A pressure *transducer* produces a voltage proportional to the pressure, as shown in Figure 8–1b. Engineers tend to refer to the voltage in Figure 8–1b as "the pressure," but it is actually an analog of the pressure. The term *analog* is also used to differentiate between analog and digital signals. As described in the next chapter, digital signals can take on only discrete values, while analog signals are continuous.

It is not always necessary to convert a parameter to an analog voltage to observe it. For instance, an aneroid barometer converts pressure to distance and amplifies it by purely mechanical means. But if the information is to be processed or carried to a remote location, it is much easier to deal with the information in an analog voltage form. This ease with which information can be handled in electrical form is what has brought electrical engineering into so many areas. Some examples of *signal processing* are amplifying, band limiting, integrating, differentiating, compressing, rectifying. By far the most common signal processing is simple amplification—raising the amplitude of some low-level signal without otherwise disturbing its waveform. The amplified signal is then large enough to do something useful, such as drive a meter.

As an example of analog instrumentation, consider the pressure monitoring system shown in Figure 8–2. A pressure transducer converts a monitored pressure into a low-level voltage v_1. This analog signal is carried by a *transmission line* (usually wires) to an amplifier where it is raised to the level v_2. This voltage is sufficient to drive a chart recorder, which plots the history of the pressure. Note that the chart

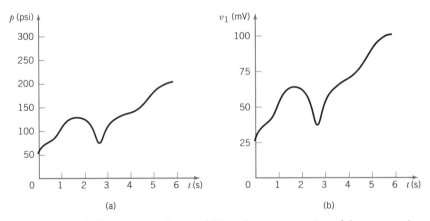

(a)   (b)

FIGURE 8–1 (a) Pressure waveform and (b) analog representation of the pressure by a voltage waveform.

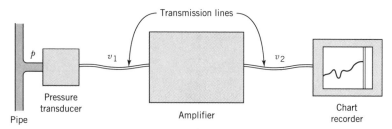

FIGURE 8–2 Instrumentation system to measure and record pressure in a pipe.

recorder is a transducer too, converting voltage levels to pen positions. The chart recorder might be replaced by a device that converts the analog signal to digital form for convenient storage and processing by computers. Such digital aspects of instrumentation will be covered in Chapter 12.

After a brief discussion of transducers, this chapter looks at the various aspects of amplification. More sophisticated processing such as multiplication, peak detection, and quantization will be taken up in Chapter 11 on communication. A practical discussion of transmission lines is given near the end of the present chapter.

8.1 TRANSDUCERS

Transducers form the interface between the world that we sense and the realm of electronics. In order for a circuit to act on information such as pressure or light intensity, these parameters must be converted into electrical signals by transducers. For the output of the circuit to be useful, it must usually be converted back from the electrical domain by other transducers. For example, the circuit might generate a sound, move a solenoid, or turn on a light.

In this section we look at *sensor* transducers that convert signals into the electrical domain. Our five senses are, in fact, such transducers, converting light, pressure, temperature, etc., into electrical signals carried by the nerves. Suppose we wish to measure temperature with an electronic circuit. Thermocouples are transducers that generate a voltage that is a function of temperature. Figure 8–3a shows a typical *transfer characteristic*—a curve that relates the input variable (temperature) to the output variable (voltage). One important property of this curve is its slope—the *sensitivity* of the transducer (42 μV/°C in this case). Another property is the specification of input signal levels for which the sensitivity is acceptable—the *range* of the transducer. The curve flattens out excessively below -200°C, the lower limit of the thermocouple's range. Sometimes the range is determined by input levels that would damage the device. The thermocouple melts above 1350°C, the upper limit of its range.

The thermocouple's transfer characteristic in Figure 8–3a assumes no current is drawn from the device. To account for reduced voltage when current is drawn, a Thevenin source and a Thevenin impedance are used to model the transducer

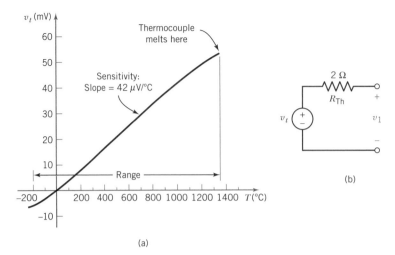

FIGURE 8–3 (a) Transfer characteristic for a Chromel-Alumel thermocouple transducer. The sensitivity is the slope of the curve. The range ends where the sensitivity becomes too small or where the transducer is harmed. (b) Typical Thevenin equivalent for a thermocouple.

(see Section 3.3). A typical impedance of a thermocouple is 2 Ω, so the Thevenin model might be that shown in Figure 8–3b.

There are a number of physical effects that generate a voltage: thermoelectric (e.g., thermocouple), photoelectric (solar cell), piezoelectric (squeezing certain crystals), and electromagnetic (moving wire through a magnetic field). These permit the conversion of temperature, pressure, light, and motion signals into voltage signals. Some devices that use these effects are listed in Table 8–1. As with the thermocouple, each device has some sensitivity, range, and impedance. The table also lists some typical values for these specifications.

Some transducers are designed to change resistance (rather than generate a voltage) in response to physical effects. For example, a thermistor changes resistance with temperature; a typical transfer characteristic is shown in Figure 8–4. A strain gauge changes its resistance when stretched. Table 8–1 lists these and other transducers that use the resistive transduction principle. Typical ranges and sensitivities are listed, where the sensitivity is now in ohms per input parameter. In practice the difference between a transducer that changes resistance and one that produces a voltage is not important. The changing resistance can be placed in a simple circuit to produce a changing current or voltage.

The *Wheatstone bridge* shown in Figure 8–5a is an example of a circuit to convert a changing resistance to a changing voltage. Here there are two voltage dividers, with the transducer in one leg of one divider. The circuit is a classic example of the principle of balance or *symmetry*. If the resistance of the transducer is $R_t = 500\ \Omega$, then the circuit is perfectly symmetrical, and $v_o = 0$. As R_t becomes greater and less than 500 Ω, v_o becomes greater and less than 0. By taking the difference of the voltage developed by the two voltage dividers (see Section 3.7), the reader should confirm that $v_o = (10\ \text{V})R_t/(500 + R_t) - 5\ \text{V}$. For $R_t \approx 500\ \Omega$, this can be ap-

TABLE 8–1 Transducers: Typical Parameters

Device	Transduction	Sensitivity	Range	Impedance
Thermocouple	Temperature-voltage	40 μV/°C	-200 to 1350°C	2 Ω
Piezoelectric crystal	Pressure-voltage	2.0 V/psi	1 to 5000 psi	1000 pF
Tachometer	Angular velocity-voltage	0.03 V/rpm	100 to 10,000 rpm	100 Ω
Hall-effect device	Magnetism-voltage	10 μV/G	1 to 10,000 G	1000 Ω
Photocell	Light-current	5μA/fc	1 to 1000 fc	10 MΩ for <0.2 V
Thermistor	Temperature-resistance	3%/°C	-50 to 300°C	100 Ω to 100 kΩ
Photoconductor	Light-resistance	3 μS/fc	0.1 to 1000 fc	300 Ω to 3 MΩ
Strain gauge	Displacement-resistance	0.05 Ω/μm	0.1 to 50 μm	200 Ω
Potentiometer	Displacement-resistance	50 Ω/mm	0.2 to 200 mm	10 to 10 kΩ
Resistive hygrometer	Humidity-resistance	10 Ω/% rel. hum.	1 to 100% rel. hum.	10 kΩ
Moving-plate capacitor	Displacement-capacitance	1 pF/mm	1 to 100 mm	1 to 100 pF
Moving-core inductor	Displacement-inductance	100 μH/mm	0.2 to 20 mm	20 μH to 2 mH
LVDT	Displacement-reluctance	1% coupling per mm	1 to 100 mm	—

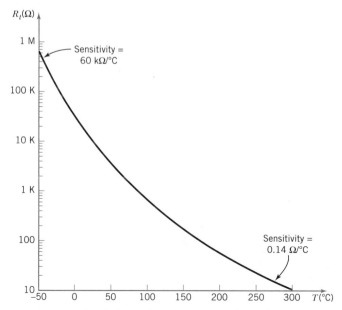

FIGURE 8–4 Typical transfer characteristic for a thermistor. Its sensitivity varies greatly over its range, but in terms of percent change in resistance, the sensitivity varies only from 8%/°C to 1.4%/°C.

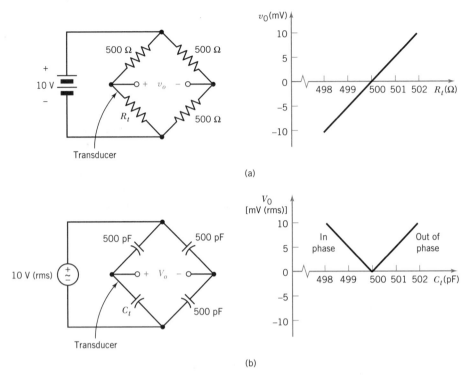

FIGURE 8–5 Wheatstone bridge converts small changes in transducer impedance to changes in voltage. (a) A resistive transducer can use a dc source. (b) A capacitive transducer needs an ac source. The source voltage and V_0 are 180° out of phase for $C_t > 500$ pF.

proximated by

$$v_o \approx (5 \text{ mV}/\Omega)(R_t - 500 \ \Omega),$$

as the graph in Figure 8–5a shows.

The advantage of symmetry is that v_o is not so sensitive to variation in the 10-V power supply so long as R_t is near 500 Ω. In fact, when $R_t = 500 \ \Omega$, v_o will be zero independent of the value of the dc power supply. However, the value of the power supply does affect the gain of the circuit (see problem D.4). The gain for the circuit in Figure 8–5a is 5 mV/Ω. If the transducer is a strain gauge with a gain of 0.05 Ω/μm, then the overall sensitivity is 5 mV/Ω × 0.05 Ω/μm = 0.25 mV/μm.

Symmetry also renders the Wheatstone bridge less sensitive to temperature. (Unless R_t is a temperature transducer!) For example, if a temperature change increases each of the four resistances by 1%, v_o will be totally unaffected, even for values of R_t far from 500 Ω (see problem D.6). We will see this principle of symmetry used to advantage again in designing differential amplifiers to boost the small voltage at v_o (see Section 14.4).

Other electrical parameters that can be varied by physical effects are capacitance, inductance, and mutual inductance. Transducers that depend on these transductive principles therefore change impedance. The impedance of a capacitor or inductor must be sensed with an ac voltage rather than a dc voltage. For example a capacitive transducer can be placed in a Wheatstone bridge, as in Figure 8–5b, with a fixed ac voltage applied. The output is now a varying ac voltage.

Figure 8–6 shows a transducer based on the mutual inductance (or variable reluctance) principle. A linear variable differential transformer (LVDT) is used with an ac voltage source to convert displacement to a proportional ac voltage. The displacement of the core varies the coupling between the windings of the transformer (see Chapter 18).

The specifications given in Table 8–1 are only typical and cover only the basic properties. To apply transducers properly, the user should also be aware of other specifications such as size, accuracy, linearity, temperature range, and frequency response. These specifications are available from manufacturers' data sheets on specific devices and from texts on transducers. See "For Further Study" at the end of this chapter.

Most of the applications for the transducers in Table 8–1 are obvious (a photoconductor measures light). But these uses can be extended. For instance, a strain gauge can be used to measure pressure by attaching it to a compliant body. The compliant body converts pressure to displacement (strain), and the resistance of the strain gauge changes. Similarly, a photoconductor together with a light source can be used to measure the thickness of some materials, a mass and a spring can be used with a displacement transducer to measure acceleration, etc.

If an electronic system is to communicate its results to a person, some visual or aural (or possibly tactile) transducer is needed to convert the electrical voltage information to something the person can perceive. The chart recorder used in the system in Figure 8–2 is one example of an *output device*. Instead of the chart recorder, a simple voltmeter calibrated in pounds per square inch (psi) might have been used. Alternatively, if the output is over a prescribed pressure, a signal might operate an alarm bell to tell an operator that something must be done, perhaps some valve turned off.

In many systems the human is eliminated as an element in the control chain for the sake of convenience, speed, or reliability. The output voltage might be fed

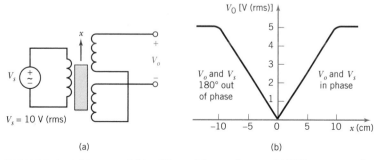

(a)   (b)

FIGURE 8–6 Linear variable differential transformer (LVDT) converts displacement of the core into changing mutual inductance. With the aid of an ac voltage source, this is converted into a changing voltage.

back to some value actuator to control the pressure being measured. Chapter 10 will consider such feedback control systems in detail.

8.2   AMPLIFIERS—INPUT AND OUTPUT IMPEDANCE

The amplifier in Figure 8–2 is an electronic circuit which increases the amplitude of a small voltage—a signal produced by a transducer in this case. A common example in the home is a stereo sound system which takes the relatively small output of a phono pickup cartridge and produces sufficient voltage and power to drive a number of large loudspeakers. As will be shown in Chapters 14 and 15, resistors, inductors, and capacitors alone are not sufficient to perform amplification; active devices such as transistors are also necessary.

Amplifier circuits are usually designed by electrical engineers, and it is generally not necessary for other engineers to deal directly with the details of the circuits. If the engineer is applying an amplifier to a system such as the one in Figure 8–2, it is usually enough to understand the overall behavior and specifications of the amplifier. In the next few sections we will look at the input, output, and transfer characteristics of amplifiers.

In this section we ignore the bandwidth restrictions and distortion of an amplifier and use the simple *two-port models* shown in Figure 8–7. The *input impedance* seen at the input terminals is modeled by a resistance R_i. The behavior at the output terminals is modeled by a Thevenin equivalent circuit—a dependent source $A_v v_1$ and a resistor R_0; A_v is called the *open-circuit voltage gain* since $v_2 = A_v v_1$ for $i_2 = 0$; R_0 is called the *output impedance*.

In Figure 8–7a the four terminals are independent; the input port may "float" at some voltage with respect to the output port. This is a *balanced input*, and the circuit is called a *differential amplifier* because it amplifies the voltage difference $v_1 = v_a - v_b$. In Figure 8–7b the two ports share a common ground; there are only three independent terminals. This is the case with stereo systems, for instance, where the input jacks share a common ground with the output terminals for the loudspeakers. We will see in Chapters 13, 14, and 15 that electronic circuits usually maintain a reference that is named "ground" to which all voltages are referred. In

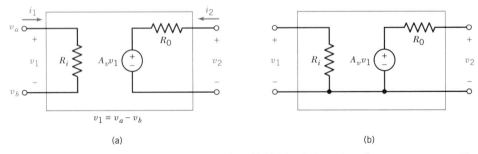

FIGURE 8–7 Two-port models for an amplifier. (a) With a balanced or *floating* input it is a differential amplifier. (b) With a common ground, the input is unbalanced.

power transmission systems this reference is physically connected to the earth, hence the name (see Chapter 17).

EXAMPLE 8.1 Consider the instrumentation system illustrated in Figure 8–2 and modeled in Figure 8–8.

Given: transducer sensitivity is 0.5 mV/psi. Recorder input voltage v_2 for full-scale chart deflection is 10.0 V. The transmission lines have practically no resistance.

Find
1. the value of A_v if full-scale deflection on the chart recorder is to correspond to 200 psi.
2. the percentage of full-scale deflection that would be caused by a pressure of 33 psi.
3. the gain v_2/v_1 of the amplifier under these conditions.

Solution
1. For a pressure $p = 200$ psi, the transducer open-circuit voltage is $v_t = 0.5$ mV $\times$ 200 psi $= 100$ mV. Not all of this voltage appears at v_1 because of the voltage divider circuit formed by R_{Th} and R_i. Then $v_1 = v_t (2000/2500) = 80$ mV. Now, $v_g = A_v(v_1) = A_v(80$ mV$)$. Looking at the voltage divider on the output of the amplifier (formed by R_o and R_L of the recorder), $v_2 = A_v(80$ mV$)(1000/1100) = A_v(72.7$ mV$)$. But this last expression must equal 10 V for full-scale chart deflection, so $A_v = \underline{137.5}$.
2. We could find v_t for 33 psi and find v_2 for the A_v of part 1. But since the system is linear, deflections will be proportional to pressures. So with 200 psi full scale, 33 psi will produce $\underline{16.5\%}$ of full scale.
3. $v_2 = v_g(1000/1100) = A_v v_1(1000/1100)$, so $v_2/v_1 = A_v(1000/1100) = \underline{125}$.

Note that the amplifier gain in the example is $v_2/v_1 = 125$, which is *not* the same as the open-circuit gain $A_v = 137.5$. This is because of the voltage divider formed by the output impedance R_o and the load R_L. The resulting reduction in v_2 is referred to as R_L *loading down* the output. Therefore the amplifier gain depends

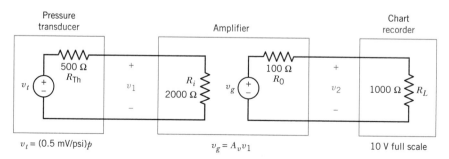

FIGURE 8–8 Model of the pressure measurement system in Figure 8–2. For $A_v = 138$, full scale on the chart recorder corresponds to $p = 200$ psi.

not only on A_v but also on R_o and R_L. In this text we will distinguish the two gains by using G_v for the voltage gain (with load) and A_v for the open-circuit voltage gain.

It is important to take into account the impedances of equipment when connecting them. Otherwise voltages can be excessively attenuated by loading. As a general rule, a load impedance should not be less than the source impedance, and it is good if it is at least a factor of 10 greater than the source impedance. Otherwise voltage amplitude is "thrown away," and more amplification is necessary. In Section 8.9 we will see that it is sometimes necessary to make the load impedance equal the source impedance to avoid reflections.

8.3   FREQUENCY RESPONSE

If a pure sinusoid is applied to an amplifier, the amplitude of the output, determined by the amplifier gain G, should be independent of the frequency of the sinusoid. In practice, though, there is always some frequency above which the gain is reduced, as shown by the *frequency response* curve of $|\mathbf{G}|$ as a function of f in Figure 8–9. This rolloff may be intentional to limit noise, or it may be forced on the designer by stray capacitance. The decade-per-decade downward slope, called a *first-*

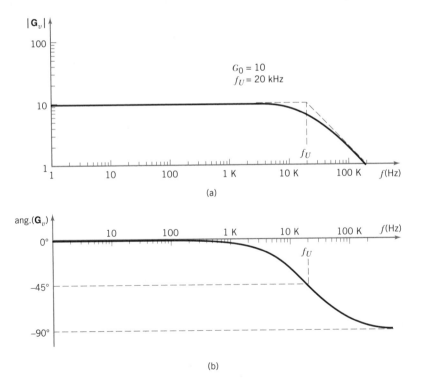

(a)

(b)

FIGURE 8–9 Frequency response of a dc-coupled amplifier with a cutoff frequency of 20 kHz. (a) Gain magnitude. (b) Gain phase.

order rolloff, corresponds to the following expression for the complex gain, which includes both magnitude and phase information:

$$\mathbf{G}_v = \frac{G_0}{1 + jf/f_U}. \tag{8.1}$$

[This response curve and gain expression are similar to the response curve in Figure 6–10 and Equation (6.21) for the *R-C* circuit in Figure 6–9.] The particular response magnitude $|\mathbf{G}_v|$ in Figure 8–9a has a *midband gain* of $G_0 = 10$, and it has a *cutoff frequency* of $f_U = 20$ kHz. The reader can show from Equation (8.1) that $|\mathbf{G}_v| = G_0/\sqrt{2}$ for $f = f_U$. Since this reduces the square of the output voltage to half of the midband value, the cutoff frequency is sometimes called the *half-power frequency*. For $f \ll f_U$, the denominator is about unity, and the response is flat with frequency. For $f \gg f_U$, the denominator is about jf/f_U, and the response falls off inversely with f. Note that with log scales on both axes, the response consists almost entirely of straight lines.

We will use the convention of showing frequency responses by just using the straight-line asymptotes—the dashed lines in Figure 8–9a. This approximation to the actual curve is called a *Bode plot*. The sharp corner, which comes at the cutoff frequency here, is called a *breakpoint*. In general, the frequency at which a breakpoint occurs is called the *breakpoint frequency* (sometimes called the *corner frequency*). The actual response can usually be recovered from the Bode plot by sketching in a curve that passes through a point a factor of 0.7 below the breakpoint.

The response in Figure 8–9a is said to be *flat* down to dc, although dc ($f = 0$) can't be shown on a log scale. A circuit with this reponse is called a *dc-coupled* amplifier.

More common are *ac-coupled* amplifiers with zero response at dc. The transfer function is

$$\mathbf{G}_v = G_0 \frac{jf/f_L}{1 + jf/f_L} \tag{8.2}$$

and the corresponding frequency response is shown in Figure 8–10a. [To emphasize this new effect—the attenuation of low frequencies—we have omitted for the moment a term as in Equation (8.1) that attenuates high frequencies.] The function of f in Equation (8.2) introduces a cutoff frequency at f_L. For $f \ll f_L$, the denominator of Equation (8.2) is about unity, and the response increases proportionally with f, due to the numerator. For $f \gg f_L$, the denominator is about jf/f_L, which cancels the numerator and makes the response flat.

All ac-coupled amplifiers actually have attenuation at high frequencies as well as low frequencies, as shown in Figure 8–10b. To distinguish the two cutoffs, f_L is called the *lower cutoff frequency*, and f_U is called the *upper cutoff frequency*. The corresponding transfer function includes a factor from Equation (8.1) and one from Equation (8.2):

$$\mathbf{G}_v = G_0 \frac{jf/f_L}{1 + jf/f_L} \frac{1}{1 + jf/f_U}. \tag{8.3}$$

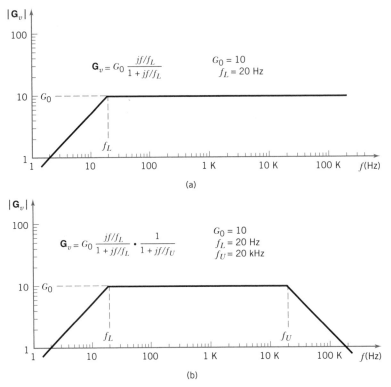

FIGURE 8–10 Frequency response of an ac-coupled amplifier (a) showing only the low-frequency cutoff, and (b) showing both high- and low-frequency cutoffs. The bandwidth is from 20 Hz to 20 kHz. The straight-line approximations here are called Bode plots.

These factors have about the same effect here as when each was alone. Unless f_L and f_U are very close, one factor is nearly unity where the other is rolling off.

Of what use is a frequency response? In audio systems such as hi-fi amplifiers the signal is rarely a pure sinusoid. Nevertheless the frequency response of the amplifier has useful meaning. Typical specifications for a high-fidelity amplifier are cutoff frequencies of $f_L = 20$ Hz and $f_U = 20$ kHz, since the ear doesn't hear frequencies beyond this range. The upper cutoff frequency can be reduced to say 5 kHz by turning down the treble control. The result is a "muffled" sound that lacks "sparkle." Likewise the lower cutoff frequency can be raised by turning down the bass control. The result is a "tinny" sound that lacks "body." These descriptive terms are not very scientific, but the point is that frequency response for audio systems conveys useful quantitative information to the user. A telephone line has lower and upper cutoff frequencies of 200 Hz and 3 kHz, and this response alters the waveform of speech. But the frequency response is more meaningful in this case than the time waveform in evaluating the system.

In many applications it *is* important to know how an amplifier affects a waveform in time. Later in this chapter we will deal with this question. First it will be useful to look at the combined effect of a number of frequency responses.

8.4   CASCADED AMPLIFIERS AND DECIBELS

Sometimes amplification is done in a number of stages, as in Figure 8–11. Signal processing circuits (amplifiers here) that follow one another are said to be in *cascade*. The first stage in Figure 8–11 may be a *preamplifier* with low noise, and the second stage may be a *power amplifier* with high output current capability. These stages could be separate units, or they could be in the same box. In any case we want to find the behavior of the overall gain V_3/V_1.

Ignoring frequency response for the moment, the amplifier stages can be characterized by loaded voltage gains* $G_{v1} = V_2/V_1 = 100$ and $G_{v2} = V_3/V_2 = 20$, respectively. The overall gain is $V_3/V_1 = G_{v1}G_{v2} = 100 \times 20 = 2000$.

Let each stage have lower and upper cutoff frequencies f_L and f_U as specified in Figure 8–11. The complex transfer functions G_{v1} and G_{v2} then have the form of Equation (8.3). The frequency responses are shown as Bode plots in Figure 8–12. In midband (the flat region) we see $|G_{v1}| = G_{01} = 20$ and $|G_{v2}| = G_{02} = 100$.

The overall gain $|V_3/V_1|$ is also plotted in Figure 8–12. The midband gain is 2000, as we have already calculated. Note that the overall bandwidth (the flat portion) is not as wide as that of the first or second stage. The response $|V_3/V_1|$ is found by multiplying the responses $|G_{v1}|$ and $|G_{v2}|$ point by point; Table 8–2 lists the gain for each stage and the overall gain for a number of frequencies. In making response plots, it is particularly helpful to select points at the breakpoint frequencies. It is clear that $|V_3/V_1|$ will have a breakpoint in its response wherever $|G_{v1}|$ or $|G_{v2}|$ has a breakpoint. The important breakpoint frequencies—the ones that determine the overall bandwidth—are the two that flank the flat portion of the $|V_3/V_1|$ curve (f_{L1} and f_{U2} in this case). Therefore the bandwidth of the overall response is from f_L to f_U, where $f_L = 100$ Hz and $f_U = 2$ kHz. The rule for cascaded amplifiers is

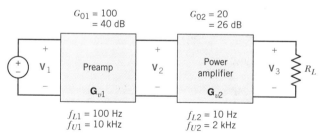

FIGURE 8–11 An example of cascaded amplifiers. The frequency responses are plotted in Figure 8–12.

---

*Our convention, again, is that **G** is the loaded gain, and **A** is the unloaded gain. Usually the input impedance of the second stage is so much greater than the output impedance of the first that the effect of loading is negligible, and **G** ≈ **A**.

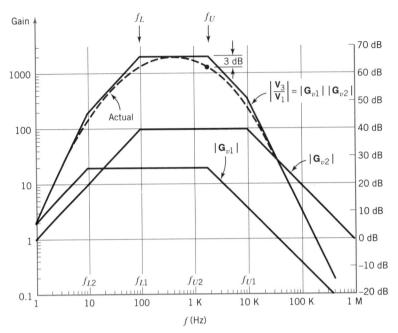

FIGURE 8–12 Frequency responses of the cascaded amplifiers in Figure 8–11. At each frequency the overall response $|\mathbf{V}_3/\mathbf{V}_1|$ is the product of the amplifier gains $|\mathbf{V}_2/\mathbf{V}_1|$ and $|\mathbf{V}_3/\mathbf{V}_2|$. When expressed in decibels, the gains add.

that *the stage with the highest lower cutoff frequency determines f_L* for the overall response. Likewise, *the stage with the lowest upper cutoff frequency determines f_U* for the overall response.

When calculating gain, engineers find it easier to add rather than multiply, especially when there are many stages. To this end the convention of specifying gain in decibels (dB) is used. This system makes use of the algebraic identity $\log(xy) = \log(x) + \log(y)$. For voltage or current gain, the expression of a gain G in decibels is

$$G_{dB} = 20 \log (G), \tag{8.4}$$

TABLE 8–2 Gain of Cascaded Stages (see Figure 8–12)

f (Hz)	$|\mathbf{G}_{v1}|$	$|\mathbf{G}_{v2}|$	$|\mathbf{V}_3/\mathbf{V}_1|$
1	2	1	2
10	20	10	200
100	20	100	2000
2000	20	100	2000
10,000	4	100	400
100,000	0.4	10	4

TABLE 8–3 Gain of Cascaded Stages (see Figure 8–12)

f (Hz)	$|G_{v1}|$ (dB)	$|G_{v2}|$ (dB)	$|V_3/V_1|$ (dB)
1	6	0	6
10	26	20	46
100	26	40	66
2000	26	40	66
10,000	12	40	52
100,000	−8	20	12

where "log" is the logarithm base ten. The same symbol is often used for the gain whether it is expressed in decibels or actual numbers. Thus we have $G_{01} = 100 = 40$ dB, and $G_{02} = 20 = 26$ dB. Therefore the overall gain is 40 dB + 26 dB = 66 dB. The reader should confirm by Equation (8.4) that a gain of 2000 corresponds to 66 dB. Note that the gains in decibels are added, not multiplied.

While the conversion to decibels requires a calculator, the addition thereafter is simple. For rough conversion, engineers remember that a voltage or current gain of ten is 20 dB, and a gain of two is 6 dB. For example a gain of 42 is about 2 × 2 × 10 = 6 dB + 6 dB + 20 dB = 32 dB. Actually 42 = 32.46 dB.

Bode plots often indicate gain on the vertical axis in decibels. Since the decibel unit already includes the logarithmic function, the scale is then linear, as shown on the right edge of the graph in Figure 8–12. Again the response $|V_3/V_1|$ is calculated point by point from the responses $|G_{v1}|$ and $|G_{v2}|$, but this time the gains (in dB) are added. Table 8–3 lists the gains in decibels at the breakpoint frequencies.

As a reminder that the response does not actually consist of straight lines, the actual overall response is plotted as a dashed curve in Figure 8–12. Note that it passes through points 3 dB (a factor of 0.7) below the breakpoints at f_L and f_U. We could have obtained this curve by evaluating Equation (8.3) at many frequencies for each stage and then multiplying, but many more points would have to be evaluated to make as accurate a plot. Because the response is down 3 dB at f_L and f_U, the bandwidth from f_L to f_U is more specifically called the *3-dB bandwidth*:

$$B_{3dB} = f_U = f_L = 2000 - 100 = 1900 \text{ Hz}.$$

Often it is sufficiently accurate to let $B_{3dB} = f_U$.

8.5   HARMONIC ANALYSIS

From its frequency response we can see how a circuit acts on a pure sinusoid: both magnitude and phase are changed. But what happens to other waveforms? If the pressure being monitored suddenly jumps, will the amplifier output jump in the same way? The answer is no. We will see in this section a connection between the frequency response of the amplifier and the shape of the waveform in time at the output.

Consider a particular class of waveforms that might appear at the input to the amplifier—*periodic* signals that have a pattern that repeats. An example is the square wave in Figure 8–13*a*. (This might have been generated by a photodiode responding to light pulses. In the next chapter there will be extensive use of binary signals that alternate between two values.) It can be shown that every periodic signal can be formed by adding together enough sine waves each with the right magnitude and phase. The square wave in Figure 8–14a can be expressed as

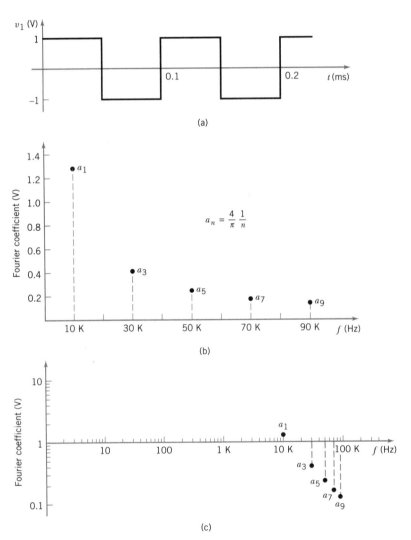

FIGURE 8–13 (a) Time waveform of a square wave. (b) Frequency components or Fourier coefficients of the square wave. (c) The Fourier coefficients plotted on log-log axes.

$$v_1(t) = (4/\pi)[\sin(2\pi ft) + \tfrac{1}{3}\sin(2\pi 3ft) + \tfrac{1}{5}\sin(2\pi 5ft) + \cdots]$$
$$= 1.27\sin(2\pi ft) + 0.42\sin(2\pi 3ft) + 0.25\sin(2\pi 5ft) + \cdots \quad (8.5)$$

where $f = 1/T$ and T is the period of repetition. In our case, $T = 0.1$ ms and $f = 10$ kHz. All the other frequencies are multiples or *harmonics* of the fundamental frequency f. Notice that the square wave has only odd harmonics (the fundamental is also the first harmonic). This infinite series of sinusoidal terms is called the *Fourier-series* expansion for $v_1(t)$.

A general form for the Fourier series is

$$v(t) = a_0 + a_1\sin(2\pi ft + \theta_1) + a_2\sin(2\pi 2ft + \theta_2)$$
$$+ a_3\sin(2\pi 3ft + \theta_3) + \cdots. \quad (8.6)$$

Given a time waveform $v(t)$, the a's, and the θ's can be calculated by hand from formulas (see problem E8.3). More often engineers rely on packaged computer programs to perform the Fourier analysis. If the actual signal $v(t)$ is available, a spectrum analyzer can be used to display the *Fourier coefficients* (the a's). The instrument displays a picture similar to that shown in Figure 8–13b, which corresponds to the square wave represented in Figure 8–13a and Equation (8.5). The same harmonics are plotted on log-log axes in Figure 8–13c, where it is clear that the harmonics fall off as $1/f$ (a straight line with these axes).

It is not obvious that something as rounded as sine waves could ever form a square wave, but it is possible with an infinite number of sine waves. As more and more terms in Equation (8.5) are included, the approximation to square corners becomes better and better. This is shown in Figure 8–14. In Figure 8–14a one cycle of the square wave is compared with the first three terms—the first, third, and fifth harmonics. The first harmonic by itself could be considered an approximation to the square wave, but it is very rounded. In the waveform in Figure 8–14b the first two terms have been added, making a better approximation. The waveform in Figure 8–14c is the sum of the first three terms. The edges are becoming steeper, and the tops are becoming flatter. With an infinite number of terms they become vertical and horizontal.

If we know the frequency response of a circuit, we can tell what it will do to the magnitude and phase of each harmonic of a periodic signal. These modified harmonics can then be added up to determine the output waveform. Consider the frequency response in Figure 8–15b for example. It either passes a frequency with a gain of ten, or it doesn't pass it at all. If the square wave in Figure 8–15a is applied to the input, only the first, third, and fifth harmonics (at 10 kHz, 30 kHz, and 50 kHz) appear at the output. As in Figure 8–14c, these harmonics add together to form the output waveform shown in Figure 8–15c. The elimination of the higher frequencies by the circuit has rounded the edges and produced some ripple in the time domain.

A cutoff as sharp as the one in Figure 8–15b is impossible to realize. Consider the more common frequency response in Figure 8–9, where the response rolls off as $1/f$. In this case the harmonics at 10 kHz, 30 kHz, 50 kHz, etc., experience gains of $8.9 \underline{/-27°}$, $5.5 \underline{/-56°}$, $3.7 \underline{/-68°}$, etc. When all the modified harmonics are

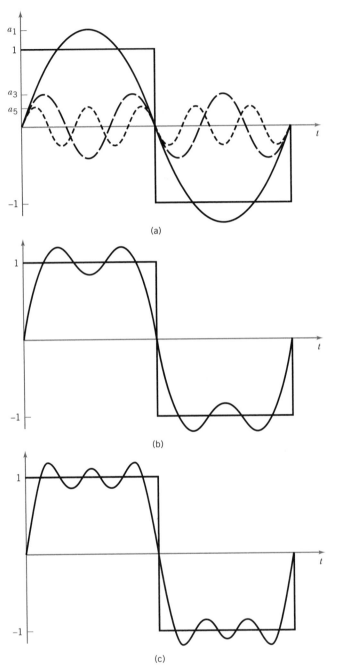

FIGURE 8–14 (a) One cycle of a square wave and its first three frequency components. (b) Square wave compared with the sum of the first two components. (c) The sum of the first three components is an even better approximation to the square wave.

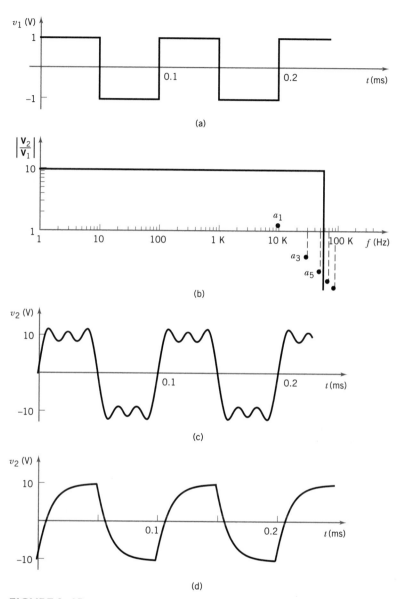

FIGURE 8–15 (a) 10-kHz square wave at the input to an amplifier. (b) Frequency response of the amplifier. Only the first three frequency components of the square wave are passed. (c) Waveform at the output of the amplifier. (d) Waveform at the output if the frequency response of the amplifier is the gradual rolloff shown in Figure 8–9.

added up, the output will appear as in Figure 8–15d. Again the edges are no longer vertical, but in this case there is no ripple. In theory we would have to add an infinite number of harmonics to get this result, but in practice about ten are sufficient to get a good approximation.

The calculations just outlined are seldom carried out by hand. A computer program usually carries out a *Fourier transform* to determine the harmonics of the input waveform. After multiplying by the frequency response, the program performs an *inverse Fourier transform* on the new harmonics to determine the output waveform. However, there are often cases where simplifying approximations allow the engineer to get some quick answers by hand.

When the input waveform is a step function, the relationship between the frequency response and the output waveform is particularly simple; a computer is not necessary. We will look at this in the next section.

8.6   TIME RESPONSE

We saw in Figure 8–15 that a reduction of the amplitude of high frequencies causes a rounding or slowness in the time response of an amplifier—a sluggishness in following change. There is also a connection between a reduction of the amplitude of low frequencies and a "sag" in a time response trying to maintain a value for a long time. In this section we will make a quantitative link between the cutoff frequencies and these behaviors of the time response.

The standard way of characterizing the time response of a circuit is to apply a step function to the input (see Figure 8–16a) and observe the *step response* at the output. We examined step responses in Chapter 7 and frequency responses in Chapter 6. Now we will show how to determine one response from the other without worrying about the specific circuit involved. The connection between the time response and the frequency response is the differential equation that relates the output voltage to the input voltage.

Let's look at the differential equation connected with the response in Figure 8–9 and the corresponding transfer function in Equation (8.1). We will show that to get this particular response, the differential equation relating input and output voltages must be of the form

$$\tau_U \frac{d}{dt} v_2 + v_2 = G_0 v_1. \tag{8.7}$$

The unit step function in Figure 8–16a has $v_1 = 1$ V. Then the solution of Equation (8.7) is

$$v_2 = G_0(1 - e^{-t/\tau_U}). \tag{8.8}$$

This step response is plotted for $G_0 = 10$ and time constant $\tau_U = 0.008$ ms in Figure 8–16b. The result is an exponential that asymptotically approaches 10 V. The initial slope is G_0/τ_U, so the time for the circuit to respond is roughly one or

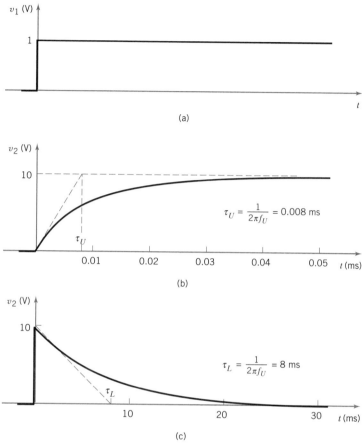

FIGURE 8–16 (a) Step applied to an amplifier input. (b) Step response at the output of an amplifier with upper cutoff frequency at $f_U = 20$ kHz (see Figure 8–9). (c) Step response of an amplifier with lower cutoff frequency at 20 Hz (see Figure 8–10a).

two time constants τ_U. A more precise specification of the time to respond is the rise time, defined near the end of this section.

To find the frequency response corresponding to Equation (8.7), let v_1 and v_2 be complex sinusoids:

$$v_1 = \mathbf{V}_1 e^{j\omega t},$$

$$v_2 = \mathbf{V}_2 e^{j\omega t},$$

where $\mathbf{V}_1$ and $\mathbf{V}_2$ are complex phasors. (This is the same approach used in deriving impedances in Section 6.3.) Substituting into Equation (8.7) we have

$$\tau_U(j\omega)\mathbf{V}_2 e^{j\omega t} + \mathbf{V}_2 e^{j\omega t} = G_0 \mathbf{V}_1 e^{j\omega t}, \tag{8.9}$$

$$\frac{\mathbf{V}_2}{\mathbf{V}_1} = \frac{G_0}{1 + \tau_U(j\omega)} = \frac{G_0}{1 + jf/f_U}, \tag{8.10}$$

where

$$\tau_U = \frac{1}{2\pi f_U}. \tag{8.11}$$

But this transfer function is the same as that for G_v in Equation (8.1). The conclusion is that the step response in Figure 8–16b corresponds to the frequency response in Figure 8–9, where $\tau_U = 0.008$ ms and $f_U = 20$ kHz. [The reader should confirm that these values satisfy Equation (8.11).] Notice that the correspondence between the responses holds independent of what circuit (or even nonelectrical device) is used to realize the responses.

We handle the case for the lower cutoff frequency in a similar way. The differential equation

$$\tau_L \frac{d}{dt} v_2 + v_2 = G_0 \tau_L \frac{d}{dt} v_1 \tag{8.12}$$

leads to a frequency response

$$\frac{\mathbf{V}_2}{\mathbf{V}_1} = G_0 \frac{jf/f_L}{1 + jf/f_L}, \tag{8.13}$$

and to a step response

$$v_2 = G_0 e^{-t/\tau_L}, \tag{8.14}$$

where

$$\tau_L = \frac{1}{2\pi f_L}. \tag{8.15}$$

A plot of Equation (8.13) (the frequency response) for $f_L = 20$ Hz is shown in Figure 8–10a, and a plot of Equation (8.14) (the step response) for $\tau_L = 8$ ms is shown in Figure 8–16c. The step response is said to "sag" since it is unable to hold the horizontal line of the step input. Equation (8.15) shows that lowering the cutoff frequency f_L increases τ_L, and the step response takes longer to sag.

When a frequency response has *both* an upper cutoff frequency f_U and a lower cutoff frequency f_L, as in Figure 8–10b and Equation (8.3), then a second-order differential equation is involved. However, if there is great enough separation between f_L and f_U, then $\tau_U \ll \tau_L$, and the fast-rising response is essentially over before any significant sag has begun. In such cases the step response is a combination of Figure 8–16b and Figure 8–16c, as the following example illustrates.

EXAMPLE 8.2 A filter for a telephone system has a transfer function of the form in Equation (8.3). It passes frequencies around 1 kHz with little attenuation, but it rejects high and low frequencies.

Given: $G_0 = 1, f_L = 300$ Hz, and $f_U = 3000$ Hz.

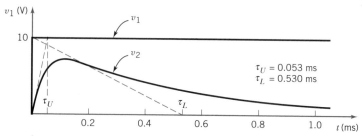

FIGURE 8–17 Input step v_1 and output step response v_2 of an amplifier with $f_L = 300$ Hz and $f_U = 3000$ Hz.

Find: the response at the output v_2 to a 1.0-V step at the input v_1.

Solution: The input step at v_1 is shown in Figure 8–17. The output v_2 doesn't follow the instantaneous rise because of the upper cutoff frequency f_U. From Equation (8.11), the resulting time constant is $\tau_U = 1/(2\pi \times 3000 \text{ Hz}) = 0.053$ ms. Therefore v_2 rises with an initial slope that would reach the final value of 1.0 V in 0.053 ms, as shown in Figure 8–17. At the same time an exponential "sag" is beginning due to the lower cutoff frequency f_L. From Equation (8.15), the time constant of this sag is $\tau_L = 1/(2\pi \times 300 \text{ Hz}) = 0.530$ ms. This is much longer than τ_U, so the rising exponential is practically complete before the sag is significant. Therefore v_2 is about 0.37 V at $t = 0.530$ ms (see Section 7.3 on first-order transients). The result is a sketch of v_2 that closely approximates the actual curve. An exact solution can be carried out with a second-order differential equation, but it involves much more work and much less insight. ▪

From Equations (8.7) through (8.10) we can see a general method for determining the step response from the frequency response. Taking the denominator of the frequency response, substitute the operator d/dt in places of $j\omega$ (or $j2\pi f$), and multiply through by v_2. The result is the left side of the governing differential equation. For a step input, the right side of the differential equation is a constant (sometimes zero). The solution of this equation yields the step response. In Chapter 10 we will apply this method to second-order differential equations. A typical frequency response and the corresponding step response for a system with a second-order rolloff are shown in Figure 8–18.

When an amplifier or a system has a cutoff steeper than -20 dB per decade, it is possible for the frequency response to exhibit *peaking*, as is the case in Figure 8–18a. The peak in the response favors frequencies in the vicinity of 10 kHz by about 4 dB here. This situation usually arises from pushing the bandwidth of a system to the point that it is on the verge of instability. (The concept of instability will be dealt with in Chapter 10.)

The corresponding step response is shown in Figure 8–18b. As might be expected from the frequency response, there is a 10-kHz component that dies out; this is called *ringing*. The response actually overshoots the final value of ten, in contrast with the exponential response in Figure 8–16b. The amount of *overshoot* here is 37%.

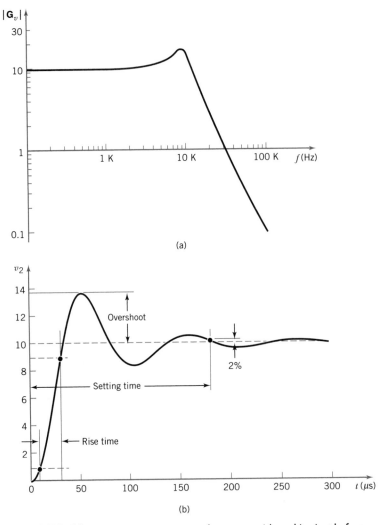

FIGURE 8–18 (a) Frequency response of a system with peaking just before a sharp cutoff. (b) Corresponding step response with ringing evident. The rise time is the time for the response to go from 10% to 90% of the final value. After the settling time, the response remains within 2% of the final value.

Another important property of the response is the length of time it takes to reach the intended level. This might be specified as the time until it first crosses the final value. But some responses, such as an exponential, very quickly reach a level near the final value, and then they take a long time (if ever) to actually reach it. Therefore the agreed-upon specification of the *rise time* is the time to go from 10% to 90% of the final value. The rise time of the waveform in Figure 8–18b is 22 μs. The reader can show that the rise time of a simple exponential response [see Figure 8–16b and Equation (8.8)] is $2.2\tau_U$.

A third property of a step response is the length of time for the ringing to cease—for the response to settle down to about the final value. The *settling time* is usually defined as the time until the response remains within 2% of the final value. In Figure 8–18b the settling time is 180 μs.

8.7   DISTORTION AND OFFSET

The simple amplifier model in Section 8.2 represents the gain by a fixed value G_v which is valid for all input levels. In practice there is some large input amplitude for which this model breaks down, and the output becomes distorted. The general term *distortion* refers to any way in which the shape of the output waveform differs from that at the input. In the previous section we saw examples of *frequency distortion*—sag, and rounding of corners. In that case the amount of distortion depended on the frequency components of the input signal. In this section we will look at *amplitude distortion*—distortion that depends on the amplitude of the signal.

The most common form of amplitude distortion is called *clipping* or *limiting*, illustrated in Figure 8–19b. The input waveform is faithfully reproduced at the output up to some level (10 V in this example). The output signal is "clipped" at this level and can't rise higher. Figure 8–19a plots the transfer characteristic (output versus input) that would produce such clipping. An input signal with amplitude greater than 0.05 V encounters a bend in the characteristic. The gain curve does not always cut off so cleanly as illustrated here; it may bend over gradually. Also it is not necessary, in general, that the positive and negative clipping take place at the same magnitude.

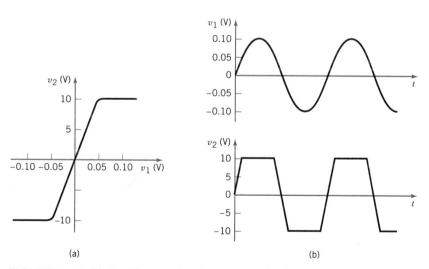

(a) (b)

FIGURE 8–19 (a) Amplifier transfer characteristic that becomes nonlinear for signals with large amplitudes. (b) Clipping distortion at the output v_2 when the input v_1 exceeds 0.05 V.

In Chapters 14 and 15 we will see that output clipping usually occurs within a couple of volts of the dc supply voltages that power an electronic circuit.* These supply voltages are typically 5–20 V.

To correct a clipping situation in a signal system, the engineer can do one of two things: arrange for a smaller voltage signal to represent the physical quantity being dealt with, or arrange for an amplifier with a larger signal capability without clipping.

In general amplitude distortion can be described as any departure of the transfer characteristic from being a straight line through the origin. The transfer characteristic in Figure 8–19a is a particular example of such a departure. There may be more subtle distortion such as a slight curvature or wiggle in the characteristic. These *nonlinearities* give rise to *harmonic distortion*—the change in the harmonic balance of a periodic signal. A manufacturer usually indicates how small these nonlinearities are by specifying the total harmonic distortion (THD). For an audio system, THD less than about 0.5% will not be noticed. THD is measured by applying a pure sinusoid to the circuit and measuring the root mean square (rms) of the new harmonic frequencies introduced at the output (ideally there should be none). The ratio of this rms to the rms of the fundamental (the original frequency) at the output is the THD. In terms of the coefficients in Equation (8.5)

$$\text{THD} = \frac{\sqrt{a_2^2 + a_3^2 + a_4^2 + \cdots}}{a_1}. \tag{8.16}$$

EXAMPLE 8.3 An amplifier with a gain of 200 clips the output at $+10$ V and at -10 V (see the characteristic in Figure 8–19a).

Given: the input is a sine wave with 1-V peak amplitude.

Find: the total harmonic distortion.

Solution: To avoid clipping, the input amplitude should be kept less than 0.05 V. Because the input amplitude for this example is 20 times too large, the output is essentially a square wave with 10-V peak amplitude. The harmonic content of a square wave with 1-V peak amplitude is given in Equation (8.5). Multiplying those coefficients by 10, we have $a_1 = 12.7$, $a_3 = 4.25$, $a_5 = 2.55$, $a_7 = 1.82$, $a_9 = 1.42, \ldots$. All the even coefficients are zero. If we truncate the numerator of Equation (8.16) at a_9, we get THD $= 43\%$. (Taking the series to infinity gives THD $= \underline{47\%}$.)

Another way in which the transfer characteristic can depart from the ideal is shown in Figure 8–20a. Here the line is simply displaced so that it doesn't pass through the origin. The result is that there is a dc voltage at the output with no

*Portable equipment operates directly off battery dc. However, most pieces of equipment are designed to plug into the 110-V ac power. These units include a "power supply" which converts the 110-V, 60-Hz ac into whatever dc voltage is needed.

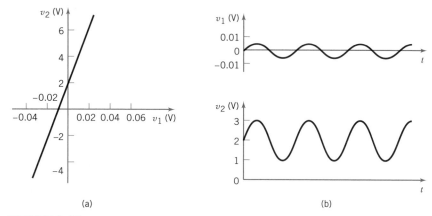

FIGURE 8-20 (a) Transfer characteristic with offset. (b) Output v_2 with dc offset. (Note that the input v_1 contains no dc.)

signal present at the input. When an input signal is present, the output signal is shifted by a *dc offset*, as in Figure 8–20b. For this example, the *output offset* is 2 V. The situation may also be described as *input offset*—the amount by which the input must be shifted to correct the output. For the example given, the input offset is 0.01 V—the point at which the characteristic intersects the input axis. A typical input offset is between 1 mV and 10 mV.

One way to correct dc offset is to use ac coupling in the circuit. This blocks the dc, but it also blocks low frequencies, causing sag as discussed in the previous section. Another way is to "trim out" the offset with an adjustment control. This correction is limited by drift with temperature and age. The most powerful method of reducing offset is to use "chopper stabilization," a technique discussed in Chapter 16.

8.8 NOISE

The term *noise* is applied to any signal present that is undesired. In audio systems noise may be 60-Hz ac hum, pops from dust on a record, or background hiss on a radio. In a video system such as TV, noise appears as "snow" or spurious lines on the screen. The presence of noise (and there is always *some* noise) limits how small a signal can be amplified practically. If there is 100 μV of noise present at the input to an amplifier, it would be futile to try to amplify a 30-μV signal from a transducer.

A common source of noise is electromagnetic interference (EMI). This includes radio waves, electric fields, and magnetic field. For example, the 60-Hz electric field that pervades man's environment will be coupled into the amplifier in Figure 8–2 through the wires (transmission line) at the input to the amplifier. If the level of interference is annoying, the noise can be reduced by shortening the wires or shielding them with a grounded metal covering (coaxial cable). In a public address system hum pickup can be reduced by placing a lower cutoff frequency at about 200 Hz—reducing the gain at 60 Hz but not in the voice range.

Stereo systems sometimes pick up police calls, and TVs sometimes pick up interference from a nearby electric drill. These are both examples of EMI in the form of radio waves. In the case of the stereo, the solution is again to shorten and shield and input wires to the amplifier. In the case of the TV, shielding can't be used since the antenna must be able to pick up the desired electromagnetic waves. Here the best solution is to eliminate the interference at the source. Capacitors across the brushes in the electric drill will cut down on the EMI from the sparks. Capacitance and resistance are also used to reduce the interference from car ignition systems.

The manufacturer or user of electrical equipment must also be concerned that he or she is not the source of excessive EMI—causing noise for someone else. The Federal Communications Commission (FCC) puts limits on the level of unlicensed radio waves that can be generated. Proper shielding and filtering must be used to meet these limits.

Even if all external noise is eliminated, there is always a small amount of noise unavoidably generated in the amplifier itself. It can be heard as the "hiss" when the gain of an audio system is turned all the way up with no signal at the input. One source of this noise is random motion of electrons in resistors. Since this motion is due to heat, the noise can be reduced by cooling the circuit. However, this is cumbersome and is used only in special situations such as the reception of signals from space. Because of its source, this noise is called *thermal noise* or sometimes *Johnson noise* after the engineer who first characterized it.

Another form of noise generated within the circuit is *shot noise*. This arises from the fact that a "continuous" current is actually the flow of discrete electrons. If the current is small enough, its "lumpiness" begins to be apparent. This is the same situation as if a "continuous" force were applied to a wall by a stream of buckshot. The discrete nature of the force would be heard as noise.

Noise can also be characterized in the frequency domain, just as we characterized a square wave by its harmonics. Both thermal noise and shot noise have components at *all* frequencies. In terms of the power associated with the noise, these components are evenly distributed in frequency. Therefore thermal and shot noise are called *white noise* by analogy to light. A convenient measure of the level of noise is the *spectral density* in terms of mean-square volts per hertz of spectrum. For example, noise with a spectral density of $N_0 = 0.02$ $(\mu V)^2$/Hz limited to a bandwidth of 60 kHz would have a mean-square voltage of $60 \times 10^3 \times 0.02$ $(\mu V)^2 = 1200$ $(\mu V)^2$. Taking the square root of this we get the rms noise voltage of 34.6 μV.

The above calculation assumes the bandwidth is limited by an abrupt cutoff as in Figure 8–15b. In the more usual case with a decade/decade rolloff as in Figure 8–9, the noise is (by comparison) gradually attenuated beyond the 3-dB bandwidth $B_{3dB} = f_U$. It can be shown that this gradual cutoff lets through the same amount of noise as if there were an abrupt cutoff at

$$B_n = (\pi/2) B_{3dB}. \tag{8.17}$$

This is called the *noise bandwidth.* The frequency response in Figure 8–9 with $B_{3dB} = 20$ kHz therefore has a noise bandwidth of $B_n = 31.4$ kHz.

In order to calculate the rms noise voltage directly, the noise spectral density is also commonly specified in terms of rms volts per $\sqrt{\text{Hz}}$. In our example, the spectral density $N_0 = 0.02$ $(\mu V)^2/\text{Hz}$ could also be specified as $\sqrt{N_0} = 0.14$ $\mu V/\sqrt{\text{Hz}}$. The rms noise voltage e_n is now found by multiplying by the square root of the noise bandwidth: $(0.14 \ \mu V/\sqrt{\text{Hz}})\sqrt{60 \text{ kHz}} = 34.6 \ \mu V$, as before. The general expression is

$$e_n = \sqrt{N_0}\sqrt{B_n}. \tag{8.18}$$

The user cares about the level of noise at the output of an amplifier. However, almost all the noise is generated in the input stage of an amplifier. So it is more appropriate to specify the equivalent noise at the input. For example, the manufacturer of an amplifier may specify that the input noise spectral density is $0.14 \ \mu V/\sqrt{\text{Hz}}$. If the (adjustable) 3-dB bandwidth of the amplifier is set to 20 kHz, then $B_n = 31.4$ kHz, and the effective noise at the input would be $e_n = (0.14 \ \mu V/\sqrt{\text{Hz}})\sqrt{B_n} = 24.8 \ \mu V$. Thus input signals of less than a millivolt would encounter a noticeable amount of noise.

The input noise spectral density $\sqrt{N_0}$ is usually a function of the source resistance R_s at the input to the amplifier. The dependence can be calculated from a more complicated noise model involving both voltage and current noise sources. The manufacturer has usually done this for the user and publishes a curve of spectral density for an effective noise voltage source e_n at the input as a function of R_s. A typical curve is shown in Figure 8–21.

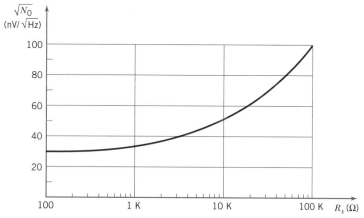

FIGURE 8–21 Spectral density of an equivalent noise voltage source e_n at the input of an amplifier. The rms voltage of the source is the spectral density times the square root of the amplifier's noise bandwidth.

EXAMPLE 8.4 The system in Figure 8–22 amplifies a 100-Hz sinusoidal signal from a transducer. The model for the amplifier includes an equivalent input noise source e_n with spectral density specified in Figure 8–21.

Given

$v_t = 100 \ \mu V$ rms. The amplifier open-circuit gain A_v has a midband gain of $A_0 = 1500$ and a 3-dB bandwidth of $f_U = 6370$ Hz.

Find

1. the level of the signal s and the level of the noise n at the output ($v_2 = s + n$);

2. a smaller bandwidth f_U so that n is only 1% of s.

Solution

1. Using superposition, set e_n to zero, and find the signal s at the output due to $v_t = 100 \ \mu V$ rms. Using the same approach as in Example 8.1, we find $s = 100 \ \mu V \times (10/14) \times 1500 \times (10/11) = $ <u>97.4 mV</u>.

   Next we set v_t to zero and find n at the output due to e_n. Since R_s is 4 kΩ, we see from Figure 8–21 that the spectral density for e_n is $\sqrt{N_0} = $ 40 nV/$\sqrt{Hz}$. The noise bandwidth is $B_n = (\pi/2)6370$ Hz $= 10$ kHz. From Equation (8.18) we find $e_n = (40 \ nV)\sqrt{10^4} = 4 \ \mu V$. Again using the approach in Example 8.1, we find $n = 4 \ \mu V \times (10/14) \times 1500 \times (10/11) = $ <u>3.9 mV</u>.

2. In order that $n = 0.01s = 0.974$ mV (a reduction by a factor of 4), we must reduce e_n by a factor of 4. But e_n is proportional to the square root of the bandwidth. Therefore we must reduce the 3-dB bandwidth by a factor of 16: $f_U = 6370$ Hz/16 $= $ <u>400 Hz</u>.

It is common to express the *signal-to-noise ratio* (SNR) in decibels. Therefore in the first part of the example SNR $= 20 \log(s/n) = 20 \log(97.4/3.9) = 28$ dB. Note that since v_t and e_n are treated in the same way by the system, the SNR remains the same if calculated at the input: SNR $= 20 \log(v_t/e_n) = 20 \log(100/4) = 28$ dB. After reduction of the bandwidth in part 2, SNR $= 20 \log(100) = 40$ dB.

Noise can always be reduced by reducing the bandwidth of the system. This will also improve the SNR *if* the signal is not affected much by the reduction in bandwidth. In part 2 of the above example, it would not pay to reduce the 3-dB bandwidth to 400 Hz if v_t were a voice signal (with significant content out to 3 kHz).

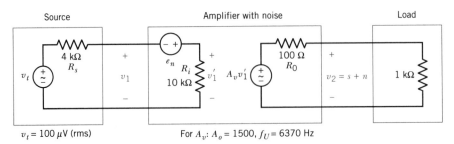

FIGURE 8–22 Amplifier model with equivalent noise voltage source e_n. The signal component s at the output is due to v_t, and the noise component n at the output is due to e_n.

8.9 TRANSMISSION LINES

The system in Figure 8–2 includes a transmission line to carry the transducer signals to the amplifier input. There is another transmission line at the output of the amplifier. Ideally these lines convey the signal unchanged; in practice they cause some loss, delay, and reflections.

The simplest form of transmission line is a pair of wires close to each other, perhaps twisted. They are usually made of copper for low resistance. For example, 10-gauge copper wire has about 1 Ω per 1000 ft. The standard way of specifying the cross-sectional area of wire, which determines the resistance per foot, is the American Wire Gauge (AWG). This is a number that increases as the cross section of the wire decreases. For each increase in the gauge, the cross section decreases by a factor of $1/1.26$, and the resistance per foot increases by a factor of 1.26. For example, 11-gauge wire has 1.26 Ω per 1000 ft, and 13-gauge wire has 2 Ω per 1000 ft (every increase of the gauge by three doubles the resistance). A wire of length d and gauge g has a resistance R given by

$$R = d \times (1\ \Omega/1000\ \text{feet}) \times 1.26^{g-10}. \qquad (8.19)$$

EXAMPLE 8.5 A pair of 26-gauge wires carries a signal 50 ft from an amplifier to an 8-Ω loudspeaker.

Find

1. the attenuation the signal undergoes from the amplifier to the loudspeaker;
2. the wire gauge needed to make the attenuation only -1 dB.

Solution

1. The wire gauge is 16 gauges higher than 10-gauge wire. Therefore the resistance is about $1.26^{16} = 40$ times as great, or 40 Ω per 1000 ft. The total length of the wire is $d = 100$ ft, so its resistance is 4 Ω. Since the speaker is 8 Ω, it gets $8/(4 + 8) = 2/3$ of the voltage from the amplifier. Therefore the attenuation due to the wire is $20 \log(2/3) = -3.5$ dB.
2. An attenuation of -1 dB corresponds to a factor of 0.89. Therefore the resistance of the wire must be $8\ \Omega(1 - 0.89)/0.89 = 0.99\ \Omega$ per 100 ft or 9.9 Ω per 1000 ft. Since 9.9 is about 1.26^{10}, we need ten gauges higher than 10-gauge wire, or 20-gauge wire.

If most of the space between the wires consists of air, a transmission line conveys signals with the velocity of light (0.3 meters, or one foot per nanosecond). If most of the space is filled with a plastic, the velocity will be about two-thirds of this. For signals with frequencies above 1 MHz, the delay in the transmission line may be important, especially if there are reflections at the ends.

To understand the nature of *reflections*, consider a rope lying on the floor. If you give one end a flick upward, a pulse travels down the rope, the far end jumps upward, and the undissipated energy tends to send a pulse back at you. If the far end of the rope is tied to a wall instead, a flick upward will produce a returning pulse that is downward. There will be no pulse reflected from the far end only if it is fastened to a sliding termination with just the right friction to completely absorb the transmitted energy.

A voltage pulse applied to a transmission line with an open circuit at the far end (the receiving end) will be completely reflected. If the receiving end is shorted, a negative pulse will be reflected. There will be no reflection if the receiving end is loaded or *terminated* with the proper impedance. This proper impedance is called the *characteristic impedance* Z_0 of the line. The reflected wave, or *reverse wave* v^- is related to the *forward wave* v^+ by $v^- = \Gamma_r v^+$, where Γ_r is called the *reflection coefficient*. It can be found from

$$\Gamma_r = \frac{R_r - Z_0}{R_r + Z_0},\tag{8.20}$$

where R_r is the terminating resistance (see Figure 8–23a). Usually the line can be considered lossless, and correspondingly Z_0 can be considered real (a pure resistance). The reader should check that Γ_r gives negative, positive, and zero reflections when R_r is equal to zero, infinity, and Z_0.

Reflections can also occur at the sending end of a transmission line. The terminating resistance at the sending end is the source impedance R_s, which determines the reflection coefficient Γ_s. When a reverse wave v^- hits the sending end, it causes a new forward wave v^+ according to $v^+ = \Gamma_s v^-$, where

$$\Gamma_s = \frac{R_s - Z_0}{R_s + Z_0}.\tag{8.21}$$

If neither R_s nor R_r equals Z_0, then multiple reflection can be set up with waves bouncing back and forth between the two ends. Consider the transmission line in Figure 8–23a. Let the source voltage v_g be a voltage step of height E. Because the line acts for a moment as an impedance Z_0, a voltage divider relation gives the initial v_s at the sending end:

$$v_s = E \frac{Z_0}{R_s + Z_0}.\tag{8.22}$$

This voltage change travels as the first forward wave

$$v^+ = v_s,\tag{8.23}$$

reaching the receiving end after a delay T, where it produces a reverse wave

$$v^- = \Gamma_r v^+.\tag{8.24}$$

The combination of the two waves produces a change in v_r:

$$\Delta v_r = v^+ + v^-.\tag{8.25}$$

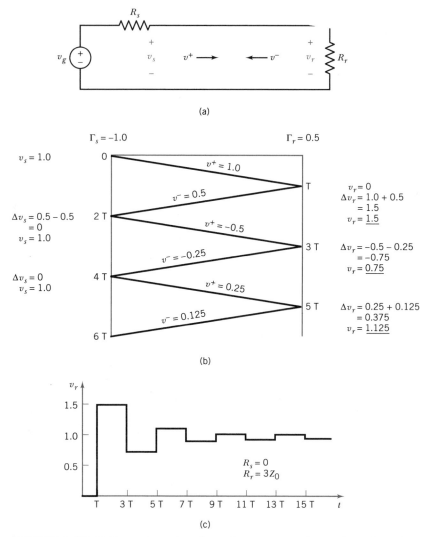

FIGURE 8–23 (a) Transmission line driven by a source with impedance R_s at the sending end and terminated by an impedance R_r at the receiving end. (b) Bounce diagram for Example 8.6. The forward waves v^+ and reverse waves v^- are caused by reflections. (c) The resulting waveform at the receiving end in the example.

Since v_r was initially zero, this is the whole of v_r now. After another delay T, the reverse wave v^- reaches the sending end, producing a new forward wave

$$v^+ = \Gamma_s v^-. \tag{8.26}$$

The combination of the two waves produces a change in v_s:

$$\Delta v_s = v^- + v^+. \tag{8.27}$$

Equations (8.24)–(8.27) are repeated every period of $2T$, making smaller and smaller changes in v_s and v_r.

EXAMPLE 8.6 A transmission line as in Figure 8–23a is misterminated at both ends.

Given: $Z_0 = 50\ \Omega$, $R_s = 0$, and $R_r = 150\ \Omega$. The source v_g is a 1.0-V step.

Find: the response v_r at the receiving end.

Solution: The application of Equations (8.20)–(8.27) is aided by the *bounce diagram* in Figure 8–23b. Since $R_s = 0$ and $R_r = 3Z_0$, Equations (8.20) and (8.21) give $\Gamma_s = -1$ and $\Gamma_r = 0.5$. Equations (8.22) and (8.23) give the initial values $v_s = v^+ = 1.0$. After a delay of T, v^+ reaches the receiving end, where it produces a reverse wave $v^- = 0.5$. The combination gives $v_r = 1.0 + 0.5 = 1.5$ (see the first value of the waveform in Figure 8–23c). The v^- gets reflected at the sending end at $t = 2T$ to produce a new forward wave $v^+ = -0.5$. At $t = 3T$ this v^+ reaches the receiving end to produce $v^- = -0.25$. According to Equation (8.25), the change in v_r is $\Delta v_r = -0.5 - 0.25 = -0.75$ for a total $v_r = 1.5 - 0.75 = 0.75$ (see the second value of v_r in Figure 8–23c). The reflections continue in a similar manner, producing a waveform that converges to $v_r = 1.0$. ▪

The "ringing" waveform in Figure 8–23c is typical of one termination greater than Z_0 and one less than Z_0. The closer R_r is to Z_0, the less the ringing. If both R_s and R_r are less than Z_0, the waveform at the receiving end is more like an exponential response; Figure 8–24 shows an example for $R_s = 0$ and $R_r = \frac{1}{3}Z_0$.

Troublesome reflections can be eliminated by making either R_s or R_r equal Z_0. Since resistors are not exact (typically 5%), it is best to make *both* nominally equal Z_0 to suppress any reflections quickly. For low-frequency applications it is not necessary to worry about proper termination of the line. A 4-m line with a delay of $T = 20$ ns would have a transient as in Figure 8–23c lasting only 150 ns or so. This would have negligible effect on an audio signal.

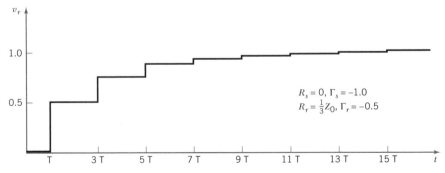

FIGURE 8–24 Waveform at the receiving end of a transmission line when there is a unit step at the sending end and when $R_s = 0$ and $R_r = \frac{1}{3}Z_0$.

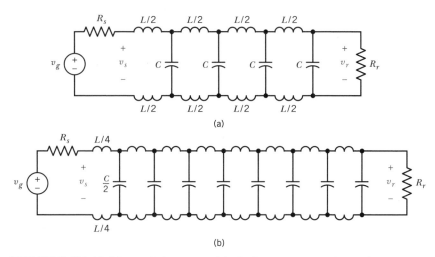

FIGURE 8–25 (a) A lumped-element model of a 4-m transmission line with capacitance *C* per meter and inductance *L* per meter. (b) A better model of the line (more nearly distributed). Z_0 is the characteristic impedance, and *T* is the delay.

The physical characteristics of a transmission line that determine its characteristic impedance Z_0 and its delay *T* are line capacitance and inductance. Because the wires of a transmission line are near each other, there is some capacitance between them, and this capacitance is distributed along the length of the line. There is also a certain amount of inductance associated with any wire; this is also distributed along the length of the line. Therefore a transmission line is called a *distributed system.* If the line has *C* capacitance per meter and *L* inductance per meter ($\frac{1}{2}L$ per meter in each wire), then a 4-m line could be approximated by the *lumped-element* model in Figure 8–25a. A better model—more nearly approaching the distributed nature of the line—can be obtained by subdividing the lumped elements as in Figure 8–25b. The characteristic impedance of the line is given by

$$Z_0 = \sqrt{L/C}, \tag{8.28}$$

and the delay of the line is given by

$$T = d\sqrt{LC}, \tag{8.29}$$

where *d* is the length of the line. For example, if $C = 100$ pF/m and $L = 250$ nH/m, then $Z_0 = 50\ \Omega$ and $T = (4\text{ m})(5\text{ ns/m}) = 20$ ns. Typical values of Z_0 range from $50\ \Omega$ to $500\ \Omega$.

A transmission line on the output of a high-impedance source looks primarily capacitive, and this may be troublesome even at relatively low frequencies. When connected to a source with $R_s = 100$ kΩ, a 4-m line with $C = 100$ pF/m looks primarily like a 400-pF capacitor. This produces a cutoff at $f_U = 1/(2\pi \times 400\text{ pF} \times 100\text{ k}\Omega) = 4$ kHz. (See Section 6.5.)

8.10   CASE STUDY

A team of engineers is assigned the task of measuring the performance of a turbine engine. A cutaway view of the engine is shown in Figure 8–26. The engine is loaded by a generator that is prevented from turning by a calibrated spring. The angle θ_3 through which this load is twisted is a measure of the engine torque. Sensor transducers are positioned to measure the inlet temperature T_1, the inlet pressure P_1, the fuel valve rotation θ_1, the combustion chamber temperature T_2, the vane rotation θ_2, the exhaust temperature T_3, the exhaust pressure P_2, the shaft rotation velocity ω_1, and the load rotation θ_3. From this data, the engineers can determine the engine efficiency and torque-speed curve under different conditions.

The temperature transducers are thermocouples. The other transducers are described below.

Rotational-Displacement Transducers

The rotational-displacement transducers θ_1, θ_2, and θ_3 are potentiometers like that pictured in Figure 8–27a. A 5-V source is placed across the ends of the resistive winding, and a wiper arm picks off some fraction of the voltage in proportion to the rotation θ of the arm. A schematic representation in Figure 8–27b shows the circuit is a voltage divider with the output voltage v_o proportional to R_1. The total resistance $R_1 + R_2$ of the winding is 10 kΩ, which determines the current drawn from the source and the Thevenin impedance of the device. When the arm is at either extreme, the Thevenin impedance is 0 Ω. When it is in the center, the Thevenin impedance is at its maximum of 2.5 kΩ. The transfer characteristic is shown in Figure 8–27c. The v_o increases linearly as θ goes from 0° to 330°; stops prevent θ from going beyond this range. The sensitivity is 5 V/330° = 15 mV/°. The *resolution* (smallest angle change that can be resolved) depends on the number of windings of the potentiometer. If there are 330 windings, then the resolution is 1°.

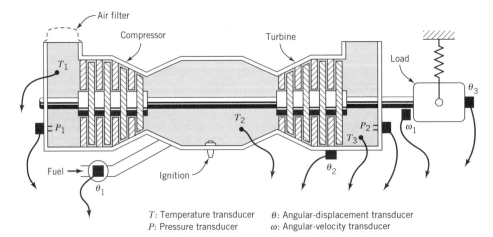

T: Temperature transducer θ: Angular-displacement transducer
P: Pressure transducer ω: Angular-velocity transducer

FIGURE 8–26 Turbine engine with transducers attached. The instrumentation in this case study is used to determine the efficiency and torque-speed curves of the engine under various conditions.

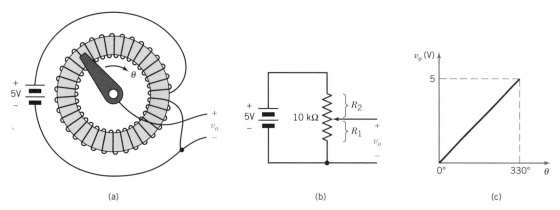

(a)   (b)   (c)

FIGURE 8–27 Rotational-displacement transducer. The output v_0 is proportional to the angle θ. (a) Mechanical representation of the potentiometer. (b) Schematic diagram of the circuit. (c) Transfer function. The range is limited to 0°–330°.

Pressure Transducers

The pressure transducers P_1 and P_2 are realized by a combination of two transducers, as shown in Figure 8–28. A Bouron C-tube converts pressure to displacement, and an LVDT converts displacement to voltage v_o. (The LVDT used here provides the Bouron C-tube with less resistance to movement than a simpler linear potentiometer would.) Recall that an LVDT requires an ac source v_s; the source here provides a 500-Hz sinusoid with 1-V peak amplitude. The output v_o is a sinusoid whose amplitude depends on the pressure. As the gauge pressure varies from -10 psi to 40 psi, the amplitude varies from -2 V to $+2$ V (a negative amplitude indicates v_o is 180° out of phase with v_s).

Some signal processing (demodulation) is necessary to convert the sinusoidal v_o to a dc signal. One simple solution might be to use a peak detector (see Section 16.5) to produce a voltage proportional to the amplitude. However, information about the phase would be lost. A *synchronous demodulator* like that in Figure 8–28b preserves the phase information. A multiplier (or mixer) circuit produces $v_1 = v_o \times v_s$, which is positive when v_o and v_s are in phase and negative when they are out of phase. (The *block diagram* in Figure 8–28b represents processes by blocks and signals by arrows; voltage transfer by two wires is indicated here by one line.) Figure 8–28c shows waveforms as the pressure increases through its range. The v_o decreases in amplitude, reverses phase, and then increases in amplitude. The v_1 correspondingly goes from 2-V negative pulses, through zero, to 2-V positive pulses. A low-pass filter (LPF) produces a smooth v_2 by removing the 1000-Hz component of v_1 (note this is twice the frequency of v_o and v_s).

The output voltage v_2 goes from -1 V to $+1$ V as the pressure goes from -10 psi to 40 psi. Therefore the sensitivity is (2 V)/(50 psi) $=$ 40 mV/psi. The resolution depends on the amount of 1000-Hz sinusoid left by the LPF. Since the sinusoidal component in v_1 has a peak amplitude equal to v_2, the LPF must reduce the sinusoidal component by a factor of 100 for 1% resolution of v_2. Then f_U of the LPF must be a factor of 100 below the frequency it is to remove: $f_U =$

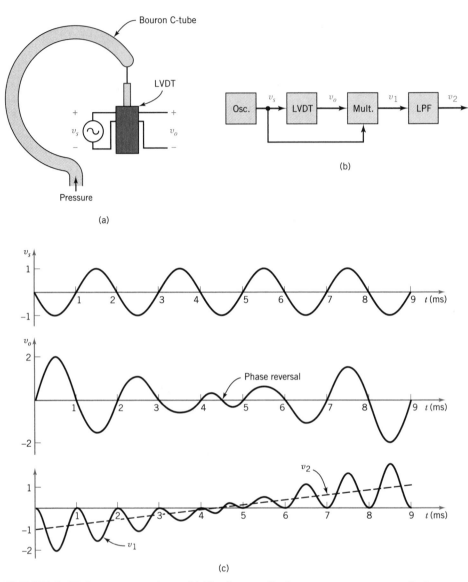

FIGURE 8-28 Pressure transducer. (a) The Bouron C-tube converts pressure to displacement, and the LVDT converts displacement to a change in v_0. (b) Block diagram of a synchronous demodulator to convert the sinusoidal v_0 to a dc v_2. (c) Waveforms in the demodulator. As pressure increases with time, v_2 increases with time.

$1000 \text{ Hz}/100 = 10 \text{ Hz}$. This implies that pressure fluctuations faster than 10 Hz will not be accurately measured.

Rotational-Velocity Transducer

One method of sensing rotational velocity depends on some irregularity in the shaft such as gear teeth. If necessary, the irregularity can be provided by connecting an iron slug to the shaft, as shown in Figure 8–29a. As the slug passes near the per-

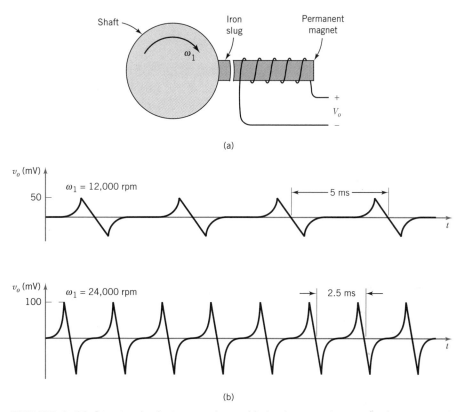

(a)

(b)

FIGURE 8–29 Rotational-velocity transducer. (a) As the iron slug passes the magnet, the change in flux causes a voltage pulse at v_0. (b) Waveforms of the voltage pulses. The pulse rate is equal to the rotational velocity in rpm.

manent magnet, it changes the flux through the windings, generating an electrical pulse at v_o (see Section 18.1). Figure 8–29 shows the form of these pulses, with one pulse per revolution of the shaft. Therefore a rotational velocity of $\omega = 12000$ rpm produces pulses at a rate of 200/sec (a pulse spacing of 5 ms). The lowest expected ω is 4800 rpm, which produces 80 pulses/sec with amplitude of 20 mV. The highest expected ω is 24000 rpm, which produces 400 pulses/sec with amplitude of 100 mV. The Thevenin impedance of the transducer is 100 Ω.

The v_o must undergo some signal processing if it is to produce an analog voltage proportional to ω. We chose instead to produce a digital signal that is a binary number proportional to ω. A binary digital counter (see Section 9.6) converts v_o directly to a digital signal by counting the pulses for 1 sec.

Signal Processing

The block diagram in Figure 8–30 shows the signal processing required by the four types of transducers in this example. The outputs of the thermocouple T and the rotational-velocity transducer ω are in the millivolt range and require amplification. The pressure transducer P requires an oscillator and a multiplier, as discussed above. The rotational-displacement transducer θ requires only a dc voltage source.

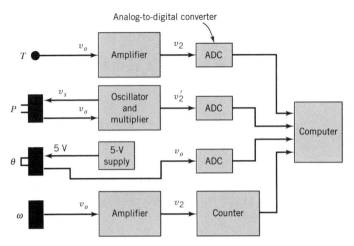

FIGURE 8–30 Block diagram of the signal processing for the four types of transducers in the case study. The measurements are finally converted to digital form for processing by a computer.

(The amplifiers and other electronic circuitry may also require dc voltage sources if they don't include their own power supplies.)

The final form for all the measurement signals is to be digital for interfacing with a computer. The computer can then make the conversion from signal levels back to temperature in degrees and pressure in psi, etc. It can also display current data, store the data, and process the data to show relations between the measured parameters (see Chapter 12). To this end, *analog-to-digital converters* (ADCs) are used to convert the analog signals from the temperature, pressure, and rotational-displacement transducers (see Section 12.6). The ADCs in this example require an input voltage in the range from 0 to 5 V. Their input impedance is 1 MΩ.

We will look, in turn, at loading, amplification, bandwidth, and noise considerations for each transducer. The outputs of the thermocouples can range from 0 to 50 mV, the Thevenin impedance is 3 Ω, and the sensitivity is 40 μV/°C. These low-level signals must be carried to their respective amplifiers by shielded wire pairs to avoid noise pickup. Twenty feet of 26-gauge wire pair will add 1.6 Ω to the Thevenin impedance of the transducers. An amplifier input impedance of at least 500 Ω is more than 100 times the total Thevenin impedance and will reduce the signal by less than 1%. An amplifier gain of 100 will raise the maximum signal level to 100 × 50 mV = 5 V. This will make use of the full range of the ADC without clipping. Temperature fluctuations are expected to be no faster than 5 Hz. An amplifier bandwidth of f_U = 50 Hz will reduce 5-Hz fluctuations by a factor of $1/[1 + (5/50)^2]$ = 0.995, or a reduction of 0.5%. Keeping the bandwidth this low will reject as much noise as possible. The amplifier must be dc coupled, and an input offset voltage of 0.2 mV will lead to an error in temperature measurement of (0.2 mV)/(40 μV/°C) = 5°C.

The pressure transducer P involves signals in the range of a couple of volts, which is well above most noise sources. Unshielded twisted pairs of wires should be adequate to conduct signals to and from the transducer. The range of the voltage

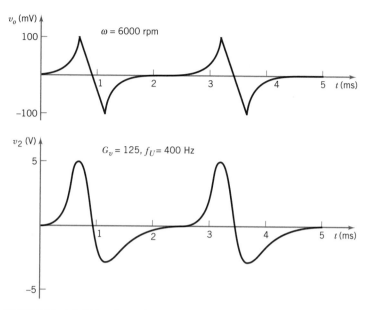

FIGURE 8–31 Waveforms from the rotational-velocity transducer before (v_0) and after (v_2) amplification. The rounding of the v_2 pulses is due to the amplifier's high-frequency cutoff at 400 Hz.

v_2 from the multiplier is from -1 V to $+1$ V (see Figure 8–28c). This must be converted to a v_2' with a range from 0 to 5 V by $v_2' = -2.5(v_2 - 1$ V$)$. The "-1 V" in this equation amounts to *level shifting*, which requires a summation amplifier (see Section 14.2).

The rotational-displacement transducer θ also involves signals in the range of a couple of volts. Unshielded twisted pairs of wires can conduct the signals to and from the transducer. The output v_o has a range of 0 to 5 V, so it is ready to be applied to the ADC. The 1-MΩ input impedance of the ADC is 400 times the 2.5-kΩ maximum Thevenin impedance of the transducer and will not load it down significantly.

The rotational-velocity transducer ω produces a signal v_o in the millivolt range, so a shielded wire pair should be used to connect it to its amplifier. The 100-Ω Thevenin impedance of the transducer calls for an amplifier with at least a 10-kΩ input impedance if the loading effects are to reduce the signal by less than 1%. The amplifier may be ac-coupled with a bandwidth from $f_L = 80$ Hz to $f_U = 400$ Hz to pass all the pulse rates. Note that v_o in Figure 8–29b actually has frequencies higher than 400 Hz because of its harmonics. These harmonics are attenuated by the roll-off at $f_U = 400$ Hz, resulting in the waveform* v_2 shown in Figure 8–31. The amplifier gain here is $G_v = 125$. For a pulse rate of 400 per second, the amplitude of v_2 is 5 V. It can be shown that for a pulse rate of 80/sec, the amplitude of v_2 is 1.8 V. These amplitudes are sufficient to drive the counter.

---

*The waveform was found by computer using a circuit analysis program called SPICE. A reference in "For Further Study" describes a version of SPICE for the personal computer.

The purpose of this example is to show how the concepts in this chapter are applied to a practical instrumentation problem. It also gives a preview of some of the more advanced signal processing covered in the following electronics chapters.

8.11   SUMMARY

Instrumentation systems monitor some physical parameter, amplify a voltage analog of it, and present it in some useful form. A transducer converts the physical parameter into the electrical domain. The signal level produced is usually small and requires some amplification. The amplifier should be able to handle the largest signal from the transducer without producing distortion.

The transducer signal may be composed of many frequencies. If the bandwidth of the amplifier is not large enough to pass some of the frequencies, the waveform will be changed. Loss of high frequencies rounds the corners and slows the response; loss of low frequencies produces sag in sustained values.

The gains of cascaded amplifiers are multiplied. When the gains are expressed in decibels, the gains are added.

Noise should be kept to a minimum by properly shielding the system and by keeping the bandwidth as narrow as possible. The amount of noise in the system determines the weakest signal level that can be amplified.

Transmission lines are intended to connect the pieces of the system without disturbing the signal. In practice long lines can attenuate the signal. Improperly terminated lines can produce reflections that are troublesome with high-frequency signals.

FOR FURTHER STUDY

J. A. Allocca and A. Stuart, *Electronic Instrumentation*, Prentice-Hall, Englewood Cliffs, NJ, 1983.

B. P. Lathi, *Signals, Systems, and Communications*, Wiley, New York, 1965.

J. D. Lenk, *Handbook of Microcomputer-Based Instrumentation and Control*, Prentice-Hall, Englewood Cliffs, NJ, 1984.

H. N. Norton, *Handbook of Transducers*, Prentice-Hall, Englewood Cliffs, NJ, 1989.

A. V. Oppenheim and A. S. Willsky, *Signals and Systems*, Prentice-Hall, Englewood Cliffs, NJ, 1983.

R. Pallas-Areny and J. G. Webster, *Sensors and Signal Conditioning*, Wiley, New York, 1991.

P. W. Tuinenga, *SPICE: A Guide to Circuit Simulation and Analysis Using PSpice*, Prentice-Hall, Englewood Cliffs, NJ, 1988.

PROBLEMS

Easy Drill Problems (answers at end of chapter)

D8.1. The transfer characteristic of a typical photocell is shown in Figure D8.1. What is its sensitivity in microamperes per footcandle, and what is its range?

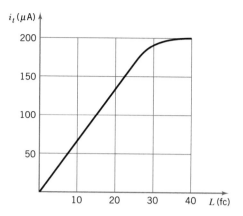

D8.2. The transfer characteristic of a typical strain gauge is shown in Figure D8.2. What is its sensitivity in ohms per millimeter, and what is its range?

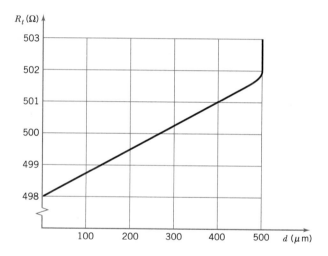

D8.3. Estimate the percent change in R_t per degree in the vicinity of 100°C for the thermistor characteristic shown in Figure 8–4. What is the sensitivity in ohms per degree there?

D8.4. For the Wheatstone bridge in Figure 8–5a, $v_0 = (10 \text{ V}) R_t/(500 + R_t) - 5 \text{ V}$. Show that the sensitivity of v_0 with respect to R_t is 5 mV/Ω for R_t close to 500. What is this sensitivity if the 10-V source is increased to 20 V?

D8.5. If R_t in Figure 8–5a is a strain gauge with the characteristic shown in Figure D8.2, what is the sensitivity of v_0 with respect to distance d in millivolts per millimeter?

D8.6. The system in problem D8.5 has $d = 0$ ($R_t = 498 \ \Omega$). Find v_o. A temperature increase causes all four resistances in the bridge to increase by 2%. Find v_o for $d = 0$ now.

D8.7. A generator with no load produces 30 mV/rpm and has a Thevenin impedance of 100 Ω. What voltage does it furnish to a 500-Ω load if it is turning at 800 rpm?

D8.8. A pressure transducer is formed by combining a strain gauge with a tube that deforms by 12 μm for every 1.0 psi of pressure. The strain gauge characteristic is that shown in Figure D8.2. What is the sensitivity of the system in ohms per psi?

D8.9. An amplifier has an open-circuit gain of 120 and an output impedance of 500 Ω. What is its effective gain if it is driving a load $R_L = 50$ Ω? How large must R_L be for the effective gain to be 80% of the open-circuit gain?

D8.10. An amplifier has $G_0 = 120$ and $f_U = 3$ kHz. Use Equation (8.1) to evaluate the magnitude of its gain at $f = 30$ kHz. (Can the 1 in the denominator be neglected?)

D8.11. An amplifier has a midband gain $G_0 = 120$ and cutoff frequencies of $f_L = 200$ Hz and $f_U = 3$ kHz. Sketch a Bode plot of its frequency response. What is its bandwidth?

D8.12. Use Equation (8.3) to evaluate the magnitude of the gain of the amplifier in D8.10 at $f = 1.8$ kHz. Repeat, neglecting the effect of the low-frequency cutoff [using Equation (8.1)]. By what percent is this approximation different from the first answer?

D8.13. Express voltage gains of 10, 30, 100, 200, 250, and 15,800 in decibels.

D8.14. A voltage divider has a voltage "gain" of 0.10. Express this gain in decibels.

D8.15. Express voltage gains of 14 dB, 70 dB, and -6 dB in regular numbers.

D8.16. An amplifier with a gain of 30 is cascaded with an amplifier with a gain of 15 for an overall gain of 450. Convert all three gains to decibels. Is the overall gain in decibels the sum of the two cascaded gains?

D8.17. Find the time constant τ_U for the step response of the amplifier in D8.10. Sketch the response to a step of height 0.1 V. [Use the initial tangent as in Figure 8–16 to make the sketch. Don't actually calculate values from Equation (8.8).]

D8.18. Find the time constants τ_L and τ_U for the step response of the amplifier in D8.11. Sketch the response to a step of height 0.1 V. (Use the initial tangents as in Figure 8–16 and Figure 8–17 to make the sketch.)

D8.19. Estimate the rise time, overshoot, and settling time of the step response in Figure D8.19.

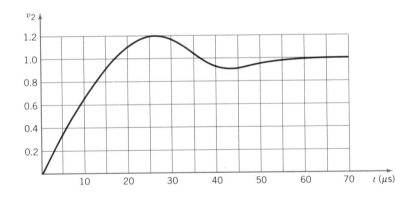

D8.20. The transfer characteristic of an amplifier is shown in Figure D8.20. What is its gain, and what is the maximum input voltage without clipping?

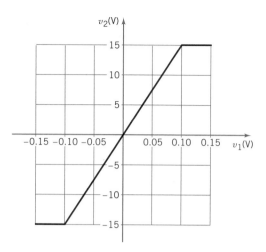

D8.21. An amplifier with a gain of 30 clips at output voltages of ± 20 V. What is the maximum input voltage without clipping?

D8.22. An amplifier with distortion has a signal $v_1 = (20 \text{ mV}) \sin(\omega t)$ applied at the input, and the output is $v_2 = 3.2 \sin(\omega t) + 0.13 \sin(2\omega t) + 0.06 \sin(3\omega t)$. What is the total harmonic distortion (THD)?

D8.23. An amplifier with gain 30 has an output voltage $v_2 = 0.5$ V when a voltage $v_1 = 0$ is applied at the input. What is the input offset voltage?

D8.24. Find the noise bandwidth B_n of the amplifier in D8.10.

D8.25. The amplifier in D8.10 has a noise spectral density of $\sqrt{N_0} = 15 \ \mu\text{V}/\sqrt{\text{Hz}}$ at the input. What is the effective rms noise e_n at the input? What is the rms noise n at the output?

D8.26. Find the resistance of 1000 ft of 18-gauge copper wire.

D8.27. One hundred feet of fine copper wire was found to have a resistance of 65 Ω. What is the gauge of the wire?

D8.28. A 2-m length of coaxial cable was measured to have a capacitance of 120 pF and a delay of 9.0 ns. What is the inductance L per meter of cable? What is the characteristic impedance Z_0 of the cable?

D8.29. A transmission line is terminated with a resistance twice its characteristic impedance. What is the reflection coefficient?

Application Problems

P8.1. In the Wheatstone bridge in Figure 8–5a R_t varies from 495 Ω to 505 Ω. How much does v_0 vary? Form a Thevenin model looking into the v_0 port of the bridge. By what percent is v_0 reduced if an amplifier with an input impedance of 10 kΩ is used to amplify v_0? Should the amplifier have a balanced or unbalanced input?

P8.2. A pressure monitoring system is the same as that in Figure 8–8 except the amplifier specifications are $A_v = 2000$, $R_i = 10$ kΩ, and $R_0 = 1000$ Ω. What pressure p will cause full-scale deflection of the chart recorder?

P8.3. An amplifier has $G_0 = 100$ and $f_U = 1.0$ kHz. Use a ruler to draw a Bode plot of the frequency response from 100 Hz to 10 kHz on log-log graph paper. Use Equation (8.1) to calculate the actual magnitude of $\mathbf{G}_v$ at $f = 100$ Hz, 200 Hz, 500 Hz, 1 kHz, 2 kHz, 5 kHz, and 10 kHz. (A programmable calculator is helpful here.) Plot these points on the same graph paper.

P8.4. Write the transfer function $\mathbf{G}_v$ for an amplifier with the Bode plot shown in Figure P8.4. Show from Figure P8.4 that the response is down by about a factor of 5 at $f = \frac{1}{5} f_L$. Check this using your expression for $\mathbf{G}_v$.

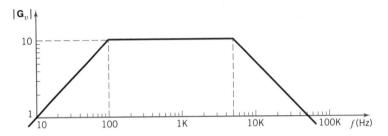

P8.5. Sketch a Bode plot for the transfer function

$$\frac{\mathbf{V}_2}{\mathbf{V}_1} = \frac{20 jf/10 \text{ kHz}}{(1 + jf/10 \text{ kHz})(1 + jf/1 \text{ MHz})}.$$

P8.6. An amplifier with a flat response down to dc has a problem with 60-Hz hum. Choose f_L for a low-frequency cutoff so the hum amplitude is reduced by a factor of $\frac{1}{8}$.

P8.7. A signal $v_1(t) = 0.1 + 0.01 \sin(2\pi ft)$, with $f = 1.0$ kHz, is applied to the input of the amplifier in D8.10. Find the signal $v_2(t)$ at the output. (Remember dc corresponds to $f = 0$.) Repeat for the amplifier in D8.11. If both amplifiers clip at output levels of ± 10 V, which amplifier is preferred for this application?

P8.8. Sketch the Bode plot for the amplifier in D8.11, and find its value at $f = 100$ Hz and $f = 5$ kHz. (*Hint:* 100 Hz is a factor of 2 below 200 Hz, and 5 kHz is a factor of $\frac{5}{3}$ above 3 kHz.) An amplifier with the response shown in Figure P8.4 is cascaded with the amplifier in D8.11. Sketch a Bode plot for the combined response, indicating the values at $f = 100$ Hz, 200 Hz, 3 kHz, and 5 kHz. What are the low- and high-frequency cutoffs of this system?

P8.9. A motor with an eccentric load is turning at 1200 rpm. A strain gauge is used to monitor the vibration in the motor mounting. The Wheatstone bridge and amplifier in Figure P8.9 produce a voltage v_2 proportional to the strain in the mounting. The capacitor C blocks the dc component from the bridge, ac-coupling the amplifier.

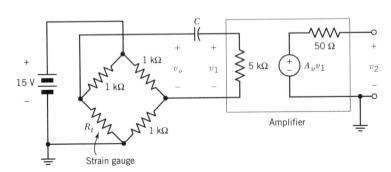

Given $R_t \approx 1000\ \Omega$, choose C so f_L is $\frac{1}{3}$ the frequency of vibration. (A Thevenin equivalent of the Wheatstone bridge is useful here.) Choose A_v so a 1% change in R_t produces a 5-V change in v_2.

P8.10. Use a programmable calculator to evaluate Equation (8.5) for $2ft = 0, 0.1, 0.2, \ldots, 1.0$. Plot a curve from these points, and compare with Figure 8–14c.

P8.11. A triangular waveform with a peak amplitude of 1.0 can be expressed by the Fourier series

$$(8/\pi^2)[\sin(x) - \tfrac{1}{9}\sin(3x) + \tfrac{1}{25}\sin(5x) - \tfrac{1}{49}\sin(7x) + \cdots]$$

where x stands for $2\pi ft$. Use a programmable calculator to evaluate this expression for $x = 0, 0.1\pi, 0.2\pi, \ldots, \pi$. Plot a curve from these points. Is it approaching a triangular form?

P8.12. Plot the coefficients of the Fourier series in P8.11 as in Figure 8–13b. How do the harmonics of a triangular wave behave compared with the harmonics of a square wave? How could you tell, from looking at the harmonics of any two waveforms, which of them was probably smoother in the time domain?

P8.13. An amplifier has the following transfer function:

$$\frac{\mathbf{V}_2}{\mathbf{V}_1} = \frac{30}{1 + jf/7\ \text{MHz}}.$$

Find the differential equation relating v_2 to v_1. Write an expression for $v_2(t)$ if $v_1(t)$ is a step of 0.1-V height.

P8.14. Find the rise time t_r in terms of τ_U for the exponential function in Equation (8.8).

P8.15. The nonlinearity of an amplifier's transfer characteristic can be represented by a squared term:

$$v_2 = 100v_1 + 40v_1^2$$

where v_1 and v_2 are in volts. What is the THD when $v_1 = 0.05\sin(2\pi ft)$? [Use a trigonometric identity to express $\sin^2 x$ in terms of $\sin(2x)$.] Repeat for v_1 with twice the amplitude.

P8.16. The nonlinearity of an amplifier can be represented by a cubic:

$$v_2 = 100v_1 + 1000v_1^3.$$

If v_1 is a sine wave, how large can its amplitude be while keeping THD < 2%?

P8.17. A thermocouple with the characteristic in Figure 8–3a is used to monitor temperature. What gain is necessary to amplify v_t so that an output voltage of $v_2 = 10$ V corresponds to $T = 1000°C$? If the amplifier has an input offset of 5 mV, what is its output offset? How many degrees does this error amount to? If the input offset remains 5 mV, can the error in degrees be reduced by reducing the gain?

P8.18. The pressure monitoring system in Figure 8–8 has an amplifier with a high-frequency cutoff of $f_U = 10$ kHz and with input noise spectral density characterized by Figure 8–21. For a monitored pressure of 2 psi (for $v_t = 1.0$ mV), what is the signal-to-noise ratio in decibels? (Assume no offset.) What must f_U be to increase the SNR to 60 dB? Do you think this bandwidth will limit the ability of the chart recorder to follow fast pressure fluctuations?

P8.19. A thermocouple characterized in Figure 8–3 is connected to the input of an amplifier by a pair of 28-gauge copper wires 100 ft long. The input impedance of the amplifier is 50 Ω. If the temperature is 1000°C, causing $v_t = 42$ mV, what is the voltage at the input to the amplifier?

P8.20. Use a bounce diagram (see Example 8.5) to confirm the v_r waveform in Figure 8–24. The step at the input has height $E = 1.0$ V.

P8.21. A 6-m transmission line has a characteristic impedance of 500 Ω and a delay of 5 ns/m. The source impedance is $R_s = 100$ Ω, and the load impedance is $R_r = 1000$ Ω. Use a bounce diagram to find the v_r waveform (out to $t = 270$ ns) in response to a step of height $E = 4$ V.

Extension Problems

E8.1. The output V_0 of the capacitance bridge in Figure 8–5 is connected to no load. The ac source has a frequency of 16 kHz. For $C_t = 510$ pF, what is the rms voltage V_0? Draw a Thevenin equivalent of the bridge looking into the V_0 port (both voltage and impedance). What is V_0 if an amplifier with an input impedance of 10 kΩ is connected to the output of the bridge? What must the frequency be so that the loaded V_0 is 99% of the unloaded V_0?

E8.2. The complex gain $\mathbf{G}_v$ in Equation (8.1) has $G_0 = 10$ and $f_U = 20$ kHz. Find an expression for the phase angle ϕ of the gain as a function of f. Show that $\phi = -27°$, $-56°$, and $-68°$ for $f = 10$ kHz, 30 kHz, and 50 kHz. What does ϕ approach as f approaches infinity?

E8.3. Equation (8.6) shows how any periodic function $v(t)$ can be expressed as the sum of sine waves. The Fourier coefficients a_1, a_2, etc., can be found from $v(t)$ by

$$a_n = \frac{2}{T} \int_0^T v(t) \, \sin(2\pi n f t + \theta_n) \, dt,$$

where the period is $T = 1/f$. Confirm the coefficients in Equation (8.5) for a square wave that is 1.0 for $0 < t < \frac{1}{2} T$ and -1.0 for $\frac{1}{2} T < t < T$. Note that all θ_n are zero. See B. P. Lathi in "For Further Study."

E8.4. Using the expression for a_n in E8.3, confirm the Fourier coefficients given for a triangular wave in P8.11.

E8.5. Given the transfer function in P8.5, find the differential equation relating v_2 to v_1.

E8.6. The thermal agitation of the free electrons in a resistor R causes noise voltage with a spectral density N_R to appear across its terminals:

$$N_R = 4kTR,$$

where k is Boltzmann's constant, equal to 1.38×10^{-23} V²/Hz-°K-Ω, and T is the absolute temperature in degrees Kelvin. The result is N_R in volts squared per hertz. Calculate $\sqrt{N_R}$ for a 100-kΩ resistor at 300° K (room temperature). The $\sqrt{N_0}$ given in Figure 8–21 for $R_s = 100$ kΩ is more than this due to additional noise density N_A from the amplifier itself. For uncorrelated noise sources the densities add: $N_R + N_A = N_0$. Find $\sqrt{N_A}$ for $R_s = 100$ kΩ.

Study Questions

8.1. List three specifications that are of interest when ordering any transducer.

8.2. How can the changing resistance of a transducer be converted to a changing voltage?

8.3. What is the advantage of symmetry in a bridge?

8.4. List three parameters that would allow an amplifier to be modeled in its region of normal operation.

8.5. List three parameters that specify limits of an amplifier—that bound its operation.

8.6. List three parameters that quantify the departure of an amplifier from ideal operation (other than those dealt with in questions 8.4 and 8.5)?

8.7. What is the 3-dB bandwidth of an amplifier?

8.8. How can the gain of an amplifier be expressed so it is easy to calculate the total gain of cascaded amplifiers?

8.9. A periodic signal in time can be represented as a sum of sinusoids. What are the amplitudes of these sinusoids called? What is the mathematical process that calculates these amplitudes from the time signal?

8.10. What is another name for the first harmonic of a signal? Which harmonics contribute to the total harmonic distortion?

8.11. What causes the sag in the step response of an amplifier? What causes the amplifier to be slow in responding to a step?

8.12. List three parameters that characterize the response of a dc-coupled amplifier to a step.

8.13. What is the difference between amplitude distortion and frequency distortion? Which category does harmonic distortion fit into?

8.14. How is the output offset related to the input offset of an amplifier?

8.15. List three sources of noise.

8.16. How is the noise bandwidth of an amplifier related to its 3-dB bandwidth?

8.17. If the bandwidth of an amplifier is doubled, more thermal noise gets through. Does the rms value of the thermal noise double?

8.18. What happens to the resistance of a wire when the area of its cross section is doubled? What gauge wire has double the cross section of a 10-gauge wire? What happens to the resistance of a wire when its length is doubled?

8.19. How do reflections in a transmission line affect the signal being carried? How can the reflections be suppressed?

8.20. List four properties of a length of lossless transmission line.

ANSWERS TO DRILL PROBLEMS

D8.1. 6.67 μA/fc, 0–27 fc

D8.2. 7.5 Ω/mm, 0–470 μm

D8.3. 3%/°C, 21.6 Ω/°C

D8.4. 10 mV/Ω

D8.5. 37.5 mV/mm

D8.6. -10 mV, -10 mV

D8.7. 20 V

D8.8. 0.09 Ω/psi

D8.9. 10.9, 2000 Ω

D8.10. 11.94 $\approx$ 12, yes

D8.11. 2800 Hz

D8.12. 102.3, 102.9, 0.6%

D8.13. 20, 29.5, 40, 46, 48, 84 dB

D8.14. -20 dB

D8.15. 5.01, 3162, 0.5

D8.16. 29.5 dB $+$ 23.5 dB $=$ 53 dB

D8.17. 53 μs

D8.18. 796, 53 μs

D8.19. 14 μs, 19%, 54 μs

D8.20. 150, 0.1 V

D8.21. 0.67 V

D8.22. 4.47%

D8.23. 16.7 mV

D8.24. 4712 Hz

D8.25. 1.03 mV, 123 mV

D8.26. 6.35 Ω

D8.27. 38 AWG

D8.28. 338 nH/m, 75 Ω

D8.29. $\frac{1}{3}$

CHAPTER 9

Digital Signals and Logic

We perceive the world with analog senses. Therefore the analog signals discussed in the last chapter have a natural appeal and are relatively easy to understand. But digital signals are rapidly superseding analog signals because (1) digital signals have several advantages and (2) digital circuits can be made smaller and more cheaply in an increasing number of applications. Sound is recorded on compact disks (CDs) as digital signals, phone conversations between cities are coded digitally, and high-definition TV (HDTV) is conveyed by digital signals.

The reader's objectives in this chapter are to:

1. Understand the advantages of digital over analog signals.
2. Review binary numbers, the basis for most digital coding.
3. Learn to express operations as Boolean functions and digital logic.

The concepts in this chapter are actually a form of mathematics, dealing with ideal variables, functions, and processes. The imperfections of digital signals and the physical means for processing them will be dealt with in Chapters 13 and 16.

We saw in the previous chapter that an analog voltage can represent a pressure. As the pressure varies, the analog voltage varies proportionately. A digital signal differs in that it takes on only a few discrete values—say 0 V, 5 V, or 10 V—and nothing else. In fact, most digital applications use only two values. Nevertheless, a digital signal can represent many different magnitudes by use of a code. For example, the sequence 0 V, 0 V, 5 V, 0 V of four digital signals could stand for the magnitude 2, while the sequence 5 V, 5 V, 5 V, 0 V could stand for the magnitude 14. This may seem an unnecessary complication, but we will show that digital signals allow more accuracy, are easier to store, and are more reliably transmitted.

The reader is already familiar with the difference between analog and digital in the form of displays. An analog watch has hands whose positions are proportional to time. A digital watch has numbers that take on discrete values; the numbers are actually a code that stands for the time. The advantages of one form of display or the other will be used to demonstrate the relative advantages of analog and digital signals.

It is possible for digital signals to stand for information other than quantities. For example, pressing a key on a computer terminal can produce a code of digital signals that stands for "A." This signal can be stored, processed, and displayed later as "A" or some other character. It would be very difficult to use an analog signal to perform the same function.

Digital signals can also stand for states or conditions. Of particular interest are situations where there are only two possible conditions—true or false, yes or no. For example, (i) the transmission is in "park" or not, (ii) the starting switch is on or not. Each of these conditions can be represented by a digital signal, and a circuit operating on the signals can apply a digital signal to the solenoid when both (i) and (ii) are true. This type of operation is called "logic," and it is the basis for all processing of digital signals.

One type of processing is conversion between digital codes. Another is computation, which leads us to the study of binary numbers and arithmetic. The processing of digital signals by logic circuits leads us to the study of Boolean algebra.

The use of digital signals is becoming increasingly common. Telephone signals, for instance, often go through a digital form if they travel any significant distance. Therefore the reader should be familiar with the concept of digital signals even if he never encounters a digital circuit directly. The discussion of digital operations in this chapter will enable the reader to design simple digital circuits. It will also prepare him for the study of digital computers in Chapter 12.

9.1   DIGITAL DISPLAYS

A digital display embodies the same properties as a digital signal; in fact digital signals are necessary to generate a digital display. Since most readers have had experience with digital displays, we will start our study of the advantages of digital signals by considering digital displays.

The readout of a voltmeter may be in either analog or digital form. Which form is better depends on the situation. Suppose the objective is to record in a notebook some observed voltage. In the case of an analog voltmeter, the engineer (getting his eye directly in front of the meter to avoid parallax) might estimate that the needle is positioned a little more than halfway between 12 and 13. It looks like either 12.6 or 12.7, and he decides to call it 12.7. Compare this with reading a digital display such as that in Figure 9–1a. The meter has already made the decision and has put the information in the form ready for copying down. One might say

FIGURE 9–1 Digital displays illustrating (a) convenient form for copying, (b) high resolution, (c) nonnumeric information.

that the meter has *coded* the information into the desired form for this situation. The same is true if a friend asks you the time. If you have a digital watch, you find the words come more quickly.

Another situation in which a digital display excels is when high *resolution* is needed. Most analog meters can be read to only three significant figures, as in the above example. However, one should be able to read an instrument with 0.01% accuracy to five significant figures. In this case a digital display such as that in Figure 9–1b is convenient. Some analog readouts try to meet the need by having very long scales (as in some signal generators and radio receivers) or by using multiple pointers (as with clocks and altimeters). These solutions are cumbersome compared with a digital readout.

Digital displays are not restricted to conveying quantitative information, as illustrated in Figure 9–1c. With digital signals we can store, transmit, and display not only voltage levels and dollar amounts, but also *items* such as "apples" and *status* such as "back order" or "on time." Some analog displays attempt to indicate status by painting the words "safe" and "danger" on a scale, but the vocabulary is limited, and borderline cases saddle the observer with the decision.

There are some situations in which an analog display is preferable. These always involve *comparisons*. For example, an analog sound-level meter easily indicates whether the signal is coming near some distortion level. An analog watch quickly gives a feeling for the time left from 2:48 until 3:15. A common use of meters is to maximize a reading while an adjustment is made. This is relatively easy to do with an analog meter, while a digital meter tends to be a confused dancing of digits.

A digital display is an interface between digital signals and people. With the use of "TV screens" and plotters, digital signals can generate complex displays in the form of pictures. The simplest form of digital display is a single light that is either on or off. In this section we will consider a form of digital display limited to numbers and some letters.

The digital display on a watch is composed of segments or dots, and each segment is controlled by an electrical signal. Since the segment is either on or off, the signal has only two states, called "true" and "false," "high" and "low," or "1" and "0." The term *digital signal* refers to the fact that the states of the signal can be equated to the digits "1" and "0."

EXAMPLE 9.1 A digital display is formed of seven segments, as illustrated in Figure 9–2a. The segments are labeled "a" through "g."

Given: There are seven wires labeled "a" through "g." A segment of the display is visible when 5 V is applied to the corresponding wire. When 0 V is applied, the segment is invisible.

Find: the voltages to produce the character "3." Express these voltages (signal states) as a pattern of digits.

Solution: The figure "3" is displayed by activating the segments shown in Figure 9–2b. This display corresponds to the digital signal in Figure 9–2c, where the voltage is "high" on wires a,b,c,d,g. A convenient way of indicating this pattern is with the digital *word* 1111001 (see Figure 9–2d).

Note that digital signals may require many wires or lines (seven here) to carry information, while an analog signal requires only one line (wire).* The seven lines in Figure 9–2c are called a *bus* when considered as a unit. The digital signal on this bus is only one way of expressing "3" with a digital signal. The binary number system, discussed in Section 9.3, has another code for "3."

Digital signals are necessary for a digital display. But digital signals are so easy to handle that they are desirable in their own right. In the next section we will look at some of the advantages of digital signals from the standpoint of signal processing.

9.2   DIGITAL SIGNALS

One advantage of a digital signal is that it can be *transmitted* perfectly. By contrast, consider the analog signal v_a in Figure 9–3a. When this signal is transmitted, it can be corrupted by additive noise n, as illustrated. In general, there is no way of extracting the original v_a from $v_a + n$ at the receiving end.

Now consider the digital signal v_d in Figure 9–3b. This is a series of 1's and 0's represented by 5 V and 0 V, respectively. If the same noise n is added to the digital signal, the original 1's and 0's are still recognizable. A circuit could be built to decide at each point in time whether $v_d + n$ is closer to 5 V or to 0 V. The only requirement for perfect recovery of v_d is that $|n| < 2.5$ V. Digital transmission will be dealt with in more detail in Chapter 11.

Another advantage of a digital signal is that it can be *stored* perfectly. The only way to store an analog voltage in a circuit is to charge a capacitor to that voltage, as in Figure 9–4a.† So long as the capacitor is isolated by high impedance from the

---

*While *line* and *wire* are almost synonymous, *wire* is a physical concept, demanding that we also count the ground wire. A *line* assumes that there is a ground somewhere to establish a reference for the voltage signal.

†Both analog and digital signals can be stored magnetically, for example on tape. We are concerned here with storage in a circuit.

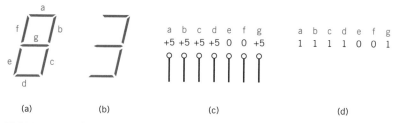

FIGURE 9–2 (a) Labeling of seven-segment display elements. (b) Display of character "3." (c) Seven-line bus with voltages to produce the character "3." (d) Seven-bit word representing the digital signals in (c).

rest of the circuit, the voltage can remain almost unchanged for perhaps a second. But after a minute or so significant charge will have leaked off the capacitor, changing the stored signal.

The circuit in Figure 9–4b illustrates how a digital signal can be stored. The switches stand for electrically operated toggle switches called *flip-flops*; we will look more at these in Section 9.6. These switches have the property that an applied signal can throw them to the left or right, and the switch will stay in that position even after the controlling signal is gone. In this way a permanent 5 V can be applied to the first line of the bus, 5 V to the second line, and 0 V to the third line. This digital signal will remain on the bus without change indefinitely.

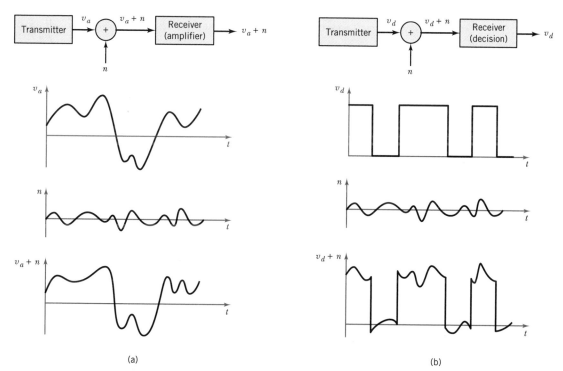

FIGURE 9–3 (a) Analog signal and corruption with noise. (b) Digital signal and corruption with the same noise. The original digital signal can be recovered, while the analog cannot.

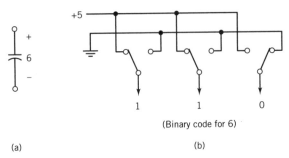

(a)   (b)

FIGURE 9–4 (a) Analog signal held on a capacitor. (b) Digital signal held with electronic switches. The digital memory does not fade, while the analog does.

A third advantage of digital signals is that they facilitate *computation*. This advantage is essentially the same feature of high resolution that was mentioned in connection with digital displays. Analog computation lacks this resolution or precision. The sliding scales in Figure 9–5a are an example of analog computation. The linear scales add, and the logarithmic scales multiply (the principle of the slide rule). The precision is limited by the ability to read the scale accurately and the care with which the scales were made. In Chapter 14 we will see addition of analog signals by a circuit that is subject to the same limitations.

Digital computation is represented by the calculations in Figure 9–5b. Multiplication by hand using this method takes longer than the sliding-scale method, but the result is much more accurate. With digital computers the many operations of digital computation can be carried out so quickly that time is no longer a deciding factor. For this reason the digital calculator has totally replaced the slide rule.

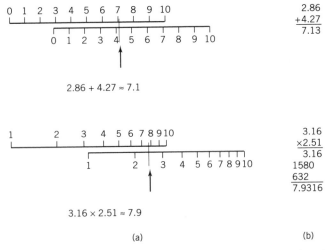

(a)   (b)

FIGURE 9–5 (a) Analog computation by combining two lengths (subject to error). (b) Digital computation by arithmetic rules. The accuracy can be as great as desired.

A person carrying out digital computation by hand uses the digits 0 through 9, as in Figure 9–5b. A digital computer goes through exactly the same process, but it uses a binary number system with only the digits 0 and 1.

9.3   BINARY NUMBERS

The *decimal* number system that we all commonly use is actually an agreed-upon code. For example the number 254 is understood to mean $2 \times 10^2 + 5 \times 10^1 + 4 \times 10^0$. Therefore we say that 2 is in the hundreds place, 5 is in the tens place, and 4 is in the units place. Because each place is a power of ten, the decimal number system is also called the base-ten number system. Also, there are ten digits—0 through 9. By contrast, the binary number system has only two digits—0 and 1.

Binary Number System

The *binary* or base-two number system has each place stand for a power of 2. For example, the binary number 110 is understood to mean $1 \times 2^2 + 1 \times 2^1 + 0 \times 2^0 = 4 + 2 + 0 = 6$. Therefore 110 base two is 6 base ten. When this is written in the form of an equation, the base is indicated by a subscript: $110_2 = 6_{10}$. When the base is obvious from context, the subscript is omitted. Each binary digit is called a *bit*. Therefore 110 is a 3-bit number. Since the left-most bit is weighted the most (by 2^2, or 4), it is called the *most significant bit* (MSB). The right-most bit is called the *least significant bit* (LSB).

EXAMPLE 9.2

1. Convert 1101_2 to decimal (base ten).
2. Convert 22_{10} to binary (base two).

Solution
1. $1101_2 = 1 \times 2^3 + 1 \times 2^2 + 0 \times 2^1 + 1 \times 2^0 = 8 + 4 + 1 = \underline{13}$.
2. The largest power of two less than 22 is 16 (or 2^4). After this is taken away, 6 remains. The largest power of two in 6 is 4 (or 2^2). Subtracting this, 2 remains (or 2^1). Then $22_{10} = 1 \times 2^4 + 0 \times 2^3 + 1 \times 2^2 + 1 \times 2^1 + 0 \times 2^0 = \underline{10110_2}$.

The table in Figure 9–6 gives more examples of decimal numbers and their binary equivalents. Notice the behavior of the digits in sequential binary numbers. The units bit alternates between 0 and 1, the twos column changes every two numbers, and the fours column changes every four numbers. The binary system is clearly less compact in terms of the number of digits required; it takes 3 bits to express the number 6 for instance. The difference is more pronounced with larger numbers: $11111000010_2 = 1986_{10}$. But the advantage of binary is the use of many simple digits rather than a few "complicated" digits. It has been shown that people find it easier to remember a few complicated things, whereas circuits are more easily designed to deal with a large number of simple things.

Decimal	Binary	Hexadecimal
0	0	0
1	1	1
2	10	2
3	11	3
4	100	4
5	101	5
6	110	6
7	111	7
8	1000	8
9	1001	9
10	1010	A
11	1011	B
12	1100	C
13	1101	D
14	1110	E
15	1111	F
16	10000	10
17	10001	11

FIGURE 9–6 Some corresponding numbers in the decimal (base-ten) number system, the binary (base-two) number system, and the hexadecimal (base-sixteen) number system. Some of the hexadecimal digits have been borrowed from the alphabet.

Binary Arithmetic

The procedures for computation in the binary system are the same as in the decimal system (see Figure 9–7). Consider the addition of seven (111) and six (110). Beginning at the right we say 1 plus 0 is 1. Then 1 plus 1 is two (10); put down 0 and carry the 1. Then 1 plus 1 plus 1 is three (11). The result is thirteen (1101). The reader might find it helpful to pencil in the carry digits at the top of the addition columns in Figure 9–7.

Because multiplying by 0 or 1 is trivial, multiplication in binary is especially simple. In multiplying seven (111) by six (110), the numbers to be added consist of seven (111) shifted to the left by various amounts. The result is 101010. As a check we can see that the binary result stands for $2^5 + 2^3 + 2^1 = 32 + 8 + 2 = 42$.

Binary division follows the usual procedure of a series of subtractions, bringing down zeros as necessary until subtraction is possible. A "1" in the quotient indicates when subtraction was possible. Note that, as with decimal division, the procedure may not terminate. The result can be truncated when sufficient accuracy has been reached.

The fourth arithmetic procedure is subtraction. Since subtraction can result in negative numbers, computers must have some way of representing negative numbers. Once negative numbers are established, it is no longer necessary to carry out

Addition:

```
  7        111
 +6      +110
 ──      ────
 13      1101
```

Multiplication:

```
  7        111
 ×6       ×110
 ──      ─────
 42        000
           111
           111
        ──────
        101010
```

Division:

```
  1.166...              1.00101...
6)7.000          110)111.00000
  6                    110
 ──                   ────
 10                    1000
  6                     110
 ──                    ────
 40                     1000
 36                      110
 ──                     ────
 40
 36
 ──
```

FIGURE 9–7 Examples of addition, multiplication, and division with binary numbers. The corresponding decimal computations are also given.

subtraction with its rules of borrowing, etc. To subtract six from seven, the computer can change the sign of six and then add: $7 - 6 = 7 + (-6)$. What is the binary representation of -6?

Negative Numbers

The definition of negative six is "that number which, when added to six (0110), results in zero." We will do this in two steps. First we form the *ones complement* of 6 by inverting all its bits (0 to 1 and 1 to 0). When the ones complement is added to the original number, the result will clearly always be all 1's (see Figure 9–8). This is the largest number that can be expressed by 4 bits. If our computer is restricted to 4 bits, then adding 1 to 1111 will cause a 1 to be lost at the left, and the result appears as zero. We can get the same result in one step if we first add the ones complement of six (1001) to unity (0001) to get the *twos complement* of six (1010). When this is added to six (0110), the result is zero. Then by definition the twos complement of six is negative six.

$$
\begin{array}{ll}
0110 & 6 \\
\underline{+1001} & \text{ones complement of 6} \\
1111 & \text{all 1's}
\end{array}
$$

$$
\begin{array}{ll}
1111 & \text{all 1's} \\
\underline{+0001} & \text{unity} \\
(1)0000 & \text{zero}
\end{array}
$$

$$
\begin{array}{ll}
1001 & \text{ones complement of 6} \\
\underline{+0001} & \text{unity} \\
1010 & \text{twos complement of 6}
\end{array}
$$

$$
\begin{array}{ll}
1010 & \text{twos complement of 6} \\
\underline{+0110} & 6 \\
(1)0000 & \text{zero}
\end{array}
$$

Therefore twos complement of 6 is -6.

FIGURE 9–8 Showing that the negative of a number is the twos complement of the number. The twos complement is formed by changing 1's to 0's and 0's to 1's and adding unity.

Now that 1010 stands for negative six, it can no longer stand for ten. This is reasonable since four bits can represent only sixteen items. These items can be either the positive integers 0 through 15, or they can be the integers -8 through 7 (see Figure 9–9). Note that all the negative numbers have 1 as the MSB, and all the positive numbers have 0 as the MSB. Therefore this is called the *sign bit*.

EXAMPLE 9.3

1. Find the 4-bit representation for -4.
2. Carry out $7 - 4 = 3$ in binary four-bit arithmetic.

Solution

1. $4_{10} = 0100_2$. Reversing each bit gives the ones complement 1011. Then adding unity gives the twos complement: $1011 + 0001 = \underline{1100}$. This represents -4.
2. $7 - 4 = 7 + (-4)$. In 4-bit binary this is $0111 + 1100 = 0011$. Because we are restricted to 4 bits, the result is 0011 (or 3) rather than 10011 because we must discard the fifth bit.

More examples of subtracting by adding negative numbers are given in Figure 9–10. Remember that this system works only if we restrict our number system to a certain number of bits—4 in this case. (In some digital computers this number may be as many as 64 bits.) With 4 bits, the correct result must lie between -8 and 7, inclusive. Therefore $6 + 7 = 13$ can't be carried out; the result appears to be -3.

Decimal	Binary
−8	1000
−7	1001
−6	1010
−5	1011
−4	1100
−3	1101
−2	1110
−1	1111
0	0000
1	0001
2	0010
3	0011
4	0100
5	0101
6	0110
7	0111
	↑—sign bit

FIGURE 9–9 Representation of both positive and negative numbers for a 4-bit code. For a 5-bit code, −4 would be 11100.

This is called *overflow*. The addition of two large negative numbers can also cause overflow [see −7 + (−6) in Figure 9–10]. A system with more than 4 bits is needed to avoid overflow in these cases.

$$
\begin{array}{ll}
\quad 0111 & \qquad 7 \\
+1010 & +(-6) \\
\hline
\quad 0001 & \qquad 1 \\
\end{array}
$$

$$
\begin{array}{ll}
\quad 0110 & \qquad 6 \\
+1001 & +(-7) \\
\hline
\quad 1111 & \qquad -1 \\
\end{array}
$$

$$
\begin{array}{ll}
\quad 0110 & \qquad 6 \\
+0111 & +7 \\
\hline
\quad 1101 \ (\text{overflow}) & -3 \ (\text{overflow}) \\
\end{array}
$$

$$
\begin{array}{ll}
\quad 1001 & \qquad -7 \\
+1010 & +(-6) \\
\hline
\quad 0011 \ (\text{overflow}) & \quad 3 \ (\text{overflow}) \\
\end{array}
$$

FIGURE 9–10 Examples of subtracting by adding a negative number. Overflow can cause erroneous results (the correct result can't be expressed by only 4 bits).

Hexadecimal Number System

It is often necessary to represent on paper (or on a display) binary numbers that are in a circuit. Binary notation is too long and hard to remember. Decimal notation is compact and familiar, but conversion between decimal and binary is tedious. A compromise between man and machine representation is the *hexadecimal* or base-16 number system. (This is popularly shortened to "hex code.") It is a compact notation and easily converted to and from binary. In this system symbols are entered into the units place to represent the numbers 0 through 15. Then 16 is represented by putting a 1 in the next place (the 16s place) to form "10." The ten decimal digits are augmented with the six letters A through F to provide the 16 symbols necessary for the hexadecimal system. The table in Figure 9–6 compares the representation in decimal, binary, and hexadecimal for some sequential numbers. Note that binary can be easily converted into hexadecimal by grouping the bits into groups of four.

EXAMPLE 9.4

1. Convert binary 1011100 to hex code.
2. Convert hexadecimal E7 to binary.

Solution

1. Beginning at the right, divide the bits into groups of 4 as far as possible: 101,1100. Taking a group at a time, $101_2 = 5_{16}$, and $1100_2 = C_{16}$. Therefore $1011100_2 = 5C_{16}$. Sometimes the notation 5CH is used to indicate base 16 (hexadecimal).
2. $E_{16} = 1110_2$ and $7_{16} = 0111_2$. Therefore E7H = $\underline{11100111_2.}$

9.4   CODING

We have already seen two ways of coding information by using 1's and 0's—binary numbers and seven-segment displays. Binary is the natural form in which to do computation. It is also an *efficient* code; it uses the fewest possible bits to represent a number. On the other hand, it is not very convenient for interfacing with people, who prefer decimal representation.

Binary-coded decimal (BCD) is used as an intermediate step between the binary machine code and the decimal input or output. For example, if a computer operator wants to enter decimal 12, he first presses the "1" key, and the computer encodes it as 0001. Then the operator presses "2," and the computer encodes this as 0010. When they are put together, the computer has the BCD 0001,0010 for decimal 12. This is a longer code than the binary 1100 for decimal 12, but this is the price of entering the number in two pieces. In BCD the bits are divided into groups of four bits (as with hexadecimal), but this time each group stands for one of ten digits—0 through 9.

▬▬▬▬ **EXAMPLE 9.5**

1. Convert decimal 65 to binary.
2. Convert decimal 65 to BCD.

Solution

1. $65_{10} = 64_{10} + 1_{10} = 1000000_2 + 1_2 = 1000001_2$.
2. $6_{10} = 0110_2$ and $5_{10} = 0101_2$. Therefore $\overline{65_{10}} = \underline{0110,0101}$ in BCD. ▬

Circuitry exists to convert between the various codes. Figure 9–12 shows a typical situation (a DVM) where there is conversion between the code listed in Figure 9–11. An analog-to-digital converter (ADC) interfaces between the world of continuous quantity and the world of integer numbers. ADCs and digital-to-analog converters (DACs) are examined in Chapter 14. The conversion from binary to BCD is not simple and often requires the use of computation or at least counters (see Section 9.6). Conversion from BCD to seven-segment display is carried out by combinational logic (see Section 9.5).

Another code that is often useful is the *one-of-n code*. In this scheme all but one of n bits are 0. For example, in a 1-of-8 code 00000001 would stand for one and 00000100 for three. If a doctor had eight waiting rooms, each with a patient, the doctor could signal to a patient that he was ready for the patient by using this code to light a lamp in the patient's room. A circuit to convert from binary code to 1-of-8 code is called a 1-of-8 decoder or sometimes a 3-to-8 decoder (3 bits can

Decimal	Binary $X_3\,X_2\,X_1\,X_0$	BCD $A_3\,A_2\,A_1\,A_0$	BCD $B_3\,B_2\,B_1\,B_0$	Seven Segment $P_a\,P_b\,P_c\,P_d\,P_e\,P_f\,P_g$	Seven Segment $Q_a\,Q_b\,Q_c\,Q_d\,Q_e\,Q_f\,Q_g$
0 0	0 0 0 0	0 0 0 0	0 0 0 0	1 1 1 1 1 1 0	1 1 1 1 1 1 0
0 1	0 0 0 1	0 0 0 0	0 0 0 1	1 1 1 1 1 1 0	0 1 1 0 0 0 0
0 2	0 0 1 0	0 0 0 0	0 0 1 0	1 1 1 1 1 1 0	1 1 0 1 1 0 1
0 3	0 0 1 1	0 0 0 0	0 0 1 1	1 1 1 1 1 1 0	1 1 1 1 0 0 1
0 4	0 1 0 0	0 0 0 0	0 1 0 0	1 1 1 1 1 1 0	0 1 1 0 0 1 1
0 5	0 1 0 1	0 0 0 0	0 1 0 1	1 1 1 1 1 1 0	1 0 1 1 0 1 1
0 6	0 1 1 0	0 0 0 0	0 1 1 0	1 1 1 1 1 1 0	1 0 1 1 1 1 1
0 7	0 1 1 1	0 0 0 0	0 1 1 1	1 1 1 1 1 1 0	1 1 1 0 0 0 0
0 8	1 0 0 0	0 0 0 0	1 0 0 0	1 1 1 1 1 1 0	1 1 1 1 1 1 1
0 9	1 0 0 1	0 0 0 0	1 0 0 1	1 1 1 1 1 1 0	1 1 1 0 0 1 1
1 0	1 0 1 0	0 0 0 1	0 0 0 0	0 1 1 0 0 0 0	1 1 1 1 1 1 0
1 1	1 0 1 1	0 0 0 1	0 0 0 1	0 1 1 0 0 0 0	0 1 1 0 0 0 0
1 2	1 1 0 0	0 0 0 1	0 0 1 0	0 1 1 0 0 0 0	1 1 0 1 1 0 1
1 3	1 1 0 1	0 0 0 1	0 0 1 1	0 1 1 0 0 0 0	1 1 1 1 0 0 1
1 4	1 1 1 0	0 0 0 1	0 1 0 0	0 1 1 0 0 0 0	0 1 1 0 0 1 1
1 5	1 1 1 1	0 0 0 1	0 1 0 1	0 1 1 0 0 0 0	1 0 1 1 0 1 1

FIGURE 9–11 Three different ways of representing numbers by using bits (1's and 0's). BCD is less efficient than binary, but it is more easily converted to a display.

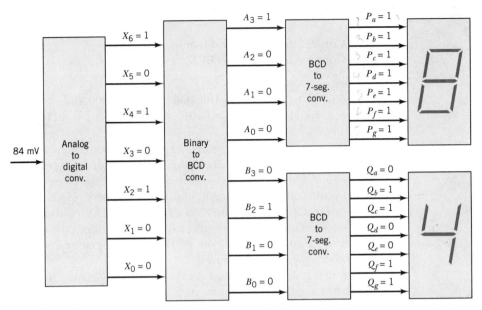

FIGURE 9–12 A digital voltmeter (DVM)—an example of conversion between several codes. (see code listings in Figure 9–11.)

represent eight numbers). Figure 9–13a shows the correspondence between the 3-bit binary codes and the 1-of-8 codes.

The circuit that performs the reverse conversion is called a *priority encoder*. Suppose that the situation was that the patients pressed a button to indicate to the doctor when *they* were ready. A priority encoder could be used to generate a binary display of the patient's waiting room number. To handle the case when two buttons

	1-of-8 Decoder		
Input		*Output*	
D_2 D_1 D_0	Q_7 Q_6 Q_5 Q_4 Q_3 Q_2 Q_1 Q_0		
0 0 0	0 0 0 0 0 0 0 1		
0 0 1	0 0 0 0 0 0 1 0		
0 1 0	0 0 0 0 0 1 0 0		
0 1 1	0 0 0 0 1 0 0 0		
1 0 0	0 0 0 1 0 0 0 0		
1 0 1	0 0 1 0 0 0 0 0		
1 1 0	0 1 0 0 0 0 0 0		
1 1 1	1 0 0 0 0 0 0 0		

(a)

Priority Encoder		
Input	*Output*	
D_7 D_6 D_5 D_4 D_3 D_2 D_1 D_0	Q_2 Q_1 Q_0	E
0 0 0 0 0 0 0 1	0 0 0	0
0 0 0 0 0 0 1 X	0 0 1	0
0 0 0 0 0 1 X X	0 1 0	0
0 0 0 0 1 X X X	0 1 1	0
0 0 0 1 X X X X	1 0 0	0
0 0 1 X X X X X	1 0 1	0
0 1 X X X X X X	1 1 0	0
1 X X X X X X X	1 1 1	0
0 0 0 0 0 0 0 0	0 0 0	1

(b)

FIGURE 9–13 (a) Conversion from binary to 1-of-8 code. (b) Conversion from 1-of-8 code to binary. If there is more than a single 1, the left-most 1 has priority. (The X's indicate "don't care.")

might be pressed at the same time, a priority would have to be assigned to the waiting rooms. The room with the higher priority would take control of the binary display. Figure 9–13b shows how the priority encoder would handle various signal patterns on the eight incoming lines. An "X" stands for "don't care"—it may be either a 1 or a 0. Note that the output for all 8 input bits 0's is distinguished by the E output going high.

The concept of a code is quite general; it does not have to represent numbers. For instance, we could agree that 00001 stands for "A," 00010 stands for "B," etc. There is a widely used ASCII code that represents numbers, letters, and punctuation in a manner similar to this. The seven-segment display is also capable of representing more than numbers, as we have seen. For example, 1100111 would generate the display "P." The seven-segment code becomes more efficient as more of its 128 possible codes are used.

9.5   COMBINATIONAL LOGIC

In the science of logic, either a thing is true or it is not true. This has an obvious correspondence to a digital (binary) signal where the signal is either a 1 or a 0. Therefore digital signals and circuitry are well suited to carry out logic operations. We will see that logic operations can be used to implement digital computation and code conversion, but first we look at a simple control application.

Basic Logic Functions

Consider a furnace control system with two thermostats, a water heater, and a pressure sensor. One thermostat is set at 70°F for the daytime, and the other is set at 60°F for the nighttime. The same burner heats both the house and the water heater. A logic statement of the system's operation might be as follows: "the furnace is on when temperature falls below 70°F during the day or below 60°F at night or when the water heater falls below 150°F. In any case the steam pressure must be below 5 psi for the furnace to be on." To make the statement briefer, we let symbols stand for the truth of the conditions and their relation:

$$F = \text{Furnace is on.}$$

$$D = \text{It is day.}$$

$$\overline{D} = \text{It is not day.}$$

$$T_1 = \text{Temperature} <70°\text{F.}$$

$$T_2 = \text{Temperature} <60°\text{F.}$$

$$H = \text{Water heater} <150°\text{F.}$$

$$P = \text{Pressure} <5 \text{ psi.}$$

$$\cdot = \text{"and"}$$

$$+ = \text{"or"}$$

Then we can write the statement as an equation:

$$F = (D \cdot T_1 + \overline{D} \cdot T_2 + H) \cdot P. \qquad (9.1)$$

This is read: "F is true when D is true and T_1 is true, or when not D is true (D is false) and T_2 is true, or when H is true; and P must be true for all."

Boolean Algebra

In the days of Socrates and Plato, logic was expressed only by sentences with words like "if . . . then," "for all," "and." Today logic is expressed by equations like (9.1), by *logic symbols*, by *truth tables*, and by *Karnaugh maps*. For example, the English statement "C is true if both A and B are true" can be expressed by $C = A \cdot B$ or by the logic symbol, truth table, or Karnaugh map in Figure 9–14a. These are all equivalent representations of the AND operation. Here a "1" stands for "true" and a "0" stands for "false." Therefore the truth table has $C = 1$ only for the row where both $A = 1$ and $B = 1$. In the Karnaugh map, the four entries in the square represent the values of C. $A = 1$ for the second column, $B = 1$ for the second row, and $C = 1$ at the intersection of these.

Figure 9–14 also shows the four ways of expressing the other logical operations. For the OR operation, $C = 1$ when either $A = 1$ or $B = 1$. For the NOT operation, $C = 1$ when $A \neq 1$.

When expressed in equation form such as $C = A \cdot B$, logic is called *Boolean algebra* after the mathematician George Boole. Boolean variables can take on only one of two values or *states*—1 or 0. By substituting values from the truth tables into

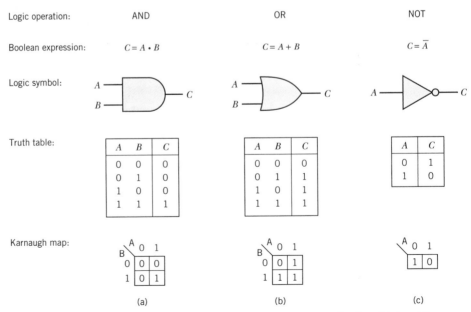

FIGURE 9–14 The basic logic operations. C is true (symbolized by 1) when (a) A and B are true, (b) A or B is true, (c) A is not true.

their corresponding Boolean expressions, we can write some Boolean identities: $1 = \bar{0}, 0 = \bar{1}, 0 = 0 \cdot 0, 0 = 0 \cdot 1, 1 = 1 \cdot 1, 0 = 0 + 0, 1 = 0 + 1, 1 = 1 + 1$. Except for the last statement, these equations look very much like standard algebra. While this similarity is tempting, the AND operation is *not* multiplication, and the OR operation is *not* addition. However, factoring does apply as with multiplication and addition:

$$A \cdot B + A \cdot C = A \cdot (B + C).$$

The logic statement in Equation (9.1) is represented by a logic diagram in Figure 9–15a. The inputs are D, T_1, T_2, H, and P; the output is F. The diagram in Figure 9–15b is a condensed version of that in Figure 9–15a.

Note that the AND operation $D \cdot T_1$ can be represented by DT_1—the same convention used with multiplication.

What will the value of F be for $D = 0$, $T_1 = 0$, $T_2 = 1$, $H = 1$, and $P = 1$? There are two methods of solution—a formal approach and an intuitive one. The formal (but more lengthy) solution is to use Boolean algebra. Substitute the given values of D, T_1, T_2, H, and P into Equation (9.1). Then use Boolean identities to simplify the expression step by step until the value of F is found:

$$F = (0 \cdot 0 + \bar{0} \cdot 1 + 1) \cdot 1 = (0 \cdot 0 + 1 \cdot 1 + 1) \cdot 1$$
$$= (0 + 1 + 1) \cdot 1 = (1) \cdot 1 = 1.$$

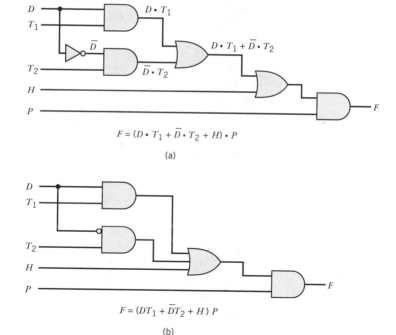

$$F = (D \cdot T_1 + \bar{D} \cdot T_2 + H) \cdot P$$

(a)

$$F = (DT_1 + \bar{D}T_2 + H) P$$

(b)

FIGURE 9–15 Logic diagram for a furnace control. The inputs are provided by a clock, three thermostats, and a pressure sensor. The output turns on the furnace. The two diagrams perform the same function.

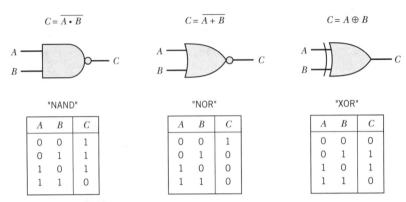

FIGURE 9-16 Three common logic gates. NAND is NOT AND, NOR is NOT OR, and XOR is exclusive OR.

Or we could go through the logic diagram in Figure 9–15a step by step. Engineers often pencil in the values in the diagram to keep track of the steps. With 0's on both its input lines, the first AND symbol will have a 0 on its output line. Continue in this way for all the symbols in Figure 9–15a and determine that $F = 1$.

Although the logic diagram in Figure 9–15b resembles a circuit diagram, it is no more than a diagrammatic representation of Equation 9.1. The lines don't represent wires, and the logic symbols don't represent logic circuits. But the resemblance allows an engineer to immediately construct an electrical circuit to carry out the logic function by viewing the logic diagram as a circuit diagram. This is the process of logic design then:

Express the verbal concept as a Boolean equation, simplify the equation with the aid of truth tables and Karnaugh maps, render the equation as a logic diagram, and construct a circuit based on it as a circuit diagram.

There are three other logic functions that are so common they are given names. These are the NAND, NOR, and XOR shown in Figure 9–16. They are not really new functions since they can be realized by the basic operations in Figure 9–14. The NAND is NOT AND, and the NOR is NOT OR, as indicated by their truth tables. XOR is an abbreviation for "exclusive OR"; it means that C is true if exclusively A or B is true (but not both). The realization of XOR by basic operations is left as an exercise.

The operations AND, OR, and NOT are the basic logic operations; any logic operation can be performed with a combination of them. Actually the AND is not essential since it can be realized by an OR and some NOTs, as shown in Figure 9–17a. A simple proof of this fact is to reverse the 1's and 0's in the truth table for OR in Figure 9–14; the result is the truth table for AND. This amounts to *DeMorgan's first theorem*:

$$\overline{A + B} = \overline{A} \cdot \overline{B}. \tag{9.2}$$

It is also possible to realize an OR by an AND and some NOTs as in Figure 9–17b. This amounts to *DeMorgan's second theorem*:

$$\overline{A \cdot B} = \overline{A} + \overline{B}. \tag{9.3}$$

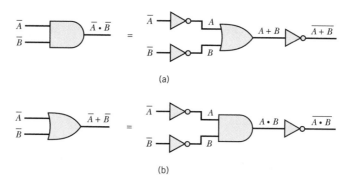

(a)

(b)

FIGURE 9–17 (a) Realizing an AND by an OR with NOTs at all inputs and outputs. (b) Realizing an OR by an AND with NOTs at all inputs and outputs. These equivalences illustrate DeMorgan's theorems.

An example of the identity in Equation 9.3 is the equivalence of the following statements: "You can go if (and only if) your homework is done and your room is clean." "You can't go if (and only if) your homework isn't done or your room isn't clean." DeMorgan's theorems allow us sometimes to simplify logic by getting rid of several NOT operations (see the diagrams on the right of Figure 9–17). Karnaugh maps also aid in simplifying logic.

Karnaugh Maps

The Karnaugh maps in Figure 9–14 can be extended to four variables, as shown in Figure 9–18. The second row corresponds to $A = 0$ and $B = 1$, and the third row corresponds to $A = 1$ and $B = 1$. Therefore, the second and third rows together correspond to simply $B = 1$. Likewise, the second and third columns together correspond to $D = 1$. Then the expression $E = BD$ is represented by a Karnaugh map with 1's at the intersection of the second and third rows with the second and third columns (see the grouping of four 1's in Figure 9–18a).

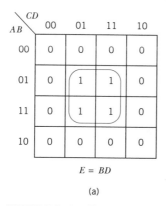

$E = BD$

(a)

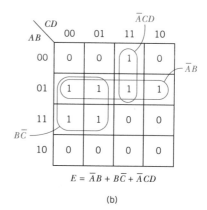

$E = \overline{A}B + B\overline{C} + \overline{A}CD$

(b)

FIGURE 9–18 Examples of grouping 1's on a Karnaugh map to reduce a logical expression. The groups may overlap, and they should be as large and as few as possible.

The expression $E = \overline{AB}\overline{C}D$ corresponds to a 1 in the second row and second column of Figure 9–18a. The other three 1's correspond to the truth of $\overline{A}BCD$, $AB\overline{C}D$, and $ABCD$. Then taken as a group, the four 1's represent $E = \overline{AB}\overline{C}D + \overline{A}BCD + AB\overline{C}D + ABCD$. But the same group represents $E = BD$. Therefore, the Karnaugh map shows the identity

$$\overline{AB}\overline{C}D + \overline{A}BCD + AB\overline{C}D + ABCD \equiv BD.$$

This simplification of the expression on the left could have been found through the use of Boolean algebra, but the Karnaugh map is usually easier.

EXAMPLE 9.6 Simplify the expression

$$E = \overline{A}\,\overline{B}CD + \overline{AB}\overline{C}\overline{D} + \overline{AB}\overline{C}D + \overline{A}BCD + \overline{AB}C\overline{D} + AB\overline{C}\overline{D} + AB\overline{C}D.$$

Solution: The first term is true when A, B, C, and D are 0, 0, 1, 1, which is the third box in the first row of the Karnaugh map in Figure 9–18b. We enter a "1" there and in the six positions corresponding to the six other terms. We then form groups of 1's following these rules: (1) A group must form a rectangle that is either one, two, or four on a side, (2) each group should be as large as possible, (3) there should be as few groups as possible, and (4) groups may overlap.

The three groups in Figure 9–18b are a solution found by following these rules. They correspond, in turn, to the expressions $\overline{A}B$, $B\overline{C}$, and $\overline{A}CD$. Then the simplification of the original expression is

$$E = \overline{A}B + B\overline{C} + \overline{A}CD.$$

Case Study

An engineer in charge of designing a digital display must develop logic to convert from BCD to seven-segment code, as given in Figure 9–11. In this case study we will follow the development of logic for driving the "e" segment; the development for the other segments is similar.

In Figure 9–19a is a portion of Figure 9–11 showing a truth table for Q_e. To complete the table for all cases, the patterns of B's not allowed by BCD (such as 1101) are also listed. For these patterns we don't care what state is assigned to Q_e; "don't care" is indicated by the entry "X." An equivalent representation of the truth table is the *Karnaugh map* in Figure 9–19b. The upper left entry indicates that Q_e is 1 when the B's are 0000. Note that the boxes are arranged so that, in moving to an adjacent box, only one B changes. This will allow us to see easily some patterns that simplify the circuit realization.

To put the logic statement for Q_e in equation form (and then in diagram or circuit form), we need to decide what to do with the X's. Let's make them all 0's (since we don't care), as shown in Figure 9–20. This is the same Karnaugh map as in Figure 9–19b but with a slightly different notation. The upper left entry says Q_e

B_3	B_2	B_1	B_0	Q_e
0	0	0	0	1
0	0	0	1	0
0	0	1	0	1
0	0	1	1	0
0	1	0	0	0
0	1	0	1	0
0	1	1	0	1
0	1	1	1	0
1	0	0	0	1
1	0	0	1	0
1	0	1	0	X
1	0	1	1	X
1	1	0	0	X
1	1	0	1	X
1	1	1	0	X
1	1	1	1	X

(a)

B_3B_2 \ B_1B_0	00	01	11	10
00	1	0	0	1
01	0	0	0	1
11	X	X	X	X
10	1	0	X	X

(b)

FIGURE 9–19 (a) Truth table showing conversion from BCD to the "e" segment of seven-segment code (taken from Figure 9–11). (b) Karnaugh map—another way of expressing the truth table in (a).

is true when $\bar{B}_3\bar{B}_2\bar{B}_1\bar{B}_0$ is true. ($\bar{B}_3$ and $\bar{B}_2$ and $\bar{B}_1$ and $\bar{B}_0$ are true, or 1; therefore all the B's are 0's.) Each of the three other 1's has a similar term associated with it. The first equation for Q_e in Figure 9–20 indicates that Q_e is true when any of these four terms are true. A circuit diagram that realizes this generation of Q_e is shown in Figure 9–21.

Q_e

	$\bar{B}_1\bar{B}_0$	$\bar{B}_1B_0$	B_1B_0	$B_1\bar{B}_0$
$\bar{B}_3\bar{B}_2$	1	0	0	1
$\bar{B}_3B_2$	0	0	0	1
B_3B_2	0	0	0	0
$B_3\bar{B}_2$	1	0	0	0

$$Q_e = \bar{B}_3\bar{B}_2\bar{B}_1\bar{B}_0 + \bar{B}_3\bar{B}_2B_1\bar{B}_0 + \bar{B}_3B_2B_1\bar{B}_0 + B_3\bar{B}_2\bar{B}_1\bar{B}_0$$
$$= \bar{B}_2\bar{B}_1\bar{B}_0 + \bar{B}_2B_1\bar{B}_0$$

FIGURE 9–20 The Karnaugh map in Figure 9–18b with the "don't care" X's replaced by 0's. The grouping of adjacent 1's reduces the complexity of the logical expression. Each 1 corresponds to a term in the first expression for Q_e. Each group of 1's corresponds to a term in the second expression for Q_e.

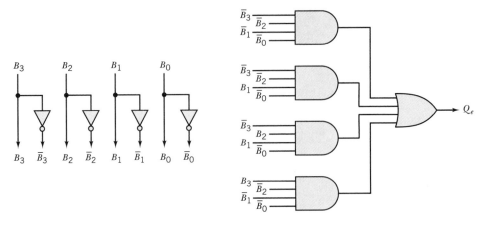

$$Q_e = \overline{B_3}\,\overline{B_2}\,\overline{B_1}\,\overline{B_0} + \overline{B_3}\,\overline{B_2}\,B_1\,\overline{B_0} + \overline{B_3}\,B_2\,B_1\,\overline{B_0} + B_3\,\overline{B_2}\,\overline{B_1}\,\overline{B_0}$$

FIGURE 9–21 Logic diagram realizing the unreduced expression for Q_e. This corresponds to the map in Figure 9–19 without grouping the 1's.

The circuit in Figure 9–21 is a brute-force approach to solving the problem. It is possible to get a simpler expression for Q_e by observing some patterns in the Karnaugh map. The technique is to look for groups of adjacent 1's. The Karnaugh map must be thought of as wrapped on a doughnut so the left and right edges meet and the top and bottom meet. For instance, the 1's in the upper and lower left corners of Figure 9–20 are adjacent. To form a group, the 1's must fill an $m \times n$ rectangle, where m and n are both powers of two. The two groups in Figure 9–21a each form a 1×2 rectangle. The left group corresponds to $\overline{B_2}\,\overline{B_1}\,\overline{B_0}$ (the states that don't change within the rectangle), and the right group corresponds to $\overline{B_3}\,B_1\,\overline{B_0}$. Therefore

$$Q_e = \overline{B_2}\,\overline{B_1}\,\overline{B_0} + \overline{B_3}\,B_1\,\overline{B_0}.$$

This is a great simplification.

Further simplification is possible if we choose the "don't cares" (the X's) so as to form larger (but not more) groups. Instead of replacing all the X's with 0's, we replace two X's in the lower right with 1's (compare Figure 9–22 with Figure 9–19b). Now we can include all 1's in two groups of *four* each. (The four corners form a group by the "wrap-around" rule.) Since the pattern in Figure 9–22 is different from that for Q_e in Figure 9–20, we are generating a new variable Q'_e. However, Q_e and Q'_e differ only in states that never occur in BCD; they will control the "e" segment in the same way in practice. The two groups in Figure 9–22 correspond to the expression $Q'_e = \overline{B_2}\,\overline{B_0} + B_1\,\overline{B_0}$. Using factoring, we can simplify the expression for Q'_e further:

$$Q'_e = (\overline{B_2} + B_1)\overline{B_0}.$$

	$\overline{B}_1\overline{B}_0$	$\overline{B}_1 B_0$	$B_1 B_0$	$B_1\overline{B}_0$
$\overline{B}_3\overline{B}_2$	1	0	0	1
$\overline{B}_3 B_2$	0	0	0	1
$B_3 B_2$	0	0	0	1
$B_3\overline{B}_2$	1	0	0	1

$$Q'_e = \overline{B}_2\overline{B}_0 + B_1\overline{B}_0 = (\overline{B}_2 + B_1)\overline{B}_0$$

FIGURE 9–22 The Karnaugh map in Figure 9–18b with the "don't care" X's replaced by two I's and the rest 0's. This allows larger groups (and a simpler expression) than in Figure 9–19. The left and right edges of the map are adjacent (picture a wrapping around), and the top and bottom edges are adjacent. Therefore the four corners form a group of adjacent 1's.

This is realized by the circuit diagram in Figure 9–23a, which is far simpler than the first try in Figure 9–21.

By using DeMorgan's theorem, the circuit can be converted into one using only NOR gates as in Figure 9–23c. The procedure consists of adding NOT "bubbles" in pairs (see Figure 23b) until the conversion in Figure 9–16b can be performed. Note that one of the NOR gates in Figure 9–23c is used strictly as an inverter.

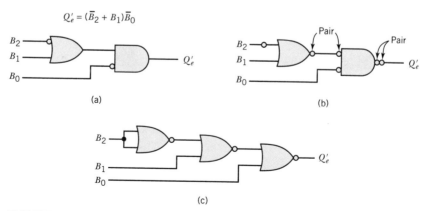

$$Q'_e = (\overline{B}_2 + B_1)\overline{B}_0$$

(a)

(b)

(c)

FIGURE 9–23 Two equivalent realizations of the expression for Q'_e (see map in Figure 9–22). Q'_e and Q_e produce the same drive for the "e" segment in practice. Note the simplicity of this logic compared with that in Figure 9–21.

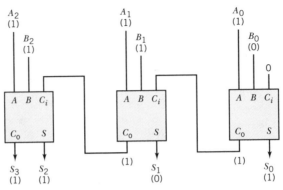

FIGURE 9–24 Using three "full adder" circuits to add a 3-bit binary number A to a number B, giving a 4-bit sum S. The C's are carries. In the example here, seven (111) plus six (110) is thirteen (1101).

Arithmetic Operations

Logic circuitry is also used to carry out the computational procedures we looked at in Section 9.3. Consider the addition of 111 and 110 in Figure 9–7. The procedure was the same in each of the three columns: the two digits in the column together with any carry from the previous column determined the digit to write under the column and whether there was any carry to the next column. Figure 9–24 shows a three-bit adder that consists of three identical circuits—one for each column. The "carry in" C_i for each stage is provided by the "carry out" C_o from the previous stage. (C_i for the LSB stage is set to 0.) The S from each stage corresponds to the digit written under each column; together these digits provide the sum.

Each of the stages in Figure 9–24 is called a *full adder*. In order to design the circuitry for a full adder, we have to set down the rules for generating S and C_o. If two or more of the inputs are true, C_o is true. If an odd number of inputs are true, then S is true. These statements are summarized by the Karnaugh maps in Figure 9–25a. In reducing the expression for C_o, we can form three groups:

$$C_o = AB + AC_i + BC_i.$$

The Karnaugh map for S has no two 1's that are adjacent, so it can't be reduced by standard procedures. But checkerboard patterns such as this are easily implemented by XOR gates. (The reader should lay out the truth table for the XOR in Figure 9–16 as a Karnaugh map to see why.) The expression is

$$S = (A \oplus B) \oplus C_i.$$

The corresponding circuitry for a full adder is shown in Figure 9–25b.

Digital Switches

It is sometimes useful to look at a logic operation as providing a switching function, much as the analog switch in Figure 9–26. When the switch is closed, $V_C = V_A$, and

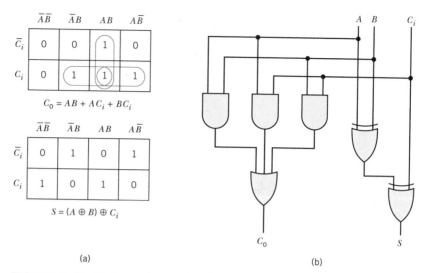

$$C_0 = AB + AC_i + BC_i$$

$$S = (A \oplus B) \oplus C_i$$

(a)

(b)

FIGURE 9–25 (a) Karnaugh maps for C_0 and S of a full adder. (b) Logic realization for one of the full adders used in Figure 9–24.

when the switch is open, $V_C = 0$. Similarly for the AND gate, when $G = 1$, $C = A$, and when $G = 0$, $C = 0$. If the digital signal at A is a series of pulses between 1 and 0, as shown in Figure 9–26, G can be used to "gate" on and off the pulses reaching C.

Two analog switches can be connected by simply joining two wires. With digital switches, the outputs must be combined with an OR gate, as in Figure 9–27. We have $E = A$ when $G = 1$ and $E = B$ when $G = 0$. This digital equivalent of a single-pole, double-throw analog switch is called a *multiplexer*.

Digital switches are used to reconfigure logic circuits. In the next section we will see such reconfiguring used in *universal shift registers*.

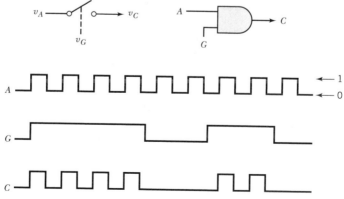

FIGURE 9–26 Analog switch compared with a logic AND gate. The gate corresponds to a closed switch ($C = A$) when $B = 1$, and to an open switch ($C = 0$) when $B = 0$.

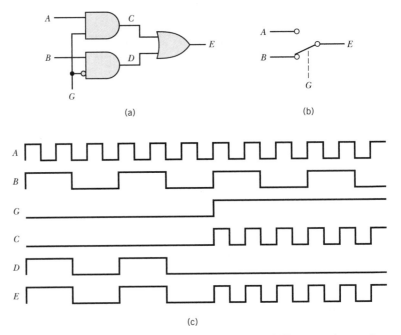

FIGURE 9–27 (a) Two-to-one multiplexer circuit. (b) The equivalent analog switch—a single-pole, double-throw switch. (c) Timing diagram showing $E = A$ when $G = 1$, and $E = B$ when $G = 0$.

9.6   SEQUENTIAL LOGIC

The logic composed of ANDs, ORs, and NOTs that we have looked at so far is called *combinational logic*; the outputs depend only on the combination of the present inputs. With *sequential logic*, the outputs depend not only on the present inputs, but also on the past inputs. This requires some kind of memory element in the circuit.

Flip-Flops and Shift Registers

The most common memory element is the *D flip-flop* shown in Figure 9–28. The output Q remains unchanged until the "clock" input C goes from a 0 to a 1. At that instant Q assumes the state of the data input D. It remembers that past state of D until the next rising edge of the clock. Because sequential logic depends on past time, truth tables are not so useful as *timing diagrams*. These are time plots of the digital signals in the circuit. It is convenient to place the plots in vertical alignment to make the order of events evident. The timing diagram in Figure 9–28 shows that the outputs change only on the rising edge of the clock signal. It is not necessary that the clock pulses be narrow; only the rising edges are important. A dot on the A waveform shows the instant at which the clock "looks" at the state of A. B then assumes that state immediately (in practice, after a small delay).

A simple use of a D flip-flop is shown in Figure 9–29. The output B is to change stage (toggle) for every rising edge of C so long as $E = 1$. When $E = 0$, B

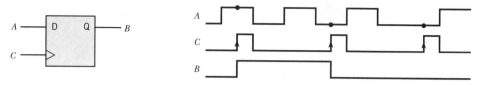

FIGURE 9–28 D flip-flop (D for "data"). The output Q can change only on the rising edge of the clock signal C, at which time it becomes the value of D. The D flip-flop serves as a memory element.

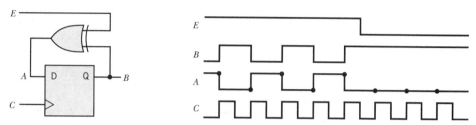

FIGURE 9–29 Simple application of a D flip-flop. B "toggles" with every clock pulse when $E = 1$. The rising edges of C sample the value of A (the dots), and B takes on this value, causing A to change immediately *after* being sampled.

is to cease toggling and hold its last value. Note that when $E = 1$, then $A = \overline{B}$, and the XOR acts as a simple NOT, or inverter (see the XOR truth table in Figure 9–17). Then A, which is the next state of B, is always the inverse of the present B. When $E = 0$, then $A = B$, and B maintains its value.

D flip-flops can be arranged to form a *serial-in, parallel-out shift register* as shown in Figure 9–30. With the first clock edge, the logic level on the input line A is shifted to E_1. With the second clock edge, this level is shifted to E_2, etc. Note that E_2 takes on the old value of E_1 even though E_1 is almost simultaneously taking on a new value. This is assured in practice by a small propagation delay from C_1 to Q. After four clock pulses, the serial data 1001 on A now appear as parallel data on outputs E_1 through E_4.

Before the parallel data on E_1 through E_4 disappears, it can be clocked into a *parallel-in, parallel-out shift register* by C_2. (The instant at which C_2 samples E_1 through E_4 is indicated by the small circles on the waveforms in Figure 9–30.) The data immediately appears on F_1 through F_4 and remains there until the next rising edge of C_2. Thus the two registers acting together form a *serial-to-parallel converter.* This serial-to-parallel conversion is necessary, for instance, when a sequence of bits has been transmitted over a single line (perhaps a phone line) and a number of them must appear together to drive a display.

A serial-in, parallel-out shift register can also be used as a *serial-in, serial-out shift register* if the first three outputs are ignored. In this mode the clock runs continuously, and the data pattern on A appears at E_4 delayed by four clock cycles. This is also called a digital delay line.

For convenience, a serial-in register is often simply called a *shift register*, and a parallel-in register is often simply called a *register*.

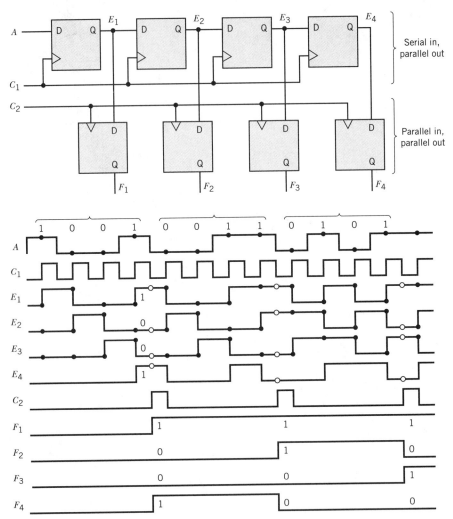

FIGURE 9–30 A serial-in, parallel-out shift register and a parallel-in, parallel-out shift register working together as a serial-to-parallel converter. C_1 clocks four data bits into the serial-in shift register from A. C_2 then clocks these four bits into the parallel-in register, where they are held until the next set of four bits is ready.

The *universal shift register* in Figure 9–31 can act as either a parallel-in, parallel-out or a serial-in, parallel-out shift register, depending on the mode control line M. When $M = 1$, the digital switches (implemented as is Figure 9–27) are up, configuring the circuit as a parallel-in, parallel-out shift register (compare Figure 9–30). The parallel data on the F lines are clocked into the register by C. The mode control then goes to $M = 0$, throwing the switches down to configure the circuit as a serial-in shift register. The next three clock pulses clock the data out serially at A_4. As this process is repeated, parallel data are converted by groups of four bits to serial data. This kind of conversion is necessary when putting parallel data gener-

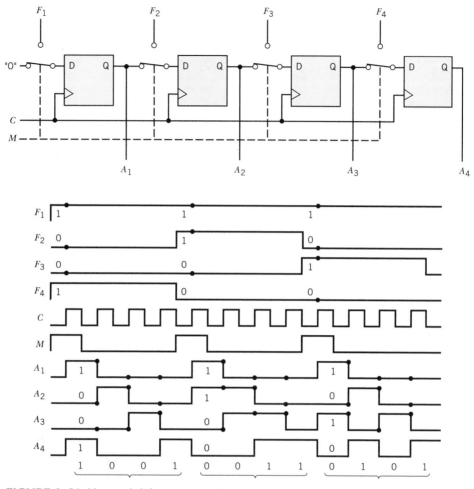

FIGURE 9–31 Universal shift register used as a parallel-to-serial converter. When $M = 1$, the switches are up, configuring the circuit as a parallel-in, parallel-out shift register. When $M = 0$, the switches are down, configuring the circuit as a serial-in shift register.

ated by a keyboard, for instance, onto a phone line. In Figure 9–31 the BCD for 935 has been converted from parallel to serial.

State Machines

A flip-flop has two states: $Q = 0$ and $Q = 1$. A circuit (or "machine") with n flip-flops can have as many as 2^n states. For example, a machine with four flip-flops might have $Q_1 = 1$, $Q_2 = 1$, $Q_3 = 0$, $Q_4 = 1$ as one of its states. With each rising clock edge the machine advances to its next state, where the next state is determined by the present state and by a number of inputs. In practical applications, each state corresponds to some event such as a step in a manufacturing process.

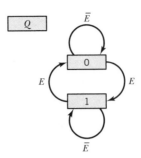

FIGURE 9–32 State diagram for the toggle circuit in Figure 9–29. The flip-flop can assume the state $Q = 0$ or the state $Q = 1$.

One of the simplest state machines is the toggle circuit shown in Figure 9–29. When the input $E = 1$, the state $Q = 0$ is followed by the state $Q = 1$, and vice-versa. When $E = 0$, the machine remains in its present state, even when a clock edge occurs. This behavior is summarized in the *state diagram* shown in Figure 9–32.

The state machine in Figure 9–33 has four flip-flops and $2^4 = 16$ states. There is one input—the variable R. The present state at any time is the values of Q_1, Q_2, Q_3, and Q_4. The next state will be the values that are currently at D_1, D_2, D_3, and D_4. The combinational logic feeding the D's makes the next state a function of the present state (the Q's) and the input (R). The following example will show how the logic is determined from the desired sequence of states.

EXAMPLE 9.7 A Gray code is a sequence of binary numbers in which each number differs from its successor in only one bit. One example of a 4-bit Gray code is the sequence 0000, 0001, 0011, 0010, 0110, 0111, 0101, 0100, 1100, 1101, 1111, 1110, 1010, 1011, 1001, 1000, 0000, and the sequence repeats.

Design: a 4-bit Gray code generator that advances when a "reset" input $R = 0$ and goes to the 0000 state when $R = 1$. The state diagram in Figure 9–34a shows the desired sequence. The inputs are a clock and R. The outputs A_1, A_2, A_3, and A_4 are the four bits of the current binary number (see Figure 9–34b).

Solution: We equate the state variables to the outputs: $A_1 = Q_1$, $A_2 = Q_2$, $A_3 = Q_3$, $A_4 = Q_4$. While it seems unnecessary to distinguish the outputs from the state variables here, we will see examples in which they are not identical. The sequence of states is given in Table 9–1. When $R = 0$, the D's give the next state (the next Q's). When $R = 1$, the next state (the D's) is 0000, regardless of the Q values.

The dependence of each D on the Q's can be expressed by the four Karnaugh maps in Figure 9–35. (The dependence on R will be taken care of simply later.) Using the rules in Section 9.5, we can group the 1's and reduce the maps to the equations for D_1, D_2, D_3, and D_4 given in Figure 9–35. Note that each D is true ($D = 1$) only when R is not true ($\overline{R} = 1$). This is simply implemented by multiplying each expression for D by $\overline{R}$.

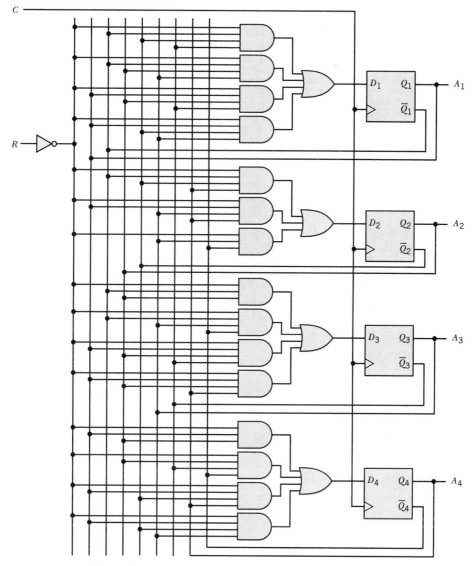

FIGURE 9–33 Circuit diagram of a typical state machine. This machine has four flip-flops and a possibility of $2^4 = 16$ states.

The combinational logic in Figure 9–33 implements the equations for the D's in Figure 9–35.

The state machine in the example could be used to activate each of 16 items, in turn, by connecting a 1-of-16 decoder to the outputs A_1 through A_4. If two or more state variables changed during a state change, we would have to be concerned that they all changed simultaneously. But with a Gray code this happens only during a reset. As we will see, this is not the case for binary counters.

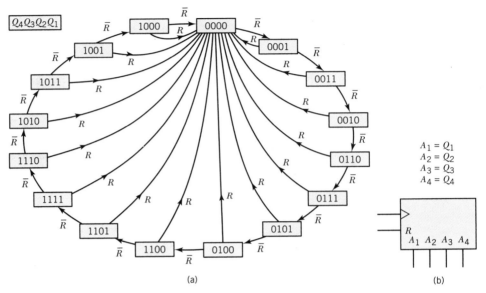

FIGURE 9–34 State diagram for a Gray code counter. Each of the 16 states differs in only one bit from the next state in the sequence. The input R is a reset.

TABLE 9–1 Sequence for Gray Counter in Figure 9–34

R	Q_4	Q_3	Q_2	Q_1	D_4	D_3	D_2	D_1
0	0	0	0	0	0	0	0	1
0	0	0	0	1	0	0	1	1
0	0	0	1	1	0	0	1	0
0	0	0	1	0	0	1	1	0
0	0	1	1	0	0	1	1	1
0	0	1	1	1	0	1	0	1
0	0	1	0	1	0	1	0	0
0	0	1	0	0	1	1	0	0
0	1	1	0	0	1	1	0	1
0	1	1	0	1	1	1	1	1
0	1	1	1	1	1	1	1	0
0	1	1	1	0	1	0	1	0
0	1	0	1	0	1	0	1	1
0	1	0	1	1	1	0	0	1
0	1	0	0	1	1	0	0	0
0	1	0	0	0	0	0	0	0
1	X	X	X	X	0	0	0	0

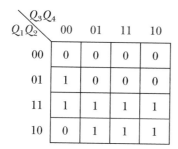

Q_1Q_2 \ Q_3Q_4	00	01	11	10
00	1	1	0	0
01	0	0	1	1
11	1	1	0	0
10	0	0	1	1

$$D_1 = (\overline{Q}_1\overline{Q}_2\overline{Q}_3 + \overline{Q}_1Q_2Q_3 + Q_1Q_2\overline{Q}_3 + Q_1\overline{Q}_2Q_3)\overline{R}$$

Q_1Q_2 \ Q_3Q_4	00	01	11	10
00	0	1	1	1
01	0	0	0	1
11	0	1	1	1
10	0	0	0	1

$$D_2 = (\overline{Q}_1\overline{Q}_2Q_4 + Q_1Q_3Q_4 + Q_3\overline{Q}_4)\overline{R}$$

Q_1Q_2 \ Q_3Q_4	00	01	11	10
00	0	0	0	1
01	1	1	1	1
11	1	1	1	0
10	0	0	0	0

$$D_3 = (\overline{Q}_1Q_2 + \overline{Q}_1Q_3\overline{Q}_4 + Q_1Q_2\overline{Q}_3 + Q_1Q_2Q_4)\overline{R}$$

Q_1Q_2 \ Q_3Q_4	00	01	11	10
00	0	0	0	0
01	1	0	0	0
11	1	1	1	1
10	0	1	1	1

$$D_4 = (Q_1Q_2 + Q_2\overline{Q}_3\overline{Q}_4 + Q_1\overline{Q}_2Q_4 + Q_1\overline{Q}_2Q_3)\overline{R}$$

FIGURE 9–35 Karnaugh maps for determining equations for the D's of the Gray code counter's flip-flops in Figure 9–33.

Counters

A binary counter is a state machine whose sequence of states is an increasing or decreasing sequence of binary numbers (as the X's in Figure 9–11). These functions are so common that they are already designed and available as integrated circuits. Figure 9–36a shows a state diagram for a 3-bit binary counter. While $E = 1$, each clock edge advances the count (the state) by one. When the *terminal* count 111 has been reached, the next clock edge takes the counter back to 000. When $E = 0$, the counter "holds"—it ceases to respond to the clock edges. The design of a circuit to implement this counter is similar to that in Example 9–7 (see problem P9.21).

The outputs A_1, A_2, and A_3 represent the binary number, with A_3 the MSB (see Figure 9–36b). They are equated to the state variables Q_1, Q_2, and Q_3. There is a fourth output TC that goes high on the terminal count 111: $TC = Q_1 Q_2 Q_3$. This output with the enable input E allows several 3-bit counters to be cascaded to form a counter that can count higher than 7.

Figure 9–37 shows three 3-bit counters cascaded to form a 9-bit counter that can count to 511. Table 9–2 illustrates how the terminal count TC and the enable E operate. When the first 3-bit counter has reached the state 111, or 7 (that is, B_1, B_2, and B_3 are 1's), then its TC generates a "carry" signal $X_1 = 1$. This enables the second 3-bit counter to count to 001 on the next clock edge, as the first 3-bit counter goes to 000. Together they then form the binary number 001000, or 8.

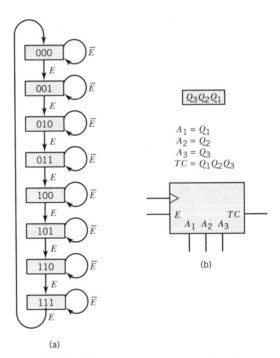

(a)

(b)

FIGURE 9–36 State diagram for a 3-bit binary counter. While the enable $E = 1$, the states count upward. While $E = 0$, the states hold.

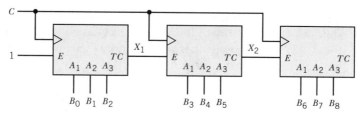

FIGURE 9–37 Three 3-bit binary counters cascaded to form a 9-bit binary counter. When a terminal count $TC = 1$, this serves as a "carry" to advance the next stage by one.

TABLE 9–2 Sequence for Binary Counter in Figure 9–37

B_8	B_7	B_6	B_5	B_4	B_3	B_2	B_1	B_0	X_1	X_2
0	0	0	0	0	0	1	0	1	0	0
0	0	0	0	0	0	1	1	0	0	0
0	0	0	0	0	0	1	1	1	1	0
0	0	0	0	0	1	0	0	0	0	0
0	0	0	0	0	1	0	0	1	0	0

One practical application of a counter might be to count cars. Each car driving over a sensor could generate a clock pulse. Another application might be to count how many times a robot arm has been extended so it can be serviced. A 20-bit counter can count to a million.

Asynchronous Logic

All of the sequential logic we have been looking at is *synchronous*; a common clock causes all flip-flops to change state simultaneously. This is a reasonable way for a digital circuit such as a computer to operate internally. But when the circuit interfaces with the outside world, events can happen at different times than the clock can expect. If a signal appeared and disappeared between two clock edges, the event would never be recognized by a synchronous circuit. To handle this situation, there are *asynchronous* logic elements; the RS flip-flop is the most common. The operation of an RS flip-flop is illustrated in Figure 9–38. Any time the "set" input

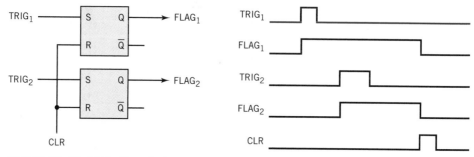

FIGURE 9–38 RS flip-flops. Each trigger signal "sets" a flip-flop, and the clear signal "resets" them. The operation is asynchronous (there is no common clock).

S goes to 1, the output Q immediately goes to 1 (or stays 1 if it was already in that state). A 1 on the "reset" input R causes Q to go to 0. In the application in Figure 9–38, two "flags" remember whether two respective trigger signals have come and gone some time in the past. A "clear" signal prepares the system to look for new triggers. A digital circuit uses RS flip-flops to hold an incoming signal until it has been clocked into the synchronous part of the circuit.

9.7 CASE STUDY

An engineer is assigned the task of designing a digital peak detector. With each clock pulse, a binary number is presented to the peak detector, and the circuit is to hold the highest value presented to it since the last reset. We will see that the design requires both combinational and sequential logic.

The engineer first blocks out the functions of his or her circuit, as shown in Figure 9–39. The input value A is a two-bit binary number—the parallel bits A_0 and A_1. (In most practical assignments there would be eight bits or more.) A register (parallel-in, parallel-out) holds the peak value B found so far—the parallel bits B_0 and B_1. A combinational logic circuit compares the values of A and B, setting $E = 1$ when $A > B$ and $E = 0$ when $A < B$. E operates a switch at the input to the register, causing the larger of A or B to appear at the register input:

$$D = A \quad \text{for} \quad E = 1,$$
$$D = B \quad \text{for} \quad E = 0.$$

D is clocked into the register to become the new value of B. (Note that if $A = B$, we don't care which value is clocked in.)

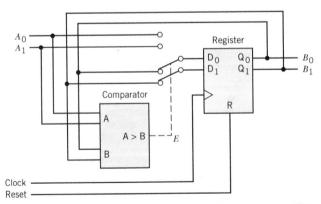

FIGURE 9–39 Peak detector design for the case study. The blocks indicate the functional operations of different pieces of the solution. The greater of the two-bit binary numbers A or B is clocked into the register.

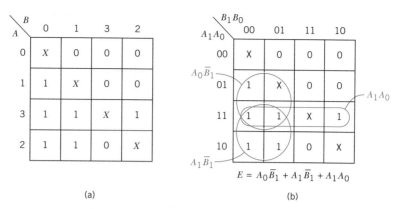

(a)   (b)

FIGURE 9–40 Karnaugh map to determine the logic to realize the Comparator in Figure 9–39. $E = 1$ when $A > B$, $E = 0$ when $A < B$, and $E = X$ (don't care) when $A = B$.

The next steps are to design the circuits inside the blocks and the circuits realizing the switches. The design of the switches is already given in Figure 9–27, and the register is two stages of the parallel-in, parallel-out register in Figure 9–30. Therefore, the main design task now is the logic to realize the Comparator.

The Karnaugh map in Figure 9–40a summarizes the desired performance of the Comparator. For convenience, the values of A and B are shown here in decimal notation. A "1" is entered when A (the row value) is greater than B (the column value), a "0" is entered when A is less than B, and an "X" (a "don't care") is entered when A equals B.

The Karnaugh map is redrawn in Figure 9–40b using the binary values for A and B. The logic is minimized by including the 1's in three groups, allowing X's as well as 1's in the groups in order to make larger and fewer groups. The groups of four entries correspond, in turn, to A_0B_1, A_1B_1, and A_1A_0. Therefore, the output of the Comparator is

$$E = A_0\overline{B}_1 + A_1\overline{B}_1 + A_1A_0. \tag{9.4}$$

The details of the circuit design for the peak detector are shown in Figure 9–41. The two switches are implemented as in Figure 9–27. Two D flip-flops realize the register. The combinational logic at the bottom realizes Equation (9.4).

An example of the circuit operation is illustrated by the timing diagram in Figure 9–42. Initially, $B = 3_{10} = 11_2$. A reset pulse causes $B = 0$. Then $A = 0$ and $B = 0$, causing $E = 0$ (see Figure 9–40b). The switch control $E = 0$ selects $D = B = 0$. On the next clock edge, this is clocked through to become $B = 0$ again. Then the input becomes $A = 1$ which, together with $B = 0$, makes $E = 1$ (since $A > B$). Therefore, the switches select $D = A = 1$, which becomes the next value of B. In a similar manner B goes on to become 2 and 3 as A reaches these values.

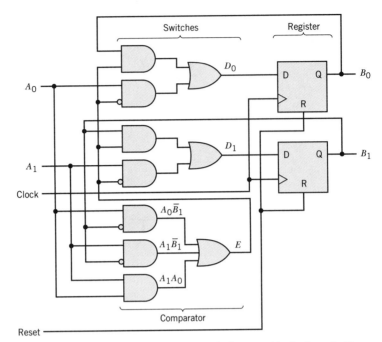

FIGURE 9–41 Circuit details of the peak detector blocked out in Figure 9–39. The logic yielding E at the bottom is a diagram of the Boolean equation determined in Figure 9–40.

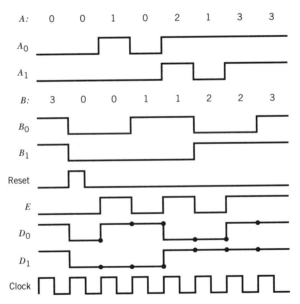

FIGURE 9–42 Timing diagram for the peak detector in Figure 9–41. The value of B advances from 0 to 1 to 2 to 3 as A (momentarily) reaches these values.

9.8 SUMMARY

Digital signals take on only discrete values—in practice two values. The binary digits "1" and "0" are represented by a high and a low voltage. Digital computation is carried out with the binary number system. For interfacing with the outside world, other digital codes for numbers are used: BCD, seven-segment display, and one-of-n are some of the common codes.

The advantages of digital signals are high precision in computation and display, flexibility in representing various kinds of information, and resistance to corruption from noise or decay with time.

Most applications of digital circuitry require both combinational logic (depending only on present inputs) and sequential logic (with memory). Combinational logic consists of AND, OR, and NOT functions. The same functions can be achieved using only NAND gates (or only NOR gates). Circuit complexity can often be reduced by the use of Karnaugh maps. Sequential logic involves the use of flip-flops, which are often configured as registers.

In synchronous logic, there is one clock signal that causes all flip-flops to change state simultaneously. Asynchronous logic, such as RS flip-flops, is necessary when interfacing with the outside world.

FOR FURTHER STUDY

W. I. Flether, *An Engineering Approach to Digital Design*, Prentice-Hall, Englewood Cliffs, NJ, 1980.

C. H. Roth, Jr., *Fundamentals of Logic Design*, West Publishing Co., St. Paul, MN, 1985.

PROBLEMS

Easy Drill Problems (answers at end of chapter)

D9.1. Which line segments are needed to display the character 5? (See Figure 9–2 for segment labels.) Give a code of seven "1" and "0" digits that represents these segments.

D9.2. Make a table of the powers of 2 from $2^0 = 1$ to $2^{10} = 1024$.

D9.3. Convert the binary (base-two) numbers, 1011, 11111, 100000, and 1110100110 to decimal (base-ten) numbers.

D9.4. Convert the decimal numbers 12, 37, 127, 128, and 2001 to binary numbers.

D9.5. Find the sum of the binary numbers 10010 and 101. Convert all three numbers to decimal, and check your addition. Do the same for the binary numbers 1011 and 11111.

D9.6. Find the product of the binary numbers 10010 and 101. Convert all three numbers to decimal, and check your multiplication. Do the same for the binary numbers 1011 and 11111.

D9.7. Show that in 4-bit binary, -8 (the twos complement of 8) is 1000.

D9.8. Using negative numbers to subtract by adding, perform $6 - 4$, $4 - 6$, and $7 - 8$. Use the table in Figure 9–9.

D9.9. Show that adding -4 and -5 in 4-bit binary results in an overflow (wrong result). Use the table in Figure 9–9. Show that adding 4 and 5 in 4-bit binary results in an overflow.

D9.10. Convert the binary numbers 10010011, 110111001001, and 110111 to hexadecimal. Convert the hexadecimal numbers 57, 3A2, FF, and 100 to binary.

D9.11. Convert the decimal numbers 37, 1986, and 9 to BCD. Which of these is the same as the binary representation?

D9.12. Write the following statement as an equation: "The stain is removed by paint thinner or by soap and water." Let R = removed, T = thinner, S = soap, and W = water.

D9.13. Fill in the missing steps, using DeMorgan's theorems, to show that $\overline{A \cdot B + C} = (\overline{A} + B) \cdot \overline{C}$.

D9.14. Convert the XOR truth table in Figure 9–16 to a Karnaugh map.

D9.15. Convert the truth table in Figure D9.15 to a Karnaugh map.

A	B	C	D	E
0	0	0	0	0
0	0	0	1	0
0	0	1	0	1
0	0	1	1	0
0	1	0	0	0
0	1	0	1	0
0	1	1	0	1
0	1	1	1	0
1	0	0	0	0
1	0	0	1	0
1	0	1	0	1
1	0	1	1	0
1	1	0	0	0
1	1	0	1	0
1	1	1	0	1
1	1	1	1	0

D9.16. Convert the expression $E = \overline{A}\,\overline{B}CD + \overline{A}BCD + A\overline{B}C\overline{D} + ABC\overline{D}$ to a Karnaugh map, as in Figure 9–19. Use the map to get a simple expression for E.

D9.17. Find simple expressions for the Karnaugh maps on p. 261, as was done in Figure 9–20. Draw the corresponding logic diagrams using AND, OR, and NOT gates.

D9.18. Given E as defined by the truth table on p. 262, use a Karnaugh map to find a simple expression for E', where $E' = E$, except possibly when $E = X$.

D9.19. Given $E = \overline{A}\,\overline{B}CD + \overline{A}B\overline{C}D + \overline{A}BC\overline{D} + ABC\overline{D}$ and don't care (X) for $\overline{A}\,\overline{B}\,CD + \overline{ABC}\,D + \overline{A}BCD + A\overline{B}\,\overline{C}D + ABCD$ true, use a Karnaugh map to find a simple expression for E', where $E' = E$ except possibly when $E = X$.

D9.20. Sketch a logic diagram to represent $E = \overline{C} + AB + ACD$. Adding pairs of "bubbles" where appropriate and using DeMorgan's theorems, sketch an equivalent logic diagram using only NAND gates.

AB \ CD	00	01	11	10
00	0	0	0	0
01	0	0	0	0
11	0	0	0	0
10	1	1	1	1

(a)

AB \ CD	00	01	11	10
00	1	1	0	0
01	0	0	0	0
11	0	0	0	0
10	1	1	0	0

(b)

AB \ CD	00	01	11	10
00	0	0	0	0
01	0	1	1	0
11	0	0	0	0
10	1	1	0	0

(c)

AB \ CD	00	01	11	10
00	1	1	0	0
01	1	1	0	1
11	0	1	0	0
10	0	0	1	0

(d)

FIGURE D9.17 Karnaugh maps for problem D9.17.

A	B	C	D	E
0	0	0	0	1
0	0	0	1	0
0	0	1	0	1
0	0	1	1	1
0	1	0	0	0
0	1	0	1	0
0	1	1	0	0
0	1	1	1	0
1	0	0	0	1
1	0	0	1	X
1	0	1	0	1
1	0	1	1	X
1	1	0	0	0
1	1	0	1	X
1	1	1	0	0
1	1	1	1	X

FIGURE D9.18
Truth table for
problem D9.18.

Application Problems

P9.1. Let $0\ \text{V} < v_C \le 1\ \text{V}$ represent the digit "1," $1\ \text{V} < v_C \le 2\ \text{V}$ represent the digit "2," etc. If the digit "9" is stored in analog form by charging a 1-μF capacitor to $v_C = 9\ \text{V}$, how long is it before a parallel 1000-MΩ resistance causes the stored digit to become "8"?

P9.2. Convert decimal 12 to a 5-bit binary number (requires a leading zero). Form the 5-bit representation of -12 (the twos complement of 12). Subtract 12 from 15 by adding your binary number to 01111. Is the result the binary for 3? (Remember to throw away the sixth bit.)

P9.3. Use the twos complement as in P9.2 to subtract 15 from 12 in the 5-bit binary number system. Is the result the twos complement of 3?

P9.4. Write an equation (Boolean expression) for "cleared for takeoff" from the following statements: "You can take off in fair weather. Foul weather requires instrument flight capability. No one takes off if the air traffic controllers are on strike." Let T = takeoff, F = fair, I = instruments, and S = strike. Reduce with a Karnaugh map.

P9.5. Realize the multiplexer function in Figure 9–27 using only NAND gates. (Add inversion bubble pairs as in Figure 9–27b, and use one of DeMorgan's theorems from Figure 9–17.)

P9.6. A seven-segment display for each of the hexadecimal digits is shown in Figure P9.6 (note lowercase "b" and "d" are used for "B" and "D"). List the segments used for each character (see segment labeling in Figure 9–2a). Make a truth table listing binary code and corresponding seven-segment code for the hexadecimal digits (as with the B's and Q's in Figure 9–11).

P9.7. For the 1-of-8 decoder in Figure 9–13, express Q_1 in terms of D_2, D_1, and D_0.

P9.8. For the BCD and seven-segment codes in Figure 9–11, make a Karnaugh map for Q_d in terms of B_3, B_2, B_1, and B_0. Use the map to find groups of 1's, and give a simplified expression for Q_d. (Converting appropriate X's to 1's, you should find three groups with four 1's, and one group with two 1's.)

P9.9. Convert the expression $Q = \overline{B_2}\overline{B_0} + \overline{B_2}B_1 + B_1\overline{B_0} + B_2\overline{B_1}B_0$ to a logic diagram with ANDs, ORs, and NOTs. Use factoring to simplify the expression first.

P9.10. To find the equivalent function of the logic function in Figure P9.10a, reverse the 1's and 0's under A and B in the XOR truth table in Figure 9–16. Can you recognize the new table? Similarly, find the equivalent of the logic function in Figure P9.10b.

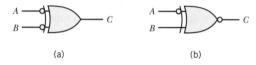

(a)   (b)

P9.11. Use four of the multiplexer shown in Figure 9–27 together with four D flip-flops to form the universal shift register shown in Figure 9–31.

P9.12. The Q's of the shift register in Figure 9–30 are initially all 0's. A is the sequence 11011011. There is a rising edge of C_1 in the center of each bit of A. Draw a timing diagram of the signals.

P9.13. The Q's in Figure P9.13 are initially all 0's. C is a sequence of seven clock pulses. Draw a timing diagram for C, Q_1, Q_2, Q_3, and D_1. (It helps to sketch 1's and 0's on the logic diagram to keep track of the states.) What function does this circuit perform?

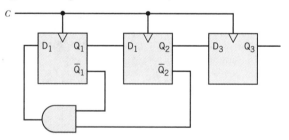

P9.14. It is desired to set a flag when any 4-bit sequence on the A line immediately repeats itself. (A clock signal C_1 accompanies the serial bit sequence A, as in Figure 9–30.) Design a circuit to do this with two 4-bit serial-in/parallel-out shift registers, four XOR gates, a four-input NOR gate, and an RS flip-flop. (The registers can delay the sequence, and the XORs can compare bits for being alike.)

P9.15. Given the sequential logic circuit in Figure P9.15, sketch the waveforms for A, B, C, and D for seven clock cycles. Initially $B = C = D = 0$.

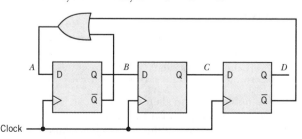

P9.16. Design a synchronous logic circuit with two-bit input A, a reset, and two-bit output B such that B is the binary sum of all A's since the last reset.

P9.17. Design a synchronous logic circuit with four-bit input A, mode control M, and eight-bit output B such that $B = A$ after a clock pulse with $M = 1$, and B grows by a factor of 2 for every clock pulse with $M = 0$.

P9.18. Design a "full subtracter" circuit (similar to the "full adder" circuit designed in Figure 9–25). The inputs are the bit A, the bit F (to be subtracted), and a borrow bit B_i. The outputs are a difference bit D and a borrow bit B_o. Use a truth table to list the outputs corresponding to each possible set of inputs. Then minimize the logic by Karnaugh maps. Arrange three full subtracters to form a three-bit subtracter.

P9.19. Design a combinational logic circuit with three-bit input F and three-bit output B such that $B = -F - 1$ (see section on Negative Numbers). Use this circuit together with the three-bit adder in Figures 9–24 and 9–25b to realize a three-bit output $S = A - F - 1$. What happens if the first C_i bit in Figure 9–24 is set to 1 rather than 0? Compare with the design in problem P9.18.

P9.20. Draw a block diagram (similar to Figure 9–39) of a sequential logic circuit with four-bit input A and four-bit output B such that B is increased by $(A - B)/2$ for every clock pulse. In the divide-by-two, the quotient is to be rounded down by discarding the fractional part. The circuit you have designed is a digital low-pass filter.

P9.21. Design a circuit to implement the state diagram of a 3-bit binary counter in Figure 9–36. Three 4-variable Karnaugh maps are needed. Produce all four outputs.

P9.22. Suppose the circuit in Figure P9.22 has just been reset. Draw a timing diagram of the outputs A_0, A_1, A_2, and A_3 for 17 clock pulses. This circuit is not synchronous because there is not a common clock for all flip-flops. What synchronous circuit is this similar to?

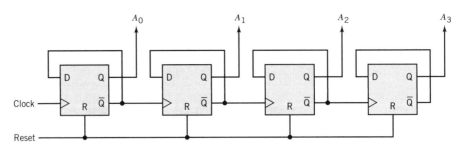

Extension Problems

E9.1. A Karnaugh map may give a simpler expression for the *complement* of a function. For example, the positions of the 0's in Figure 9–19b give an expression for $\overline{Q_e}$. Convert the appropriate X's to 0's to get as simple an expression for $\overline{Q_e}$ as possible by forming groups of 0's. Draw the corresponding logic diagram.

E9.2. Simplify the expression $E = \overline{A}\,\overline{B}C\overline{D} + \overline{A}BC\overline{D} + A\overline{B}C\overline{D} + ABC\overline{D}$ by factoring out the $C\overline{D}$. Then factor out $\overline{A}$ and A, and use the Boolean identities $\overline{A} + A \equiv 1$ and $1 \cdot A \equiv A$. Is the final result the same as the solution to D9.16?

E9.3. The logic circuit in Figure E9.3 has $Q_1 = 0$ and $Q_2 = 1$ initially. For the R and S signals shown, draw a timing diagram for R, S, Q_1, and Q_2. Is this combinational or sequential logic? (Do Q_1 and Q_2 depend only on the current states of R and S?)

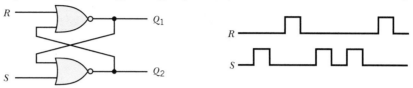

Study Questions

9.1. Give three advantages of digital displays over analog displays. Give one advantage of analog displays over digital displays.

9.2. Give three advantages of digital signals over analog signals.

9.3. How are the ones complement, the twos complement, and the negative of a number related?

9.4. How many numbers can be represented by four bits? How does your answer change if one of the bits is a sign bit?

9.5. If computers prefer binary numbers and people prefer decimal numbers, why are hexadecimal numbers ever used?

9.6. What kind of encoder converts a one-of-n code to a binary code? What happens if a three-of-n code (with three 1's) is applied to this encoder?

9.7. What feature does combinational logic lack that sequential logic has?

9.8. What functions do the "plus" and "times" symbols represent in Boolean algebra? What is the solution to $X = 1 + 1$?

9.9. What is the difference between a Boolean equation and a logic diagram? Between a logic diagram and a circuit diagram?

9.10. Which gives more information, a truth table or a Karnaugh map? Which is more compact?

9.11. How is a universal shift register related to a serial-in shift register and a parallel-in shift register?

9.12. At what instants do the variables of a synchronous circuit change? Of an asynchronous circuit?

ANSWERS TO DRILL PROBLEMS

D9.1. a,c,d,f,g, 1011011

D9.2. 1, 2, 4, 8, 16, 32, 64, 128, 256, 512, 1024

D9.3. 11, 31, 32, 934

D9.4. 1100, 100101, 1111111, 10000000, 11111010001

D9.5. $10010 + 101 = 10111$, $18 + 5 = 23$; $1011 + 11111 = 101010$, $11 + 31 = 42$

D9.6. $10010 \times 101 = 1011010$, $18 \times 5 = 90$; $1011 \times 11111 = 101010101$, $11 \times 31 = 341$

D9.7. 2's comp(1000) = 1's comp(1000) + 1 = 0111 + 1 = 1000

D9.8. $0110 + 1100 = 0010, 0100 + 1010 = 1110, 0111 + 1000 = 1111$

D9.9. $1100 + 1011 = 0111$ (7), $0100 + 0101 = 1001$ (−7)

D9.10. 93, DC9, 37; 1010111, 1110100010, 11111111, 100000000

D9.11. 0011,0111; 0001,1001,1000,0110; 1001

D9.12. $R = T + S \cdot W$

D9.13. $\overline{A \cdot \overline{B} + C} = \overline{(A \cdot \overline{B})} \cdot \overline{C} = (\overline{A} + B) \cdot \overline{C}$

D9.14.

	$\overline{B}$	B
$\overline{A}$	0	1
A	1	0

D9.15.

AB \ CD	00	01	11	10
00	0	0	0	1
01	0	0	0	1
11	0	0	0	1
10	0	0	0	1

D9.16. Map same as for D9.15; $E = C\overline{D}$

D9.17. $A\overline{B}, \overline{B}\,\overline{C}, \overline{A}BD + A\overline{B}\,\overline{C}, \overline{A}\,\overline{C} + \overline{A}B\overline{D} + B\overline{C}D + A\overline{B}CD$

D9.18. $E' = \overline{B}\overline{C}\,\overline{D} + \overline{B}D$

D9.19. $E' = \overline{A}D + BC$

D9.20. A NAND with inputs A and B, a NAND with inputs A, C, and D, and a NAND with inputs from C and the other two NANDs.

CHAPTER 10

Feedback Control Systems

A control system uses electronics to adjust some parameter such as the position of a cutting tool, the temperature of a chemical bath, or the speed of a motor. Control systems are the opposite of instrumentation systems in the sense that motor nerves are the opposite of sensor nerves in the body. A *feedback* control system combines the two, controlling a parameter, measuring the parameter to see if the adjustment was done correctly, and changing the control, if necessary.

To see the importance of feedback, suppose that a motor is specified to achieve a speed of 3000 rpm when 60 V is applied. (In Chapters 19 and 20 it is shown that the speed of a dc motor is proportional to the voltage applied to the armature.) Figure 10–1a diagrams this arrangement with a *signal flow graph*, so called because it traces signals or parameters through the system. One might simply apply 60 V and trust the motor speed to be 3000 rpm. However, the actual speed will depend on the loss in the line delivering the voltage to the motor and on the particular load on the motor. If these parameters are uncertain, then this *open-loop control* will give an uncertain motor speed.

A better method of control would be to monitor the motor speed with a tachometer, compare this with the desired speed, and adjust the applied voltage until the error is reduced to zero. Figure 10–1b diagrams this *closed-loop control* system. The comparison function is carried out by subtraction, represented here by the circle with a " + " inside and a " − " outside. A signal flow graph that comes back on itself, as here, is called a *feedback loop*.

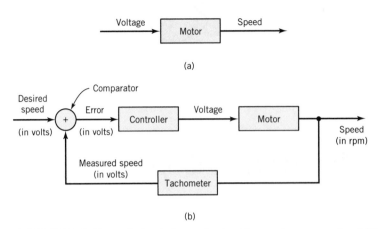

FIGURE 10–1 Control of a shaft speed using (a) open-loop control and (b) feedback control.

The reader's objectives in this chapter are to:

1. learn to diagram analog equations by signal flow graphs (just as Boolean equations are expressed by logic diagrams),
2. understand the tradeoffs between accuracy, stability, and speed of response in a feedback control system,
3. be able to design an electronic Controller to achieve the desired accuracy, stability, and speed.

Most engineering students take a course in control theory that treats the topic in depth. This chapter is only an introduction to the topic, but it is complete enough that the reader will be able to meet the above objectives. From the standpoint of electrical engineering, the thrust of the chapter is to understand the design of the Controller, which is an electronic circuit. The feedback theory here will also provide a foundation for studying operational amplifiers in Chapter 14.

10.1   IDEAL FEEDBACK SYSTEM

We start with an open-loop control system—the motor example suggested above. Figure 10–2a shows a more detailed flow graph of the system in Figure 10–1a. A small control voltage v_1 is applied to a power amplifier,* which in turn applies a voltage v_2 to the motor. Suppose that $v_2 = 60$ V produces a shaft speed $y = 3000$ rpm and that the relationship is linear. Then we can write $y = (50 \text{ rpm/V})v_2$, as represented by the second box of the flow graph. Let the gain of the power amplifier

---

*A power amplifier is capable of supplying significant current as well as increased voltage.

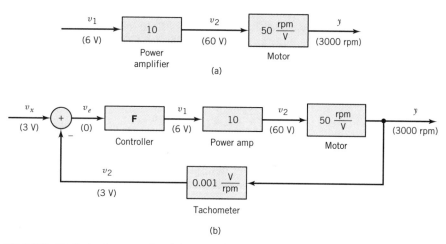

FIGURE 10–2 (a) An example of the open-loop system in Figure 10–1a. (b) An example of the feedback control system in Figure 10–1b. The controller adjusts v_1 until $v_e = 0$. The gain from input to output is determined primarily by the feedback gain.

be 10, as represented by the first box. Then $v_1 = 6$ V would give $v_2 = 60$ V and the desired speed $y = 3000$ rpm. As discussed before, this open-loop system gives only rough control because of uncertainty about the gains in the two boxes. If a load caused the motor transfer function to become 40 rpm/V, the speed would drop to 2400 rpm.

We can overcome this lack of precise control by adding a feedback path as in Figure 10–2b. This is a more detailed flow graph of the system in Figure 10–1b. The tachometer has a gain here of 0.001 V/rpm. For example $y = 3000$ rpm produces $v_z = 0.001$ V/rpm $\times$ 3000 rpm $= 3$ V.

The basic philosophy of the loop's operation is the following. Suppose we desire $y = 3000$ rpm, or, equivalently, $v_z = 3$ V. Then we should apply $v_x = 3$ V to the system. The comparator takes the difference between v_x and v_z to produce the error voltage v_e. If y is the desired speed, then $v_e = 0$. Otherwise an error voltage is applied to the Controller block labeled "F." For now we can think of the Controller as adjusting the speed up or down according to the sign of v_e until v_e is reduced (ideally) to zero. This means that v_z is brought equal to $v_x = 3$ V, which corresponds to the condition $y = 3000$ rpm. The numbers for this example are in parentheses in Figure 10–2b.

The gain from the input v_x to the output y when feedback is present is called the *closed-loop gain* y/v_x. By considering the ideal behavior of the loop just described, we can find its closed-loop gain. Then we will generalize to any feedback loop. The feedback gain provides $v_z = (0.001 \text{ V/rpm})y$, or writing this backwards, $y = (1000 \text{ rpm/V})v_z$. But ideally we end up with $v_z = v_x$, and we get $y = (1000 \text{ rpm/V})v_x$. It is clear that the closed-loop gain of 1000 rpm/V came from inverting the feedback gain of 0.001 V/rpm. In general,

The ideal closed-loop gain is the reciprocal of the feedback gain.

This result is formalized in the next section.

Notice that in determining the closed-loop gain we almost ignored the forward path. As long as the "Controller" box is doing its job of adjusting v_1 so the error v_e is reduced almost to zero, the relation of y to v_x is determined entirely by the feedback gain (ideally). This is the advantage of feedback control. All the poorly defined and uncertain elements of the system are put in the forward path, and the one or two (often passive) elements constituting the feedback path are kept very accurate. The system in Figure 10–2a depends on the (poorly controlled) forward gain, while that in Figure 10–2b depends on the (accurate) feedback gain.

We do care something about what we put into the forward path since it must do a reasonable job of reducing v_e to zero. A satisfactory solution is to simply make the "Controller" box be a large gain F, so $v_1 = Fv_e$. A differential amplifier, discussed in Section 8.2, usually performs the function of both the comparator and the Controller. With v_x and v_z as inputs, the differential amplifier has an output $v_1 = F(v_x - v_z)$. (See Figure 8–7a.)

How much gain is sufficient? Ideally we need $F = \infty$ so that the v_1 necessary to maintain the desired shaft speed corresponds to $v_e = 0$. Such a value of F is not practical, of course. In fact, if F gets too large, the system will be unstable. A finite value of F requires $v_e \neq 0$; there is some error or lack of precision in the control. The next section looks at the relationship between F and the precision of the closed-loop gain.

10.2   CLOSED-LOOP GAIN ERROR

In Chapter 8 we presented the concept of an analog world in which voltages stand for other (usually nonelectrical) parameters. This is just what is going on at the comparator. As suggested in Figure 10–1b, the input voltage v_x is proportional to the desired output, the fed-back voltage v_z is proportional to the actual output, and the error voltage v_e is proportional to the output error. Since a practical feedback control system doesn't reduce v_e to zero, there will be some output error. Now, if this error were predictable, there would be no problem; we could simply compensate by applying a slightly higher v_x. The problem is that the exact error isn't known because of uncertainty in the forward gains. Let's look at an example to get a feel for the relationships involved.

EXAMPLE 10.1 The motor control system in Figure 10–2 has a finite gain F for the controller.

Given: $v_x = 3$ V is applied to get a desired output (ideally) of 3000 rpm.

Find: the departure from 3000 rpm
1. for $F = 100$,
2. for $F = 200$,
3. for the motor transfer function changed from 50 rpm/V to 40 rpm/V due to a load.

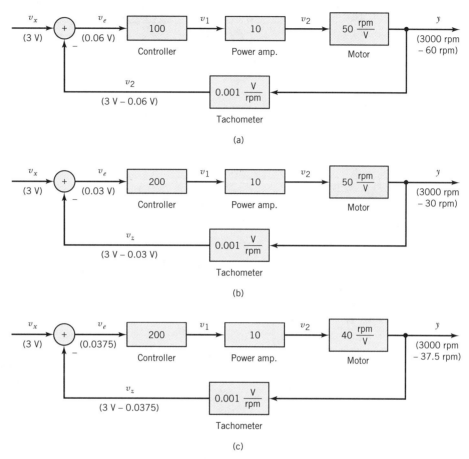

FIGURE 10–3 Examples showing that the output error decreases when the controller gain is increased. The system also becomes less sensitive to changes in the motor response due to loading as in (c).

Solution

1. Considering the top three blocks in Figure 10–3a, we write the output $y = (100 \times 10 \times 50 \text{ rpm/V})v_e = (50{,}000 \text{ rpm/V})v_e$. This means v_e must be 0.06 V in order to support the desired $y = 3000$ rpm. But 0.06 V is 2% of the applied $v_x = 3$ V. Then the output speed must actually differ from the desired $y = 3000$ rpm by 2%. That is, the error in speed is $\underline{60 \text{ rpm}}$, and the actual speed is 2940 rpm. (At this point we should recalculate v_e using $y = 2940$ rather than 3000 rpm, but it would make little difference in the result.)

2. For $F = 200$ (see Figure 10–3b), the output in terms of the error voltage is $y = (100{,}000 \text{ rpm/V})v_e$. Then a v_e of only 0.03 V is necessary to give the desired $y = 3000$ rpm. Since the error voltage is 1% of the applied 3 V, the speed error is also 1% of 3000 rpm, or $\underline{30 \text{ rpm}}$.

3. A load changes the motor transfer function to 40 rpm/V (see Figure

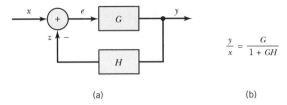

(a)   (b)

FIGURE 10–4 (a) General signal flow graph for any feedback system. (b) An expression for the closed-loop gain of the system.

10–3c). Now $y = (200 \times 10 \times 40 \text{ rpm/V})v_e = (80{,}000 \text{ rpm/V})v_e$, and $v_e = 0.0375$ V is needed to give $y = 3000$ rpm. Since 0.0375 V is 1.25% of the applied 3 V, the speed error is 1.25% of the desired 3000 rpm, or 37.5 rpm.

There are three points to be noted from this example: (i) Increasing F decreased v_e and therefore decreased the output error. (ii) Changes in the output due to loading are simply changes in the output error. (iii) Since the input v_x was constant, changes in the output were due to changes in the closed-loop gain. The conclusion is that if the error is small to begin with, then any changes in the closed-loop gain will be correspondingly small, and this can be improved by increasing F.

In the previous section we found the ideal closed-loop gain to be the reciprocal of the feedback gain, which we assume to be constant. We define the closed-loop gain error to be the difference between this ideal and the actual closed-loop gain. To find the exact relationship between this error and F, we need to look more carefully at the mathematics of the loop.

Figure 10–4 gives a general representation of a feedback control system. All the blocks in the forward path are combined into one block with gain G. The system in Figure 10–3b, for example, would have $G = 200 \times 10 \times 50 \text{ rpm/V} = 100{,}000$ rpm/V. All the blocks of the feedback path are combined in the block with gain H.* For the system in Figure 10–3b, H is simply 0.001 V/rpm. Since the parameters in the general system may be voltage, speed, temperature, phase, or any other quantity, we have used x, y, z, and e to represent them. Now, a signal flow graph is simply an engineer's way of representing a number of interrelated equations. The flow graph in Figure 10–4 represents the equations

$$e = x - z, \qquad y = Ge, \qquad z = Hy.$$

Solving these equations for the closed-loop gain we have

$$\frac{y}{x} = \frac{G}{1 + GH}. \qquad (10.1)$$

*In control theory the symbols G and H are usually used to represent forward gain and feedback gain. The symbols A and β are often used in electrical engineering, particularly when the system is entirely electronic.

This last equation is a useful summary of the flow graph in Figure 10–4; it is probably the most-used formula in control engineering. If you ever forget the equation, it is easily derived from the flow graph.

Equation 10.1 gives us a formal analysis tool to apply to the system in Figure 10–3b. There $G = 100,000$ rpm/V and $H = 0.001$ V/rpm, and from Equation (10.1) we have $y/x = 990.1$ rpm/V. If we apply an input voltage $x = 3$ V, then the shaft speed is $y = 2970$ rpm, which is 1% less than the desired result. This agrees with the informal result we found in part 2 of Example 10.1, and also indicates that variations in the forward path can change the closed-loop gain by about 1%.

In order to get an even "tighter" control, we can make $F = 2000$. This leads to $y/x = 999$—within 0.1% of the ideal $y/x = 1000$. The system is correspondingly less subject to variations.

The ideal control system would have $G = \infty$. This is not realistic, but let G approach infinity in Equation (10.1) to find the *ideal closed-loop gain*. The 1 in the denominator becomes insignificant, and G in the numerator and denominator cancel, leaving

$$y/x = 1/H \qquad \text{for} \qquad G = \infty.$$

This is the reciprocal of the feedback gain—the informal result we found in the previous section. Ideally the system in Figure 10–3 has $y/x = 1/(0.001 \text{ V/rpm}) = 1000$ rpm/V. To the extent that the system approaches the ideal, the closed-loop gain depends only on the feedback gain of the tachometer. And to the extent that the tachometer is accurate, the motor speed is accurate and insensitive to variations in the forward gain G.

To emphasize the difference between the ideal gain $1/H$ and the actual closed-loop gain, we can rewrite Equation (10.1) as

$$\frac{y}{x} = \frac{1/H}{1 + 1/GH} \approx \frac{1}{H}\left(1 - \frac{1}{GH}\right) \tag{10.2}$$

From Equation (10.2) it is clear that $GH = 100$ leads to $y/x = 0.99(1/H)$, or 1% error, as in part 2 of Example 10.1. In general,

The percentage error of the system gain from the ideal is given by $1/GH$.

Clearly, increasing F increases G, which increases GH, which decreases the error. Notice that it is not sufficient to keep G large; GH must be kept much greater than unity. This important parameter GH is called the *loop gain*; it is the product of the gains all the way around the loop.

It may seem that the error in the system gain is no problem; if the gain is 1% too low, just decrease H or increase x by 1% to compensate. The problem is that 1% may not always be the right amount. The G that gave $1/GH = 0.01$ was only the nominal G. Under other conditions at a later time the actual G may be half of the nominal, giving $1/GH = 0.02$. To argue that this error could also be compensated is to invoke another feedback loop—one with a human in it. The point of a

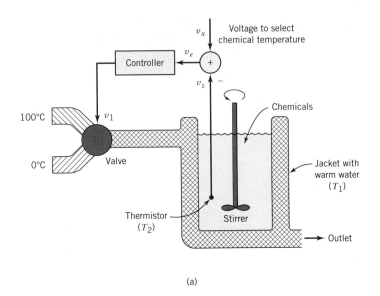

(a)

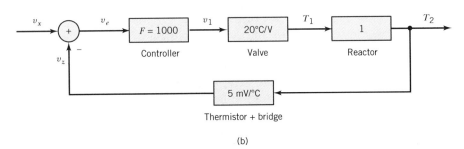

(b)

FIGURE 10–5 (a) Feedback system for controlling the temperature of a jacketed chemical reactor. (b) Signal flow graph for the system.

control system is that, once the input is applied and the system is initially adjusted, the system is to handle changing conditions without further attention.

EXAMPLE 10.2 The system in Figure 10–5a is a jacketed reactor for controlling the temperature of a chemical solution during a reaction. An electrically controlled valve mixes together two sources of water, one at 0°C and one at 100°C. The mixed water with temperature T_1 flows through a jacket and eventually brings the chemical solution to the same temperature ($T_2 = T_1$). A thermistor, together with a Wheatstone bridge (see Section 8.1), monitors the temperature T_2. The voltage v_z it develops is compared with a reference voltage v_x to provide an error voltage v_e to the Controller with gain F.

Given: Corresponding to $v_1 = 0$ the mixed water temperature is 0°C, and for $v_1 = 5$ V the temperature is 100°C. The valve's response is about linear, so we can model the valve by a transfer function of 20°C/V. The transfer function of the thermistor-bridge combination is 5 mV/°C.

Find: the ideal closed-loop gain T_2/v_x, and find the value of F that will keep the error in the closed-loop gain to 1%.

Solution: For this system, $G = F(20°\text{C/V})$, and $H = 5 \text{ mV/°C}$. For $F = \infty$, the ideal closed-loop gain would be $T_2/v_x = 1/H = \underline{200°\text{C/V}}$. To come within 1% of this, we need $GH = 100$. This requires $G = 20{,}000°\text{C/V}$ and $F = \underline{1000}$. Because of the 1% error, applying $v_x = 0.5 \text{ V}$ to the system results in $T_2 = 99$ rather than 100°C. [This can also be found directly from Equation (10.1), where $x = v_x$ and $y = T_2$.]

In Example 10.2 it was stated that the chemical solution "eventually" came to a certain temperature. We were not concerned with the time behavior of the system there, only the final steady-state condition. In the terms of Chapter 7, we assumed "the transient was over." In the next section we will be concerned with the time it takes a feedback system to respond.

10.3 TIME AND FREQUENCY RESPONSE WITH FEEDBACK

Nothing happens instantly. In the mechanical world momentum and filling processes cause delay, and in the electrical world inductance and capacitance cause delay. In a control system, it is usually the mechanical delays that dominate. The flow graph in Figure 10–3b in fact gives the *static* or dc behavior of the motor control system because it doesn't take these delays into account. The flow graph will now be modified to include the time response of each block so that the *dynamic* behavior of the system can be obtained.

Consider the block in Figure 10–3 representing the dc motor. It has a steady-state transfer function 50 rpm/V. But the dynamic behavior is governed by the differential equation $\tau(d/dt)y + y = cv_2$, where $c = 50 \text{ rpm/V}$. The *time constant* τ is determined by the moment of inertia and resistance of the armature (see Chapter 20). A typical value is $\tau = 200 \text{ ms}$ for a fractional horsepower motor. Note that the differential equation has the same form as that for a first-order electrical circuit. Then, as developed in Section 8.6, the step response of the motor to a suddenly applied voltage is the exponential $y = 50v_2(1 - e^{-t/\tau})$ plotted in Figure 10–6b. It comes up to essentially full speed in about a second (5τ). We found further in Section 8.6 that the corresponding frequency response is $\mathbf{Y}/\mathbf{V}_2 = c/(j\omega/\omega_1 + 1)$, where $\omega_1 = 1/\tau = 5 \text{ rad/s}$ (see plot in Figure 10–6a). This means that the motor speed won't fully follow fluctuations in the applied voltage faster than about $\omega_1/2\pi \approx 1 \text{ Hz}$.

With $c = 50$ and $\omega_1 = 5$, the transfer function for the motor is $\mathbf{Y}/\mathbf{V}_2 = 50/(j\omega/5 + 1)$. In accordance with standard control notation, we let s stand for $j\omega$

$$s \equiv j\omega,$$

and write the transfer function as $50/(s/5 + 1)$. The flow graph in Figure 10–7 includes the frequency dependence (and therefore the time dependence) of the

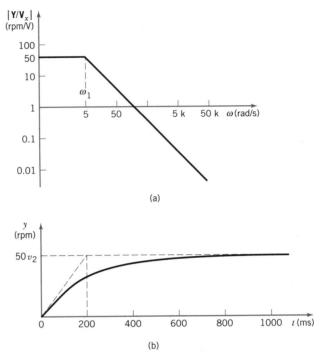

(a)

(b)

FIGURE 10–6 Responses of the open-loop motor control system in Figure 10–2a. (a) Frequency response. (b) Step response.

system in Figure 10–2 by replacing each dc gain with the transfer function as a function of s (of $j\omega$).

The controller F and the power amplifier in the system of Figure 10–2b also have cutoff frequencies, but we assume them to be high enough that they have negligible effect. The power amplifier has a gain of 10, and let the controller have a gain of $F = 200$.

The forward gain $\mathbf{G}$ is now a function of s (frequency), and it is sometimes represented by $\mathbf{G}(s)$ as a reminder. This complex gain is the product of the three transfer functions it comprises:

$$\mathbf{G}(s) = 200 \times 10 \times \frac{50}{s/5 + 1} = \frac{100,000}{s/5 + 1}. \tag{10.3}$$

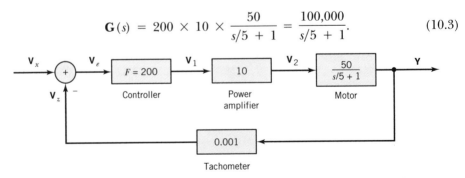

FIGURE 10–7 The system in Figure 10–3b with frequency responses included in the transfer functions. The bandwidth of the motor is 5 rad/s.

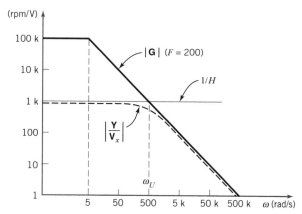

FIGURE 10–8 Frequency response curves for $|\mathbf{G}|$ and $1/H$ of the system in Figure 10–7. The closed-loop gain $|\mathbf{Y}/\mathbf{V}_x|$ follows the lower of these two curves.

The magnitude of $\mathbf{G}(s)$ is plotted as a function of ω in Figure 10–8 (remember that $s \equiv j\omega$). On the same axes we have also plotted $1/H = 1000$, assuming that the feedback has effectively unlimited frequency response (often a good assumption). We will show that the closed-loop frequency response can be obtained from these two curves.

The expression

$$\frac{\mathbf{Y}}{\mathbf{X}} = \frac{\mathbf{G}}{1 + \mathbf{G}H} \qquad (10.1')$$

for the closed-loop transfer function still holds when $\mathbf{G}$ is complex and a function of frequency. Consider two ranges of frequencies in Figure 10–8: the range well below 500 rad/s where $1/H << |\mathbf{G}|$, and the range well above 500 rad/s where $|\mathbf{G}| << 1/H$. It is easy to show from Equation (10.1) that in the first range $|\mathbf{Y}/\mathbf{X}| \approx 1/H$ and in the second range $|\mathbf{Y}/\mathbf{X}| \approx |\mathbf{G}|$. The rule, then, is that $|\mathbf{Y}/\mathbf{X}|$ follows the lower of the two curves. This is illustrated in Figure 10–8. In the range around $|\mathbf{G}| = 1/H$ the rule gives a poorer approximation, but it is as good as a Bode plot. Then the closed-loop response can be expressed as

$$|\mathbf{Y}/\mathbf{X}| = \text{lesser of } \{1/H, |\mathbf{G}|\}. \qquad (10.4)$$

Compare the closed-loop response $|\mathbf{Y}/\mathbf{V}_x|$ in Figure 10–8 with the open-loop response of the motor in Figure 10–6a. The cutoff frequency is no longer 5 rad/s but now 500 rad/s—an increase by a factor of 100. Let ω_1 be the bandwidth without feedback, and let ω_U be the closed-loop bandwidth (the frequency at the intersection of $|\mathbf{G}|$ and $1/H$). It can be shown that

$$\omega_U = (1 + G_0 H)\omega_1, \qquad (10.5)$$

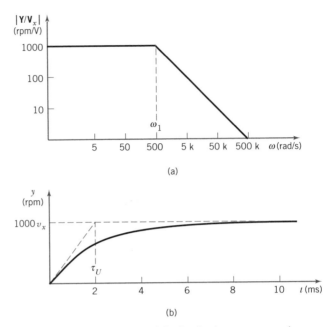

FIGURE 10-9 Responses of the feedback motor control system in Figure 10-7. (a) Frequency response. (b) Step response. Because $G_0H = 100$, these responses are 100 times faster than those in Figure 10-6.

where G_0 is the dc gain of $\mathbf{G}(s)$:

$$G_0 \equiv \mathbf{G}(0).$$

In our case $1 + G_0H = 1 + 100{,}000 \times 0.001 = 101$. This rule holds when the second corner frequency of $|\mathbf{G}|$ is high enough that it doesn't affect the bandwidth. The best rule is always to draw a picture rather than remember a formula.

What is the effect of feedback on the step response of the system? The frequency response curve of $|\mathbf{Y}/\mathbf{V}_x|$ in Figure 10-8 is a first-order response with a high-frequency cutoff $\omega_U = 500$ rad/s. A Bode plot of the response is shown in Figure 10-9a. The corresponding step response (see Section 8.6) is the exponential shown in Figure 10-9b. The time constant of the exponential is $\tau_U = 1/\omega_U = 1/(500$ rad/s$) = 2$ ms, which is 100 times faster than the response of the open-loop motor.*

In the previous section we found that the percentage error of the closed-loop gain is given by $1/GH$. Now that $\mathbf{G}$ is a function of frequency, we need to specify the frequency for which the error is being found. Usually we are interested in the percentage error of the static (or dc) closed-loop gain, given by $1/G_0H$.

In this section we have learned another reason to keep the dc loop gain G_0H

---

*This speed of response holds if the step at v_x is not too large. Otherwise it is possible for the feedback control to drive the power amp into saturation in its attempt to follow a large change in v_x quickly.

high: to increase the speed and bandwidth of the system. But there is a limit; with too much gain, the system can become unstable.

10.4   STABILITY

A system is defined as unstable if it supports sustained oscillation. This happens when a public address system "squeals." Oscillation is clearly unacceptable, but in most applications performance is unacceptable even if the system has ringing that dies out. We saw this effect in Section 7.4 for an underdamped second-order system. If the loop gain of a feedback control system is increased too far, it can become an underdamped system, as in the following example.

Let us decrease the closed-loop gain error and increase the bandwidth of the motor system in Figure 10–7 by increasing the dc loop gain G_0H. We do this by increasing the dc gain of the controller from 200 to 2000. For a gain of 2000, a typical bandwidth for the controller is 5000 rad/s. (In Chapter 14 we will see that the bandwidth often varies inversely with the dc gain. We assume a *gain-bandwidth product* of 10 Mrad/s here.) The transfer function for the controller is then $\mathbf{F} = 2000/(s/5000 + 1)$. The frequency response of the controller is plotted in Figure 10–10a, along with the responses of the other blocks in the forward path. The forward gain is the product of these three responses:

$$\mathbf{G} = \frac{2000 \times 10 \times 50}{(s/5000 + 1)(s/5 + 1)} = \frac{25 \times 10^9}{s^2 + 5000s + 25000}. \qquad (10.6)$$

The magnitude of $\mathbf{G}$ is plotted as a function of frequency in Figure 10–10b along with $1/H$. As before, the closed-loop response $|\mathbf{Y}/\mathbf{V}_x|$ follows the lower of the two curves. However, this time we will have to be a little more careful in the vicinity of $|\mathbf{G}| = 1/H$. Because there is a second corner frequency in $|\mathbf{G}|$ low enough to be important, the closed-loop response actually has a little *peaking* in the vicinity of 5000 rad/s. This means $|\mathbf{Y}/\mathbf{V}_x|$ goes above its dc level (see the dashed curve in Figure 10–10b). The closed-loop bandwidth is not so clearly defined for this *second-order system*. For the *first-order system* in Figure 10–8, the closed loop has a 3-dB bandwidth at the intersection of $|\mathbf{G}|$ and $1/H$. For second-order systems we will find that this continues to be a good working definition:

> The closed-loop bandwidth ω_U is the frequency at the intersection of $|\mathbf{G}|$ (*its Bode plot*) with $1/H$.

What does this peaking in the frequency response indicate about the transient response of the system? As was done in Section 8.6, we will obtain the transient response from the frequency response. From $H = 0.001$ and from Equation (10.5) we have

$$\frac{\mathbf{Y}}{\mathbf{V}_x} = \frac{\mathbf{G}}{1 + \mathbf{G}H} = \frac{25 \times 10^9}{s^2 + 5000s + 25 \times 10^6} = \frac{25 \times 10^9}{s^2 + 2\alpha s + \omega_0^2}$$

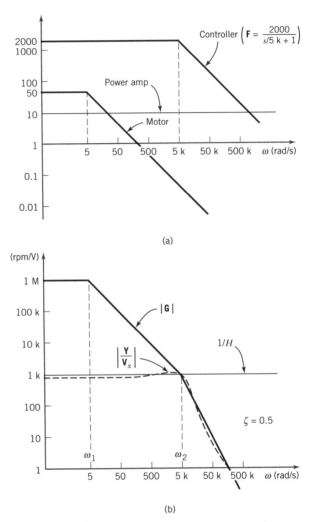

FIGURE 10–10 (a) Frequency response curves for the components of the forward path in Figure 10–7 when the controller transfer function is 2000/(s/5k + 1). (b) Frequency response curves for |G| and 1/H of the system in Figure 10–7. Because the curves intersect at the second break in |G|, the system is on the edge of stability.

where

$$\alpha = 2500, \qquad \omega_0 = 5000.$$

We get the differential equation for the transient response by substituting the differential operator d/dt for $j\omega$ (or s, in the present notation) in the denominator, and then multiplying through by the output variable y:

$$\left(\frac{d}{dt}\right)^2 y + 2\alpha \left(\frac{d}{dt}\right) y + \omega_0^2 y = 0.$$

This is the same form that was used to study the transient response of second-order systems in Section 7.4. The first thing to notice is that $\alpha < \omega_0$ for our case, so the system is underdamped. The transient response is therefore given by Equation (7.10).

For convenience, transient responses to a step input are plotted for various amounts of damping in Figure 10–11. The *damping ratio* is

$$\zeta \equiv \alpha/\omega_0,$$

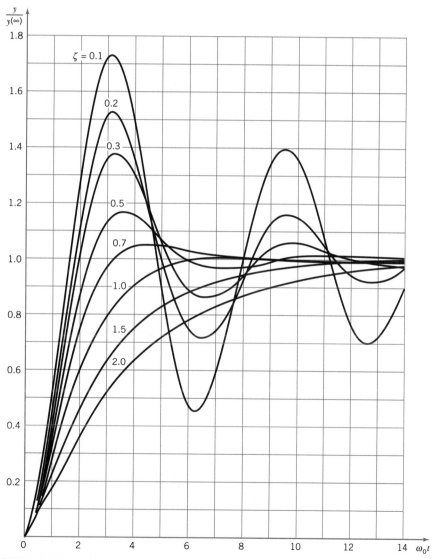

FIGURE 10–11 Normalized step responses of second-order systems. ζ is the damping ratio. The ringing for $\zeta < 0.5$ is usually considered excessive.

in accordance with standard notation. For our case $\zeta = 0.5$, so the step response overshoots* the final value by about 17% here. The rise time (10% to 90%) takes from $\omega_0 t = 0.5$ to $\omega_0 t = 2.2$. Since ω_0 is 5000, this corresponds to a rise time of 0.34 ms. The settling time (after which the response remains within 2% of the final value) is $\omega_0 t = 8.5$, or $t = 1.7$ ms.

Is this step response acceptable? There is no definite rule since what is acceptable depends on the application. An overshoot of 17% is probably just starting to get excessive. Engineers often use

$$\zeta \geq 0.5 \tag{10.7}$$

as the condition for acceptable response. It can be shown that $\zeta = 0$ corresponds to a sinusoidal step response—a sustained oscillation, which is our definition of instability. Therefore, the damping ratio ζ is a measure of the margin against instability. The greater ζ is, the more stable the system is.

To illustrate a system that is clearly unacceptable, we will look at a case where the second breakpoint in $|\mathbf{G}|$ occurs above the intersection of the curves. Suppose for the motor system in Figure 10–7 the dc gain of the controller is increased to 10,000, and the bandwidth of the controller is (10 Mrad/s)/10,000 = 1000 rad/s. Then the transfer function of the controller is $10,000/(s/1000 + 1)$. Figure 10–12

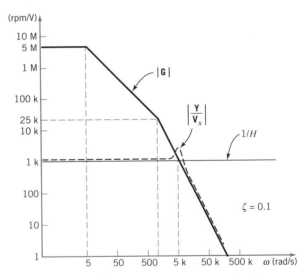

FIGURE 10–12 Frequency response curves for the system in Figure 10–7 when the controller transfer function is 10,000/(s/1k + 1). Because the second break in $|\mathbf{G}|$ comes before the intersection of the two curves, the system is unstable. This is seen in the peaking in the closed-loop response.

---

*See Section 8.6 for definitions of the step response parameters.

compares a plot of the new $|\mathbf{G}|$ with the plot of $1/H$. The second corner frequency for $|\mathbf{G}|$ is 1000 rad/s—well before the crossing of the two curves at 5000 rad/s. The result is that $|\mathbf{Y}/\mathbf{V}_x|$ has considerable peaking around 5000 rad/s, as shown in Figure 10–12. The reader can show that the corresponding differential equation for the transient has $\alpha = 500$ rad/s and $\omega_0 = 5000$ rad/s. Then $\zeta = 0.1$, and the overshoot of the step response is 73%, as can be seen from Figure 10–11. This is clearly unacceptable.

In the following example we analyze the dynamics of another system.

EXAMPLE 10.3 We will take into account the response times of the components of the jacketed reactor system in Figure 10–5. For a step change in the valve position, it takes 1 s for the new-temperatured water to flush out the old. This can be approximated by an exponential time constant of $\tau_U = 1$ s and a cutoff frequency of $\omega_U = 1/\tau_U = 1$ rad/s. Therefore, the transfer function of the valve is

$$\frac{\mathbf{T}_1}{\mathbf{V}_1} = \frac{20°\text{C/V}}{s/1 \,+\, 1}.$$

Because of the thermal mass of the chemical solution, there is a 10-s time constant for the response of the chemical temperature $\mathbf{T}_2$ to a step change in the water temperature $\mathbf{T}_1$. This corresponds to a cutoff frequency of 0.1 rad/s and a transfer function of

$$\frac{\mathbf{T}_2}{\mathbf{T}_1} = \frac{1}{s/0.1 \,+\, 1}.$$

for the reactor (see Figure 10–13a).

The controller has a bandwidth so great that it has essentially no effect on the analysis. Its gain F can be considered independent of frequency.

Find: the gain F for the controller that will result in a damping ratio of $\zeta = 0.5$. Find the closed-loop bandwidth ω_U and the closed-loop static gain error.

Solution: The forward gain is given by

$$\mathbf{G} = F(\mathbf{T}_1/\mathbf{V}_1)(\mathbf{T}_2/\mathbf{T}_1) = F(\mathbf{T}_2/\mathbf{V}_1).$$

We don't know F, but we can plot the portion we do know—the product of the two transfer functions:

$$\frac{\mathbf{T}_2}{\mathbf{V}_1} = \frac{20°\text{C/V}}{(s/1 \,+\, 1)(s/0.1 \,+\, 1)}.$$

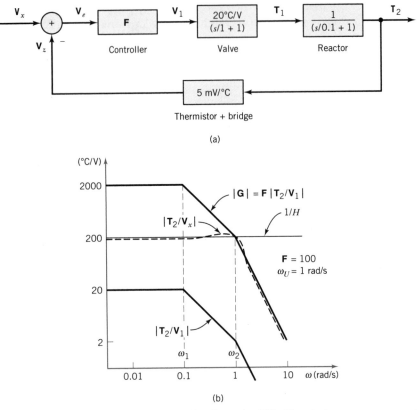

FIGURE 10–13 (a) Feedback system for Example 10.3. Chemical reactor system of Figure 10–5 with frequency cutoffs for the valve and reactor. (b) Frequency responses with F selected for $\omega_U = \omega_2$, or for $\zeta = 0.5$. ω_U is the frequency at which $|\mathbf{G}|$ (its Bode plot) intersects $1/H$.

The magnitude of this function, plotted in Figure 10–13b, begins with a dc gain of 20°C/V. At $\omega = 0.1$ rad/s the magnitude begins falling a decade per decade until $\omega = 1$ rad/s, where it has reached 2°C/V. Then it descends at two decades per decade.

The reciprocal of the feedback gain is $1/H = 200$°C/V, which is also plotted in Figure 10–13b. We saw in Figure 10–10b that the desired $\zeta = 0.5$ corresponds to $|\mathbf{G}|$ crossing $1/H$ at the second breakpoint—at 1 rad/s in this case. Thus we must choose F to raise $|\mathbf{T}_2/\mathbf{V}_1|$ by a factor of 100 to the position $|\mathbf{G}| = F|\mathbf{T}_2/\mathbf{V}_1|$ shown in Figure 10–13b. The solution, then, is $F = \underline{100}$, and the forward transfer function is

$$\mathbf{G} = \frac{2000}{(s/1 + 1)(s/0.1 + 1)} = \frac{200}{s^2 + 1.1s + 0.1}.$$

The dc forward gain is $G_0 = 2000$°C/V, the dc loop gain is $G_0 H =$

$(2000°C/V)(5 \text{ mV}/°C) = 10$, and the closed-loop static gain error is $1/G_0H = 0.1 = \underline{10\%}$. The closed-loop transfer function is

$$\frac{T_2}{V_x} = \frac{G}{1 + GH} = \frac{200}{s^2 + 1.1s + 1.1} = \frac{200}{s^2 + 2\alpha s + \omega_0^2},$$

where $\alpha = 0.55$ and $\omega_0 = 1.05$ rad/s. Therefore $\zeta = \alpha/\omega_0 = 0.524$, which is close to the design value of 0.5.

The dashed curve in Figure 10–13b is the closed-loop response, approximately following the lower of $|G|$ or $1/H$. The closed-loop bandwidth is $\omega_U = 1$ rad/s, at the intersection of $|G|$ with $1/H$. This coincides with the second breakpoint of $|G|$ at $\omega_2 = 1$ and approximately with $\omega_0 = 1.05$ rad/s. For other dampings ($\zeta \neq 0.5$), we will find $\omega_U \neq \omega_0$. ∎

The dc loop gain G_0H in this example was only 10, compared with 100 in Example 10.2. As a result, the static gain error is only 10% rather than 1%. Also, in approximate agreement with Equation (10.5), the bandwidth was increased by a factor of only 10—from $\omega_1 = 0.1$ rad/s to $\omega_U = 1$ rad/s. But any attempt to increase G_0H by increasing F would reduce the stability. It is possible to improve the dc loop gain while maintaining the stability, but the solution requires a more complicated Controller function. The Controller must include compensation to modify the breakpoints that restrict our design.

10.5   COMPENSATION

The design problem in Example 10.3 lends some insight to the relationship between the breakpoint frequencies and the minimum static gain error. The dc loop gain G_0H can be seen in Figure 10–13b as the factor of 10 separating $G_0 = 2000°C/V$ from $1/H = 200°C/V$. Because $|G|$ descends at a decade per decade from G_0 at ω_1 to $1/H$ at ω_U, we also have

$$G_0H = \omega_U/\omega_1 \qquad (10.8)$$

[compare Equation (10.5)]. For stability we desire $\zeta \geq 0.5$, which requires

$$\omega_2 \geq \omega_U. \qquad (10.9)$$

Then in designing for a stability of $\zeta \geq 0.5$ we are restricted to a dc loop gain of

$$G_0H \leq \omega_2/\omega_1, \qquad (10.10)$$

where ω_1 is the first breakpoint frequency and ω_2 is the second breakpoint frequency of $|G|$. If we are to increase the dc loop gain in Example 10.3 to more than 10, we must first increase the ratio of the breakpoint frequencies to more than 10. The technique of compensation is to move one of the breakpoint frequencies

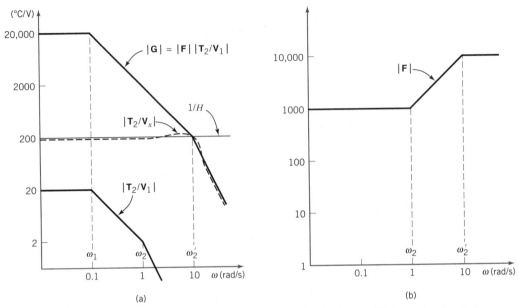

FIGURE 10-14 Frequency responses for Example 10.4. (a) $|\mathbf{T_2}/\mathbf{V_1}|$ and $1/H$ from the chemical reactor system in Figure 10-13. The desired $|\mathbf{G}|$ gives $G_0H = 100$, $\omega_U = 10$ rad/s, and $\zeta = 0.5$. (b) The $|\mathbf{F}|$ to achieve the desired $|\mathbf{G}|$. It effectively moves the original breakpoint in $|\mathbf{G}|$ at ω_2 out to ω_2'.

by choosing the Controller transfer function to cancel a breakpoint and introducing a new one at a different frequency. We illustrate this in the following example.

EXAMPLE 10.4

Given: the chemical reactor in Example 10.3 with breakpoints in the forward gain at $\omega_1 = 0.1$ rad/s and $\omega_2 = 1$ rad/s.

Find: a transfer function for the Controller so the dc loop gain is 100, the closed-loop bandwidth is $\omega_U = 10$ rad/s, and the damping ratio is $\zeta = 0.5$.

Solution: According to Equation (10.10), a dc loop gain of 100 requires that $\omega_2/\omega_1 = 100$ if we are to maintain $\zeta = 0.5$. Also, to achieve the desired bandwidth, Equation (10.9) requires $\omega_2 \geq 10$ rad/s. Therefore, we must raise ω_2 to $\omega_2' = 10$ rad/s.

The desired forward gain $|\mathbf{G}|$ is plotted in Figure 10-14a. The dc gain G_0 is a factor of 100 above $1/H$, it has breakpoints at $\omega_1 = 0.1$ rad/s and $\omega_2' = 10$ rad/s, and the second breakpoint lies on $1/H$ (or $\omega_2' = \omega_U$) so that $\zeta = 0.5$. This desired transfer function is

$$\mathbf{G} = \frac{20,000°\text{C/V}}{(s/0.1 + 1)(s/10 + 1)}.$$

The portion of the forward gain that we already know is

$$\frac{\mathbf{T}_2}{\mathbf{V}_1} = \frac{20°C/V}{(s/0.1 + 1)(s/1 + 1)},$$

whose magnitude is also plotted in Figure 10–14a. Since $\mathbf{G} = \mathbf{F}(\mathbf{T}_2/\mathbf{V}_1)$, the portion to be designed is

$$\mathbf{F} = \frac{\mathbf{G}}{\mathbf{T}_2/\mathbf{V}_1} = \frac{1000(s/1 + 1)}{(s/10 + 1)},$$

whose magnitude is plotted in Figure 10–14b. Notice that $|\mathbf{F}|$ is the ratio (distance on a log plot) between the $|\mathbf{G}|$ and $|\mathbf{T}_2/\mathbf{V}_1|$ plots in Figure 10–14a. The s function in the numerator causes the break upward in the $|\mathbf{F}|$ plot, and the s function in the denominator causes the break downward.

The dashed curve in Figure 10–14a is the closed-loop response—the lower of $|\mathbf{G}|$ or $1/H$. As desired, the bandwidth is 10 rad/s. ▨

10.6   DIFFERENTIATION AND INTEGRATION

Many control systems include integration, whose effect is similar to that of a low-pass transfer function. Just as velocity u is the time-derivative of position y, it is equivalent to say that position is the time-integral of velocity. Starting with y, differentiation yields u, and starting with u, integration yields y:

$$u = dy/dt,$$

$$y = \int u \, dt.$$

Substituting s for the d/dt operator gives the transfer functions for differentiation and integration in the frequency domain:

$$\mathbf{U} = s\mathbf{Y} = j\omega\mathbf{Y},$$

$$\mathbf{Y} = (1/s)\mathbf{U} = (1/j\omega)\mathbf{U}.$$

The magnitudes of the functions are simply

$$|s| = \omega, \qquad \text{(differentiation)}$$
$$|1/s| = 1/\omega. \qquad \text{(integration)}$$

Recall that a low-pass transfer function has the form $1/(j\omega + \omega_1)$, where ω_1 is the breakpoint frequency. Therefore, integration can be treated as a low-pass transfer function with a breakpoint at $\omega_1 = 0$. When a control system includes integration, then, the first "breakpoint" is at zero frequency (dc), although it doesn't appear as a break in the plot. If we agree to this nomenclature, then the position of the "second" breakpoint ω_2 still defines stability and damping.

EXAMPLE 10.5 An *X-Y* plotter is controlled by two electrical signals—one for each axis. A feedback control system, shown in Figure 10–15a is used to increase the accuracy and speed. We will look at the control for the *Y* axis.

Given: A motor drives the pen at a velocity u of 10 cm/s per volt. Because of the mass of the pen and its transport system, the frequency response of the velocity cuts off at 40 rad/s. Therefore the transfer function of the pen drive is $10/(s/40 + 1)$. The position y is the integral of the speed u. A linear potentiometer senses the position y and develops 0.2 V/cm as a feedback signal v_z.

Find: a transfer function **F** for the Controller that gives a closed-loop bandwidth $\omega_U = 200$ rad/s and a damping ratio of $\zeta = 0.5$. Find the percentage error of the static gain.

Solution: The transfer function $\mathbf{U}/\mathbf{V}_1 = 10/(s/40 + 1)$ of the pen drive has the magnitude plot shown in Figure 10–15b. The magnitude of the integra-

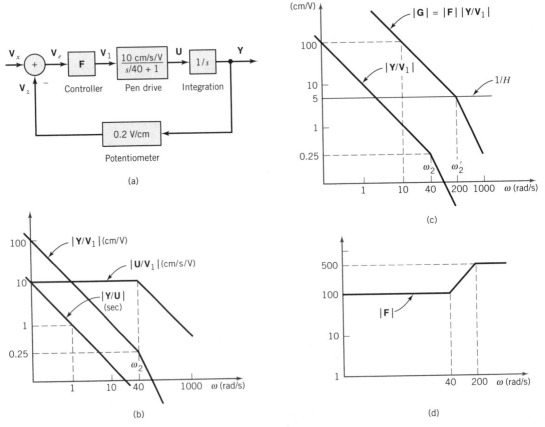

FIGURE 10–15 X-Y plotter system for Example 10.5. (a) Signal flow graph of the Y-drive system. The Integration box represents a mathematical relationship, not the transfer function of a physical device. (b) Components of the forward transfer function. The $|\mathbf{Y}/\mathbf{V}_1|$ found in (b). (c) The desired $|\mathbf{G}|$ compared with the $|\mathbf{Y}/\mathbf{V}_1|$ found in (b). The necessary $|\mathbf{F}|$ is the ratio of the two curves in (c). On a log plot the ratio is seen as the vertical separation between the curves.

tion function $Y/U = 1/s$ is simply a straight line (on a log plot) descending a decade per decade and passing through unity gain at $\omega = 1$ rad/s. The product of these two functions is

$$\frac{Y}{V_1} = \frac{10}{s(s/40 + 1)},\qquad (10.11)$$

whose magnitude is also plotted in Figure 10–15b. Note that it tends to infinity as ω approaches zero ($\omega = 0$ can't be shown on a log plot). We can think of the first breakpoint being there: $\omega_1 = 0$. The product of this function with F will give the forward gain G.

What is the desired G? Figure 10–15c shows the position of $|G|$ relative to $1/H$ to give the desired bandwidth and stability. The intersection of the two plots is at $\omega_U = 200$ rad/s, and $\omega_2' = \omega_U$ for a damping of $\zeta = 0.5$. The corresponding transfer function is

$$G = \frac{1000}{s(s/200 + 1)}.$$

We can check the value of the numerator as follows. Pick a frequency well below $\omega_2' = 200$, say at $\omega = 10$. Then the term $s/200$ is negligible, and $|G|$ becomes $1000/|s| = 1000/\omega = 1000/10 = 100$, which is the value of $|G|$ plotted for $\omega = 10$ in Figure 10–15c.

Since $G = F(Y/V_1)$, the controller function F provides the missing factor between Y/V_1 and the desired G:

$$F = \frac{G}{Y/V_1} = \frac{100(s/40 + 1)}{(s/200 + 1)},$$

whose magnitude is plotted in Figure 10–15d.

The percentage static gain error is given by $1/G_0 H$, where G_0 is the dc gain of G. But G is infinite at dc. Therefore $1/G_0 H = 0$. ▪

We see from this example an important result of integration:

There is no static gain error for a control system with integration in the forward gain.

In practice there is position error because of dc offset (see Section 8.7), and there is some gain error because of inaccuracy of the potentiometer. But there is no error due to loading on the pen or changes in the pen drive's transfer function.

10.7   OTHER DAMPING RATIOS

So far, we have been designing systems for a damping ratio of $\zeta = 0.5$ by making $\omega_2' = \omega_U$; the second breakpoint lies on the intersection of $|G|$ with $1/H$. For more stable designs with less overshoot in the step response, the breakpoint must lie below the $1/H$ line. An equivalent statement is: the second breakpoint frequency ω_2' must be greater than the intersection frequency ω_U. It can be shown that, for $\omega_1 \ll \omega_U$,

$$\zeta = 0.5\sqrt{\omega_2'/\omega_U}, \qquad (10.12)$$

$$\omega_0 = \sqrt{\omega_2'\omega_U}. \qquad (10.13)$$

(See problem P10.6.) We see the familiar result

$$\zeta = 0.5 \qquad \text{for} \qquad \omega_2' = \omega_U.$$

Other common dampings are

$$\zeta = 0.7 \qquad \text{for} \qquad \omega_2' = 2\omega_U,$$

$$\zeta = 1.0 \qquad \text{for} \qquad \omega_2' = 4\omega_U.$$

The damping $\zeta = 0.7$ is the smallest damping with no peaking in the frequency response. The damping $\zeta = 1.0$ is the smallest damping with no overshoot in the step response.

EXAMPLE 10.6 As an example of the tradeoffs involved in achieving greater damping, we will redesign the system in Example 10.5 for $\zeta = 1$.

Find: a Controller transfer function to give a closed-loop bandwidth $\omega_U = 200$ rad/s and a damping ratio $\zeta = 1$. Find the step response overshoot, rise time, and settling time.

Solution: Equation (10.12) says that $\zeta = 1.0$ requires $\omega_2' = 4\omega_U = 800$ rad/s. The desired forward gain $|\mathbf{G}|$ passes through $1/H$ at $\omega_U = 200$ rad/s, and it breaks at $\omega_2' = 800$ rad/s (see plot in Figure 10–16a). The corresponding transfer function is

$$\mathbf{G} = \frac{1000}{s(s/800 + 1)},$$

The given part of $\mathbf{G}$ is still the transfer function $\mathbf{Y}/\mathbf{V}_1$ in Equation (10.11). Then $\mathbf{F}$ supplies the missing factor:

$$\mathbf{F} = \frac{\mathbf{G}}{\mathbf{Y}/\mathbf{V}_1} = \frac{100(s/40 + 1)}{(s/800 + 1)},$$

whose magnitude is plotted in Figure 10–16b.

The $\zeta = 1.0$ curve in Figure 10–11 shows the step response of the system. This curve has an overshoot of 0%. It intersects 0.1 at $\omega_0 t = 0.5$, 0.9 at $\omega_0 t = 3.9$, and 0.98 at $\omega_0 t = 6.5$. Therefore, the rise time is $(3.9 - 0.5)/\omega_0$,

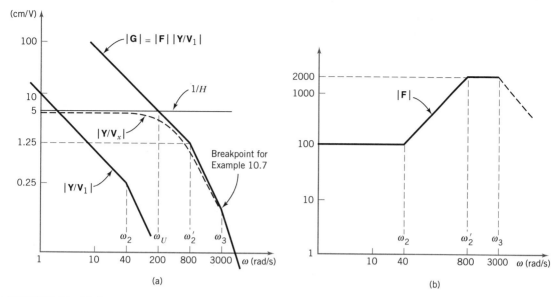

FIGURE 10–16 Frequency responses for Examples 10.6 and 10.7. The X-Y plotter system in Figure 10.15 is redesigned for $\zeta = 1.0$. (a) The second breakpoint of $|\mathbf{G}|$ is now at ω_2' a factor of 4 greater than ω_U. (b) The $|\mathbf{F}|$ to achieve the desired $|\mathbf{G}|$. By moving ω_2' out to 800 rad/s, we have required a gain of 2000 from the controller. The breakpoint at ω_3 applies to Example 10.7.

and the settling time is $6.5/\omega_0$. From Equation (10.13), $\omega_0 = \sqrt{200 \times 800}$ = 400 rad/s. Then the rise time is <u>8.5</u> ms, and the settling time is <u>16.3 ms</u>. ■

Note that the maximum gain of $|\mathbf{F}|$ in Example 10.6 is 2000 rather than the 500 in Example 10.5. This higher gain is more difficult to achieve, especially at higher frequencies. Therefore, the general rule is to

> *Keep the damping ratio as small as possible, consistent with design restrictions on overshoot and peaking.*

10.8   PHASE MARGIN

The *Barkhausen criterion* for oscillation (instability) of a feedback system is that the loop gain be unity. Then the signal v_e will produce (sustain) itself with no input at v_x. We have been speaking of $\mathbf{G}(s)\,H$ as the loop gain. If we actually go all the way around the loop (see Figures 10–1 and 10–4), there is a minus sign connected with the comparator. Therefore, the Barkhausen criterion says that a feedback system will oscillate if there is some $s = j\omega$ for which

$$\mathbf{G}(s)\,H = -1 = 1\,\underline{/-180°}. \qquad (10.14)$$

The magnitude of $\mathbf{G}(s)H$ is unity where the plot of $|\mathbf{G}|$ intersects $1/H$. Since this point is difficult to solve for, we will approximate the point by ω_U, where the *Bode plot* of $|\mathbf{G}|$ intersects $1/H$:

$$|\mathbf{G}(j\omega_U)H| \approx 1.$$

Half of the Barkhausen criterion in Equation (10.14) is satisfied at ω_U. The other half is that the angle of $\mathbf{G}(s)H$ be $-180°$:

$$\text{ang}[G(j\omega_U)H] = -180°. \tag{10.15}$$

We will find this angle for a second-order system (two breakpoints in $\mathbf{G}$) and extend it to higher-order systems in the next section.

A second-order system has $\mathbf{G}(s)H$ of the form

$$\mathbf{G}(s)H = \frac{G_0 H}{(s/\omega_1 + 1)(s/\omega_2' + 1)} = \frac{\omega_U/\omega_1}{(s/\omega_1 + 1)(s/\omega_2' + 1)},$$

where we have used Equation (10.8) in the numerator. We want the angle of this expression for $s = j\omega_U$:

$$\mathbf{G}(j\omega_U)H = \frac{\omega_U/\omega_1}{(j\omega_U/\omega_1 + 1)(j\omega_U/\omega_2' + 1)}.$$

For $\omega_1 \ll \omega_U$,

$$\mathbf{G}(j\omega_U)H \approx \frac{\omega_U/\omega_1}{(j\omega_U/\omega_1)(j\omega_U/\omega_2' + 1)} = \frac{1}{j(j\omega_U/\omega_2' + 1)},$$

$$\text{ang}[\mathbf{G}(j\omega_U)H] = -90° - \arctan(\omega_U/\omega_2'). \tag{10.16}$$

(See Appendix B for determining the angle of a complex number.) Since Equation (10.8) holds only for $\omega_2' \gtrsim \omega_U$, Equation (10.16) also holds only for $\omega_2' \gtrsim \omega_U$, which means that $|\mathbf{G}|$ crosses $1/H$ as it descends at a decade per decade. Since $|\mathbf{G}|$ must be descending at *two* decades per decade for the angle of $\mathbf{G}$ to be $-180°$, the system is not unstable. But how far is the angle from $-180°$ at ω_U? This is the *phase margin* ϕ_m against instability:

$$\phi_m \equiv \text{ang}[\mathbf{G}(j\omega_U)H] - (-180°)$$

$$= 90° - \arctan(\omega_U/\omega_2') \tag{10.17}$$

$$= \arctan(\omega_2'/\omega_U).$$

Using Equation (10.12), we can relate the phase margin to the damping ratio:

$$\phi_m = \arctan(4\zeta^2), \tag{10.18}$$

$$\zeta = 0.5\sqrt{\tan \phi_m}. \tag{10.19}$$

For example, $\zeta = 0.5$ corresponds to $\phi_m = 45°$, and $\zeta = 1$ corresponds to $76°$.

If we know the phase margin ϕ_m and the bandwidth ω_U, we can find the natural frequency ω_0. From Equations (10.13) and (10.17),

$$\omega_0 = \omega_U \sqrt{\tan \phi_m}. \tag{10.20}$$

10.9   HIGHER-ORDER SYSTEMS

So far, we have looked at systems with two breakpoints in the forward gain $\mathbf{G}$. The resulting closed-loop transfer function has the form

$$\frac{\mathbf{Y}}{\mathbf{V}_x} = \frac{A}{s^2 + 2\zeta\omega_0 s + \omega_0^2},$$

with a second-order polynomial in the denominator (see the form in Example 10.3). Therefore, they are called *second-order systems*. The natural frequency ω_0 is defined in terms of the constant term in the polynomial, and the damping coefficient ζ is defined in terms of the coefficient of the s term.

When $\mathbf{G}$ has three breakpoints, the denominator of the closed-loop transfer function becomes a third-order polynomial, and the feedback control is a third-order system. There is no ω_0 or ζ defined for a third-order polynomial. But a third-order system does have a phase margin, and Equations (10.19) and (10.20) can provide some indication of how the system will behave.

An nth-order system has a forward gain of the form

$$\mathbf{G}(s) = \frac{G_0}{(s/\omega_1 + 1)(s/\omega_2' + 1)(s/\omega_3 + 1) \cdots (s/\omega_n + 1)},$$

and in a development similar to that of Equation (10.17),

$$\phi_m \approx \arctan(\omega_2'/\omega_U) - \arctan(\omega_U/\omega_3) - \cdots - \arctan(\omega_U/\omega_n) \tag{10.21}$$

This is a conservative estimate; the actual phase margin will be greater because $|\mathbf{G}|$ actually crosses $1/H$ to the left of ω_U.

EXAMPLE 10.7 The control system designed in Example 10.6 assumed that the controller's transfer function $\mathbf{F}$ had no upper cutoff frequency. That is, the gain $|\mathbf{F}|$ = 2000 was assumed to extend from $\omega = 800$ rad/s to infinity. But, in practice, all amplifiers are bandlimited. If the gain-bandwidth product of the amplifier used for the controller is 6 Mrad/s, then the gain of 2000 will encounter a cutoff at $\omega_3 = 3000$ rad/s (see dashed portion of $|\mathbf{F}|$ in Figure 10–16b). This introduces a third breakpoint in $|\mathbf{G}|$ at $\omega_3 = 3000$ rad/s (see Figure 10–16a).

Given: a third-order system with breakpoints for $|\mathbf{G}|$ at $\omega_1 = 0$, $\omega_2' = 800$, and $\omega_3 = 3000$. The Bode plot of $|\mathbf{G}|$ crosses $1/H$ at $\omega_U = 200$.

Find: the phase margin ϕ_m. Find the damping ratio of a second-order system with the same phase margin. Assuming similar behavior for the third-order system, estimate the rise time, overshoot, and settling time of the system's step response.

Solution: From Equation (10.21) the phase margin is approximated by

$$\phi_m \approx \arctan(800/200) - \arctan(200/3000)$$

$$= \arctan(4) - \arctan(0.067)$$

$$= 76° - 3.8° = \underline{72.2°}.$$

From Equation (10.19) a second-order system with this phase margin would have a damping ratio of $\zeta = 0.5\sqrt{\tan 72.2°} = \underline{0.88}$. This damping is between the curves for $\zeta = 0.7$ and $\zeta = 1.0$ in Figure 10–11. The reader should sketch a curve for $\zeta = 0.88$ on the figure (about halfway between the two, but a little closer to the $\zeta = 1.0$ curve). This curve has an overshoot of about $\underline{1\%}$. It intersects 0.1 at $\omega_0 t = 0.5$, 0.9 at $\omega_0 t = 3.2$, and 0.98 at $\omega_0 t = 4.3$. Therefore the rise time is $(3.2 - 0.5)/\omega_0$ and the settling time is $4.3/\omega_0$. From Equation (10.20), $\omega_0 = 200\sqrt{\tan 72.2°} = 353$ rad/s. Then the rise time is $\underline{7.65}$ ms, and the settling time is $\underline{12.2 \text{ ms}}$.

Compare the results of Example 10.7 with those of Example 10.6. The location of a third breakpoint in **G** a factor of 15 beyond ω_U has caused the overshoot to increase from 0% to 1%, the rise time to decrease by 10%, and the settling time to decrease by 25%. If the design intent is to maintain the stability of Example 10.6 (0% overshoot) in the face of the third breakpoint in Example 10.7, the Controller transfer function **F** could be redesigned to increase ω_2' so the phase margin is again 76°.

10.10   CASE STUDY

A submarine is loaded with weights so that it has about neutral buoyancy. Its weight is then 8 million pounds. Ballast tanks are filled or emptied of water to change this weight, making the boat either sink or rise. If a submarine is at rest in the water, it is difficult to make it hover at some fixed depth. A feedback control system is needed to keep the boat from slowly rising or sinking. A depth sensor, based on water pressure, senses any depth change and causes pumps to either fill or empty the ballast tanks to correct the error. The task is to design a Controller transfer function to make the system bandwidth be $\omega_U = 0.1$ rad/s and the damping be $\zeta = 1.0$.

The control system is diagrammed in Figure 10–17a. A voltage v_1 controls the speed of the pumps, which pump 10 (ft^3/s)/V, or 640 (lb/s)/V. The net force from the weight of water in the ballast tanks is the integral of the flow rate:

$$w = \int A \, v_1 \, dt, \quad \text{or} \quad \mathbf{W} = (A/s) \, \mathbf{V}_1,$$

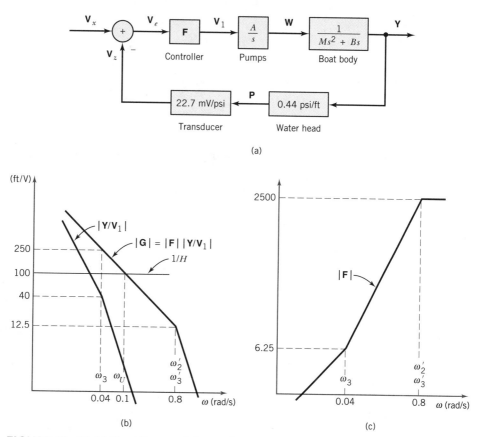

FIGURE 10-17 (a) Signal flow graph for a submarine depth control system. Pumps increase the net weight **W** by filling ballast tanks. The boat body affects the dynamics through mass and drag. (b) The desired $|\mathbf{G}|$ gives a bandwidth of $\omega_U = 0.1$ rad/s and an effective damping ratio of $\zeta = 1.0$. (c) Controller design to move $\omega_2 = 0$ out to $\omega_2' = 0.8$ rad/s and $\omega_3 = 0.04$ rad/s out to $\omega_3' = 0.8$ rad/s.

where $A = 640$ (lb/s)/V. This force can be either positive or negative as the total weight of the boat becomes either greater or less than that of the water it displaces.

The effect of the force w is to accelerate the mass of the boat downward and to overcome the friction of the hull passing downward through the water. The force required to accelerate the mass is $M\,(d/dt)^2 y$, where y is the depth in feet and M is the mass of 250,000 slugs (32 lb have a mass of one slug). The force required to overcome friction is assumed proportional* to velocity: $B\,(d/dt)y$, where $B = 10,000$ lb/(ft/s). Therefore, the total force to move the boat body is

$$w = M\,(d/dt)^2 y + B\,(d/dt)y,$$

*Frictional resistance is actually dependent on the 1.825 power of velocity. We assume a simpler relationship to keep the differential equation linear and subject to our means of analysis.

or in complex notation,

$$\mathbf{W} = Ms^2\mathbf{Y} + Bs\mathbf{Y}.$$

Solving for the depth in terms of the force,

$$\mathbf{Y} = \mathbf{W}/(Ms^2 + Bs),$$

which gives the "Boat Body" transfer function in Figure 10–17a.

The forward gain without the Controller is the product of the pump and boat body transfer functions:

$$\frac{\mathbf{Y}}{\mathbf{V}_1} = \frac{A}{s(Ms^2 + Bs)} = \frac{0.064 \ (\text{ft/V})/s^2}{s^2(s/\omega_3 + 1)},$$

where $\omega_3 = B/M = 0.04$ rad/s. The first two breakpoints are at $\omega_1 = 0$ and $\omega_2 = 0$. The Bode plot of this transfer function is plotted in Figure 10–17b. Note that the magnitude at ω_3 is $0.064/\omega_3^2 = 40$ ft/V, and the slope to the left of ω_3 is -2 decades per decade.

The water pressure p is 0.44 psi/ft, and a transducer produces 22.7 mV/psi. The product is the feedback gain $H = 10$ mV/ft. The reciprocal $1/H = 100$ ft/V is also plotted in Figure 10–17b.

The desired forward gain $|\mathbf{G}|$ passes through $1/H = 100$ at $\omega_U = 0.1$ rad/s with a slope of a decade per decade (see Figure 10–17b). This means that $\mathbf{F}$ must provide compensation to cancel the breakpoints at $\omega_2 = 0$ and $\omega_3 = 0.04$, which are to the left of ω_U. To achieve a damping of 1.0, Equation (10.18) says we need a phase margin of $\phi_m = 76°$. The position of the new ω_2' and ω_3' determine this phase margin. Keeping ω_3' as close as possible to ω_2' will minimize the gain required of the Controller. Let us make $\omega_2' = \omega_3'$. Then solving Equation (10.21) for $\phi_m = 76°$ yields $\omega_2' = \omega_3' = 8\omega_U = 0.8$ rad/s, as shown in Figure 10–17b. The corresponding transfer function is

$$\mathbf{G} = \frac{10 \ (\text{ft/V})/s}{s(s/\omega_2' + 1)(s/\omega_3' + 1)}. \qquad (10.22)$$

At $\omega = \omega_U = 0.1$ rad/s, the second two factors in the denominator can be neglected since $\omega_U/\omega_2' \ll 1$ and $\omega_U/\omega_3' \ll 1$. Then $|\mathbf{G}(j\omega_U)| \approx 10/\omega_U = 100 = 1/H$, as desired.

The Controller transfer function necessary to realize this $\mathbf{G}$ is

$$\mathbf{F} = \frac{\mathbf{G}}{\mathbf{Y}/\mathbf{V}_1} = \frac{(156 \ \text{sec}) \ s(s/\omega_3 + 1)}{(s/\omega_2' + 1)(s/\omega_3' + 1)}.$$

The magnitude is plotted in Figure 10–17c. The s factor in the numerator causes $|\mathbf{F}|$ to rise a decade per decade at low frequencies. At $\omega_3 = 0.04$ rad/s, the magnitude is about $(156 \ \text{sec})\omega_3 = 6.25$. From there it rises at two decades per decade to a magnitude of $(0.8/0.04)^2 6.25 = 2500$ at 0.8 rad/s. The double breakpoint at

$\omega'_2 = \omega'_3 = 0.8$ rad/s causes the curve to go flat for $\omega > 0.8$ rad/s. (We assume any breakpoint introduced by the Controller is much greater than 10 rad/s.)

The phase margin of this third-order system is 76°, which is the same as that of a second-order system with damping $\zeta = 1.0$. From Equation (10.20) the second-order system would have $\omega_0 = 0.2$ rad/s. Then Figure 10–11 gives the approximate step response of the third-order system. The $\omega_0 t = 2$ corresponds to $t = 10$ sec, the $\omega_0 t = 4$ corresponds to $t = 20$ sec, etc. This response is plotted in Figure 10–18 (dashed curve).

Our method of approximation has allowed us to avoid the analysis of third-order systems. To gain some confidence in this method, we will present the results of an exact analysis and compare them with the approximation in Figure 10–18. From Equation (10.22) and $\omega'_2 = \omega'_3 = 0.8$ rad/s, we have

$$\mathbf{G} = \frac{10}{1.56s^3 + 2.5s^2 + s},$$

$$\frac{\mathbf{Y}}{\mathbf{V}_x} = \frac{\mathbf{G}}{1 + \mathbf{G}\mathbf{H}} = \frac{10}{1.56s^3 + 2.5s^2 + s + 0.1}$$

$$= \frac{100}{(s/0.155 + 1)(s/0.394 + 1)(s/1.05 + 1)}$$

$$= 100 \left[\frac{1.94}{s/0.155 + 1} - \frac{1.04}{s/0.394 + 1} + \frac{0.1}{s/1.05 + 1} \right].$$

Thus this overdamped system has a transfer function that is the sum of three first-

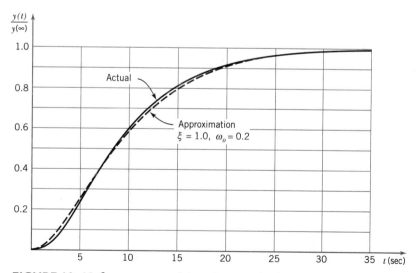

FIGURE 10–18 Step response of the submarine depth control system in Figures 10–17. The approximation assumes the third-order system behaves the same as a second-order system with the same phase margin. It differs from the actual response by 0.02 at most.

order transfer functions. By superposition, the step response is the sum of three first-order step responses (see Sections 7.2 and 7.3):

$$y(t)/y(\infty) = 1 - 1.94e^{-0.155t} + 1.04e^{-0.394t} - 0.1e^{-1.05t}.$$

This exact response is plotted in Figure 10–18 (solid curve). It differs from the approximate response by 0.02 at most.

Suppose that the captain sets v_x to descend an additional 50 ft. The step response shows that the submarine will descend 45 ft (90% of the distance) after 20 s, and it doesn't overshoot the 50 ft. If v_x is set for a step increase of 500 ft, the boat will *not* go 90% of the distance in 20 s; some part of the system will hit a limit. Either the pumps will not be able to pump as fast as required, or the ballast tanks will not hold as much water as required. The system becomes nonlinear, and its behavior is best analyzed by a computer simulation.

The approach to this design problem was typical of most engineering design. Approximations are made to simplify the problem, lending insight into the system behavior. This allows a first-cut design based on familiar procedures. Then a detailed computer analysis checks the design, and small changes can be made to the design to compensate for undesirable discrepancies. Once the engineer gets in the ball park, it is not difficult to see how to achieve small corrections.

10.11   SUMMARY

The primary purpose of feedback is to establish the closed-loop gain precisely in the face of variations in the forward path of the control. If the feedback gain H is accurate, then the closed-loop gain is accurate and is nearly $1/H$. This is true only if the loop gain $|\mathbf{G}H|$ is kept much greater than unity.

Feedback also increases the bandwidth and speed of the system. In most cases the closed-loop bandwidth is greater than the "open-loop bandwidth" (the bandwidth of the forward gain $\mathbf{G}$) by a factor of about $G_0 H$, where G_0 is the value of $\mathbf{G}$ at dc. The duration of the transient response is correspondingly reduced by a factor of $G_0 H$.

If $G_0 H$ is made too large, the system can become unstable, having a ringing transient or even going into sustained oscillation. A measure of the stability is the damping ratio $\zeta \equiv \alpha/\omega_0$. For $\zeta \geq 1$ there is no ringing, and for a value of 0.7 or 0.5 the amount of ringing may be acceptable. A phase margin $\phi_m \geq 45°$ generally implies $\zeta \geq 0.5$.

The damping ratio is determined by selecting the gain of the Controller (the error amplifier) to position the breakpoints in the Bode plot of $|\mathbf{G}|$. If the second breakpoint is at or below $1/H$, then the damping ratio is generally more than 0.5, and the stability is considered acceptable. The closed-loop response follows the lower of the $|\mathbf{G}|$ or $1/H$ responses.

The technique of compensation shapes the frequency response of the Controller to cancel some breakpoint in $|\mathbf{G}|$. This increases the closed-loop bandwidth while maintaining stability.

The precision, the bandwidth, and the stability of a feedback system are determined (within limits) by the design of the Controller. Realization of the desired transfer function for the Controller is dealt with in Chapter 14.

FOR FURTHER STUDY

W. L. Brogan, *Modern Control Theory*, Prentice-Hall, Englewood Cliffs, NJ, 1982.

J. R. Leight, *Applied Control Theory*, Peter Peregrinus, Stevenage, U.K. (Institution of Electrical Engineers), 1982.

G. J. Thaler, *Automatic Control Systems*, West Publishing Co., St. Paul, 1989.

PROBLEMS

Easy Drill Problems (answers at end of chapter)

D10.1. The feedback element in a system is a potentiometer that serves as a position-to-voltage transducer with a gain of 0.25 V/cm. What is the *ideal* closed-loop gain y/v_x of the system? What is the output position y (ideally) for an input voltage $v_x = 3$ V?

D10.2. The feedback system in D10.1 has a forward gain of $G = 1000$ cm/V. What is the percent difference between the actual and the ideal closed-loop gains? What is the *actual* gain y/v_x?

D10.3. Choose a gain F for the motor control system in Figure 10–2b so the closed-loop gain y/v_x differs from the ideal gain by 0.5%.

D10.4. The closed-loop transfer function of a system is

$$\frac{\mathbf{G}}{1 + \mathbf{G}H} = \frac{10^5}{s^2 + 40s + 10^4}$$

Find ω_0, α, and ζ. Is the system stability acceptable?

D10.5. For

$$\mathbf{G} = \frac{1000}{(s/1 + 1)(s/50 + 1)}, \qquad H = 0.01,$$

sketch Bode plots of $|\mathbf{G}|$ and $1/H$. Estimate ω_U where the plots cross. Do you expect $\zeta > 0.5$?

D10.6. Find ω_0, α, and ζ for the $\mathbf{G}$ and H in D10.5.

D10.7. For the $\mathbf{G}$ and H in D10.5, find the phase margin ϕ_m of the system. (Assume $\omega_1 \ll \omega_U$.)

D10.8. For $\zeta = 0.7$ and $\omega_0 = 100$ rad/s, estimate the percent overshoot and the settling time of a system's step response (see Figure 10–11). Repeat for $\zeta = 0.4$ and $\omega_0 = 100$ rad/s. (Settling time is defined in Figure 8–18b.)

D10.9. For the Bode plot in Figure D10.9, what are the values of A and B? Write the transfer function that has this magnitude plot.

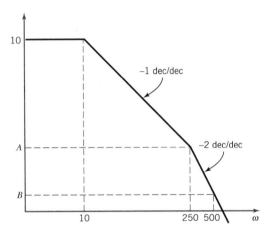

D10.10. Write the transfer function that has the magnitude plot represented by the Bode plot in Figure D10.10. (The first breakpoint is at $\omega = 0$.)

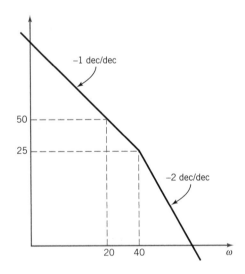

D10.11. Sketch the Bode plot for the magnitude of $400/(s/2 + 1)(s/8) + 1)(s/40 + 1)$. What is the value of the Bode plot at the breakpoints?

D10.12. Sketch the Bode plot for the magnitude of $1600/s^2(s/40 + 1)$. What is the value of the Bode plot for $\omega = 10$, for $\omega = 40$?

D10.13. Sketch the Bode plot for the magnitude of $10(s/50 + 1)/(s/2000 + 1)$. What is the maximum value?

D10.14. Sketch the Bode plot for the magnitude of $s^2/(s/5 + 1)(s/50 + 1)$. What is the value of the Bode plot at the breakpoints?

D10.15. Plot $|\mathbf{Y}/\mathbf{V}_1|$ and $1/H$ for the system in Figure D10.15. What constant value for $\mathbf{F}$ will achieve $\zeta = 0.5$? What $\mathbf{F}$ will achieve $\zeta = 1.0$? (Use the fact that $\omega_1 \ll \omega_U$.)

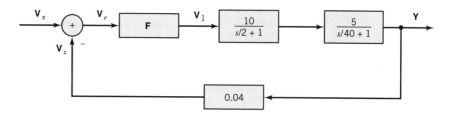

D10.16. Plot $|\mathbf{Y}/\mathbf{V}_1|$ and $1/H$ for the system in Figure D10.15. Plot the desired $|\mathbf{G}|$ for $\omega_U = 120$ rad/s and $\zeta = 0.5$ (keep $\omega_1 = 2$ rad/s). Write the corresponding transfer function $\mathbf{G}$. What $\mathbf{F}$ will achieve this $\mathbf{G}$? Plot $|\mathbf{F}|$. What is the maximum gain of $|\mathbf{F}|$?

Application Problems

P10.1. A lamp produces 5 lumens/mA, and a photocell monitoring its output produces 0.5 mA per lumen of light falling on it. Only 0.1% of the lamp's light falls on the photocell. What is the *ideal* gain y/i_x of the control system in P10.1? What current gain F is necessary to make the actual gain y/i_x differ from the ideal by 2%?

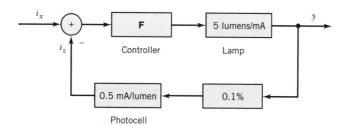

P10.2. For the system in P10.1, $F = 10,000$. What is the percent error from the ideal of the gain y/i_x? For $i_x = 2.5$ mA, find y. Suppose that the lamp response decreases to 4 lumens/mA; what is y for $i_x = 2.5$ mA now? Compare the percent change in y with the percent change in the lamp response.

P10.3. The pressure control system in Figure P10.3 has a pump with a response bandwidth of 100 rad/s. For a flat Controller response $\mathbf{F} = 1000$, sketch Bode plots of $|\mathbf{G}|$, $1/H$, and $|\mathbf{Y}/\mathbf{V}_x|$ as a function of ω. What is the bandwidth of $|\mathbf{Y}/\mathbf{V}_x|$? Calculate $G_0 H$ and compare with the increase in bandwidth.

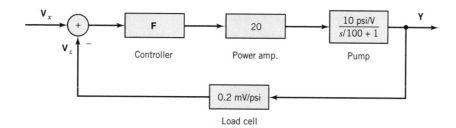

P10.4. Repeat P10.3 for a Controller bandwidth of 5000 rad/s: $\mathbf{F} = 1000/(s/5000 + 1)$. Considering the position of the second breakpoint in $\mathbf{G}$, is $\zeta \geq 0.5$?

P10.5. Repeat P10.3 for $\mathbf{F} = 2000/(s/2500 + 1)$. Considering the position of the second breakpoint in $\mathbf{G}$, is $\zeta \geq 0.5$? Find ω_0, α, and the damping ratio ζ of the closed-loop response. Estimate the percent overshoot of the system step response (see Figure 10–11).

P10.6. Equations (10.12), (10.13), and (10.17) through (10.21) hold for $\omega_1 << \omega_U$. In this problem, you will prove Equations (10.12) and (10.13) for $\omega_1 = 0$. Given $\mathbf{G} = A/s(s/\omega_2' + 1)$ whose Bode plot intersects $1/H$ at ω_U (see Figure P10.6), find A in terms of H and ω_U. Show that the closed-loop response has $\zeta = 0.5\sqrt{\omega_2'/\omega_U}$, and $\omega_0 = \sqrt{\omega_2'\omega_U}$.

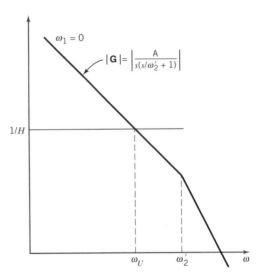

P10.7. The jets controlling the roll of a rocket provide 100 N-m of torque per volt of control voltage v_1. The moment of inertia of the rocket is $J = 250$ kg-m^2. The angle of roll θ is governed by $d^2\theta/dt^2 = T/J$, where T is the torque. θ is monitored by a gyroscope with pickoff that produces 1.0 V per radian of θ. Sketch the signal flow graph of a feedback system to control θ with an input voltage v_x. The closed-loop dc gain is to be $\theta/v_x = 1$ rad/V. Design $\mathbf{F}$ for a bandwidth $\omega_U = 4$ rad/s and a damping ratio $\zeta = 0.5$. Redesign for $\zeta = 1.0$. Draw the Bode plot for each $|\mathbf{F}|$. Do you think the Controller cutoff frequency will be a concern in either case?

P10.8. The speed control system for a steam turbine is shown in Figure P10.8. The valve passes a flow q_v of steam that is related to a control voltage v_1 by $A = 5$ (m^3/s)/V. Due to its mass, the valve has a cutoff frequency of $\omega_3 = 20$ rad/s. The turbine bowl volume has a filling time constant of 1 s, for a cutoff frequency of $\omega_2 = 1$ rad/s. For a given load, the turbine develops a shaft speed y that is related to the steam flow q_t by $B = 200$ rpm/(m^3/s). The turbine inertia results in a cutoff frequency of $\omega_1 = 0.2$ rad/s. A transducer monitors the shaft speed and produces 1.0 mV/rpm of feedback voltage.

Choose a constant value $\mathbf{F}$ for the Controller so the damping is $\zeta = 0.5$ (neglect the effect of the breakpoint at ω_3). What is the percent static gain error? What is the bandwidth ω_U?

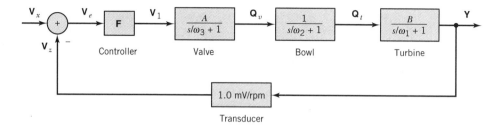

V_x ⊕ → V_e → **F** → V_1 → $\dfrac{A}{s/\omega_3 + 1}$ → Q_v → $\dfrac{1}{s/\omega_2 + 1}$ → Q_t → $\dfrac{B}{s/\omega_1 + 1}$ → Y

V_z

Controller — Valve — Bowl — Turbine

1.0 mV/rpm

Transducer

P10.9. Design a Controller function **F** that compensates the system in P10.8 to achieve a damping of $\zeta = 0.7$ and a bandwidth of $\omega_U = 2$ rad/s. Leave ω_1 unchanged, and neglect the effect of ω_3. What is the percent static gain error?

Now taking the effect of ω_3 into account, find the phase margin ϕ_m. What is the damping ratio of a second-order system with the same ϕ_m? Estimate the overshoot and rise time of the system step response.

P10.10. Design a Controller function **F** that compensates the system in P10.8 to achieve a phase margin of $\phi_m = 76°$ and a bandwidth of $\omega_U = 5$ rad/s. Assume the controller has unlimited bandwidth. Leave ω_1 unchanged, and make $\omega_2' = \omega_3'$. What is the percent static gain error? Estimate the rise time of the system step response.

Suppose the Controller has a gain-bandwidth product of 200,000 rad/s, introducing a breakpoint at ω_4. For your designed **F**, what is ω_4? What does this do to the phase margin and to the overshoot of the step response?

P10.11. Design a control system so v_x controls the forward velocity of u of a helicopter. The mass of the helicopter is $M = 1500$ kg. The vertical thrust of the blades is $w_v = Mg = 14{,}700$ N, where $g = 9.8$ m/s². Forward thrust is gained by rotating the body forward by an angle θ. This gives the blades a forward thrust component $w_f = w_v \tan \theta \approx w_v \theta$ for $\theta < 0.75$ rad. The torque to rotate the body is achieved by increasing the pitch of the blades as they approach the back and decreasing the pitch as they approach the front. The pitch difference determines the torque T, which is controlled by voltage v_1 according to $T = Av_1$, where $A = 1000$ N-m/V. The moment of inertia of the body to rotation is $J = 20{,}000$ kg-m². The air resistance to forward motion is Bu, where u is the forward velocity and $B = 73.5$ N/(m/s). A Pitot tube measures the wind speed u and produces 0.02 V/(m/s) of feedback voltage v_z. (See related systems in Section 10.10 and problem P10.7.)

Design a Controller function **F** so the system bandwidth is $\omega_U = 0.4$ rad/s and the damping is $\zeta = 0.7$ (or $\phi_m = 54°$). Make the frequency of the second and third breakpoints after compensation be the same.

Extension Problems

E10.1. Phase margin ϕ_m is defined as the difference between ang$[GH]$ (or ang$[G]$) and $-180°$ at the frequency where $|GH| = 1$, or $|G| = 1/H$. In this chapter we have approximated $|G|$ by its Bode plot, making ϕ_m easier to find but less accurate. For example, Figure 10–10b shows the Bode plot of $|G|$ crossing $1/H = 1000$ at $\omega = 5000$ rad/s, leading to $\phi_m = 45°$. But from Equation (10.6), $|G| = 1000$ for $s = j\omega = j3930$ rad/s (the reader should show this). At this frequency, ang$[G] = -121°$, and $\phi_m = 52°$, a significant difference from 45°.

The accuracy of ϕ_m is not so important if we are consistent in its definition when we compare second-order systems with higher-order systems. We have used ϕ_m

in this way as a design tool. But when the design is done, the engineer should be able to give an accurate estimate of the ϕ_m of his or her control system to others.

By trial and error, find the $s = j\omega$ for which $|\mathbf{G}| = 1/H = 200$, where $\mathbf{G}$ is given in Equation (10.3). (A programmable calculator is useful here.) Start with the value indicated by the Bode plot in Figure 10–13b. For this ω, find ang[$\mathbf{G}$] and ϕ_m. Compare with the ϕ_m we have been associating with $\zeta = 0.5$.

By trial and error, find the $s = j\omega$ for which $|\mathbf{G}| = 1/H = 5$, where $\mathbf{G}$ is given in Example 10.6. Start with the value indicated by the Bode plot in Figure 10–16a. For this ω, find ang[$\mathbf{G}$] and ϕ_m. Compare with the ϕ_m we have been associating with $\zeta = 1.0$.

E10.2. Suppose the closed-loop transfer function of a feedback system is

$$\frac{\mathbf{G}}{1 + \mathbf{G}H} = \frac{1000}{(s + 10)(s + 30)}.$$

For what values of s does the function go to infinity? These are called the "poles" of the function. Place X's on a complex plane at these values of s. This is a "pole plot" in the "s plane."

E10.3. Suppose the closed-loop transfer function of a feedback system is

$$\frac{\mathbf{G}}{1 + \mathbf{G}H} = \frac{K}{(s^2 + 2\alpha s + \omega_0^2)} = \frac{K}{(s - a)(s - b)}.$$

Since the function goes to infinity for $s = a$ and $s = b$, these are the poles of the function. Show that $a = -\alpha + j\omega_n$, and $b = -\alpha - j\omega_n$, where ω_n is related to α and ω_0 as shown in Figure E10.3. Express the damping ratio ζ in terms of the angle θ. For fixed ω_0, sketch pole plots for $\zeta = 0$, for $\zeta = 0.5$, and for $\zeta = 1.0$.

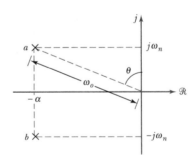

Study Questions

10.1. Name two improvements that feedback provides for a control system.

10.2. Is the closed-loop gain of a control system governed more by the forward gain G or the feedback gain H?

10.3. What is the symbol for multiplication by a constant in a signal flow graph? What is the symbol for addition? For subtraction?

10.4. Express the "closed-loop gain" in terms of other gains in the feedback control system.

10.5. Express the "loop gain" in terms of other gains in the feedback control system. What must the loop gain be for a closed-loop gain error of 0.5%?

10.6. By what factor does feedback increase the bandwidth of a control system? If the bandwidth is doubled, how does the transient response of the system change?

10.7. What relationship between $|\mathbf{G}|$ and $1/H$ defines ω_U? What property of the closed-loop system does ω_U represent?

10.8. Name three measures of stability other than overshoot of the system step response.

10.9. Is ω_0 the bandwidth of the system? What property of the system does it represent? How is it used in determining the system step response?

10.10. Generally speaking, how large must the damping ratio be for acceptable system stability? How large must the phase margin be?

10.11. Which component in the system does the engineer design to determine the accuracy, stability, and bandwidth of the system?

10.12. How many breakpoints in $|\mathbf{G}|$ should be to the left of ω_U for the system to be stable? If there are too many breakpoints to the left, how are they moved to the right?

10.13. What is the static gain error of a system with integration in the forward gain? What is the frequency response of integration?

10.14. What is the lowest damping ratio with no step response overshoot? What is the lowest damping ratio with no peaking in the frequency response?

10.15. Which measure of stability—damping ratio or phase margin—is defined for systems higher than second order? How can the behavior of a second-order system be compared with that of a higher-order system?

ANSWERS TO DRILL PROBLEMS

D10.1. 4.0 cm/V, 12 cm

D10.2. 0.4%, 3.984 cm/V

D10.3. 400

D10.4. $\omega_0 = 100$, $\alpha = 40$, $\zeta = 0.2$, no ($\zeta < 0.5$)

D10.5. $\omega_U = 10$, yes ($\omega_2 > \omega_U$)

D10.6. $\omega_0 = 23.5$, $\alpha = 25.5$, $\zeta = 1.09$

D10.7. $\phi_m = 78°$

D10.8. 5%, 63 ms; 25%, 110 ms

D10.9. $A = 4$, $B = 1$, $100/(s/10 + 1)(s/250 + 1)$

D10.10. $1000/s(s/40 + 1)$

D10.11. 400, 100, 4

D10.12. 16, 1

D10.13. 400

D10.14. 25, 250

D10.15. $\mathbf{F} = 10$, $\mathbf{F} = 5$

D10.16. $\mathbf{G} = 1500/(s/2 + 1)(s/120 + 1)$, $\mathbf{F} = 30(s/40 + 1)/(s/120 + 1)$, 90

CHAPTER 11

Communications

Communications deals with the transmission of information from one place to another. This can be done with smoke signals, written letters, telegraph, telephone, or computer data links, for example. Many of the concepts in this chapter apply to any form of communication, but we are most interested in communication systems that involve electrical signals. As we saw in Chapters 8 and 9, these signals can be in either analog or digital form.

The simplest form of communication system is shown in Figure 11–1a. An information source (a voice signal or some other analog signal) is amplified by a *transmitter* and carried some distance over a *transmission medium* (a pair of wires, for example). In such a simple system the communication problem is one of distance, distance great enough that the transmission medium significantly attenuates the signal. Therefore a *receiver* must amplify the signal again at the far end. Most of the concepts involved in this simple system have been covered in Chapter 8. Both the amplifiers and the transmission medium have bandwidth limitations which distort the signal, and noise and reflections tend to mask the signal.

Sometimes the information to be transmitted is in digital code. This is shown explicitly in the simple digital communication system in Figure 11–1b. As mentioned in Chapter 9, such a system has the advantage that the original signal can be recovered perfectly if the noise and distortion are not too great. A decision is made in the receiver as to whether a "1" or a "0" was transmitted.

An important feature of some communication systems is the ability to have many information sources share the same transmission medium, as shown in Figure

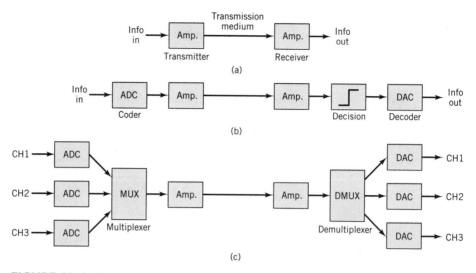

FIGURE 11–1 Communication systems involving (a) distance, (b) digital coding, and (c) multiplexing a number of channels onto one system. The system in (c) may also appear without the coder (ADC) and decoder (DAC) blocks.

11–1c. This feature, called *multiplexing*, is a principal topic of this chapter. A transmission medium is either expensive (as with a long cable of glass fiber) or limited in availability (as with band space for radio waves). Multiplexing makes efficient use of the medium by using it to carry as many channels of information as possible.

A later section of the chapter looks at the limits of a communication system—how much information it can carry. We will also see that particular coding schemes may be necessary to take full advantage of this capacity.

The reader's objectives in this chapter are to:

1. be able to specify a communication system to meet his or her needs in terms of number of users and capacity per user,
2. understand the advantages and limitations of space-division, frequency-division, and time-division multiplexing.
3. keep up with the growing use of digital communication,
4. be able to estimate the amount of information conveyed by written words, spoken words, pictures, music, and data.

11.1 MULTIPLEXING

When a communication system is set up between two points, there is usually more than one information source that wants to use the system. Two towns linked by a telephone system want to have several simultaneous conversations possible. A remote power station may have to continuously send back status information about several pieces of equipment. Any process by which several users are able to share one communication system without interfering with one another is called *multi-plexing*. In this way each user is provided with a *channel*—what appears to the user to be a private communication system, although the user is actually sharing one larger system with many others.

A common example where multiplexing is needed is at a cocktail party. Suppose that you are trying to hear a friend talk to you, but there is a loud conversation going on nearby. You have several options to help sort out the information your friend is providing. If you and your friend decide to go to a far corner of the room by yourselves, this is an example of *space-division multiplexing*. If you agree with the interfering speaker to alternate sentences (first he talks, then your friend, then he, etc.), this is an example of *time-division multiplexing*. It may be that the interfering speaker has a very high voice and your friend has a very low voice. You could probably mentally filter out the high voice and listen for the low voice; this would be an example of *frequency-division multiplexing*. If all else fails you might try listening for the right topic. Suppose the interfering speaker is talking about fishing and your friend is talking about music. You would be able to pick out your friend's conversation just because his words fit in with the topic of music. This is an example of *code-division multiplexing*.

The same methods are used to multiplex several channels onto one electrical communication system. The system may provide many lines in one cable—each line devoted to a separate channel. This is space-division multiplexing. Figure 11–2a shows an example of space-division multiplexing with three senders of Morse code operating simultaneously. The transmission medium can be reduced to one line if

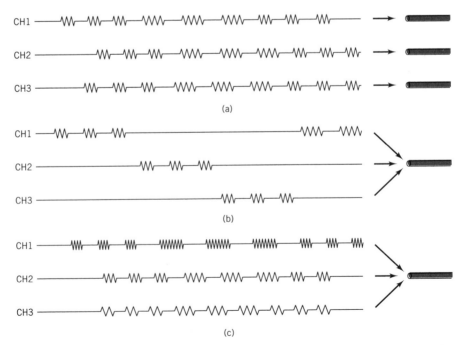

FIGURE 11–2 Combining three channels carrying Morse code (a) by space-division multiplexing, (b) by time-division multiplexing, and (c) by frequency-division multiplexing.

the senders time-share the line, as in Figure 11–2b. In this case each sender in turn gets to transmit a piece of his or her message (one character here). At the far end the messages are separated and re-formed without gaps. This is time-division multiplexing. Another method of combining the three signals on one line without interference is to use three different tones, as in Figure 11–2c. At the receiving end three filters tuned to the tones separate the messages. This is frequency-division multiplexing.

The most interesting forms are time-division and frequency-division multiplexing. Space-division multiplexing might be considered a trivial case since it amounts to several separate communication systems that are only loosely associated. Code-division multiplexing is used little because of the complex signals and wide bandwidth it requires. The "spread-spectrum" techniques of code-division multiplexing are usually justified only in military applications where secrecy and jamproofing are needed.

Most of this chapter is devoted to describing ways of implementing time-division and frequency-division multiplexing. We have already been introduced to the concepts of the time domain and the frequency domain in Chapter 8, where a square wave was expressed either as a time waveform or as the sum of sinusoids with certain frequencies and amplitudes. Multiplexing in the time domain or frequency domain amounts to dividing the domain into nonoverlapping *slots* or *bands* as illustrated in Figure 11–3. In the case of time-division multiplexing in Figure 11–3a, slots 1, 4, 7, etc., are devoted to channel 1; slots 2, 5, 8, etc., to channel 2; and slots 3, 6, 9, etc., to channel 3. The amount of information in each time slot may be very small, perhaps amounting to only one data point or even only one bit. Frequency-division multiplexing is illustrated in Figure 11–3b, with each channel occupying one band. The method by which the frequency content of a message is moved to a certain band is the technique of carrier modulation. We look next at this topic.

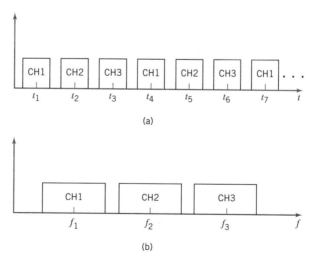

(a)

(b)

FIGURE 11–3 Multiplexing by interleaving channels in either (a) the time domain or (b) the frequency domain.

11.2 AMPLITUDE MODULATION

Let the message to be transmitted be a sinusoidal signal

$$m(t) = 0.6 \cos(\omega_m t), \qquad \omega_m = 2\pi \times 10 \text{ kHz},$$

as depicted in Figure 11–4a. This is not a very exciting message, but it will serve to illustrate multiplexing. Suppose this is to be channel 3 in Figure 11–3b, with $f_3 = 120$ kHz at the center of the channel. The message is shifted up to the vicinity of 120 kHz by using $m(t)$ to modulate a sinusoidal *carrier*

$$v_c = \cos(\omega_c t), \qquad \omega_c = 2\pi \times 120 \text{ kHz}$$

(see Figure 11–4c).

Amplitude modulation (AM) involves varying the amplitude of the carrier, as shown in Figure 11–4d. This was done by passing v_c through an amplifier whose gain $m^+(t)$ varies according to $m(t)$. The modulated carrier is

$$v_t = m^+(t)v_c \tag{11.1}$$

where

$$m^+(t) = m(t) + 1 = 0.6 \cos(\omega_m t) + 1.$$

Note that a constant (unity in our case) was added to $m(t)$ to keep $m^+(t)$ always positive. This ensures that the outline, or *envelope*, of v_t has the shape of the original message $m(t)$. If the amplitude of $m(t)$ becomes too great (greater than 1.0 here), the top and bottom envelopes can cross in the middle, causing distortion in the receiver.

An AM transmitter consists of an AM modulator and an RF (radio frequency) amplifier. Figure 11–5 shows a block diagram of an AM modulator, which implements Equation (11.1). The heart of the modulator is a multiplier circuit, which is essentially an amplifier with a voltage-controlled gain. An emitter-coupled transistor pair, described in Section 15.7, can perform this function (see problem E15.2).

Using the trigonometric identity $\cos(A) \cos(B) = 0.5 \cos(A - B) + 0.5 \cos(A + B)$, we can write Equation (11.1) as

$$v_t = 0.3 \cos(\omega_c - \omega_m)t + 0.3 \cos(\omega_c + \omega_m)t + \cos(\omega_c t). \tag{11.2}$$

For our case, the frequencies of these three sinusoids are 110 kHz, 130 kHz, and 120 kHz. Now that the transmitted signal is expressed as the *sum* of sinusoids, it can be represented by a plot in the frequency domain. Figure 11–6d shows the three frequency components of v_t. The carrier is in the center, and the modulation has caused *sideband* components on each side of it.

Figure 11–6 also represents $m^+(t)$ and v_c in the frequency domain for comparison. The original message $m(t)$ is evident in Figure 11–6b as the component

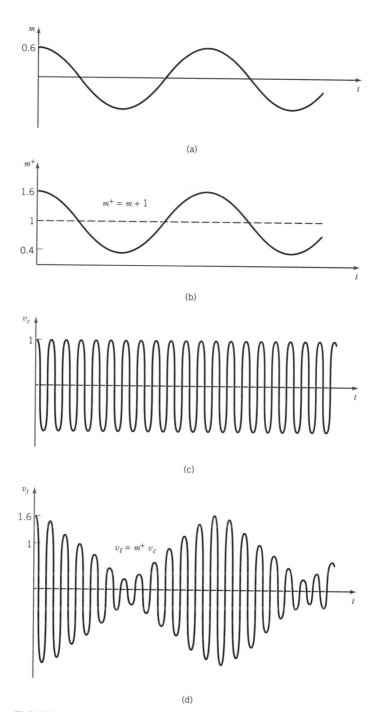

FIGURE 11–4 Amplitude modulation, showing in the time domain (a) the message or modulating signal, (b) the raised (always positive) message, (c) the carrier, and (d) the modulated carrier which is transmitted.

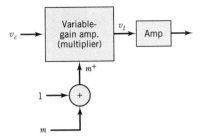

FIGURE 11–5 AM transmitter. The modulator consists of a multiplier—an amplifier with gain $m^+(t)$ controlled by the message signal $m(t)$.

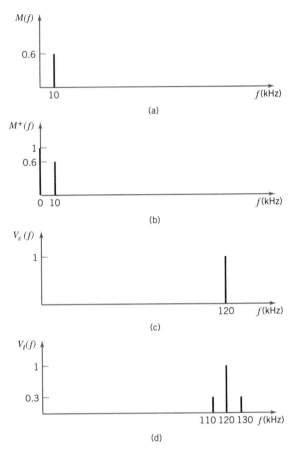

FIGURE 11–6 Amplitude modulation, showing in the frequency domain (a) the message, (b) the raised message, (c) the carrier, and (d) the modulated carrier. The effect of carrier modulation is to shift the message to a higher frequency.

of $M^+(f)$ at 10 kHz. The component at $f = 0$ is the dc component added to keep $m^+(t)$ positive. The effect of the carrier in Figure 11–6c can be thought of as shifting both these components up in frequency by 120 kHz (compare Figure 11–6b and Figure 11–6d). Note that a lower sideband at 110 kHz also appears.

The reader can confirm that if the dc component of $m^+(t)$ is eliminated, then the carrier component of v_t at 120 kHz is eliminated. This is sometimes done to save transmitted power. It is called *suppressed-carrier* (SC) modulation and requires a special receiver to recover the original message.

Either one of the sidebands tells all there is to know about the original message. The bandwidth of the transmitted signal can be halved by eliminating one sideband (usually removed by filtering). This is called *single-sideband* (SSB) transmission. It usually also involves partial or complete suppression of the carrier.

Figure 11–7a shows the functions of an AM receiver. After some amplification, the received signal v_r looks about the same as the transmitted signal v_t (compare

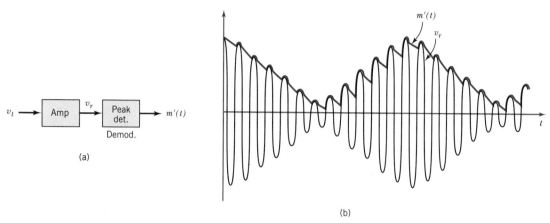

FIGURE 11–7 Demodulation of AM. A peak detector follows the envelope of the modulated carrier. The output $m'(t)$ is an approximation to the original message $m(t)$.

Figure 11–7b and Figure 11–4d). A peak detector (see Section 16.6) follows the envelope to get a good approximation $m'(t)$ to the original message $m(t)$. (The "bumps" on $m'(t)$ are at a frequency beyond the range of human hearing.) This process of recovering the original message from the modulated carrier is called *demodulation*.

So far we have considered only a purely sinusoidal message. A more typical message signal $m(t)$ is shown in Figure 11–8a. It is composed of many frequencies, and the highest frequency present (with period $1/B_m$) determines the bandwidth B_m of the signal. This is illustrated in Figure 11–8b by a plot of the *spectrum* $M(f)$ of the message. The height of the curve for each value of f indicates the relative amount of that frequency component present. The spectrum $V_t(f)$ of the transmitted signal shows that, because of the upper and lower sidebands, the transmitted bandwidth B_t for AM is double the message bandwidth B_m:

$$B_t = 2B_m. \tag{11.3}$$

The carrier component at $f_c = \omega_c/2\pi$ is also evident.

The purpose of modulating was to move the message spectrum to a different frequency band, one centered on the carrier frequency. If several messages each modulate a different carrier frequency, they can be added together without interfering with each other in the frequency domain (see Figure 11–9). This is an example of frequency-division multiplexing. The messages can now share a common transmission medium such as a pair of wires or a region of space (for radio transmission). Note in Figure 11–9b that a *guard band* has been left between adjacent channels. This simplifies the separation (demultiplexing) of the channels by filtering at the receiving end.

EXAMPLE 11.1 Four messages are to be multiplexed by AM.

Given: The bandwidth of each message is $B_m = 4$ kHz, and a guard band of 2 kHz is needed for demultiplexing.

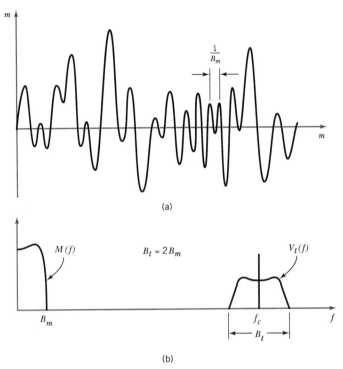

FIGURE 11–8 (a) Typical message waveform. (b) Corresponding spectra $M(f)$ for the message and $V_t(f)$ for the modulated carrier. The transmission bandwidth is twice the message bandwidth.

Find: the lowest carrier frequencies f_{c1}, f_{c2}, f_{c3}, and f_{c4} that can be used. Also find the highest-frequency component the transmission medium must carry.

Solution: From Figure 11–9 it can be seen that the spacing between the carrier frequencies is $2B_m$ plus the guard band, or 10 kHz. f_{c1} can be <u>zero</u> (m_1 is transmitted "as is"). Then $f_{c2} = $ <u>10 kHz</u>, $f_{c3} = $ <u>20 kHz</u>, and $f_{c4} = $ <u>30 kHz</u>. The highest-frequency component the transmission medium must carry is $f_{c4} + B_m = $ <u>34 kHz</u>.

Modulation has another purpose independent of the need to multiplex; it makes radio transmission practical. To efficiently radiate at a given frequency, the length of the antenna should be at least a tenth of the wavelength of the radio wave. The wavelength λ is given by

$$\lambda = c/f, \tag{11.4}$$

where c is the speed of light $(10^9$ ft/s$)$* and f is the frequency. To radiate low

---

*The wavelength λ may be found in terms of meters by expressing c as 3×10^8 m/s.

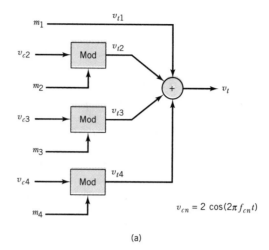

(a)

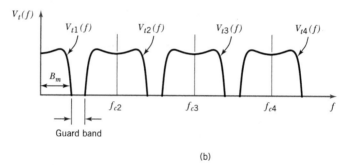

(b)

FIGURE 11-9 Four channels sharing a transmission line by AM frequency-division multiplexing. The guard bands between channels allow the channels to be more easily separated by filters at the receiver.

frequencies (around 1 kHz) such as voice, the antenna would have to be many miles long to have any reasonable efficiency. If this same voice information is modulated up to, say, 1 MHz (as with commercial AM radio), the antenna length need only be 100 ft or so.

11.3   PHASE MODULATION

In phase modulation (PM) the message varies the phase of a carrier v_c to produce the transmitted wave v_t. An example for a sinusoidal message is shown in Figure 11-10. When $m(t) = 0$, $v_t = v_c$; when $m(t)$ is positive, v_t is ahead of v_c in phase. The modulated wave can be expressed as

$$v_t = \cos(\omega_c t + \theta), \tag{11.5}$$

where

$$\theta = \beta m(t) = \beta \cos(\omega_m t). \qquad (11.6)$$

It can be shown that, for small phase deviation, the spectrum of a PM signal is practically the same as that of an AM signal (see problem E11.1). One advantage of PM is that the transmitted power is constant, simplifying the design of the power amplifier in the transmitter. Another is that the message is immune to gain variation in transmission.

PM is used in commercial radio to provide so-called AM stereo. It takes advantage of the fact that the same carrier can be modulated by both AM and PM at the same time. The AM provides the information for the L + R channel, and the PM provides the information for the L − R channel. Both channels occupy the same frequency band around the carrier, but they can be separated from each other at the receiver by AM and PM demodulation.

PM modulation and demodulation are usually performed by a phase-locked loop. This is a control loop that includes a phase comparator and a voltage-controlled oscillator. The availability of phase-locked loops in integrated circuits has simplified the process of PM modulation and demodulation. Just as the AM demodulator in Figure 11–7 is not sensitive to PM, a properly designed PM demodulator is not sensitive to AM.

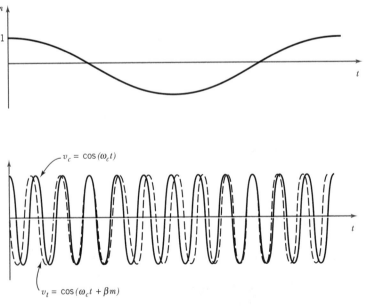

FIGURE 11–10 Phase modulation. When the message signal m is positive, the phase of the modulated carrier v_t is earlier than the unmodulated carrier v_c. Here $\beta = \pi/2$.

11.4 FREQUENCY MODULATION

In frequency modulation (FM) the message signal modulates the frequency of the carrier. The transmitted frequency is

$$f_t = f_c + \Delta f m(t), \tag{11.7}$$

where f_c is the carrier frequency, $m(t)$ is the message, and Δf determines the amount of frequency deviation. Figure 11–11 shows an example for a sinusoidal $m(t)$. Note that f_t is centered on f_c. Equation (11.7) is usually implemented by applying $m(t)$ to a voltage-controlled oscillator (VCO). The VCO is often placed in a phase-locked loop to provide better control of the carrier frequency.

The transmitted signal v_t with this varying frequency f_t is shown in Figure 11–11c, and its spectrum is shown in Figure 11–11d. As f_t swings back and forth from $f_c - \Delta f$ to $f_c + \Delta f$, it generates an almost continuous spectrum between those two frequencies. Because the sinusoidal modulation $m(t)$ spends more time near its extremes, the spectrum has its greatest values at its extremes.* The transmission bandwidth is about twice the peak frequency deviation Δf. But a more accurate approximation is given by *Carlson's rule*:

$$B_t \approx 2\Delta f + 2B_m, \tag{11.8}$$

where B_m is the message bandwidth. A typical message spectrum $M(f)$ and a typical transmission spectrum $V_t(f)$ are shown in Figure 11–12. The shape of $V_t(f)$ indicates that $m(t)$ spends more time near zero than near its extremes (see Figure 11–8a, for example).

EXAMPLE 11.2 Three messages are to be frequency-division multiplexed using FM.

Given: The carrier frequencies allotted are 96.1 MHz, 96.5 MHz, and 96.9 MHz. The required guard band between channels in a listening area is 220 kHz. The message bandwidths are $B_m = 15$ kHz.

Find: the peak deviation Δf allowed.

Solution: Figure 11–13 illustrates the situation in the frequency domain. B_t is the spacing between carrier frequencies less the guard band: $B_t = 400$ kHz $- 220$ kHz $= 180$ kHz. From Equation (11.8) the peak deviation is $\Delta f = 0.5(B_t - 2B_m) = 0.5(180 \text{ kHz} - 30 \text{ kHz}) = \underline{75 \text{ kHz}}$. Note that Δf is proportional to the peak amplitude of the message. ∎

In AM the transmission bandwidth was twice the message bandwidth: $B_t = 2B_m$; in FM it can be much more than this: $B_t = 12B_m$ in Example 11.2. In the next

---

*A rigorous quantitive analysis of the spectrum involves Bessel functions.

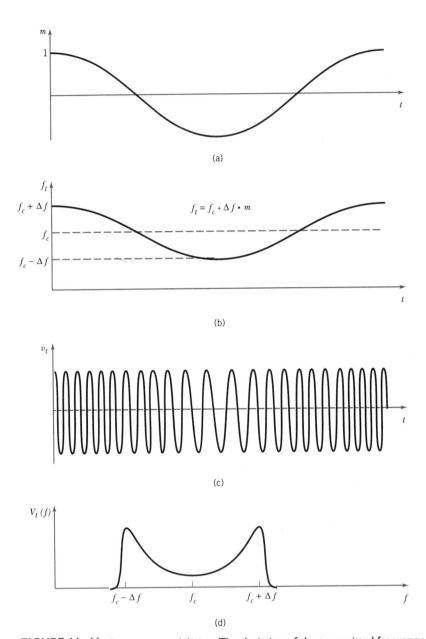

FIGURE 11–11 Frequency modulation. The deviation of the transmitted frequency f_t from the center frequency f_c is proportional to the message signal m. The transmitted signal v_t (with frequency f_t) is shown (c) in the time domain and (d) in the frequency domain.

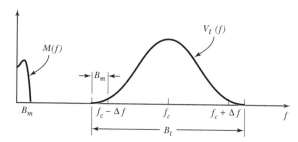

FIGURE 11–12 Spectra $M(f)$ of the message and $V_t(f)$ of the transmitted signal. The bandwidth B_t of the transmitted signal can be much greater than the message bandwidth B_m, depending on Δf.

section we will see that this "wasting" of the frequency domain is the price paid for improving the demodulated signal-to-noise ratio.

11.5   SIGNAL-TO-NOISE RATIO

Sources of noise in electronic systems were discussed in Section 8.8. The standard measure of the noise level in a system is the ratio of the signal *power* to the noise *power*, or simply the signal-to-noise ratio (SNR). For example, a SNR of unity would be extreme degradation, while a SNR of 3000 is typical of AM radio reception.

The SNR can be expressed in terms of signal and noise *voltage*. The power associated with the received signal v_r is $(v_r)^2_{\mathrm{rms}}/R$, where R is the resistance v_r appears across. Received with v_r is some rms noise voltage n with power n^2/R. Then the received signal-to-noise ratio is

$$\mathrm{SNR}_r = \left(\frac{(v_r)_{\mathrm{rms}}}{n}\right)^2. \qquad (11.9)$$

This power ratio is often expressed in decibels (dB), where a *bel* is a ratio of ten. The number of factors of ten in a ratio is found by taking the log base ten. Since

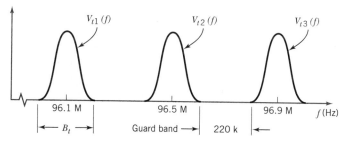

FIGURE 11–13 Three channels frequency-division multiplexed by FM (see Example 11.2). The parameters are those for commercial FM radio.

there are ten decibels in a bel, the definition is

$$\text{SNR}_r \text{ (in dB)} \equiv 10 \log(\text{SNR}_r). \qquad (11.10)$$

Substituting Equation (11.9) into (11.10), we have

$$\text{SNR}_r \text{ (in dB)} = 20 \log \left(\frac{(v_r)_{\text{rms}}}{n} \right). \qquad (11.11)$$

But this last expression is the convention for *voltage* ratio in terms of decibels [see Equation (8.4)]. Therefore a power ratio of 10,000 and a voltage ratio of 100 are both 40 dB. This tends to be confusing, even to practicing engineers.*

EXAMPLE 11.3 The level $(v_r)_{\text{rms}}$ of a received signal is 3 V (rms), and the noise level n at the same point is 15 mV (rms).

Find: the signal-to-noise ratio in both real numbers and in decibels.

Solution: $\text{SNR}_r = \{[3 \text{ V (rms)}]/[0.015 \text{ V (rms)}]\}^2 = \underline{40,000}$. In terms of decibels, $\text{SNR}_r = 10 \log(40,000) = \underline{46 \text{ dB}}$.

Some of the noise accompanying the received signal v_r can be removed by filtering, but noise that occupies the same band of frequencies as v_r can't be removed. This remaining noise is demodulated with v_r and produces noise with the recovered message $m'(t)$. Let the signal-to-noise ratio after demodulation be SNR_m. It can be shown that for amplitude modulation,

$$\text{SNR}_m(\text{AM}) = (\tfrac{2}{3})\text{SNR}_r(\text{AM}), \qquad (11.12)$$

where the SNRs are in real numbers, not decibels. [The relationship actually depends on $m(t)$; Equation (11.12) holds for modulation by a maximum-amplitude sinusoid.]

Suppose that an FM signal and an AM signal of the same power are received, that the message bandwidth B_m is the same, and that the received noise power density N_0 (see Section 8.8) is the same. It can be shown that after demodulation, the FM receiver gives better performance:

$$\text{SNR}_m(\text{FM}) = 3(\Delta f/B_m)^2 \text{SNR}_m(\text{AM}). \qquad (11.13)$$

This improvement in signal-to-noise ratio can be made as great as desired by increasing Δf. However, this also increases the transmission bandwidth [see Equation (11.8)], and greater transmission bandwidth per channel allows fewer channels in

---

*The confusion could be avoided by adhere:.ce to the strict definition of the decibel as applying only to power ratios. However, common usage has extended the use of the term to voltage ratios and even to phase ratios (e.g., in phase-locked loops).

a multiplexed system. So an FM system offers a tradeoff between signal-to-noise ratio and bandwidth.

EXAMPLE 11.4 Suppose the FM system in Example 11.2 has a signal-to-noise ratio after demodulation of $\text{SNR}_m = 50$ dB.

Find: The SNR_m if an AM system had been used under the same conditions [same $(v_r)_{\text{rms}}$, B_m, and N_0].

Solution: Reversing Equation (11.10) to get SMR_m in real numbers, SNR_m (FM) $= 10^{50 \, \text{dB}/10} = 10^5$. From Example 11.2, $\Delta f = 75$ kHz, and $B_m = 15$ kHz, giving $3 \, (\Delta f/B_m)^2 = 75$. Then from Equation (11.13), SNR_m (AM) $= 10^5/75 =$ 1333, or 31.3 dB. That is, the FM provided a signal-to-noise ratio improvement of $50-31.3 = 18.7$ dB over the AM. This was obtained at the expense of six times greater transmission bandwidth B_t.

11.6   PULSE MODULATION

If we plan to time-division multiplex several messages, the messages can't be continuous in time; they must be sampled so that their samples can be interleaved in time (see Figure 11–3a). This sampling process is shown for a message signal v_m in Figure 11–14. Every interval T_s the signal v_m is sampled to produce the signal v_{ms} in Figure 11–14c. This process can be described mathematically by using the series of sampling pulses $s(t)$ in Figure 11–14b. If we express v_{ms} as the product $v_m s(t)$, then v_{ms} has the value of v_m at instants T_s apart and the value zero elsewhere. The spaces leave room to interleave samples from other messages. (We are assuming for now that the pulses are very narrow.) Because it is the amplitude of the pulses in v_{ms} that contains the information about v_m, this process is called *pulse amplitude modulation* (PAM).

To understand the effect of sampling, it will help to look at the frequency spectra of the signals. Fourier analysis of the sampling waveform $s(t)$ shows that it is composed of many sinusoids with frequencies at multiples of f_s, where $f_s = 1/T_s$ is the sampling frequency [see the spectrum $S(f)$ in Figure 11–14b].

When v_m multiplies $s(t)$, it can be thought of as modulating many carriers—the sinusoidal components of $s(t)$. This is similar to the situation in Figure 11–9. The spectrum $V_{ms}(f)$ in Figure 11–14c therefore looks much the same as the spectrum $V_t(f)$ in Figure 11–9. There we had to observe $\omega_{c2}/2\pi > 2B_m$ to avoid overlap. Similarly, with sampling we have to observe

$$f_s > 2B_m \tag{11.14}$$

for the same reason. Any overlap that does occur is called *aliasing*, which causes distortion in the time domain. Inequality (11.14) is called the *Nyquist criterion*. It says, in effect, that if a signal is sampled often enough, no information about the original signal is lost. This is because Equation (11.14) assures that the original spectrum $V_m(f)$ is still available in $V_{ms}(f)$ without being "messed up" by overlap.

The process for recovering v_m in its entirety is simply to filter v_{ms} with a low-pass filter of bandwidth B_m. Since the resulting spectrum is identical to $V_m(f)$, the

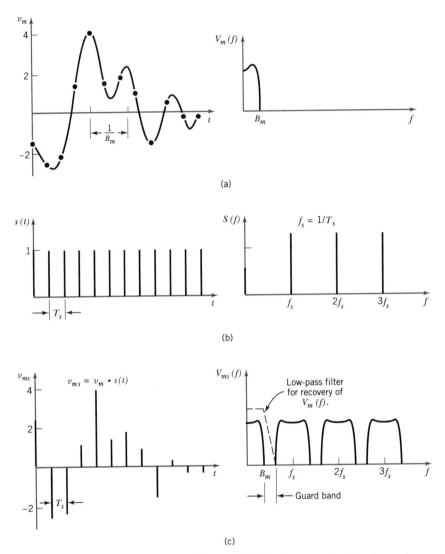

FIGURE 11–14 Pulse amplitude modulation (PAM) involves sampling. Signals are shown in both the time and frequency domains. (a) The message signal. (b) The sampling waveform. (c) The sampled message—mathematically the product of the message and the sampling waveform s(t). For $V_m(f)$ to be recoverable, f_s must be greater than $2B_m$.

recovered signal is identical to v_m in the time domain also. If Equation (11.14) is barely satisfied, the filter must have a very sharp cutoff to separate out the desired part of the spectrum. The guard band in Figure 11–14c allows a simpler low-pass filter to do the job. Therefore a more realistic bound in practice is perhaps $f_s \geq 2.5\,B_m$.

The PAM communication system in Figure 11–15a achieves time-division multiplexing by sampling each of the three messages at an interval of T_s and interleaving

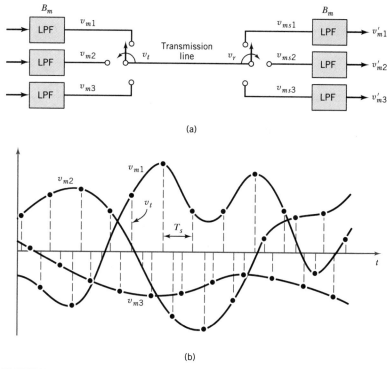

(a)

(b)

FIGURE II–15 (a) PAM system that time-division multiplexes three channels together. (b) The transmitted pulses v_t are the values of the three message signals sampled in rotation.

the samples. The low-pass filters at the inputs limit the bandwidth B_m of v_{m1}, v_{m2}, and v_{m3} to prevent aliasing. The filtered messages are sampled in turn, as represented by the rotating switch. (In practice the sampling is actually done electronically by transistor switches, as discussed in Chapter 16.) The transmitted signal v_t is a series of pulses, as in the example shown in Figure 11–15b. Since there are three messages, the spacing between the transmitted pulses is $\frac{1}{3} T_s$.

At the receiver the samples are separated by a synchronized electronic switch, and low-pass filters recover the messages. For a system with no noise or distortion, the v_m and the v'_m are the same. While the system does change the frequency content of the messages in the process of sampling them (see Figure 11–14), frequency multiplexing is not involved here. All the sampled signals occupy the same range of frequencies; they avoid each other only in the time domain.

All transmission systems have a limited bandwidth. How small can the transmission bandwidth B_t be without affecting the received messages? It can be shown from the Nyquist criterion that the messages can be fully recovered if

$$B_t > 0.5Nf_s, \tag{11.15}$$

where N is the number of messages multiplexed by time division and f_s is the sampling rate for each message. Equation (11.14) together with Equation (11.15) gives

$$B_t > NB_m. \tag{11.16}$$

EXAMPLE 11.5 Five messages are to be time-division multiplexed by PAM.

Given: The message bandwidth for each channel is $B_m = 20$ kHz. The guard band to ease the filter requirements is 10 kHz.

Find: the necessary sampling frequency f_s and the necessary transmission bandwidth B_t.

Solution: According to the Nyquist criterion, $f_s > 2B_m = 40$ kHz. A sampling frequency of $f_s = \underline{50\ kHz}$ allows a guard band of 10 kHz. Then $T_s = 1/f_s = 20$ ms. Once the pulses from the five channels are interleaved, the pulse rate is $5f_s = 250$ kHz. According to Equation (11.15), the minimum transmission bandwidth is $B_t = 0.5 \times 5 \times 50$ kHz $= \underline{125\ kHz}$.   ■

11.7   DIGITAL TRANSMISSION

As we saw in Chapter 9, there are several advantages to having information in digital form. It can stand for a definite integer (dollars, for example) or for a string of letters. It is more immune to noise that would cause the information to be read incorrectly at the receiver. Also, digital signals are more easily switched and stored by electronic circuits. These features often make it desirable to encode even analog information in digital form. All long-distance telephone links installed today use digital transmission. HDTV in the United States uses digital transmission. In fact, the world is moving toward a system in which telecommunications and data communications are combined in a common digital communications network.

Time-division multiplexing is possible with digital information; the pulses representing 1's and 0's of several messages can be interleaved. Since multiplexing is usually associated with some form of modulation, digital transmission is often referred to as *pulse code modulation* (PCM).

Quantization

Before a continuous analog message v_m can be digitally encoded, it must be converted into a sequence of voltage levels $v_0, v_1, v_2, \ldots$. This sampling operation, represented in Figure 11–16a, is the same as in PAM. The message v_m in Figure 11–16b is sampled at a rate of $f_s = 1/0.1$ ms $= 10$ kHz; so the bandwidth must be limited to $B_m < 5$ kHz to prevent aliasing. The resulting sequence of voltages v_i here is approximately 1.7 V, 2.3 V, 2.5 V, 3.2 V, etc.

Digital information takes the form of integers—usually in binary form (see Chapter 9). Because integers don't form a continuum, they can't represent the voltages v_i exactly. So one must decide how much resolution is necessary; do 0.1-V

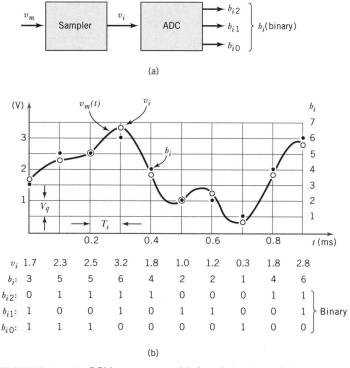

FIGURE 11-16 PCM transmitter. (a) Sampled values of the message signal are converted to binary numbers b_i by an analog-to-digital converter (ADC). (b) V_q is the quantization interval; b_i is the nearest integer to the corresponding v_i/V_q.

differences need to be resolved, or are 0.5-V differences good enough? Suppose a *quantization interval* of $V_q = 0.5$ V is decided on, as in Figure 11–16b. Then the analog samples v_i are converted to the numbers (digits) b_i according to

$$b_i = v_i/V_q \quad \text{(rounded to nearest integer).} \quad (11.17)$$

For our case the sequence of v_i/V_q is 3.4, 4.6, 5.0, 6.4, etc. Rounding these to the nearest integer, we get the sequence of digits b_i: 3, 5, 5, 6, etc., or in binary code (as we choose to transmit it): 011, 101, 101, 110, etc. The conversion in Equation (11.17) is performed by the *analog-to-digital converter* (ADC) in Figure 11–16a. These two operations—sampling and ADC—constitute the transmitter of a digital transmission system.

The three bits necessary to represent each binary number b_i are transmitted in parallel on three lines. For example, the sequence on one line is b_{i2}—the most significant bits. Let $+3$ V stand for a "1," and let -3 V stand for a "0."* Then the

---

*In Chapter 9, we let 5 V stand for "1" and 0 V stand for "0." This is representative of logic circuits. In communication systems of some length it is more common to use voltage levels symmetric about zero. This eliminates dc and requires less power for the same spacing between levels.

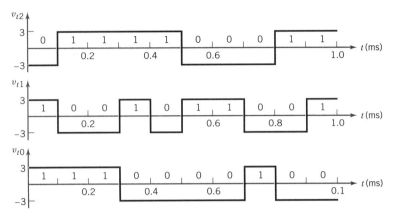

FIGURE 11–17 Non-return-to-zero (NRZ) voltages represent in parallel the bits of successive binary numbers; v_{t0} represents the least significant bit (LSB).

sequences b_{i2}, b_{i1}, and b_{i0} are represented by the transmitted voltage waveforms v_{t2}, v_{t1}, and v_{t0} in Figure 11–17. A waveform holds its value (say $+3$ V corresponding to a "1") for the complete $T_s = 0.1$ ms until it is ready to take on a new value corresponding to the next bit. This is called a *non-return-to-zero* (NRZ) format. As mentioned before, this form of transmission, with pulses representing a binary code, is called *pulse code modulation* (PCM).

The functions performed in a PCM receiver are shown in Figure 11–18a. The numbers b_i represented by the transmitted voltages are converted to a sequence of proportional voltages v_i' according to

$$v_i' = b_i V_q. \tag{11.18}$$

The circuit that performs this conversion is called a *digital-to-analog converter* (DAC).* These v_i' are estimates of the original samples v_i; they differ from the original in that their values are *quantized* as a result of the rounding by the ADC [see Equation (11.17)]. At this point the sequence v_i' could be called a quantized PAM signal (illustrated by the dots in Figure 11–18b). As in the previous section, the pulses v_i' are converted to a continuous signal by a low-pass filter with bandwidth B_m to produce the v_m' shown in Figure 11–18b. Note that, at intervals of $T_s = 0.1$ ms, v_m' passes through the values v_i'.

Because of the quantization, the recovered signal v_m' is, in general, not the same as the original message v_m'. The difference,

$$v_e \equiv v_m' - v_m,$$

is called the *quantization error* (see Figure 11–18c). The magnitude $|v_e|$ is never

---

*The realization of analog-to-digital and digital-to-analog converters is discussed in Chapter 14.

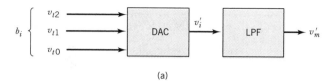

(a)

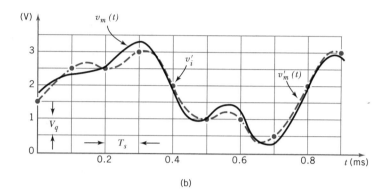

(b)

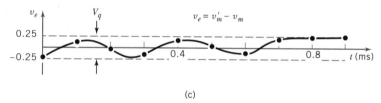

(c)

FIGURE 11-18 PCM receiver. (a) A digital-to-analog converter (DAC) converts the binary numbers b_i back to quantized voltages v_i'. (b) The recovered signal v_m' differs from the original v_m because of quantization. (c) The peak quantization error is $\frac{1}{2}V_q$.

greater than $\frac{1}{2} V_q = 0.25$ V, and the values of v_e are evenly distributed on the interval -0.25 V to 0.25 V. Therefore $(v_e)_{\text{rms}} = 0.14$ V. In general the rms error or *quantization noise* is given by

$$n_q \equiv (v_e)_{\text{rms}} = V_q/\sqrt{12}, \tag{11.19}$$

where V_q is the quantization interval. Except for the effect of occasional reception errors, n_q determines the signal-to-noise ratio SNR_m after decoding:

$$\text{SNR}_m \text{ (PCM)} = \left[\frac{(v_m)_{\text{rms}}}{n_q}\right]^2 \tag{11.20}$$

[This SNR_m can also be expressed in decibels by using Equation (11.10).] If SNR_r is high enough that reception errors are negligible, *the SNR_m for PCM is essentially unrelated to the received SNR_r.* This is not true of AM and FM [see Equations (11.12) and (11.13)].

EXAMPLE 11.6 A sinusoidal "message" $v_m(t)$ is to be transmitted by PCM.

Given: The message signal is $v_m(t) = (5V) \sin \omega_m t$. The signal-to-quantization-noise ratio of the recovered v'_m is to be SNR$_m \geq 48$ dB.

Find: the quantization interval V_q and the number of bits necessary to represent the binary numbers.

Solution: From Equation (11.10), SNR$_m$ in real numbers is 63,100. The rms of v_m is $5V/\sqrt{2} = 3.53$ V. Then from Equation (11.20), $n_q = 3.53V/\sqrt{63,100} = 14$ mV. From Equation (11.19) this requires that $V_q = \sqrt{12} \times 14$ mV = 48.5 mV. In order to code v_m into positive numbers b_i, it is necessary to first raise it by adding 5 V to produce a v_m^+ that is always positive. Then the largest voltage in v_m^+ is 10 V. Since there is a quantization level every $V_q = 48.5$ mV from 0 V to 10 V, the number of quantization levels required is $10V/V_q + 1 = 206$. Now, 7 bits can represent $2^7 = 128$ levels and 8 bits can represent $2^8 = 256$ levels. Therefore we require 8 bits.

In the example, the waveform v_m used only 206 of the 256 available levels. The greatest SNR$_m$ is obtained when the signal to be coded is as large as possible, "filling" the levels. The SNR$_m$ also depends on the waveform of v_m. Typical voice or music waveforms have peaks that only occasionally fill the levels and have less power than a sinusoid of the same peak-to-peak amplitude.

There is no standard waveform for rating the SNR$_m$ of a PCM system. But if the waveform v_m in Figure 11–19 is used, a simple expression for the SNR$_m$ results. The waveform is a sawtooth that fills the levels (5 levels in the example here).

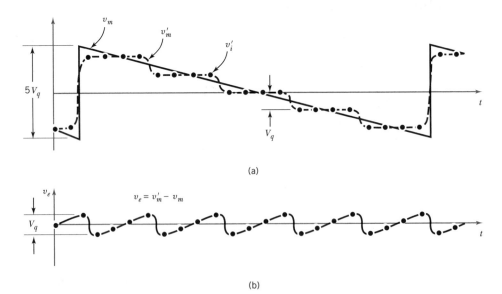

(a)

(b)

FIGURE 11–19 (a) Message signal v_m with sawtooth waveform and corresponding quantized waveform v'_m with M levels. (b) The error (noise) waveform v_e is also a sawtooth but with $1/M$th the amplitude. Therefore the SNR is M^2.

Therefore its peak-to-peak amplitude is $5V_q$, where V_q is the quantization interval. The waveform v'_m recovered at the receiver is shown passing through the quantized voltage levels v_i. The error voltage v_e in Figure 11–19b is the difference between v'_m and v_m. It is also a sawtooth but with a peak-to-peak amplitude of V_q. Since the waveforms of v_m and v_e are both sawtooths, the ratio of their rms values is the same as the ratio of their peak-to-peak values—5 in this case. In general,

$$\frac{(v_m)_{\mathrm{rms}}}{(v_e)_{\mathrm{rms}}} = \frac{(v_m)_{\mathrm{rms}}}{n_q} = M,$$

where M is the number of quantization levels. Then together with Equation (11.20) we have the signal-to-noise ratio

$$\mathrm{SNR}_m(\mathrm{PCM}) = M^2. \tag{11.20'}$$

This holds when v_m is a sawtooth that fills the levels. In the absence of any other information about v_m, this provides as good an estimate of SNR_m as any.

Since N bits can represent 2^N levels, the minimum number of bits necessary to represent M levels is

$$N = \log_2 M \qquad \text{(round up to integer).} \tag{11.21}$$

To make full use of the bits, M is usually chosen to be a power of 2; that is, $M = 2^N$. In this case $\mathrm{SNR}_m(\mathrm{PCM}) = 2^{2N}$, and the signal-to-noise ratio in decibels is

$$\mathrm{SNR}_m(\mathrm{PCM})\mathrm{dB} = 10 \log(2^{2N}) = 20N \log(2) = 6N. \tag{11.22}$$

The results in Example 11.6 agree well with this rule of thumb.

Serial PCM

When the transmission distance is greater than about 20 ft, it is inconvenient and expensive to run a bus of multiple lines. In this case the parallel binary numbers are usually converted to a serial format so they can be transmitted on just one line. (Parallel-in, serial-out registers are discussed in Section 9.6.) Such a transmission system is shown in Figure 11–20a. The binary numbers (or *words*) are shown with the corresponding transmitted waveform v_t in Figure 11–20b. Note that the bits "1" and "0" and the corresponding *symbols* + 3 V and − 3 V now come at three times the rate that they did in Figure 11–19. This is the price paid for reducing from three transmission lines to one. The interval T between symbols is called the *unit interval*, equal here to $\frac{1}{3} T_s = 0.033$ ms. In general,

$$T = T_s/N = 1/Nf_s, \tag{11.23}$$

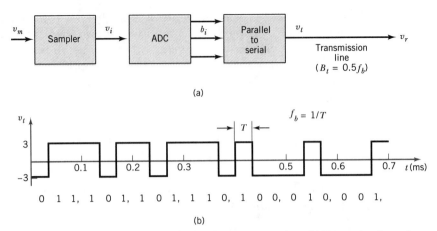

(a)

(b)

FIGURE 11–20 (a) Serial PCM over a single transmission line. (b) Transmitted waveform after interleaving the bits of the parallel signals in Figure 11–17. Check that the figures agree.

where N is the number of bits for each binary number b_i, and f_s is the sampling frequency. The symbol rate f_b is called the *baud*.* The relationship is

$$f_b \equiv 1/T = Nf_s. \tag{11.24}$$

In our case $f_b = 1/0.033$ ms $= 30$ kHz.

What transmission bandwidth B_t is necessary for this serial PCM system? In effect we have a system with a pulse rate $f_b = Nf_s$, as in Figure 11–15, so the minimum bandwidth is

$$B_t = 0.5f_b \tag{11.25}$$

to assure $v_r = v_t$ at intervals of T. These instants when v_r has the correct value will allow the recovery of the number sequence b_i.

EXAMPLE 11.7 A voice signal is to be encoded and transmitted by a serial PCM system.

Given: The signal has a bandwidth $B_m = 10$ kHz. The SNR_m after decoding is to be 60 dB.

Find: the minimum bandwidth B_t of the transmission line.

Solution: The Nyquist criterion [Equation (11.14)] says the voice signal must be sampled at a rate f_s greater than $2B_m = 20$ kHz. Let $f_s = 25$ kHz. Then the system generates a binary number b_i every 40 μs. How many bits must the number have? Using the rule of thumb in Equation (11.22), a signal-to-noise ratio of 60 dB requires $N = 10$ bits. From Equation (11.24), the frequency of

---

*The phrase "baud rate" is frequently encountered, but this is redundant, meaning "symbol rate rate."

the serial bits is $f_b = 10 \times 25$ kHz $= 250$ kHz. From Equation (11.25), the necessary bandwidth is $B_t = 0.5 \times 250$ kHz $= \underline{125 \text{ kHz}}$. ▨

Several serial PCM channels can be time-division multiplexed onto one line (see Figure 11–1c). This is done by interleaving the bits from the channels, as indicated in Figure 11–3b. For three channels, the total baud is three times the baud f_b for each channel. This requires three times the transmission bandwidth. The receiver takes every third bit and reconstructs the original channel 1; it does the same for channels 2 and 3. The interleaving and separation are done with a switching arrangement much the same as the PAM time-division multiplexing and demultiplexing in Figure 11–14a.

Error Rate

In any transmission system there is noise n present at the input to the receiver (see Section 8.8). The attenuation of the transmission medium requires that the receiver include amplification to bring the signal back up to normal level, and this amplification increases the level of the noise also. In the block diagram in Figure 11–21, the amplification and any filtering to limit noise are not shown explicitly; they are assumed to be part of the transmission path.

When n is added to the received signal v_r, the waveform $v_r + n$ no longer passes through the desired value v_t at the center of each unit interval (see the example waveform in Figure 11–21). The difference between $v_r + n$ and v_t at those instants is the noise n. The task of the receiver circuitry is to recover the transmitted waveform v_t perfectly except for an occasional error.

The first step in recovering v_t is to compare $v_r + n$ with a *threshold* halfway between the symbol levels (0 V here). The comparator produces $v_c = 3$ V when $v_r + n > 0$ and $v_c = -3$ V when $v_r + n < 0$. The output v_c is trying to take on the same values as the original v_t.

The instants when v_c stands the best chance of being the same as v_t are at the center of each unit interval. A clock signal* s_2 samples v_c at these instants with a D flip-flop (see Section 9.6). If v_c has the correct value at these instants, the flip-flop output v'_t will be the same as v_t (with a small delay). If $n(t) > 3$ V at a sampling instant when $v_r = -3$ V, then v_c has the wrong value, and v'_t is in error for that unit interval. In the example in Figure 11–21, $n(t) > 3$ V at the eighth symbol, causing v_c to be above 0 V. As a result v'_t is $+3$ V when it should have been -3 V. This error changes the binary number, and the DAC produces a wrong value of v'_m at that instant. For an audio signal this would be heard as a small click. For data transmission it might change the dollar amount of an account.

If the error rate is small enough, its effects will hardly be noticed. A good rule of thumb for PCM transmission of analog information is that one error every million bits (on average) is practically unnoticeable.[†] This is called an *error probability* of

---

*Sampling pulses are produced by a clock recovery circuit. This is usually a phase-locked loop that generates the pulses s_2 from the received signal v_r.

†For transmission of data where errors are critical, a *parity bit* is transmitted periodically to allow checking for most errors.

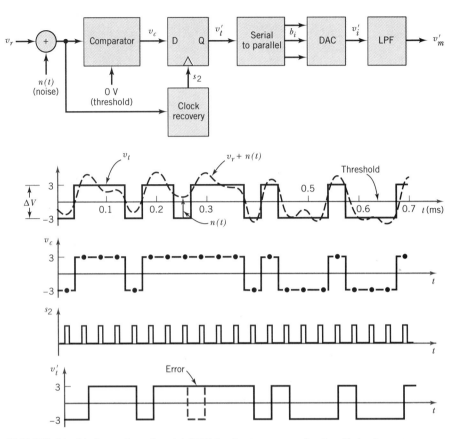

FIGURE 11–21 Reception of serial PCM in the presence of noise. If the instantaneous noise $n(t)$ is greater than $\frac{1}{2}\Delta V$, the comparator makes an incorrect decision (an error) as to the value of the original v_t.

$p_e = 10^{-6}$. The error probability is determined by how large the noise n is relative to the received signal v_r. Let the difference between the high and low symbol levels of v_r be ΔV. (In our case $\Delta V = 6$ V.) It can be shown that for Gaussian noise, the error probability p_e is related to the ratio $\Delta V/n$ by the curve in Figure 11–22, where n is the rms noise voltage. This curve can't be expressed by common functions, but it is closely approximated by

$$p_e \approx 0.8(x^{-1} - 4x^{-3})\exp(-x^2/8) \tag{11.26}$$

where

$$x = \Delta V/n.$$

For example, we need $\Delta V = 9.5n$ in order that $p_e = 10^{-6}$.

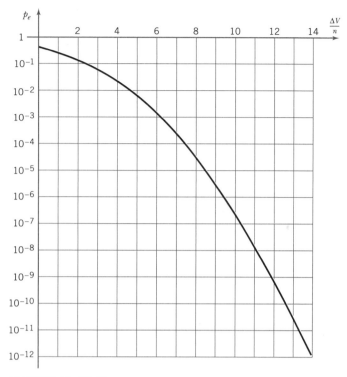

FIGURE 11–22 The error probability p_e as a function of the rms noise n and the distance ΔV between the symbol levels.

EXAMPLE 11.8 Three PCM channels, each with a baud of $f_b = 200$ kHz as in Example 11.7, are time-division multiplexed by interleaving their bits. Noise in the system produces an unacceptably high error rate.

Given: The baud for each channel is $f_b = 200$ kHz. The transmitted symbol difference is $\Delta V = 6$ V. After amplifying and filtering, the received signal v_r again has a symbol difference of $\Delta V = 6$ V, and the received noise is $n = 1$ V (rms).

Find: the error probability p_e. Choose a new transmitted symbol difference $\Delta V'$ such that $p_e = 10^{-5}$. Find the corresponding *error rate* (errors per second) for each channel.

Solution: We have $\Delta V/n = 6$. According to Figure 11–22, $p_e = \underline{1.6 \times 10^{-3}}$. In order that $p_e = 10^{-5}$, the curve shows we need a ratio $\Delta V'/n = 8.5$. But $n = 1$ V (rms). Therefore $\Delta V' = \underline{8.5 \text{ V}}$. From Example 11.7, the baud for each channel is $f_b = 200$ kHz. Then the baud for the multiplexed signal (interleaving the bits from the three channels) is $3f_b = 600$ kHz. That is, there are 600,000 bits transmitted per second. Then $p_e = 10^{-5}$ says there is an average of $(600,000) \, 10^{-5} = 6$ errors per second. On average only one-third of these errors occur on any given channel, so the error rate for one channel is $6/3 = \underline{2 \text{ errors per second}}$.

Would two errors per second be noticeable in music? It would be like a record needle hitting a small bit of dust about every half second. Aside from this there is some background quantization noise, a ''fuzziness'' about 48 dB below the signal level for systems with 8-bit encoding. Note that the quantization noise is not related to the noise at the receiver input; only the error rate is.

In practice a PCM system is usually designed by making ΔV at the transmitter as large as is convenient (perhaps 10 V) and then increasing the length of the transmission path until the maximum allowable p_e is reached. Such a design involves a study of how $\Delta V/n$ decreases with transmission length, which is beyond the scope of this chapter.

Combined Modulation Schemes

PCM signals such as v_t in Figure 11–20b are necessarily carried by transmission lines or waveguides—wires, coaxial cables, or glass fibers. They can't be transmitted in this form by radio because of the low frequencies in the spectrum (it is possible to have 100 successive ''1'' bits, for example). As with analog signals, carrier modulation can be used to raise the spectrum into a broadcast band. The only difference is that now the modulation has only two values rather than a continuum of values. When FM is used to transmit PCM, f_t shifts back and forth between two frequencies as either a ''1'' or a ''0'' is transmitted. Therefore this combination of PCM and FM is called *frequency-shift keying* (FSK). FSK is used by *modems* (*mo*dulator-*dem*odulators) that allow computers and terminals to communicate data over telephone lines.

A combination of PCM and PM is called *phase-shift keying* (PSK). Here the carrier phase is $+90°$ or $-90°$ depending on whether a ''1'' or a ''0'' is transmitted. It can be shown that the transmitted signal has suppressed carrier. Therefore special techniques are necessary to recover the carrier in order to demodulate v_t and recover the PCM signal. The telephone system uses PSK with its microwave radio links for long distance service.

Multiple-Level PCM

As we have seen, PCM has a much greater transmission bandwidth B_t than the bandwidth B_m of an analog message. In Example 11.7 we had $B_m = 10$ kHz and $B_t = 100$ kHz. One thing gained was greater immunity to noise. It is possible to make a tradeoff, sacrificing some noise immunity for less bandwidth, by using multiple levels. Rather than having two possible *symbols*, -3 V and $+3$ V, consider having four possible symbols: -3 V, -1 V, $+1$ V, and $+3$ V. This is called *quaternary* PCM, while the more common two-level signal is called *binary* PCM. With quaternary code, each symbol can represent a block of 2 bits: 00, 01, 10, or 11 corresponding to the four levels. With 2 bits per symbol, the bit rate is twice the symbol rate f_b (the baud). Compare the binary PCM in Figure 11–20b with the quaternary PCM in Figure 11–23 of the same bit sequence. Note that for the same peak v_t, the ΔV must be smaller.

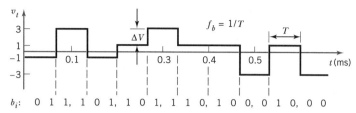

FIGURE 11–23 Quaternary PCM, with each symbol using one of four levels to represent two bits. The advantage is a lower symbol rate (baud) f_b. The baud here is half that in Figure 11–20b.

EXAMPLE 11.9 The encoded voice with a bit rate of 200,000 bits per second (200 kbits/s) in Example 11.7 is to be transmitted by PCM over a band-limited, noisy system.

Given: The maximum transmitted voltage is limited to ± 4.5 V. The noise at the receiver (after the signal is filtered and amplified back to its original level) is 1 V (rms) for a transmission bandwidth of 100 kHz. For a bandwidth of 50 kHz, the noise is 0.4 V (rms).

Find: the bandwidth B_t and the error probability p_e for binary and for quaternary coding.

Solution: For binary coding, the symbol rate (the baud) equals the bit rate: $f_b = 200$ kHz. From Equation (11.25), $B_t = 0.5 f_b = \underline{100 \text{ kHz}}$. The maximum distance ΔV between symbol levels is achieved by choosing the levels -4.5 V and $+4.5$ V, making $\Delta V = 9$ V. Then $\Delta V/n = 9$ V$/1$ V $= 9$, and from Figure 11–22, $p_e = \underline{3 \times 10^{-6}}$.

For quaternary coding, the baud is half the bit rate; $f_b = 100$ kHz, and $B_t = \underline{50 \text{ kHz}}$. For maximum ΔV we choose the four symbols levels -4.5 V, -1.5 V, $+1.5$ V, and $+4.5$ V. Then $\Delta V = 3$ V, $\Delta V/n = 3$ V$/0.4$ V $= 7.5$, and $p_e = \underline{10^{-4}}$. This is a higher error rate, so binary coding is preferred here.

Binary PCM is not always optimum. Sometimes the reduction in bandwidth allowed by quaternary PCM reduces the noise so much that p_e is less for quaternary than for binary code.

In general m-ary coding (with m transmission levels per symbol) can be used. The number of bits each symbol represents is

$$\text{bits per symbol} = \log_2(m). \qquad (11.27)$$

For $m = 8$, there are $\log_2 (8) = 3$ bits per symbol, and the baud is one-third of that for binary coding. The tradeoff is that ΔV is one-seventh as large.

As m is increased, the point will be reached where $\Delta V < n$, and adjacent levels become essentially indistinguishable. Any further increase in m past this point results in no more bits per symbol. This suggests that the bit rate is limited by the system bandwidth and the noise. In Section 11.9 we quantify this relationship. First we need to extend the concept of bit rate to the more general concept of information rate.

11.8   INFORMATION THEORY

What measure is there of the amount of information that has been communicated? You obviously convey less information if you tell your name than if you tell what you did on your summer vacation; different numbers of words are involved. But how can we compare the information content of two different forms of communication, say pictures and words, or words and numbers? There is a saying that a picture is worth 10,000 words,* which might be true if you had to describe a picture so completely that someone else could reproduce it exactly. What is needed is a common measure of information that can be applied to all forms of communication. Then we can compare. This is similar to decision-making in the business world where all factors can be reduced to dollars in order to compare alternatives.

The measure of *information* is not dollars but the number of *bits* it would take to represent the information. In fact the bit is the smallest unit of information—the indication that something is true or false, yes or no, on or off, high or low, "1" or "0". In the game of "twenty questions" all items in the universe are reduced to a string of yeses and nos, hopefully twenty or fewer. By substituting "1" for "yes" and "0" for "no," we have effectively assigned a unique binary number from 00000000000000000000 to 11111111111111111111 to each item. It is apparent that with twenty questions we are limited to $2^{20} = 1,048,576$ items. Put another way, to tell which of these items you are thinking of is to convey $\log_2 (1,048,576) = 20$ bits of information. In general, one of M equally likely items conveys (or is worth) I bits of information where

$$I = \log_2(M). \tag{11.28}$$

EXAMPLE 11.10 A message of ten characters is sent from a computer keyboard.

Given: The characters are drawn from a set of the 86 characters found on a standard typewriter and 42 special control characters, making a total of 128 equally likely characters.

Find: The information conveyed by the ten characters.

Solution: What is the number M of different messages that can be sent? Since each of the ten characters has 128 possibilities, there are $M = 128^{10}$ different (and equally likely) patterns of characters possible. Then the information is $I = \log_2 (128^{10}) = 10 \log_2 (128) = 10 \times 7 = \underline{70 \text{ bits}}$.   ▰

An equivalent way of finding the information of a message is to encode the message in PCM using the least numbers of bits. (This is true if all possible messages are equally likely.) In the above example a 7-bit ASCII code would be used to represent the characters. (ASCII stands for American Standard Code for Information Interchange.) For example, the binary number 0101000 stands for "(";

---

*The number 10,000 in the original Chinese is often converted to 1000 in translation.

1000001 stands for "A"; and so on. There are $2^7 = 128$ such numbers—one for each of the 128 characters. Seven bits must be sent for each character, and $10 \times 7 = 70$ bits must be sent for ten characters.

In the following example the information content is found by encoding the elements of the message with the least number of bits, as if it were to be transmitted by PCM.

EXAMPLE 11.11. Compare the information of a 4×4 in. black-and-white newspaper picture with the information of 10,000 words.

Given: The picture is made up of 75,000 dots, and each dot is one of 16 sizes. Each word has, on average, six characters drawn from a set that consists of the 26 capital letters and the symbols comma, period, question mark, semicolon, apostrophe, and space. All possible pictures are equally likely, and all possible words are equally likely.

Find: which conveys the greater information, the picture or the 10,000 words.

Solution: Since there are 16 dot sizes, it takes $\log_2 (16) = 4$ bits to represent each dot. Since there are 75,000 dots, it takes at least $75,000 \times 4 = 300,000$ bits to encode the picture in PCM. Therefore $I = \underline{300,000 \text{ bits}}$ of information.

The 10,000 words comprise 60,000 characters. Since there are 32 possible characters, it requires $\log_2 (32) = 5$ bits to encode each character. Then the total number of bits is $60,000 \times 5 = \underline{300,000 \text{ bits}}$. This is the same information as the picture. ▬

If the picture is larger or the dot spacing is smaller, then the picture is worth more bits. A TV picture has about 500 lines with about 500 resolvable points along each line. Then the total number of picture elements (called *pixels*) is $500 \times 500 = 250,000$. Assuming again that 16 shades of gray are satisfactory, each element corresponds to 4 bits, and the TV picture (one frame) is worth $I = 250,000 \times 4 = 1,000,000$ bits.

What is the information content of an analog signal such as music? To represent an analog signal perfectly (with infinite resolution) would take an infinite number of bits. But an analog signal with noise added can be represented by a finite number of bits—if the number of bits is chosen so that the quantization noise is the same as the added noise.

EXAMPLE 11.12 Consider a 1-s segment of modern music, where this "music" can be any audible sound. Noise is also present.

Given: The bandwidth of the music is $B_m = 15$ kHz. The music waveform v_m has values evenly distributed over the interval -4 V to 4 V. The added noise (sound not part of the original information) is $n = 4.51$ mV (rms).

Find: the information of the segment of music (with noise). Find the signal-to-noise ratio SNR_m.

Solution: We encode the music (without noise) so that the resulting quanti-zation noise n_q equals n. From Equation (11.19), the quantization interval must be $V_q = [4.51 \text{ mV (rms)}] \sqrt{12} = 15.62 \text{ mV}$. Then for each sample the number of quantization levels from -4 V to 4 V is $8 \text{ V}/15.62 \text{ mV} = 512$. Since $512 = 2^9$, this requires 9 bits per sample. According to Equation (11.14), the minimum sampling rate is $f_s = 2B_m = 30 \text{ kHz}$, or 30,000 samples in 1 s. Then the minimum number of bits to encode the music (with the same noise) is $I = 9 \times 30,000 = \underline{270,000 \text{ bits}}$. This is the information content of the 1-s segment of music.

Since v_m is evenly distributed over an 8-V interval, its rms value is $(v_m)_{rms} = 8 \text{ V}/\sqrt{12} = 2.31 \text{ V (rms)}$. Then $\text{SNR}_m = [2.31 \text{ V}/4.51 \text{ mV (rms)}]^2 = 262,000 = \underline{54.2 \text{ dB}}$. This is the signal-to-noise ratio for both the original signal and for the encoded signal with quantization noise. ▄

In the above example, the ear may not be able to discern all the information available in the music. However, a properly designed electronic device would be able to distinguish 2^I different segments of music. [Compare Equation (11.26)].

The *information rate* of a message is the number of bits of information conveyed per second, whether or not the information is actually digitally encoded as bits. For example, an analog TV signal has 30 complete frames (pictures) per second. Then if each frame is worth 1,000,000 bits, the information rate is 30,000,000 bits per second, or 30 Mbits/s. (Actually this is high because all pictures are not equally likely; consecutive pictures are usually similar.)

In Example 11.12 the information of the 1-s segment of music was $I = 270,000$ bits. Therefore the information rate was 270,000 bits per second (or 270 kbits/s).

11.9   SYSTEM CAPACITY

The maximum rate at which a communication system can transmit information depends on whether or not the message is modulated or encoded and on what type of modulation or encoding is used. The *capacity* of a system is the maximum infor-mation rate, regardless of what type of modulation or encoding is used. We will see that the capacity is determined by two properties of the communication system: the system bandwidth B_s and the received signal-to-noise ratio SNR_r. Both of these parameters are limited in any system.

It can be shown by an analysis similar to that in Example 11.12 that the max-imum information rate for a communication system is

$$C = 2B_s \log_2(\sqrt{\text{SNR}_r + 1}) = B_s \log_2(\text{SNR}_r + 1), \tag{11.29}$$

where SRN_r is in real numbers (not decibels). It was shown by the first information theorist, Claude Shannon, that no modulation or encoding scheme can provide a greater information rate for the same B_s and SNR_r. Therefore Equation (11.29) gives the information-rate capacity or *Shannon capacity* of a system.

EXAMPLE 11.13 Consider two noisy communication systems with limited transmission bandwidths.

Given:

$$\text{System 1: } B_s = 15 \text{ kHz} \quad \text{and} \quad \text{SNR}_r = 54.2 \text{ dB.}$$

$$\text{System 2: } B_s = 45 \text{ kHz} \quad \text{and} \quad \text{SNR}_r = 18.0 \text{ dB.}$$

Find: the Shannon capacity C for each system.

Solution: System 1: Converting $\text{SNR}_r = 54.2$ dB to real numbers we have $10^{54.2/10} = 262{,}000$. From Equation (11.29), $C = 15 \text{ kHz} \times \log_2 (262{,}000) = 15 \text{ kHz} \times 18 = 270$ kbits/s, or 270,000 bits per second.

System 2: $\text{SNR}_r = 18.0 \text{ dB} = 63.1$. Then $C = 45 \text{ kHz} \times \log_2 (63.1) = 270$ kbits/s, or 270,000 bits per second (the same as for system 1). ▬

The message in Example 11.12 had a bandwidth of 15 kHz and an information rate of 270,000 bits per second. Therefore either system 1 or system 2 in Example 11.13 has the capacity to transmit that message, given the proper modulation or encoding. Since system 1 has the same signal-to-noise ratio desired for the message bandwidth, it can transmit the message directly; no modulation is necessary.

How can system 2 transmit the message in Example 11.12? If the message is transmitted directly (baseband), the signal-to-noise ratio of the message will be the same as that of the system: $\text{SNR}_m = \text{SNR}_s = 18.0$ dB. Thus it is not the same message; it has less information because of the increased noise. The problem is that the message bandwidth $B_m = 15$ kHz occupies only one-third of the system bandwidth $B_s = 45$ kHz, wasting two-thirds of the system capacity.

In the following example, the message of Example 11.12 is transmitted over system 2 by PCM encoding. This essentially expands the message bandwidth by a factor of 3 to fully use the capacity of the system.

EXAMPLE 11.14 Music with an information rate of 270 kbits/s is to be transmitted over a communication system with a capacity of 270 kbits/s by m-ary PCM.

Given: The message (music) has a bandwidth of $B_m = 15$ kHz, and the signal-to-noise ratio after decoding is to be $\text{SNR}_m = 54.2$ dB. The system has the specifications $B_s = 45$ kHz and $\text{SNR}_r = 18.0$ dB (or 63.1 in real numbers).

Find: the number m of transmission levels and the symbol rate f_b for the m-ary PCM. Also find the error probability p_e.

Solution: Let the bandwidth of the transmitted signal B_t equal B_s. This together with Equation (11.25) gives the maximum symbol rate (or baud) $f_b = 2B_s = 90$ kHz, or 90,000 symbols per second. To transmit 270,000 bits per second, this requires $270{,}000/90{,}000 = 3$ bits per symbol. Equation (11.27) gives the number of transmission levels $m = 2^3 = 8$ for the m-ary PCM.

As shown in Example 11.12, when the message is coded with 270 kbits/s, the quantization noise gives $\text{SNR}_m = 54.2$ dB as desired.

To find p_e, we need $\Delta V/n$, which is determined by $m = 8$ and $\text{SNR}_r = 63.1$. The received signal spans a range of $(m - 1)\Delta V$. Since this signal is about evenly distributed over the range, its rms value is $(v_r)_{\text{rms}} = (m - 1)\Delta V/\sqrt{12} = 2.02\Delta V$. From Equation (11.9), $n = (v_r)_{\text{rms}}/\sqrt{\text{SNR}_r} = 0.254\Delta V$, and $\Delta V/n = 3.93$. From Figure 11–22 this yields $p_e = \underline{0.025}$, or one error every 40 bits on average. ∎

This example shows that PCM can realize nearly the full system capacity given by Equation (11.27). This is also true for SSB (single-sideband AM), but FM and standard AM are less efficient in their use of capacity.

One concern with the results of Example 11.14 is that an error probability of $p_e = 0.025$ is rather high—so high that it can't be ignored in calculating SNR_m. However, Shannon showed that with sufficiently long error-correction codes, PCM can realize the capacity in Equation (11.29) with a p_e arbitrarily close to zero. By this method, the bound given by Equation (11.29) can be approached as closely as desired.

If error-correcting codes are not used, p_e can be reduced only by sacrificing some capacity. For $p_e = 10^{-6}$, the maximum transmission rate for a PCM system is about

$$C' = B_s \log_2(\text{SNR}_r/7.5 + 2). \tag{11.30}$$

This holds for $\text{SNR}_r \geq 16$, or 12 dB. Otherwise $p_e = 10^{-6}$ can't be attained even for binary PCM.

11.10   TRANSMISSION MEDIA

The bandwidth B_s of a communication system is usually determined by the frequency response of the transmission medium—copper lines or glass fiber. (The bandwidth of air to radio waves is virtually unlimited if it is not raining.) Figure 11–24 shows the frequency response of two copper media—a twisted wire pair and a coaxial cable. In both cases the response in decibels is proportional to $-\sqrt{f}$.

Because the rolloff with frequency is relatively slow, the bandwidth of a copper transmission line can be extended by equalization, as shown in Figure 11–15. A filter whose loss has the opposite shape of the cable response is placed in cascade with the cable. The equalized cable response is the sum of the two, resulting in a flat response out to $B_s = 130$ MHz. The tradeoff is that the response now has a loss of 40 dB for all frequencies within the band, not just at 130 MHz. But frequency distortion (see Section 8.5) has been eliminated for signals within B_s, and the flat loss can be made up by amplification.

The cable response in Figure 11–25 could have been equalized out to $B_s = 300$ MHz, but then the flat loss would have to be 60 dB. The necessary increase in amplification from 40 dB to 60 dB would increase the noise at the receiver. Depending on the application, this level of noise may be unacceptable, and the cable bandwidth couldn't be stretched to 300 MHz.

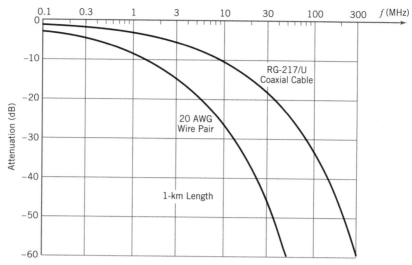

FIGURE 11–24 Frequency response of copper transmission media 1 km in length. Because of the skin effect, the resistance increases with frequency, and the loss in decibels grows as $\sqrt{f}$.

The loss in copper transmissions lines is caused by the resistance of the copper. The reason the loss increases with frequency is that the current is confined to an increasing thin layer at the copper surface as the frequency increases. This is called the *skin effect.* The effective thickness of the conducting skin is inversely proportional to $\sqrt{f}$.

The cross section of a *coaxial cable* is shown in Figure 11–26a. In this case the

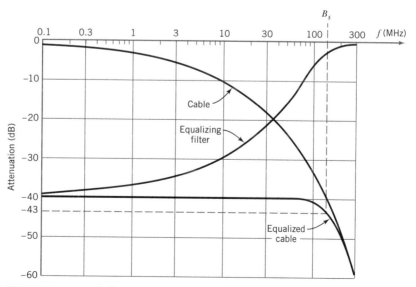

FIGURE 11–25 Cable response made flat by an equalizing filter, resulting in a 3-dB bandwidth of $B_s = 130$ MHz. The response below B_s is made to have the same loss that the cable has at $f = B_s$.

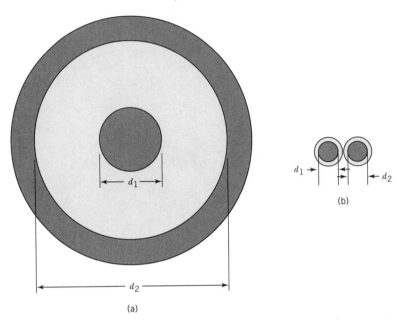

FIGURE 11–26 (a) Cross section of a coaxial cable. For RG-217/U cable, d_1 = 0.106 in. and d_2 = 0.370 in. (b) Cross section of a twisted wire pair. For 20-gauge wire, $d_1 = d_2 = 0.032$ in. At high frequencies, loss is inversely proportional to the diameter of a conductor.

conducting skin is on the inner surface of the cylindrical shield (outer conductor) and on the outer surface of the central wire (inner conductor). Therefore the resistance of each is inversely proportional to its circumference or diameter. The loss of a cable is given by

$$A = -14.2 \text{ dB} \times \frac{\ell}{1 \text{ km}} \times \frac{1''/d_1 + 1''/d_2}{Z_0/1\Omega} \times \sqrt{\frac{f}{1 \text{ MHz}}} \qquad (11.31)$$

where ℓ is the length of the cable, d_1 is the diameter of the inner conductor, d_2 is the inner diameter of the outer conductor, and Z_0 is the characteristic impedance of the cable (see Section 8.9). For example, an RG-217/U coaxial cable has $d_1 = 0.106''$, $d_2 = 0.370''$, and $Z_0 = 50 \ \Omega$. Then 1 km of the cable has an attenuation of $A = -3.44 \text{ dB} \times \sqrt{f/1 \text{ MHz}}$. This response is graphed in Figure 11–24.

The cross section of a twisted wire pair is shown in Figure 11–26b. The same Equation (11.31) gives the loss for the wire pair, where $d = d_1 = d_2$ is the diameter of each conductor. The plastic insulation around each conductor holds them at a center-to-center spacing about $1.7d$, resulting in a characteristic impedance of about $Z_0 = 100 \ \Omega$. For 20-gauge wire, $d = 0.032''$, and the attenuation of a 1-km twisted pair is $A = -8.9 \text{ dB} \times \sqrt{f/1 \text{ MHz}}$. This response is graphed in Figure 11–24.

More information on transmission lines is available in Chapter 29 of *Reference Data for Radio Engineers* (see "For Further Study" at the end of the chapter).

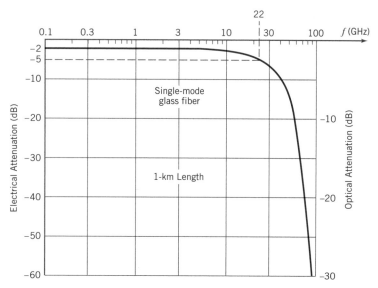

FIGURE 11–27 Frequency response of an optical fiber to modulated light of wavelength 1.3 μm. There is a flat attenuation (2 dB here) due to light loss, and an attenuation that increases with frequency due to dispersion (pulse widening). The attenuation in decibels grows as f^2, where f is the frequency of the modulating signal. The effective electrical attenuation is twice that of the optical attenuation in decibels.

Glass fiber is a far better transmission medium than copper. Light amplitude modulated with a signal is transmitted over the fiber and is converted to an electrical signal at the receiving end. Figure 11–27 shows the response of 1 km of single-mode glass fiber to the signal frequency. The flat attenuation is only 2 dB, and the 3-dB bandwidth is 22 GHz (or 22,000 MHz)! This is a recent development made possible by the production in the 1970s of glass pure enough to have low optical attenuation.

A careful distinction needs to be made between the optical attenuation and the effective electrical attenuation. Each photon of light carries energy, and the number of photons per second determines the optical power. Since an attenuation of 3 dB is a halving of power, it is natural to say that a fiber that halves the number of photons per second has 3 dB of (optical) loss. But at the receiving end, each photon is converted to an electron, and a halving of the electrons per second (current) is 6 dB of (electrical) loss. This is because electrical power is proportional to the square of current. As a result, the effective electrical attenuation of a fiber is twice the decibels of its optical loss. When comparing the performance of a glass fiber with that of copper wire, we must be consistent in looking at the electrical attenuation.

Another possible confusion is that the light itself has a frequency. This confusion is avoided by speaking of the *wavelength* of the light—typically 1.3 μm. This corresponds to 230 THz, or 230,000,000 MHz. But in speaking of the frequency response of a fiber, it is always the frequency of the modulating signal that is referred to.

There are two loss mechanisms in glass. One is the actual reduction of the number of photons due to scattering and absorption. This loss in decibels is proportional to the length of the fiber and is independent of the frequency of the signal. The other loss is due to dispersion. A pulse of light transmitted over the fiber is dispersed so that it is about 1 cm longer after traveling 1 km. This causes the cycles of a high-frequency signal to overlap and reduce the amplitude of the modulation. Photons are not lost, but amplitude of the light fluctuation is lost. This loss in dB is proportional to the square of the signal frequency and the square of the fiber length.

The expression for optical attenuation in a glass fiber is

$$A_{\text{opt}} = -3 \text{ dB } (f \ \ell/B)^2 - C \ \ell, \tag{11.32}$$

where B is the range-bandwidth product, C is the loss in decibels per kilometers, and ℓ is the fiber length in kilometers. The first term is due to dispersion, and the second term is due to scattering and absorption. The effective electrical attenuation of $A_{\text{elec}} = 2 \times A_{\text{opt}}$. Therefore

$$A_{\text{elec}} = -3 \text{ dB } (f \ \ell/B')^2 - C' \ \ell, \tag{11.32'}$$

where $B' = 0.7B$ and $C' = 2C$. Typical values for a good single-mode fiber are $B' = 22$ GHz-km and $C' = 2$ dB/km. The corresponding response for a 1-km length of fiber is shown in Figure 11–27.

The f^2 dependence of fiber attenuation in dB results in the sudden rolloff of the response in Figure 11–27. This makes it impractical to equalize the fiber to increase its bandwidth much. In practice, the bandwidth of the fiber is already so great that it is not usually the limiting factor.

11.11   CASE STUDY

An engineer is to design a communication system for a large commercial airliner that will let each passenger listen to one of 127 music channels. All of the channels are to be multiplexed onto one wire pair that loops back and forth among the seats—a total length of 440 m. At each seat, a number 1 through 127 typed into a key pad makes one of the channels available through a pair of earphones. The sound is to be in stereo with 20 kHz bandwidth and 96 dB signal-to-noise ratio (the quality of a CD player).

Some technical constraints are known. The EMI (electromagnetic interference) environment in the plane leads to a noise density of $\sqrt{N_0} = 100$ nV/$\sqrt{\text{Hz}}$ at each seat's receiver (see Section 8.8 on noise). The signal voltage on the wire pair is not to exceed 10 V.

The engineer decides to try a 20-gauge wire pair—a lightweight and flexible transmission medium. If this proves too lossy, a heavier gauge can be used. Since one wire pair is specified, the choices for multiplexing are frequency-division and time-division. Both will be investigated, starting with AM, the simplest form of frequency-division multiplexing.

Figure 11–9 shows the scheme for multiplexing with AM. B_m = 20 kHz, and with a guard band of 5 kHz, each AM signal requires 45 kHz of bandwidth. Each of the 127 channels requires two of these AM signals for stereo, or 90 kHz per channel. Then the total system bandwidth is B_s = 127 × 90 kHz = 11.4 MHz. To test the feasibility of the scheme, the engineer calculates the transmitted signal level necessary to achieve the 96 dB signal-to-noise ratio.

The 3-dB bandwidth of each AM signal is $2B_m$ = 40 kHz. From Equation (8.17) the noise bandwidth of each receiver is $B_n = (\pi/2) \times 40$ kHz = 63 kHz. From Equation (8.18) the noise at the receiver is $n = 100$ nV/$\sqrt{\text{Hz}} \times \sqrt{63\text{ kHz}}$ = 25 μV rms. For a message signal-to-noise ratio SNR_m = 96 dB, with AM the received signal-to-noise ratio SNR_r must be $\frac{3}{2}$ (or 3.5 dB) greater than 96 dB [see Equation (11.12)]. From Figure 11–24, the attenuation of the wire pair is 30 dB/km at 11.4 MHz. Then the seat at a distance of 440 m would see a loss of 0.44 × 30 dB = 13.2 dB. Then the transmitted signal must be 96 + 3.5 + 13.2 dB = 112.7 dB greater than 25 μV rms. Since 112.7 dB is a factor of 432,000 [see Equation (8.4)], the transmitted signal for each channel must be 432,000 × 25 μV rms = 10.8 V rms. The transmitted power for all channels is 127 times that for one channel. Therefore the transmitted voltage for all channels is $\sqrt{127}$ × 10.8 V rms = 122 V rms. But this exceeds the specified maximum of 10 V.

The calculations show that AM requires a larger signal-to-noise ratio than is available. If we are to transmit the required information rate, Shannon's capacity theorem [Equation (11.29)] says reducing the SNR_r must be accompanied by increasing the transmission bandwidth B_s. One method of trading off greater B_s for smaller SNR_r is to use FM [see Equation (11.13)]. But a more effective tradeoff can be achieved by using PCM.

PCM allows time-division multiplexing—interleaving the 127 channels in time as in Figure 11–3a. The total bit rate will determine the necessary B_s, and this together with a reasonable error probability p_e will determine the necessary transmitted signal voltage.

From Equation (11.22), a SNR_m of 96 dB requires 16 bits per sample. The Nyquist sampling theorem [Equation (11.14)] requires a sampling rate of at least twice 20 kHz. An f_s = 44 kHz is chosen to leave some margin. This gives 16 × 44 kHz = 704 kb/s for each of the left and right stereo signals, or a total bit rate of 1.408 Mb/s for each of the 127 channels. One additional channel will be needed to provide framing information. Then the transmitted bit rate is 128 × 1.408 Mb/s = 180.224 Mb/s. For binary serial PCM, Equation (11.25) gives the minimum system bandwidth as half the bit rate, or 90 MHz. To leave some margin, B_s = 100 MHz is chosen. From Equation (11.31), the loss of a 20-gauge wire pair for f = 100 MHz and ℓ = 0.44 km is 40 dB. To keep the frequency response flat, the cable must be equalized out to 100 MHz, resulting in a flat loss of 40 dB for the farthest seat.

The noise bandwidth of the equalized line is $(\pi/2) \times 100$ MHz = 157 MHz. Then the noise at the receiver is $n = 100$ nV/$\sqrt{\text{Hz}} \times \sqrt{157\text{ MHz}}$ = 1.25 mV rms. A reasonably small error probability is $p_e = 10^{-10}$. From Figure 11–22 this requires

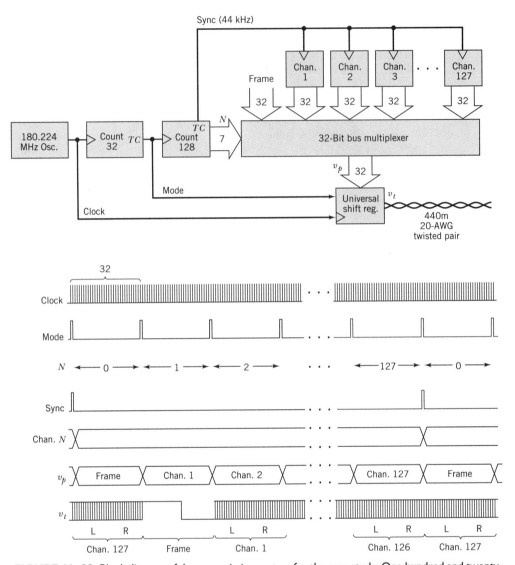

FIGURE 11–28 Block diagram of the transmission system for the case study. One hundred and twenty-seven channels of digitized stereo sound are multiplexed together onto one twisted wire pair. A 180.224-MHz master clock times the operations.

$\Delta V_r/n = 12.8$, or $\Delta V_r = 16$ mV. With a transmission loss of 40 dB or a factor of $1/100$, the transmitted symbol difference must be 100×16 mV $= 1.6$ V. This is conveniently implemented by differential emitter-coupled logic (see Chapter 13). The requirement that the signal on the wire pair be less than 10 V has been met.

Figure 11–28 shows one possible implementation of the PCM transmitter. The 127 channels each produce 32 bits (16 bits for the left signal and 16 bits for the right) 44,000 times a second. The multiplexer takes each of these blocks of 32 bits in turn and delivers them to the universal shift register for conversion to a serial format (see Figure 9–31).

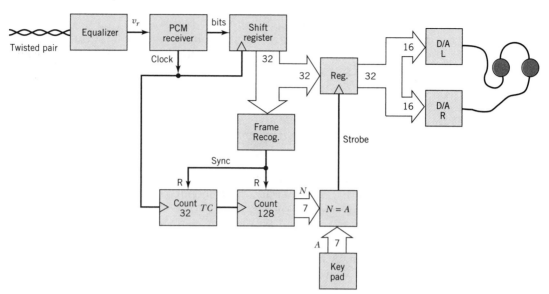

FIGURE 11–29 Block diagram of the receiver for the case study. By choosing a number *A* with a key pad, the listener selects which blocks of code to capture and convert to analog sound.

The result is a series of bits appearing on the twisted wire pair (see the v_t waveform in Figure 11–28). Sixteen bits for the L (left) signal of Channel 1 are followed by 16 bits for the R signal of Channel 1 and so on through the 32 bits for Channel 127. In order to mark the position of Channel 1 for the receiver, a special block of 32 framing bits is inserted just before Channel 1. These consist of 16 1's followed by 16 0's.

The timing of the various signals is controlled by a clock signal from a 180.224-MHz oscillator, which shifts the serial bits out of the universal shift register. A mode signal causes the universal shift register to accept 32 bits in parallel from the multiplexer every 32 clock cycles. This is generated by the terminal count (TC) of the "Count 32" counter. The mode signal also increments the count of the "Count 128" counter by one. This count is brought out as the 7-bit binary number *N*. As *N* increments from 0 to 127, it causes the multiplexer to select the corresponding channel for the universal shift register (the frame signal is Channel 0). At the terminal count of the "Count 128" counter, a sync pulse tells the channels to prepare new 32-bit blocks for the multiplexer.

The receiver at one of the seat locations is shown in Figure 11–29. The frequency response of the wire pair is equalized to be flat (see Figure 11–25); the amount of equalization depends on the distance of the seat from the transmitter. A PCM receiver (see Figure 11–21) decides whether each bit is a 1 or a 0 and recovers a 180.224-MHz clock signal. The bits are shifted into a shift register so they can be examined 32 at a time. When the 32 bits are 16 1's followed by 16 0's, the Frame Recognizer generates a sync pulse to reset the counters to zero. Then the number *N* indicates which channel the current block of 32 bits represents. If *N* agrees with the number A selected by the key pad, a strobe pulse is generated to latch the 32-bit block in the register. One digital-to-analog converter (D/A) looks at 16 of the bits and generates the left signal for the headphones. Another D/A generates the right signal.

11.12 SUMMARY

A communication system usually involves multiplexing. Many channels can share the system by interleaving in either the frequency domain or the time domain. Carrier modulation schemes such as AM, SSB, PM, and FM are ways of implementing frequency-division multiplexing. The transmission bandwidth for SSB is the same as the message bandwidth. For AM it is double the message bandwidth. For FM it can be many times as great, but this is traded off for a higher signal-to-noise ratio for the recovered message.

Time-division multiplexing involves sampling of each message often enough that essentially none of the message is lost. The Nyquist criterion says the sampling rate must be at least twice the bandwidth of the message. The sampled values can modulate the amplitude, width, or phase of a sequence of pulses, producing PAM, PWM, or PPM.

PCM is a digital form of time-division multiplexing in which the sampled values are digitally encoded. Numbers are assigned to quantization levels of the values; the more levels, the finer the resolution and the less the quantization noise. The numbers are transmitted by binary, ternary, or m-ary symbols depending on whether two, three, or more transmission levels are used. (Transmission levels are not to be confused with the quantization levels for the message.) The number m of transmission levels together with the symbol rate f_b determines the bit rate.

The measure of information in a message is the minimum number of bits required to encode it. The capacity (the maximum information rate) of a communication system is determined by its bandwidth and its signal-to-noise ratio. The total information rate for all messages on a system can't exceed the Shannon capacity of the system. PCM and SSB use the capacity more efficiently than AM or FM do. PCM offers an effective tradeoff of bandwidth for signal-to-noise ratio— better than that of FM.

FOR FURTHER STUDY

P. F. Panter, *Modulation, Noise, and Spectral Analysis*, McGraw-Hill, New York, 1965.

Transmission Systems for Communications, 5th ed., AT&T Bell Telephone Laboratories, Holmdel, NJ, 1982.

Reference Data for Radio Engineers, 7th ed., Howard W. Sams, Indianapolis, 1985.

B. P. Lathi, *Modern Digital and Analog Communication Systems*, Holt, Rinehart, and Winston, New York, 1983.

PROBLEMS

Easy Drill Problems (answers at end of chapter)

D11.1. An AM modulator transmits $v_t = [1 + m(t)] \cos(\omega_c t)$, where $\omega_c = 2\pi \times 1$ MHz. Let $m(t) = 0.8 \sin(\omega_m t)$, where $\omega_m = 2\pi \times 1$ kHz. Use the identity $\cos(A)\sin(B) \equiv 0.5 \sin(A + B) - 0.5 \sin(A - B)$ to express v_t as the sum of three sinusoids. Sketch the components of v_t in the frequency domain, as in Figure 11–6d. (The sketch need not be to scale, but label the frequencies of the components.)

D11.2. Repeat D11.1 for $m(t) = 0.8 \sin(\omega_m t) + 0.4 \sin(2\omega_m t)$, where $\omega_m = 2\pi \times 1$ kHz. Sketch $m(t)$ in the time domain; does it exceed unity? Sketch v_t in the time domain.

D11.3. Repeat D11.1 for $m(t)$ the square wave v_1 in Figure 8–15. The Fourier expansion of v_1 is given in Equation (8.5), where $f = 10$ kHz (use only four terms). Sketch v_t in the time domain.

D11.4. Commercial AM stations broadcast at about 1 MHz, and commercial FM at about 100 MHz. What are the respective wavelengths? What are the respective minimum antenna lengths for efficient transmission?

D11.5. The received signal level at an antenna is $(v_r)_{\text{rms}} = 100$ μV, and the received rms noise is $n = 3$ μV. Find the received signal-to-noise ratio in both real numbers and in decibels.

D11.6. The antenna impedance in D11.5 is 300 Ω. What is the received signal power? What is the noise power? Does the power ratio agree with the answer to D11.5?

D11.7. The bandwidth of a message is 3 kHz. If the signal is to be sampled for PAM,
(a) what is the minimum sampling frequency to avoid aliasing, and
(b) what is the sampling frequency if there is to be a guard band of 2 kHz?

D11.8. Twenty-four channels like that in D11.7 (b) are to share a system by time-division multiplexing. What is the minimum system bandwidth?

D11.9. A message to be transmitted by PCM is sampled to produce the following v_i sequence: 2.21, 0.98, 0.47, 2.85, 2.02, 1.20. The quantization interval is $V_q = 0.1$ V. What is the sequence of transmitted numbers (in 5-bit binary)? What is the recovered sequence v_i' in the receiver?

D11.10. What is the quantization noise for $V_q = 0.1$ V? Compare this with the root mean square (rms) of v_e in D11.9, where the v_e sequence is $v_i' - v_i$.

D11.11. For a received symbol difference of $\Delta V = 5$ V, estimate from Figure 11–22 the error probability for rms noise $n = 0.42$ V, 0.63 V, 1.1 V, and 1.7 V. Compare with p_e estimated by Equation (11.24).

D11.12. Represent the binary sequence 1, 0, 0, 0, 1, 1, 0, 1, 1, 0 by a quaternary signal with the levels -3 V, -1 V, 1 V, and 3 V, as in Figure 11–23.

D11.13. What is the information conveyed by a four-digit decimal display? (All numbers in its range are equally likely.)

D11.14. What is the information conveyed by a 10-V full-scale analog meter? There is a 0.1-V (rms) needle vibration superimposed on the reading, and all readings are equally likely.

D11.15. What is the Shannon capacity of a system with a 3-kHz bandwidth and a 48-dB signal-to-noise ratio?

D11.16. **(a)** What is the capacity of a 30-km single-mode glass fiber with a 12-dB signal-to-noise ratio at the receiver? The range-bandwidth product is $B' = 22$ GHz-km.
(b) What is the maximum information rate for PCM over this fiber with $p_e = 10^{-6}$ and no error correction code?

D11.17. A telephone-grade voice channel requires about 64 kbits/s. How many voice channels can the system in D11.16(b) carry? It takes about 20,000 bits to transmit the characters on the page of a book. How long will it take the system in D11.16(b) to transmit the contents of a 700-page book?

D11.18. What is the attenuation of 2 km of 20-gauge wire equalized out to 10 MHz?

Application Problems

P11.1. Commercial AM radio is allowed carrier frequencies from 540 kHz to 1600 kHz. The message bandwidth for each channel is limited to 7.5 kHz, and the guard band between channels in a listening area is 5 kHz. With these constraints, what is the minimum spacing between AM channels in a listening area? What is the maximum number of channels that can be received in a listening area?

P11.2. Several FM channels have a message bandwidth 7.5 kHz, a peak frequency deviation of 75 kHz, and a minimum guard ban of 45 kHz. Find the minimum spacing between carrier frequencies of adjacent channels.

P11.3. An AM signal from P11.1 and an FM signal from P11.2 are received with the same strength, and the received noise density is the same for both channels. What is the difference in decibels between the SNR_m (after demodulation) for the AM and for the FM?

P11.4. A 1-kHz sinusoid is sampled at a 7-kHz rate and quantized so that the signal-to-quantization-noise ratio is at least 30 dB. For serial binary PCM transmission, find the minimum baud f_b and the minimum transmission bandwidth.

P11.5. A message signal with 3-kHz bandwidth is sampled so there is a 2-kHz guard band, and the samples are encoded with 8-bit binary PCM. Twenty-four such channels are time-division multiplexed into one serial binary PCM signal. What is the baud of each channel? What is the baud of the transmitted signal?

P11.6. The transmitted signal in P11.5 is received with a symbol difference of 3 V, and the received rms noise is 270 mV. Find the error probability, the error rate for each channel, and the mean time between errors for each channel.

P11.7. Repeat P11.5 for quaternary PCM (four transmission levels).

P11.8. Spy A has arranged to pass information to a confederate by releasing one of six colors of smoke into the air. Spy B plans to release a sequence of three puffs of smoke; each puff is either black or white. By what factor does the information of spy B exceed the information of spy A? (For each scheme, all signals are equally likely.)

P11.9. A typist can type 60 words per minute.
(a) If each word is one of 10,000 equally likely English words, what is the information rate?
(b) With an average of 6 characters per word, the character rate is 360 per minute. If each character were one of 32 equally likely characters (counting punctuation), what would the information rate be? What is the efficiency of coding information by spelling English words?

P11.10. An m-ary PCM system has m equally likely levels evenly distributed between -5 V and 5 V. The signal-to-noise ratio is $\mathrm{SNR}_r = 48$ dB. What is the rms noise n? What is the number of levels m for an error probability $p_e = 10^{-6}$? What is the information (in bits) conveyed by one symbol?

P11.11. The system in P11.10 has a bandwidth of 3 kHz. What is the maximum symbol rate it can transmit? For the m determined in P11.10, what is the information rate of the system? Compare this with the system capacity given by Equation (11.30) (for $p_e = 10^{-6}$). How much less than the capacity determined in D11.15 is this?

P11.12. A binary digital signal with $\Delta V = 1.6$ V at the transmitter has a data rate of 125 Mb/s. What bandwidth does the signal require? If the noise level at the receiver is $n = 0.1$ mVrms, what transmission attenuation leads to a 10^{-8} error rate? A twisted wire pair with impedance $Z_0 = 100\ \Omega$ and wire diameter 0.0403 in. is equalized to carry the signal. What is the maximum transmission distance for a 10^{-8} error rate?

P11.13. A digital signal with a data rate of 125 Mb/s is conveyed over a glass fiber. With no optical attenuation, the received signal (after optical-to-electrical conversion) would have $\Delta V = 1.6$ V. The noise level at the receiver is 0.1 mVrms. The glass fiber has optical attenuation of $C = 1$ dB/km and range-bandwidth product of $B = 32$ GHz-km. What is the maximum transmission distance for an error rate of 10^{-8}?

P11.14. Eight strain gauges are used to measure the vibration of a part in a wind tunnel. The vibration frequencies are 2 kHz or less. The analog signals are converted to binary digital signals so that the signal-to-quantization-noise ratio is 72 dB after conversion back to analog. The signals must be transmitted 200 m with an attenuation not more than 3 dB. Design a system like that in Figure 11–1c using a 100-Ω twisted wire pair as the transmission medium. How many bits must the ADCs and DACs use to convert each sample? What must the wire diameter be?

Extension Problems

E11.1. Consider the phase modulation described by Equations (11.5) and (11.6) for the case $\beta \ll 1$. Show that

$$v_t \approx \cos(\omega_c t) + 0.5\beta \sin(\omega_c + \omega_m)t + 0.5\beta \sin(\omega_c - \omega_m)t.$$

Use the identities $\cos(A + B) \equiv \cos(A)\cos(B) - \sin(A)\sin(B)$, $\sin(A)\cos(B) \equiv 0.5\sin(A + B) + 0.5\sin(A - B)$, and the facts that $\cos\theta \approx 1$ for $\theta \ll 1$, and $\sin\theta \approx \theta$ for $\theta \ll 1$.

Sketch the components of v_t in the frequency domain for the case $\omega_c = 2\pi \times 120$ kHz, $\omega_m = 2\pi \times 10$ kHz, and $\beta = 0.2$, and compare with the spectrum of an AM signal (see Figure 11–6d).

E11.2. An FSK modem transmits a "1" (a *mark*) as 2300 Hz and a "0" (a *space*) as 1700 Hz. The message (data) has a baud of 1200 bits/s. What is the minimum bandwidth of the message? What is the minimum transmission bandwidth? (FSK is a form of FM.)

E11.3. The spectrum $V_m(f)$ of a message with bandwidth $B_m = 3$ kHz is shown in Figure E11.3. The spectrum $V_t(f)$ of the same message transmitted by SSB (single-sideband AM) is also shown. The carrier frequency here is 12 kHz. A communication system has a bandwidth of $B_s = 95$ kHz and a signal-to-noise ratio of $\mathrm{SNR}_r = 40$ dB. Using a guard band of 1 kHz, how many messages with the spectrum shown can be transmitted over the system by frequency multiplexing with SSB? Given that $\mathrm{SNR}_m = \mathrm{SNR}_r = 40$ dB for each demodulated SSB channel, what is the maximum information rate for each channel? Compare the total information rate for all channels with the Shannon capacity of the system. How efficiently does SSB use the available capacity (for the given guard band)?

E11.4. The probability density that Gaussian noise will have the value x is given by

$$p(x) = \frac{1}{n\sqrt{2\pi}} \exp\left(\frac{-x^2}{2\,n^2}\right),$$

where n is the rms value of the noise. For PCM transmission, the probability of an error is the probability that the noise exceeds $\Delta V/2$:

$$p_e = \int_{\Delta V/2}^{\infty} p(x)\,dx = \int_{\Delta V/2n}^{\infty} \frac{1}{\sqrt{2\pi}} \exp\left(\frac{-y^2}{2}\right)\,dy,$$

where $y = x/n$. Use math tables or numerical methods to evaluate p_e for a few values of $\Delta V/n$, and compare with Figure 11–22.

Study Questions

11.1. Name four domains in which signals can be multiplexed.

11.2. Name three types of modulation that can be used for frequency-division multiplexing.

11.3. What advantage does FM have over AM? What advantage does AM have over FM?

11.4. What happens to an FM signal when the amplitude of the message signal increases?

11.5. How often must you look at a (band-limited) signal so that you can guess perfectly its values between samples? What is the name of this theorem?

11.6 PAM signals are quantized in time. What *two* parameters are quantized in PCM?

11.7. In serial binary PCM, what happens to the bit rate if the voltage quantization interval is reduced by a factor of 8? What happens to the signal-to-noise ratio of the received message? Be quantitative in your answers.

11.8. For AM and FM the signal-to-noise ratio of the received message is proportional to the signal-to-noise ratio of the received signal. This is not true for PCM. What does the received signal-to-noise ratio determine for PCM?

11.9. What is the measure of information in a message? How much information is there in a symbol that has one of four levels? What, in practice, keeps a continuous signal with an infinite number of possible levels from conveying an infinite amount of information?

11.10 What two parameters of a transmission channel determine its information capacity? What is the name of the theorem stating this relationship?

11.11. Name three materials that commonly serve as transmission media.

11.12. Why does the loss of a copper wire increase as the frequency of the signal increases? Does the loss in dB increase as $\sqrt{f}$ or as f^2?

11.13. What name is given to the filter that compensates for the increasing loss as f increases?

11.14. Does the attenuation of the message signal in dB increase as $\sqrt{f}$ or as f^2 for glass fiber?

11.15. If the optical loss of a glass fiber increases by 3 dB, by how many dB does the received signal decrease after conversion to an electrical signal?

ANSWERS TO DRILL PROBLEMS

D11.1. $v_t = -0.4 \sin(2\pi f_1 t) + \cos(2\pi f_2 t) + 0.4 \sin(2\pi f_3 t)$, where $f_1 = 0.999$ MHz, $f_2 = 1.000$ MHz, $f_3 = 1.001$ MHz

D11.2. $v_t = -0.2 \sin(2\pi f_1 t) - 0.4 \sin(2\pi f_2 t) + \cos(2\pi f_3 t) + 0.4 \sin(2\pi f_4 t) + 0.2 \sin(2\pi f_5 t)$, where $f_1 = 0.998$ MHz, $f_2 = 0.999$ MHz, $f_3 = 1.000$ MHz, $f_4 = 1.001$ MHz, $f_5 = 1.002$ MHz, $m_{(max)} = 1.04$

D11.3. $v_t = -0.091 \sin(2\pi f_1 t) - 0.127 \sin(2\pi f_2 t) - 0.212 \sin(2\pi f_3 t) - 0.637 \sin(2\pi f_4 t) + \cos(2\pi f_5 t) + 0.637 \sin(2\pi f_6 t) + 0.212 \sin(2\pi f_7 t) + 0.127 \sin(2\pi f_8 t) + 0.091 \sin(2\pi f_9 t)$, where $f_1 = 0.96$ MHz, $f_2 = 0.97$ MHz, $f_3 = 0.98$ MHz, $f_4 = 0.99$ MHz, $f_5 = 1.00$ MHz, $f_6 = 1.01$ MHz, $f_7 = 1.02$ MHz, $f_8 = 1.03$ MHz, $f_9 = 1.04$ MHz

D11.4. AM: $\lambda \approx 1000$ ft, antenna ≥ 100 ft; FM: $\lambda \approx 10$ ft, antenna ≥ 1.0 ft

D11.5. 1111, 30.5 dB

D11.6. 33.33 pW, 0.03 pW, SNR = 1111

D11.7. 6 kHz, 8 kHz

D11.8. 96 kHz

D11.9. 10110, 01010, 00101, 11101, 10100, 01100; 2.2, 1.0, 0.5, 2.9, 2.0, 1.2

D11.10. 28.9 mV, 26.8 mV

D11.11. 1.0×10^{-9}, 4×10^{-5}, 1.2×10^{-2}, 7×10^{-2}; 1.24×10^{-9}, 4.50×10^{-5}, 1.27×10^{-2}, 6.31×10^{-2}

D11.12. 1 V, -3 V, 3 V, -1 V, 1 V

D11.13. 13.3 bits

D11.14. 4.9 bits

D11.15. 47.8 kbits/s

D11.16. 3 Gbits/s, 1.5 Gbits/s

D11.17. 23,437; 9.3 ms

D11.18. -54 dB

CHAPTER 12

Microcomputers

A computer is a collection of simple digital circuits—the OR gates, AND gates, and flip-flops studied in Chapter 9. What gives a computer its tremendous power and flexibility is the fact that there are very many of these circuits, and they are under the control of a set of instructions. These instructions cause the circuitry to perform (i) arithmetic functions, (ii) logic operations, (iii) decisions based on comparing data, and (iv) data storage and retrieval. These functions are not new; the simple circuits in Chapter 9 perform them. But with the computer's organization these operations can be performed rapidly and in limitless combinations and orders.

Computers take the form of large mainframes, personal computers, and engineering workstations. But computers are far more widely used in less obvious applications: controlling microwave ovens, VCRs, digital radios, digital watch displays, printers, digital scale displays, car engines, and many more. In these applications they are called *embedded controllers, microprocessors,* and *microcomputers.* All computers operate basically in the same way, independent of their size. In this chapter we will study a simple microcomputer as representative of all computers.

The data on which a computer operates are provided by a variety of input interface devices (see Figure 12–1). An analog voltage representing temperature or pressure must be converted to digital code by an analog-to-digital converter (ADC). Digital signals indicating the status of a system can be provided by switch contacts. People enter data to a computer through a keyboard or thumbwheel switches. As progress is made in speech recognition, computers will be able to respond to many spoken commands.

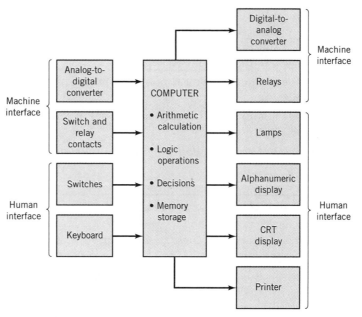

FIGURE 12–1 A computer with peripheral devices that allow it to interface with the rest of the world. Sources of data are shown on the left, and destinations of data from the computer are on the right.

Several output interface devices are illustrated in Figure 12–1. A computer can provide analog signals to a system by means of a digital-to-analog converter (DAC), and it can provide binary information or control to a system by means of relays. Interfacing with people is usually through visual devices. A lamp conveys binary information, alphanumeric displays (e.g., seven-segment displays) present text and numbers, a cathode-ray tube (CRT) displays pictures, and a printer gives a "hard" copy. A computer can also produce music and speech through a DAC.*

The applications of a computer are control, calculation, and data management. In industry a computer with machine interfaces can become part of a feedback loop, monitoring and controlling manufacturing processes. Businesses use a computer to calculate payroll data, and engineers use computer-aided design to handle complicated or repetitive calculations. Examples of data management are word processing, inventory records, and airline reservations. The physical devices that constitute a computer and the peripheral devices that can be connected to it are called *hardware*, while the set of instructions and the programs written with them are called *software*. Many of the hardware concepts have been covered in Chapter 9. We will devote most of this chapter to investigating the power of software.

---

*The announcement, "The number you have dialed has been disconnected," is produced by a computer from its digital memory.

The reader's objectives in this chapter are to:

1. understand the capabilities of microcomputers and be able to recognize situations where they should be applied,
2. be able to interface with a microcomputer—provide the inputs and handle the outputs,
3. be able to understand simple microcomputer programs and to write simple programs of your own,
4. be able to assemble the hardware for simple microcomputer configurations.

12.1 MICROCOMPUTER ARCHITECTURE

Some microcomputers, such as the Intel 8048, are contained completely in one 40-pin integrated circuit (IC). Therefore it is called a *single-chip microcomputer*. A diagram of the pin connections to the 8048 is shown in Figure 12–2. The pins provide power, permit the flow of data in and out, and control the data flow of

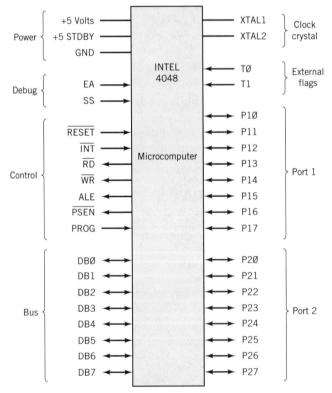

FIGURE 12–2 Pin connections of a typical single-chip microcomputer, the Intel 8048. A 40-pin DIP (dual in-line package) measures about 2in. $\times$ $\frac{3}{4}$in.

peripherals. Within the IC are several blocks of circuitry which perform the functions of data storage, computation, etc.

A typical arrangement (or *architecture*) of the functional blocks is shown in Figure 12–3. The *Accumulator* is an 8-bit parallel-in parallel-out shift register through which most of the data being processed passes. The input-output (I/O) *ports* and the I/O *bus* are also registers, but they are directly accessible from the "outside world." (We will distinguish between a bus and a port in Section 12.5.) An 8-bit word in I/O Port 1 can be read (sensed) at the eight pins P10 through P17 on the package (see Figure 12–2). Similarly, Port 2 is connected to pins P20 through P27, and the bus is connected to pins DB0 through DB7. Data can also be written (entered) into the I/O ports or bus by these pins.

There are two blocks of *memory* indicated in Figure 12–3—one to store data and one to store instructions (the program). Each memory is a set of registers that can be accessed one at a time by applying an address. A *Control Unit* interprets the instructions and applies the appropriate data memory address, routes the data from place to place, and tells the *Arithmetic and Logic Unit* (ALU) what to do. The ALU can add, perform logic functions, and shift bits left or right. All of the complex capability of a computer can be reduced to these few, simple operations.

A single-chip microcomputer includes all the functions in Figure 12–3 in a single 40-pin device called *a dual in-line package* (DIP). When the functions are distributed among several DIPs, the DIP containing the Control Unit, ALU, and Accumulator is called a *microprocessor,* reflecting the fact that it is not a complete computer by itself.

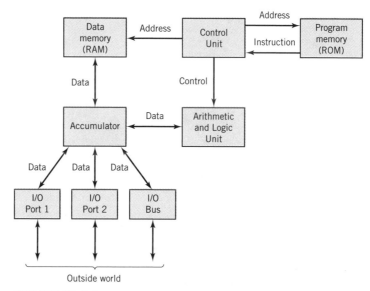

FIGURE 12–3 Internal architecture of a typical microcomputer, the Intel 8048. The Control Unit responds to instructions from program memory and controls the flow of data and the operations of the Arithmetic Logic Unit (ALU).

A simple example will show the function of each block in Figure 12–3. Suppose that we want to add two 8-bit numbers—one applied to Port 1 and the other to Port 2. For the moment these are being used as input ports, and the output (the sum) will appear on the bus. A set of instructions to perform the addition has been stored in the program memory. These instructions are coded in the form of 1's and 0's, but their rendering in English is the following:

0. Port 1 data to Accumulator.

1. Accumulator data to memory.

2. Port 2 data to Accumulator.

3a. Memory and Accumulator data to ALU. ⎫
3b. ALU performs addition.   ⎬ (combine)
3c. ALU data (sum) to Accumulator.   ⎭

4. Accumulator data to bus.

When the microcomputer is turned on, it automatically goes to instruction 0, which, in this case, tells the Control Unit to allow data to pass from Port 1 to the Accumulator. The Control Unit then goes to the next instruction and performs that. After instruction 4, the sum appears at the bus. Usually just one instruction handles instructions 3a, 3b, and 3c above: "Add memory and Accumulator data, and load sum in Accumulator."

The statement of the instructions can be shortened by using special notation. Let [x] mean "the data contained at x," and let → mean "replaces." Then the seven instructions above can be represented in shorthand by

0. $[P1] \rightarrow [A]$

1. $[A] \rightarrow [MEM]$

2. $[P2] \rightarrow [A]$

3. $[MEM] + [A] \rightarrow [A]$

4. $[A] \rightarrow [BUS]$

When a microcomputer programmer is actually designing an instruction sequence, he or she usually uses an even briefer (but not so self-explanatory) code for the instructions called an *assembly code:*

0. IN A,P1

1. MOV RØ,A

2. IN A,P2

3. ADD A,RØ

4. OUTL BUS,A

In instruction 1, RØ is register zero in memory. Throughout this chapter we use the common convention of putting a slash through zero to differentiate it from the letter O. The first part of the assembly code tells what happens, and the second part tells what it happens to. For example, MOV RØ,A means "move data; memory register RØ gets the contents of the Accumulator register." In cases like this, the comma is the reverse of our arrow; it can be read "gets the contents of."

The actual 8048 instructions in program memory are in binary form:

0. ØØØØ1ØØ1
1. 1Ø1Ø1ØØØ
2. ØØØØ1Ø1Ø
3. Ø11Ø1ØØØ
4. ØØØØØØ1Ø

This is called the *machine code*. It is more handily represented (from a human standpoint) in hexadecimal form (see Section 9.3):

0. Ø9
1. A8
2. ØA
3. 68
4. Ø2

In this form it is called *object code,* although machine code and object code are essentially the same. The reader should confirm that the binary and hexadecimal forms agree.

The particular instruction codes used in this example hold only for the Intel 8048 microcomputer. Unfortunately, each manufacturer has its own set of codes, but it is easy to learn others once you have learned one set. There are about 60 different types of instructions in the 8048 instruction set. Table 12–1 lists both the assembly code (under "Instruction") and the object code for most of the instructions. We will become familiar with more of these in Section 12.3.

Microcomputers include their own ROM and RAM with the expectation that these small memories are usually all that will be required for the task. That is, they are designed for simple tasks. For more complex tasks, they shed the in-chip memory (they become microprocessors) to make room for larger busses, more instructions, and greater speed. Microprocessors have busses ranging from 8 to 32 bits, instruction execution times ranging from 0.01 to 2 μs, and costs ranging from $2 to $700.

The instruction execution time is usually specified in terms of the inverse— millions of instructions per second (MIPS). Therefore the range is 0.5–100 MIPS, with the cost roughly proportional to the speed. The time required for the 8048 to execute an instruction is either 1.36 μs or 2.72 μs, depending on the complexity of the instruction. For a program with an equal mix of the two types of instructions, the speed of the 8048 is 0.5 MIPS—among the slowest.

TABLE 12–1

Instruction	Object Code	Operation Performed
MOV A,#MM	23 MM	$[A] \leftarrow MM$
MOV A,RN	F8 to FF	$[A] \leftarrow [RN]$
MOV A,@R1	F1	$[A] \leftarrow [[R1]]$
MOV RN,#MM	B8 to BF MM	$[RN] \leftarrow MM$
MOV RN,A	A8 to AF	$[RN] \leftarrow [A]$
MOV @ R1,#MM	B1 MM	$[[R1]] \leftarrow MM$
MOV @ R1,A	A1	$[[R1]] \leftarrow [A]$
MOVX A,@R1	81	$[A] \leftarrow [[R1]]$
MOVX @ R1,A	91	$[[R1]] \leftarrow [A]$
MOVP A,@A	A3	$[A] \leftarrow ([A])$
XCH, A,RN	28 to 2F	$[A] \leftrightarrow [RN]$
XCH A,@R1	21	$[A] \leftrightarrow [[R1]]$
INS A,BUS	Ø8	$[A] \leftarrow [BUS]$
IN A,P1	Ø9	$[A] \leftarrow [P1]$
IN A,P2	ØA	$[A] \leftarrow [P2]$
OUTL BUS,A	Ø2	$[BUS] \leftarrow [A]$
OUTL P1,A	39	$[P1] \leftarrow [A]$
OUTL P2,A	3A	$[P2] \leftarrow [A]$
ADD A,#MM	Ø3 MM	$[A] \leftarrow [A] + MM$
ADD A,RN	68 to 6F	$[A] \leftarrow [A] + [RN]$
ADD A,@R1	61	$[A] \leftarrow [A] + [[R1]]$
ADDC A,#MM	13 MM	$[A] \leftarrow [A] + MM + [C]$
ADDC A,RN	78 to 7F	$[A] \leftarrow [A] + [RN] + [C]$
ADDC A,@R1	71	$[A] \leftarrow [A] + [[R1]] + [C]$
ANL A,#MM	53 MM	$[A] \leftarrow [A]$"AND"MM
ANL A,RN	58 to 5F	$[A] \leftarrow [A]$"AND"$[RN]$
ANL A,@R1	51	$[A] \leftarrow [A]$"AND"$[[R1]]$
ANL BUS,#MM	98 MM	$[BUS] \leftarrow [BUS]$"AND"MM
ANL P1,#MM	99 MM	$[P1] \leftarrow [P1]$"AND"MM
ANL P2,#MM	9A MM	$[P2] \leftarrow [P2]$"AND"MM
ORL A,#MM	43 MM	$[A] \leftarrow [A]$"OR"MM
ORL A,RN	48 to 4F	$[A] \leftarrow [A]$"OR"$[RN]$
ORL A,@R1	41	$[A] \leftarrow [A]$"OR"$[[R1]]$
ORL BUS,#MM	88 MM	$[BUS] \leftarrow [BUS]$"OR"MM
ORL P1,#MM	89 MM	$[P1] \leftarrow [P1]$"OR"MM
ORL P2,#MM	8A MM	$[P2] \leftarrow [P2]$"OR"MM
INC A	17	$[A] \leftarrow [A] + 1$
INC RN	18 to 1F	$[RN] \leftarrow [RN] + 1$
INC @R1	11	$[[R1]] \leftarrow [[R1]] + 1$
DEC A	Ø7	$[A] \leftarrow [A] - 1$
DEC RN	C8 to CF	$[RN] \leftarrow [RN] - 1$
DJNZ RN,#MM	E8 to EF MM	$[RN] \leftarrow [RN] - 1$, go to MM if $[RN] \neq \emptyset$
CLR A	27	$[A] \leftarrow \emptyset$
CPL A	37	$[A] \leftarrow \overline{[A]}$

TABLE 12–1 *(continued)*

Instruction	Object Code	Operation Performed
RR A	77	Rotate right Accumulator
RL A	E7	Rotate left Accumulator
CALLØ #MM	14 MM	Call subroutine at ØMM
CALL1 #MM	34 MM	Call subroutine at 1MM
CALL2 #MM	54 MM	Call subroutine at 2MM
CALL3 #MM	74 MM	Call subroutine at 3MM
RET	83	Return from subroutine
RETR	93	Return, replace status bits
EN I	Ø5	Enable interrupts
DIS I	15	Disable interrupts
JMPØ #MM	Ø4 MM	Go to address ØMM
JMP1 #MM	24 MM	Go to address 1MM
JMP2 #MM	44 MM	Go to address 2MM
JMP3 #MM	64 MM	Go to address 3MM
JBØ #MM	12 MM	Go to MM if [AØ] = 1
JB1 #MM	32 MM	Go to MM if [A1] = 1
JB2 #MM	52 MM	Go to MM if [A2] = 1
JB3 #MM	72 MM	Go to MM if [A3] = 1
JB4 #MM	92 MM	Go to MM if [A4] = 1
JB5 #MM	B2 MM	Go to MM if [A5] = 1
JB6 #MM	D2 MM	Go to MM if [A6] = 1
JB7 #MM	F2 MM	Go to MM if [A7] = 1
JC #MM	F6 MM	Go to MM if [C] = 1
JNC #MM	E6 MM	Go to MM if [C] = Ø
JZ #MM	C6 MM	Go to MM if [A] = Ø
JNZ #MM	96 MM	Go to MM if [A] ≠ Ø
DJNZ RN,#MM	E8 to EF MM	$[RN] \leftarrow [RN] - 1$, go to MM if $[RN] \neq Ø$

12.2 MEMORY

A memory unit contains many registers that store data. A small memory with only four registers is shown in Figure 12–4. In response to an address code, the memory can connect any of these registers for access by the Accumulator or ALU. The largest computers have 64-bit registers, while some microcomputers have only 4-bit registers. Microcomputers most commonly have memories with 8-bit registers, as in the case of the 8048. The 8 bits in any given register are called a *byte* for short.*

Any of the 4 bytes in Figure 12–4 is addressed by closing the switches connecting that register with the eight data lines. The signal closing the switches is

---

*Four bits are sometimes called a ''nibble.''

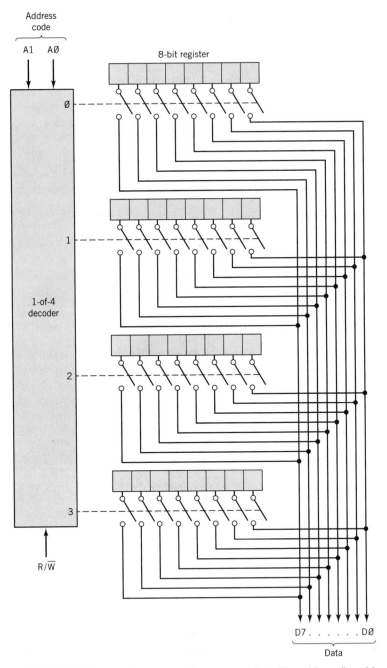

FIGURE 12–4 Internal operation of a memory block. The address (bits A1 and AØ) determines which of the four registers has its eight bits connected to the data lines. Read-write memory (RWM or RAM) connects either the input or output, depending on the signal on the R/$\overline{W}$ line. Read-only memory (ROM), which has no R/$\overline{W}$ line, connects only the register output to the data lines. The number of registers and the number of bits per register can vary.

produced by decoding the two address bits A_1A_0. (See 1-of-n decoders in Section 9.4). For example, the address 10 (binary) connects register 2 to the data lines. The 8048 has 1024 bytes of program memory, so it requires 10 address bits ($2^{10} = 1024$). In computer jargon this is often rounded off to 1000 and called a "kilobyte." Computers with a 32-bit address bus can have as much as 2^{32} (or 4 billion!) memory locations, and each location may have as many as 64 bits.

The $R/\overline{W}$ line connected to the memory determines whether the input or the output of the addressed register is connected to the data line. If the input is connected, data can be *written* into the register. If the output is connected, data can be *read* from the registers. Such a memory is called a *read-write memory* (RWM) or a *random access memory* (RAM). The first name is more descriptive, but the latter is more commonly used (probably because the acronym is easily pronounced). The Control Unit causes $R/\overline{W}$ to be "1" or "∅" depending on whether a program instruction moves data from the Accumulator to memory or from memory to the Accumulator.

Some memories have no provision for writing into them; only the register outputs can be connected for reading. These are called *read-only memories* (ROMs). Data are originally stored in a ROM at the time of manufacture (as with the 8048) or by the user electrically melting fusable metal links in his ROM. The latter are called *programmable ROMs* (PROMs) Some PROMs can be programmed nondestructively and later erased by exposure to ultraviolet light. These are *erasable PROMs* (EPROMs). EPROMs are more expensive than ROMs, but they are very useful in system development. For example the 8748 is identical to the 8048 except it has an EPROM rather than a ROM. The design engineer can test out his prototype with an 8748 and switch to the 8048 for production quantities.

The principal feature of a ROM is that it is *nonvolatile*—it doesn't lose its data when the power is turned off. This makes it useful for program memory since the computer usually has to know what to do (have a program) the instant it is turned on. Therefore ROM, nonvolatile, and program memory are almost synonymous in usage.

Data memory is necessarily RAM so that data can be both stored and retrieved. However, RAM is volatile; any data stored in it are lost when power is turned off. Any data to be saved must be loaded onto a magnetic tape or disk.

The program memory for the 8048 is diagrammed in Figure 12–5. There are 1024 bytes of ROM addressed by ten address bits. For the reader's convenience the address bits are grouped in fours and represented in hexadecimal. For example, address 1∅,1111,111∅ is represented by 2FE. We will see that a program can not only address data memory, but it can also address itself. Since the program is stored in bytes (8 bits) it is often convenient for the program to change only the last 8 bits of a 10-bit program memory address. In this case all addresses with the same first 2 bits (same first hex character) are called a *page* of memory. A page has 256 bytes, and the 8048 has four pages.

The data memory for the 8048 is diagrammed in Figure 12–6. Since it has only 64 bytes of RAM, the computer is not well suited for data management application. The small data memory here is used mostly as a place to hold data temporarily while making calculations, much as an engineer uses a scratchpad. Any one of the registers can be accessed through its address (hex ∅∅ to 3F) for data storage

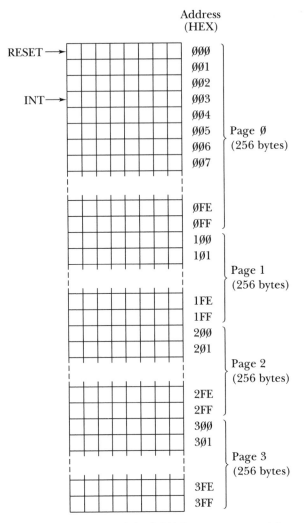

FIGURE 12–5 Map of the ROM (program memory) for the Intel 8048. There are 1024 bytes of memory (addresses given in hex). A RESET signal sends the computer to ∅∅∅, and an interrupt (INT) signal sends it to ∅∅3.

or retrieval. The first eight registers are special in that they can also be accessed by a shortened address.* We will use the symbols R∅ to R7 to refer to these *general-purpose* registers.

Another section of data memory, the 16 registers following the general-purpose registers, is called the *stack*. The computer uses this area to automatically store some data in connection with CALL and Interrupt instructions. If these instructions

---

*This shortened addressing is similar to a feature provided on some telephones whereby a few commonly called numbers can be dialed by just one digit.

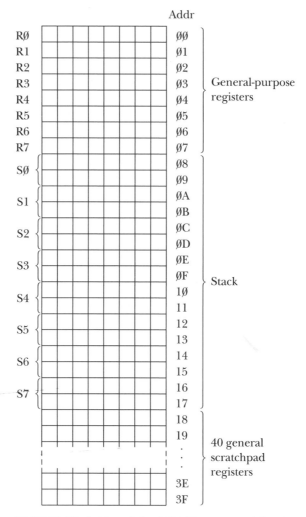

FIGURE 12–6 Map of the RAM (data memory) for the Intel 8048. The registers RØ–R7 are addressed more easily than the others. The Stack is used by subroutines and interrupts.

are not used in a program, the stack space is available for use as general scratchpad memory.

12.3 INSTRUCTION SET

The collection of instructions a computer is designed to recognize and execute is called its *instruction set*. Throughout the rest of this chapter we will be learning about and applying some of the instructions for the 8048. After completing this chapter, the reader will be able to use all the instructions listed in Table 12–1.

The simple program in Section 12.1 gave an example of some of the instructions an 8048 can perform. Instructions fall into one of three categories: (i) move data from place to place, (ii) perform arithmetic or logic operation on the data; or (iii) go to some particular place in the program. When executing a program, the computer normally goes to the next instruction in sequence, but some instructions tell the computer to jump ahead or behind in the sequence.

We begin by studying just a few instructions—those listed in Figure 12–7. The instruction MOV R3,A moves the data in the Accumulator (A) into register number 3 (R3). The list in Figure 12–7a puts this more generally: MOV RN,A where N can be any integer from $\emptyset$ to 7. Figure 12–7b shows that the object codes (in hexadecimal) for these instructions are A8 through AF. Here we are using the short form of addressing the general-purpose registers. The reader should confirm that the last three bits in each code are the last three bits of the corresponding register's address. (See Figure 12–6 for the addresses of the general-purpose registers.)

The instruction ORL A,R3, with object code 4B, performs logic OR on the data in the Accumulator and in R3. This is done bit by bit; the first bit in the Accumulator is ORed with the first bit in R3, and the result is put in the first bit of the Accumulator, etc. For example if the Accumulator contains $1\emptyset11\emptyset1\emptyset\emptyset$ and R3 contains $\emptyset\emptyset111\emptyset1\emptyset$, then after execution of ORL A,R3, the Accumulator would con-

Instruction (Intel 8048)	Object Code	Operation Performed
MOV RN,A	A8 to AF	Move data [RN] ← [A]
MOV A,RN	F8 to FF	Move data [A] ← [RN]
ORL A,RN	48 to 4F	OR Logic [A] ← [A]"OR"[RN]
ANL A,RN	58 to 5F	AND Logic [A] ← [A]"AND"[RN]
CPL A	37	[A] ← $\overline{[A]}$
RR A	77	Rotate Right the Accumulator
IN A,P1	$\emptyset$9	Input Port 1 [A] ← [P1]
OUTL P2,A	3A	Output Port 2 [P2] ← [A]
JMP$\emptyset$ #MM	$\emptyset$4 MM	Jump to program memory location $\emptyset$MM (requires 2 bytes of code)

(a)

Instruction	R$\emptyset$	R1	R2	R3	R4	R5	R6	R7
MOV RN,A	A8	A9	AA	AB	AC	AD	AE	AF
MOV A,RN	F8	F9	FA	FB	FC	FD	FE	FF
ORL A,RN	48	49	4A	4B	4C	4D	4E	4F
ANL A,RN	58	59	5A	5B	5C	5D	5E	5F

(b)

FIGURE 12–7 (a) Partial listing of instructions for the 8048. Instructions with "RN" are expanded upon in (b). The notation "#MM" indicates an 8-bit number, usually given in hexadecimal code.

tain 1Ø111110. Similarly, ANL A,R3 performs logic AND bit by bit, and CPL A complements (logic NOTs) each bit in the Accumulator (see ones complement in Section 9.3).

The instruction RR A rotates the bits in the Accumulator to the right. The second bit becomes the first bit, the first bit becomes the zeroth bit, and the zeroth bit becomes the seventh bit (rotates around the end).

The instruction JMPØ #3DH jumps the program to location Ø3DH (the final H simply means that Ø3D is in hexadecimal). After execution of this instruction, the next program step to be executed is that in the location Ø3D, where Ø is the page number and 3D is the location on that page. The generalized notation #MM in Figure 12–7a indicates that MM is a two-digit hexadecimal number (8 bits). Note that this instruction takes 2 bytes of object code, Ø4 and 3D.

Let's look at an example that makes use of the instructions in Figure 12–7. We will use the same furnace control problem that was used in Section 9.5. In that example the solution was in the form of logic gates—few enough gates to fit in two simple ICs. In the following example, the same function is provided by a microcomputer.

EXAMPLE 12.1 The status F of a furnace (on or off) is to be determined from the logic $F = (D \cdot T_1 + \overline{D} \cdot T_2 + H)P$, where D, T_1, T_2, H, and P are the status for a clock; three thermostats, and a pressure gauge, respectively. Recall that " + " means OR and "·" means AND. See Example 9.1 for the origin of this logic function.

Given: The logic signals D, T_1, T_2, H, and P are connected to lines Ø, 1, 2, 3, and 4 of Port 1, respectively.

Objective: Write a program to generate the furnace status F on line 7 of Port 2.

Solution: A straightforward program is shown in Figure 12–8. The figure also shows the bits in the Accumulator and the registers, where "X" means "don't care" and ″ means "unchanged." Only the least significant bit (bit Ø) of each register is shown.

When the computer is turned on, it is *reset*. This causes it to look in program memory location ØØØ for its first instruction. For the purpose of handling interrupts (discussed in Section 12.5), it is necessary to leave locations ØØ2 to ØØ9 free. Therefore the first instruction jumps over these to location ØØA, where the program continues. An "input" instruction loads the Port 1 data into the Accumulator. The first logic operation to be performed is $D \cdot T_1$ (D ANDed with T_1). But D and T_1 are bits Ø and 1 in the Accumulator. In order to AND D and T_1, they must be at the same bit location (say bit Ø) in two separate registers. Therefore we get D, T_1, T_2, H, and P into bit Ø of registers R2, R3, R4, R5, and R6, respectively. This is done by a series of "rotate right" and "move" instructions. Figure 12–8 shows the contents of the registers during this procedure. Then the logic operations are carried out to generate F. Finally a "rotate right" instruction takes F from bit Ø to bit 7, and an "output" instruction places it on the Port 2 line. (The last instruction in

Loc.	Obj.	Instruction	Accumulator	R2	R3	R4	R5	R6	R7	
000	04	JMP0 #0AH	× × × × × × × ×	×	×	×	×	×	×	
001	0A		" " " " " " " "	"	"	"	"	"	"	
00A	09	IN A,P1	× × × P H T_2 T_1 D	"	"	"	"	"	"	
00B	AA	MOV R2,A	" " " " " " " "	D	"	"	"	"	"	
00C	77	RR A	D × × × P H T_2 T_1	"	"	"	"	"	"	
00D	AB	MOV R3,A	" " " " " " " "	"	T_1	"	"	"	"	
00E	77	RR A	T_1 D × × × P H T_2	"	"	"	"	"	"	
00F	AC	MOV R4,A	" " " " " " " "	"	"	T_2	"	"	"	
010	77	RR A	T_2 T_1 D × × × P H	"	"	"	"	"	"	
011	AD	MOV R5,A	" " " " " " " "	"	"	"	H	"	"	
012	77	RR A	H T_2 T_1 D × × × P	"	"	"	"	"	"	
013	AE	MOV R6,A	" " " " " " " "	"	"	"	"	P	"	
014	FA	MOV A,R2	× × × × × × × D	"	"	"	"	"	"	
015	5B	ANL A,R3	" " " " " " " •	"	"	"	"	"	"	$D \cdot T_1$
016	AF	MOV R7,A	" " " " " " " "	"	"	"	"	"	•	$D \cdot T_1$
017	FA	MOV A,R2	" " " " " " " D	"	"	"	"	"	"	
018	37	CPL A	" " " " " " " $\bar{D}$	"	"	"	"	"	"	
019	5C	ANL A,R4	" " " " " " " •	"	"	"	"	"	"	$\bar{D} \cdot T_2$
01A	4F	ORL A,R7	" " " " " " " •	"	"	"	"	"	"	$D \cdot T_1 + \bar{D} \cdot T_2$
01B	4D	ORL A,R5	" " " " " " " •	"	"	"	"	"	"	$D \cdot T_1 + \bar{D} T_2 + H$
01C	5E	ANL A,R6	" " " " " " " F	"	"	"	"	"	"	$(D \cdot T_1 + \bar{D} T_2 + H) \cdot P$
01D	77	RR A	F × × × × × × ×	"	"	"	"	"	"	
01E	3A	OUTL P2,A	" " " " " " " "	"	"	"	"	"	"	
01F	04	JMP0 #0AH	" " " " " " " "	"	"	"	"	"	"	
020	0A		" " " " " " " "	"	"	"	"	"	"	

FIGURE 12–8 Program for Example 12.1. The control of a furnace is determined by the function $F = (D \cdot T_1 + \bar{D} \cdot T_2 + H)P$ on five of the data bits at Port 1. The "#0AH" in an instruction indicates hexadecimal number 0A.

the program takes the computer back to the beginning of the program for repeated execution.)

A relay connected to line 7 of Port 2 controls the furnace, turning it on when $F = 1$.

This example helps answer the question, "When should I use a microcomputer?" The answer is that a microprocessor can be used in place of almost any logic circuitry. A rough rule of thumb is that a circuit needing more than 30 gates and flip-flops will be cheaper if replaced by a microcomputer.

Sometimes it pays to go to a microcomputer even if it appears to be more expensive in the short run. A design change at some later time can be costly with a pure hardware solution; the printed circuit board will have to be redesigned, and new ICs will have to be ordered. With a microcomputer solution, a redesign usually involves only a change in software; a new program must be written and a new ROM obtained. In the case of a PROM or EPROM, the design change can take only a few hours. For Example 12.1 a redesign to have the furnace go on to heat water only at night would only require that a couple of instructions be added to substitute $\bar{D} \cdot H$ for H.

Another reason for using a microcomputer is that for a little more money you can have many more features added. The microcomputer usually has unused capability just sitting there that could be used to add some "bells and whistles." Consider how little strain Example 12.1 puts on the microcomputer. The 8048 takes 1.36 μs to execute most program bytes. Therefore the 23-byte program in Example 12.1 takes 31.3 μs to execute, and the furnace control is updated about 32,000 times a second. Since 32 times a second would be more than adequate, the microcomputer only has to be active 0.1% of the time. With the remaining time, it could read a humidistat and calculate a more comfortable temperature T_1 for the furnace control, for example. We will add features like this in Section 12.5.

Yet another advantage of a microcomputer-based system is lower maintenance cost. With fewer devices, it is easier to locate a faulty device. Also, the microcomputer can execute a special test program that exercises all the devices in the system. Thus the power of the microcomputer can be utilized to help automate fault location.

12.4   PROGRAMMING TECHNIQUES

This section introduces some programming techniques: masks, subroutines, loops, indirect addressing, lookup tables, and conditional jumps. To introduce these through example, we will modify and embellish the furnace control problem used in the previous section.

Masks

Suppose we wanted to preserve the 4 middle bits of the byte 1Ø1Ø1Ø1Ø and set the others to "Ø." This can be done by ANDing the byte with the *mask* ØØ1111ØØ. The reader can show that the result is ØØ1Ø1ØØØ. If we wanted to preserve the 4 middle bits but set the others to "1," we can OR the byte with the mask 11ØØØØ11. The result is 111Ø1Ø11. In each case a 4-bit-wide "hole" in the mask allows the middle bits to come through. Let's look at a practical application.

In Example 12.1 it was assumed that only line 7 on Port 2 was being used. We didn't have to worry about feeding "garbage" (the X's) to the other seven lines. Suppose the microcomputer is being time shared for other purposes, and lines Ø through 6 of Port 2 have meaningful data on them. The OUTL P2,A instruction in Figure 12–8 would destroy those data. The following example uses masking to solve the problem.

EXAMPLE 12.2 Revise the program in Figure 12–8 so that sending the *F* bit to line 7 of Port 2 doesn't change the bits P2Ø through P26 on lines Ø through 6 of Port 2.

Solution: The revised portion of the program is shown in Figure 12–9b. (The program through memory location Ø1D stays the same.) The new instructions used here are listed for convenience in Figure 12–9a.

The first step is to make all bits other than F in the Accumulator be "Ø." This is done by ANDing the Accumulator data with the mask 1ØØØØØØØ (8Ø in

Instruction (Intel 8048)	Object Code	Operation Performed
ORL A, #MM	43 MM	[A] ← [A] "OR" MM
ANL A, #MM	53 MM	[A] ← [A] "AND" MM
IN A, P2	∅A	[A] ← [P2]
OUTL P1,A	39	[P1] ← [A]

(a)

Loc.	Obj.	Instruction	Accumulator							
.										
∅1D	77	RR A	F	×	×	×	×	×	×	×
∅1E	53	ANL A, #8∅H	F	∅	∅	∅	∅	∅	∅	∅
∅1F	8∅									
∅2∅	A	MOV R7,A	"	"	"	"	"	"	"	"
∅21	∅A	IN A,P2	P27	P26	P25	P24	P23	P22	P21	P2∅
∅22	53	ANL A,#7FH	∅	P26	P25	P24	P23	P22	P21	P2∅
∅23	7F									
∅24	4F	ORL A, R7	F	P26	P25	P24	P23	P22	P21	P2∅
∅25	4A	OUTL P2,A	F	"	"	"	"	"	"	"
∅26	∅4	JMP∅ #∅AH	F	"	"	"	"	"	"	"
∅27	∅A									

Note: 8∅H = 1∅∅∅∅∅∅∅$_2$, 7FH = ∅1111111$_2$

(b)

FIGURE 12–9 (a) Instructions introduced in Example 12.2. (b) Program for Example 12.2. This is a modification to the end of the program in Figure 12–8 so that the bits P2∅–P26 at Port 2 are not changed. The bits patterns 1∅∅∅∅∅∅∅ and ∅1111111 here are called *masks*.

hex). The next step is to mask the bits already at Port 2. This time we want to preserve bits P2∅ through P26, so the mask is ∅1111111. Now the two bytes can be combined by ORing. The reader can show that the ∅'s in one byte don't affect the corresponding bits in the other byte when the bytes are ORed. The byte sent to Port 2 returns the same bits P2∅ through P26, and only P27 is updated to the current F.   ▬

Subroutines

Consider a program that needs to change a bit at Port 2 several times. Each time the instructions in locations ∅1E through ∅25 of Figure 12–9 (or similar instructions) would have to be duplicated. This would be wasteful of program memory space. A better solution would be to use a *subroutine*—a group of instructions set aside that can be executed as many times as needed.

When the main program accesses a subroutine, it is said to *call* the subroutine. Instructions that perform this are listed in Figure 12–10a. Each instruction accesses a different page of memory. For example, if the subroutine is at address Ø7 (hex) on page 2, CALL2 #Ø7H is used. The second byte of the instruction gives the address on the page. The "call" instructions cause the computer to jump to a certain program memory location, just as the "jump" instructions do. The difference is that with a "call" instruction, the computer remembers where it was in the program when the subroutine was called. Then it can return to that spot when it is done with the subroutine.

Instruction (Intel 8048)	Object Code	Operation Performed
CALLØ #MM	14 MM	Go to subroutine addr ØMM
CALL1 #MM	34 MM	Go to subroutine addr 1MM
CALL2 #MM	54 MM	Go to subroutine addr 2MM
CALL3 #MM	74 MM	Go to subroutine addr 3MM
RET	83	Return to addr following call

(a)

Loc.	Obj.	Instruction	Comment
Ø1D	77	RR A	; F IN BIT 7
Ø1E	B9	MOV R1,#8ØH	; STORE MASK IN R1.
Ø1F	8Ø		; FOR BIT 7.
Ø2Ø	34	CALL 1 #ØØH	; GO TO SUBROUTINE AT
Ø21	ØØ		; LOCATION 1ØØ.
Ø22	Ø4	JMPØ #ØAH	; REPEAT LOOP.
Ø23	ØA		
			; SUBROUTINE:
1ØØ	59	ANL A,R1	; MASK THE BIT.
1Ø1	AF	MOV R7,A	; STORE BIT IN R7.
1Ø2	F9	MOV A,R1	; LOAD MASK IN ACCUMULATOR.
1Ø3	37	CPL A	; MASKBAR = COMPLEMENT OF MASK.
1Ø4	A9	MOV R1,A	; STORE MASKBAR IN R1.
1Ø5	ØA	IN A,P2	; INPUT FROM PORT 2.
1Ø6	59	ANL A,R1	; MASKBAR OTHER BITS.
1Ø7	47	ORL A,R7	; COMBINE BIT WITH OTHER BITS.
1Ø8	4A	OUTL P2,A	; OUTPUT TO PORT 2.
1Ø9	83	RET	; RETURN TO MAIN PROGRAM.

(b)

FIGURE 12–10 (a) Instructions introduced in Example 12.3. (b) Program for Example 12.3. This is a subroutine version of the program portion in Figure 12–9. The subroutine occupies program memory locations 1ØØ through 1Ø9.

EXAMPLE 12.3 Write the output instructions in Figure 12–9b in the form of a subroutine.

Given:　The main program will go to location 1∅∅ (hex) for the subroutine. The subroutine will look in the Accumulator for the datum to go to the output, and it will look in R1 for the first mask.

Solution:　The subroutine and the revised portion of the main program are given in Figure 12–10b. The instruction at location ∅1D puts the datum in bit 7 of the Accumulator, and the next instruction puts the first mask in R1, as the subroutine expects. Then the subroutine is called, causing the instruction at location 1∅∅ to be executed next. The subroutine performs the same function as the program portion in Figure 12–9. The only difference is that the second mask is generated by complementing the first mask. This saves a couple of steps in the main program—steps that will be saved every time the subroutine is called. The final instruction in a subroutine is RET. This causes the computer to return to the location in the main program following the "call" instruction (to ∅22 in this case).

Since the subroutine was called only once, there was no savings; the 10-byte subroutine in Figure 12–10 allowed the main program to be reduced by 4 bytes (compare Figure 12–9). The advantage comes with a program that calls the subroutine many times. If it were called five times, there would be a savings of 10 bytes.

It is possible for a subroutine to call a subroutine—a situation called *nested subroutines.* An example of nesting to a depth of two subroutines is shown in Figure 12–11. The main program (on page ∅ of memory) calls the first subroutine on page 3, and it in turn calls a second subroutine on page 2. When the second subroutine is done, the computer returns to the first, and when that is done, the computer returns to the main program.

The "call" instruction automatically causes the computer to remember where

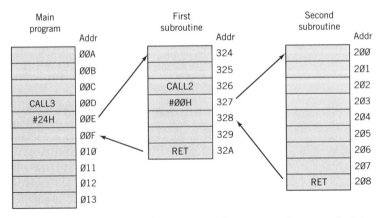

FIGURE 12–11 Two nested subroutines. The computer keeps track of the address of each instruction following a CALL instruction so it can return to the proper place when each subroutine has been completed.

it was at the time the subroutine was called so it can return to the proper point in program memory. The programmer doesn't have to worry about this feature except to allow room in data memory for the computer to do the work. The space in memory the computer uses is called the *stack* (see Figure 12–6). The depth of nested subroutines is limited by the size of the stack. Because the 8048 stack has eight pairs of bytes, there can be a nesting depth of eight. If a program is to handle interrupts, 2 to 4 bytes of the stack will be used to remember the program address to return to after the interrupt has been handled.

Loops

The program in Figure 12–8 has a "move" and "rotate right" instruction sequence repeated five times. This suggests that a technique such as a subroutine could be used to save program memory space. However the "call" and "return" instructions alone would use more bytes than the instructions they were replacing. A better solution here is to use a *loop*—a portion of the program that cycles through itself a given number of times (five in this case).

A program involving loops and decisions can become complicated and difficult to hold in mind. A useful method for keeping track of the program is to diagram it with a *flow chart*. The following example, which includes a loop, makes use of a flow chart.

EXAMPLE 12.4 Replace the instructions in location $\emptyset\emptyset B$ through $\emptyset13$ in Figure 12–8 with a loop to store the data in R2 through R6.

Solution: A flow chart for this portion of the program is shown in Figure 12–12. To store in R2 through R6, we need to go through the loop five times. Let the number M be a cycle counter. If we initially set $M = 5$ and decrease M by 1 each cycle, we can tell we're done when $M = \emptyset$. In each cycle the store instruction is the same: move data into R N, but the address N will change from 2 up to 6. Therefore an address N must be initially set to 2. With each execution of the loop, N is incremented and M is decremented. When $M = \emptyset$, we're done.

In Figure 12–13b is a program with each instruction corresponding to a box of the flow chart. The initial values of M and N (5 and 2) are stored in registers R7 and R1, respectively. This is called *initializing* the counters. The data word is moved from Port 1 into the Accumulator. Then the program enters the loop, which occupies memory locations $\emptyset\emptyset F$ through $\emptyset13$.

The first instruction in the loop is the powerful *indirect addressing* command MOV @R1,A. This moves the contents of the Accumulator into the register *whose address* is found in R1. Remember that R1 was initialized with the number 2, which is the short-form address for R2. R1 is said to be "pointing at" R2. So far we have stored the status D in the least significant bit (LSB) of R2.

Next the Accumulator is rotated to get T_1 into the LSB position. The address N in register R1 is then incremented by 1 (from 2 to 3). The instruction DJNZ decrements M by 1, checks to see if $M = \emptyset$, and jumps to the start

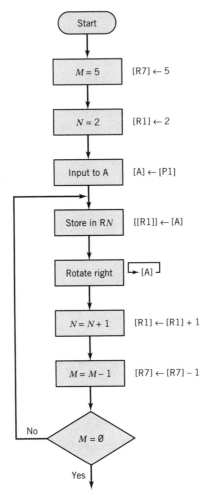

FIGURE 12–12 Flow chart of the program for Example 12.4. The loop stores data in registers R2 through R6 as N goes from 2 to 6.

of the loop if it is not zero. (The acronym for the instruction stands for "decrement and jump if not zero.") The DJNZ instruction corresponds to the last two boxes of the flow chart in Figure 12–12.

The second time through the loop, T_1 is stored in the LSB of R3, and the loop is repeated until all the data are stored. Compare these operations with those of the program in Figure 12–8.

The new program portion in Example 12.4 happens to occupy the same amount of space in program memory as the instructions it replaces—9 bytes. If

Instruction (Intel 8048)	*Object Code*	*Operation Performed*
MOV RN,#MM	B8 to BF MM	[RN] ← MM
MOV A,#MM	23 MM	[A] ← MM
MOV @ R1,A	A1	[[R1]] ← [A]
INC RN	18 to 1F	[RN] ← [RN] + 1
INC A	17	[A] ← [A] + 1
DJNZ RN,#MM	E8 to EF MM	[RN] ← [RN] − 1, go to MM if [RN] ≠ Ø

(a)

Loc.	*Obj.*	*Instruction*	*Comment*	
ØØØ	Ø4	JMPØ #ØAH	; START	
ØØ1	ØA			
ØØA	3F	MOV R7,#05H	; INITIALIZE LOOP CYCLE	
ØØB	Ø5		; COUNTER WITH 5.	
ØØC	39	MOV R1,#Ø2H	; INITIALIZE ADDRESS POINTER	
ØØD	Ø2		; WITH 2.	
ØØE	09	IN A,P1	; INPUT FROM PORT 1.	
ØØF	A1	MOV @ R1,A	; STORE BIT. (LOOP START) ←	
Ø1Ø	77	RR A	; ROTATE ACCUMULATOR RIGHT.	
Ø11	19	INC R1	; INCREMENT ADDRESS POINTER.	[R7] ≠ Ø
Ø12	EF	DJNZ R7,#ØF	; IF CYCLE COUNTER ≠ Ø,	
Ø13	ØF		; JUMP TO LOOP START.	
				[R7] = Ø
Ø14	FA	MOV A,R2	; BEGIN LOGIC OPERATIONS. ←	
⋮				

(b)

FIGURE 12–13 (a) Instructions introduced in Example 12.4. (b) Program for Example 12.4. These instructions replace the first part of the program in Figure 12–8. Each instruction corresponds to a box in the flow chart in Figure 12–12; the DJNZ instruction corresponds to the last two boxes.

there had been six registers to load, there would have been a savings since the loop can be made to load any number of registers just by changing the initial M (the Ø5 at location ØØB).

In Example 12.1 (see Figure 12–8), register R2 was addressed directly by the instruction AA at location ØØB; the last three bits of AA give the address 2. This is called *direct addressing*. In Example 12.4 (see Figure 12–13), the address of R2 is given indirectly by the instruction A1; it says to look in R1 for the address of R2. This indirect addressing allows addresses to be calculated, rather than being unalterably specified in the program.

Lookup Tables

The function $F = (D \cdot T_1 + \overline{D} \cdot T_2 + H)P$ assigns a value to F for each possible combination of the input values D, T_1, T_2, H, and P. A brute-force, but very flexible, way of assigning values to F is to use a *lookup table*. Let the binary number whose bits are D, T_1, T_2, H, and P be the address of a register, and store the proper value of F in that register. Then the input values can be used to look up the corresponding value of F.

EXAMPLE 12.5 Write a program to evaluate the function $F = (D \cdot T_1 + \overline{D} \cdot T_2 + H)$ P by using a lookup table.

Given: The input values are available as the following bits of Port 1: P1∅ = D, P11 = T_1, P12 = T_2, P13 = H, P14 = P. The output F is to be bit 7 of Port 2 (P27 = F).

Solution: The program is shown in Figure 12–14. We set the unused input bits P15, P16, and P17 to "∅" by ANDing the data from Port 1 with the mask ∅∅∅11111 (or 1F in hex). Depending on the input values, this gives us an address somewhere from ∅∅∅ (hex) to ∅1F. Suppose we want the lookup table located from ∅20 to ∅3F. Then we have to add 2∅ (hex) to the Accumulator. The instruction MOVP A, @ A looks in the program register specified by the Accumulator and moves the contents of that program register into the Accumulator. (The instruction looks on the same page the program is currently on.)

Now we have to determine the proper contents for each register in the table. If $F = 1$, then the contents 1∅∅∅∅∅∅∅ (or 8∅ hex) will put a 1 in bit 7, as desired. Otherwise the contents are ∅∅. For example if the input data are $D = 1$, $T_1 = 1$, $T_2 = \emptyset$, $H = \emptyset$, $P = 1$, the masked number in the Accumulator is ∅∅∅1∅∅11 (or ∅13 hex). After 2∅ is added to this, the address is ∅33. Since $F = 1$ for these input values, location ∅33 must contain 8∅.

Usually data are stored in data memory, but in the above example the data were stored in program memory. This was because program memory is ROM (nonvolatile), and the data will not be lost when power is turned off. Parentheses are used to indicate the contents of *program* memory, while square brackets indicate the contents of a *data* register. Therefore the instruction MOVP A,@A can be expressed as [A] ← ([A]), which can be read, "The contents of the program memory whose address is in the Accumulator are moved into the Accumulator." With two types of memory in a computer, the programmer has to keep in mind whether it is data memory or program memory he or she is addressing.

Implementing the program with a lookup table took more program memory space. But with this structure the function determining F can be changed simply by changing some of the table entries. Also, other functions can easily be added by making use of the unused bits in each table register. For example, one of the bits could signal an error such as the occurrence of $T_2 = 1$ when $T_1 = \emptyset$ (temperature shouldn't be below the nighttime setting if it is above the daytime setting).

Loc.	Obj.	Instruction	Comment
∅∅∅	∅4	JMP∅ #OAH	; START.
∅∅1	∅A		
∅∅A	∅9	IN A,P1	; INPUT PORT 1; P1∅ = D, P11 = T1,
			; P12 = T2, P13 = H, P14 = P.
∅∅B	53	ANL A,#1FH	; SET UNUSED PORT BITS TO "∅."
∅∅C	1F		
∅∅D	∅3	ADD A,#2∅H	; [A] ← [A] + 2∅H, STARTS
∅∅E	2∅		; TABLE AT LOCATION ∅2∅.
∅∅F	A3	MOVP A, @ A	; [A] − ([A]) GET DATA FROM TABLE.
∅1∅	3A	OUTL P2,A	; OUTPUT TO PORT 2; P27 = F.
∅11	∅4	JMP∅ #∅AH	; REPEAT LOOP.
∅12	∅A		
			; TABLE LOCATED ∅2∅ TO ∅3F.

Loc.	Obj.	Loc.	Obj.	Loc.	Obj.	Loc.	Obj.
∅2∅	∅∅	∅28	∅∅	∅3∅	∅∅	∅38	8∅
∅21	∅∅	∅29	∅∅	∅31	∅∅	∅39	8∅
∅22	∅∅	∅2A	∅∅	∅32	∅∅	∅3A	8∅
∅23	∅∅	∅2B	∅∅	∅33	8∅	∅3B	8∅
∅24	∅∅	∅2C	∅∅	∅34	8∅	∅3C	8∅
∅25	∅∅	∅2D	∅∅	∅35	∅∅	∅3D	8∅
∅26	∅∅	∅2E	∅∅	∅36	8∅	∅3E	8∅
∅27	∅∅	∅2F	∅∅	∅37	8∅	∅3F	8∅

FIGURE 12–14 Program for Example 12.5. This program realizes the function $F = (D \cdot T_1 + \overline{D} \cdot T_2 + H)P$ by a lookup table. The instruction MOVP A,@A replaces the address in the Accumulator with the data at that address in *program* memory.

Conditional Jumps

One way of implementing logic functions is with series and parallel switches (see Section 16.11). For example, current can flow through two series switches if switch 1 AND switch 2 are closed. Current can flow through two parallel switches if switch 1 OR switch 2 is closed. Figure 12–15 shows a circuit of switches that implements the logic $F = (D \cdot T_1 + \overline{D} \cdot T_2 + H) P$. The switches correspond to the input variables, and the state of each switch corresponds to "∅" or "1." A lit lamp corresponds to $F = 1$.

The implementation of logic by switches is analogous in computer programming to the *conditional jump*—a jump that depends on the result of a test. (We have already seen one example, the instruction DJNZ.) Just as with a switch, a conditional jump routes in one of two directions depending on the state of a variable. The following example illustrates programming with conditional jumps.

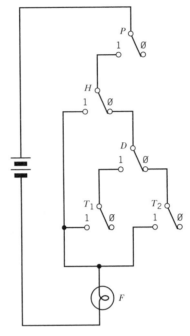

$$F = (D \cdot T_1 + \overline{D} \cdot T_2 + H)P$$

FIGURE 12–15 Realization of a logic function with switches. Switch D in position "1" corresponds to $D = 1$, and the lamp on corresponds to $F = 1$.

EXAMPLE 12.6 Write a program to evaluate $F = (D \cdot T_1 + \overline{D} \cdot T_2 + H)\ P$ by using conditional jumps.

Given: The input values D, T_1, T_2, H, and P will be bits 0, 1, 2, 3, and 4 of the Accumulator, respectively. The output F is to be bit 7 of Port 2 (P27 = F).

Solution: A flow chart for the program is shown in Figure 12–16. Note that it has the same topology as the logic circuit with switches in Figure 12–15. The corresponding program is shown in Figure 12–17. After the input values have been loaded into the Accumulator, the instruction JB4 #ØFH tests whether bit 4 in the Accumulator (which has the value P) is "1" and causes the computer to jump to location ØØF if the answer is "Yes." (Conditional jumps stay within the same page of memory.) At location ØØF is the test to see if bit 3 is "1" ($H = 1$). If the proper tests are passed, the computer executes the instruction ORL P2,#8ØH. This causes binary 1ØØØØØØØ to be ORed with the data at Port 2, setting P27 to "1." Otherwise instruction ANL P2,#7FH sets P27 to "Ø."

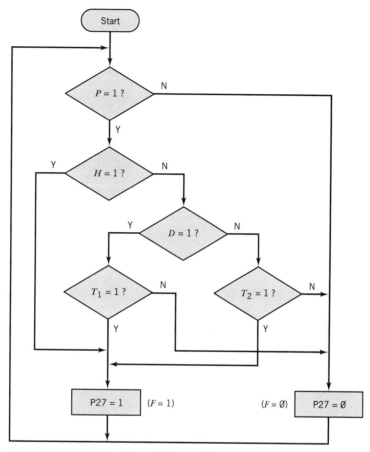

FIGURE 12–16 Flow chart for Example 12.6. The chart is isomorphic to the switching circuit in Figure 12–15. It realizes the logic function $F = (D \cdot T_1 + \overline{D} \cdot T_2 + H)P$ by conditional jumps.

12.5 ARITHMETIC OPERATIONS

The instruction set of most computers doesn't include the operations "multiply," "cube," "square root," or "sine," and some (such as the 8048) don't have a "subtract" instruction. High-level languages such as Pascal, C, and BASIC have these commands, but they don't communicate directly with the computer's Control Unit. A complex mathematical operation must first be broken down to simpler instructions in object code (sometimes called *machine code*). We will see that the instructions ADD, CPL, RR, and RL are sufficient to realize any mathematical operation or function.

As shown in Section 9.3, we can subtract a number by changing its sign (twos complementing it) and adding. Recall that the twos complement is obtained by complemeting the bits and then adding unity. For example, decimal 22 is

Loc.	Obj.	Instruction	Comment
ØØØ	Ø4	JMPØ #ØAH	; START.
ØØ1	ØA		
ØØA	Ø9	IN A,P1	; BITØ = D,BIT1 = T1,BIT2 = T2,BIT3 = H,BIT4 = P.
ØØB	92	JB4 #ØFH	; TEST BIT4, P = 1? YES, GO TO TEST H = 1?
ØØC	ØF		
ØØD	Ø4	JMPØ #19H	; NO, GO TO SET F = Ø.
ØØE	19		
ØØF	72	JB3 #1DH	; TEST BIT3, H = 1? YES, GO TO SET F = 1.
Ø1Ø	1D		
Ø11	Ø2	JBØ #17H	; TEST BITØ, D = 1? YES, GO TO TEST T1 = 1?
Ø12	17		
Ø13	52	JB2 #1DH	; TEST BIT2, T2 = 1? YES, GO TO SET F = 1.
Ø14	1D		
Ø15	Ø4	JMPØ #19H	; NO, GO TO SET F = Ø.
Ø16	19		
Ø17	32	JB1 #1DH	; TEST BIT1, T1 = 1? YES, GO TO SET F = 1.
Ø18	1D		
Ø19	9A	ANL P2,#7FH	; SET F = P27 = Ø. (7FH = Ø1111111B)
Ø1A	7F		
Ø1B	Ø4	JMPØ #ØAH	; REPEAT LOOP.
Ø1C	ØA		
Ø1D	8A	ORL P2,#8ØH	; SET F = P27 = 1. (8ØH = 1ØØØØØØØB)
Ø1E	8Ø		
Ø1F	Ø4	JMPØ #ØAH	; REPEAT LOOP.
Ø2Ø	ØA		

FIGURE 12–17 Program for Example 12.6. Each instruction corresponds to a box in the flow chart in Figure 12–16. The instruction JB4 #ØFH jumps to location ØØF if bit 4 in the Accumulator is a "1."

ØØØ1Ø11Ø in binary. Complementing gives 111Ø1ØØ1, and adding unity gives 111Ø1Ø1Ø, which represents −22.

Section 9.3 also covers binary multiplication. A simple case of multiplication is when we want to multiply by a power of two. For example, to multiply ØØØ1Ø11Ø by two, simply add a zero on the right to get ØØ1Ø11ØØ. (This is the same rule as for multiplying by ten in decimal.) The instruction that adds a zero on the right is RL A (rotate left the Accumulator). This only works if the MSB is not "1"; otherwise the "1" would get rotated out from the left side and appear on the right. Two applications of RL multiplies by four (ØØØ1Ø11Ø times four is Ø1Ø11ØØØ). A number N can be multiplied by six (not a power of two) through $N \times 2 + N \times 4$; for example ØØØ1Ø11Ø times six is ØØ1Ø11ØØ + Ø1Ø11ØØØ = 1ØØØØ1ØØ. This last example actually resulted in an overflow, since "1" in the MSB indicates a negative number (1ØØØØ1ØØ represents decimal −124, not 132).

Dividing by a power of two is also easy. Applying an RR A (rotate right the Accumulator) instruction to ØØØ1Ø11Ø divides it by two to give ØØØØ1Ø11 (in decimal, $22 \div 2 = 11$). In effect we have moved the decimal point (actually the binary point) one place to the left. Two applications of RR should divide by four: $\emptyset\emptyset\emptyset1\emptyset11\emptyset$ becomes $1\emptyset\emptyset\emptyset\emptyset1\emptyset1$. Unfortunately, a "1" has rotated off the right end and appeared on the left. The troublesome "1" can be wiped out by first ANDing $\emptyset\emptyset\emptyset1\emptyset11\emptyset$ with the mask $111111\emptyset\emptyset$ to give $\emptyset\emptyset\emptyset1\emptyset1\emptyset\emptyset$. Now when we RR twice (divide by four), the result is $\emptyset\emptyset\emptyset\emptyset\emptyset1\emptyset1$ (or 5 in decimal). The result should have been (in decimal) $22 \div 4 = 5.5$, but the masking resulted in roundoff error.

For an example of applying arithmetic operations, we will add some frills to the furnace control program in Figure 12–8. As it now stands, the input bit T_1 is "1" when room temperature falls below some setting, say 70°F. Suppose we want to adjust this setting to take account of humidity. It is roughly true that an 8% rise in relative humidity is equivalent in terms of comfort to a 1°F rise in temperature. Then a "comfort index" could be formed by HUM/8 + TEM, where HUM is the relative humidity in percent, and TEM is the temperature in degrees Fahrenheit. If the comfort index is maintained at 78, then TEM = 76°F for HUM = 16%, TEM = 70°F for HUM = 64%, and TEM = 66°F for HUM = 96%. This is maintained if T_1 is set to "1" (the furnace is brought on during the day) when HUM/8 + TEM < 78. Since HUM or TEM will not exceed 100 (or $\emptyset11\emptyset\emptyset1\emptyset\emptyset$ in binary), they can be represented by 8 bits, where the MSB is the sign bit.

EXAMPLE 12.7 Write a subroutine to set T_1 (bit 1 of the input data) to "1" when HUM/8 + TEM < 78.

Given: The number TEM is available in data memory address 19, and HUM is available at address 1A.

Solution: The subroutine and a portion of the main program are shown in Figure 12–18. After the main program puts the input data in the Accumulator, it calls the subroutine. After storing the input data (by indirect addressing), the subroutine gets the value HUM and divides it by 8. Since the Accumulator is rotated three times, the three least significant bits had to be first set to "$\emptyset$" by a mask. The address in R1 is decremented to 19, where the value TEM is, and TEM is added to HUM/8 to determine the comfort index. If it is below 78, subtracting 78 will give a negative number, and T_1 should be set to "1." The reader should confirm that 78 in binary is $\emptyset1\emptyset\emptyset111\emptyset$, and -78 in binary is $1\emptyset11\emptyset\emptyset1\emptyset$ or B2 in hex. Therefore B2 is added to the comfort index, and bit 7 (the sign bit) is tested for a "1" (a minus sign). If it is minus, the computer jumps to the instructions that OR $\emptyset\emptyset\emptyset\emptyset\emptyset\emptyset1\emptyset$ with the input data we stored at address 1B (where R1 now points). This sets bit 1 (T_1) of the input data to "1," as desired.

We have seen how subtraction, multiplication, and division can be achieved using 8048 instructions. Exponentiation, x^n, is achieved by multiplying x by itself n times. Other functions can then be evaluated by a power series—for example $\sin(x) = x - x^3/6 + x^5/120 - x^7/5040 + \cdots$. The series is easily evaluated by a loop that derives the next term in the series from the last. For $\sin(x)$, the nth term is $-x^2/(2n-1)(2n-2)$ times the previous term. The series is infinite, but

Loc.	Obj.	Instruction	Comment
ØØA	Ø9	IN A,P1	; INPUT DATA.
ØØB	54	CALL2 #ØØH	; CALL SUBROUTINE AT 2ØØ.
ØØC	ØØ		
ØØD	AA	MOV R2,A	; CONTINUE MAIN PROGRAM.
.			
.			
2ØØ	B9	MOV R1,#1BH	; POINT AT ADDR 1B.
2Ø1	1B		
2Ø2	A1	MOV @ R1,A	; STORE INPUT DATA AT 1B.
2Ø3	C9	DEC R1	; POINT AT ADDR 1A (HUM).
2Ø4	F1	MOV A, @ R1	; LOAD HUM FROM 1A.
2Ø5	53	ANL A,#F8H	; ROUND OFF WITH MASK
2Ø6	F8		; 11111ØØØ.
2Ø7	77	RR A	; HUM/2.
2Ø8	77	RR A	; HUM/4.
2Ø9	77	RR A	; HUM/8.
2ØA	C9	DEC R1	; POINT AT ADDR 19 (TEM).
2ØB	61	ADD A, @ R1	; HUM/8 + TEM.
2ØC	Ø3	ADD A,#B2H	; HUM/8 + TEM + (-78),
2ØD	B2		; (-78 IS B2 IN HEX).
2ØE	19	INC R1	; POINT AT ADDR 1A.
2ØF	19	INC R1	; POINT AT ADDR 1B (INPUT DATA).
21Ø	F2	JB7 #16H	; IS BIT 7 = 1? THAT IS,
211	16		; HUM/8 + TEM $-$ 78 $<$Ø?
212	23	MOV A,#FDH	; NO, PREPARE MASK, BINARY
213	FD		; 111111Ø1 (OR FD IN HEX).
214	51	ANL A, @ R1	; SET T1 = Ø IN INPUT DATA.
215	83	RET	; RETURN TO MAIN PROGRAM.
216	23	MOV A, #Ø2H	; YES, PREPARE MASK, BINARY
217	Ø2		; ØØØØØØ1Ø (OR Ø2 IN HEX).
218	41	ORL A, @ R1	; SET T1 = 1 IN INPUT DATA.
219	83	RET	; RETURN TO MAIN PROGRAM.

FIGURE 12–18 Program for Example 12.7. When added to the program in Figure 12–8, this subroutine causes the bit corresponding to T_1 to be a "1" when HUM/8 + TEM $<$78, where HUM is the relative humidity in percent, and TEM is the temperature in Fahrenheit.

it can be terminated when the last term changes the answer by some small amount, say 0.1%.

We have limited our binary numbers so far to 8 bits, which restricts us to numbers from 0 to 255 or from -128 to 127. This is convenient because the 8048 has only 8-bit registers. But it is possible to deal with larger (say 16-bit) numbers 8 bits at a time, and an 8-bit microcomputer will have instructions that allow this.

Suppose we want to add the decimal numbers 13141 and 4095. Figure 12–19 shows the addition in decimal, hex, and binary. If the addition is to be done on an 8048, the binary numbers ØØ11ØØ11Ø1Ø1Ø1Ø1 and ØØØØ111111111111 must be added 8 bits at a time. Taking the eight lowest bits of each first:

DECIMAL:	HEX:	BINARY:
13141	3355	ØØ11ØØ11Ø1Ø1Ø1Ø1
+ 4Ø95	+ØFFF	+ØØØØ1111111111111
17236	4354	Ø1ØØØØ11Ø1Ø1Ø1ØØ

(a)

INSTRUCTION	C	ACCUMULATOR	R2	R3	R4
MOV A, #55H		Ø1Ø1Ø1Ø1			
MOV R2, #FFH		Ø1Ø1Ø1Ø1	11111111		
ADD A,R2	1	Ø1Ø1Ø1ØØ	11111111		
MOV R4,A	1	Ø1Ø1Ø1ØØ	11111111		Ø1Ø1Ø1ØØ
MOV A,#33H	1	ØØ11ØØ11	11111111		Ø1Ø1Ø1ØØ
MOV R2,#ØFH	1	ØØ11ØØ11	ØØØØ1111		Ø1Ø1Ø1ØØ
ADDC A,R2	Ø	Ø1ØØØØ11	ØØØØ1111		Ø1Ø1Ø1ØØ
MOV R3,A	Ø	Ø1ØØØØ11	ØØØØ1111	Ø1ØØØØ11	Ø1Ø1Ø1ØØ

(b)

FIGURE 12–19 (a) Addition of two large numbers in decimal, in hex, and in binary (the same two numbers in each case). (b) Program implementing 16-bit addition to add the two numbers in (a). The final sum is stored in registers R3 and R4.

$$\begin{array}{r} Ø1Ø1Ø1Ø1 \\ + 11111111 \\ \hline (1)\ Ø1Ø1Ø1ØØ \end{array}$$

This gives us 8 bits of the result and a ninth bit—a carry bit we will need in the second half of the problem. Now taking the 8 highest bits of the numbers, we add them plus the carry bit:

$$\begin{array}{r} ØØ11ØØ11 \\ + ØØØØ1111 \\ (1) \\ \hline Ø1ØØØØ11 \end{array}$$

These 8 bits joined with the 8 bits of the first result give the 16-bit sum Ø1ØØØØ11Ø1Ø1Ø1ØØ, which is 17236 in decimal.

Instructions to carry out the above procedure are shown in Figure 12–19b. When an ADD instruction results in a 9-bit sum, the *carry bit* (C bit) is stored in a special *status register*. (This register also has other status bits, which aren't discussed in this chapter.) When adding the higher-order bits, an "add with carry" instruction ADDC is used to include the carry bit in the sum. The final 16-bit sum is stored in registers R3 and R4.

There are also special "rotate" instructions involving the carry bit that permit the programmer to conveniently multiply 16-bit numbers. (See problems E12.1 and E12.2.)

12.6   PERIPHERAL DEVICES

Some of the peripheral devices that interface with computers were shown in Figure 12–1. Other common peripherals are external memory (both ROMs and RAMs), universal asynchronous receiver-transmitters (UARTs), direct memory access units (DMAs), and timers. UARTs permit the computer to receive and send data over a single line such as a telephone line. DMAs can store data in memory more quickly than the microprocessor using the memory. Timers can provide elapsed-time information for control purposes and time documentation. Since all these devices communicate with a microcomputer in basically the same way, we will look at just a couple of examples—ADCs and RAMs—to illustrate the techniques.

A peripheral device communicates with the microcomputer through either an I/O port or an I/O bus. The difference is that when the microcomputer outputs data at a *port*, it is automatically latched (held there). A *bus* is time shared; data appear for a short time, and the peripheral must latch it internally, or a separate latch must be provided. Then the bus data can be changed and a different peripheral can latch on to that. Therefore a bus is a more natural "party line" than a port is.

Analog-to-Digital Converters

Suppose the microcomputer needs two pieces of analog information such as temperature and humidity (see Example 12.7). Transducers provide analog voltages v_1 and v_2 representing these two parameters. Then two ADCs convert these analog voltages to digital signals, say 8 bits each. Figure 12–20 shows the ADCs' connection with the microcomputer over the bus. In this case the bus is used only to input data from the ADCs. Which of the ADCs does the talking is determined by a chip-select pin $\overline{CS}$ on each device. When the $\overline{CS}$ of one ADC is low (a "$\emptyset$"), that ADC is connected to the bus and can provide data. When $\overline{CS}$ is high (a "1"), the ADC is internally disconnected from the bus, and the other ADC can control the bus if its $\overline{CS}$ is low. See Figure 13–13, where $\overline{OE}$ acts as $\overline{CS}$.

Note the convention used here for the pin label: a bar over the abbreviation indicates that action (described by the abbreviation) takes place when the logic level is low. This is called an *active low* signal. Another example is $\overline{WR}$, which when low, causes data to be WRitten into a device. An example of an *active high* signal is ALE, which when high, Enables an Address Latch (see Figures 12–23 and 12–24).

An ADC takes some time to make a conversion; a typical time is 70 μs. It is possible to wire the ADC so that it continually performs conversions, updating its digital output every 70 μs or so. However, this is not a good mode if it is communicating with a computer. Because the microcomputer and ADC are asynchronous (have different clocks), the microcomputer might ask for data from the ADC while

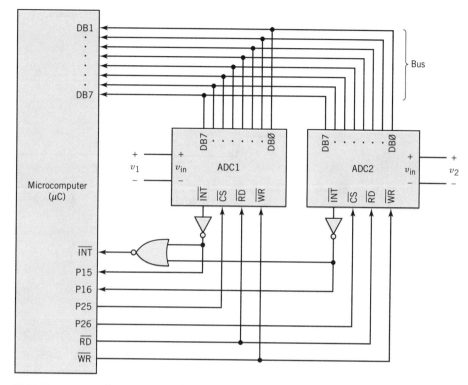

FIGURE 12–20 Connection of two analog-to-digital converters to a microcomputer. When ADC 1 has data ready, it puts an interrupt signal on $\overline{\text{INT}}$ and on P15. The μC selects ADC 1 with a signal on its $\overline{\text{CS}}$. A signal on $\overline{\text{RD}}$ causes ADC 1 data to appear on the bus.

it is updating. This would result in half old data and half new data—in other words, "garbage."

One solution is to have the microcomputer tell the ADC to make one conversion and stop. This is done by putting a low pulse on the ADC $\overline{\text{WR}}$ line. Then the microcomputer must wait until it is sure the ADC is done converting (perhaps 80 μs) and then ask for the data by putting a low pulse on the $\overline{\text{RD}}$ line. But during 80 μs the microcomputer could have been carrying out as many as 50 instructions instead of wasting its time waiting.

A better solution is for the microcomputer to start the ADC and then go about its business while the conversion is being carried out. (See the timing diagram in Figure 12–21.) When the conversion is done, the ADC *interrupts* the microcomputer by a low signal on the microcomputer's interrupt pin $\overline{\text{INT}}$. The NOR gate allows either ADC to do the interrupting. The microcomputer stops whatever it was doing and goes to a special subroutine—an interrupt service routine. This routine finds out which device interrupted it and puts a low signal on that device's chip-select pin $\overline{\text{CS}}$. When the microcomputer asks for the data by a low signal on the $\overline{\text{RD}}$ line, only the selected device responds with data on the bus. Before leaving the routine, the microcomputer starts the ADC on another conversion by a low signal on the

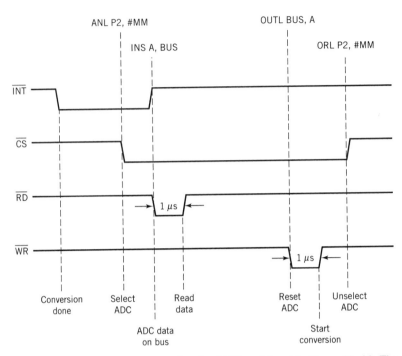

ANL P2, #MM   OUTL BUS, A

INS A, BUS   ORL P2, #MM

$\overline{INT}$

$\overline{CS}$

$\overline{RD}$
→| 1 µs |←

$\overline{WR}$
→| 1 µs |←

Conversion   Select   Read   Reset   Unselect
done   ADC   data   ADC   ADC

ADC data   Start
on bus   conversion

FIGURE 12–21 Timing diagram for the ADC peripheral in Figure 12–20. The ADC initiates a cycle with an interrupt when a conversion is done. The cycle concludes with a WR signal starting another conversion.

$\overline{WR}$ line and then "unselects" the ADC so it won't respond to further $\overline{RD}$ and $\overline{WR}$ signals.

EXAMPLE 12.8 Write an interrupt service routine to store data from ADC 1 at address 19 or store data from ADC 2 at address 1A.

Given: The interrupt signal from ADC 1 puts a low on the microcomputer's $\overline{INT}$ line and a high on the P15 line (line 5 of Port 1). The interrupt signal from ADC 2 puts a low on $\overline{INT}$ and a high on P16. ADC 1 and ADC 2 chip-selects are connected to P25 and P26, respectively (see Figure 12–20).

Solution: An interrupt service routine is shown in Figure 12–22. During its regular operation, the microcomputer checks often to see if the $\overline{INT}$ line is low. If it is, the microcomputer automatically goes to ∅∅3 in program memory (see Figure 12-5). The first instruction there disables further interrupts so the microcomputer won't be bothered while it is handling this one. The next instruction jumps to the location of the interrupt service routine—3∅∅ in this case.

Loc.	Obj.	Instruction	Comment
ØØ3	15	DIS I	; DISABLE FURTHER INTERRUPTS.
ØØ4	64	JMP3 #ØØH	; GO TO INT SERVICE ROUTINE.
ØØ5	ØØ		
.			
.			
3ØØ	B8	MOV RØ,#18H	; POINT AT ADDR 18.
3Ø1	18		
3Ø2	AØ	MOV @RØ,A	; SAVE A AT ADDR 18.
3Ø3	Ø9	IN A,P1	; INPUT FROM PORT 1.
3Ø4	D2	JB6 #1ØH	; TEST FOR ADC 2 INTERRUPT,
3Ø5	1Ø		; GO TO 31Ø IF YES (IF P16 = 1).
3Ø6	9A	ANL P2,#DFH	; SET P25 = Ø, SELECTS ADC 1.
3Ø7	DF		
3Ø8	Ø8	INS A,BUS	; INPUT FROM BUS.
3Ø9	18	INC RØ	; POINT AT ADDR 19.
3ØA	AØ	MOV @RØ,A	; STORE ADC 1 DATA AT 19.
3ØB	Ø2	OUTL BUS,A	; WR PULSE STARTS ADC 1.
3ØC	8A	ORL P2,#2ØH	; SET P25 = 1, UNSELECTS ADC 1.
3ØD	2Ø		
3ØE	64	JMP3 #19H	; GO TO EXIT STEPS.
3ØF	19		
31Ø	9A	ANL P2,#BFH	; SET P26 = Ø, SELECTS ADC 2.
311	BF		
312	Ø8	INS A,BUS	; INPUT FROM BUS.
313	18	INC RØ	; POINT AT ADDR 19.
314	18	INC RØ	; POINT AT ADDR 1A.
315	AØ	MOV @RØ,A	; STORE ADC 2 DATA AT 1A.
316	Ø2	OUTL BUS,A	; WR PULSE STARTS ADC 2.
317	8A	ORL P2,#4ØH	; SET P26 = 1, UNSELECTS ADC 2.
318	4Ø		
319	B8	MOV RØ,#18H	; POINT AT ADDR 18.
31A	18		
31B	FØ	MOV A,@RØ	; RESTORE ORIGINAL A.
31C	Ø5	EN I	; ENABLE INTERRUPTS.
31D	93	RETR	; RETURN, RESTORE STATUS BITS.

FIGURE 12–22 Program for Example 12.8. The interrupt service routine, which starts at location 3ØØ, stores ADC 1 data at address 19 or ADC 2 data at address 1A, depending on which ADC caused the interrupt.

The microcomputer is usually interrupted in the middle of some procedure that uses a number of registers. If the service routine messes up the contents of any of those registers, the microcomputer will be confused when it returns to the main program. Therefore the first task of the service routine is to save the contents of any registers it is going to use that the main program also uses. In this case only the Accumulator needs to be saved.

The next task is to find out which device did the interrupting by looking at bit 5 and bit 6 from Port 1. If bit 6 is a "1," then it was an ADC 2 interrupt, and the microcomputer jumps to the block of instructions from 31∅ to 318 that handle that ADC. Otherwise it goes to the block of instructions from 3∅6 to 3∅F that handle ADC 1.

Suppose that the interrupt was from ADC 2. Then the instruction at 31∅ ANDs 1∅111111 (or BF in hex) to Port 2, setting P26 = ∅ and selecting ADC 2. The INS A,BUS instruction applies an $\overline{RD}$ pulse that transfers data from ADC 2 to the microcomputer via the bus (see Figure 12–21). After the data have been stored at 1A, a $\overline{WR}$ pulse is applied to start ADC 2 doing another conversion. ORing ∅1∅∅∅∅∅∅ (or 4∅ in hex) to Port 2 unselects ADC 2.

The exit steps restore the original contents of the Accumulator (held at 18), and enable the microcomputer to respond to future interrupts. The instruction RETR is a special "return" instruction that restores the carry bit and other status bits that were present before the interrupt.

In the example each ADC required two port lines—one for its $\overline{CS}$ and one for its $\overline{INT}$. With eight ADCs (or other devices that generate interrupts), all 16 port lines on the 8048 would be used up. A more efficient use of the port lines would be to use only three port lines to specify a binary number from ∅ to 7 and use a 1-of-8 decoder to create the corresponding eight $\overline{CS}$ signals. Similarly, the eight $\overline{INT}$ signals from the devices can be coded into only three lines by a priority encoder (see Section 9.4). In this way the eight devices can be serviced by six port lines. The tradeoff is that two extra packages—the coder and decoder—are required.

External Memory

The 8048 microcomputer has only 64 bytes of RAM on the chip, so it is necessary to add some external data memory (RAM) for applications requiring larger data storage. As we saw in Figure 12–4, a memory device has two busses—a data bus and an address bus. Some microcomputers also have both a data bus and an address bus, and the interfacing is straightforward. However, the 8048 and some other microcomputers and microprocessors have only one bus in order to reduce the number of pins on the package. In this case the one bus must be time shared to handle both data and addresses.

Figure 12–23 shows how a microcomputer such as the 8048 can be connected with RAM devices. Figure 12–24 shows the timing sequence of the signals. The bus DB∅–DB7 first outputs an address, which is held by a parallel-in, parallel-out register or latch (see Section 9.6). An ALE (address latch enable) signal is automatically generated by the microcomputer. While the latch holds the address on A∅–A7, the bus can be used to apply data on D∅–D7. When a WR pulse is applied, the RAM that has been selected by a low $\overline{CS}$ stores the data.

If the external RAM has more than 256 bytes, its addresses won't fit on an eight-line bus. The RAMs in Figure 12–23 each have 2048 bytes, so they need a bus

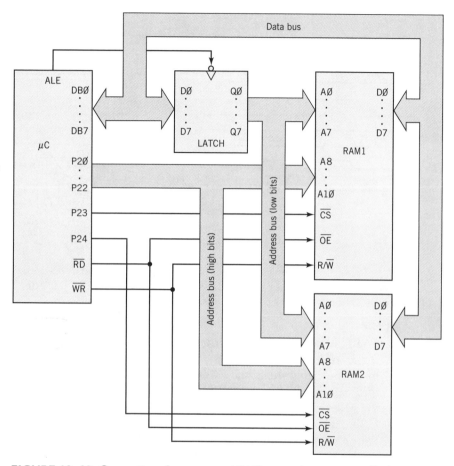

FIGURE 12-23 Connection of two external RAMs to a microcomputer. Both memory address and data appear on the μC bus. A latch holds the address on the RAMs while the data are sent over the bus. $\overline{RD}$ is connected to $\overline{OE}$ (output enable), and $\overline{WR}$ is connected to R/$\overline{W}$ (read/write).

with 11 lines. The additional three lines (forming an address bus for the higher-order bits) are provided by P2Ø–P22 of Port 2. The chip-select signals are also provided by Port 2; a "Ø" on P23 selects RAM 1, and a "Ø" on P24 selects RAM 2.

Suppose we want to store the data 1Ø1Ø1Ø1Ø (or AA in hex) at external address 111ØØ11ØØ11 (or 733 in hex) in RAM 1. A sequence of instructions that does this is

MOV A,#F7H
OUTL P2,A
MOV R1,#33H
MOVX @ R1,#AAH
ORL P2,#Ø8H

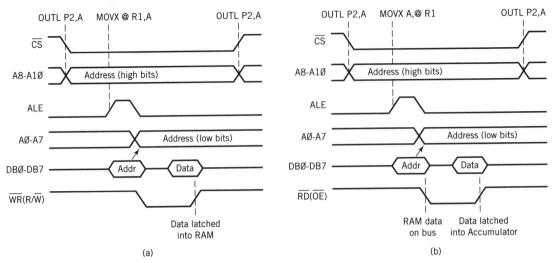

FIGURE 12–24 Timing diagrams for the RAM peripheral in Figure 12–23. (a) Write cycle. The lower-order bits of the address are latched to AØ–A7 when ALE goes low. Data are latched into the selected RAM when $\overline{WR}$ goes high. (b) Read cycle. Data are latched into the Accumulator when $\overline{RD}$ goes high.

The first two instructions write 1111Ø111 to Port 2. This not only provides the higher-order address bits 111, it also makes P23 = Ø, selecting RAM 1. The MOVX instruction does the actual writing of data to the external memory (the "X" stands for "external"). It puts the lower-order address bits on the bus, latches them with the ALE signal, puts the data bits on the bus, and latches them with the $\overline{WR}$ signal. ORing ØØØØ1ØØØ (or Ø8 in hex) to Port 2 unselects RAM 1. Here the address and data are supplied immediately by the instructions. Usually they are supplied as the contents of some register (see Example 12.9 below).

The sequence for reading data from an external RAM is about the same, except that the microcomputer pulses the $\overline{RD}$ line rather than the $\overline{WR}$ line (see Figure 12–24b). This signals the addressed RAM register to place its data on the bus, and the data are latched into the Accumulator at the end of the pulse. The following sequence of instructions reads the data at external memory address 377 (hex) in RAM 2.

MOV A,#EBH
OUTL P2,A
MOV R1,#77H
MOVX A, @ R1
ORL P2,#1ØH

The first two instructions write 111Ø1Ø11 to Port 2. The last three bits are 3 (hex), the higher-order bits of the address. P24 is "Ø," selecting RAM2. Note that the unused bits are kept high so no other devices are selected. The MOVX instruction again provides the rest of the address on the bus and generates the necessary latching signals.

EXAMPLE 12.9 Write a subroutine to store data in the next available byte of external memory and set P27 = $\emptyset$ when memory is full.

Given: The data to be stored are in the Accumulator. General scratchpad memory addresses 2$\emptyset$ and 21 point to the last-used byte of external memory. The lower-order address bits (LOBITS) are at 2$\emptyset$, and the higher-order address bits (HIBITS) are at 21. RAM 1 in Figure 12-23 is the only external memory. The subroutine is to be located starting at 38$\emptyset$.

Solution: The subroutine is shown in Figure 12–25. Since LOBITS and HIBITS are stored in General Scratchpad Memory, indirect addressing must be used. They give the address of the last-used external memory byte, so this address must be incremented. Unity is added to LOBITS, and any carry is added to HIBITS. If a carry causes HIBITS to become 1$\emptyset\emptyset\emptyset$ (bit 3 = 1), the memory is full, and the microcomputer jumps to the instruction that sets P27 = $\emptyset$. If the new address is valid, its new LOBITS and HIBITS are stored in address 2$\emptyset$ and 21.

The actual storage sequence begins with the instruction ORL A,#F$\emptyset$H, which keeps the three higher-order bits (HIBITS) intact, keeps bit 3 ($\overline{\text{CS}}$) a "$\emptyset$," and sets the remaining bits to "1" so other chips won't be selected. This is output to Port 2, selecting the RAM. The LOBITS are set up in R1 for indirect addressing, and the data to be stored are loaded into the Accumulator. Upon execution of MOVX @R1,A, then LOBITS appears on the bus, ALE latches it to the RAM, the data appear on the bus, and a $\overline{\text{WR}}$ pulse writes it in RAM (see Figure 12–24a).

Some RAMs are compatible with a single-bus microcomputer and include the latch that holds addresses that appear momentarily on the bus. This reduces the number of packages, which is always desirable. Further RAMs can be connected to the bus, each requiring its own chip-select line. As with the ADCs, the port lines are more efficiently used if a 1-of-8 decoder is used to produce the $\overline{\text{CS}}$ signals. The decoder is, of course, an additional package which may not be necessary if enough port lines are available.

The 8048 also has provision for adding external ROM for longer programs. The limit is an additional 3 kilobytes for a total of 4 kilobytes. The reader is referred to user manuals on the 8048 for information on external ROMs.

If more than one or two external memory devices are used, the designer has lost the advantage of a single-chip microcomputer. For applications requiring large memory, microprocessors with more extensive instruction sets should be used. This allows programs to be written with fewer instructions, so less ROM is required.

12.7 DEVELOPMENT SYSTEMS

The design of a microcomputer system starts with designing the interconnection of the microprocessor with memory and I/O ports (if it is not a single-chip microcomputer) and the interconnection with peripheral devices. Some port lines are dedicated to control and some to data, and addresses are established (see Figure

Loc.	Obj.	Instruction	Comment
38Ø	AB	MOV R3,A	; HOLD DATA IN R3.
381	B9	MOV R1,#2ØH	; POINT AT ADDR 2Ø.
382	2Ø		
383	81	MOV A,@ R1	; GET LOBITS.
384	AA	MOV R2,A	; HOLD LOBITS IN R2.
385	19	INC R1	; POINT AT ADDR 21.
386	81	MOV A,@ R1	; GET HIBITS.
387	2A	XCH A,R2	; EXCHANGE A AND R2.
388	Ø3	ADD A,#Ø1	; INCREMENT LOBITS TO
389	Ø1		; GET NEW LOBITS.
38A	2A	XCH A,R2	; RE-EXCHANGE A AND R2.
38B	E6	JNC #9ØH	; IF NO CARRY FROM LOBITS,
38C	9Ø		; THEN GO TO 39Ø.
38D	17	INC A	; CARRY, INCREMENT HIBITS.
38E	72	JB3 #9FH	; IF HIBITS = 1ØØØ (MEMORY
38F	9F		; OVERFLOW), GO TO 39F.
39Ø	A1	MOV @R1,A	; STORE NEW HIBITS AT 21.
391	2A	XCH A,R2	; EXCHANGE A AND R2.
392	C9	DEC R1	; POINT AT ADDR 2Ø.
393	A1	MOV @R1,A	; STORE NEW LOBITS AT 2Ø.
394	2A	XCH A,R2	; RE-EXCHANGE A AND R2.
395	43	ORL A,#FØH	; SET BIT 3 TO "Ø" AND
396	FØ		; HIGHER BITS TO "1."
397	3A	OUTL P2,A	; CS AND HIBITS TO RAM.
398	FA	MOV A,R2	; LOBITS TO ACCUMULATOR.
399	A9	MOV R1,A	; POINT AT LOBITS.
39A	FB	MOV A,R3	; DATA TO ACCUMULATOR.
39B	91	MOVX @ R1,A	; LOBITS AND DATA TO RAM.
39C	8A	ORL P2,#Ø8H	; P23 = 1, UNSELECT RAM.
39D	Ø8		
39E	83	RET	; RETURN.
39F	9A	ANL P2,#7FH	; P27 = Ø, MEMORY FULL.
3AØ	7F		
3A1	83	RET	; RETURN.

FIGURE 12–25 Program for Example 12.9. The subroutine stores data in the next available address in external RAM. The last-used address is kept in addresses 2Ø and 21.

12–20 and Figure 12–23). This is the *hardware* design. Next the program must be written to communicate with the peripherals and to perform the calculations and data moving. This is the *software* design.

In order to implement the program so the hardware and software can be tested, the program must be reduced to machine language (the binary code we have been representing by the object code in hex), and the program must be put in PROM. A few iterations will undoubtedly be necessary to *debug* the program—locate and correct its errors. The designer is aided in these steps by a "development system'' that can be purchased from the manufacturer of the microcomputer being used or from some manufacturers of test equipment.

A development system performs one or more of three functions, depending on its cost. (i) It *assembles* the program, converting symbols such as MOVX @R1,A into machine code 1ØØ1ØØØ1 (or A1 in hex). (ii) It *emulates* the operation of the microcomputer in the user's system, allowing the whole system to be tried out before the PROM is available. (iii) It *programs* the PROM or EPROM,* permanently storing the program in memory.

Figure 12–26 shows a block diagram of a development system. The heart of the system is a computer (usually a single board) that has programs in ROM to assemble, monitor, and edit the user's program, and to emulate the microcomputer and to program the PROM. The user communicates with it through a keyboard and a CRT. When the development system is emulating the microcomputer, a cable with a 40-pin plug at the end plugs into the microcomputer socket in the *target system*—the user's system that will eventually receive the microcomputer. To program the PROM, there is a socket to receive the PROM (or microcomputer with internal PROM such as the Intel 8748). The user's program can be stored on a disk or a cassette tape.

Assembler

For the simple programs in this chapter, it is reasonable to convert to machine language by looking up the object code for each instruction (see the instruction set listing in Table 12–1). This is called *assembling by hand.* The object code (or machine code) can then be programmed into a PROM by an inexpensive PROM programmer unit. However, the conversion by hand to object code can be tedious for longer programs. Also, it is subject to error, and the iteration time during debugging is long.

The assembler program in a development system automatically converts the instructions to object code. More than this, it takes care of assigning instruction addresses. Consider the following program to multiply a number in R2 by decimal number 25.

```
Ø1Ø   CLR A
Ø11   MOV R4,#25D
Ø13   ADD A,R2
Ø14   DJNZ R4,#13H
```

To assemble this, the user would enter the following *source code* through the keyboard.

---

*This is popularly referred to as "blowing" a PROM since a PROM has fusable links that are melted, just as a fuse is blown.

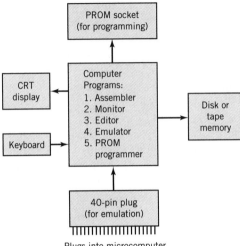

FIGURE 12–26 Block diagram of a microcomputer development system. The Assembler converts source code to object code. The Monitor observes the user's program while it is running, and the Editor changes it. The Emulator mimics the microcomputer in the target system. The PROM programmer stores the user's program in a PROM.

```
          ORG #Ø1ØH
          CLR A
          MOV R4,#25D
LOOP:     ADD A,R2
          DJNZ R4,LOOP
          END
```

The first and last lines are *directives* or *pseudoinstructions* that give information to the assembler; they don't produce an object code as instructions do. The directive ORG establishes the origin for the program—the address of the first instruction. The END directive is always the last line of source code. The *label* LOOP has been assigned to the ADD instruction. After the assembler has assigned the location Ø13 to the ADD instruction, it will substitute "#13H" for "LOOP" in the DJNZ instruction. This frees the programmer from having to change the "#13H" whenever instructions are inserted above the ADD.

From this source code the assembler produces the following object code.

LOC	OBJ
Ø1Ø	27
Ø11	BC
Ø12	19
Ø13	6A
Ø14	EC
Ø15	13

The reader should hand-assemble the program to check that this is correct. The object code, which we represent in hex, is actually the 1's and Ø's that will be stored in ROM. At this point the development system can program a PROM or store the object code on tape or disk.

Another feature of an assembler is to allow a name to be used in place of a number in the source code. For example, the source code for the program above could be written as follows:

```
COST    EQU #25D
        ORG #Ø1ØH
        CLR A
        MOV R4,COST
LOOP:   ADD A,R2
        DJNZ R4,LOOP
        END
```

the directive EQU equates the name COST with the number 25. If at some future time the cost is 35 instead of 25, only the EQU directive has to be changed. This is convenient, especially if several instructions involve COST. The EQU directive could alternatively have been expressed in hexadecimal as "COST EQU #19H." An EQU directive can also be specified in binary, such as "MASK EQU #11111ØØØB."

EXAMPLE 12.10 Write the source code for the program in Figure 12–17. When assembled, the object code produced will be that listed in Figure 12–17.

Solution: The source code is shown in Figure 12–27. The binary numbers that will be used to set P27 to "1" or "Ø" are specified in two EQU directives. Numbers in source code must start with "#" (note that numbers in hex code might not start with a number). The conditional jump instructions here include labels that stand for addresses. A label can have up to six characters; the first character must be a letter. Comments should always accompany the source code so that other people (or the author some months later) can easily follow the program.

We have mentioned only the EQU, ORG, and END directives here. Assemblers have other directives to facilitate program assembly. These may be found in user's manuals for the assemblers.

BIT	EQU #1∅∅∅∅∅∅∅B	; P27
NBIT	EQU #∅1111111B	; NOT P27
	ORG #∅∅∅H	
	JMP∅ LOOP	; START
	ORG #∅∅AH	
LOOP:	IN A,P1	; BIT∅ = D,BIT1 = T1,BIT2 = T2,BIT3 = H,BIT4 = P.
	JB4 HTEST	; TEST BIT4, P = 1? YES, GO TO TEST H = 1?
	JMP∅ NF	; NO, GO TO SET F = ∅.
HTEST:	JB3 F	; TEST BIT3, H = 1? YES, GO TO SET F = 1.
	JB∅ T1TEST	; TEST BIT∅, D = 1? YES, GO TO TEST T1 = 1?
	JB2 F	; TEST BIT2, T2 = 1? YES, GO TO SET F = 1.
	JMP∅ NF	; NO, GO TO SET F = ∅.
T1TEST:	JB1 F	; TEST BIT1, T1 = 1? YES, GO TO SET F = 1.
NF:	ANL P2,NBIT	; SET F = P27 = ∅.
	JMP∅ LOOP	; REPEAT LOOP.
F:	ORL P2,BIT	; SET F = P27 = 1.
	JMP∅ LOOP	; REPEAT LOOP.
	END	

FIGURE 12–27 Source code for Example 12.10. From this input the Assembler produces the object code in Figure 12–17. The programmer can use labels in place of addresses in the source code. The pseudoinstructions EQU, ORG, and END are directives to the Assembler, not program instructions.

It is often convenient to write a program in pieces—main program, subroutines, tables, etc. When assigning ORG directives, the programmer has to be careful not to have the pieces overlap in memory. Most assemblers can automatically relocate and link together the program pieces. Also, some assemblers have libraries of commonly used subroutines that can be made part of the user's program. Library subroutines are usually available to multiply, to convert from BCD to binary, and to evaluate the sine function, for example. The user's manual on the particular assembler will give further information on relocating, linking, and libraries.

Monitor

Once the user's program is assembled, the development system can execute it to see if it works as intended. It is helpful in diagnosing bugs to be able to stop the program at various points and examine the contents of some of the registers. For this purpose the development system has a *Monitor* program that will *single-step* through the program (one instruction at a time) or establish *breakpoints*. The program will run until it hits a breakpoint, stop for registers to be examined, and continue at the command of the user.

When the Monitor is testing the program, there are no peripherals to provide input data to the program. Therefore an instruction such as INS A,BUS has to be temporarily replaced by a MOV A,#37D instruction to simulate input data. If the program performs satisfactorily, the original input instructions can be put back in.

EXAMPLE 12.11 Suppose the program in Figure 12–8 has an error: the MOV R3,A instruction (at location ØØD) was entered as "MOV R4,A." Show how the monitor would catch the bug by single-stepping.

Solution: In place of the IN A,P1 instruction we temporarily put the following block of instructions:

CLR A

MOV R2,A

MOV R3,A

MOV R4,A

MOV R5,A

MOV R6,A

MOV R7,A

MOV A,#ØØØ11111B

This clears R2 through R7 (makes their contents zero) and enters data equivalent to $D = T_1 = T_2 = H = P = 1$ into the Accumulator. At this point the Monitor program will let us display the contents of R2 and see that it is ØØØØØØØØ. After single-stepping executes the next instruction MOV R2,A, then R2 will have ØØØ11111, as expected. After RR A, the Accumulator will have 1ØØØ1111. After the next instruction, which should be MOV R3,A, we see ØØØØØØØØ in R3 instead of 1ØØØ1111. This would lead us to examine the instruction and see that "R4" had been entered instead of "R3."

The error in the program would be corrected by using the *Editor*—a program in the development system that allows the user to delete portions of the program, insert code, and move blocks of code around. After "R4" is changed to "R3," the program should produce the expected results. Then the block of code shown above can be deleted and the original IN A,P1 inserted.

Emulator

The substitution of "move" for "input" instructions and the use of the Monitor allows the user to catch most of the bugs in the program. However, the user's system isn't fully tested until the microcomputer and programmed PROM are actually put in the target system. This will check that the proper chip-select and interrupt lines are activated and that the timing of signals is correct. When the development system is in the Emulator mode, it can create this condition.

Suppose that a program has been assembled for an Intel 8048 microcomputer that is to be used in the furnace control example we have been studying. The development system's 40-pin plug is inserted into the socket for the 8048 in the target system (the furnace control). When the Emulator begins execution of the user's program (still in the Emulator's memory), the target system will operate just as if it had the actual programmed 8048. The thermostat, pressure, and clock status signals can be changed to see if the furnace responds correctly.

If an operation error is found, it may be the fault of either hardware or software. To track down software bugs, the Emulator has monitor features to execute single-stepping, set breakpoints, and examine register contents. When the bug has been corrected, the emulation is repeated until the target system operates properly.

PROM Programming

After the user's program is fully tested, it is ready to be stored in a PROM. Most development systems program EPROMs—PROMs that can be erased and reprogrammed. The EPROM is inserted in the socket on the development system, and the user's program is written into the EPROM.

In the case of a single-chip microcomputer, the EPROM is in the microcomputer package. In the 8048 family of microcomputers, the 8748 has the EPROM program memory. A development system for the 8048 will have a socket to receive the 8748 to program it.

Since an EPROM is more expensive than a ROM, the programmed EPROM is usually just used for the prototype target system. If experiments with the prototype turn up needs for program changes, the EPROM can be reprogrammed on the development system.

When the final program is decided on, it is sent to the semiconductor manufacturer for the development of a custom ROM (or 8048) with the user's program incorporated at the time of manufacture. The manufacturer makes a *mask* that determines the pattern of metal paths on the ROM—a pattern that permanently encodes the user's program in the ROM. Although there is a charge for development of the mask, the ROM is cheaper than the EPROM for large enough quantities—for lots of about 500 or more.

12.8   CASE STUDY

The task is to design an echo box for an electric guitar. Each strum on the guitar is to produce a series of echoes that die out. The echoes are to be spaced about $\frac{1}{7}$ of a second apart, each echo being half the amplitude of the previous one. The signal flow graph in Figure 12-28a expresses this mathematically. The signal v_1 from the guitar passes immediately to the output v_2. After a delay of τ and attenuation by a factor of 0.5, the signal again appears at the output as an echo. This repeats until the echo is too soft to be heard (see Figure 12-28b). Note that while the echo is dying out, another can start and continue on top of it. Since the system is linear, superposition holds.

The delay of $\frac{1}{7}$ of a second could be achieved with a cable, but the length would have to be about 30,000 kilometers (a signal travels two-thirds the speed of light in a cable). The most practical means of delay is to convert the analog signal to PCM (see Section 11.7), store the digital codes in memory, read them out $\frac{1}{7}$ of a second later, and convert them back to an analog signal. A block diagram of such a system is shown in Figure 12–29. The low-pass filter (LPF) at the input prevents aliasing (see Section 11.6), the analog-to-digital converter (ADC) digitally encodes

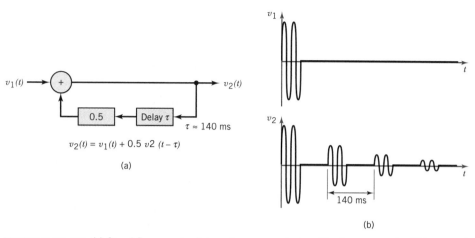

FIGURE 12–28 (a) Signal flow graph of the echo system designed in the case study. (b) Input and output waveforms for a brief sinusoidal burst (meant to model a guitar strum).

the signal, the microcomputer (μC) and memory (RAM) delay the signal, and the digital-to-analog converter (DAC) and LPF return the signal to analog form.

We would like to pass all the audio spectrum up to 20 kHz, but this would require a sampling rate of at least 40 kHz according to the Nyquist criterion. This would give the 8048 only $1/(40 \text{ kHz}) = 25 \mu s$ to process each sample. Since each instruction takes 1.36 or 2.72 μs, there would be time for only 9 to 18 instructions. This will not be enough. We will reduce the LPF cutoff frequency to 12 kHz (about the bandwidth of AM radio), allowing a sampling frequency as low as 24 kHz and increasing the number of possible instructions by 67%.

The size of the RAM depends on the length of code for each sample and the number of samples that must be stored. The 8048 naturally handles 8-bit codes, which give a SNR of about 48 dB (about that of the telephone). Suppose we choose a sampling frequency of 28 kHz—17% higher than the Nyquist frequency. Then for a delay of $\frac{1}{7}$ of a second, we need to store $28,000/7 = 4000$ bytes of digitally

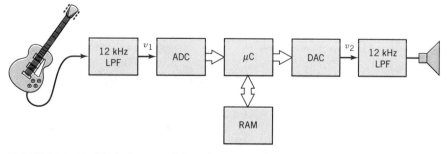

FIGURE 12–29 Block diagram of the echo system. An analog-to-digital (ADC) converter prepares the sound for digital signal processing, and a digital-to-analog converter (DAC) brings it back to an analog signal.

encoded samples. The nearest memory size to this is 4096 bytes, requiring 12 address lines ($2^{12} = 4096$).

The hardware configuration is shown in Figure 12–30. The lower 8 bits of the RAM address are provided by the data bus and a latch. The upper 4 bits are provided by bits $\emptyset$ through 3 of Port P2. Bits 4, 5, and 6 of Port P1 provide the chip select signals for the RAM, ADC, and DAC, respectively. The design of the analog low-pass filters (see Figure 12–29) could be either passive (Section 6.5) or active (Section 15.6).

The design of the software begins with the flow chart shown in Figure 12–31. When the power is first turned on, the RAM should have no signal in it. When v_1 into the ADC is 0 V, the corresponding ADC output is half of its maximum integer 255, or 128. So the RAM is cleared by storing 128 (80 in hexadecimal) in each of the 4096 locations. The rest of the instructions form a loop to read a sample that has been stored in the RAM for $\frac{1}{7}$ of a second, attenuate it by halving it, add it to a new sample read from the ADC, and write the signal with echo to the DAC. The next RAM location is addressed, and the loop repeats.

Since the ADC represents 0 V by the number 128, halving the echo in volts corresponds to halving the distance between the ADC's ECHO number and 128:

$$\begin{aligned} \text{ECHO}' &= (\text{ECHO} + 128)/2 \\ &= \text{ECHO}/2 + 64. \end{aligned}$$

This explains the "ADD 64" step following the "DIVIDE ECHO BY 2" step in Figure 12–31.

The general-purpose registers (see Figure 12–6) are assigned the following contents.

R1: lower 8 bits (line) of RAM address

R2: upper 4 bits (page) of RAM address

R3: temporary store of echo after retrieval from RAM

R4: RAM chip select 111$\emptyset$1111 (or EF in hex)

R5: ADC chip select 11$\emptyset$11111 (or DF in hex)

R6: DAC chip select 1$\emptyset$111111 (or BF in hex)

R7: 64 to be added to ECHO/2

The RAM address in R1 and R2 is changed once each cycle of the loop. When the contents of R4, R5, or R6 are moved into P1, the corresponding chip is selected by making its $\overline{\text{CS}}$ pin low.

The instructions to carry out the functions in the flow chart are shown in Figure 12-32. The block of instructions located in ROM addresses $\emptyset\emptyset$A through $\emptyset 2\emptyset$ clear the external RAM. The DJNZ instruction is the quickest way to cycle through the 4096 RAM addresses and stop when the clearing is complete. Since the DJNZ counts downward, we would logically start at the highest address—FF in R1 and $\emptyset$F in R2. But DJNZ leaves the loop when first it finds $\emptyset\emptyset$ in R1 or R2, and

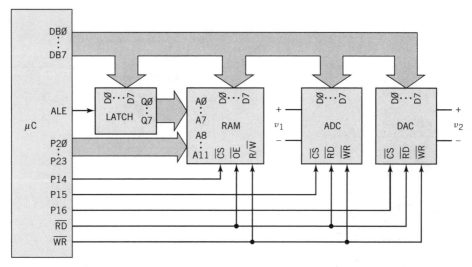

FIGURE 12–30 Hardware arrangement for the echo system, showing pin assignments for address busses and chip selects. The memory chip (RAM) provides the delay for the echo.

these addresses would not be cleared. Therefore we start with ∅∅ in R1 and X∅ in R2. The contents of R1 and R2 in sequential cycles is as follows:

R2	R1	R2 (binary)	R1 (binary)	RAM Address
1∅	∅∅	∅∅∅1∅∅∅∅	∅∅∅∅∅∅∅∅	∅∅∅∅∅∅∅∅∅∅∅∅
1∅	FF	∅∅∅1∅∅∅∅	11111111	∅∅∅∅11111111
1∅	FE	∅∅∅1∅∅∅∅	1111111∅	∅∅∅∅1111111∅
⋮	⋮	⋮	⋮	⋮
1∅	∅1	∅∅∅1∅∅∅∅	∅∅∅∅∅∅∅1	∅∅∅∅∅∅∅∅∅∅∅1
∅F	∅∅	∅∅∅∅1111	∅∅∅∅∅∅∅∅	1111∅∅∅∅∅∅∅∅
∅F	FF	∅∅∅∅1111	11111111	1111111111111
∅F	FE	∅∅∅∅1111	1111111∅	111111111111∅
⋮	⋮	⋮	⋮	⋮
∅F	∅1	∅∅∅∅1111	∅∅∅∅∅∅∅1	1111∅∅∅∅∅∅∅1
∅E	∅∅	∅∅∅∅111∅	∅∅∅∅∅∅∅∅	111∅∅∅∅∅∅∅∅∅
⋮	⋮	⋮	⋮	⋮
∅1	∅1	∅∅∅∅∅∅∅1	∅∅∅∅∅∅∅1	∅∅∅1∅∅∅∅∅∅∅1

Note that R2 starts out with ∅∅∅1∅∅∅∅ rather than ∅∅∅∅∅∅∅∅. However, the RAM is connected to only the four lowest bits of P2 (see Figure 12–30), which gets the contents of R2. Therefore we only have to worry about the four lowest bits of R2. With the final decrements of R1 and R2, they both contain zero, and the program falls out of the clear routine into the next block of instructions.

The block of instructions beginning at location ∅21 executes the delay. The use of the RAM begins at this address and follows the same sequence shown above. This time, however, we don't want the loop to stop when the contents of R2 reach

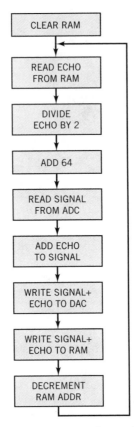

FIGURE 12–31 Flow chart for the echo system software. These functional blocks execute the mathematical operations indicated in the signal flow graph in Figure 12–28. Data read from the RAM were stored there 140 ms before.

zero. The final instruction in locations $\emptyset38$ and $\emptyset39$ throws the program back to the start of the delay routine when this happens.

In Section 12.5 we divided by 2 by masking the LSB to be zero and rotating the Accumulator to the right with the RR A instruction. To save the masking time, we use the instruction RRC A, which rotates the accumulator so the carry bit (which is zero at the time of the instruction) goes into the MSB, and the LSB goes into the carry bit. (See the similar RLC A instruction in problem E12.1.)

The instructions in locations $\emptyset29$ and $\emptyset2A$ read the ADC and immediately start it making another conversion. This takes less time than answering an interrupt, as

FIGURE 12–32 Software for the echo system. The program begins with instructions to clear the RAM, as outlined in Figure 12–30. Then the second half of the program operates in a loop to implement the delay, attenuation, and addition of the echo.

Loc.	Obj.	Instruction	Comment	Selected
∅∅∅	∅4	JMP∅ #∅AH		XXX
∅∅1	∅A			XXX
:				
∅∅A	B9	MOV R1,#∅∅H	; INITIALIZE R1	XXX
∅∅B	∅∅			XXX
∅∅C	BA	MOV R2,#1∅H	; INITIALIZE R2	XXX
∅∅D	1∅			XXX
∅∅E	BC	MOV R4,#EFH	; RAM SELECT (P14=∅) IN R4	XXX
∅∅F	EF			XXX
∅1∅	BD	MOV R5,#DFH	; ADC SELECT (P15=∅) IN R5	XXX
∅11	DF			XXX
∅12	BE	MOV R6,#BFH	; DAC SELECT (P16=∅) IN R6	XXX
∅13	BF			XXX
∅14	BF	MOV R7,#4∅H	; 64 (DECIMAL) IN R7	XXX
∅15	4∅			XXX
∅16	FA	MOV A, R2	; GET RAM PAGE	XXX
∅17	3A	OUTL P2,A	; ESTABLISH RAM PAGE	XXX
∅18	FC	MOV A,R4	; GET RAM SELECT	XXX
∅19	39	OUTL P1,A	; SELECT RAM	RAM
∅1A	23	MOV A,#8∅H	; 128 (DECIMAL) IN ACCUMULATOR	RAM
∅1B	8∅			RAM
∅1C	91	MOVX @R1,A	; CLEAR RAM LINE	RAM
∅1D	E9	DJNZ R1,#1CH	; NEW RAM LINE	RAM
∅1E	1C			RAM
∅1F	EA	DJNZ R2,#16H	; NEW RAM PAGE	RAM
∅2∅	16			RAM
∅21	FA	MOV A,R2	; BEGIN DELAY ROUTINE	RAM
∅22	3A	OUTL P2,A	; ESTABLISH RAM PAGE	RAM
∅23	81	MOVX A,@R1	; GET ECHO FROM RAM	RAM
∅24	67	RRC A	; ECHO/2	RAM
∅25	6F	ADD A,R7	; ECHO/2 + 64	RAM
∅26	AB	MOV R3,A	; TEMP STORE ECHO	RAM
∅27	FD	MOV A,R5	; GET ADC SELECT	RAM
∅28	39	MOV P1,A	; SELECT ADC	ADC
∅29	∅8	INS A,BUS	; READ SIGNAL FROM ADC	ADC
∅2A	∅2	OUTL BUS,A	; WRITE TO ADC (START CONV)	ADC
∅2B	6B	ADD A,R3	; SIGNAL + ECHO	ADC
∅2C	2E	XCH A,R6	; GET DAC SELECT	ADC
∅2D	39	OUTL P1,A	; SELECT DAC	DAC
∅2E	2E	XCH A,R6	; GET SIGNAL + ECHO BACK	DAC
∅2F	∅2	OUTL BUS,A	; WRITE SIGNAL + ECHO TO DAC	DAC
∅3∅	2C	XCH A,R4	; GET RAM SELECT	DAC
∅31	39	OUTL P1,A	; SELECT RAM	RAM
∅32	2C	XCH A,R4	; GET SIGNAL + ECHO BACK	RAM
∅33	4∅	MOVX @R1,A	; STORE SIGNAL + ECHO IN RAM	RAM
∅34	E9	DJNZ R1,#23H	; NEW RAM BYTE	RAM
∅35	23			RAM
∅36	EA	DJNZ R2,#21H	; NEW RAM PAGE	RAM
∅37	21			RAM
∅38	∅4	JMP∅ #∅21	; GO TO START OF DELAY ROUTINE	RAM
∅39	21			RAM

was done in Example 12.8. However, we must make sure that the ADC can complete the conversion in less time than it takes the loop to go through one cycle. The shortest loop is the block of instructions from location $\emptyset23$ to $\emptyset35$. Each instruction that includes INS, OUTL, MOVX, or # takes 2.72 μs to execute (there are 8 such instructions). The remaining instructions (ten of them) each take 1.36 μs to execute. Therefore the total time for the short loop is 35.36 μs. ADCs are available with conversion times less than this; the cost increases with the speed.

For every 256 cycles of the short loop we enter a new RAM page, requiring 6.8 μs for the three additional instructions. This is negligible compared with 256 $\times$ 35.36 μs = 9052 μs, and we can say the sampling rate is $1/(35.36 \mu s)$ = 28 kHz. This is the sampling rate we chose at the start of the design. (This was not luck, of course, but the result of a couple iterations through the problem.)

The main effort throughout the software design was to minimize the time required for the instructions that must be executed for every sample (the "short loop"). Instructions such as DJNZ and XCH reduce the number of instructions by performing multiple operations. The 8048 is generally not intended for real-time signal processing because of its lower speed and small bus. Larger microcomputers and microprocessors have 16- and 32-bit busses that can handle larger numbers (for greater SNR) and longer instructions (for fewer instructions and shorter loop times). They also have faster clocks for shorter execution time of each instruction.

12.9 SUMMARY

Computers can input data, store data, perform mathematical and logic operations on data, and output data. Mainframe computers operate at the highest speed and can store the most data. Microcomputers are small enough to be included in such equipment as intelligent microwave ovens. In this application, "intelligent" refers to the ability to monitor and store many data inputs and respond in complex ways to the data.

The performance of a microcomputer in a given application could be duplicated by a custom logic circuit, but the cost would be very high. The genius of the computer is that it has a standard architecture—a number of data registers and an ALU under the direction of a Control Unit. The customization comes in the set of instructions that the Control Unit follows. In other words, the computer hardware is universal, allowing high volume and relatively low cost; the software allows the computer to be specialized for a particular application.

There are two kinds of memory in a computer—RAM, which is volatile, for data, and ROM, which is nonvolatile, for program instructions. The memory is arranged so that a number of bits—4, 8, 16, or 32—are addressed at one time. A 4-kilobyte memory can address 4096 bytes (4096 $\times$ 8 bits). A typical microcomputer program might have 3000 instructions, requiring about 4K of ROM. Usually the RAM doesn't have to be as large as the ROM.

The machine language for computer instructions consists of binary codes that tell the Control Unit to perform simple operations such as move data, add, OR, jump to some instruction, etc. The software designer usually writes the program for a microcomputer in assembly language—mnenomics such as ADD A,R2. These are converted to machine language by the assembler of a microcomputer development system. Programs for minicomputers and mainframes are written in high-level languages such as BASIC, C, or COBOL. These too must be converted to machine language—by a compiler.

Programs for microcomputers use many of the same techniques as high-level languages—loops, subroutines, conditional jumps. Unlike high-level languages, assembly language does not have statements such as "$X = A*B$" or "IF $A < B$;" these must be built up from simpler instructions. Also, the programmer of a microcomputer must decide specifically where data are to be stored to avoid overlap and to address the proper data. The technique of indirect addressing allows an address to be calculated at the time of program execution.

A computer communicates data to and from the outside world by I/O ports and busses. Most data transfer is over the data bus, which is time shared among several peripheral devices such as external memory, ADCs, DACs, and latches. The address bus and the control bus indicate which device is being addressed and whether it is to talk or listen. Sometimes a peripheral device will indicate that it wants to talk with the microcomputer by generating an interrupt signal.

Development systems are available to help the user program and debug a microcomputer system. They make the conversion from assembly code to object code, run the program, emulate the electrical behavior of the microcomputer, and program PROMs.

FOR FURTHER STUDY

T. C. Bartee, *Digital Computer Fundamentals,* McGraw-Hill, New York, 1985.

J. D. Lenk, *Handbook of Microcomputer-Based Instrumentation and Control,* Prentice-Hall, Englewood Cliffs, NJ, 1984.

Embedded Controllers and Processors Handbook (No. 270645), Intel Corp., 1992.

Macroassembler Users Guide (No. 980937), Intel Corp., 1992.

PROBLEMS

Easy Drill Problems (answers at end of chapter)

D12.1. What is the binary code in program memory corresponding to the instruction RL A (rotate left the Accumulator)?

D12.2. What is the actual number of bytes referred to as "8 kilobytes?"

D12.3. How many address bus lines are necessary to address 64 kilobytes of memory? There is one address for every byte (8 bits).

D12.4. The complete 8-bit address of register R5 is $\emptyset\emptyset\emptyset\emptyset\emptyset1\emptyset1$ (see Figure 12–6). Its 3-bit abbreviated address is $1\emptyset1$. Convert the object codes for "MOV R5,A" and "MOV R3,A" to binary, and show that the last 3 bits of code are the abbreviated address of the register involved (see Figure 12–7b).

D12.5. Suppose [A] = $1\emptyset\emptyset1\emptyset11\emptyset$ and [R5] = $\emptyset1\emptyset1\emptyset\emptyset11$. After the instruction ORL R5,A is executed, what is [A] (the contents of the Accumulator)?

D12.6. Suppose [A] = $1\emptyset\emptyset1\emptyset11\emptyset$. After the instruction RR A is executed, what is [A]? After RR A is executed again, what is [A]?

D12.7. If it takes 1.36 μs to execute each byte of instructions, how long does it take to complete one cycle of the program loop in Figure 12–14?

D12.8. Design a mask to preserve bits $\emptyset$ and 1 in the Accumulator and set the other six bits to $\emptyset$. Should [A] be ANDed or ORed with this mask? Repeat the problem, but set the other 6 bits to 1.

D12.9. Give a sequence of two instructions that will jump to location $\emptyset A\emptyset$ if [A] is odd and to location $\emptyset B\emptyset$ if [A] is even.

D12.10. Suppose that [R5] = FF and [R1] = $\emptyset5$. Give three different single instructions that will each load the number FF into the Accumulator. Give the object code for each instruction. Which instruction uses indirect addressing?

D12.11. Express decimal -90 in hexadecimal. (See "Negative Numbers" in Section 9.3.) This is to be done in the context of an 8-bit number system.

D12.12. Give a sequence of two instructions that replaces the number N in the Accumulator with $-N$.

D12.13. Give a sequence of four instructions that multiplies the number in the Accumulator by 6.

D12.14. Suppose [A] = $\emptyset1111111$ and [R2] = $\emptyset\emptyset\emptyset\emptyset\emptyset\emptyset\emptyset1$. What is the C bit (carry bit) in the status register after execution of the ADD A,R2 instruction? Repeat for [A] = 11111111.

D12.15. Which instruction causes a pulse on the $\overline{RD}$ pin of the Intel 8048? Which three instructions cause a pulse on the $\overline{WR}$ pin?

D12.16. Which instructions cause a pulse on the ALE pin?

D12.17. An interrupt causes an instruction at a certain location to be executed. What is that location?

D12.18. For the external memory arrangement in Figure 12–23, write a sequence of instructions to store the contents of the Accumulator at address 25F of RAM 1.

D12.19. The following program produces a number at Port 2 that is a function of the number at Port 1. What is the function?

$\emptyset1\emptyset$	IN A, P1
$\emptyset11$	JB7 #15H
$\emptyset13$	JMP$\emptyset$ #17H
$\emptyset15$	CPL A
$\emptyset16$	INC A
$\emptyset17$	OUTL P2,A
$\emptyset18$	JMP$\emptyset$ #1$\emptyset$H

D12.20. Give two different single instructions that will each load the number $\emptyset\emptyset$ into the Accumulator. Use no other register.

Application Problems

P12.1. Replace the instructions at locations $2\emptyset5$ to $2\emptyset9$ (inclusive) in Figure 12–18 with a CALL to a divide-by-8 subroutine that begins at location 1A8 (hex). Give the CALL instruction and its object code. What depth of subroutine nesting is involved here?

P12.2. Replace the three RR A instructions at locations $2\emptyset7$ to $2\emptyset9$ in Figure 12–18 with a loop that is executed three times, performing RR A once each time. Give instructions and object codes. Is the number of program bytes increased or reduced?

P12.3. Suppose that R2–R7 each contains the square of its address. Write a three-instruction subroutine (including RET) to replace the number N in the Accumulator with N^2, where $2 \leq N \leq 7$. (*Suggestion:* Use indirect addressing.)

P12.4. Suppose that *memory* locations $\emptyset A\emptyset$-$\emptyset F9$ contain 127 sin 0°, 127 sin 1°, . . . , 127 sin 89° (rounded to the nearest integer). Write a three-instruction subroutine (including RET) to replace the number N in the Accumulator with 127 sin N, where $0° \leq N \leq 89°$. The subroutine is to start at program memory location $\emptyset9D$. Give the contents in hexadecimal of the memory location $\emptyset B4$.

P12.5. Write a program starting at location $\emptyset\emptyset A$ to evaluate 127 sin N, for $0° \leq N \leq 179°$, by the following approach. If $N \leq 89°$, call the subroutine in P12.4. Otherwise use the fact that sin $N = -\sin (N - 90°)$, assured that now $0° \leq (N - 90°) \leq 89°$. [*Suggestion:* Convert the test for $N \leq 89°$ to a test for a negative number, which is indicated by the "sign bit" (see "Negative Numbers" in Section 9.3).]

P12.6. By following the multiplication method in Figure 9–7, write a program starting at location $\emptyset\emptyset A$ to multiply a number N_1 in the Accumulator by a number N_2 in R2, where N_1 and N_2 are each less than 16. (*Suggestion:* Keep the running total in R3, and use XCH instructions to get numbers into and out of the Accumulator. Use a loop that is cycled four times, since N_1 and N_2 are 4 bits or less.)

P12.7. Suppose that 8 bits represent numbers from $\emptyset$ to 255; negative numbers are not represented. Write a program starting at location $\emptyset\emptyset A$ that adds the numbers in R3 through R7 and puts the sum in the Accumulator. If there is an overflow (sum > 255), set bit $\emptyset$ at Port 1 to a "1." (*Suggestion:* Use the instruction JC.)

P12.8. Draw a diagram as in Figure 12–20 of four ADCs connected to the μC. Use a priority encoder for the ADC $\overline{\text{INT}}$ lines, and use a 1-of-4 decoder for the $\overline{\text{CS}}$ lines (see Section 9.4).

P12.9. In the figure on p. 408 asynchronous data are interfaced with an Intel 8048 through a latch. "Data ready" is indicated by the Strobe line going high, causing $\overline{Q}$ to go low. The latch output is connected to the bus when its $\overline{\text{OE}}$ is low. Write an interrupt service routine to read the data into register R7. The routine must also remove the low signal on the $\overline{\text{INT}}$ pin by resetting the D flip-flop with a high pulse at least 2 μs wide on R. Otherwise the μC will answer the same interrupt as soon as the routine is done.

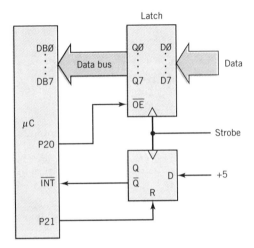

P12.10. Assemble the instructions in D12.19 by hand—that is, convert them to object code. Write source code for a development system assembler that will generate the same object code. Include all necessary directives.

P12.11. Write source code for the program portion in Figure 12–13b. Use an EQU directive so the five data entries can easily be reduced to four or three at some later time.

P12.12. For the purpose of debugging with a development system Monitor, modify the program in D12.19 so the decimal number −86 (10101010 in binary) takes the place of the Port 1 number. What are the contents of the Accumulator after the execution of each instruction?

Extension Problems

E12.1. The instruction RLC A rotates the Accumulator left through the carry bit, as shown in Figure E12.1. Suppose [R1] = ∅∅∅∅∅∅∅∅ and [R2] = ∅1111111 at the beginning of execution of the following program loop.

∅1∅   XCH A,R2

∅11   RLC A

∅12   XCH A,R2

∅13   XCH A,R1

∅14   RLC A

∅15   XCH A,R1

∅16   JMP∅ #1∅H

The XCH instruction doesn't affect the C bit. What are the contents of the C bit, the Accumulator, R1, and R2 after the execution of each instruction for two cycles of the loop?

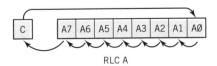

RLC A

E12.2. The RRC A instruction rotates the Accumulator to the right through the carry bit (arrows reversed in Figure E12.1). Modify the multiplication program in P12.6 by using the RLC A and RRC A instructions to multiply a 16-bit number N_1 in R4 and R5 by a 16-bit number N_2 in R6 and R7, given $N_1 \times N_2 < 65536$.

E12.3. After power is applied to the microcomputer, it can be reset by applying a low level (less than 1.0 V) to the $\overline{\text{RESET}}$ pin for at least 10 ms. The circuit in Figure E12.3 does this automatically by keeping $\overline{\text{RESET}}$ low for a short time after power is applied with the switch. How large must C be? Assume no current into the $\overline{\text{RESET}}$ pin. (The Intel 8048 includes the 80-kΩ resistor internally.)

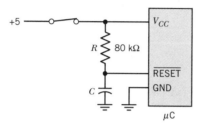

Study Questions

12.1. What are the four principal types of operations (instructions) a microcomputer performs on data?

12.2. Why does a computer need two separate areas of memory?

12.3. The Accumulator is one of many registers. What is its special role?

12.4. What is the difference between a microcomputer and a microprocessor?

12.5. What is the relationship among the terms *data memory, volatile memory, RAM, nonvolatile memory, program memory,* and *ROM*?

12.6. Define *assembly code, object code,* and *machine code.*

12.7. How many address bits does 16 megabytes of memory require?

12.8. Suppose the Accumulator contains the number 237. After an instruction that moves data from the Accumulator to R2, what data are left in the Accumulator?

12.9. The OR and AND are supposed to be operations on binary numbers. How can the ORL and ANL instructions operate on two numbers such as 237 and 128?

12.10. What action is taken as a result of instructions that make decisions?

12.11. What is the last instruction in a subroutine?

12.12. Which instruction treats the contents of the program memory as if it were data?

12.13. It is especially easy for a computer to divide by a number if that number comes from what class of numbers?

12.14. It is possible for several peripherals to talk to the computer on the same data bus. What keeps them from talking at the same time?

12.15. What are the three main functions of a development system for a computer chip?

ANSWERS TO DRILL PROBLEMS

D12.1. 1110̸0̸111

D12.2. 8192 bytes

D12.3. 16 lines

D12.4. 1Ø1Ø11Ø1 ends in 1Ø1, 1Ø1Ø1Ø11 ends in Ø11

D12.5. 11Ø1Ø111

D12.6. Ø1ØØ1Ø11, 1Ø1ØØ1Ø1

D12.7. 12.24 μs

D12.8. ØØØØØØ11, AND; 111111ØØ, OR

D12.9. JBØ #AØ; JMPØ #BØ

D12.10. MOV A,#FFH or MOV A,R5 or MOV A,@R1

D12.11. 1Ø1ØØ11Ø

D12.12. CPL A; INC A

D12.13. RL A; MOV R1,A; RL A; ADD A,R1

D12.14. Ø; 1

D12.15. INS A,BUS; OUTL BUS,A and ANL BUS#MM and ORL BUS,#MM

D12.16. MOVX @ R1,A and MOVX A,@R1

D12.17. ØØ3

D12.18. MOV R1, #5FH
MOV R3,A
MOV A,#F2H
OUTL P2,A
MOV A,R3
MOVX @R1,A
ORL P2,#Ø8H

D12.19. Absolute value

D12.20. CLR A or MOV A,#ØØH

CHAPTER 13

Digital Integrated Circuits

In this chapter we look at practical considerations in implementing the logic functions introduced in Chapter 9. Logic functions deal with logic states—mathematical symbols of "1" and "0." The logic circuits that perform these functions deal with logic levels—voltages that are high (H) or low (L). In the physical implementation of the mathematics, it is arbitrary whether H stands for "1" and L stands for "0" or the reverse. Engineers usually think and implement in *positive logic*, in which H is "1," but sometimes it is useful to use *negative logic*, in which L is "1." Throughout this chapter we use positive logic.

The waveforms in Figure 13–1 illustrate the distinction between an ideal logic function and its physical realization by a logic circuit. Consider the logic AND function in Figure 13–1a. As an input variable A goes through a sequence of "1" and "0" states, the output C ideally follows instantaneously. Essentially the same symbol (see Figure 13–1b) is used to represent the logic circuit that performs this function. (The power supply connections are V_{CC} and V_{EE}.) Here a "1" is represented by a high voltage level—about 5 V here—and a "0" by a low voltage level—about 0 V. These voltage levels might be far from the nominal. For example, the circuit might provide a high level of only 4 V. This may still be adequate to represent a "1," but would 2.5 V represent a "1" or a "0"? Another departure from the ideal is that logic circuits have "propagation delay" from input to output—the t_{pd} in Figure 13–1b. They also require some "transition time" t_T to go from high to low

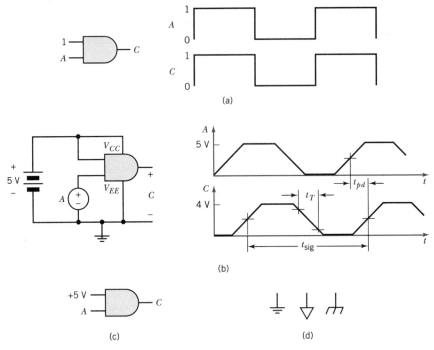

FIGURE 13–1 (a) Logic AND function. The inputs and output are logic states. (b) Logic AND circuit. The inputs and output are voltages with propagation delay and transition time. The signal frequency is limited by the transition time. V_{CC} and V_{EE} are power supply connections. (c) Circuit diagram with voltages referenced to ground. (d) Ground symbols.

or from low to high. Devices for which these times are long are said to be "slow" and can't be used with high-frequency signals.

The circuit diagram in Figure 13–1b can be simplified by hiding the voltage supplies for power. When the intent of the diagram is to clarify the design, not to show all wiring connections, the simpler diagram in Figure 13–1c is usually used. The voltage symbols +5V, A, and C no longer have the + and − reference associated with them; the voltages are assumed to be relative to ground. This concept was introduced in nodal analysis (see Figure 3–3a). Ground is usually (but not always) the negative terminal of the power supply. Various ground symbols are shown in Figure 13–1d. Throughout these four chapters on electronic circuits we will use the convention of hiding the power supplies, only indicating its connections by symbols such as ground and "+5."

In this chapter we examine the practical limitations in realizing logic functions—limitations of voltage level tolerance, speed or frequency, current driving capability, and power dissipation. There are a number of *logic families*—CMOS, TTL, and ECL—each with its own specifications for these limitations. Manufacturers of logic circuits publish the specifications in data books for each of the families, and they often include a discussion of what the specifications mean, how they are measured, and how they affect a circuit design. The material in this chapter is a compendium of such information; it gives the reader an overview of the available options in integrated circuits and provides the basics of practical digital circuit design.

The reader's objectives in this chapter are to:

1. Appreciate the practical limits of digital circuits;
2. Be able to choose the logic family that provides the best tradeoffs among power consumption, speed, and cost for the task at hand;
3. Understand the interfacing problems between the logic voltage levels of two logic families;
4. Be able to drive relays and lamps with logic devices;
5. Understand "tri-state" logic and bussing concepts frequently encountered with computers;
6. Be able to estimate the power consumed by a circuit, estimate the resulting temperature rise of the circuit, and use air flow to control the temperature.

13.1 INTEGRATED CIRCUITS

Logic circuits are usually integrated circuits (ICs)—circuits that are small enough to fit in packages such as those shown in Figure 13–2. The "dual in-line packages" or *DIPs** shown at the top and middle are typically from three-quarters to two inches long. The more compact package at the bottom is an example of a "small-outline integrated circuit" (SOIC). A package may include thousands of components such as resistors, capacitors, and transistors. Many useful logic functions are available in ICs. (We refer to only a limited selection in this chapter; for complete and current information on available functions, refer to manufacturers' catalogs.) Only the most popular functions that will sell many thousands of DIPs are integrated because of the high cost of designing and preparing to manufacture this kind of circuit. However, ICs are very cheap to produce once these "front-end" costs have been covered.

Figure 13–3 shows a photograph of an IC—an MC10104 "quad AND gate" manufactured by Motorola, Inc. The dimensions of this *chip* are about 1 mm $\times$ 1 mm. It is made of silicon with impurities added at certain spots to form resistors and transistors. Then a metal pattern is deposited on the chip to form capacitors and fine "wires" to interconnect the components. Finally the IC is packaged in a DIP for ease in handling and in making connections.

A *logic diagram* of the MC10104 is shown in Figure 13–4a. This provides the user with a description of its logic function and the pin number of each connection. As shown in Figure 13–4b, the pins are numbered counterclockwise (looking at the top of the device) starting at the "notch" in one end. In this case two V_{CC} pins

---

*"Dual in-line" refers to the two rows of connection pins on the package.

FIGURE 13–2 (Top to bottom) 40-pin DIP houses a complex integrated circuit (IC); 16-pin DIP is typical of packaging for most logic circuits; 16-pin small-outline integrated circuit (SOIC) occupies less space. (Photograph by Peter Dreyer Photography)

connect to the positive supply voltage, the V_{EE} pin connects to the negative supply voltage, and the 13 other pins provide connections to the four AND gates. The total number of connections to a digital IC varies from 14 to about 64, depending on the complexity of the circuit.

The digital IC's number "MC10104" tells that the manufacturer is Motorola, that the IC is in the ECL family, and that it performs the function of four AND gates (called a "quad AND"). Table 13–1 compares the numbers used by several manufacturers to designate ECL quad AND ICs. The prefix letter(s) indicate the manufacturer, and the "10" indicates the 10K series of ECL. The last three digits "104" indicate the function—quad AND. It is common to refer to an IC by its last digits when it is not important to specify the manufacturer. In this chapter we use the generic designation "ECL '104" for an ECL quad AND, for example.

Table 13–1 also lists the numbers several manufacturers use to designate the quad AND function in the TTL and CMOS families. Note that the function code "08" is different from that used by ECL, but TTL and CMOS use the same function code. The TTL numbers are specifically for the advanced Schottky series. In this chapter we use "TTL '08" to designate a TTL quad AND. The CMOS numbers are specifically for the high-speed CMOS or HCMOS series. In this chapter "CMOS '08" is used for a CMOS quad AND.

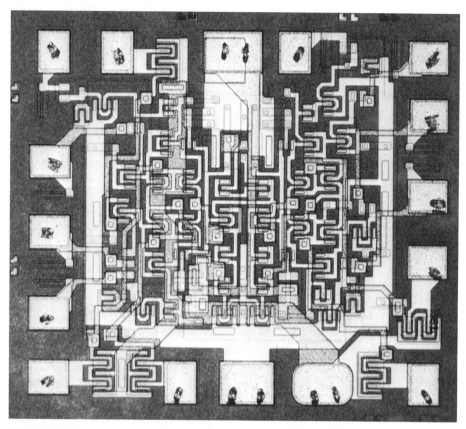

FIGURE 13–3 Integrated circuit (IC) on a 1 × 1 mm silicon chip. The circuit is an MC10H104 quad AND gate. (Photograph courtesy of Motorola, Inc.) Larger ICs may have as many as 100,000 gates.

The standard measure of the complexity of a digital IC is the number of gates composing the circuit. ICs with from 1 to 10 gates are referred to as *small-scale integration* (SSI). *Medium-scale integration* (MSI) has up to 500 gates, and *large-scale integration* (LSI) has up to 10,000 gates. Beyond 10,000 gates is *very large-scale integration* (VLSI). (These boundaries are very rough; there are no official definitions.) The quad AND in Figure 13–4 would be classified as SSI, and the data multiplexer with 14 gates in Figure 13–5 is MSI. Microprocessor ICs are an example of VLSI.

TABLE 13–1 IC Numbers for "Quad AND" Function

Manufacturer	CMOS	TTL	ECL
Motorola	MC74HC08	MC74F08	MC10104
Texas Instruments	SN74HC08	SN74AS08	—
National	MM74HC08	DM74AS08	—
Signetics	74HC08	74F08	10104
Fairchild	—	74F08	F10104

ECL '104

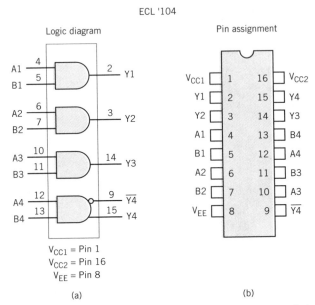

FIGURE 13–4 Quad AND gate from the emitter-coupled logic (ECL) family. (a) The logic diagram describes the function of the IC. (b) The pin assignment diagram shows a top view of the DIP.

The multiplexer in Figure 13–5 is the digital equivalent of two rotary switches, each connecting one of four inputs to an output. (See Figure 11–15a for an example of an analog multiplexer.) A binary number N on the address lines A0 and A1 causes an input data line DN to be connected to the output Y. For example, a high (H) on A1 and a low (L) on A0 (binary for 2) causes $Y_a = D2_a$ and $Y_b = D2_b$. These relationships are summarized in the *function table* in Figure 13–5. ICs with this function are available in both TTL and CMOS. For example, Texas Instruments makes a TTL SN74AS153, and Motorola makes a CMOS MC74HC153. The same function is available in ECL (the MC10174, for example), but it has a different pin assignment.

13.2 LOGIC FAMILIES

The three commonly used logic families are *complementary MOS logic* (CMOS), *transistor-transistor logic* (TTL), and *emitter-coupled logic* (ECL). The major reason for choosing one logic family over another is the speed of circuit operation. The frequency that a logic gate can pass is limited by its *transition time* t_T, as shown in Figure 13–1b. We will see in Section 13.5 that t_T depends on the output load. A good rule of thumb is that the period T_{sig} of the signal should not be any shorter than $2t_T$. Then the signal frequency $f_{sig} = 1/T_{sig}$ is limited by

$$f_{sig} < 0.5/t_T. \tag{13.1}$$

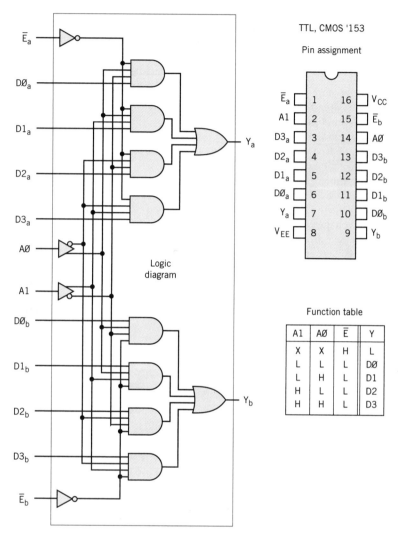

TTL, CMOS '153

Pin assignment

$\overline{E}_a$	1	16	V_{CC}
A1	2	15	$\overline{E}_b$
D3$_a$	3	14	AØ
D2$_a$	4	13	D3$_b$
D1$_a$	5	12	D2$_b$
DØ$_a$	6	11	D1$_b$
Y$_a$	7	10	DØ$_b$
V_{EE}	8	9	Y$_b$

Logic
diagram

Function table

A1	AØ	$\overline{E}$	Y
X	X	H	L
L	L	L	DØ
L	H	L	D1
H	L	L	D2
H	H	L	D3

FIGURE 13–5 Dual four-input data multiplexer from the transistor-transistor logic (TTL) family or the complementary MOS (CMOS) logic family. With 14 gates, this IC is an example of medium-scale integration (MSI).

Values of t_T for minimum loading are listed in Table 13–2. This condition allows the maximum f_{sig}, which is listed as f_{max} in Table 13–2. ECL can operate at speeds up to an f_{max} of 250 MHz, while TTL goes up to 100 MHz and CMOS up to 30 MHz.* Some gallium arsenide (GaAs) logic, with ECL-compatible logic levels, is available at speeds in the gigahertz. At present the high cost of this logic restricts it to special applications.

---

*Family specifications such as maximum signal frequency are constantly improving as technology advances. The specifications given in this chapter are for the most advanced series of each family at the time of this writing—high-speed CMOS, advanced Schottky TTL, and 10KH ECL.

TABLE 13–2 Logic Family Specifications
High-speed CMOS, Advanced Schottky TTL, 10KH ECL

	Symbol	CMOS	TTL	ECL	
Typical power per gate	P_D	0.001	5.0	25	mW
Guaranteed maximum frequency	f_{max}	30	100	250	MHz
Maximum transition time	t_T	15	5.0	2.0	ns
Maximum propagation delay (gate)	t_{pd}	15	4.5	1.5	ns
Guaranteed toggle frequency	f_{tog}	20	75	250	MHz
Maximum propagation delay (flip-flop)	t_{pd}	30	9	2.0	ns
Minimum setup time	t_{su}	20	4.5	1.5	ns
Minimum hold time	t_h	5	0	1.0	ns
Typical output voltage (H)	V_{OH}	5	3.3	−0.9	V
Typical output voltage (L)	V_{OL}	0	0.25	−1.8	V
Minimum noise margin	NM	1.0	0.4	0.15	V
Positive supply voltage (tolerance)	V_{CC}	5(±20%)	5(±10%)	0	V
Negative supply voltage (tolerance)	V_{EE}	0	0	−5.2(±5%)	V
Maximum output current	I_O	10	−2 (H)	−30 (H)	mA
			20 (L)		mA
Maximum input current	I_I	0.001	0.02 (H)	0.3 (H)	mA
			−0.5(L)		mA
Maximum input capacitance	C_{in}	10	5	5	pF

The reason for using a slower logic family when possible is that it dissipates less power and is cheaper. The plots in Figure 13–6 show the amount of power dissipated by each logic family as a function of frequency. Since the power also depends on the size or complexity of the IC, the power is given in milliwatts per gate (mW/gate). For example, ECL at 1 MHz dissipates 25 mW/gate, so a '104 with four gates would dissipate 100 mW. As the frequency of operation increases, the power also increases as the result of charging stray capacitance during switching (see Section 13.7). The end of the curve indicates the maximum operating frequency. A device dissipates minimum power when not switching ($f = 0$); this is called the *static power dissipation*.

The static power for each family is listed in Table 13–2. CMOS has a typical static power of 0.001 mW/gate, which is too low to be seen in Figure 13–6. TTL has a static power of 5 mW/gate, and ECL has a static power of 25 mW/gate (compare with the plot in Figure 13–6).

EXAMPLE 13.1 A data multiplexer (MUX) is needed to convert 4 bits of parallel data into a stream of serial bits, as shown in Figure 13–7.

Given: The data rate of the serial data is 140 Mbits/s (the spacing between the bits is 7 ns). The multiplexer comprises 14 gates.

Find: the proper logic family for an integrated circuit to do the multiplexing, and find the power dissipated by the IC.

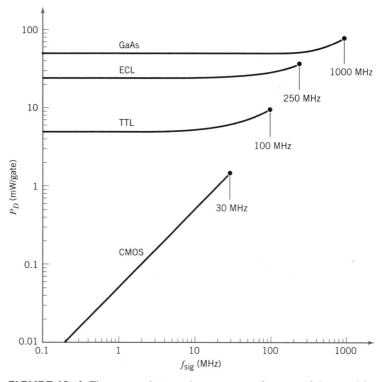

FIGURE 13–6 The power dissipated per gate as a function of the signal frequency. Logic families that dissipate more power have a higher limit on the signal frequency. The power at dc is the static power. (For CMOS, the static power is about 0.001 mW/gate.)

Solution: The bit spacing of 7 ns corresponds to half a period T_{sig} (compare the waveform of $A0$ in Figure 13–7 with the waveform of C in Figure 13–1b). Therefore $T_{sig} = 14$ ns, and $f_{sig} = 1/T_{sig} = 70$ MHz. The multiplexing function can be provided by a CMOS '153, a TTL '153, or an ECL '332. However, the CMOS is not fast enough since it only goes up to $f_{sig} = 30$ MHz. Either TTL or ECL will handle the 70-MHz application here. Of the two, <u>TTL</u> dis-

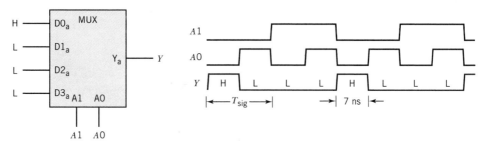

FIGURE 13–7 Logic function for Example 13.1. The signal frequency is $1/T_{sig} = 1/14$ ns $= 70$ MHz. Therefore TTL should be used (the lowest-power family that can handle the frequency).

sipates less power, so a TTL '153 is our choice. From Figure 13–6, the dissipation is 8 mW per gate at 70 MHz, so the TTL '153 with 14 gates will dissipate about 112 mW. This is actually pessimistic since not all the gates will be switching at 70 MHz. ■

Another difference between logic families is the pair of voltage levels they use to represent logic "1" and "0." From the standpoint of distinguishing between high and low levels, it is best to make these levels as far apart as possible. As shown in Figure 13–8, CMOS does the best in this regard; the nominal high and low levels

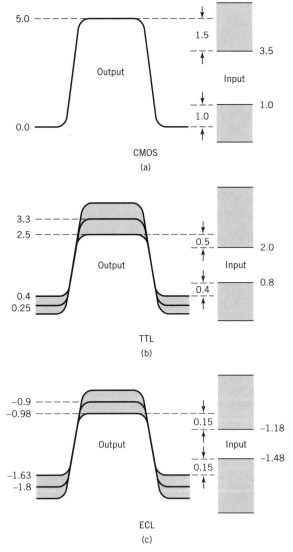

FIGURE 13–8 Logic levels used by the three logic families. A high voltage level represents a logic "1," and a low voltage level represents a logic "0." The highest low output level must be less than the highest acceptable low input level. The difference is the noise margin (e.g., 0.4 V for TTL).

are 5 V and 0 V. TTL has about a 3-V swing, going from a nominal high level of 3.3 V to a low level of 0.25 V. ECL has almost a volt swing, going typically from a high level of -0.9 V to a low level of -1.8 V. This smaller voltage swing is an advantage in achieving higher speed; it takes less time to make a smaller voltage change.

As shown in Figure 13–8b, the high level for TTL is not tightly controlled—possibly going down to 2.5 V under some conditions. Is 2.5 V still a high enough voltage to be interpreted as high by the input of another TTL device? In fact a TTL device is guaranteed to treat any input voltage above 2 V as high, so there is still a margin of 0.5 V (see Figure 13–8b). This *noise margin* assures that any interference that couples into a TTL logic line won't cause a high level to appear as low if the amplitude of the interference is less than 0.5 V. The difference between the highest low output level and the highest low input level for TTL is 0.4 V (see Figure 13–8b). This is the margin against a low level appearing high. Choosing the lower of these two margins, the overall noise margin for TTL is 0.4 V.

The noise margins for CMOS and ECL are determined in the same way as for TTL. From Figure 13–8a and Figure 13–8c it can be seen that these are 1.0 V for CMOS and 0.15 V for ECL. The difference isn't as extreme as it appears since transmission lines for ECL signals have about one-quarter the characteristic impedance of those for CMOS signals. Therefore interference introduced through stray coupling has less amplitude on ECL lines than on CMOS lines. Noise margin in connection with reflections will be discussed in Section 13.5.

Because the logic levels are different for each family, a circuit design using ICs from two different families must use *translator* ICs to convert from one set of levels to another. Under some circumstances CMOS and TTL are compatible without translators.

Table 13–2 lists some other specifications for the logic families. The time specifications are discussed in Section 13.4, and the current and capacitance specifications are discussed in Section 13.5.

13.3   WIRED LOGIC

In this section we look at ways that logic functions can sometimes be realized without using additional ICs. First we mention a practical detail. Most digital ICs are complete in themselves, requiring only a dc power supply to operate. But ECL requires a *pull-down resistor* at each output, as shown in Figure 13–9. The output of an ECL IC can pull a line high, but the current to pull the line low must be provided by a resistor to the negative supply. These resistors are external, not included in the package, so the user can eliminate them at unused outputs. This saves power. External pull-down resistors also permit the use of "wired logic."

The logic function realized by the circuit in Figure 13–9 is

$$Y = D0 \cdot G0 + D1 \cdot G1 + D2 \cdot G2 + D3 \cdot G3,$$

where "·" stands for AND, and "+" stands for OR. There is a simpler way to realize the OR function performed by the ECL '109 device here; it is to use *wired-OR* logic, as in Figure 13–10. Here the OR gate in Figure 13–9 has been replaced by a simple

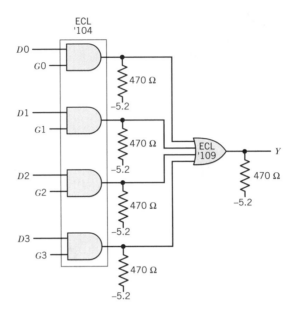

$$Y = D0 \cdot G0 + D1 \cdot G1 + D2 \cdot G2 + D3 \cdot G3$$

FIGURE 13–9 A logic function Y realized by ECL. ("·" and "+" represent AND and OR.) ECL requires a pull-down resistor at each output.

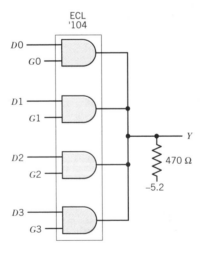

$$Y = D0 \cdot G0 + D1 \cdot G1 + D2 \cdot G2 + D3 \cdot G3$$

FIGURE 13–10 An example of wired-OR logic. This is a simpler way of realizing the logic function in Figure 13–9; the ECL '109 OR has been replaced by a wired connection. Any one of the ECL '104 outputs can pull Y high.

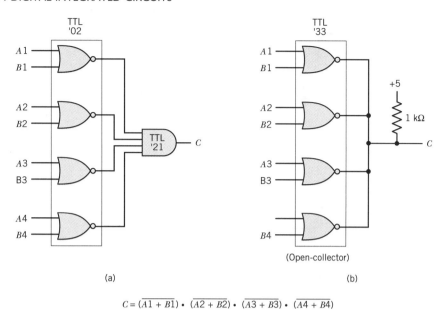

$$C = (\overline{A1 + B1}) \cdot (\overline{A2 + B2}) \cdot (\overline{A3 + B3}) \cdot (\overline{A4 + B4})$$

FIGURE 13–11 An example of wired-AND logic. The TTL '21 AND in (a) is replaced by a wired AND in (b). Any one of the TTL '33 outputs can pull C low (they must be open-collector outputs).

connection of the outputs of all the AND gates. If any one of the AND gate outputs in Figure 13–10 goes high, then Y goes high. Wired-OR logic can always be used with ECL unless there is a problem with output loading (see Section 13.5).

In a similar way, *wired-AND* logic can sometimes be used with TTL. Figure 13–11 shows two ways of realizing the function

$$C = (\overline{A1 + B1}) \cdot (\overline{A2 + B2}) \cdot (\overline{A3 + B3}) \cdot (\overline{A4 + B4}).$$

In Figure 13–11a the AND function is realized by a TTL '21 device. (Note that TTL usually requires no pull-down or pull-up resistors.) In Figure 13–11b the outputs of the quad NOR are connected together to achieve the AND function. The TTL '33 is a special quad NOR with *open-collector* outputs that can only pull a line low; an external *pull-up resistor* is needed to pull the line high. Therefore any of the outputs can pull C low, and C is high only when all the outputs are high. Open-collector TTL devices are necessary to implement wired-AND logic, but not all logic functions are available in an open-collector version. To avoid the need for pull-up resistors, open-collector TTL is not used unless there is a specific need for it.

EXAMPLE 13.2 The circuit in Figure 13–12a realizes the function $F = A \cdot B + C \cdot D$.

Given: The gates are of the TTL family.

Objective: Reduce the number of gates by using wired logic.

Solution: In Figure 13–12b the function is preserved by adding pairs of "bubbles" (inversions). In Figure 13–12c DeMorgan's second law (see Section 9.5)

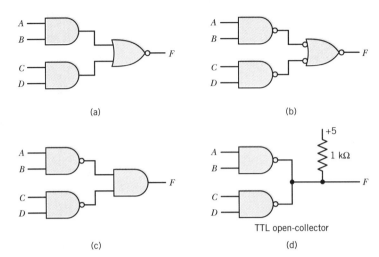

FIGURE 13–12 Circuit simplification in Example 13.2. The function in (a) is unaffected by adding pairs of bubbles (inversions) in (b). In (c) DeMorgan's rule is used to replace the OR and bubbles by an equivalent AND. In (d) the AND gate is replaced by a wired AND.

allows us to replace the OR and the inversions at its inputs and output by an AND gate. Finally in Figure 13–12d we replace the AND gate with a wired AND.

Logic devices with *tri-state* outputs can achieve the equivalent of wired logic without the bother of external resistors. Figure 13–13 shows a functional representation of a TTL or CMOS '374 parallel-in, parallel-out shift register, or edge-triggered latch. The eight D-type flip-flops have tri-state outputs; each output can be either a high or a low state, or they can all be a *high-impedance state* (off). When a high level is applied to the *output enable pin* $\overline{\text{OE}}$, all the outputs are disconnected as by opening a switch. Thus two or more tri-state outputs can be connected to one line, as in Figure 13–13, and only the enabled output controls the line. All the other devices on the line must be in the high-impedance state. This arrangement is usually used when several devices are "talking" over one *bus* (only one device at a time can talk). When $\overline{\text{OE}}_1$ is low, Data 1 (latched by the CLK) appears on the bus in Figure 13–13.

13.4   PROPAGATION DELAY

As with any physical system, it takes a digital IC time to respond to an input. For example, the NAND gate in Figure 13–14a has the signals A and B applied to its inputs. When A and B both go high at the same time, the output D doesn't fall until after a *propagation delay* t_{pd}. For the ECL gates here t_{pd} is only about 1.5 ns, but even a small propagation delay may cause problems.

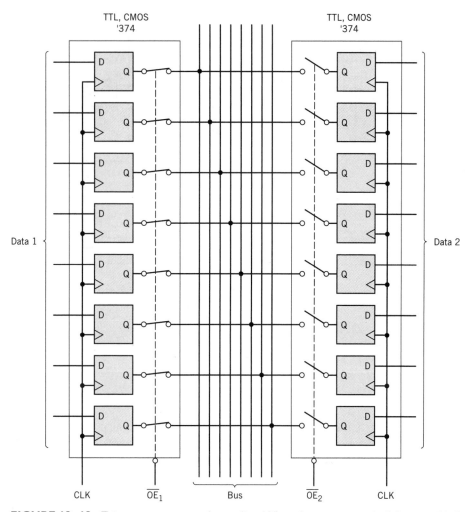

FIGURE 13–13. Tri-state outputs can share a line. When the outputs on the left are enabled ($\overline{OE}_1$ is low), they control the bus lines. Meanwhile the outputs on the right must be in the "high-impedance" state ($\overline{OE}_2$ is high) so they are disconnected from the lines.

Consider the effect of delay in the circuit in Figure 13–14a. (These three gates realize the exclusive-OR function $C = A \oplus B$.) Note that both A and B pass through two gates in reaching C. Therefore the only effect is that C is shifted in time by $2t_{pd} = 3.0$ ns here. A shift is no bother if we are not comparing C with some other signal resulting from A and B. This is the same situation as when a television signal is received from a transmitter some 40 miles away. We are not bothered by the 200-μs delay as long as we are not comparing the reception with some other reference. But if the signal arrives by two paths—say also bouncing off an airplane—then it can arrive with two different delays. In this case we see a "ghost," and delay does cause a problem.

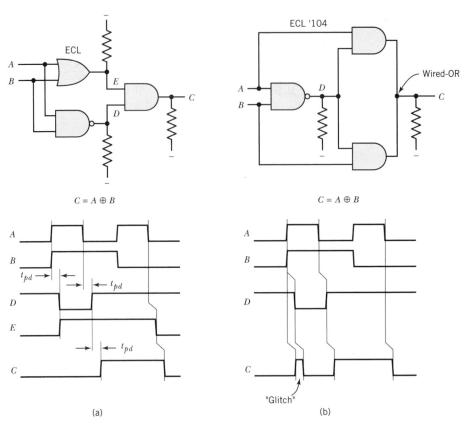

FIGURE 13–14 An example of a race condition. (a) All paths from inputs to output have the same number of gates (same number of propagation delays). (b) Paths with different propagation delays result in a "glitch" in the output. Both circuits realize an exclusive-OR function.

The circuit in Figure 13–14b is a case where there are two different delays from input to output. (This is another way of realizing the XOR function.) The signal A goes through one AND gate to reach the output, experiencing a delay of t_{pd}. It also goes by a different path through a NAND gate and then an AND gate, experiencing a delay of $2t_{pd}$. This results in a "glitch" in the output C when A and B go high together, where C should have stayed low. In general, any situation sensitive to the differences of two delays is called a *race condition*. Depending on the application, a delay difference may be short enough that it causes no problems, but it is usually good to keep the same number of propagation delays in all parallel paths.

Maximum propagation delays for each of the logic families are given in Table 13–2. Propagation delay is about inversely proportional to the maximum operating frequency, so CMOS has the longest t_{pd}, ECL has the shortest, and TTL is in between. Note that t_{pd} for flip-flops is about twice the t_{pd} for gates in the same family.

The circuit in Figure 13–15 illustrates the effects of propagation delay with flip-flops. The first D-type flip-flop has input A and output B. After A goes low, B remains high until the next rising edge of CLK_1. Then B goes low after a propa-

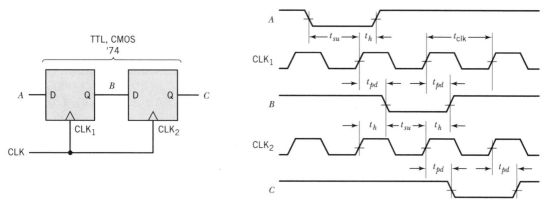

FIGURE 13–15 The propagation delay of the first flip-flop provides the hold time for the second flip-flop. However, it also reduces the setup time for the second flip-flop. This limits the minimum clock period and the maximum clock frequency.

gation delay t_{pd}. If the rising edge of CLK_1 comes too soon after A goes low, there is not enough time for the flip-flop to recognize that A is now low, and B will not go low. Therefore the new data at the input (low in this case) must be present for a minimum *setup time* t_{su} before the rising CLK_1 edge. Similarly, the data must remain for a minimum *hold time* t_h after the rising CLK_1 edge. This assures that the flip-flop has enough time to set the output equal to the input before the input changes again. Minimum setup times and hold times for each logic family are listed in Table 13–2.

Since the two flip-flops in Figure 13–15 share the same CLK signal, the rising edge of CLK_2 is reading B as high at the same time CLK_1 is reading A as low. Therefore we must be concerned that B hold the high long enough for the second flip-flop to read it before the first flip-flop changes it to low. As can be seen from the waveforms, the propagation delay of the first flip-flop (t_{pd} from CLK_1 to B) provides the hold time for the second (t_h from B to CLK_2). Therefore we must always have $t_{pd} > t_h$ for flip-flops in this configuration. Table 13–2 doesn't list the minimum values of t_{pd}, but they are always greater than the minimum t_h.

We must also be concerned that the second flip-flop have enough setup time. The waveforms in Figure 13–15 show that the setup time for CLK_2 is given by $t_{su} = T_{clk} - t_{pd}$, where T_{clk} is the period of the clock (from one rising edge of CLK to the next). For CMOS, maximum t_{pd} for a flip-flop is 30 ns, and the minimum t_{su} is 20 ns. Therefore we must assure that $T_{clk} > 50$ ns, corresponding to a limit on the clock frequency of $f_{clk} = 1/T_{clk} < 20$ MHz (see Table 13–2). In general, the clock frequency is limited by

$$f_{clk} < 1/(t_{pd} + t_{su}),$$ (13.2)

where t_{pd} is the maximum propagation delay of a flip-flop and t_{su} is the minimum setup time. For the lowest attainable propagation delay (t_{pd} the propagation delay of one flip-flop, as here) manufacturers list the highest attainable f_{clk} for a logic family as the *toggle frequency* f_{tog}. (See the specifications in Table 13–2.)

For some circuit configurations, the total propagation delay is greater than the t_{pd} of one flip-flop, and f_{clk} can't be as high as f_{tog}. The following example is an illustration of this.

EXAMPLE 13.3 The divide-by-3 counter in Figure 13–16 provides three clock phases A, B, and C, each at one-third the frequency f_{clk} of the signal CLK.

Given: The D-type flip-flops and the AND gate are from the TTL family.

Find: the maximum f_{clk}.

Solution: After a rising edge of CLK, there is a propagation delay $t_{pd\ 1}$ through the flip-flops before A, B, and C respond. Then there is another propagation delay $t_{pd\ 2}$ through the AND gate before D responds. Therefore the total propagation delay in setting up the next input for the first flip-flop is $t_{pd} = t_{pd\ 1} + t_{pd\ 2}$. From Table 13–2, the maximum $t_{pd\ 1}$ and $t_{pd\ 2}$ are 9 ns and 4.5 ns, and the minimum setup time is $t_{su} = 4.5$ ns. Therefore maximum $t_{pd} = 13.5$ ns, and from Equation (13.2) the maximum clock frequency is given by $f_{clk} \leq 1/(13.5 + 4.5 \text{ ns}) = \underline{55 \text{ MHz}}$.

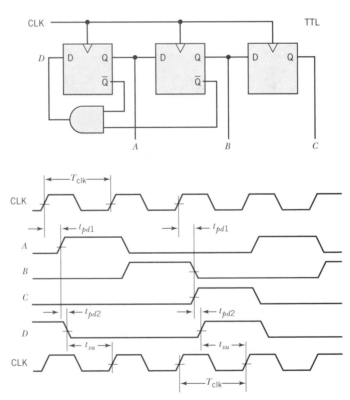

FIGURE 13–16 Divide-by-3 counter in Example 13.3. The total delay t_{pd} from a rising clock edge to D being ready is $t_{pd\ 1} + t_{pd\ 2}$. Then the maximum clock frequency is $1/(t_{pd} + t_{su})$.

13.5   FANOUT

In Figure 13–14b, the output of the NAND gate has two inputs connected to it. This is called a *fanout* of 2. Fanouts are often much greater than this, especially in the case of a clock signal that needs wide distribution. Figure 13–17a shows a case of clock distribution with a fanout of 16. The limit on fanout is determined by the maximum current the output can provide. In steady state, logic inputs require some dc current; this leads to a *dc fanout* specification. Inputs also have some input capacitance that must be charged by current when the logic level changes; this leads to an *ac fanout* specification.

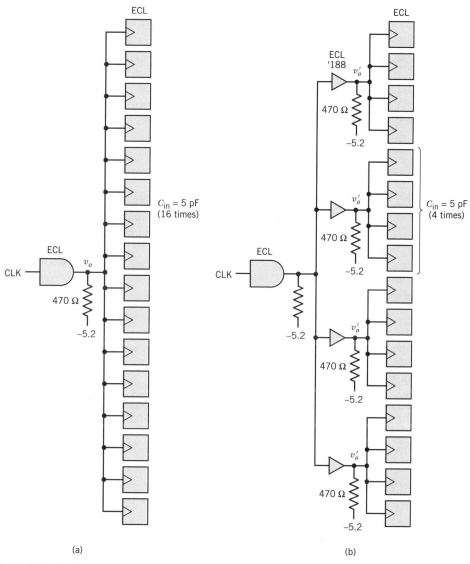

(a)   (b)

FIGURE 13–17 (a) A clock fanout of 16. (b) Buffers reduce the fanout to 4. This reduces the transition times.

Table 13–2 lists the maximum current I_O that an output can provide. It also lists the maximum dc current I_I that a logic input will require. If an output feeds n inputs, we must be sure that $nI_I < I_O$. Then the maximum dc fanout is given by

$$n \le I_O/I_I. \tag{13.3}$$

For example, TTL has for a low logic level $I_O = 20$ mA and $I_I = 0.5$ mA. Therefore the maximum dc fanout for this family is 40. This considers the worst case since TTL requires very little input current with a high logic level.

In calculating the fanout capability of ECL, the current of the pull-down resistor must be taken into account. In Figure 13–17a the 470-Ω resistor to -5.2 V draws about 9 mA for a high logic level, so of the $I_O = 30$ mA, only 21 mA is available for feeding inputs. Since the maximum dc input current is $I_I = 0.3$ mA, there is no problem feeding the 16 inputs in Figure 13–17a in the steady state. The maximum dc fanout is $n = 21$ mA/0.3 mA $= 70$.

For CMOS the dc fanout is $I_O/I_I = 10$ mA/0.001 mA $= 10,000$. There is probably no situation that requires such a high fanout, so the dc fanout can be considered essentially unlimited for this family. However, there are other considerations such as capacitive loading.

When a capacitor is charged or discharged (its voltage level is changed), it draws current according to the relationship $I = C \, dv_0/dt$. Then when v_0 in Figure 13–17a goes from low to high, the output current must not only supply the dc current, it must also supply I to charge the input capacitance of the 16 devices. The more limiting situation for ECL is a high-to-low transition. Since the output of ECL can't "sink" current, all of the current $-I$ to discharge the input capacitance must be provided by the pull-down resistor. This situation is modeled in Figure 13–18a.

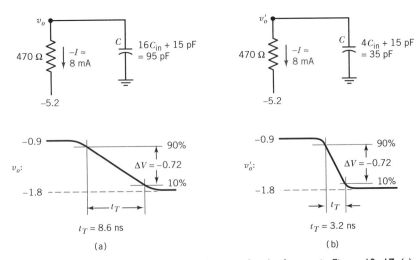

FIGURE 13–18 Analysis of the transition times for the fanouts in Figure 13–17. (a) Sixteen input capacitances require 8.6 ns to be discharged by $-I$. (b) Four input capacitances require 3.2 ns to be discharged by $-I$.

The total capacitance to be discharged is

$$C = nC_{in} + 15 \text{ pF,} \tag{13.4}$$

where the 15 pF accounts for the output capacitance of the driving device. For our case, $n = 16$, $C_{in} = 5$ pF, and $C = 95$ pF. Although v_0 is changing, the voltage across the 470-Ω resistor is about 3.8 V, and the current through it is $-I \approx 8$ mA. The *transition time* t_T is defined as the time for v_0 to traverse the middle 80% of its swing from high to low or from low to high (compare with "rise time" in Section 8.6). The total voltage swing is $V_{OL} - V_{OH} = -1.8 + 0.9 = -0.9$, and 80% of this is $\Delta V = -0.72$ (see Figure 13–18a). Then $dv_0/dt = \Delta V/t_T$, and from $I = C \, dv_0/dt$ the transition time is

$$t_T = C \, \Delta V/I, \tag{13.5}$$

where

$$\Delta V = 0.8(V_{OH} - V_{OL}) \tag{13.6}$$

For our case, Equation (13.5) gives $t_T = 95$ pF $\times$ 0.72 V/8 mA = 8.6 ns. This is considerably slower than what is required in applications using ECL.

Longer transition times limit the maximum operating frequency of a circuit [see Equation (13.1)]. Therefore a circuit design may have to be modified to reduce t_T to accommodate high-frequency signals. One way of reducing t_T is to reduce the fanout, as shown in Figure 13–17b. Instead of one gate feeding 16 inputs, it feeds four *buffers*, and each of the buffers handles four inputs. The transition time of v_0' at the inputs can be determined from the model in Figure 13–18b. Now the capacitance to be discharged is only $C = 4C_{in} + 15$ pF = 35 pF, and from Equation (13.5), $t_T = 35$ pF $\times$ 0.72/8 mA = 3.2 ns. If this is still not fast enough, the 470-Ω pull-down resistor could be reduced to 270 Ω to increase I.

The transition times for CMOS and TTL can be determined from the same Equations (13.4)–(13.6). For CMOS use $I = I_O = 10$ mA, and for TTL use $I = I_O = 2$ mA. Table 13–2 provides the other necessary variables.

EXAMPLE 13.4 A CMOS logic circuit is to handle signals with frequencies as high as $f_{sig} = 3$ MHz.

Given: From Table 13–2, $V_{OH} = 5$, $V_{OL} = 0$, $I_O = 10$ mA, and $C_{in} = 10$ pF.

Find: the maximum ac fanout n.

Solution: To have $f_{sig} = 3$ MHz, Equation (13.1) allows a maximum of $t_T = 1/(6 \text{ MHz}) = 167$ ns. From Equation (13.6), $\Delta V = 0.8 (5 - 0) = 4$ V. Then with $I = I_O = 10$ mA, we have from Equation (13.5) $C = 167$ ns $\times$ 10 mA/4 V = 418 pF. Since $C_{in} = 10$ pF, Equation (13.4) gives $n = (418 - 15 \text{ pF})/10$ pF = <u>40</u>.

13.6   DRIVERS

There are occasions when more than the usual output current is required from a logic gate. These are the cases when the output is driving something other than another logic gate—things such as transmission lines, relays, and light-emitting diodes. Logic devices with special capability to handle the extra current are called *drivers*, *buffers*, and *interface gates*.

Figure 13–19 shows a TTL '760 driver feeding a 150-Ω transmission line. This device has an open-collector output and a special current-sinking capability of 60 mA. To avoid reflections, the line must be terminated with its characteristic impedance of $Z_0 = 150 \ \Omega$ at either the source or load end or both (see Section 8.9). Usually it is easiest to terminate the load end with a pull-up resistor (for TTL) or a pull-down resistor (for ECL). In Figure 13–19a the pull-up resistor is $R_L = 150 \ \Omega$ at the load end. We need to check that the driver can handle the maximum current through R_L. This occurs when v_2 is low—about 0.4 V. The current is $(5 - 0.4)/R_L = 30$ mA, which is within the 60-mA capability of the driver. (Compare with the standard 20 mA for TTL listed in Table 13–2.)

A 50-Ω line driven by ECL is shown in Figure 13–19b. The pull-down resistor has been reduced to $R_L = 50 \ \Omega$ to properly terminate the line. Note that R_L doesn't go to -5.2 V; it goes to -2 V instead. This is to keep the current through R_L low enough that the ECL output can handle it. When v_2 is high—about -0.9 V—the current through R_L is $(2 - 0.9)/R_L = 22$ mA. This is within the $I_O = 30$ mA capability of ECL.

An alternative arrangement that doesn't require a -2-V power supply is shown in Figure 13–19b. Here the -5.2-V supply and the two resistors form a Thevenin voltage of -2 V and a Thevenin impedance of 50 Ω. (The reader should show this.)

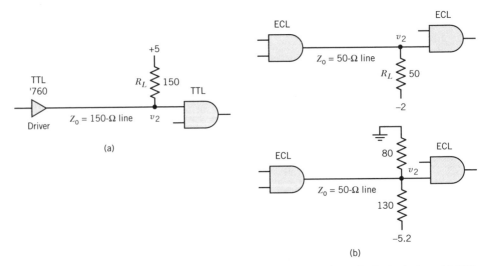

FIGURE 13–19 Proper termination of transmission lines being driven by logic devices. (a) TTL uses an open-collector driver with the pull-up resistor at the receiving end. (b) ECL uses a pull-down resistor to -2 V at the receiving end. The two ECL circuits are equivalent.

How long does a transmission line have to be before reflections due to mis-termination become a problem? A good rule is to terminate the line in Z_0 if the round-trip delay 2τ of the line is greater than the transition time of the signal; that is, if

$$\tau > \tfrac{1}{2}\, t_T, \tag{13.7}$$

where the delay τ of a line of length d is given by

$$\tau = (0.05 \text{ ns/cm})\, d. \tag{13.8}$$

EXAMPLE 13.5 The transmission line in Figure 13–20 is driven by ECL and terminated in 470 Ω to -5.2 V.

Given: The characteristic impedance of the line is $Z_0 = 100$ Ω, and its length is $d = 20$ cm. The source impedance of the driving ECL gate is $R_s = 0$. The signal has a transition time of $t_T = 1.4$ ns.

Find
1. whether the line is short enough that it need not be terminated in Z_0;
2. the response of v_2 to a rising transition of v_1 for $R_L = 470$ Ω, and whether the response stays within the noise margin.

Solution
1. From Equation (13.8), the delay of the line is $\tau = 1.0$ ns, but Equation (13.7) requires $R_L = Z_0$ if $\tau > 1.4 \text{ ns}/2 = 0.7$ ns. We must therefore reduce R_L, but first we will look at the response when $R_L = 470$ Ω.
2. From Section 8.9, the reflection coefficient at the source is

$$\Gamma_s = \frac{R_s - Z_0}{R_s + Z_0} = \frac{0 - 100}{0 + 100} = -1, \tag{13.9}$$

and the reflection coefficient at the load is

$$\Gamma_r = \frac{R_L - Z_0}{R_L + Z_0} = \frac{470 - 100}{470 + 100} = 0.65. \tag{13.10}$$

For a step of $\Delta V = 0.9$ V at v_1, the first reflection at v_2 is $\Gamma_r \Delta V = 0.65 \times 0.9 \text{ V} = 0.59$ V, causing an overshoot by this amount. After a delay of τ, this reflection gets returned at the source by $\Gamma_s = -1$ to become -0.59 V. After another delay of τ this arrives at the load to wipe out the first reflection and generate a second reflection $0.65 \times (-0.59 \text{ V}) = -0.38$ V. This can be seen in Figure 13–20 as v_2 "ringing" under 0.9 V by 0.38 V. The reflections continue, with the third being $-0.65 \times (-0.38) = 0.25$ V, and the fourth being $-0.65 \times 0.25 \text{ V} = -0.16$ V, etc. The reader should make a bounce diagram of this process, as in Section 8.9.

The first couple of negative reflections -0.38 V and -0.16 V are large enough to exceed the ECL noise margin of 0.15 V. The result is some

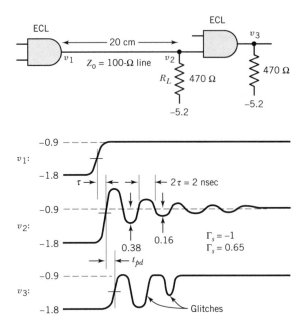

FIGURE 13–20 Example of an improperly terminated line (R_L should equal Z_0). Ringing due to reflections cause glitches at v_3. (See Example 13.5.)

glitches 2τ wide in v_3. Therefore we need $R_L = Z_0 = 100\ \Omega$. As in Figure 13–19b, it is best to connect a 100-Ω resistor to -2 V rather than -5.2 V to avoid drawing large currents from the output. ∎

Figure 13–21 shows how an open-collector TTL can be used to drive interface devices. In Figure 13–21a the relay gets about 4.5 V across its coil when the output of the '760 goes low. Since the resistance of the coil is 100 Ω, the current is 45 mA. This is more current than an ordinary open-collector TTL could handle, so a driver was chosen. The diode protects the TTL output from a high voltage when the output is turned off. (The inductance of the coil keeps its current going for a while,

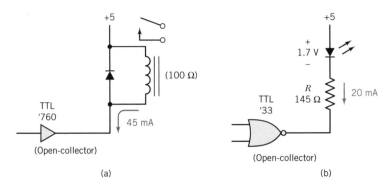

FIGURE 13–21 Logic devices driving (a) a relay and (b) a light-emitting diode (LED).

and this current circulates through the diode rather than developing a high voltage.)

In Figure 13–21b an open-collector TTL is to drive a light-emitting diode (LED) with 20 mA when the output goes low. In order to drive the LED with a well-controlled current, a resistor R is placed in series. Since the voltage drop of the LED is rather constant at 1.7 V, the resistor has about 2.9 V across it when the output of the gate goes low—to about 0.4 V. Therefore the resistor value is $R = 2.9$ V/20 mA $= 145$ Ω.

13.7   POWER AND HEAT

The amount of power that a logic circuit dissipates is important for two reasons. It determines how much current the dc power supply must be able to provide, and it is equal to the amount of heat that is generated and must be disposed of.

Static Power Dissipation

In Section 13.2 we determined the power dissipated by a device by multiplying the number of gates composing it by the power per gate given in Figure 13–6. Since the number of gates is not always known, a more accurate method is to use the manufacturer's specification for the supply current I_{CC} or I_{EE}.

The currents and voltages for a digital IC are defined in Figure 13–22: V_{CC} is the positive "rail" of the power supply, and V_{EE} is the negative rail. Usually one or the other is defined as ground (zero volts). Table 13–2 gives the nominal values for these voltages; for example, TTL has $V_{CC} = 5.0$, and $V_{EE} = 0$. In calculating the static power, we must also take into account the output current I_O (the total for all outputs of the IC). This will be significant if there are pull-up or pull-down resistors. Since I_{CC} goes through a voltage drop of $V_{CC} - V_{EE}$ in the device, and I_O goes through a drop of $V_O - V_{EE}$ in the device, the power dissipated is

$$P_D = I_{CC}(V_{CC} - V_{EE}) + I_O(V_O - V_{EE}). \qquad (13.11)$$

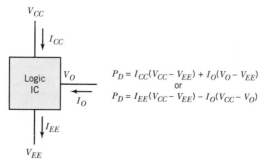

FIGURE 13–22 Analysis of power dissipated by a logic IC.

By using the fact that $I_{CC} = I_{EE} - I_O$, this can be put in the form

$$P_D = I_{EE}(V_{CC} - V_{EE}) - I_O(V_{CC} - V_O), \tag{13.12}$$

which is more convenient when working with ECL, for which I_O is negative.

EXAMPLE 13.6 Consider the circuit using a TTL '33 in Figure 13–11b.

Given: $I_{CC} = 5.6$ mA (from the '33 data sheet), $V_{CC} = 5$, $V_{EE} = 0$, and $V_O = 0.25$ (from Table 13–2).

Find: the power P_D dissipated by the '33. Also find the total current I_S drawn from the +5-V supply by this circuit.

Solution: For $V_O = 0.25V$, the 1-kΩ pull-up resistor causes $I_O = (5-0.25)/1$ kΩ $= 4.75$ mA. From Equation (13.11), $P_D = 5.6$ mA (5V) + 4.75 mA (0.25V) = <u>29.2 mW</u>. The total current draw by the IC and the pull-up resistor is $I_S = I_{CC} + I_O = $ <u>10.35 mA</u>.

A system typically includes dozens of ICs like that in Example 13.6. The power supply selected for the system must be able to provide the total current found by summing the I_S for each IC with its associated resistors.

Temperature

As the dissipated power in the form of heat leaves the IC package (the DIP), it generates a temperature difference from the inside of the DIP to the outside. Let T_J be the temperature of the IC inside the DIP (called the *junction temperature*), and let T_A be the temperature of the air surrounding the DIP (called the *ambient temperature*). Then

$$T_J - T_A = \theta_{JA}P_D, \tag{13.13}$$

where θ_{JA} is the *thermal resistance** and P_D is the power dissipated by the IC. A typical value of θ_{JA} for a 16-pin DIP is 100°C/W. The temperature T_J should never exceed 165°C, but in practice a limit lower than 165°C is selected to prolong the life of the IC.

One way to reduce T_J is to reduce T_A by providing better ventilation in the box containing the circuitry. If a room houses many electronic units, it may be necessary to air condition the room to keep down the T_A. Another way is to reduce θ_{JA} by blowing air over the DIP with a fan. The plot in Figure 13–23 shows θ_{JA} as a function of the airflow in linear feet per minute (lfpm). As the airflow is increased, θ_{JA} is reduced from 100°C/W to near 40°C/W. Airflow above 800 lfpm achieves little further reduction in θ_{JA}.

---

*The term *thermal resistance* comes from an analogy to $V = RI$, where temperature drop is like V, thermal resistance is like R, and power is like I.

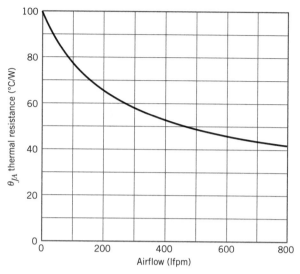

FIGURE 13-23 The thermal resistance (and therefore the junction temperature) of a DIP can be reduced by blowing air across the DIP ("lfpm" stands for linear feet per minute).

EXAMPLE 13.7 The IC with the highest power dissipation in a piece of equipment is an ECL '136.

Given: The '136 has four outputs with 470-Ω pull-down resistors, and the ambient air temperature is 85°C. From the manufacturer's data sheet the '136 has $I_{EE} = 120$ mA, and from Table 13-2 $V_{CC} = 0$ and $V_{EE} = -5.2$ V.

Find: the power P_D dissipated by the '136, and find the airflow necessary to maintain a junction temperature of $T_J = 140°$C.

Solution: The worst-case power dissipation is for $V_O = -1.8$ V. Then the pull-down resistors cause a current of $(1.8 \text{ V} - 5.2 \text{ V})/470 \text{ }\Omega = -7.23$ mA at each of the four outputs. The total output current is $I_O = 4\ (-7.23 \text{ mA}) = -29$ mA. From Equation (13.12), $P_D = 120$ mA (5.2 V) + 29 mA (1.8 V) = 676 mW.

From Equation (13.13), $\theta_{JA} = (T_J - T_A)/P_D = (140° - 85°)/676 \text{ mW} = 81.4°$C/W. From the plot in Figure 13-23, this requires an airflow of about 80 lfpm.

The thermal impedance θ_{JA} can be further reduced by connecting a *heat sink* to the DIP. This is a metal device with fins to increase the surface area. The result is to lower the curve in Figure 13-23 by 20–40%.

CMOS Power Dissipation

Since the static power dissipated by a CMOS IC is negligible, we must find its power by an analysis of its dynamic operation. The result will be a power that increases proportionally with the signal frequency, as shown in Figure 13-6.

When there is a low-to-high transition, the output of a CMOS device charges a capacitive load C. The charge transferred into C is $\Delta Q = C(V_{OH} - V_{OL})$. For a signal frequency of f_{sig}, the charge transferred per second (or current) is $I = \Delta Q f_{sig}$. Since I_{CC} provides the charging current and I_{EE} provides the discharging current, each is equal to I. Therefore the average supply current is

$$I_{CC} = \Delta Q f_{sig} = C(V_{OH} - V_{OL})f_{sig}. \tag{13.14}$$

The same expression for C can be used here as was used for fanout:

$$C = nC_{in} + 15 \text{ pF}, \tag{13.4}$$

where n is the fanout and C_{in} is the input capacitance.

Since I_O provides both the charging and discharging current, its average is zero. Then Equation (13.11) becomes

$$P_D = I_{CC}(V_{CC} - V_{EE}). \tag{13.15}$$

Note that for CMOS, $V_{OH} - V_{OL} = V_{CC} - V_{EE} = 5 \text{ V}.$

EXAMPLE 13.8 A CMOS IC has one output fanning out to three inputs ($n = 3$).

Given: The signal frequency is $f_{sig} = 3 \text{ MHz}$, and the typical input capacitance is $C_{in} = 10 \text{ pF}$.

Find: the typical current I_{CC} supplied to the IC and the typical power P_D dissipated by the IC.

Solution: From Equation (13.4), $C = 3 \times 10 \text{ pF} + 15 \text{ pF} = 45 \text{ pF}$. From Equation (13.14), $I_{CC} = 45 \text{ pF} \times 5 \text{ V} \times 3 \text{ MHz} = \underline{0.67 \text{ mA}}$. From Equation (13.15), $P_D = 0.67 \text{ mA} \times 5 \text{ V} = \underline{3.35 \text{ mW}}$. ■

In the example, the IC had only one output. For an IC with more than one output, the contribution to I_{CC} from each output must be taken into account.

13.8   PROGRAMMABLE LOGIC DEVICES

Large-scale integrated circuits save board space when realizing a digital design, and if the circuit is produced in large numbers, LSI saves cost also. But if only a few hundred circuits are to be made, the up-front costs of custom LSI are prohibitive. For this level of production a *programmable logic device* (PLD) is the most cost-effective solution. A PLD is an LSI circuit with possibly hundreds of logic gates and flip-flops that can be configured quickly and simply by programming it as one would a *programmable read-only memory* (PROM). Since PLDs are based on CMOS logic, the power requirements are modest, but the operating frequency is limited to 30 or 60 MHz.

A PLD is basically a combination of a logic circuit with a PROM. Figure 13–24a shows a very simple PLD that can be programmed to make D a combinational logic function of A, B, and C. By storing a pattern of 1's and 0's in the PROM, certain inputs to the AND gate can be "turned on." In the example here, the two 0's turn on the $\overline{A}$ and C inputs; the other inputs are inhibited by the 1's. A simplified diagram to indicate this programming is shown in Figure 13–24b. The dots indicate inputs that are effectively connected to the AND. The resulting logic function is $D = \overline{A}C$, as shown in Figure 13–24c.

Figure 13–25 shows a PLD that can be used to realize a four-variable state machine with one input. Sixteen programmable AND gates have been connected with four OR gates and four flip-flops to make a very flexible circuit. The potential inputs to each AND gate are R, $\overline{R}$, A_1, $\overline{A}_1$, A_2, $\overline{A}_2$, A_3, $\overline{A}_3$, A_4, and $\overline{A}_4$. The dots here indicate the programmed connections to realize the Gray code counter designed in Example 9.7.

Commercially available PLD's are larger than the simple examples used here. One measure of the size of a PLD is the number of logic gates it has (where a flip-flop comprises about four gates). For example, Altera Corporation manufactures the EP320 with about 300 gates (see Figure 13–26) and the EPM71024 with about 20,000 gates. These devices are also available as *erasable* programmable logic devices

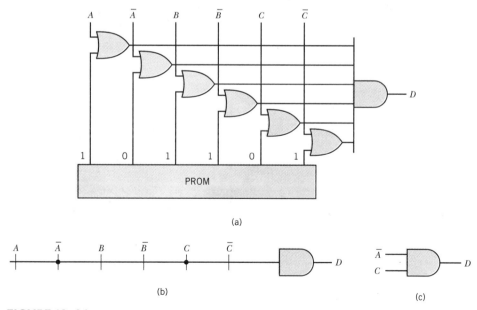

(a)

(b)

(c)

FIGURE 13–24 (a) A programmable logic device (PLD) combines logic circuitry with a programmable read-only memory (PROM) to let the user customize an IC. (b) Abbreviated symbol for the circuit in (a). Dots represent enabled inputs to the AND gate. (c) The logical equivalent of the PLD programmed as shown.

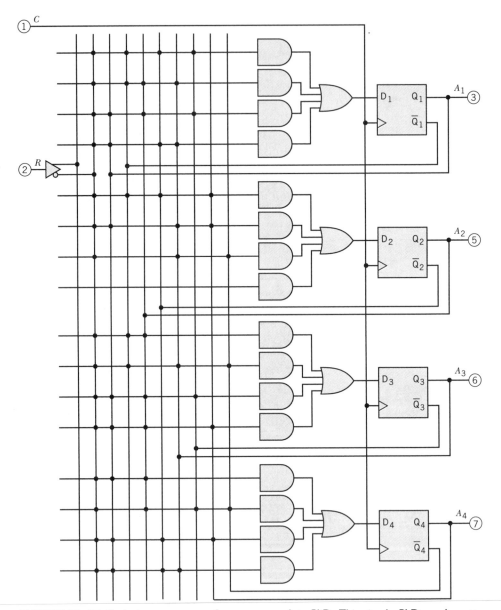

FIGURE 13-25 Typical arrangement of a more complete PLD. This simple PLD can be programmed to realize any 16-state state machine with four product terms for each flip-flop and one input. The dots here indicate programming to realize the Gray counter in Example 9.7.

(EPLDs) to allow the designer to experiment with many configurations using just one device over and over. When the final design is settled on, it can be transferred to the less expensive PLD. The manufacturers of PLDs sell development software that allow the user to enter a design as Boolean equations or as logic diagrams drawn on a PC screen with a "mouse."

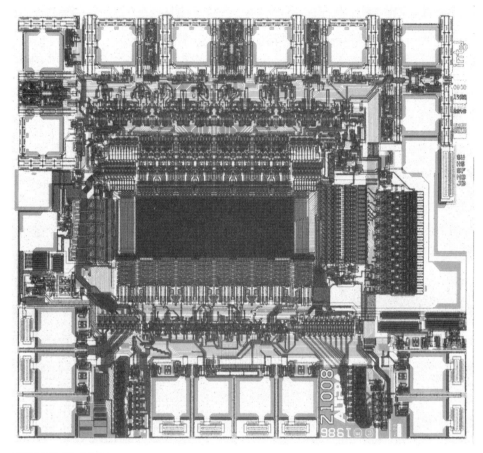

FIGURE 13–26 Microphotograph of the Altera EP320 programmable logic device. This tiny chip has about 300 gates, 8 flip-flops, 8 product terms for each flip-flop, and 10 inputs. Can you identify some of the components? The 20 white squares around the edge are bonding pads for connections to the package. (Photograph courtesy of Altera Corporation.)

13.9   SUMMARY

Logic functions are available in miniature integrated circuits (ICs). The three logic families—CMOS, TTL, and ECL—have different specifications for speed, power, current driving, and noise margin. CMOS is the slowest (up to 30 MHz) but consumes the least power; its static power is negligible. TTL is faster (100 MHz) and consumes more power. ECL is fastest (250 MHz) and consumes the most power.

For each logic family there is a range of high voltages that stand for logic "1" and a range of low voltages that stand for logic "0." These ranges include a noise margin so a logic level will remain within its range even when some interference is added. The noise margin is 1.0 V for CMOS, 0.4 V for TTL, and 0.15 V for ECL.

ECL outputs always require pull-down resistors, and TTL sometimes uses pull-up resistors. A logic OR can be realized by having several ECL outputs share one pull-down resistor, and logic AND can be realized by having several TTL outputs share a pull-up resistor.

If the propagation delays of parallel signal paths are not matched, unwanted "glitches" can appear in the output. In a chain of flip-flops, the propagation delay of one stage reduces the setup time of the next. This limits the maximum clock frequency.

The output of a logic device must provide enough current to drive the load properly. The inputs to the driven logic device require dc current and current to charge their input capacitance during level changes. The number of inputs an output feeds is its fanout. An output may also drive a transmission line, a relay, or an LED.

A transmission line with round-trip delay greater than the transition time of a signal must be properly terminated to avoid reflections. This is usually done at the far end with a pull-down or pull-up resistor equal to the characteristic impedance of the line. Any reflections due to mistermination must not exceed the noise margin.

CMOS and TTL require a 5-V power supply, and ECL requires a -5.2-V supply. The current requirement of the supply is determined by summing the supply currents drawn by each logic device (see manufacturers' specifications). For CMOS, the supply current is calculated as a function of the signal frequency. Power dissipation by a device causes a temperature rise within the device. In order to keep the junction temperature less than 165°C (or some other limit) it may be necessary to cool the device with a fan.

FOR FURTHER STUDY

The TTL Data Book, Vol. 3, Texas Instruments, 1984.
High-Speed CMOS Logic Data Book, Texas Instruments, 1989.
Motorola MECL Device Data, Motorola, 1989.
Altera Data Book, Altera Corp., San Jose, CA, 1992.

PROBLEMS

Easy Drill Problems (answers at end of chapter)

D13.1. Sketch the waveform of an L,H,L,H,L pattern (as in Figure 13–1b)
(a) for the case $T_{sig} = 4t_T$ and
(b) for the case $T_{sig} = 2t_T$. Which case corresponds to $f_{sig} = f_{max}$?

D13.2. Would the '374 octal flip-flop in Figure 13–13 be considered small-, medium-, or large-scale integration? (It takes at least two gates to realize each flip-flop.)

D13.3. What logic family will handle data signals with $f_{sig} = 20$ MHz and dissipate the least power? Use Figure 13–6 to estimate the total power dissipated if the circuit consists of 100 gates (assuming that all 100 gates are switching at 20 MHz). Repeat for $f_{sig} = 150$ MHz.

D13.4. Table 13–2 reports the *minimum* noise margin. What is the noise margin of a TTL gate that has a high output level of 3.5 V and a low output level of 0.2 V?

D13.5. Suppose the circuit in Figure 13–16 is realized with ECL. Redraw the circuit diagram, adding 470-Ω pull-down resistors to -5.2 V where necessary.

D13.6. Use two open-collector TTL gates with wired logic to realize the function $E = (\overline{AB}) \cdot (\overline{CD})$.

D13.7. Suppose the logic in Figure 9–15a is realized with CMOS gates and inverters. Initially D, T_1, and P are high, and T_2 and H are low. When D goes low, what is the maximum delay before F goes low?

D13.8. Suppose the logic in Figure 9–15a is realized with CMOS gates and inverters. Initially D, T_1, T_2, and P are high, and H is low. When D goes low, F should remain high. What actually happens due to propagation delay?

D13.9. For the circuit in Figure 13–16, let $f_{clk} = 30$ MHz. What is the setup time for the second flip-flop? What is it for the first flip-flop?

D13.10. What is the maximum dc fanout for ECL if no pull-down resistor is used? What is it if a 50-Ω pull-down resistor to -2 V is used?

D13.11. Find the fanout for each output in Figure P13.4. What is the maximum input capacitance each output must drive?

D13.12. If a CMOS circuit is to be used with signal frequencies up to 10 MHz, what is the maximum fanout allowed?

D13.13. Suppose the transmission line in Figure 13–19b has $Z_0 = 100 \ \Omega$. Change the values of the 80-Ω and 130-Ω resistors so R_L is 100 Ω to -2 V.

D13.14. Suppose the transmission line in Figure 13–19a has $Z_0 = 100 \ \Omega$. For $R_L = 100 \ \Omega$, how much current must the driver handle? (It can handle up to 60 mA.)

D13.15. What is the longest transmission line for which proper termination is not necessary for CMOS? Assume the transition time (maximum) given in Table 13–2. Repeat for TTL and for ECL.

Application Problems

P13.1. The ECL circuit in Figure P13.1 produces a similar sequence to that in Figure 13–16. Draw a timing diagram for D_0, Q_0, and Q_1 for five CLK cycles, starting with Q_0 and Q_1 high. Include propagation delays. Using the t_{pd} (different for gate and flip-flop) and the t_{su} listed in Table 13–2, find the maximum clock frequency.

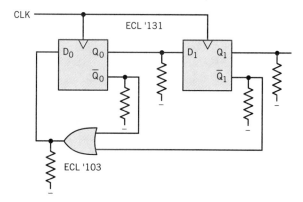

P13.2. Redraw the circuit in Figure P13.1 using wired logic to eliminate the OR gate. Repeat problem P13.1 for this new realization.

P13.3. The wired-AND in Figure P13.3 performs the same function as the tri-state switches in Figure 13–13. Show this by making a truth table for Y_a as a function of $\overline{OE}_1$, $\overline{OE}_2$, Q_{1a}, and Q_{2a}. (The inputs $\overline{OE}_1$ and $\overline{OE}_2$ are never both low together.) Why is tri-state preferable to wired-AND for this bus application?

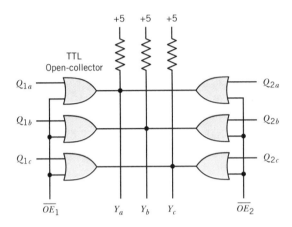

P13.4. The circuit in Figure P13.4 is designed to make each E go high sequentially as the CLK is pulsed. Because the two flip-flops form a ripple counter (see "Counters" in Section 9.6), propagation delay affects the operations. Draw a timing diagram for the Q's and E's for five CLK pulses, starting with Q_0 and Q_1 low. Include propagation delays. Show that E_0 and E_2 each have a "glitch" that occurs out of sequence. What is the maximum f_{clk}? (Both flip-flops toggle.)

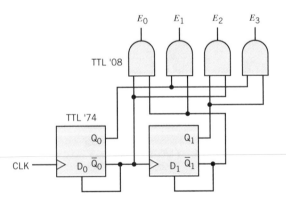

P13.5. The circuit in Figure P13.5 (on p. 445) is a synchronous counter using TTL logic. Draw a timing diagram for D_0, D_1, Q_0, and Q_1 for five CLK cycles, starting with Q_0 and Q_1 both low. Include propagation delays. For $f_{\text{clk}} = 50$ MHz, find the setup times for the first and second flip-flops. Are these adequate? What is the maximum f_{clk}?

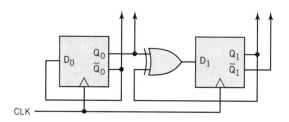

P13.6. Substitute the synchronous counter in Figure P13.5 for the ripple counter (the two flip-flops) in Figure P13.4, and repeat problem P13.4. Which circuit has less trouble with glitches? Which circuit is faster?

P13.7. The greatest fanout in Figure P13.4 is 4. What is the low-to-high transition time for this output? Use nominal TTL levels of 3.3 V for high and 0.25 V for low. What is the maximum signal frequency for this case?

P13.8. Let the transmission line in Figure 13–19a have $Z_0 = 75\ \Omega$ and a length of 100 cm. Initially the driver voltage v_s and the receiving-end voltage v_2 are both 5 V. Then the driver goes low to 0.4 V, causing a step of $\Delta v_s = -4.6$ V, and the output impedance of the driver is essentially zero. Use a bounce diagram to find the waveform of v_2 at the receiving end for the first 45 ns. Does v_2 ever exceed the maximum low input level (see Figure 13–8b) after initially going low?

P13.9. The two flip-flops in Figure P13.1 constitute a complete ECL '131 IC. All pull-down resistors are 470 Ω to -5.2 V. The '131 draws $I_{EE} = 45$ mA, and the '103 draws $I_{EE} = 21$ mA. Find the power dissipated by the '131 when Q_0 and $Q_1 = L$. Find the junction temperature if the ambient air is 30°C and still. Find the current that the -5.2 V supply must provide to the whole circuit.

P13.10. Let the circuit in Figure P13.4 be CMOS rather than TTL. Suppose $f_{clk} = 20$ MHz so $f_{sig} = 10$ MHz for the first flip-flop, and $f_{sig} = 5$ MHz for the second flip-flop. Find the power dissipated by the two flip-flops. Remember that each output has a different capacitive load.

P13.11. The combinational logic for the variable D_1 is given by the Karnaugh map in Figure P13.11a. Figure 13.11b shows a portion of a PLD diagram. Place dots on the appropriate intersections in the diagram to implement the logic in the Karnaugh map.

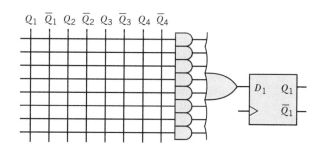

(a)   (b)

Extension Problems

E13.1. Figure E13.1b shows the transfer curve for the logic inverter in Figure E13.1a. $V_{IL\,max}$ and $V_{IH\,min}$ are the chosen limits for low and high input voltages. Draw similar horizontal lines for $V_{OL\,max}$ and $V_{OH\,min}$, the resulting limits for the low and high output voltages. Estimate the noise margin. Suggest new $V_{IL\,max}$ and $V_{IH\,max}$ lines that would maximize the noise margin for this transfer curve. *Hint:* Make high and low margins equal. (In practice a band of possible transfer curves must be used.)

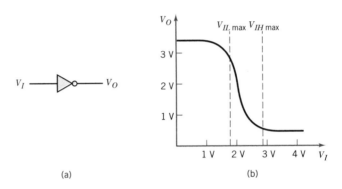

(a)   (b)

E13.2. Use DeMorgan's theorems to show that with negative logic ($L = 1$ and $H = 0$) AND gates become OR gates, and OR gates become AND gates. What do NANDs, NORs, XORs, inverters, and D-type flip-flops become?

Study Questions

13.1 Name three ways in which the response of a logic gate differs from the ideal.

13.2. How does positive logic differ from negative logic?

13.3 What does VLSI stand for?

13.4. When viewed from the top, does an IC have its pins numbered clockwise or counterclockwise? How is pin 1 identified?

13.5. Of the logic families, what is the principal advantage of CMOS logic? What is the principal advantage of ECL?

13.6. What is the power supply voltage for CMOS logic? For TTL? For ECL?

13.7. What is the distance between a logic level and the logic threshold called?

13.8. Which type of logic can be used in a wired AND? Which type of logic can be used in a wired OR?

13.9. How are the outputs from two or more ICs connected to the same logic bus kept from interfering with each other?

13.10. Name two problems caused by excessive propagation delay.

13.11. What limits the fanout of ECL? What limits the fanout of CMOS logic?

13.12. Name three types of loads (other than another logic gate) that a logic gate may be required to drive.

13.13. Name the three factors that determine the junction temperature within an IC. What steps can be taken to lower the junction temperature?

13.14. One technology permits logic circuits to be designed and configured with a PC rather than by soldering devices together with wires. What are these ICs called?

ANSWERS TO DRILL PROBLEMS

D13.1. Case (b)

D13.2. Medium-scale integration

D13.3. CMOS, 100 mW; ECL, 3.0 W

D13.4. 1.5 V (H), 0.6 V (L)

D13.5. At each output (Q's are outputs)

D13.6. Two NANDs with wired AND

D13.7. 60 ns

D13.8. A low pulse as much as 15 nsec wide at F

D13.9. 24.3 ns, 19.8 ns

D13.10. 100, 26

D13.11. Q_0: 2, 10 pF; $\overline{Q}_0$: 4, 20 pF; Q_1: 2, 10 pF; $\overline{Q}_1$: 3, 15 pF

D13.12. 11

D13.13. 160 Ω, 260 Ω

D13.14. 46 mA

D13.15. CMOS: 150 cm, TTL: 50 cm, ECL: 20 cm

CHAPTER 14

Operational Amplifiers

An operational amplifier (op amp) is basically an amplifier with a very high gain. In Chapter 10 we saw the value of high-gain amplifiers in feedback control systems. In this chapter we use them in electronic feedback circuits to realize precision amplifiers, filters, oscillators, wave shapers, and other functions. There is such a demand for op amps that they are available in integrated circuits. In that form they can be thought of and used as basic building blocks themselves. In this chapter we discuss the properties of op amps and look at a number of their applications.

The reader's objectives in this chapter are to:

1. Learn to design simple amplifiers and filters using op amps,
2. Learn to design simple oscillators and rectifiers using the op amp as a nonlinear device,
3. Appreciate the limitations of op amps and how they affect circuit performance.

Ideally an op amp should have such properties as infinite gain and infinite bandwidth. In practice manufacturers provide a wide selection of op amps, with the more expensive types more nearly approaching the ideal. In 90 percent of the applications an inexpensive "general-purpose" op amp such as the μA741 is adequate. Therefore the example circuits in this chapter use the specifications for the μA741. The analysis for other op amps is the same but uses different numbers.

14.1   IDEAL OP AMP MODEL

An op amp is a differential amplifier (see Section 8.2.), which can be modeled by the two-port circuit shown Figure 14–1a. What characterizes it as an op amp is its very high voltage gain—an A of 100,000 or higher. Therefore only 50 μV or so is needed at v_i to produce $v_0 = 5$ V.

The symbol for an op amp is shown in Figure 14–1b. Unlike the model, the op amp is not actually connected to ground. However, it is connected to a positive and a negative voltage supply, effectively establishing ground as about halfway between the two supplies. They are typically $+15$ V and -15 V, but they may be as low as $+5$ V and -5 V. The "plus" and "minus" terminals at the input of the op amp indicate the polarity of v_i that makes v_0 positive. The "minus" terminal is called the *inverting input terminal* because raising its voltage lowers the voltage at the output v_0. The "plus" terminal is called the *noninverting input terminal* because raising its voltage raises the voltage at v_0.

The power supply connections are often not shown on the op amp symbol. The exact supply voltages are not important to the op amp's operation so long as the signal voltages don't try to exceed the supplies.

An ideal op amp would have infinite gain and no input current, and the output v_0 would be unaffected by any load. These properties can be summarized as

$$A = \infty, \qquad R_{\text{in}} = \infty, \qquad R_{\text{out}} = 0.$$

An ideal op amp has other properties such as infinite bandwidth and infinite voltage range at the input and output. But the most important property for simplifying design ideas is $A = \infty$. In practice this can't be achieved, but analysis based on a model with infinite gain leads to excellent agreement with actual performance in most cases. In Sections 14.7, 14.8, and 14.9 we will look at ways in which op amps depart from the ideal. Otherwise, we will assume ideal op amps when studying their applications.

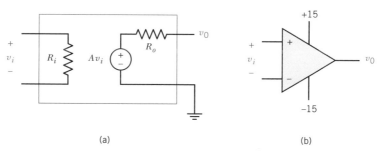

(a)   (b)

FIGURE 14–1 (a) Two-port model for a differential amplifier. An op amp is a differential amplifier with very high gain: $A \geq 100,000$. (b) Op amp symbol. The power supply connections are often not shown.

14.2   INVERTING CONFIGURATION

An amplifier with infinite gain is not very useful by itself. The op amp's usefulness comes from imbedding it in other elements which determine the response of the circuit. In Figure 14–2a an op amp is used together with two resistors to form an amplifier with a gain of -3.

In all linear op amp applications the output is somehow connected back to the inverting input terminal—through R_2 in this case. This results in negative feedback, as studied in Chapter 10. Under this condition the op amp acts so as to make its input voltage v_i be always zero. Suppose that v_1 goes negative, taking the inverting terminal slightly below ground (see Figure 14–2a). Then v_i is positive, causing v_0 to go positive—positive enough to raise the inverting terminal so that v_i is again zero. Because the (ideal) gain is infinite, not even a small voltage can remain at v_i.

This is most important in understanding an op amp circuit: *the voltage between the two op amp input terminals is zero.* It is a nontrivial accomplishment because it is done without drawing any current at the op amp input. For instance, in the circuit of Figure 14–2a we could make zero volts from ground to the junction of R_1 and R_2 by merely shorting that junction to ground. But this would draw current through the short, and the circuit operation would be disturbed. The op amp, on the other hand, draws no current (ideally) into its input terminals.

The analysis of an op amp circuit is easy once we make the assumptions (1) that it is ideal and (2) that it is achieving the task of keeping 0 V between its input terminals. For the circuit in Figure 14–2a, R_2 is three times the value of R_1. Since the same current i flows through them, the voltage v_{R2} across R_2 is three times the voltage v_{R1} across R_1: $v_{R2} = 3v_{R1}$. But since the inverting terminal (with the minus) is virtually grounded by the op amp, $v_{R1} = v_1$, and $v_{R2} = -v_0$. Therefore, $v_0 = -3v_1$. In general,

$$v_0 = -(R_2/R_1)\, v_1. \tag{14.1}$$

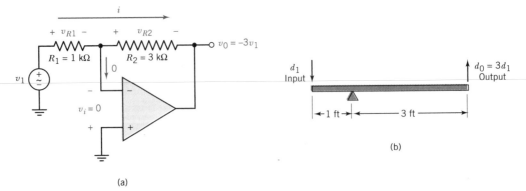

(a)

(b)

FIGURE 14–2 (a) Inverting configuration amplifier with gain of -3. (b) First-class lever analogy of the inverting configuration. Height corresponds to voltage, length to resistance, and angle to current.

Because this amplifier design yields a negative gain, it is called the *inverting config-uration*. Note that the gain $-R_2/R_1$ depends only on the values of the resistors, and these can be made very accurate (0.1% if necessary).

A mechanical analogy is useful in visualizing the operation of the inverting op amp configuration. Consider the first-class lever in Figure 14–2b. The lengths of the arms are proportional to the values of the resistors in Figure 14–2a; that is, a ratio of 1 : 3. In this analogy, height is the analog of voltage. Therefore the fulcrum fixes a point on the lever, just as the op amp fixes the voltage between the resistors at ground. As the left end of the lever is moved down a distance d_1, the right end of the lever moves up a distance $d_2 = 3d_1$. (This assumes the two arms change by the same angle, just as the circuit has the same current in the two resistors.) The opposite direction of the motions reflects the fact that this mechanical amplifier is inverting (has negative gain).

Another use for the inverting configuration is the *summing amplifier* shown in Figure 14–3. Such a circuit might be used in combining or "mixing" audio signals. The sources v_1, v_2, and v_3 establish currents i_1, i_2, and i_3 through their respective resistors to ground (this is a *virtual ground* established by the op amp's maintaining zero volts at its input). Each current is proportional to its corresponding voltage; for example $i_1 = v_1/1 \text{ k}\Omega$. By KCL, the current i_4 through the 3-kΩ feedback resistor is the sum of these currents:

$$i_4 = i_1 + i_2 + i_3 = (v_1 + v_2 + v_3)/1 \text{ k}\Omega.$$

But $v_0 = -3 \text{ k}\Omega \times i_4$. Therefore

$$v_0 = -3(v_1 + v_2 + v_3).$$

The output voltage is the sum of the input voltages with, in addition, some gain.

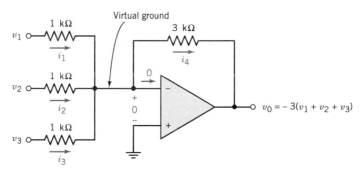

FIGURE 14–3 Summation circuit with a gain of -3. Summation occurs at the virtual ground node, where the input currents sum by KCL to produce i_4.

The following example shows the use of a summing amplifier to convert digital signals to an analog signal.

EXAMPLE 14.1   A *digital-to-analog converter* (DAC) is to produce an analog output voltage v_0 equal to -1 V times the 4-bit binary number at the input.

Given: The bits (least significant to most significant) are represented by v_1, v_2, v_3, and v_4. Logic "1" is represented by 5 V and logic "0" by 0 V.

Find: a summation circuit design to convert the digital signals v_1 through v_4 to an analog voltage v_0 according to the table in Figure 14–4a.

Solution: One possible design is shown in Figure 14–4b. Consider the second row in the table when $v_1 = 5$ V and the other inputs are 0 V. The 5 V across the 40-kΩ resistor causes $\frac{1}{8}$ mA to flow through the 8 kΩ, developing $v_0 = -1$ V. Similarly, when $v_2 = 5$ V in the third row of the table, $v_0 = -2$ V. By superposition, $v_0 = -3$ V when both v_1 and v_2 are 5 V (see the fourth row). The largest output occurs when all inputs are 5 V, resulting in $v_0 = -15$ V.

Note that in the summation, each input receives a different gain or "weighting;" $v_0 = -1.6v_4 - 0.8v_3 - 0.4v_2 - 0.2v_1$.

In this example the lowest value of v_0 is -15 V. Since v_0 can't go below the negative supply voltage, the negative supply must be at least as low as -15 V. In practice it should be even a couple of volts lower—say -17 V.

A DAC like that in Example 14.1 can be used in the construction of an analog-to-digital converter (ADC), which performs the reverse operation of a DAC. Problem E14.3 at the end of the chapter examines this further.

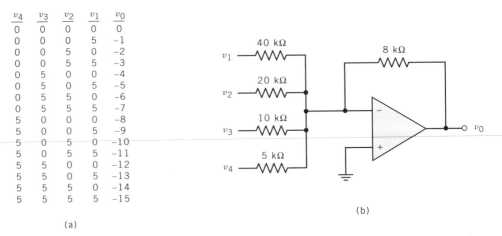

v_4	v_3	v_2	v_1	v_0
0	0	0	0	0
0	0	0	5	-1
0	0	5	0	-2
0	0	5	5	-3
0	5	0	0	-4
0	5	0	5	-5
0	5	5	0	-6
0	5	5	5	-7
5	0	0	0	-8
5	0	0	5	-9
5	0	5	0	-10
5	0	5	5	-11
5	5	0	0	-12
5	5	0	5	-13
5	5	5	0	-14
5	5	5	5	-15

(a)

(b)

FIGURE 14–4   (a) Table giving desired relationship (see Example 14.1) between digital input signals v_1 through v_4 and analog output voltage v_0. The input signals represent a binary number, where 5 V and 0 V represent "1" and "0," respectively. (b) Digital-to-analog conversion circuit implementing the table in (a).

14.3   NONINVERTING CONFIGURATION

As the name implies, the noninverting configuration of an op amp realizes an amplifier with positive gain. More importantly, the input impedance of the amplifier is very high. Because of this it is useful in *buffer* applications—where the source is protected from having to provide a high load current.

Figure 14–5a shows an example of the noninverting configuration. Note again the polarity of the input; the inverting terminal is connected to the output through R_2. Since there is a virtual short at the op amp input, the source voltage v_1 appears at the junction of R_1 and R_2. The voltage v_1 establishes a current $i = v_1/R_1$ through R_1, and this same current flows through R_2 (no current flows into the inverting terminal). Ground is now at the left, so the output voltage is the voltage across both resistors: $v_0 = (R_1 + R_2)i = (R_1 + R_2)v_1/R_1 = 4v_1$. In general the gain for the noninverting configuration is

$$\frac{v_0}{v_1} = \frac{R_1 + R_2}{R_1}. \tag{14.2}$$

The analogy in this case is a third-class lever as shown in Figure 14–3b. The fulcrum corresponds to ground, and the $1 : 4$ ratio of the lever arms corresponds to the resistance from v_1 to ground and from v_0 to ground. The motion amplification by the lever is clearly a factor of 4. Since the input and output move in the same direction, the gain is positive (noninverting).

A special case of the noninverting configuration, shown in Figure 14–5c, finds frequent application. In this case $R_1 = \infty$, and $R_2 = 0$. The input v_1 is virtually

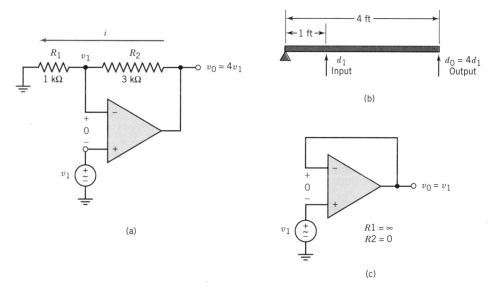

FIGURE 14–5 (a) Noninverting configuration amplifier with a gain of 4. (b) Third-class lever analogy to the noninverting configuration. (c) Noninverting configuration with a gain of 1, called a "voltage follower."

shorted to the inverting terminal, and the inverting terminal is actually shorted to v_0. Therefore $v_0 = v_1$. Because the output follows the input exactly, this configuration is called a *voltage follower*. A gain of unity is not very impressive; the circuit's purpose is solely one of buffering. The source v_1 sees infinite impedance, and any load connected to v_0 sees zero impedance—the output impedance of an ideal op amp. In practice, an op amp with feedback has an output impedance of less than an ohm. Buffering is important for sources with high Thevenin impedance—that suffer a reduced voltage if required to provide much load current. The noninverting amplifier in Figure 14–5a provides buffering and amplification too.

EXAMPLE 14.2 A microphone needs to have its voltage amplified to a level of 4 V (rms).

Given: The source impedance of the microphone is $R_s = 10$ kΩ, and its (unloaded) output voltage is $v_s = 20$ mV (rms).

Find: an amplifier circuit design to provide the needed gain. Compare designs using a noninverting op amp configuration and an inverting configuration.

Solution: A gain of 4 V/20 mV = 200 is needed. This is provided by the noninverting configuration in Figure 14–6a with a gain of $v_0/v_1 = (199$ kΩ $+ 1$ kΩ$)/1$ kΩ $= 200$. Note that since $i_1 = 0$, there is no drop in R_s, and $v_1 = v_s$.

The inverting configuration in Figure 14–6b would provide a gain $v_0/v_1 = -200$ kΩ$/1$ kΩ $= -200$ for $R_1 = 1$ kΩ and $R_2 = 200$ kΩ. But the 1 kΩ to virtual ground would load down the microphone so that $v_1 = v_s/11$ (the 10 kΩ and 1 kΩ form a voltage divider). Therefore v_0/v_s is only $-200/11 = -18.2$. To get $v_0/v_s = -200$, one solution would be $R_1 = 0$ and $R_2 = 2$ MΩ. This solution relies excessively on the value of R_s, which may not be well controlled. The better configuration is the noninverting one in this situation, because of the high source impedance.

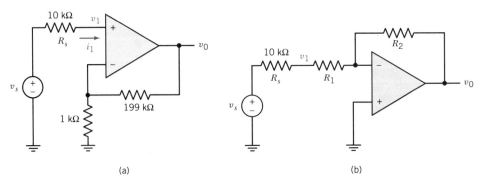

(a)   (b)

FIGURE 14–6 (a) Solution to the design problem in Example 14.2. A noninverting configuration (with high input impedance) is used here to give $|v_0/v_s| = 200$. (b) Solution using an inverting configuration. $R_1 = 0$ and $R_2 = 2$ MΩ give the same gain magnitude of 200 as in (a), but this gain is dependent on R_s.

14.4   DIFFERENTIAL CONFIGURATION

Some applications require that the *difference* between two voltages be amplified. This is the case in a control system where the comparator takes the difference between the input voltage and the feedback voltage (see Chapter 10). Another example is in biomedical applications where a measurement is made of the voltage difference between two points on the patient's body. In these cases the differential op amp configuration in Figure 14–7a is needed. This is basically a combination of the two configurations already considered. If v_1 is grounded, v_2 sees an inverting configuration. If v_2 is grounded, v_1 (through a voltage divider) sees a noninverting configuration. An analysis based on this superposition approach (see problem D14.3) leads to the result

$$v_o = (R_2/R_1)(v_1 - v_2). \tag{14.3}$$

The important point here is that the output depends only on the *difference* between the input voltages. If these voltages are moved up and down together, there is no effect on the output. This is called *common-mode rejection.*

Another lever analogy makes this clear. In Figure 14–7b the voltage divider (connected to v_1) is likened to a second-class lever with a fulcrum anchoring its right end. At its effort point is a pivot connecting it to a first-class lever. This corresponds to the virtual short connecting the lower resistors to the upper resistors. The whole arrangement resembles a pair of scissors. The distance of the tip above the fulcrum depends only on the spacing between the "handles." If both "handles" are moved up the same amount, the height of the upper right tip essentially doesn't change (the analog breaks down for large common-mode input).

EXAMPLE 14.3 An *X–Y* plotter uses a linear potentiometer to sense the *y* position of the pen, as shown in Figure 14–8. A differential amplifer is to compare the voltage v_z from the potentiometer with the input voltage v_x. The output v_1 powers

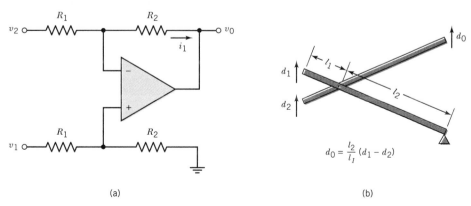

(a)   (b)

FIGURE 14–7 (a) Differential configuration amplifier. The output is responsive only to the *difference* of v_1 and v_2. (b) Lever analogy to the differential configuration. The arrangement resembles a pair of scissors.

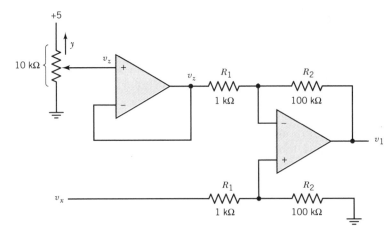

FIGURE 14–8 Solution to the design problem in Example 14.3 requiring $v_0 = 100\,(v_x - v_z)$. The voltage follower serves as a buffer to keep the current through R_1 from affecting v_z.

a motor that controls the y position of the pen. (See Figure 10–15a for a flow graph of the system.)

Given: The total impedance of the potentiometer is 10 kΩ, and its length is 25 cm.

Find: a differential amplifier design with a gain of 10 so that $v_1 = 100(v_x - v_z)$. The differential amplifier is not to affect v_z by loading down the potentiometer.

Solution: The differential op amp configuration shown in Figure 14–8 provides $v_1 = (R_2/R_1)(v_x - v_z)$, which should equal $100(v_x - v_z)$. This is satisfied by any reasonable resistor values with the ratio of 100—say $R_1 = 1$ kΩ and $R_2 = 100$ kΩ. Other resistors with a ratio of 100 could be used—for example, 80 kΩ and 8 MΩ. For an ideal op amp it makes no difference, but we will see that for practical op amps, resistors in the range from about 500 Ω to 500 kΩ are best.

The unity-gain voltage follower provides buffering so that no current (ideally) is drawn from the potentiometer. Otherwise current as great as 1 mA through R_1 would change v_z by as much as 2.5 V. (The Thevenin impedance of the potentiometer is 2.5 kΩ when the wiper is in the center.)

14.5   INTEGRATION AND DIFFERENTIATION

Electronic circuits are often called on to carry out mathematical operations. We have already seen a circuit that adds (Figure 14–3) and one that multiplies by a constant (Figure 14–2).* We will now look at a circuit to perform integration.

_____

*Variable-gain amplifiers can multiply two voltages, but an ordinary op amp cannot do this with fixed resistors.

Actually, a capacitor by itself does this if the input is a current and the output is a voltage. From Chapter 4 we have the $v-i$ relationship for a capacitor:

$$v = \frac{1}{C} \int i \, dt.$$

If we wish to represent all variables by *voltages*, then the input voltage must first be converted into a proportional current i, and a capacitor does the rest. This is essentially what the integrator circuit in Figure 14–9a does. Since R_1 is connected to virtual ground, v_1 appears directly across it, and a proportional current is produced: $i = v_1/R_1$. The capacitor is also connected to virtual ground, so the output voltage is the voltage across the capacitor—the integral of the same current:

$$v_o = -\frac{1}{R_1 C} \int v_1 \, dt.$$

Both input and output variables are now voltages.

We could carry out differentiation by simply substituting an inductor for the capacitor in Figure 14–9a. However, inductors are expensive, lossy, and sometimes bulky. Therefore capacitors are used instead whenever possible. Consider the alternative form of the $v-i$ relationship for a capacitor: $i = C \, dv/dt$. If we can turn i into a proportion voltage, then we have built a differentiator. In Figure 14–9b the op amp takes the current from the capacitor and pulls it through R_2 to produce the output voltage:

$$v_o = -R_2 C \frac{dv_1}{dt}.$$

The circuit analysis so far has assumed ideal op amps. When op amp limitations are considered, the result is usually that the whole circuit is limited in voltage

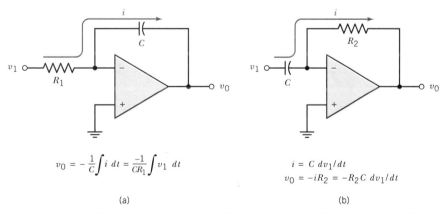

$$v_0 = -\frac{1}{C} \int i \, dt = \frac{-1}{CR_1} \int v_1 \, dt$$

(a)

$$i = C \, dv_1/dt$$
$$v_0 = -iR_2 = -R_2 C \, dv_1/dt$$

(b)

FIGURE 14–9 (a) Integrator circuit. (b) Differentiator circuit. This circuit is actually unstable. See Figure 14–15c for a practical circuit.

or frequency range. Worse yet, a circuit design that looks fine with an ideal op amp can oscillate when built in practice; this is true for the differentiator in Figure 14–9b. In Section 14.8 we will add a "Band-Aid" to the differentiator circuit to make it stable.

14.6   FILTERS

A filter is, in general, any circuit with a gain that is a function of frequency:

$$\frac{\mathbf{V}_0}{\mathbf{V}_1} = \frac{a_n(j\omega)^n + \cdots + a_1(j\omega) + a_0}{b_m(j\omega)^m + \cdots + b_1(j\omega) + b_0}.$$

In Section 6.4 we saw examples of passive filters (only resistors, capacitors, and inductors) that gave a desired frequency response. When op amps are also used, the filter design often becomes simpler, and a wider variety of responses is possible. Some configurations with op amps allow inductors to be eliminated from a filter design. Filter functions with m and n within two of each other can be realized by the general inverting configuration in Figure 14–10. The transfer function is

$$\mathbf{V}_0/\mathbf{V}_1 = -\mathbf{Z}_2/\mathbf{Z}_1, \tag{14.4}$$

where $\mathbf{Z}_1$ has replaced R_1 in Figure 14–2a and Equation (14.1) and $\mathbf{Z}_2$ has replaced R_2.

A low-pass filter is often used in voice communications to limit noise (see Section 8.8) or to prevent aliasing (see Section 11.6). The configuration in Figure 14–10 can be used to realize a low-pass filter that rolls off at either one or two decades per decade.

EXAMPLE 14.4 A low-pass filter is needed to attenuate noise at high frequencies but still let through enough voice frequencies so the speaker can be recognized.

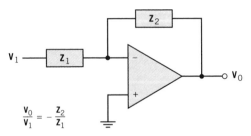

$$\frac{\mathbf{V}_0}{\mathbf{V}_1} = -\frac{\mathbf{Z}_2}{\mathbf{Z}_1}$$

FIGURE 14–10 General inverting configuration. Various filter functions can be realized by constructing $\mathbf{Z}_1$ and $\mathbf{Z}_2$ of resistors, inductors, and capacitors.

Given: The in-band gain magnitude is to be unity, and the response is to roll off at one decade per decade starting at 3 kHz (see Bode plot in Figure 14–11a). That is, the response is 3 dB down at 3 kHz.

Find: an active filter design with an input impedance of 10 kΩ to meet the specifications. The transfer function will be $\mathbf{V}_0/\mathbf{V}_1 = -3 \text{ kHz}/(jf + 3 \text{ kHz})$.

Solution: For a low-pass, $\mathbf{Z}_1$ in Figure 14–11 is a resistor R_1, and $\mathbf{Z}_2$ is a resistor R_2 in parallel with a capacitor, as shown in Figure 14–11b. The input impedance is R_1 to virtual ground, so $R_1 = 10$ kΩ. At low frequencies the capacitor is essentially an open circuit, so the gain magnitude there is $R_2/R_2 = 1$. Therefore $R_2 = 10$ kΩ also. The frequency where the response starts rolling off is the frequency for which the capacitor starts shorting out R_2. Specifically, the cutoff frequency f_U is that f for which the magnitude of the capacitor's impedance equals R_2. Then $|\mathbf{Z}_C| = 1/(2\pi f_U C) = R_2 = 10$ kΩ. For $f_U = 3$ kHz, solving for C yields $C = 5300$ pF.

As a check let us find the transfer function for our design. $\mathbf{Z}_2$ is the parallel combination of $\mathbf{Z}_C$ and R_2:

$$\mathbf{Z}_2 = \frac{R_2 \mathbf{Z}_C}{R_2 + \mathbf{Z}_C} = \frac{R_2/(2\pi jfC)}{R_2 + 1/(2\pi jfC)} = \frac{R_2/(2\pi R_2 C)}{jf + 1/(2\pi R_2 C)} = \frac{R_2 f_U}{jf + f_U},$$

where

$$f_U = 1/(2\pi R_2 C) = 1/(2\pi 10 \text{ kΩ} \times 5300 \text{ pF}) = 3 \text{ kHz}.$$

Then from Equation (14.4), $\mathbf{V}_0/\mathbf{V}_1 = -\mathbf{Z}_2/R_1 = -f_U/(jf + f_U)$, as desired.   ▪

As a second example of filter design we will realize the compensation response incorporated in the controller for the *X–Y* plotter system in Example 10.7.

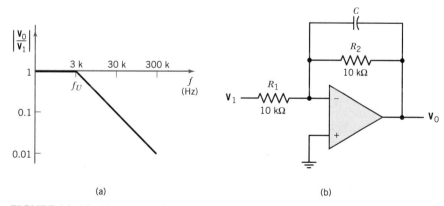

(a)   (b)

FIGURE 14–11 (a) Low-pass filter response desired in Example 14.4. The Bode plot here has axes with logarithmic scales. (b) Op amp circuit realizing the response in (a). The capacitor starts shorting out R_2 at $f = 3$ kHz.

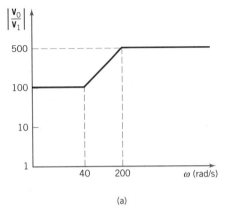

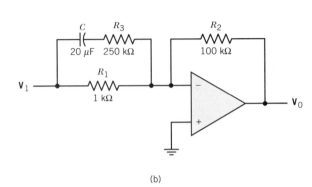

(a)   (b)

FIGURE 14–12 (a) Compensation filter response desired in Example 14.5. (b) Circuit realizing the desired response. At low frequencies the capacitor is an open circuit, and $|\mathbf{V_0}/\mathbf{V_1}| = R_2/R_1 = 100$. At high frequencies the capacitor is a short circuit, and $|\mathbf{V_0}/\mathbf{V_1}| = R_2/(R_3 \| R_1) = 500$.

EXAMPLE 14.5 A controller is needed with the response shown in Figure 14–12a. [See the desired $\mathbf{F}(s)$ in Figure 10.15d.]

Given: The corresponding transfer function is

$$\frac{\mathbf{V}_0}{\mathbf{V}_1} = \frac{100(j\omega/40 + 1)}{j\omega/200 + 1},$$

Find: a filter design with the configuration in Figure 14–10 to realize the specified response. The input impedance at low frequencies is to be 1 kΩ.

Solution: The final filter design is shown in Figure 14–12b. The design starts at low frequencies, where the capacitor is essentially an open circuit. Then R_1 = 1 kΩ to satisfy the input impedance requirement. The low-frequency gain magnitude is R_2/R_1, and from the response in Figure 14–12a, this must equal 100. Therefore R_2 = 100 kΩ. At 40 rad/s the capacitor starts shorting out R_1 to increase the gain. When the capacitor's impedance becomes less than R_3, it can be considered a short circuit, and we essentially have R_3 in parallel with R_1 (or $R_3 \| R_1$) at the input for high frequencies. The gain magnitude is $R_2/(R_3 \| R_1)$, which, according to Figure 14–12a, is equal to 500. Solving for R_3 yields R_3 = 250 Ω. The break in the curve at 200 rad/s is where $|\mathbf{Z}_C| = R_3$. Therefore $1/\omega C = 1/200C = 250$, and C = 20 μF.
      The reader can use Equation (14.4) to confirm that these values give the desired transfer function.

14.7   FREQUENCY AND GAIN LIMITATIONS

An ideal op amp would have both infinite gain and infinite bandwidth. In practice both are limited, as shown by the frequency response of a typical "general-purpose" op amp such as the μA741 (see Figure 14–13d). At low frequencies the gain $|\mathbf{A}|$ is about 100,000 (called the *dc gain*), and at high frequencies $|\mathbf{A}|$ rolls off until it is

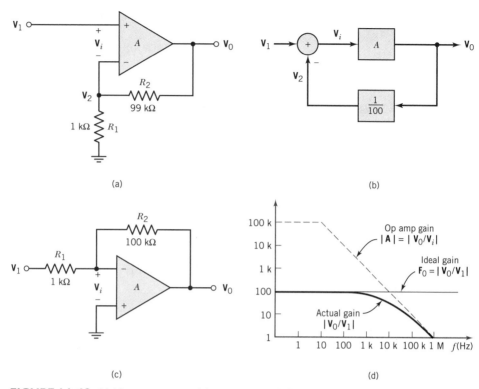

(a)

(b)

(c)

(d)

FIGURE 14–13 (a) Noninverting amplifier with gain of 100 (ideally for all frequencies). (b) Signal flow graph of the circuit in (a). (c) Inverting amplifier with gain of 100. (d) Typical op amp response $|\mathbf{A}|$ with a dc gain of 100,000 and a decade per decade rolloff to unity gain at 1 MHz. The actual gain $|\mathbf{V}_0/\mathbf{V}_1|$ is the lower of the op amp gain and the ideal $|\mathbf{V}_0/\mathbf{V}_1|$. The intersection determines the bandwidth of 10 kHz here.

unity at about 1 MHz. Notice that the gain rolls off in a regular way—one decade (factor of 10) decrease in gain for each decade increase in frequency. To find the gain of an op amp at a particular frequency, it is better to figure backward from the unity-gain frequency than to figure forward from the dc gain. This is because the unity-gain frequency is better controlled than the dc gain. For example, if you wish to know the gain at 10 kHz, figure that 10 kHz is two decades down from 1 MHz, and the gain is therefore two decades—or $10^2 = 100$.

The limited frequency response of an op amp will of course limit the frequency response of the circuit in which it is used. It is usually true that the gain of the whole circuit at a particular frequency cannot exceed the gain of the op amp at that frequency. Consider the noninverting configuration in Figure 14–13a, for example. Ideally the gain is given for all frequencies by Equation (14.2): $\mathbf{V}_0/\mathbf{V}_1 = (R_1 + R_2)/R_1 = 100$. This is represented by the horizontal line in Figure 14–13d. We can find the actual gain $\mathbf{V}_0/\mathbf{V}_1$ by using feedback theory from Chapter 10. The flow graph in Figure 14–13b relates the variables of the circuit in Figure 14–13a. If the feedback function is denoted H, then $H = 1/100$, and $|\mathbf{V}_0/\mathbf{V}_1|$ is the lesser of $|\mathbf{A}|$ and $1/H$ (see Section 10.3). But $1/H$ is the ideal gain, so the actual gain has a cutoff where $|\mathbf{A}|$ crosses the ideal gain at 10 kHz in Figure 14–13d. This cutoff frequency is called the bandwidth.

The analysis for the inverting configuration in Figure 14–13c is not as simple as for the noninverting. But it can be shown that the same result roughly holds: *the actual gain follows the lower of the ideal gain and* $|\mathbf{A}|$. This holds more closely the higher the ideal gain; it does not hold very well for gains less than about 10. In any case, the ideal gain given by Equation (14.1) or (14.2) holds only when it is significantly less than $|\mathbf{A}|$.

Let F_0 represent the ideal gain of a flat-gain amplifier. Because the op amp gain $|\mathbf{A}|$ rolls off at a decade per decade, any increase in F_0 results in a reduction of the bandwidth by the same factor. For example, changing the 99-kΩ resistor in Figure 14–13a to 999 kΩ would increase the ideal gain from $F_0 = 100$ to $F_0 = 1000$. The reader can sketch a line at 1000 on Figure 14–13d and find that it intersects $|\mathbf{A}|$ at 1 kHz rather than 10 kHz. There has been a tradeoff of one-tenth the bandwidth for ten times more gain. It is easy to show that the *gain-bandwidth product* is constant and equal to the unity-gain frequency—1 MHz for this op amp. Therefore the 3-dB bandwidth is given in terms of the ideal gain F_0 and the gain-bandwidth product GBP by

$$B_{3\text{dB}} = \frac{\text{GBP}}{F_0}. \tag{14.5}$$

For example, the controller in Figure 10–10a has $F_0 = 2000$ and $B_{3\text{dB}} = 5$ krad/s, corresponding to GBP = 10 Mrad/s, or about 1.6 MHz.

In considering an op amp for an application, one should calculate the available gain for the bandwidth that the application requires. For an audio application a bandwidth of $B_{3\text{dB}} = 20$ kHz may be desired. For an op amp with a GBP of 1 MHz, the greatest gain possible with this bandwidth is $F_0 = 1$ MHz/20 kHz = 50. This is a useful gain for most purposes, so a general-purpose op amp is sometimes called a *voice-frequency op amp*. If a higher gain is needed, two stages can be cascaded. In some situations it may be necessary to use an op amp with a higher gain-bandwidth product. High-performance op amps with GBP as high as 200 MHz are available. Op amps with *external compensation* allow the user to adjust the rolloff of $|\mathbf{A}|$ to get a greater bandwidth at high gains.

Let's look at the effect of an op amp's frequency response on the integrator circuit in Figure 14–9a. In Section 14.5 we analyzed the integrator's operation in the time domain; here we analyze it in the frequency domain. Since the integrator circuit comes under the general class of circuits in Figure 14–10, the transfer function of the integrator is given by Equation (14.4), where $\mathbf{Z}_1 = R_1$ and $\mathbf{Z}_2 = \mathbf{Z}_C = 1/(j2\pi f C)$. Then the gain magnitude is $|\mathbf{V}_0/\mathbf{V}_1| = |\mathbf{Z}_C|/R_1 = 1/(2\pi f R_1 C) = f_1/f$, where $f_1 = 1/(2\pi R_1 C)$. (For the values in the circuit in Figure 14–14a, $f_1 = 16$ kHz.) Thus the ideal gain descends inversely with f, and it goes to unity at $f = 16$ kHz, as shown in Figure 14–14b.* At frequencies lower than 0.16 Hz the ideal analysis calls for gains greater than 100,000—greater than the op amp can provide.

---

*It is no coincidence that an op amp by itself has the same sort of rolloff. Most op amps (specifically, *internally compensated* op amps) include a feedback capacitor to achieve the rolloff for stability purposes.

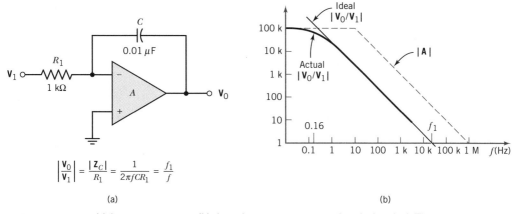

FIGURE 14-14 (a) Integrator circuit. (b) Actual response compared with the ideal. The circuit is a good integrator so long as the response is falling off at a decade per decade.

The rule again applies that, at each frequency, the circuit gain is the lesser of the ideal gain and the op amp gain. The curves in Figure 14-14b illustrate this, showing the actual gain flattening off below 0.16 Hz. This means that the circuit is a good integrator only for frequencies above 0.16 Hz. It can also be shown that it is a good integrator only for frequencies below the unity-gain frequency of the op amp— 1 MHz here.

14.8 STABILITY

We saw in Chapter 10 that it is possible for a feedback system to become unstable and oscillate. Linear applications of op amps always involve feedback, so the possibility exists of instability due to too little phase margin. A simple example of an op amp circuit that can be unstable is the differentiator in Figure 14-9b. Since the differentiator has the general configuration shown in Figure 14-10, we apply Equation (14-4) to get $|\mathbf{V}_0/\mathbf{V}_1| = R_2/|\mathbf{Z}_C| = 2\pi f R_2 C$. For the values in the circuit in Figure 14-15a this gives $|\mathbf{V}_0/\mathbf{V}_1| = f/100$ Hz. This ideal gain rises at a rate of a decade per decade, as shown in Figure 14-15b, crossing the decreasing op amp gain $|\mathbf{A}|$ at 10 kHz. We might expect to simply say that the circuit acts as a differentiator up to 10 kHz and be done with the analysis. However, feedback theory shows that if the two curves cross at that steep an angle, the phase margin goes to zero, and the circuit will be unstable (see Section 10.8). Therefore the response of the op amp must be made to go flat (horizontal) in the vicinity of the crossing.

If a 320-Ω resistor is put in series with the capacitor as in Figure 14-15c, it will dominate the impedance of the capacitor at frequencies above 5.0 kHz. Then above 5.0 kHz the capacitor can be neglected, and the ideal gain is a flat 16 kΩ/ 320 Ω = 50, which crosses the op amp gain $|\mathbf{A}|$ at 20 kHz. The curves cross at a decade/decade, and the circuit is stable (see curves in Figure 14-15d). The circuit looks like a differentiator only so long as the gain is rising a decade per decade. Therefore it is a good differentiator only up to about 5.0 kHz.

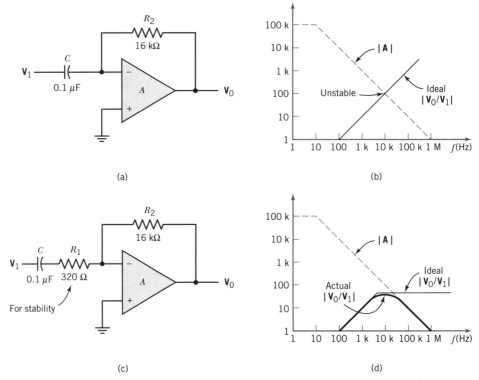

FIGURE 14-15 (a) Differentiator circuit as in Figure 14–9b. (b) Ideal response $|V_0/V_1|$ = $R_2/|Z_C|$ crosses op amp response $|A|$ at too steep an angle; the circuit is unstable. (c) Differentiator circuit with resistor R_1 added for stability. (d) Ideal response is now flat as it crosses the op amp response, making the circuit stable.

14.9   dc OFFSET

In Section 14.2 we characterized the input of an ideal op amp as having zero voltage maintained across it and zero current through it. For real op amps, neither is actually zero; there is a small dc *input offset voltage* V_{IO}, and there is a small dc *input bias current* I_B. For a general-purpose op amp typical values for these parameters are V_{IO} = 5 mV dc and I_B = 0.1 μA dc.* The effect of each is to produce dc offset at the output (see Section 8.7). We will look first at the effect of V_{IO}.

The input offset voltage V_{IO} is defined as that input voltage which will produce zero volts at the output. Because of the op amp's high gain, any moderate voltage at the output will correspond to about V_{IO} at the input. Therefore, in a negative feedback situation, V_{IO} always appears at the op amp input. Depending on what

---

*Some op amps include a trim adjustment to reduce V_{IO} to as low as 10 μV. High-performance op amps with FET inputs are available with I_B as low as 10 pA at room temperature.

circuit the op amp is used in, the input offset voltage will be amplified by some amount and appear as an offset at the output.

EXAMPLE 14.6 A noninverting configuration has a gain of 11; ideally $v_o = 11v_1$. The circuit is shown in Figure 14–16.

> *Given:* The op amp has $V_{IO} = 5$ mV and $I_B = 0$.

> *Find:* the offset voltage at the output.

> *Solution:* Since 5 mV appears at the op amp input, the voltage at the junction of the resistors is $v_1 + 5$ mV. Both these voltage sources are amplified 11-fold to produce $v_O = 11(v_1 + V_{IO}) = 11v_1 + 55$ mV. The offset at the output is <u>55 mV</u>.

From the example it is clear the input offset voltage should be kept much less than the input signal:

$$V_{IO} << v_1. \tag{14.6}$$

This same rule applies for the inverting and differential configurations. If the amplifier doesn't have to have a flat response down to dc, the output offset can be blocked with a large coupling capacitor. Too small a capacitor can lead to "sag" (see Section 8.6).

The input bias current I_B has a similar effect. There are effectively current sources within the op amp that pull I_B into each input terminal. It can be shown (see problem P14.13) that, as I_B is pulled through the resistors at the input, it effectively produces an additional input offset voltage V_{eq}:

$$V_{eq} = R_{eq}I_B, \tag{14.7}$$

where R_{eq} is the Thevenin equivalent resistance seen by the inverting input terminal.

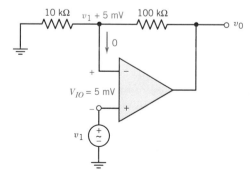

FIGURE 14–16 (a) Amplifier for Example 14.6. The op amp has a 5-mV input offset.

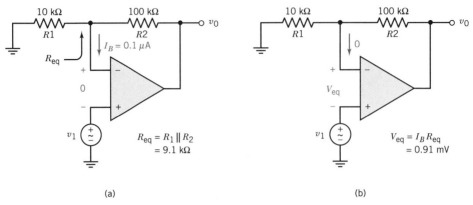

(a)   (b)

FIGURE 14–17 (a) Amplifier for Example 14.7. The op amp has a 0.1-μA input bias current. (b) Equivalent circuit showing that the input bias current has the same effect as an equivalent input offset voltage V_{eq}.

EXAMPLE 14.7 Consider the same amplifier circuit as in Example 14.6, but this time the offset is due to I_B (see Figure 14–17a).

Given: $V_{IO} = 0$ and $I_B = 0.1$ μA.

Find: the offset voltage at the output.

Solution: The op amp draws $I_B = 0.1$ μA into each of its input terminals. The I_B drawn through the v_1 source has no effect. The I_B drawn into the inverting terminal encounters an equivalent resistance R_{eq} equal to R_1 in parallel with R_2, or $R_{eq} = 9.1$ kΩ. (Thevenin impedance is found by setting all voltage and current sources to zero; this grounds v_0.) Then $V_{eq} = 0.1$ μA $\times$ 9.1 kΩ = 0.91 mV. As in Example 14.6, this equivalent input offset voltage experiences a gain of 11 to produce an output offset of <u>10 mV</u>.

If the V_{IO} of Example 14.6 and the I_B of Example 14.7 were present together, by superposition the output offset would be the sum of the two separate results: 55 mV + 10 mV = 65 mV. Here the effect of V_{IO} dominates, and I_B interacting with the resistors has not significantly degraded the performance. A good rule of thumb is that the resistors should be chosen so that

$$V_{eq} < V_{IO}. \qquad (14.8)$$

For general-purpose op amps, this makes it desirable to keep R_{eq} less than about 50 kΩ for both noninverting and inverting configurations.

14.10   NONLINEAR APPLICATIONS

When using an op amp in a linear application, the basic assumption is that negative feedback causes the voltage between the input terminals of the op amp to be kept at zero. In nonlinear applications there is either no feedback or there is positive feedback, and the input voltage is not kept at zero. In such cases the high gain of the op amp causes the output to become saturated ("pinned") at either its positive or negative extreme—about a volt short of the supply voltages. In the examples that follow, we will assume that the supplies are $+15$ V and -15 V, and the output is either $+14$ V or -14 V.

Comparator

An op amp by itself as in Figure 14–1 can be used as a *comparator*. The input terminals compare two voltages, and the output is either $+14$ or -14, depending on which of the input voltages is higher.

EXAMPLE 14.8 An alarm is to be set off when room temperature exceeds 150°F.

Given: a temperature-to-voltage transducer for which 150°F produces a voltage $v_1 = 5$ V. The alarm sounds when -14 V is applied, and it is silent when $+14$ V is applied.

Find: a circuit design to monitor the transducer voltage and apply the proper voltage to the alarm.

Solution: We need a circuit that compares v_1 with a 5-V reference and changes its output suddenly when v_1 exceeds 5 V. Such a circuit is shown in Figure 14–18a, and its transfer characteristic is shown in Figure 14–18b. When v_1 exceeds 5 V, v_o falls from 14 V to -14 V. For an op amp with dc gain of 100,000, a change of only 28 V/100,000 $= 0.28$ mV in v_1 is needed to cause this fall.

Schmitt Trigger

One problem with a comparator is that it can "chatter." If the temperature in the example is advancing slowly, it can reach the point just at 150°F where v_0 is at neither 14 V nor -14 V but banging back and forth due to noise. What is needed in this case is a little *hysteresis*. A wall switch is an example of mechanical hysteresis. An over-the-center mechanism causes the contacts to snap closed when the switch is pushed upward past a certain point. The switch must be pushed downward con-

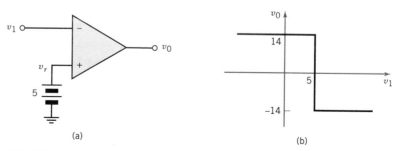

FIGURE 14–18 (a) Op amp used as a comparator in Example 14.8. (b) The output is at either the high or the low limit depending on whether v_1 is lower or higher than the reference voltage v_r.

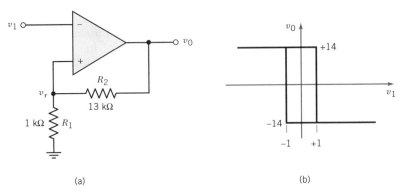

(a)   (b)

FIGURE 14–19 Schmitt trigger circuit. Positive feedback provides hysteresis by moving the reference voltage v_r.

siderably below that point before the contacts open. This avoids the chattering and arcing that can occur with a switch that has no hysteresis (see Figure 17–4c).

Hysteresis is implemented in electronics by a Schmitt trigger circuit, such as the one in Figure 14–19a. This configuration is similar to the noninverting configuration in Figure 14–5, but the resistors connect to the noninverting terminal, resulting in positive feedback. Now the input v_1 is being compared with a reference voltage v_r that depends on the output voltage v_0. When $v_0 = 14$ V, the resistor voltage divider makes $v_r = 1$ V. Then v_1 must go upward past 1 V to cause v_0 to go to -14 V. Now $v_r = -1$ V, and v_1 must go below -1 V to send v_0 back to 14 V. (See the characteristic in Figure 14–19b.)

EXAMPLE 14.9 Noise in the signal v_1 requires that hysteresis be added to the circuit in Figure 14–18a.

Given: The peak-to-peak noise is 1 V. The output is to go to -14 V when the average of v_1 exceeds 5 V, that is, when v_1 exceeds 5 V $+ (1$ V$)/2 = 5.5$ V.

Find: a Schmitt trigger circuit design with the minimum hysteresis to avoid chatter due to the noise. Compare the performance of the original circuit in Figure 14–18a with that of the new design.

Solution: The hysteresis must be at least as wide as the peak-to-peak noise. Therefore the minimum hysteresis is the 1-V "window" shown in Figure 14–20b. The output v_0 goes to -14 V when v_1 exceeds 5.5 V, as desired, and v_0 goes to 14 V when v_1 falls back past 4.5 V. The circuit in Figure 14–20a realizes this characteristic if it meets two conditions: $v_r = 5.5$ when $v_0 = 14$, and $v_r = 4.5$ when $v_0 = -14$. Using superposition and voltage division, v_r is determined by

$$v_r = \frac{R_1}{R_1 + R_2} v_0 + \frac{R_2}{R_1 + R_2} V_x.$$

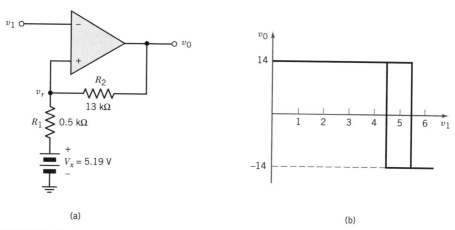

(a)

(b)

FIGURE 14–20 Schmitt trigger solution to the design problem in Example 14.9. Hysteresis has been added to the comparator in Figure 14–18 to avoid "chatter" due to noise.

For the two conditions, this gives us two equations in three unknowns: R_1, R_2, and V_x. Then we have a free pick of a third equation; we choose $R_1 + R_2 = 14$ kΩ. Solving these three equations gives $R_1 = 0.5$ kΩ, $R_2 = 13.5$ kΩ, and $V_x = 5.19$ V.

In Figure 14–21 is an example of a noisy v_1 whose average increases with time. For the circuit in Figure 14–18a without hysteresis, the output v_o chatters as v_1 passes through the 5-V reference, as shown in Figure 14–21a. The performance of the circuit with hysteresis is shown in Figure 14–21b. The output switches when v_1 exceeds 5.5 V and can't switch again until v_1 falls below 4.5 V.

Oscillator

An oscillator is a circuit that produces a periodic waveform such as a sine wave or a square wave. Oscillators find many applications such as tone generators for electronic music, clocks for computers or timers, carriers for communication systems, and switching control for power circuits.

We can modify the Schmitt trigger in Figure 14–19 to make an oscillator to generate a square wave. Consider the *astable multivibrator* in Figure 14–22a. As with the Schmitt trigger, the reference voltage keeps changing due to positive feedback. But there is also negative feedback that produces v_1 by charging a capacitor. As shown in Figure 14–22b, v_1 no sooner reaches the 1-V reference v_r than the reference switches to -1 V. When v_1 finally reaches -1 V, the reference goes to 1 V, and the cycle is repeated. During this cycle v_0 is switching from 14 V to -14 V and back, producing a square wave.

In order to design for a specific frequency, we need to be quantitative about the rate at which v_1 changes. Since v_1 stays close to zero compared with v_0, the voltage $v_0 - v_1$ across R is either about 14 V or about -14 V. Correspondingly the capacitor current i is either $14/R$ or $-14/R$. From the v–i relationship for a ca-

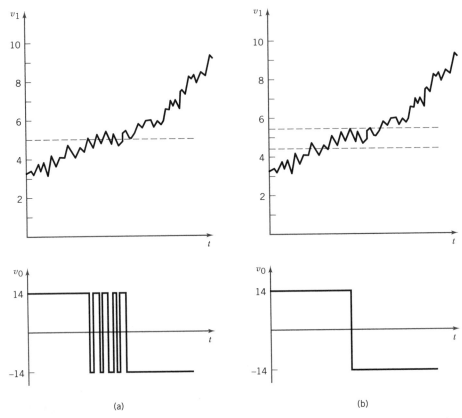

(a)   (b)

FIGURE 14–21 (a) Response of the comparator in Figure 14–18 to a noisy input. Note the chattering of v_0 as v_1 crosses the threshold. (b) Response of the Schmitt trigger in Figure 14–20 to the same input: v_0 switches as v_1 crosses the upper threshold and won't switch again until v_1 crosses the lower threshold.

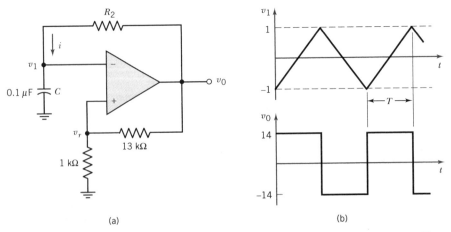

(a)   (b)

FIGURE 14–22 Astable multivibrator—an oscillator that produces a square wave. The frequency depends on the value of the RC product. For $R = 35$ kΩ, $f = 1/2T = 1.0$ kHz.

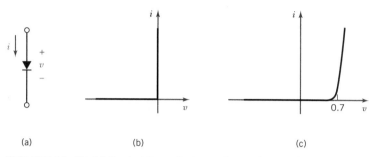

(a)   (b)   (c)

FIGURE 14–23 (a) Symbol for a diode—a device that conducts in only one direction. (b) The v–i characteristic of an ideal diode. (c) The v–i characteristic of a typical diode. The voltage when conducting is close to 0.7 V rather than zero.

pacitor, this gives a rate of change $dv_1/dt = i/C = \pm 14/RC$. The length of time for v_1 to rise 2 V (a half cycle) is $T = 2/(dv_1/dt) = RC/7$. For example, $f = 1/2T = 1$ kHz requires $T = 0.5$ ms and $RC = 3.5$ ms. Then for $C = 0.1$ μF, $R = 35$ kΩ.

Ideal Diode

The diode is a nonlinear element that conducts current in only one direction. It was introduced in Section 3.9, and Chapter 16 looks at some diode applications such as peak detection and rectification. We look briefly at diodes here to show how an op amp can be used to make a diode's characteristic more nearly ideal.

The symbol for a diode is shown in Figure 14–23a. Ideally it has no voltage drop across it when it is conducting (in the direction of the arrow). This is indicated by the v–i characteristic with $v = 0$ for $i > 0$ in Figure 14–23b. When the voltage is reversed, there is no conduction: $i = 0$ for $v < 0$. Actual diodes have about 0.7 V drop when conducting rather than the desired ideal of zero volts. (See the v–i characteristic in Figure 14–23c.) By using a diode together with an op amp as in Figure 14–24a, a more nearly ideal diode characteristic can be realized, as shown

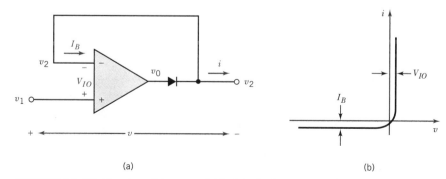

(a)   (b)

FIGURE 14–24 Circuit realizing a nearly ideal diode characteristic. For v positive, the circuit is a voltage follower. For v negative, the output is an open circuit.

in Figure 14–24b. We will assume that the op amp has the specifications $V_{IO} = 5$ mV and $I_B = 0.1$ μA.

When $v < 0$ (v_1 below v_2), the op amp acts like a comparator, and v_0 limits at about -14 V. Assuming that -14 V is below v_2, this puts a reverse voltage across the diode, and it doesn't conduct. Then $i = -I_B = -0.1$ μA (the horizontal line in the v–i characteristic of Figure 14–24b).

When $i > 0$, the diode conducts, and the op amp acts like a voltage follower (compare Figure 14–5c). Therefore $v_2 = v_1 - V_{IO} = v_1 - 5$ mV, and $v = 5$ mV (the vertical line in Figure 14–24b). Because of the diode's drop, v_0 is about 0.7 V above v_2, but this is incidental. The resulting v–i characteristic is much closer to the ideal than that of a diode alone. With a high-performance op amp it can be made even closer by reducing V_{IO} and I_B.

14.11   SUMMARY

An op amp is a high-gain differential amplifier that is available as an integrated circuit. Its linear applications involve negative feedback, and as a result the voltage at its input is held at about zero. The noninverting configuration has a high input impedance, and it doesn't load down a signal source driving it. The inverting configuration has a lower input impedance, but it can provide a wider variety of transfer functions.

Some of the linear applications of an op amp are amplifiers, summers, integrators, differentiators, and filters. Some nonlinear applications are comparators, Schmitt triggers, oscillators, and ideal diodes. These involve positive feedback or no feedback.

Ideally we would like an op amp to have infinite gain, infinite bandwidth, zero input current, zero offset voltage, and an unlimited range of input and output voltages. In practice a general-purpose op amp has a dc gain of about 100,000, and its gain rolls off so that it reaches unity at about 1 MHz. An amplifier built with such an op amp has a gain-bandwidth product of 1 MHz. The input bias current is typically 0.1 μA, and the input offset voltage is typically 5 mV. These cause significant output offset if they are nearly as large as the input signal. With supply voltages of $+15$ V and -15 V, the input and output voltages are restricted to lie between about $+13$ V and -13 V.

FOR FURTHER STUDY

R. G. Irvine, *Operational Amplifier Characteristics and Applications*, Prentice-Hall, Englewood Cliffs, NJ, 1981.

W. G. Jung, *IC Op Amp Cookbook*, Howard W. Sams, Indianapolis (ITT), 1974.

A. P. Malvino, *Electronic Principles*, McGraw-Hill, New York, 1984.

P. R. Gray and R. G. Meyer, *Analysis and Design of Analog Integrated Circuits*, Wiley, New York, 1977.

PROBLEMS

Easy Drill Problems (answers at end of chapter)

D14.1. **(a)** Using an ideal op amp, design an amplifier with a gain of -10 and an input impedance of 10 kΩ.
   (b) Design an amplifier with a gain of 10. What is the input impedance?

D14.2. Using an ideal op amp, design a differential amplifier with a gain of 100; that is, $v_0 = 100(v_1 - v_2)$.

D14.3. Prove the result of D14.2 by superposition. First set $v_1 = 0$, effectively grounding the noninverting input terminal, and analyze the inverting configuration. Then set $v_2 = 0$, and analyze the noninverting configuration (fed by a voltage divider).

D14.4. Sketch a Bode plot of the response of the high-pass filter shown in Figure D14.4.

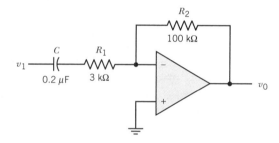

D14.5. Suppose the op amp used in D14.1 has a gain $|\mathbf{A}|$ that rolls off to unity at 1.0 MHz. What is the 3-dB bandwidth of the amplifier in part **(a)** and in part **(b)**?

D14.6. Choose a new value of R_1 in the differentiator in Figure 14–15c so the break in the ideal $|\mathbf{V}_0/\mathbf{V}_1|$ curve occurs at 10 kHz (where it crosses $|\mathbf{A}|$).

D14.7. **(a)** The op amp in Figure P14.9 has an input offset voltage $V_{IO} = 3$ mV. Find the resulting output offset voltage at v_0. (Since the offset is dc, the capacitor acts as an open circuit.)
   (b) Repeat for the circuit in Figure D14.4.

D14.8. **(a)** The op amp in Figure P14.9 has an input bias current $I_B = 0.2$ μA. Find the resulting output offset voltage at v_0. Assume the input offset voltage is zero. (Since the bias current is dc, the capacitor acts as an open circuit.)
   (b) Repeat for the circuit in Figure D14.4.

D14.9. Design a Schmitt trigger with the transfer characteristic shown in Figure D14.9.

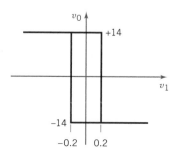

D14.10. For the oscillator in Figure 14–22, select R for a frequency of 200 Hz.

D14.11. For the component values in Figure 14–12b, find the impedance $\mathbf{Z}_1$ at the op amp input and the impedance $\mathbf{Z}_2$ in the feedback path. Using Eq. (14.5), show the transfer function $\mathbf{V}_0/\mathbf{V}_1$ is that given in Example 14.5.

Application Problems

P14.1. Use an op amp and potentiometers (variable resistors) to design a mixing amplifier with output $v_0 = -c_1 v_1 - c_2 v_2 - c_3 v_3$, where each c can be varied from 1 to 20 independent of the others. Give the necessary resistance range for the potentiometers.

P14.2. Use a differential configuration to shift an input voltage v_1 by 5 V; that is, $v_0 = v_1 + 5$ V. Assume $+15$ V, -15 V, $+5$ V, and -5 V supply voltages are available. Repeat the design without the $+5$V and -5V supply voltages (Use a resistor divider to obtain a lower dc voltage, but take its Thevenin impedance into account.)

P14.3. Using three op amps, design a circuit to amplify the voltage difference between two electrodes pasted to someone's chest. The gain is to be 100, and the input impedance seen by both electrodes is to be very high. (This arrangement is sometimes called an "instrumentation amplifier.")

P14.4. The reverse-biased photodiode in Figure P14.4 passes 0.5 μA of current i_d for each 1 μW of light falling on it. Express v_0 in terms of the input light power p.

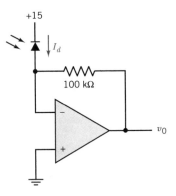

P14.5. Design an integrator to produce a voltage v_0 that rises at 2 V/s, given the input v_1 is a -15-V step (see Figure P14.5). The input impedance of the circuit is to be 100 kΩ.

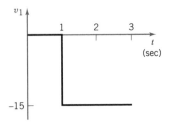
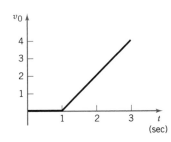

P14.6. Design a low-pass filter with a gain of 30 at low frequencies and a cutoff at 3 kHz (see Figure P14.6). The input impedance is to be 10 kΩ.

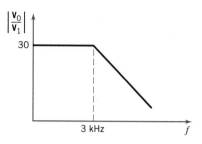

P14.7. A feedback control system needs a compensation filter. Design a filter to have the response shown in Figure P14.7. (Compare the response in Figure 10–14b.) Make the resistors large enough that the capacitor is only 100 μF.

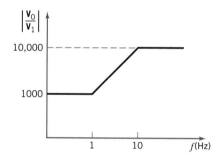

P14.8. Suppose the op amp used in P14.7 has unity gain at 1.0 MHz. Where will the actual filter response start to roll off (beyond 10 Hz)? Sketch the actual response out to 1 kHz.

P14.9. For the "bass-boost" filter shown in Figure P14.9, sketch the response from 10 Hz to 100 kHz. Take into account the op amp gain rolloff to unity gain at 1.0 MHz.

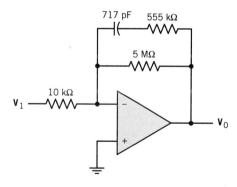

P14.10. Design a differentiator to convert the triangular waveform v_1 in Figure P14.10 (on p. 476) into the square wave v_0 shown. Use a 0.1-μF capacitor. Use a stabilizing resistor R_1 (as in Figure 14–15c) so the ideal frequency response breaks just as it crosses the op amp response $|\mathbf{A}|$. The unity-gain frequency of the op amp is 1.0 MHz.

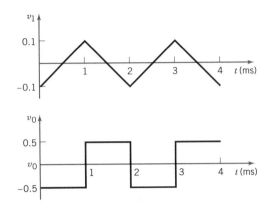

P14.11. A temperature transducer produces 410 mV corresponding to 70°F and 530 mV corresponding to 75°F. Design a Schmitt trigger to switch when the temperature increases to 75°F and switch back when the temperature falls to 70°F.

P14.12. Design an oscillator to generate the 500-Hz triangle wave shown in Figure P14.10. Use a 0.1-μF capacitor.

P14.13. The circuits in Figures P14.13a and P14.13b represent the circuits seen by the inverting input terminals in Figures 14–17a and b. Find the Thevenin equivalents looking into x in each case. (Use superposition to find the Thevenin voltage, setting first v_0 to zero and then either I_B or V_{eq} to zero.) If the Thevenin equivalents are to be the same, what must be the relationship between V_{eq} and I_B? Does this agree with Equation (14.7)?

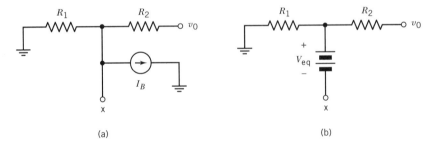

(a)   (b)

Extension Problems

E14.1. For the *half-wave rectifier* circuit in Figure E14.1, the input voltage is $v_1 = 2.0 \sin 2\pi f t$, where $f = 500$ Hz. Sketch the waveforms for v_1, v_2, and v_0. The op amp output v_0 is limited to lie between $+14$ V and -14 V.

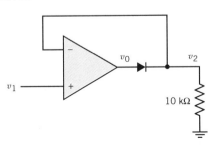

E14.2. The amplitude of any actual op amp output is limited. The *rate* at which the output changes is also limited. For a general-purpose op amp, the maximum rate of change, called the *slew rate*, is typically 0.5 V/μs (both rising and falling). Suppose the output is trying to produce a 10-V peak-to-peak square wave at 10 kHz. Sketch the actual waveform, distorted by the slew-rate limit. What is the largest-amplitude *sine wave* at 10 kHz that the op amp can produce without distortion?

E14.3. The *analog-to-digital converter* (ADC) in Figure E14.3 converts a negative voltage v_{in} to a binary number, where $v_1 = 5$ V is a "1" and $v_1 = 0$ V is a "0" for the least significant bit. The circuit incorporates a digital-to-analog converter whose output v_0 is -1 V times the binary number at its input (see Figure 14-4). The 1-MHz oscillator causes the counter (see Figure 9–32) to count upward in binary, and v_0 correspondingly decreases. When v_0 just passes v_{in}, the op amp comparator turns off the AND gate, and the counter holds that binary number. The process begins with a Reset signal that makes the binary count go to zero. What binary number will be held for $v_{in} = -9.3$ V? How long does the conversion take? Sketch a timing diagram of all voltage waveforms.

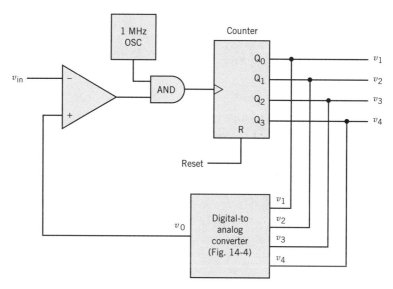

Study Questions

14.1. What three parameters characterize an ideal op amp?

14.2. When an op amp is used in a linear mode (with negative feedback), what is the voltage across its input terminals ideally? In practice?

14.3. What is the advantage of making an amplifier with the noninverting configuration rather than the inverting configuration?

14.4 What is the minimum gain you can design a noninverting configuration to have?

14.5 Which configuration is used to sum two voltages? Which configuration is used to subtract one voltage from another?

14.6. If a capacitor by itself integrates current, why is an op amp usually used in making an integrator?

14.7. In designing a filter, we are concerned with frequencies for which a capacitor is essentially an open circuit or essentially a closed circuit. What is the impedance of the capacitor compared to in determining this?

14.8. Name four ways in which an op amp differs from the ideal.

14.9. When the design of an amplifier (using an op amp) is changed to double the gain, what happens to the bandwidth?

14.10. What determines whether I_B or V_{IO} will have the greater effect in an op amp circuit?

14.11. What is the advantage of a Schmitt trigger over a comparator? What is its disadvantage?

14.12. A diode ideally has an ''on'' voltage of zero and an ''off'' current of zero. An op amp can help a diode more nearly realize one of these parameters. Which one? By how much is the other parameter degraded?

ANSWERS TO DRILL PROBLEMS

D14.1. Figure 14–2a, $R_1 = 10$ kΩ, $R_2 = 100$ kΩ; Figure 14–5a, $R_2 = 9R_1$, $R_{in} = \infty$

D14.2. Figure 14–7a, $R_2 = 100R_1$

D14.4. $\mathbf{V}_0/\mathbf{V}_1 = 33.3$ for $f > 265$ Hz

D14.5. $B_{3dB} = 100$ kHz

D14.6. 160 Ω

D14.7. (a) 1.503 V, (b) 3 mV

D14.8. (a) 1.0 V, (b) 20 mV

D14.9. Figure 14–19, $R_2 = 69R_1$

D14.10. 175 kΩ

CHAPTER 15

Transistors as Amplifiers

A transistor is an *active* circuit element, meaning that it can amplify the power of a signal. None of the elements introduced in the first five chapters—resistors, capacitors, inductors, and transformers—can do this. (While a step-up transformer does amplify the voltage of a signal, at the same time it reduces the current, and the power remains the same.) These are *passive* elements; they can't amplify power.

Active elements are necessary to build the amplifiers discussed in Chapter 8. Before 1960 the active elements were vacuum tubes; today tubes have been almost entirely replaced by transistors. Because of their small size (about 0.1 mm across), transistors have made possible very complex functions in small integrated circuits (ICs) such as the digital ICs of Chapter 13 and the op amps of Chapter 14. Also, individual transistors are available in packages like those pictured in Figure 15–1. In this chapter transistors are used as linear elements in amplifiers (analog circuits). In Chapter 16 we will look at their application as switches in nonlinear circuits (including logic circuits).

A transistor has three terminals, as shown in Figure 15–2a. The basic function of a transistor is to use small changes in the voltage and current at one terminal to induce a large change in the current through the other two terminals. This is analogous to a hydraulic lift, where a small force and a small motion of a valve lever control a large change in fluid flow (see Figure 15–2b). The fluid flow, in turn,

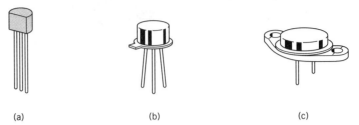

(a)   (b)   (c)

FIGURE 15–1 Transistor packages: (a) plastic package, (b) hermetically sealed metal can, (c) power transistor.

provides a large force to lift the load. In this analogy, the valve acts very much like a transistor.*

There are two classes of transistors—bipolar transistors (usually referred to simply as transistors) and field-effect transistors (FETs)—which are based on different physical principles. These physical principles are described briefly in Section 15.10. Most of this chapter is devoted to circuits using one or two bipolar transistors. This reflects the more common use of bipolar transistors and the fact that they are easier to design with. Of the bipolar transistors, the most popular are the *npn* types made of silicon. Therefore a treatment of *pnp* types is left to the extension problems at the end of the chapter.

The reader's objectives in this chapter are to:

1. Be able to model transistors by resistors, capacitors, and dependent current sources.

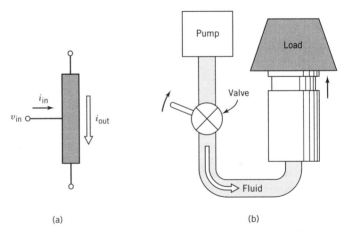

(a)   (b)

FIGURE 15–2 (a) Transistor action. Small input voltage and current control a large output current. (b) Hydraulic lift. The valve is analogous to the transistor.

---

*In fact, a vacuum tube is called a "valve" in England.

2. Design simple transistor amplifier circuits.

3. Be able to model simple transistor amplifier circuits with the two-port model (input impedance, output impedance, and voltage gain) introduced in Chapter 8.

4. Analyze the high- and low-frequency cutoffs of simple transistor amplifier circuits.

In these objectives we can see the top-down structure of engineering. In Chapter 8 we designed instrumentation systems based on amplifiers. We didn't know details of how the amplifiers were constructed; instead we had a simple model that summarized their behavior as seen from the outside. Now in this chapter we will design amplifiers based on transistors. We won't have to know the details of how transistors work; instead we will learn a simple model that summarizes their behavior as seen from the outside. Finally the last section gives a nonnumerical description of how transistors work.

15.1   BIPOLAR TRANSISTOR MODEL

Figure 15–3a shows the symbol for an *npn* bipolar transistor. The three terminals are called collector (C), base (B), and emitter (E). The arrow on the emitter shows the direction in which current can flow—out of the terminal. This symbol is used to represent the transistor in a circuit schematic, but the behavior of the transistor can be understood better if the symbol is replaced by a model consisting of a few basic elements—resistors, capacitors, inductors, and sources. Some advanced models for transistors have as many as 15 elements, but the 4-element model in Figure 15–3b is adequate for most applications. We begin with an even simpler model—the *ideal transistor* model in Figure 15–3c, which reduces the model in Figure 15–3b

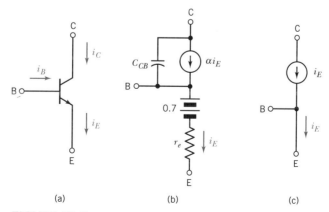

FIGURE 15–3 (a) Symbol for *npn* bipolar transistor. (b) A transistor model called the T model. (c) Simplified transistor model. This models the ideal behavior of a transistor.

to just one element. Even such a stripped-down model is adequate for many applications. As we take into account subtler properties of the transistor, we will increase the complexity of the model until we have the model in Figure 15–3b.

The ideal transistor model has one element—a *dependent current source*, which depends on the value of another current or voltage. In this case the dependent source causes the collector current i_C to be equal to the emitter current i_E. The reader should confirm by KCL that this implies that the current i_B into the base terminal must be zero. It is also worth noting that in this ideal model the voltage from base to emitter is zero.

15.2   COMMON-COLLECTOR CONFIGURATION

One frequent application of a transistor is to handle the current that a voltage source by itself would otherwise have to provide. For example, the unaided voltage source v_1 in Figure 15–4 must provide a peak current of $i_{max} = v_{1max}/R_L = 10 \text{ V}/100 \text{ }\Omega = 100 \text{ mA}$ to the load R_L. In Figure 15–5a a transistor is inserted between the source and the load. (It is customary with electronic circuits to not show the dc voltage source providing the $+15 \text{ V}$ to the collector.) Because the collector is connected to a dc voltage (*ac ground*), it serves as a common terminal for the input port and the output port. Therefore the transistor is said to be in the *common-collector configuration*.

In Figure 15–5b the transistor has been replaced by the ideal transistor model. Because of the zero voltage drop from base to emitter, the load voltage v_2 is still the same as the source v_1.* Therefore a common-collector amplifier doesn't amplify voltage; its voltage gain is unity. The useful feature is that the source v_1 no longer has to supply any current (ideally) since the base current is zero for the ideal transistor model. All the load current (emitter current here) is provided by the collector. The power gain of the transistor is now evident; the source provides zero power, while the load dissipates significant power—about 580 mW. A more complete tran-

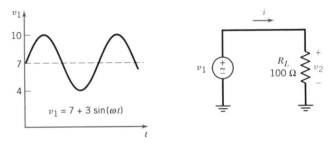

FIGURE 15–4 An unaided voltage source driving a load R_L. The source must provide all the current drawn by the load.

---

*The common-collector configuration is sometimes called an *emitter follower* because the emitter voltage follows the base voltage.

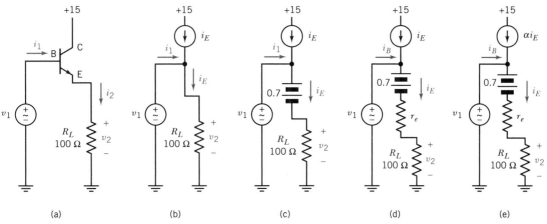

FIGURE 15–5 (a) Common-collector transistor helping the source drive the load. (b) Transistor replaced by simple (ideal) transistor model. Here $v_2 = v_1$ as in Figure 15–4, but the source handles no current. (c) $V_{BE} = 0.7$ V in the transistor model shows that actually $v_2 = v_1 - 0.7$ V. (d) Emitter resistance r_e in the transistor model shows that the ac voltage gain is not quite unity. (e) Current gain α (typically 0.99) shows that the transistor provides most of the load current, but the source v_1 must handle a small part.

sistor model will show that the source must actually provide a small amount of current.

Two rules must be followed in using an *npn* transistor in any linear application:

$$v_B \leq v_C, \qquad i_E > 0, \tag{15.1}$$

where v_B is the base voltage and v_C is the collector voltage with respect to ground. If these conditions are satisfied, the transistor is said to be in its *active state* of operation, and our transistor models are valid. For the circuit in Figure 15–5b, the first condition requires $v_1 \leq 15$ V, and the second requires $v_1 > 0$. Note that it was necessary that v_1 include a dc component as well as the sinusoid to meet these conditions (see waveform in Figure 15–4).

Base-Emitter Drop V_{BE}

One feature of the ideal transistor model is that the voltage drop V_{BE} from base to emitter is zero. In a real transistor there is a small, nearly constant drop of about 0.7 V. This can be accounted for in the circuit analysis by augmenting the transistor model as in Figure 15–5c. The battery symbol in the model indicates a base-emitter voltage drop $V_{BE} = 0.7$ V. The actual voltage may vary from 0.5 V to 0.9 V in different applications, but this small difference from 0.7 V can usually be ignored.

The relationship between input and output voltages for the common-collector configuration is now

$$v_2 = v_1 - 0.7 \text{ V}. \tag{15.2}$$

We will see in Section 15.6 that the dc portion of the signals is usually blocked by capacitors, and a dc *level shift* such as this makes little difference in circuit performance. What is of more interest is the *incremental voltage gain* or *ac voltage gain* defined by

$$A_v \equiv \Delta v_2 / \Delta v_1, \tag{15.3}$$

where Δv_1 is a small increment in v_1 and Δv_2 is the corresponding increment in v_2. Differentiating Equation (15.2), we find the ac gain to be $A_v = 1$.

Emitter Resistance r_e

To a first-order approximation the base-emitter voltage drop v_{BE} is a constant. In fact the voltage depends on the emitter current i_E as shown in Figure 15–6. The actual curve is an exponential, but a good second-order approximation is the straight line shown fitted to the actual curve. The best fit is to make the line tangent to the curve at $i_E = I_E$, where I_E is the current on which the transistor operation is centered. The equation for this straight-line approximation is

$$v_{BE} = 0.7 + i_E r_e,$$

where r_e depends on the slope of the line. The last term of this equation is accounted for by an emitter resistance r_e in the transistor model (see Figure 15–5d). It is shown in Section 15.10 that

$$r_e = 26 \text{ mV} / I_E, \tag{15.4}$$

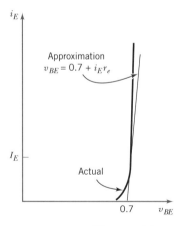

FIGURE 15–6 The actual base-emitter characteristic is an exponential. A linear approximation to this is a straight line tangent to the exponential at $i_E = I_E$ (the average of i_E).

where I_E is the average emitter current. The 26 mV is a universal constant for semiconductors at room temperature, so Equation (15.4) holds for all bipolar transistors—silicon or germanium, *npn* or *pnp*.

What is the effect of r_e on circuit performance? Analyzing the circuit with the more complete model in Figure 15–5d, we find

$$v_2 = \frac{R_L}{R_L + r_e}\,(v_1 - 0.7). \tag{15.5}$$

From this it follows that the ac voltage gain for the common-collector configuration is actually a little less than unity:

$$A_v \equiv \frac{\Delta v_2}{\Delta v_1} = \frac{R_L}{R_L + r_e} \approx 1. \tag{15.6}$$

To see that A_v is still close to unity, we need to evaluate r_e. The waveform for v_1 in Figure 15–4 has an average of 7 V. Then, neglecting the drop across r_e in Figure 15–5d, the average voltage across R_L is $7 - 0.7 = 6.3$ V, and the average emitter current is $I_E = 6.3\text{ V}/100\ \Omega = 63$ mA. From Equation (15.2), $r_e = 26\text{ mV}/63\text{ mA} = 0.41\ \Omega$. Then with $R_L = 100\ \Omega$, Equation (15.6) gives the voltage gain $A_v = 100/100.41 = 0.996$, which is certainly very close to unity. Therefore r_e can usually be ignored for the common-collector configuration, and the voltage gain is almost always taken as unity.

Current Ratios α and β

The ratio of a transistor's collector current to its emitter current is defined as α:

$$\alpha \equiv i_C/i_E. \tag{15.7}$$

In the ideal transistor model (and in the models in Figure 15–5c and Figure 15–5d), α is unity because the dependent current source in the collector equals i_E. A more accurate model would reflect the fact that i_C is actually about 99% of I_E, and the remaining 1% comes from the base. Therefore α is about 0.99. This is incorporated in the transistor model by making the dependent current source be αi_E, as in Figure 15–5e.

Although α doesn't vary much from transistor to transistor, a small change in α has a significant effect on the base current. To make this more evident, another parameter β is defined as the ratio of the collector current to the base current:

$$\beta \equiv i_C/i_B \quad \text{or} \quad i_B = i_C/\beta. \tag{15.8}$$

A little algebra shows that

$$\beta = \alpha/(1 - \alpha) \quad \text{and} \quad \alpha = \beta/(\beta + 1). \tag{15.9}$$

A typical β is 100, corresponding to $\alpha = 0.99$. But β can be as low as 25 or as high as 300. The corresponding range of α is much smaller—from 0.96 to 0.997. When a manufacturer gives the properties of a transistor, it is β rather than α that is specified, and α is found by means of Equation (15.9). In circuit calculations we usually use β, finding the base current by dividing the collector current by β. Since the collector and emitter currents are the same within a percent or two, we will find i_B from either Equation (15.8) or

$$i_B \approx i_E/\beta. \tag{15.8'}$$

In Figure 15–5e the common-collector configuration is analyzed using the transistor model that includes α. Equation (15.5) still holds; the only part of the analysis that is changed is that i_B is no longer zero. Suppose $\beta = 100$, corresponding to an α of 0.99. Then from Equation (15.8'), $i_B = i_E/100$. Since $i_1 = i_B$ and $i_2 = i_E$, the input and output currents are related by

$$i_1 = i_2/\beta = i_2/100. \tag{15.10}$$

For example, the peak value of i_2 is $(v_{1\max} - V_{BE})/R_L = (10 - 0.7)/100 \ \Omega = 93$ mA. Then the peak input current is $93 \ \text{mA}/100 = 0.93$ mA. So the source v_1 does have to handle some current, but it is only a hundredth of what it would have to handle if the transistor were not helping.

The ac current gain is defined as

$$A_i \equiv \Delta i_2/\Delta i_1, \tag{15.11}$$

where Δi_2 is an increment in the output current and Δi_1 is the corresponding increment in the input current. From Equation (15.10), this gain is

$$A_i = \beta = 100 \tag{15.12}$$

for this common-collector configuration. This gain is almost always used in reverse—finding i_1 by dividing i_2 by A_i. On rare occasions a current source is applied to the input of a common-collector configuration, and A_i is used in the forward direction. These occasions must always involve feedback because β varies considerably from transistor to transistor.

dc Circuit Analysis

The transistor model in Figure 15–5e is fairly complete. Our circuit analysis of that figure gives correct results, but engineers seldom do the analysis that way. Instead it is done in two parts, once with only dc sources present, and once with ac sources. The complete solution is the sum of the two parts. This approach, based on superposition, is possible because the transistor model is linear in the active state [see Equation (15.1)].

For the dc circuit analysis, all ac sources and the ac portion of sources are set to zero. In Figure 15–5e we had the source $v_1 = 7 + 3 \sin(\omega t)$. The second term here is the only ac term in the circuit. So we set it to zero and find the response of the circuit to $V_1 = 7$ V (the dc component of v_1). By convention,

The dc components of voltage and current are labeled with uppercase letters.

Since any dc drop across r_e is negligible compared with the 0.7 V, r_e can be eliminated from the transistor model for the purposes of the dc analysis. The resulting model (see Figure 15–7a) is called the *dc transistor model*.

To find the dc voltages and currents in the circuit of Figure 15–5a, we analyze the *dc circuit model* in Figure 15–7b, which has all ac sources set to zero and uses the dc transistor model. The output voltage is $V_2 = 7 - 0.7 = 6.3$ V. The output current I_2 equals the emitter current $I_E = V_2/R_L = 6.3$ V/100 Ω = 63 mA. The input current is $I_1 = I_E/\beta = 63$ mA/100 = 0.63 mA. These dc voltages and currents, when the ac source is set to zero, are called the *quiescent* or *bias* voltages and currents.

ac Circuit Analysis

The second part of the circuit analysis is to set all dc sources to zero and find the response to the ac sources. Setting the 0.7-V dc source in the transistor model to zero leaves the *ac transistor model* shown in Figure 15–8a. Setting the dc portion of v_1 to zero leaves $v_1' = 3 \sin(\omega t)$. Setting the +15-V dc supply to zero effectively grounds the collector.

The result of eliminating all dc sources is the *ac circuit model* in Figure 15–8b. By convention,

The ac components of voltage and current are given lowercase subscripts,

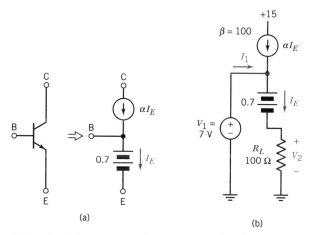

(a)   (b)

FIGURE 15–7 (a) *npn* bipolar transistor and its dc transistor model. (b) dc circuit model of the common-collector configuration in Figure 15–5a. The ac portion of v_1 is set to zero. Analysis of this circuit finds the bias (or quiescent) voltages and currents.

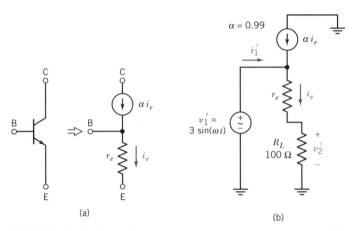

FIGURE 15–8 (a) *npn* bipolar transistor and its ac transistor model. (b) ac circuit model of the common-collector configuration in Figure 15–5a. All dc voltages and currents are set to zero. Analysis of this circuit finds the ac current gain and ac voltage gain of the configuration.

as with i_e. In the case of numerical subscripts, we have added a prime to indicate the ac component, as with v'_1. The value of r_e is obtained from Equation (15.4): $r_e = 26$ mV$/I_E = 26$ mV$/63$ mA $= 0.41$ Ω, where I_E was found in the dc circuit analysis. Analyzing the circuit in Figure 15–8b for the ac voltage gain, we have

$$A_v \equiv \frac{v'_2}{v'_1} = \frac{R_L}{R_L + r_e} = 0.996 \approx 1.$$

Comparing this with Equation (15.6), we see that the ac analysis gives the same result as the definition using incremental voltages.

The ac transistor model in Figure 15–8b shows that α (or 99% in our case) of i_e comes from the collector. Then $1 - \alpha$ (or 1% in our case) of i_e comes from the base. This calculation is usually done using β, as in Equation (15.8′): $i_b = i_e/\beta = i_e/100$. (Some transistor models show this β relationship explicitly, but they obscure other insights. It is not difficult to use the model in Figure 15–8a together with the knowledge that $i_b = i_e/\beta$.) Since $i'_1 = i_b$ and $i'_2 = i_e$, the ac current gain is

$$A_i \equiv i'_2/i'_1 = \beta = 100.$$

Comparing this with Equation (15.12), we see again that the ac analysis gives the same result as the definition using incremental currents.

If the dc circuit analysis in Figure 15–7b and the ac circuit analysis in Figure 15–8b give the same results as the analysis of the circuit model in Figure 15–5e, why make two problems out of one? Because each of the two problems becomes much simpler. The dc analysis gets the less interesting dc parameters out of the way so the engineer can concentrate on the real purpose of the circuit in the ac analysis.

The dc analysis has two objectives: it determines I_E so r_e can be found for the ac analysis, and it checks that the transistor is properly biased to keep it operating in its active state.

Operation in the Active State

The linear transistor models we have been using are valid when the transistor is in the active state—when it can amplify power. Equation (15.1) gives the conditions for operation in the active state. Since these conditions apply to the *total* voltages and currents, we must sum the results of the dc circuit analysis and the ac circuit analysis.

From dc circuit analysis (Figure 15–7b), $V_B = V_1 = 7$ V, $V_C = 15$ V, and $I_E = 63$ mA. From ac circuit analysis (Figure 15–8b), $v_b = v_1'$, $= 3\sin(\omega t)$, $v_c = 0$ V, $v_2' = A_v v_1' = 2.99\sin(\omega t)$, and $i_e = v_2'/R_L = (29.9 \text{ mA})\sin(\omega t)$. Then the total voltages and currents are the sums of the dc and ac components:

$$v_B = V_B + v_b = 7 + 3\sin(\omega t) \leq 10 \text{ V},$$

$$v_C = V_C + v_c = 15 + 0 = 15 \text{ V},$$

$$i_E = I_E + i_e = 63 \text{ mA} + (29.9 \text{ mA})\sin(\omega t) \geq 33.1 \text{ mA}.$$

These clearly satisfy the requirements of Equation (15.1) that $v_B \leq v_C$ and $i_E > 0$. If these were not satisfied, the dc component of v_B would have to be lowered or raised to satisfy the first or second requirement, respectively.

EXAMPLE 15.1 A voltage regulator is to provide a constant dc voltage $v_L = 10$ V to a load R_L from a nominal 15-V supply voltage V_{CC}. The regulator comprises a 10-V reference V_r (which can supply very little current), an op amp, and a common-collector transistor (see Figure 15–9). The reader should draw a flow graph of this system, as is done in Chapter 10, to understand how the regulator operates to keep v_L nearly constant at 10 V.

Given:　The load R_L can vary from 20 Ω to 1000 Ω. The supply voltage V_{CC} can vary from 13 V to 16 V. The op amp can provide a maximum i_o of 20 mA.

Find:　the β necessary for the common-collector transistor to provide the needed current. Find the maximum power the transistor must dissipate.

Solution:　Since the load voltage is kept constant at $V_L = 10$ V, the maximum load current is $i_L = 500$ mA when $R_L = 20$ Ω. Neglecting the current into the op amp, $i_E \approx i_L$, and $i_B = i_E/\beta \approx i_L/\beta \leq 500 \text{ mA}/\beta$. The op amp must provide $i_o = i_B \leq 500 \text{ mA}/\beta$. Since we must keep $i_o \leq 20$ mA, this requires $\beta \geq \underline{25}$, which is a reasonable specification for a power transistor.

The power dissipated by a transistor is

$$p = (v_C - v_E)i_E, \tag{15.13}$$

which is a maximum here for $v_C = 16$ V, $v_E = 10$ V, and $i_E = 500$ mA. Then $p_{\max} = \underline{3 \text{ W}}$. Power transistors that handle 3 W have a case like that shown in

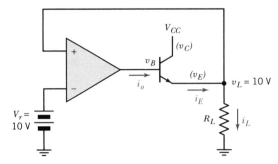

FIGURE 15-9 Voltage regulator circuit studied in Example 15.1. The common-collector transistor boosts the current-handling capability of the op amp. The circuit provides a nearly constant 10 V to the load R_L at up to 500 mA.

Figure 15–1c. Bolting this case to a metal chassis helps the transistor dissipate the power without excessive heat. A thin mica insulator is sometimes necessary to keep the case from shorting to ground.

We need to check that the base voltage is always below the collector voltage: $v_B = 10 + 0.7 < V_{CC} \geq 13$ V. Also, if the op amp is operated from V_{CC}, it requires that v_B be always at least 2 V below V_{CC}, which is satisfied here.   ■

15.3   COMMON-BASE CONFIGURATION

A transistor in the common-base configuration can help a current source deliver power to a load. Consider first the unaided current source i_1 in Figure 15–10. It is providing current to a 1-kΩ load resistor R_L, which happens to be connected to a dc voltage supply. Using KVL, we find the voltage across the current source to be $v_2 = 10 - 1$ kΩ $\times i_1$. For example, $v_2 = 9$ V when $i_1 = 1$ mA, and $v_2 = 3$ V when $i_1 = 7$ mA. Some current sources have trouble handling voltages or voltage variations as large as these.

In Figure 15–11a a transistor has been inserted between the source and the load. Because the base is connected to a dc voltage—is common to the input and output ports—the transistor is in the *common-base configuration*. In Figure 15–11b the transistor has been replaced by the ideal transistor model to illustrate the basic operation of this configuration. The current from the source i_1 passes straight through the transistor to the load, making $i_2 = i_1$. So the current gain A_i of the common-base configuration is unity (ideally). Unlike the unaided source in Figure 15–10, however, the voltage across the source i_1 is now fixed at only 2 V. Therefore the source i_1 still determines the load current, but it handles much less voltage and sees no voltage variation (ideally).

A more exact analysis involves the use of the dc transistor model in Figure 15–7a and the ac transistor model in Figure 15–8a. Solving first for the emitter bias

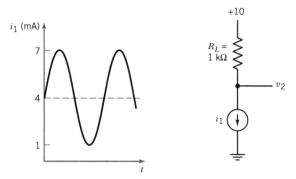

FIGURE 15-10 An unaided current source driving a load R_L. The source must provide all the voltage at the load.

current I_E, we use the dc circuit model in Figure 15–12b. The source i_1 has been replaced by its dc component $I_1 = 4$ mA (see waveform in Figure 15–10). But $I_E = I_1$; therefore $I_E = 4$ mA. The quiescent voltage across the source is $V_1 = 2.0 - 0.7 = 1.3$ V, and the quiescent collector voltage is $V_2 = 10 - \alpha I_E R_L = 10 - 0.99 \times 4$ mA $\times 1$ k$\Omega = 6.04$ V.

In the ac circuit model in Figure 15–12c, the source i_1 has been replaced by its ac component $i_1' = (3$ mA$) \sin(\omega t)$. The emitter resistance is $r_e = 26$ mV$/I_E = 6.5$ Ω. The output current is $i_2' = \alpha i_e = \alpha i_1'$. Therefore the ac current gain for a common-base configuration is

$$A_i \equiv i_2'/i_1' = \alpha \approx 1. \tag{15.14}$$

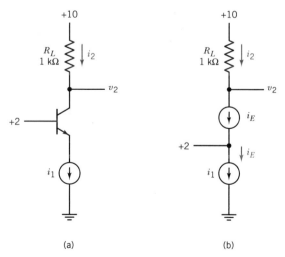

(a)   (b)

FIGURE 15-11 (a) Common-base transistor aiding the source i_1. (b) Transistor replaced by ideal transistor model. The same current is delivered to the load, but now the source provides a small, constant voltage.

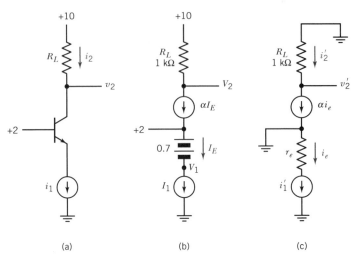

FIGURE 15–12 (a) Common-base configuration; (b) dc circuit model, I_1 is the dc component 4 mA of i_1; (c) ac circuit model, where i_1' is the ac component (3 mA)sin(ωt) of i_1.

The input voltage is $v_1' = -i_e r_e$, and the output voltage is $v_2' = -\alpha i_e R_L = -(2.97\,\text{V})\sin(\omega t)$. Therefore the ac voltage gain is

$$A_v \equiv v_2'/v_1' = \alpha R_L/r_e \approx R_L/r_e. \tag{15.15}$$

For our case, $R_L/r_e = 1000/6.5 = 154$. So the source has to handle only 1/154 of the ac load voltage—that is, about 0.04 V peak-to-peak here.

To check for operation in the active state, we find the total voltages and currents to see if Equation (15.1) is satisfied. First, $v_C = V_2 + v_2' = 6.04 - (2.97\,\text{V})\sin(\omega t) \geq 3.07\,\text{V}$, which is greater than $v_B = 2.0\,\text{V}$. Next, $i_E = I_1 + i_1' = 4\,\text{mA} + (3\,\text{mA})\sin(\omega t) \geq 1\,\text{mA}$, which is never negative.

15.4   COMMON-EMITTER CONFIGURATION

In the common-emitter configuration a transistor is able to provide at the same time both voltage gain and current gain greater than unity. Therefore it is the most popular configuration for amplification. The basic form of the configuration is shown in Figure 15–13a (disregard the capacitor for now). The input voltage is applied to the base, the output voltage is at the collector, and the emitter is grounded through a resistor R_E. We will analyze the specific case for $v_1 = 2\,\text{V} + (1.0\,\text{V})\sin(\omega t)$, $R_C = 3\,\text{k}\Omega$, $R_E = 1\,\text{k}\Omega$, and $\alpha = 0.99$.

The dc circuit model is shown in Figure 15–13b. The dc component of v_1 is $V_1 = 2\,\text{V}$. So $I_E = (2 - 0.7)/R_E = 1.3\,\text{mA}$, and the quiescent output voltage is $V_2 = 12\,\text{V} - \alpha I_E R_C = 8.14\,\text{V}$.

The ac circuit model is shown in Figure 15–13c (disregard the dashed line for now). The ac component of v_1 is $v_1' = (1.0 \text{ V}) \sin(\omega t)$. The ac voltage gain is found by the following line of analysis. The source v_1' produces a current through $r_e + R_E$, α of this current flows through the collector and produces a voltage drop across R_C, and the negative of this drop is the output v_2'. The reader should confirm that the resulting relationship is

$$A_v \equiv \frac{v_2'}{v_1'} = \alpha\frac{-R_C}{r_e + R_E} \approx \frac{-R_C}{r_e + R_E}. \tag{15.16}$$

Since $I_E = 1.3$ mA, $r_e = 26$ mV/1.3 mA $= 20$ Ω. Then for $R_C = 3$ kΩ and $R_E = 1$ kΩ, Equation (15.16) gives $A_v = 2.94$. For $v_1' = (1.0 \text{ V}) \sin(\omega t)$, the ac output component is $v_2' = -(2.94 \text{ V}) \sin(\omega t)$.

The reader should show that the transistor stays in the active state for the case we have been analyzing.

Emitter Bypass Capacitor

The gain of the common-emitter configuration in Figure 15–13 could be greatly increased by eliminating R_E. In that case Equation (15.16) would become

$$A_v = -\alpha R_C/r_e \approx -R_C/r_e. \tag{15.17}$$

For $R_C = 3$ kΩ and $r_e = 20$ Ω, this would give $A_v = 150$—a sizable gain! But there is a problem with completely eliminating R_E; a dc analysis with $R_E = 0$ shows that it would be very difficult to establish an I_E that is well determined. Instead, the solution is to short out R_E with a capacitor, as is shown dashed in Figure 15–13a.

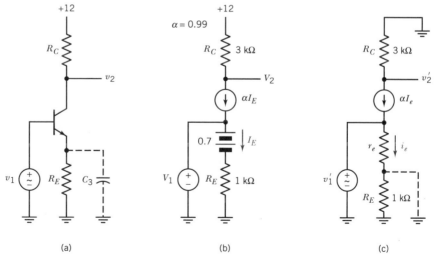

(a)   (b)   (c)

FIGURE 15–13 (a) Common-emitter configuration, which provides both voltage gain and current gain. The voltage gain is greatly increased by bypassing R_E with a large C_3. (b) The dc circuit model. (c) The ac circuit model. When C_3 is present, R_E is effectively shorted out.

The capacitor doesn't conduct dc current, so the dc analysis is the same as if there were no capacitor (see Figure 15–13b). Then $I_E = 1.3$ mA, and $r_e = 20\ \Omega$, as before.

For the ac analysis, the capacitor is such a low impedance that it can be considered a short circuit (see the dashed line in Figure 15–13c). This is true if the capacitor value C_3 is large enough that its impedance magnitude is much less than r_e for all signal frequencies; that is, $1/\omega C_3 << r_e$. Generally, a factor of 3 is sufficient, making the requirement

$$1/\omega C_3 \leq r_e/3. \tag{15.18}$$

Then R_E is effectively eliminated from the ac circuit, and Equation (15.17) holds.

With an ac voltage gain as great as 150, the amplitude of the ac input voltage must be reduced; otherwise the transistor will not stay in the active state. Both the conditions in Equation (15.1) can be seen in terms of the collector voltage v_2. The condition $i_E \geq 0$ corresponds to $v_2 \leq 12$ V, and $v_C \geq v_B$ corresponds to $v_2 \geq 2$ V. The dc analysis gave $V_2 = 8.14$ V, which is 3.86 V below 12 V and 6.14 V above 2 V. Then the maximum amplitude allowed for v_2' is 3.86 V, and the input v_1' is restricted to an amplitude of 3.86 V/150 = 26 mV. For example, the total input could be $v_1 = 2$ V + (26 mV) $\sin(\omega t)$. The reader should sketch the corresponding $v_2(t)$.

Biasing and ac Coupling

Signal sources usually don't have a dc component such as the 2 V in $v_1 = 2$ V + (26 mV) $\sin(\omega t)$. But if the dc component were missing, the condition $i_E > 0$ could not be met. The solution is to take an input signal such as $v_1' = $ (26 mV) $\sin(\omega t)$ and add a dc component to it by biasing and ac coupling, as in Figure 15–14a. We will show by superposition that this arrangement adds 2.0 V to v_1'. To make the analysis easier we first form the Thevenin equivalent of R_1, R_2, and the +12 supply as in Figure 15–14b. Recall that V_{Th} is the open-circuit voltage of the voltage divider, and R_{Th} is equal to R_1 in parallel with R_2.

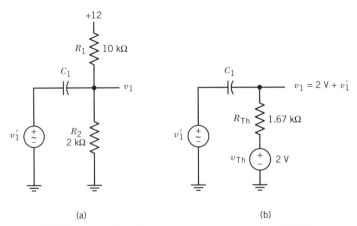

(a)   (b)

FIGURE 15–14 (a) Use of ac coupling and biasing to add 2 V dc to v_1'. (b) R_1, R_2, and the 12-V supply modeled by a Thevenin equivalent.

To find by superposition the v_1 produced, we first set $V_{Th} = 0$ and find the portion of v_1 due to v_1'. The *coupling capacitor* C_1 together with R_{Th} form a high-pass filter (see Section 6.5). If we choose C_1 large enough, the filter will pass all frequency components of v_1' (see Section 15.8), and $v_1 = v_1'$. Now set $v_1' = 0$ to find the dc portion of v_1 due to $V_{Th} = 2.0$ V. Since C_1 passes no dc current, there is no dc voltage drop across R_{Th}, and $v_1 = V_{Th} = 2.0$ V. Then by superposition, the total is

$$v_1 = 2.0 \text{ V} + v_1'. \tag{15.19}$$

This v_1 can be applied to the base of the common-emitter amplifier in Figure 15–13a, as shown in Figure 15–15. R_1, R_2, and R_E are called *biasing resistors* since they establish the emitter bias current I_E. Assuming i_B is so small it can be neglected, v_1 will remain that given in Equation (15.19). Since this v_1 is the same used in our analysis of the circuit in Figure 15–13a, the solution will be the same: $v_2 = 8.14$ V $- 150v_1'$. The ac portion $v_2' = -150v_1'$ is usually separated from the dc portion by a coupling capacitor C_2 at the output, as in Figure 15–15. This blocks the dc and, if C_2 is large enough, passes the ac portion unchanged (see Section 15.8).

The assumption was made that i_B is small enough to be negligible in the analysis. We will use the dc circuit model in Figure 15–16 to show that the dc component I_B has only a small effect on the biasing. Note that R_{Th} in Figure 15–16 models R_1 and R_2 in Figure 15–15. Assuming at first that $I_B = 0$, the drop across R_{Th} is zero, and $V_1 = 2$ V. $I_E = (V_1 - 0.7)/1$ k$\Omega = 1.3$ mA, and $I_B = I_E/\beta = 1.3$ mA$/100 = 13$ μA. Then, in fact, the drop across R_{Th} is 13 μA $\times$ 1.67 k$\Omega = 0.022$ V, which is negligible (about 1%) compared with $V_{Th} = 2.0$ V.

Note that if R_1 and R_2 in Figure 15–15 are chosen too large, the drop across the larger R_{Th} may not be negligible. A good rule of thumb is to keep R_{Th} less than twice R_E.

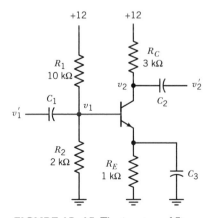

FIGURE 15–15 The input v_1 of Figure 15-13a provided by the biasing circuit in Figure 15–14. The input v_1' can go both positive and negative.

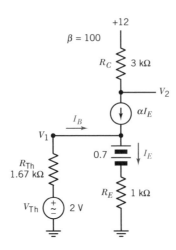

FIGURE 15–16 A dc circuit model of the circuit in Figure 15–15. Analysis of this circuit shows that I_B causes a negligible drop of 0.022 V across R_{Th}.

EXAMPLE 15.2 A common-emitter amplifier is to be designed to amplify a signal $v_1' = (0.3 \text{ V}) \sin(\omega t)$ by a gain of -10 to give $v_2' = -(3.0 \text{ V}) \sin(\omega t)$.

Given: The transistor has $\beta = 200$, and the dc power supply is a 9-V battery. The collector resistor R_C is to be 5 kΩ (output impedance is discussed in Section 15.6). The lowest frequency to be amplified is 100 Hz, or $\omega = 628$ rad/s.

Find: C_3 and the values of the resistors in Figure 15–17a to provide the necessary gain and output amplitude.

Solution: In following the design solution, the reader is encouraged to write in the dc voltages and the resistor values on Figure 15–17a as they are chosen. First we choose a base bias voltage V_1. Too low a V_1 makes the drop across $R_{E1} + R_{E2}$ too small to establish a stable I_E bias. The total v_1 must stay below 3 V to allow the 6-V "swing" of v_2 while maintaining $v_1 \le v_2 \le 9$ V. About the highest V_1 satisfying this $V_1 = 2.5$ V, making the maximum v_1 equal $2.5 + 0.3 = 2.8$ V. The values $R_1 = \underline{6.5 \text{ k}\Omega}$ and $R_2 = \underline{2.5 \text{ k}\Omega}$ form a voltage divider that provides $V_1 = 2.5$ V (neglecting I_B). These also yield $R_{Th} = R_1 \| R_2 = 1.8$ kΩ. (We need to check later that $R_{Th} \le 2R_E$ so that I_B *can* be neglected. If not, R_1 and R_2 must be reduced, keeping their ratio the same.)

Next we choose the emitter bias current I_E. For maximum voltage swing at V_2, the quiescent V_2 should be about centered between V_1 and the 9-V supply. Let $V_2 = 6$ V. Then we require $I_E = (9 - 6)/R_C = 0.6$ mA. The quiescent emitter voltage is $V_E = V_1 - 0.7 = 1.8$ V. Then we get the necessary I_E by making $R_E = V_E/I_E = 1.8 \text{ V}/0.6 \text{ mA} = 3 \text{ k}\Omega$, where $R_E = R_{E1} + R_{E2}$. (Note that $R_{Th} \le 2R_E$ is satisfied.)

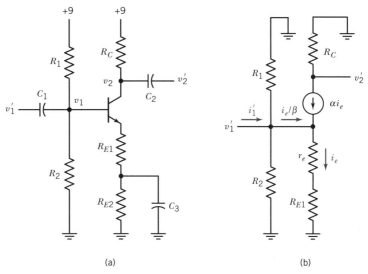

FIGURE 15–17 Common-emitter amplifier for Example 15.2. The total R_E = R_{E1} + R_{E2} is only partially bypassed, moderating the gain.

For the ac analysis, $r_e = 26\text{ mV}/I_E = 43\ \Omega$. This value is incorporated in the ac circuit model in Figure 15–17b. C_3, C_1, and C_2 have been replaced by short circuits. Note that part of the total R_E is shorted by C_3, leaving R_{E1}. For this case Equation (15.16) becomes $A_v = -R_C/(r_e + R_{E1})$. But $R_C = 5\text{ k}\Omega$, $r_e = 43\ \Omega$, and we want $A_v = -10$. Therefore $R_{E1} = \underline{457\ \Omega}$, and $R_{E2} = 3\text{ k}\Omega - 457\ \Omega = \underline{2540\ \Omega}$.

Equation (15.18) was based on making the impedance of C_3 much less than r_e. Here R_{E1} is in series with r_e in the ac circuit model. Therefore the equation becomes

$$1/\omega C_3 \le (r_e + R_{E1})/3. \qquad (15.18')$$

For $\omega = 628$, this requires $C_3 \ge 9.55\ \mu\text{F}$. Choose $C_3 = \underline{10\ \mu\text{F}}$. ∎

15.5 HYBRID-π TRANSISTOR MODEL

There are several popular ac models for the transistor. One is the T model in Figure 15–8a that we have been using. Another is the hybrid-π model shown in Figure 15–18b.* This model is particularly useful in seeing the input impedance of the transistor in the common-emitter configuration. With the emitter grounded (at

---

*This is a simplified version; a more complete hybrid-π model has about six elements.

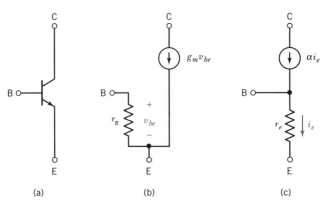

FIGURE 15–18 Two ac transistor models: (a) Transistor symbol, (b) hybrid-π model, and (c) T model. For most configurations the T model is simpler. The hybrid-π model is useful when the emitter is grounded.

least through a bypass capacitor), the impedance looking into the base is simply r_π. The model also expresses the collector current in terms of the base-emitter voltage: $i_c = g_m v_{be}$, where g_m is called the *transconductance* of the transistor.

The T model is shown for comparison in Figure 15-18c. These two models are completely equivalent, both giving the same result in analyzing a circuit. Therefore the parameters of one model can be related to those of the other. For example, when v_{be} is applied across r_e in the T model, the result is $i_c = \alpha v_{be}/r_e$. For this to be the same result as for the hybrid-π model, it must be true that

$$g_m = \alpha/r_e \approx 1/r_e. \qquad (15.20)$$

Similarly, since $i_b = i_c/\beta$, $i_b = \alpha v_{be}/\beta r_e$ in terms of the T model parameters. From the hybrid-π model it can be seen that $i_b = v_{be}/r_\pi$. Therefore it must be true that

$$r_\pi = \beta r_e/\alpha \approx \beta r_e. \qquad (15.21)$$

Why have two transistor models if they are equivalent? Because each lends different insights in different applications. For common-collector and common-base configurations, the T model is easier to understand. It can also be used to analyze the common-emitter configuration, but it doesn't show the input impedance explicitly. With the emitter of the hybrid-π model grounded, it is clear that the impedance seen looking into the base is r_π. When using the T model one must remember that the impedance looking into the base is not r_e as it might appear. Rather it is βr_e because the αi_e source is providing most of the current through r_e.

15.6   INPUT AND OUTPUT IMPEDANCE

In Chapter 8 all the details of amplifier design were hidden inside a two-port model—a box like that in Figure 15–19. The elements in the box model three

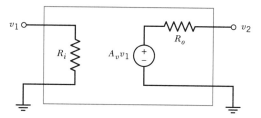

FIGURE 15–19 Two-port model used to represent the properties of an amplifier circuit.

parameters of the amplifier: voltage gain A_v, input impedance R_i, and output impedance R_o. From Equation (15.16) or (15.17) we already have A_v for the common-emitter amplifier. R_i and R_o can be determined from the ac circuit model.

EXAMPLE 15.3 A two-port model is to be developed for the common-emitter amplifier circuit in Figure 15–15.

Given: $\beta = 100$, and (from an earlier analysis of the dc circuit model) $I_E = 1.3$ mA and $r_e = 20\ \Omega$.

Find: R_i, R_o, and A_v.

Solution: First form the ac circuit model as in Figure 15–20. C_3 shorts out R_E, grounding the emitter. The hybrid-π transistor model was chosen as more convenient in this case with a grounded emitter. From Equation (15.20), $g_m \approx 1/r_e = 50$ mS, and from Equation (15.21), $r_\pi \approx \beta r_e = 2$ kΩ.

Looking into the input terminal, we see only resistors to ground. Therefore R_i is simply their parallel combination: $R_i = R_1 \| R_2 \| r_\pi = 10\text{k}\Omega \| 2\text{k}\Omega \| 2\text{k}\Omega = \underline{909\ \Omega}$.

To find R_o we set $v_1 = 0$ and look into the output terminal. With no input, the current source in the transistor model is $g_m v_1 = 0$, which is an open circuit. Then all we see at the output is R_C, and $R_o = R_C = \underline{3\ \text{k}\Omega}$.

The output voltage is $v_2 = -i_c R_C = -g_m v_1 R_C$. Therefore the ac voltage gain is $A_v \equiv v_2/v_1 = -g_m R_C = -50$ mS $\times$ 3 k$\Omega = \underline{-150}$. This is the same result found from Equation (15.17) by the T model.

In general, the two-port parameters for a common-emitter amplifier (with R_E bypassed) are given by

$$A_v = -R_C/r_e = -g_m R_C,$$
$$R_i = R_1 \| R_2 \| \beta r_e = R_1 \| R_2 \| r_\pi, \qquad (15.22)$$
$$R_o = R_C.$$

When a portion of R_E is not bypassed, as in Figure 15–17, an analysis of the ac circuit model in Figure 15–17b yields the expressions

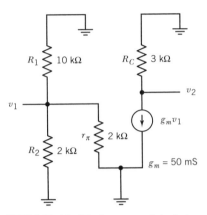

FIGURE 15–20 An ac model of the circuit in Figure 15–15. The ac analysis in Example 15.3 shows $A_v = 150$, $R_i = 909\ \Omega$, and $R_o = 3\ \text{k}\Omega$.

$$A_v = -R_C/(r_e + R_{E1})$$
$$R_i = R_1 \,\|\, R_2 \,\|\, \beta(r_e + R_{E1}), \qquad (15.23)$$
$$R_o = R_C,$$

where the expression for R_i is found from the definition

$$R_i \equiv v_1'/i_1', \qquad (15.24)$$

where i_1' is the input current in response to v_1'. Note that the T transistor model is used in Figure 15–17b; the hybrid-π model is not so convenient when the emitter is not directly grounded (through a bypass capacitor).

The input impedance of a common-emitter stage can be increased by putting a common-collector stage in front of it, as in Figure 15–21. This doesn't affect the voltage gain, but the current gain of the common-collector stage [see Equation (15.12)] reduces the base current at the input by a factor of β, thereby increasing R_i. Similarly, a common-collector stage following the common-emitter stage reduces R_o by a factor of β. The two-port parameters for the three-stage amplifier in Figure 15–21 are given by

$$A_v = -R_C/r_{e2} = -g_{m2}R_C,$$
$$R_i = R_1 \,\|\, R_2 \,\|\, \beta^2 r_{e2} = R_1 \,\|\, R_2 \,\|\, \beta r_{\pi 2}, \qquad (15.25)$$
$$R_o = (R_C/\beta) \,\|\, R_4,$$

where r_{e2} and g_{m2} relate to the common-emitter transistor, and β is assumed the same for all the transistors.

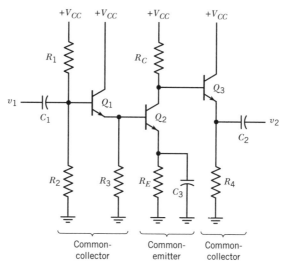

FIGURE 15–21 Three-stage amplifier. The common-collector stages at the input and output increase the input impedance and decrease the output impedance of the amplifier.

15.7   DIFFERENTIAL CONFIGURATION

The transistor amplifier stage in Figure 15-22a is called a *differential amplifier, emitter-coupled pair,* or *long-tail pair.* Like the common-emitter configuration, it has high voltage gain and high current gain. In addition, it lends itself to dc coupling (eliminating coupling capacitors), it has better linearity, and it responds only to the voltage *difference* between the two input terminals.

While most amplifiers have one input terminal grounded, a differential amplifier is characterized by having its two input terminals able to "float" with respect to ground. All operational amplifiers have a differential amplifier stage at the input.

Part of the circuit biasing is provided by a dc current source, 2 mA in this case. While this is shown as a single element, it is actually realized by a biased transistor, such as the circuit in Figure 15–23. (The analysis of this circuit is the same as the dc analysis of the circuit in Figure 15–15.) With $v_i = 0$ at the input to the differential amplifier in Figure 15–22a (the two bases tied together), circuit symmetry tells us that the 2 mA splits equally between the two transistors. Therefore the bias current I_E in each transistor is 1 mA, and r_e for the transistor model is $r_e = 26\ \text{mV}/I_E = 26\ \Omega$.

Figure 15–22b shows an ac circuit model of the differential amplifier using the T model for the transistors. Notice that setting the dc current source to zero turned it into an open circuit. Since the emitters are in series in an ac sense, $i_{e2} = -i_{e1}$. Approximating $\alpha \approx 1$, it is useful to visualize a single current i_{e1} flowing counterclockwise around the loop (joining the grounds at the top).

First we calculate the voltage gain v_2/v_i. The input v_1 appears across r_{e1} and r_{e2} in series, so $i_{e1} = v_i/(r_{e1} + r_{e2})$. The output is taken across R_C, so $v_2 =$

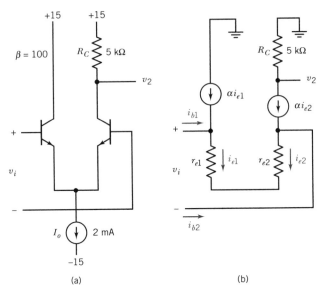

FIGURE 15-22 (a) Differential configuration. The circuit responds to the *difference* between the base voltages. (b) An ac circuit model.

$-\alpha i_{e2} R_C \approx i_{e1} R_C$. Therefore the voltage gain for the differential amplifier* is given by

$$A_v \equiv v_2/v_i = R_C/2r_e. \tag{15.26a}$$

For $R_C = 5$ kΩ and $r_e = 26$, the gain is $A_v = 96$.

Notice that v_2 depends only on the difference v_i between the base voltages. If a common voltage were added to each of the base voltages, v_2 would be unchanged. This is called *common mode rejection*.

The input current for this circuit is i_{b1} $(= -i_{b2})$. Then by Equation (15.24) the input impedance is $R_i = v_i/i_{b1}$. Now, $v_i = i_{e1}(r_{e1} + r_{e2})$, and $i_{b1} = i_{e1}/\beta$. Therefore

$$R_i = \beta(r_{e1} + r_{e2}). \tag{15.26b}$$

For $\beta = 100$ and $r_e = 26$ Ω, $R_1 = 5.2$ kΩ.

We find the output impedance R_o by first setting the input to zero. Then $i_{e1} = i_{e2} = 0$, and the current source is $\alpha i_{e2} = 0$. This makes the current source an open circuit, and looking into the output we see

$$R_o = R_C. \tag{15.26c}$$

---

*Some differential amplifiers include another R_C in series with the first collector and take a "balanced" output voltage between the two collectors. This results in twice the voltage gain.

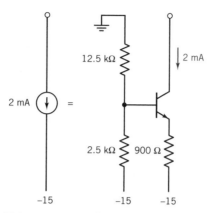

FIGURE 15–23 Realization of the 2-mA source in the circuit of Figure 15-22a.

These parameters apply to the two-port model of a differential amplifier shown in Figure 15–24. Note that neither input terminal need be grounded.

15.8   FREQUENCY RESPONSE

All *ac-coupled amplifiers,* such as the common-emitter stage in Figure 15–15, have a gain rolloff at both low frequency and high frequency. The low-frequency rolloff is due to the coupling capacitors C_1 and C_2. [The bypass capacitor C_3 can also cause rolloff, but it should not dominate if Equation (15.18) is satisfied.] Figure 15–25 shows the two-port model in Figure 15–19 expanded to include the coupling capacitors in series with the input and output terminals. The values of A_v, R_i, and R_o are found from Equation (15.22). The load on the output of the stage and the Thevenin equivalent of the source driving the stage are also shown; they affect the cutoff frequency. Analysis of this model yields the transfer function from source to load;

$$\mathbf{V}_L/\mathbf{V}_S = G_o \frac{jf/f_{L1}}{1 + jf/f_{L1}} \frac{jf/f_{L2}}{1 + jf/f_{L2}}, \tag{15.27}$$

where

$$G_o = A_v \frac{R_i}{(R_s + R_i)} \frac{R_L}{(R_o + R_L)}, \tag{15.28}$$

$$f_{L1} = \frac{1}{2\pi(R_S + R_i)C_1}, \text{ and } f_{L2} = \frac{1}{2\pi(R_o + R_L)C_2}. \tag{15.29}$$

It can be seen that each coupling capacitor contributes a cutoff frequency. One of them will dominate by being higher—usually f_{L2} because of the lower values of R_o

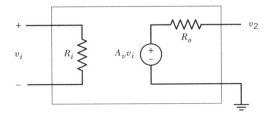

FIGURE 15–24 Two-port model of a differential amplifier. Unlike the amplifier model in Figure 15–19, either input terminal (or neither) can be grounded.

and R_L. The corresponding frequency response is shown in Figure 15–26. (Compare the response for one lower cutoff frequency in Figure 8–10a.)

The differential amplifier in Figure 15–22 is a *dc-coupled amplifier* because it has no coupling capacitors. None is needed at the input. If the output is level shifted to cancel the dc component of v_2, none is needed at the output (this is done in op amps). Such a dc-coupled amplifier has no lower cutoff frequencies; its gain is G_o all the way down to $f = 0$ (dc).

Rolloff of the gain at high frequency is due to stray capacitance that unintentionally bypasses the signal to ground. One source of such stray capacitance is in the transistors, as shown in the ac circuit model for a differential amplifier in Figure 15–27. Here a more complete transistor model that includes collector-base capacitance C_{CB} has been used* (compare with Figure 15–22). If the v_3 terminal of the input is grounded, then C_{CB2} bypasses v_2 to ground, and C_{CB1} bypasses v_1 to ground. Because C_{CB} is very small—perhaps 20 pF—this bypassing doesn't occur except at high frequencies—in the megahertz.

A high-frequency two-port model of the differential amplifier in Figure 15–27 is shown in Figure 15–28. The values of A_v, R_i, and R_o are found from Equation (15.26). The model has again been expanded, this time showing the bypassing by the C_{CB}'s. Analysis of this model yields the following transfer function from source to load:

$$\mathbf{V}_L/\mathbf{V}_S = G_o \frac{1}{1 + jf//f_{U1}} \frac{1}{1 + jf//f_{U2}}, \tag{15.30}$$

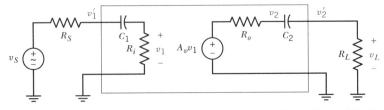

FIGURE 15–25 Two-port model of the ac-coupled amplifier in Figure 15–15. In Figure 15–19, C_1 and C_2 were assumed to be effectively short circuits; here they are shown explicitly.

*In the hybrid-π model, the collector-base capacitance is called C_μ.

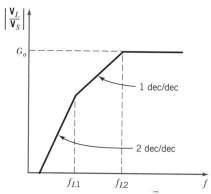

FIGURE 15–26 Gain rolloff at low frequencies due to the coupling capacitors in Figure 15–25.

where

$$f_{U1} = \frac{1}{2\pi(R_S \| R_i)\; C_{CB1}} \quad \text{and} \quad f_{U2} = \frac{1}{2\pi(R_o \| R_L)\; C_{CB2}}, \quad (15.31)$$

and G_o is given by Equation (15.28). Again each of the capacitances leads to a cutoff frequency. The corresponding frequency response is shown in Figure 15–29. (Compare the response for one upper cutoff frequency in Figure 8–9a.)

The effect of stray capacitances in a common-emitter amplifier is similar. The expression for f_{U2} is about the same as in Equation (15.31), but f_{U1} is lowered by a factor of A_v. The reason for this involves feedback and is called the *Miller effect*. A detailed treatment of this topic is beyond the scope of this text, but see problem E15.4.

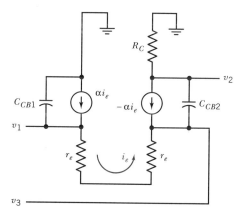

FIGURE 15–27 More complete ac circuit model of the differential amplifier in Figure 15–22. The collector-base capacitance C_{CB} is included in the transistor models.

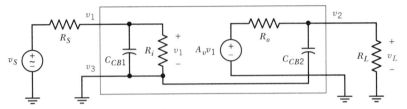

FIGURE 15–28 Two-port model of the differential amplifier in Figure 15–27. In Figure 15–24, C_{CB1} and C_{CB2} were assumed to be effectively open circuits; here they are shown explicitly.

15.9   FIELD-EFFECT TRANSISTORS

Field-effect transistors (FETs), like bipolar transistors, are made of semiconductor material—usually silicon. The physical principle is different (see Section 15.10), but its behavior is very similar. In fact the models for the FET are practically the same as those used for the bipolar transistor. The chief advantage of an FET is its extremely high input impedance—greater than $10^9 \ \Omega$. As we will see, this property comes at the expense of reduced voltage gain.

There are two types of FETs: insulated-gate FETs (IGFETs or MOSFETs) and junction FETs (JFETs). In this section we look at an n-channel JFET as representative of amplifying (or linear) field-effect transistors. The equations describing the behavior of MOSFETs are similar to those for JFETs.

The symbol for an n-channel JFET is shown in Figure 15–30a. The drain (D), the gate (G), and the source (S) are similar to the collector, the base, and the emitter of a bipolar transistor. The high input impedance of an FET is reflected in a small gate current $i_G < 1$ nA at room temperature. A voltage v_{GS} applied from gate to source controls the current i_D flowing from drain to source. The relationship is

$$i_D = I_{DSS}(1 - v_{GS}/V_P)^2, \qquad (15.32)$$

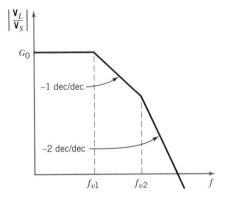

FIGURE 15–29 Gain rolloff at high frequencies due to the stray capacitances in Figure 15–28.

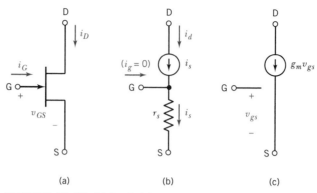

FIGURE 15–30 (a) Symbol for an n-channel junction FET. (b) T model for the FET. (c) Transconductance model for the FET. These are ac models suitable for ac analysis. Note that the gate current is zero.

where the *saturation current* i_{DSS} and the negative *pinch-off voltage* V_P are constants that are properties of the FET.* Note that $I_D = 0$ for $v_{GS} = V_P$, and $i_D = I_{DSS}$ for $v_{GS} = 0$. Unlike bipolar transistors where we can take $V_{BE} = 0.7$ V, $V_T = 26$ mV, and $\beta \approx 100$ as representative of all bipolar transistors, for an FET the designer must have a data sheet that specifies I_{DSS} and V_P for the particular FET being used. Even then these parameters can vary from device to device. In general I_{DSS} is between 1 mA and 20 mA, and V_P is between -0.5 V and -10 V.

ac Model

Once an FET has been biased, an ac model can be established for it. One ac model is shown in Figure 15–30b. This is essentially the same as the T model for the bipolar transistor in Figure 15–11 except that $\alpha = 1$. Therefore the gate current for the model is zero:

$$i_g = 0. \tag{15.33}$$

Zero is a very good approximation since the actual i_g is less than a nanoampere. The drain current i_d, which equals the source current i_s, is related to v_{gs} by

$$i_d = v_{gs}/r_s, \tag{15.34}$$

where the source resistance is given by

$$r_s = -V_P/2\sqrt{I_D I_{DSS}}. \tag{15.35}$$

This expression for r_s is developed in Section 15.10. From Equation (15.35), a bias current of $I_D = 0.44I_{DSS}$ results in $r_s = 0.75V_P/I_{DSS}$.

---

*The symbol $V_{GS\,(\text{off})}$ is sometimes used in place of V_P.

Another ac model for the FET, the *transconductance model*, is shown in Figure 15–30c. With this model it is clear that the gate current is zero since there is an open circuit looking into the gate. (Note the similarity to the hybrid-π model in Figure 15–18b with $r_\pi = \infty$.) The dependent current source makes

$$i_d = g_m v_{gs}, \tag{15.36}$$

where g_m is the *transconductance* of the FET. Comparing Equations (15.34) and (15.36), we see

$$g_m = 1/r_s. \tag{15.37}$$

Either ac model can be used; the T model is simpler to apply except perhaps when the source is grounded (usually through a bypass capacitor).

Biasing

The conditions for an FET to be in its active state are

$$0 < i_D < I_{DSS}, \qquad v_D > v_G + |V_P|. \tag{15.38}$$

From Equation (15.32) it can be shown that the first condition is equivalent to

$$V_P < v_{GS} < 0. \tag{15.39}$$

The drain bias current I_D and the drain bias voltage V_D are chosen so the requirements of Equation (15.38) are satisfied when the ac signals are also present. In problems P15.14 and E15.3 the reader can investigate tradeoffs in selecting I_D. In this section we look at just one of the possible biasings: $I_D = 0.44 I_{DSS}$, which leads to some simple relations for component values.

The *common-source* configuration for an FET (see Figure 15–31a) is the counterpart of the common-emitter configuration for a bipolar transistor. The *self-biasing* method has been used here to establish the drain bias current I_D. Since the gate current is zero, there is no voltage drop across the 10-MΩ resistor, and the bias voltage at the gate is zero. Then the voltage drop across R_S causes the source voltage to be above the gate, giving a negative $V_{GS} = -I_D R_S$, as required by Equation (15.39). The design procedure is to solve Equation (15.32) for V_{GS} corresponding to the desired I_D and then choose $R_S = -V_{GS}/I_D$. For example, if we desire $I_D = 0.44 I_{DSS}$, then $V_{GS} = (1 - \sqrt{0.44}) V_P = 0.33 V_P$, and $R_S = (0.33/0.44)|V_P|/I_{DSS} = 0.75|V_P|/I_{DSS}$.

ac Analysis

The r_s or g_m for the ac analysis is determined from the bias current I_D by Equation (15.35) or (15.37). A convenient biasing is $I_D = 0.44 I_{DSS}$ because it yields $r_s = R_S$. The reader can show from Equation (15.35) that

$$r_S = R_S = 0.75|V_P|/I_{DSS} \qquad \text{for} \qquad I_D = 0.44 I_{DSS}. \tag{15.40}$$

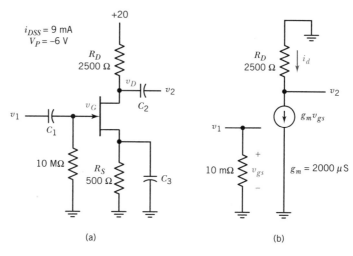

FIGURE 15–31 (a) Common-source configuration designed in Example 15.4. (b) An ac model of the circuit. The ac voltage gain is −5.

EXAMPLE 15.4

The common-source amplifier in Figure 15–31a is to be biased and designed for a voltage gain of −5.

Given: The FET's parameters are $V_P = -6$ V and $I_{DSS} = 9$ mA.

Find: values for R_S and R_D so $I_D = 0.44 I_{DSS}$, and $v_2/v_1 = -5$.

Solution: From Equation (15.40), $R_S = 0.75$ (6 V/9 mA) = 500 Ω. Then for our ac model Equation (15.40) gives $r_s = 500$ Ω, or from Equation (15.37) $g_m = 1/r_s = 2000$ μS.

Figure 15–31b shows an ac circuit model. The FET has been replaced with the ac model in Figure 15–30c, R_S bypassed by C_3 has been shorted out, and the dc supply voltage has been set to zero (ground). Working back from the output,

$$v_2 = -R_D i_d = -R_D g_m v_{gs} = -R_D g_m v_1.$$

Then for $v_2/v_1 = -5$, we need

$$R_D = 5/g_m = 5/2000 \ \mu S = \underline{2500 \ \Omega}.$$

The selection for I_D has satisfied the first condition of Equation (15.38) for being in the active state. Now we need to check the second condition $v_D > v_G + |V_P| = 0 + 6$ V. The quiescent drain voltage is $V_D = 20 - I_D R_D = 20 - 4$ mA $\times$ 2500 Ω = 10 V (see Figure 15–31a). This leaves 4 V of margin for the ac signals to swing. Specifically, v_1 (and v_G) can swing up 0.67 V, while v_2 (and v_D) swing down by 5(0.67 V) = 3.33 V.

From the example it is clear that the voltage gain for a common-source FET amplifier is

$$A_v = -g_m R_D = -R_D/r_s. \qquad (15.41)$$

This is similar to Equation (15.22) for the gain of a common-emitter amplifier, with r_s replacing r_e. For the same current levels (and therefore $R_D \approx R_C$), r_s is generally much larger than the corresponding r_e [compare Equation (15.35) with (15.4)]. Therefore the voltage gain with an FET is usually much less than that possible with a bipolar transistor. This is the penalty paid for getting a higher input impedance. (R_i for the FET amplifier in Figure 15–31a is 10 MΩ.) Another disadvantage of FETs is that the biasing is much more sensitive to variations in transistor parameters (V_P and I_{DSS}). More complicated biasing schemes can reduce this sensitivity.

15.10   SEMICONDUCTOR PHYSICS OF TRANSISTORS

The transistor models used in this chapter describe the transistors' behavior well enough for most design work. This section is included for those who want to deepen their understanding of the models. It provides additional details such as non-linearities and the effects of temperature.

Semiconductors

Transistors are made of semiconductors, which, as the name implies, lie somewhere between insulators and conductors. Metals are good conductors because they have *conduction electrons* that are not tightly bound to the atoms. With the application of an electric field, these electrons can be convinced to move rather easily. Insulators are poor conductors because they have almost no conduction electrons. Semiconductors can be made to have some conduction electrons and become fairly good conductors. They can also have another means of conduction that involves *holes* or electron vacancies.

The most popular semiconductor material is silicon, which has almost no conduction electrons in its pure state; it is a very poor conductor. However, it can be *doped* by adding some impurity atoms of phosphorus, arsenic, or antimony. These are called *donors* because they have electrons that are not tightly bound; they donate conduction electrons. A semiconductor doped with a donor is called an *n-type* material because the charge carriers are electrons (negative charges). Figure 15–32a shows the conduction process for *n*-type material. Application of a voltage across the bar of material sets up an electric field that causes the conduction electrons to drift from the negative to the positive end. Since the electrons are negative, the conventional current goes in the reverse direction, as in metal.

Semiconductors can also be doped with *acceptor* impurities such as boron, gallium, or indium. When an acceptor atom is introduced into the lattice structure of silicon, it leaves an electron vacancy or *hole* in the lattice structure. In the presence of an electric field, that vacancy robs an electron from a silicon atom, and that

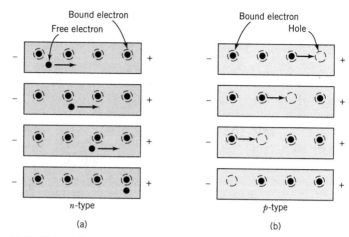

FIGURE 15–32 Conduction in semiconductors by (a) electrons and (b) holes.

silicon atom in turn robs an electron from another silicon atom. This successive robbing goes on in a consistent direction, and conduction by holes results, as depicted in Figure 15–32b. As the bound electrons successively move to the right, a hole appears to move to the left. From a modeling standpoint, it serves to think of the hole as a positive charge moving to the left and constituting a current from right to left. A semiconductor doped with acceptors is called a *p-type* material because the carriers are holes (positive charges).

Semiconductor Junctions

If an *n*-type and a *p*-type material are joined together, a *p-n* junction results. We will see that this junction has the property of conducting current in only one direction. Suppose that a voltage is applied to the junction with the positive terminal on the *p*-type, as in Figure 15–33a. Electrons and holes are driven in opposite directions across the junction where they recombine, electrons and holes canceling each

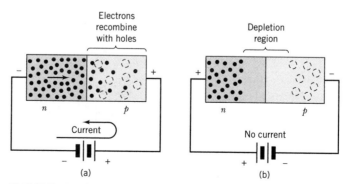

FIGURE 15–33 *p-n* junctions that are (a) forward biased (conducting) and (b) reverse biased.

other. The total current is the sum of the electron current and the hole current crossing the junction. If one of the materials, say the n-type material, is much more heavily doped than the other, then its carriers dominate the total current. This is the situation in Figure 15–33a, where many more electrons cross the junction from left to right than holes cross from right to left. As the electrons enter deeper and deeper into the p-type material, there are fewer and fewer of them as they recombine with holes.

In Figure 15–33b the voltage is applied to the p-n junction in the opposite direction. Now both electrons and holes are pulled back from the junction, leaving a *depletion region* with no carriers. Since there are no conduction electrons on the left nor holes on the right to flow across the junction, the current stops. The junction is said to be *reversed biased*, and it won't conduct.

A p-n junction is represented in Figure 15–34a by a diode symbol with the arrow pointing toward the n-type material. Figure 15–34b graphs the current that flows in response to a voltage across the p-n junction. For $v < 0$ there is almost no current. For $0 < v < 0.6$ V there is a small amount of current. At about $v = 0.7$ V the current finally becomes significant, and its rapid growth with v becomes evident. The derivation of this curve is beyond the scope of this text, but its mathematical expression is

$$i = I_s \left(e^{v/V_T} - 1 \right). \tag{15.42}$$

Both constants I_s and V_T in the expression are dependent on temperature. Specifically,

$$V_T = (87 \ \mu\text{V/}^\circ\text{K}) \, T,$$

which gives

$$V_T = 26 \text{ mV} \tag{15.43}$$

at room temperature of $T = 300^\circ$K. For a typical silicon junction I_s is about 2×10^{-14} A at room temperature, but this value depends on the cross-sectional area of the junction. I_s doubles for about every 10°C increase in temperature.

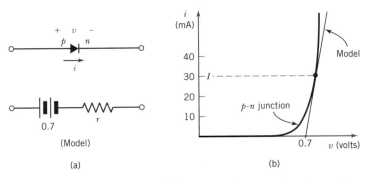

FIGURE 15–34 (a) Diode symbol representing a semiconductor junction and linear model of the junction. (b) Exponential characteristic of the junction.

Rather than deal with the nonlinear expression in Equation (15.42), it is usually convenient to approximate the exponential junction characteristic with a linear characteristic in the vicinity of operation. If the bias current is I, the exponential can be approximated by a tangent to the curve at $i = I$, as in Figure 15–34b. This tangent corresponds to a junction model consisting of a dc source and a small resistance r (see Figure 15–34a). The source has a nominal value of 0.7 V, but this value obviously depends on the bias current I through the junction. A good rule to remember is that the dc drop increases by 60 mV for every factor of 10 increase in I. The dc drop decreases by about 2 mV for every degree centigrade increase in temperature.

If the junction is operating in the vicinity of $i = I$, the slope of the characteristic is about equal to the slope of the tangent there. Then the *incremental resistance* r in the junction model is given by $1/r = di/dv$. Taking the derivative of Equation (15.42), $1/r = (1/V_T) e^{v/V_T} \approx i/V_T$. If the variation in i is small, then we can approximate it by its average value I, and $1/r = I/V_T$, or

$$r = V_T/I \approx 26 \text{ mV}/I. \qquad (15.44)$$

Bipolar Transistors

An *npn* transistor consists of two junctions sharing a very thin *p*-type material in the middle, as in Figure 15–35. In order to get the desired effect, the emitter junction must be forward biased (conducting), and the collector junction must be reverse biased. As in Figure 15–33a, electrons are *injected* from the emitter into the base, where they would normally recombine with holes. However, the base region is very thin, and the holes are few, owing to light doping. Therefore about 99% of the injected electrons escape recombination and find themselves at the collector junction. There they are swept into the collector by the reverse bias across the collector junction. Normally the reverse bias would produce no current because of lack of conduction electrons in the *p*-type base material. However, the base finds itself in the unusual situation of having many conduction electrons as a result of injection from the emitter.

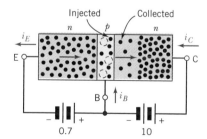

FIGURE 15–35 *npn* transistor consisting of two junctions. Almost all of the electrons injected from the emitter into the thin base region are collected by the collector.

The basic transistor action is this: the base-emitter voltage determines the emitter current according to the characteristic in Figure 15–34b. Almost all of the emitter current flows through the collector, and the base current makes up the remaining 1% or so. The *Ebers-Moll transistor model* in Figure 15–36b summarizes this action with a diode having an exponential characteristic and with a dependent current source. The fraction of emitter current passing through the collector is indicated by the current source's α, which is typically 0.99. For linear analysis, the diode in the Ebers–Moll model can be replaced by the diode model in Figure 15–34a to form the T model for the transistor in Figure 15–36c. The resistance r_e in the model is given by Equation (15.44).

Field-Effect Transistors

A field-effect transistor (FET) is made of a bar of doped semiconductor (*n*-type in Figure 15–37b) which provides a *channel* for the flow of current from drain (D) to source (S). A semiconductor junction connected to the gate (G) is introduced near the center of the bar to control this current.

The control achieved by the gate voltage v_{GS} is illustrated in Figures 15–37c and 37d. In Figure 15–37c a positive voltage applied to the drain causes a drain current $i_D = 9$ mA to flow downward (electrons to flow upward). With zero volts applied to the gate, the junction does nothing; it neither passes current itself nor does it affect the drain current. In Figure 15–37d the junction is reverse biased by applying $v_{GS} = -6$ V to the gate. The resulting depletion region eliminates the conduction electrons from the channel in the vicinity of the junction. In fact, the reverse bias here is so great that the depletion region extends completely across

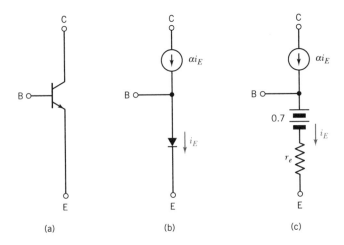

FIGURE 15–36 Transistor models: (a) Transistor symbol, (b) Ebers-Moll, (c) linearized model with $r_e = 26$ mV/I_E.

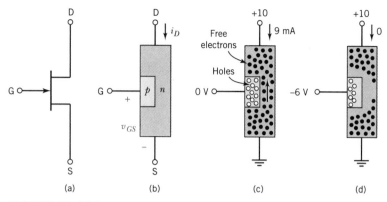

FIGURE 15–37 Principle of FET operation. (a) *n*-channel FET. (b) Construction of *n*-type and *p*-type semiconductor. (c) With zero volts on the gate, a large i_D can flow. (d) With -6 V on the gate, the reverse-biased junction forms a depletion region that inhibits i_D. For the pinch-off condition here i_D is reduced to zero.

the channel, stopping all drain current. The value of v_{GS} for which i_D is just zero is called the *pinch-off voltage* V_P.

For values of v_{GS} between zero and V_P the depletion region inhibits i_D to a greater or lesser degree. The relationship, given by Equation (15.32), is reproduced here:

$$i_D = I_{DSS}(1 - v_{GS}/V_P)^2. \tag{15.32}$$

Note that i_D is independent of the drain voltage v_D. This holds if v_D is at least $|V_P|$ above v_G [see Equation (15.38)]. A plot of Equation (15.32) is shown in Figure 15–38. This parabola shows that for $v_{GS} = V_P$, $i_D = 0$, and for $v_{GS} = 0$, $i_D = I_{DSS}$. The curve does not extend to v_{GS} positive because a forward bias on the junction would cause current to flow in the gate. Biasing the FET amounts to selecting a point on the curve around which the device will operate. For example, bias values of $V_{GS} = 0.33 V_P$ and $I_D = 0.44 I_{DSS}$ are shown in Figure 15–38.

An ac model for an FET indicates how much i_D increases for a small increment in v_{GS}. This transconductance g_m is given by the slope of the curve in Figure 15–38 at the bias point $v_{GS} = V_{GS}$. Taking the derivative of Equation (15.32),

$$g_m \equiv \frac{di_D}{dv_{GS}} = \frac{2 I_{DSS}}{-V_P}\left(1 - \frac{V_{GS}}{V_P}\right). \tag{15.45}$$

Using Equation (15.32) to solve for V_{GS} in terms of I_D, this can be expressed as

$$g_m = 2\sqrt{I_D I_{DSS}}/(-V_P). \tag{15.46}$$

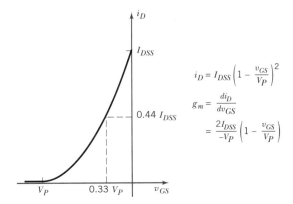

FIGURE 15–38 Characteristic curve for an FET. The transistor is biased at some point on the curve, and the slope of the curve at that point is the transconductance g_m.

Equation (15.35) for r_s follows from $r_s = 1/g_m$. It would appear from Equation (15.46) that g_m should be larger for an FET with a large I_{DSS}. In making an FET, I_{DSS} is increased by widening the channel. Unfortunately it takes more reverse bias to pinch off a wider channel, so the magnitude of V_P is also increased. This counters the increased I_{DSS} in Equation (15.46), and g_m actually changes very little. As a result g_m (for $I_D = I_{DSS}$) is typically between 1500 and 6000 μS. This is much less than the typical 20 : 1 range found for both I_{DSS} and V_P.

15.11   SUMMARY

A transistor is a small semiconductor device that amplifies electrical signals. It can be modeled by a dependent current source that responds to small input voltage changes and small input currents. For a bipolar transistor the required input voltage change is especially small, and for an FET the input current is especially small.

The three configurations for a single-transistor amplifier are common collector, common base, and common emitter. A common-collector stage (or emitter follower) has only unity voltage gain but performs buffering by providing a current gain of β. A common-base stage has only unity current gain but provides a low input impedance. This is a desirable load for signal current sources. A common-emitter stage has both high voltage gain and high current gain. The voltage gain is greatest when an emitter bypass capacitor is used.

Two common models for the transistor are the T model and the hybrid-π model. The emitter resistance r_e in the T model relates the collector current to the base-emitter voltage: $i_c = v_{be}/r_e$, where r_e is given by $r_e = 26$ mV$/I_E$ and I_E is the bias current. The base current is related to the collector current by $i_b = i_c/\beta$, where β is typically within a factor of 2 of 100. The parameters of the hybrid-π model are related to those of the T model by $g_m \approx 1/r_e$ and $r_\pi \approx \beta r_e$.

For proper operation as an amplifier (operation in the active state), an *npn* transistor must always have current flowing *out* of the emitter and the collector voltage *above* the base voltage. This corresponds to forward-biasing the emitter-base junction and reverse-biasing the collector-base junction. This is assured by applying dc bias voltages and currents to the transistor along with the ac signals. By using ac coupling capacitors, these bias voltages are blocked from appearing at the input and output, while the ac signals are easily passed.

A differential configuration with two transistors has properties similar to the common-emitter configuration. In addition, fewer coupling capacitors are needed, and the circuit responds to the difference between two input voltages.

The ac-coupled amplifiers exhibit a rolloff of gain at low frequencies due to the coupling capacitors. All amplifiers exhibit a rolloff at high frequencies due to stray capacitances. C_{CB} in transistors is an important source of stray capacitance.

FETs have a similar ac model to bipolar transistors, and they can be used in similar configurations. The FET is characterized by having a very small gate current, but it also realizes a rather low voltage gain. It is more difficult to make the biasing of an FET be independent of the parameters that characterize it: the saturation current I_{DSS} and the pinch-off voltage V_P. The drain current is related to the gate-source voltage by $i_D = I_{DSS} (1 - v_{GS}/V_P)^2$.

FOR FURTHER STUDY

A. P. Malvino, *Electronic Principles*, McGraw-Hill, New York, 1984.

P. R. Gray and R. G. Meyer, *Analysis and Design of Analog Integrated Circuits*, Wiley, New York, 1977.

PROBLEMS

Easy Drill Problems (answers at end of chapter)

D15.1. For the circuit in Figure D15.1 (p. 518) let R_E and $R_C = 2.5$ kΩ. Redraw the circuit diagram with the transistor replaced by the ideal transistor model (Figure 15–3c). Solve for V_E, I_E, and V_C. Are the conditions in Equation (15.1) satisfied?

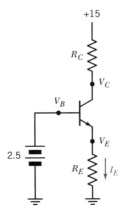

D15.2. Repeat D15.1, this time replacing the transistor by the dc model (Figure 15–7a) with $\alpha = 0.99$.

D15.3. Redraw the circuit diagram in Figure D15.1, replacing the transistor with the dc model (Figure 15–7a) with $\alpha = 0.99$. Choose R_E so $I_E = 1$ mA. What is the largest value of R_C for which Equation (15.1) is satisfied?

D15.4. A transistor has $\beta = 50$. What is its α? Repeat for $\beta = 200$. If a transistor has $\alpha = 0.993$, what is its β?

D15.5. For the common-collector configuration in Figure D15.5, assume that I_B is so small that V_B is essentially 2.5 V (there is negligible drop across R_{Th}). Using a dc transistor model with $\alpha = 0.99$ (or $\beta = 100$), find I_E. For this value of I_E find I_B and a more accurate value of V_B (including the drop across R_{Th}). For this value of V_B, find V_E and I_E. By what percent does this I_E differ from the first estimate?

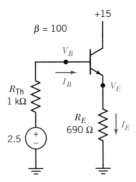

D15.6. Repeat D15.5 for $R_{\text{Th}} = 7$ kΩ. Does this value of R_{Th} satisfy the rule of thumb under "Biasing and ac Coupling" in Section 15.4? What is the result?

D15.7. A dc analysis of the circuit in Figure D15.7 shows $I_E = 2.6$ mA (see problem D15.5). Find r_e for the ac transistor model (Figure 15–8a). Draw the ac circuit model (assume C_1 acts as an ac short) and solve for the gain v_e/v_1. Solve for the gain v_c/v_1.

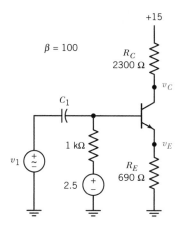

D15.8. For the circuit in Figure D15.7, suppose that $r_e = 10\ \Omega$ for the transistor, and the lowest frequency component of v_1 is 200 Hz. Choose a suitable value C_3 of a capacitor to bypass R_E.

D15.9. Suppose that R_E in Figure D15.7 has been bypassed by a suitable capacitor. Draw the ac circuit model using the T model for the transistor (Figure 15–18c) with $r_e = 10\ \Omega$. C_1 acts as an ac short. Solve for the gain v_c/v_1.

D15.10. Repeat D15.9 using the hybrid-π model for the transistor (Figure 15–18b). (What g_m and r_π correspond to $\beta = 100$ and $r_e = 10\ \Omega$?)

D15.11. Replace the 2.5-V source and the 1-kΩ resistor in Figure D15.7 by an equivalent voltage divider from the $+15$-V supply (similar to Figure 15–15).

D15.12. The common-base configuration in Figure 15–11a is driven by a current source $i_1 = 6\ \text{mA} + (1\ \text{mA}) \sin(\omega t)$. What is r_e for the ac transistor model? Draw the ac circuit model, and solve for the ac voltage v_1' at the emitter and the ac voltage v_2' at the collector.

D15.13. In Figure D15.13 the base bias voltage for the common-base configuration in Figure 15–11a has been supplied by a voltage divider. Redraw Figure D15.13, replacing the voltage divider with its Thevenin equivalent. The capacitor bypasses all ac current. Find the (dc) base voltage V_B for $\beta = 100$ and an average i_1 of 4 mA.

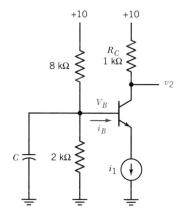

D15.14. The T model for a transistor has $\beta = 150$ and $r_e = 50\ \Omega$. Find the corresponding g_m and r_π for the hybrid-π model.

D15.15. The hybrid-π model for a transistor has $g_m = 40$ mS and $r_\pi = 2500\ \Omega$. Find the corresponding β and r_e for the T model.

D15.16. For the common-emitter amplifier in Figure P15.5, $R_1 = 40$ kΩ, $R_2 = 10$ kΩ, and $I_E = 0.5$ mA. Draw the ac circuit model using the hybrid-π model for the transistor. Find the input impedance of the amplifier.

D15.17. Find the input impedance R_i and the gain v_2/v_i for the differential configuration in Figure 15–22a. Repeat for $I_o = 0.2$ mA and $R_C = 50$ kΩ.

D15.18. Confirm Equations (15.27)–(15.29) by analyzing the circuit in Figure 15–25 to find $\mathbf{V}_L/\mathbf{V}_S$.

D15.19. The amplifier modeled in Figure 15–25 has $A_v = -120$, $R_i = 18$ kΩ, $R_o = 7\ \Omega$, $C_1 = 1\ \mu$F, and $C_2 = 50\ \mu$F. The source and load impedances are $R_S = 2$ kΩ and $R_L = 100\ \Omega$. Sketch the Bode plot of $|\mathbf{V}_L/\mathbf{V}_S|$ as in Figure 15–26.

D15.20. Confirm Equations (15.30), (15.31), and (15.28) by analyzing the circuit in Figure 15–28 to find $\mathbf{V}_L/\mathbf{V}_S$.

D15.21. The common-emitter amplifier in Figure 15–15 is driven by a source with impedance $R_S = 0$ and has a load $R_L = 10$ kΩ from v_2' to ground. Choose C_1, C_2, and C_3 for a flat response down to 300 Hz. (See the analysis in Example 15.3.)

D15.22. The amplifier modeled in Figure 15–28 has $A_v = 96$, $R_i = 5.2$ kΩ, $R_o = 5.0$ kΩ, and $C_{CB1} = C_{CB2} = 20$ pF. The source and load impedances are $R_S = 500\ \Omega$ and $R_L = 10$ kΩ. Sketch the Bode plot of $|\mathbf{V}_L/\mathbf{V}_S|$ as in Figure 15–29.

D15.23. As shown in D15.17, the ac voltage gain of a circuit stays the same if all resistors are increased and all currents are decreased by the same factor. Repeat D15.22 with R_i, R_o, R_S, and R_L all decreased by a factor of 10.

D15.24. Suppose the FET in Figure 15–31 has the parameters $I_{DSS} = 3$ mA and $V_p = -3$ V. What value of R_S will provide the bias $I_D = 0.44 I_{DSS} = 1.32$ mA? What are the resulting values of r_s and g_m for the ac transistor models?

Application Problems

P15.1. The voltage divider consisting of R_1 and R_2 in Figure P15.1a is designed to provide a nominal $V_L = 5$ V to the load R_L. Show by means of a Thevenin equivalent that V_L varies from 5 V to 4.5 V as the load current I_L varies from 0 to 15 mA.

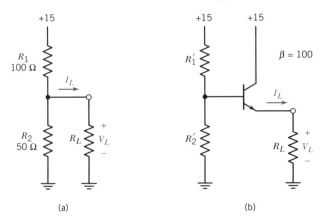

(a)   (b)

Choose R_1' and R_2' in Figure P15.1b to provide the same performance. (Use the dc transistor model in working out your design.) Compare the power dissipated in R_1 and R_2 with that dissipated in R_1' and R_2' when $I_L = 0$.

P15.2. The v_o and the R_o in Figure P15.2a are the Thevenin equivalent for the output of the amplifier stage in Figure 15–15 (without the capacitor C_2). Given $v_o = 8.14$ V $+ (0.5$ V$) \sin(\omega t)$, find the output voltage v_2 when a 2-kΩ load resistor is connected to the amplifier, as shown.

In Figure P15.2b a common-collector transistor is added. Do a dc analysis neglecting I_B, and find I_E. Do an ac analysis (including i_b) to find v_2' (the ac component of v_2).

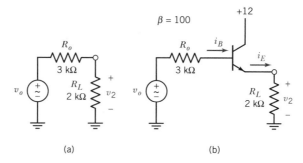

(a)   (b)

P15.3. The output impedance of the common-collector configuration in Figure P15.2b is defined as $R_o' \equiv (v_o' - v_2')/i_e$, where v_o' and v_2' are the ac components of v_o and v_2. Do an ac analysis of the circuit to show that $R_o' = r_e + R_o/\beta$.

P15.4. A solar cell produces a current i_1 of 5μA per foot-candle of light so long as the voltage across the cell is no more than 0.2 V (see Table 8–1). Design a common-base amplifier to produce an output voltage that changes by -10 mV per foot-candle. Assume a $+10$-V dc supply is available. For what range of i_1 is the transistor in its active state [is Equation (15.1) satisfied]?

P15.5. For the common-emitter amplifier in Figure P15.5, choose values of R_1 and R_2 so that the base bias voltage V_1 is 3 V. (Neglect the base bias current I_B.) Choose a value of R_E so that $I_E = 10$ mA. At this point revise the values of R_1 and R_2, if necessary, so that $R_{Th} < 2 R_E$. Choose R_C so that the ac gain is $v_2'/v_1' = -200$. Find the collector bias voltage V_2. What is the largest amplitude of voltage swing at the collector while satisfying Equation (15.1)?

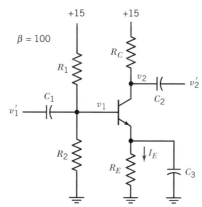

P15.6. Carry out the amplifier design in Example 15.2 with the same specifications except that $R_C = 10$ kΩ.

P15.7. Carry out the amplifier design in Example 15.2 with the same specifications except that $v_1' = (0.15 \text{ V}) \sin(\omega t)$, requiring an ac voltage gain of -20.

P15.8. For the common-emitter amplifier in Figure P15.5, let $R_1 = 40$ kΩ and $R_2 = 10$ kΩ. Choose R_E and R_C for a voltage gain of 350 and an input impedance of 5 kΩ. How much is the quiescent v_2 above the quiescent v_1? [See Equation (15.1).]

P15.9. Repeat P15.8 for a voltage gain of 600. To what voltage must the dc supply be raised to satisfy Equation (15.1) for a 2-V peak-to-peak output signal?

P15.10. The input impedance of a common-emitter stage with partially bypassed emitter resistor (Figure 15–17a) is $R_i = R_1 \| R_2 \| \beta(r_e + R_{E\ 1})$, where R_i is defined as the ratio of the ac input voltage to the ac input current: $R_i \equiv v_1'/i_1'$. Use the ac circuit model in Figure 15–17b to prove this result.

P15.11. For the three-stage amplifier in Figure 15–21, let $V_{CC} = 9$ V and the resistor values be $R_1 = 60$ kΩ, $R_2 = 30$ kΩ, $R_3 = 10$ kΩ, $R_E = 600$ Ω, $R_C = 1000$ Ω, and $R_4 = 200$ Ω. Solve a dc circuit model for all bias voltages and currents, assuming $\beta = \infty$ (all base currents are negligible). Equation (15.25) is an approximation that holds when the I_E of each stage is about ten times that of the previous stage. Is that true here? Solve Equation (15.25) for $\beta = 150$ to find the voltage gain and the input and output impedances. Note that the r_{e2} in Equation (15.25) is that of the common-emitter stage.

P15.12. Redraw the differential configuration in Figure 15–22a, replacing the current source I_o by the circuit in Figure 15–23. Let the bases of both upper transistors be at a dc voltage V_{cm} (the "common-mode voltage"). For $V_{cm} = 0$, do a dc analysis to find all bias voltages and currents. What is the highest V_{cm} for which $v_B \le v_C$ is satisfied for all three transistors? [See Equation (15.1) for operation in the active state.] What is the lowest V_{cm}? This is called the "input common-mode voltage range."

P15.13. Draw an ac circuit model for the differential amplifier in Figure P15.13. Solve for the ac gain v_2/v_i. Solve for the input impedance $R_i \equiv v_i/i_{b1}$. [Equation (15.26) does not hold.]

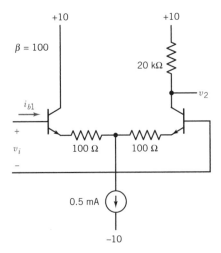

P15.14. Suppose the FET in Figure 15–31 has the parameters $I_{DSS} = 5$ mA and $V_P = -2$ V. Choose R_S so that $I_D = 0.44 I_{DSS} = 2.2$ mA. What is the largest value of R_D for which Equation (15.38) is satisfied quiescently? Using half this value for R_D, find the ac voltage gain and the output impedance of the circuit.

Extension Problems

E15.1. Figure E15.1 shows an *npn* common-emitter stage followed by a *pnp* common-emitter stage. Note that the *pnp* stage is essentially the same as the *npn* stage turned upside down. The arrow in the *pnp* transistor symbol marks the emitter and shows the direction in which current can flow. The ac model for a *pnp* transistor is the same as that for an *npn* (see Figure 15–18). The dc model for a *pnp* transistor is that shown in Figure 15–7a but with the 0.7-V source's polarity reversed.

Draw the dc circuit model for Figure E15.1, and find the bias voltages and currents for $V_1 = 2.1$ V. Neglect the base currents. Draw the ac circuit model, using hybrid-π models for the transistors. Use $\beta = 100$ for both transistors. The capacitors act as ac shorts. Solve for the ac voltage gain v_3/v_1. (Analyze the gain and impedances of the second stage first.)

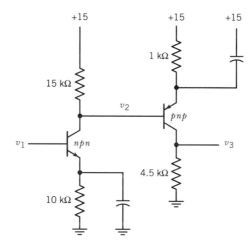

E15.2. Do an ac analysis of the differential amplifier in Figure E15.2 (p. 524) to find the ac gain v_z/v_x as a function of i_o. Do a dc analysis to find i_o as a function of v_y. Express the ac gain v_z/v_x as a function of v_y. An amplifier with a voltage-controlled gain such as this is used for *automatic gain control* (AGC) applications.

Express v_z in terms of v_x and v_y. This expression shows that the circuit can be used as a *two-quadrant multiplier*. Find v_z when $v_x = (20$ mV$) \cos(\omega_c t)$ and $v_y = -8$ V $+ (5$ V$) \cos(\omega_m t)$. Compare this v_z with the transmitted signal for amplitude modulation in Section 11.2.

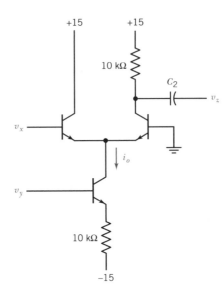

E15.3. An FET can be biased for any I_D that satisfies Equation (15.38) by solving Equation (15.32) for the V_{GS} corresponding to that I_D. Then for the biasing in Figure 15–31, $R_S = -V_{GS}/I_D$. Repeat P15.14, but choosing R_S so $I_D = 0.2$ mA. Note that the ac voltage gain has increased at the expense of higher output impedance.

E15.4. It was shown in the text that the amplifier in Figure 15–15 has an ac voltage gain of -150. Then the collector-to-base capacitance C_{CB} of the transistor has an ac voltage of $151v_1'$ across it. Find the current through this capacitance for $C_{CB} = 20$ pF and $v_1' = (20 \text{ mV}) \sin(2\pi ft)$, where $f = 2$ MHz. This current must be provided by the source at the input.

The current that the source must provide can be reduced by adding a common-base stage, as in Figure E15.4. This combination is called a *cascode* configuration. Draw the ac circuit model of the cascode amplifier, and solve for the ac voltage gain v_2'/v_1' to show it is still -150. Show that the current through C_{CB} has been reduced to $(10 \ \mu\text{A}) \cos(2\pi ft)$.

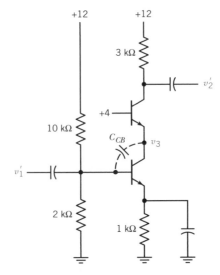

Study Questions

15.1. What is the difference between active devices and passive devices?

15.2. How many elements does it take to model an ideal transistor? How many parameters does it take to describe it?

15.3. If the common-collector configuration has a voltage gain less than unity, why is it called an amplifier?

15.4. How does the ac gain of a common-collector configuration depend on the base-emitter drop V_{BE}?

15.5. According to Equation (15.17) the gain of a common-emitter configuration can be made infinite by setting $r_e = 0$. Why can't this be done in practice?

15.6. If β is an important transistor parameter, why can't it be seen in the T model? How can it be found from the T model?

15.7. The analysis of a transistor circuit is broken into two parts. What are they called? What is the advantage of dealing with two problems rather than one?

15.8. Why is it necessary to bias a transistor?

15.9. In which configuration is the hybrid-π transistor model easier to use than the T model?

15.10. Which transistor configuration is used at the input of an op amp?

15.11. For what kind of analysis is it necessary to include the capacitor in the transistor model?

15.12. Which has a higher current ratio β, a bipolar transistor or an FET?

15.13. Which will provide greater voltage gain, a bipolar transistor or an FET?

ANSWERS TO DRILL PROBLEMS

D15.1. $V_E = 2.5$ V, $I_E = 1.0$ mA, $V_C = 12.5$ V

D15.2. $V_E = 1.8$ V, $I_E = 0.72$ mA, $V_C = 13.22$ V

D15.3. $R_E = 1.8$ kΩ, $R_C \leq 12.6$ kΩ

D15.4. $\alpha = 0.98$, $\alpha = 0.995$, $\beta = 142$

D15.5. $I_E = 2.61$ mA, $I_B = 26.1$ μA, $V_B = 2.474$ V, $I_E = 2.57$ mA

D15.6. $V_R = 2.32$ V, $I_E = 2.35$ mA (down 10% from 2.61 mA)

D15.7. $r_e = 10$ Ω, $v_e/v_1 = 0.986$, $v_c/v_1 = -3.28$

D15.8. $C_2 = 140$ μF

D15.9. $v_c/v_1 = -228$

D15.10. $g_m = 0.1$ S, $r_\pi = 1000$ Ω, $v_c/v_1 = -228$

D15.11. $R_1 = 6$ kΩ, $R_2 = 1.2$ kΩ

D15.12. $r_e = 4.33$ Ω, $v_1' = -(4.33$ mV$)\sin(\omega t)$, $v_2' = -(1.0$ V$)\sin(\omega t)$

D15.13. $V_B = 1.936$ V

D15.14. $g_m = 20$ mS, $r_\pi = 7500$ Ω

D15.15. $\beta = 100$, $r_e = 25$ Ω

D15.16. $R_i = 3125 \ \Omega$

D15.17. $R_i = 5.2 \ \text{k}\Omega$, $v_2/v_i = 96$; $R_i = 52 \ \text{k}\Omega$, $v_2/v_i = 96$

D15.19. $G_o = 101$, $f_{L1} = 7.96 \ \text{Hz}$, $f_{L2} = 29.7 \ \text{Hz}$

D15.21. $C_1 = 0.584 \ \mu\text{F}$, $C_2 = 0.041 \ \mu\text{F}$, $C_3 = 80 \ \mu\text{F}$

D15.22. $G_o = 58.4$, $f_{U1} = 17.4 \ \text{MHz}$, $f_{U2} = 2.39 \ \text{MHz}$

D15.23. $G_o = 58.4$, $f_{U1} = 174 \ \text{MHz}$, $f_{U2} = 23.9 \ \text{MHz}$

D15.24. $R_S = 750 \ \Omega$, $r_s = 750 \ \Omega$, $g_m = 1333 \ \mu\text{S}$

CHAPTER 16

Diodes and Transistors as Switches

Some circuit functions can't be realized from linear amplifiers, resistors, and capacitors alone; they require switches or operations equivalent to switching. These are nonlinear functions such as ac-to-dc conversion, dc-to-ac conversion, modulation, demodulation, and digital logic functions. This chapter examines the behavior of diodes and transistors as switches and their application in a number of nonlinear circuits.

Ideally a switch is a short circuit at one time and an open circuit at another. Semiconductor devices approximate this behavior fairly well, going from tens of ohms to billions of ohms. A diode can be viewed as a switch that is either an open or a short circuit depending on the direction current tries to pass through it. The transistor, used as an amplifying device in Chapter 15, can also be used as a switch. If the base (or gate) is driven "hard" enough, the path from collector to emitter (or drain to source) is either an open or a short circuit (approximately).

Switching circuits don't involve new mathematics, just the basic KVL, KCL, and Ohm's law of circuit analysis. The new discipline is learning how to think about circuits that change during the analysis. Sometimes it isn't clear what state of the circuit should be analyzed until the solution is known. The way out of this seeming dilemma is to develop a sense of the interaction of switches with other circuit elements. Analogies are helpful in this.

The reader's objectives in this chapter are to:

1. be able to analyze circuits with nonlinear elements such as diodes,
2. understand the basic operation of modulators and demodulators used in communications,
3. understand the basic operation of digital logic circuits,
4. be able to design simple switching circuits using bipolar transistors and FETs.

16.1 DIODES

A diode is a device that passes current in only one direction. It is usually realized by a semiconductor junction,* whose characteristic was described in Section 15.10. The exponential *V-I* characteristic of a junction diode is reproduced here in Figure 16–1a. As in the analysis of transistor circuits, it will be useful to start with a simplified version of the characteristic so the basic ideas aren't clouded by detail.

The *ideal diode characteristic* is shown in Figure 16–1b. The vertical line corresponds to a switch that is *on* (closed); there is no voltage across the device. The horizontal line corresponds to a switch that is *off* (open); there is no current through the device. Although it ignores the forward voltage drop of about 0.7 V, this characteristic is a good model to begin with. In Section 16.7 the 0.7-V voltage drop will be brought into the analysis.

The symbol for a diode is shown in Figure 16–1c. Note that the arrow portion of the symbol indicates the direction in which current can flow. When analyzing a diode circuit, the symbol represents the characteristic in Figure 16–1a. but in preliminary analysis it can also be used to represent the ideal diode characteristic in Figure 16–1b. In such cases the symbol is said to represent an *ideal diode* (the word "ideal" may be written next to the symbol to emphasize this). However, any such

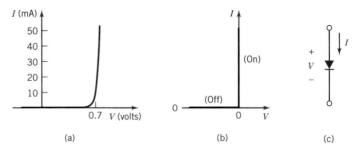

FIGURE 16–1 (a) Exponential *V-I* characteristic for a diode. (b) Ideal diode characteristic—a simple model of the characteristic in (a). (c) Diode symbol.

---

*Diodes can also be realized by selenium rectifiers, "cat's whisker" crystals, and vacuum tubes, but these are seldom used today.

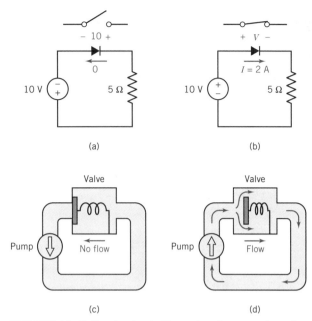

FIGURE 16–2 Simple circuit illustrating the operation of a diode. In (a) the diode won't conduct current with reverse bias (*V* negative). The diode acts like an open switch. In (b) the diode conducts current in the direction of its arrow with (ideally) no voltage drop. The diode acts like a closed switch. (c) and (d) show hydraulic analogs of the circuits in (a) and (b).

reference to an ideal diode should not suggest that such a device exists; it is simply a modeling concept.

The circuits in Figure 16–2 show a diode in some simple contexts. To emphasize the basic behavior, we assume here that it is an ideal diode. In Figure 16–2a the polarity of the source is trying to force current backward through the diode. As a result, no current flows, and the diode is equivalent to an open switch. In Figure 16–2b the source is oriented to force current through the diode in the forward direction. As a result, current does flow, and there is a voltage drop of zero (ideally) across the diode, as for a closed switch. Therefore all the 10 V is dropped across the 5-Ω resistor, and the current is $I = 10 \text{ V}/5 \ \Omega = 2 \text{ A}$.

The hydraulic systems in Figure 16–2c and 2d are mechanical analogs of the electrical circuits in Figures 16–2a and 2b. In Figure 16–2c the valve is held closed (or *off*) by the pressure of the pump, and no fluid flows. In Figure 16–2d the pressure of the pump opens the valve (the valve is *on*), and there is fluid flow determined by the size (resistance) of the pipe.

The next few sections look at some circuit applications of diodes. In each case the analysis assumes the ideal-diode model. In Section 16.7 the analyses are revised in line with an improved diode model.

16.2   HALF-WAVE RECTIFIERS AND CLIPPERS

When an ac voltage source is applied to the diode circuit in Figure 16–3b, the diode conducts only on the positive half cycles, as shown in Figure 16–3c. During the time $v_1 > 0$, the current is $i = v_1/R_L$; otherwise $i = 0$. This is called half-wave rectification. In the example shown, $v_1 = (160 \text{ V}) \sin(\omega t)$ and $R_L = 40 \ \Omega$. Therefore the peak current is $I_m = 160 \text{ V}/40 \ \Omega = 4 \text{ A}$. The important point is that the current waveform now has a dc component (an average) that is not zero. Although the waveform is not smooth, such a current is useful in applications such as battery charging or electroplating. R_L is the load—the resistance of the electrolyte in an electroplating process for example.

The average of a periodic waveform $i(t)$ is defined as

$$I_{\text{dc}} \equiv \frac{1}{T} \int_0^T i(t) \, dt, \tag{16.1}$$

where T is the period. For the particular case of the half-wave-rectified sine wave in Figure 16–3c, the reader should confirm that

$$I_{\text{dc}} = I_m/\pi, \tag{16.2}$$

where I_m is the peak current. For our example, the dc component of the current is $I_{\text{dc}} = 4 \text{ A}/\pi = 1.27 \text{ A}$ (the dashed line in Figure 16–3c).

A similar application of a diode is to limit or "clip" a signal—eliminate the portion of a waveform above or below a certain voltage. Figure 16–4b shows a circuit to clip a signal voltage at 10 V. The purpose may be to protect some delicate electronic components from damage if greater than 10 V is applied to them. In analyzing the circuit behavior, it is helpful to visualize the circuit distorted so the position of each node is proportional to its voltage. This has been actually drawn out in Figure 16–5, showing the behavior for successively higher v_1. In the first two drawings the diode points upward, indicating that it is *reverse biased*. No current flows, there is no drop across the resistor, and $v_2 = v_1$. In Figure 16–5c the diode points downward, "catching" v_2 at 10 V. The diode conducts with zero volts across it (ideally), and the resistor drops the difference between v_1 and v_2. Therefore $v_2 =$

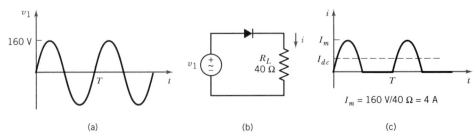

FIGURE 16–3 A half-wave rectifier produces dc current from an ac voltage source. The current waveform is not smooth, but it has a nonzero dc component (average).

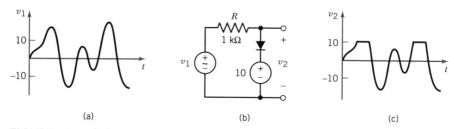

FIGURE 16–4 A clipper circuit eliminates the input waveform above or below some voltage—above 10 V here.

v_1 when $v_1 < 10$ V, and $v_2 = 10$ V when $v_1 > 10$ V. The waveform in Figure 16–4c reflects this result.

By reversing the orientation of the diode in Figure 16–5c, the portion of the waveform above 10 V is preserved, and the portion below 10 V is eliminated.

16.3   ZENER DIODES

Zener diodes are useful in simplifying the realization of clipper circuits and other circuits. These are diodes with a carefully controlled *breakdown voltage*. When a large enough reverse voltage is applied to any diode, it breaks down—starts to conduct current in the reverse direction.* For ordinary diodes, this occurs somewhere be-

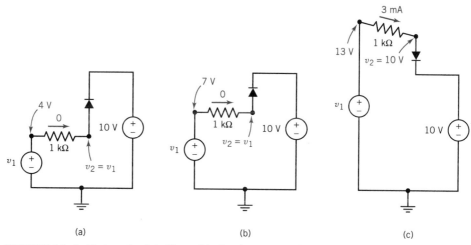

FIGURE 16–5 Limiter circuit in Figure 16–4 redrawn so node positions are proportional to their voltage. The diode is off when pointing up, and it is on when pointing down. v_2 is effectively "caught" at 10 V as v_1 increases past 10 V.

---

*Breakdown does not of itself damage a diode, but it may produce excessive current that destroys the diode. In a properly designed Zener diode circuit, the current is limited to a safe level by the rest of the circuit.

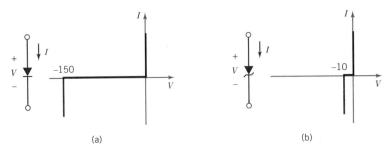

FIGURE 16–6 Characteristics showing the breakdown voltage of (a) a typical standard diode and (b) a 10-V Zener diode.

tween 50 V and 1000 V, as in the typical characteristic in Figure 16–6a. *Zener diodes* are designed to break down at lower voltages—as low as a few volts. Figure 16–6b shows the *V-I* characteristic and symbol for a 10-V Zener diode. When current flows in the forward direction, the voltage drop is nearly zero, as with a regular diode. When current flows in the reverse direction, the voltage drop is -10 V.

A Zener diode can be modeled by a dc source together with standard diodes, as shown in Figure 16–7a. The reader should confirm that for this model, also, the voltage can't exceed 10 V in one direction, and it can't exceed zero volts in the other direction. For voltages in between, the model doesn't conduct any current.

The circuit in Figure 16–7b uses a Zener diode to limit v_2 to the range $0 < v_2 < 10$ V. This is not quite the same performance as the circuit in Figure 16–4b, which clips only at 10 V. By adding a diode in series with the Zener diode (see Figure 16–8a), conduction of the Zener diode in the forward direction can be prevented. The model for this diode pair is simpler: just one diode and a dc source as in Figure 16–8a. Then the diode and 10-V source in Figure 16–4b can be replaced by a diode and a 10-V Zener diode, as has been done in Figure 16–8b. This is a more practical solution than using a source in a clipper.

A similar use for Zener diodes is in dc *voltage regulation*, as illustrated in the following example. (Compare the design problem with the same specifications in Example 15.1.)

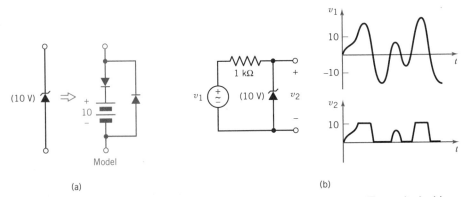

FIGURE 16–7 (a) Zener diode and its model. (b) Clipper circuit using a Zener diode. Note that v_2 is clipped at both 10 V and at 0 V.

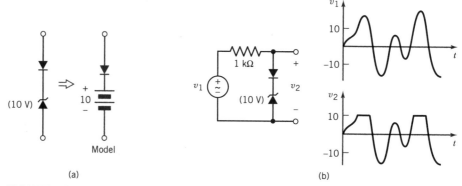

(a)   (b)

FIGURE 16–8 Clipper circuit formed by replacing the diode and the 10-V source in Figure 16–4 with a diode and a 10-V Zener diode. Now v_2 is clipped only at 10 V (compare Figure 16–7).

EXAMPLE 16.1 A voltage regulator is to provide a constant dc voltage $v_L = 10$ V to a load R_L from a nominal 15-V supply voltage V_{CC}. The simple regulator comprises a Zener diode and a dropping resistor R_d (see Figure 16–9).

Given: The load R_L can vary from 20 Ω to 1000 Ω. The supply voltage V_{CC} can vary from 13 V to 16 V. The Zener diode maintains 10 V across its terminals so long as at least 36 mA flows through it.

Find: the value of R_d that will satisfy the specifications. Find the maximum power dissipated in the Zener diode.

Solution: R_d must be chosen so that $i_Z = 36$ mA when $V_{CC} = 13$ V (its minimum) and $R_L = 20$ Ω (its minimum). The reader should confirm that i_Z is a minimum for these conditions. The load current is then $i_L = 10 \text{ V}/20 \text{ Ω} = 500$ mA, and $i_{CC} = i_L + i_Z = 536$ mA. Since there is $13 - 10 = 3$ V across R_d, its value is $R_d = 3 \text{ V}/536 \text{ mA} = \underline{5.6 \text{ Ω}}$.

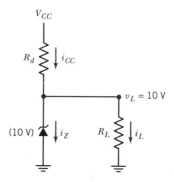

FIGURE 16–9 Voltage regulator designed in Example 16.1. The load voltage v_L is nearly constant despite changes in V_{CC} and R_L.

The maximum i_Z is for the conditions $V_{CC} = 16$ V and $R_L = 1000\ \Omega$. Then $i_L = 10$ mA, $i_{CC} = (16\ \text{V} - 10\ \text{V})/5.6\ \Omega = 1.07$ A, and $i_Z = i_{CC} - i_L = 1.06$ A. The maximum power dissipated in the Zener diode is therefore $p = 10\ \text{V} \times 1.06\ \text{A} = \underline{10.6\ \text{W}}$.

This regulator is simple, but it is also inefficient; the maximum Zener diode dissipation is greater than the maximum load dissipation. Also, v_L varies somewhat from 10 V because of resistance in the Zener diode. (A more complete Zener diode model includes this resistance.)

16.4   dc POWER SUPPLY

The electronic circuits in Chapter 15 require one or more dc voltage sources for their operation. For portable circuits this is provided by a battery, but usually it is provided by a *dc power supply* that converts 115-V ac power to dc power. A dc power supply can be constructed by adding a capacitor to a half-wave rectifier circuit (such as that in Figure 16–3) to hold the peak value of the voltage. When a transformer is included to reduce the voltage, the supply has the form shown in Figure 16–10.

EXAMPLE 16.2 Design a dc power supply to convert 115 V ac at 60 Hz into 14 V dc.

Given: The supply is to provide a maximum current of 1.4 A (to a load $R_L = 14\ \text{V}/1.4\ \text{A} = 10\ \Omega$), and the variation (or *ripple*) in the output voltage must be 2% or less.

Find: the turns ratio of the transformer and the size of the capacitor in Figure 16–10 to satisfy the specifications.

Solution: First the 115 V ac must be reduced to a smaller ac voltage whose peak is 14 V. A transformer with 115 V ac primary and 10 V ac secondary will do this. (The rms voltage is the peak voltage over $\sqrt{2}$, or 10 V = 14 V/$\sqrt{2}$.) As the secondary voltage v_2 rises to its peak of 14 V (see waveform in Figure 16–10), the diode conducts current, charging the capacitor. Because we as-

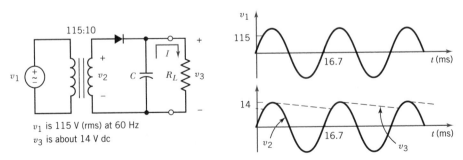

FIGURE 16–10 A dc power supply. In Example 16.2 the supply is designed to convert 115 V ac to 14 V dc. For a load current $I = 1.4$ A and for $C = 83,500\ \mu\text{F}$, the output ripple is 2%.

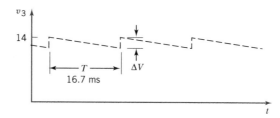

FIGURE 16–11 Simplified waveform for v_3 in Figure 16–10, assuming the diode is off for nearly the entire period T. The voltage drops ΔV as the load current discharges the capacitor.

sume no voltage drop across the diode, the capacitor voltage v_3 follows v_2 until it reaches 14 V. Then as v_2 decreases, the diode turns off, preventing the capacitor from discharging back into the transformer. The capacitor holds v_3 at about 14 V, causing the diode to become reverse biased (the voltage across it is opposed to current flow). The diode remains off for almost a complete cycle of 16.7 ms (the reciprocal of 60 Hz). When v_3 again reaches 14 V, it brings v_3 back up to a full 14 V, replacing the charge lost through the load resistor R_L during the 16.7 ms.

Capacitor size is chosen to make the "sag" in v_3, due to the lost charge, less than the specified 2%. Figure 16–11 shows a simplified waveform for v_3 that assumes the diode is off for the entire period T while the capacitor supplies the load current I. Once a cycle, the diode conducts for an instant to replace the voltage ΔV lost by the capacitor. This simplification is valid if the ripple is small. For a load current $I = 1.4$ A, the change ΔQ in capacitor charge during the period $T = 16.7$ ms is

$$\Delta Q = IT = 1.4 \text{ A} \times 16.7 \text{ ms} = 23.4 \text{ mC}. \qquad (16.3)$$

From the relation $Q = CV$ for a capacitor, the corresponding change in voltage is

$$\Delta V = \Delta Q / C. \qquad (16.4)$$

For a ripple of 2%, this change is $\Delta V = 0.02 \times 14 \text{ V} = 0.28$ V. Solving Equation (16.4) for the size of capacitor: $C = \Delta Q / \Delta V = 23.4 \text{ mC} / 0.28 \text{ V} = 83,500 \ \mu\text{F}$.

If a full-wave rectifier were used, C would only have to be half as large for the same ripple (see problem E16.2). ▪

Another hydraulics analogy may give a better feel for the holding function of the capacitor in Example 16.2. Figure 16–12 shows a system that converts a varying fluid level in a "source" tank to a nearly constant level in a "holding" tank. Let the bottom of the holding tank be the zero reference for fluid level. A pump drives the level h_2 in the source tank up and down sinusoidally. When h_2 tries to exceed

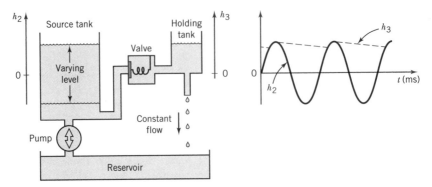

FIGURE 16–12 Hydraulic analog of a dc power supply. An alternating fluid level in the source tank is converted to a nearly constant level in the holding tank.

the level of h_3 in the holding tank, the greater pressure on the left opens the valve, and h_3 seeks the same level as h_2. When h_2 decreases, the pressure on the right is greater, the valve closes, and h_3 holds the peak value of h_2. During the time it is waiting for the next peak of h_2, h_3 decreases slightly as a result of the small output flow from the bottom of the holding tank. If the resistance of the output tube to flow is increased, or the diameter of the holding tank is increased, then the rate of decline of h_3 is slower. If the period between the peaks of h_2 is known, then the diameter of the holding tank can be chosen to keep the fluctuation in h_3 as small as desired. As an exercise, calculate the diameter of the cylindrical holding tank so the peak-to-peak fluctuation of h_3 is 5% of its maximum. It is given that the peak value of h_2 is 5 ft every 10 s, and that the output flow is 0.25 ft^3/s. Compare this design with the selection of the capacitor in Example 16.2. (*Answer:* 3.57 ft diameter.)

16.5   PEAK DETECTORS AND ENVELOPE DETECTORS

The power supply circuit in Figure 16–10 also finds applications in signal processing (as compared with power processing). The similar circuit in Figure 16–13a charges a capacitor to the peak voltage of the signal v_2. Therefore it is called a *peak detector.* The RC product is large enough that the output v_3 has the value of the widely spaced largest peaks, as shown in Figure 16–13b. One application is in radio receivers, where a peak detector monitors the strength of the amplified radio signal. If the amplitude is so great that it would cause distortion, the gain is automatically reduced. This is called an *automatic gain control* (AGC).

By reducing the RC product in Figure 16–13a, the output can be made to follow the peak of every cycle, as illustrated in Figure 16–13c. In this case the peak detector is called an *envelope detector.* It is used in AM radio receivers to demodulate the received signal (see Section 11.2). The design of an envelope detector requires that the RC product not be so large that v_3 misses some peaks as in Figure 16–13b. On the other hand, if RC is too small, v_3 starts looking more like v_2 than its envelope.

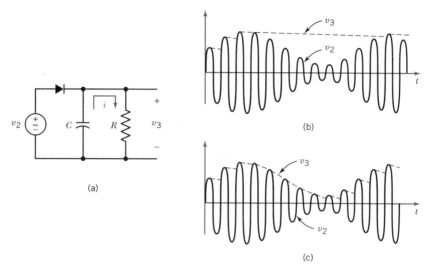

FIGURE 16–13 (a) Peak detector circuit. If the RC product is very large, v_3 equals the highest peaks in v_2, as in (b). If the RC product is small enough, v_3 will be able to follow the envelope of v_2, as in (c). The circuit is then an envelope detector and can perform AM demodulation.

What is the bound on the RC product so that the envelope falls no faster than v_3? Between peaks the diode is off, and the current $i = v_3/R$ discharges the capacitor. The rate of change of the output voltage is

$$\frac{dv_3}{dt} = -\frac{i}{C} = -\frac{v_3}{RC} \qquad (16.5)$$

Then if v_3 is to follow the envelope voltage v_e, the rate of change of v_e must satisfy $|dv_e/dt| < v_e/RC$, or

$$\frac{|dv_e/dt|}{v_e} < \frac{1}{RC} \qquad (16.6)$$

EXAMPLE 16.3 A 600-kHz AM radio carrier with the envelope v_e shown in Figure 16–14 is to be demodulated.

Given: The envelope is $v_e = 4\sin(\omega t) + 6$, where $\omega = 2\pi \times 16$ kHz $= 100$ krad/s. The value of R in the peak detector of Figure 16–13 is 10 kΩ.

Find: the maximum value of C in Figure 16–13 for which v_3 will follow v_e.

Solution: For the specified v_e, the left-hand side of Equation (16.6) is

$$\frac{|dv_e/dt|}{v_e} = \frac{4\omega|\cos(\omega t)|}{4\sin(\omega t) + 6}.$$

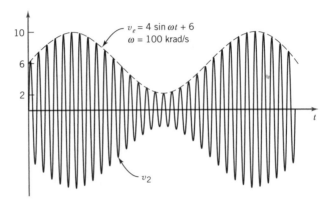

FIGURE 16–14 AM signal. The envelope v_e is to be detected in Example 16.3. For the values $R = 10$ kΩ and $C = 1000$ pF in Figure 16–13, v_3 will follow this envelope.

A little calculus shows that the maximum value for this expression occurs for $v_e = 3.33$. The value there is $0.895\omega = 89.5$ krad/s. Then Equation (16.6) is satisfied if $RC < (89.5 \text{ krad/s})^{-1} = 11.2$ μs, or $C < 11.2$ μs/10 kΩ = $\underline{1120}$ $\underline{\text{pF.}}$ Any smaller value than necessary increases the ripple in v_3. ▪

16.6   DIODE VOLTAGE DROP

In applications where voltages are large—20 V and greater—analysis using the ideal-diode model may be adequate, as in Figure 16–2. When smaller signals are involved, the assumption of zero diode voltage drop is probably not acceptable. In such cases the improved diode model in Figure 16–15 is used. The 0.7-V source in the model approximates the forward drop of the diode, as shown by the characteristics in Figure 16–15b. The actual drop may range from 0.5 V to 0.9 V depending on the current and the size of the diode junction, but 0.7 V is a better approximation than zero volts.*

Consider the effect of including the diode voltage drop in the design of the clipper circuit in Figure 16–4. If v_2 is to be clipped at 10 V, the dc voltage source should be 9.3 V. Then v_2 will be "caught" at $9.3 + 0.7 = 10$ V when the diode conducts.

The design of the dc power supply in Figure 16–10 should also be modified slightly to account for the 0.7-V diode drop. The peak voltage to which the capacitor is charged is actually only $v_2 - 0.7$ V. If the output voltage is to be $v_3 = 14$ V dc, than v_2 must have a peak value of $14 + 0.7 = 14.7$ V. This corresponds to an rms voltage $14.7/\sqrt{2} = 10.4$ V at the transformer secondary.

---

*Schottky diodes have forward voltage drop as low as 0.3 V. These can be used in applications where a large reverse breakdown voltage is not required.

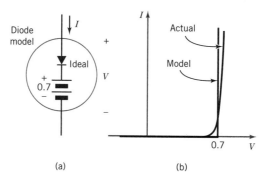

FIGURE 16-15 (a) Diode model that includes the 0.7-V drop. (b) Characteristic of the diode model compared with actual diode characteristic.

The diode drop has a similar effect in peak detectors (see Figure 16-13); the output v_3 will actually be 0.7 V below the peak voltage of v_2. Either this can be taken into account in interpreting v_3, or an op amp can be used to help the diode approach the ideal-diode characteristic (see Section 14.10). It should be remembered that op amps generally have useful gain only up to 100 kHz or so; RF amplifiers must be used at radio frequencies.

16.7   TRANSISTORS AS SWITCHES

Both bipolar transistors and FETs can be used as switches, where a signal at the base (or gate) determines whether the switch is on or off. This is a different kind of switch than that realized by a diode, where the switch is either on or off depending on the direction of the voltage across the switch. A third terminal is necessary to control a switch if the voltage doesn't change direction; the switch must be realized by a transistor.

Models for the bipolar transistor used as a switch are shown in Figure 16-16. The transistor is "off" when $i_B = 0$. This causes $i_E = 0$, and the transistor is said to be in the *cutoff* state. The corresponding model is shown in Figure 16-16b, indicating that no current flows in any of the terminals. From a voltage standpoint, cutoff is caused by applying

$$v_{BE} \leq 0.3 \text{ V} \qquad (16.7)$$

from base to emitter. If v_{BE} is made more negative than about -5 V, the base-emitter junction may break down.

The bipolar transistor is "on" when

$$i_B \geq i_C/\beta, \qquad (16.8)$$

where i_C is determined by other elements in the circuit. This causes v_{CE} to be 0.2 V, and the transistor is said to be in the *saturation* state (see the model in Figure

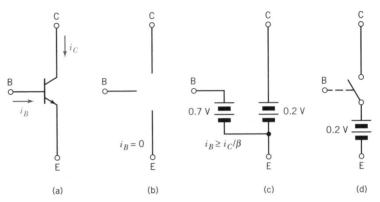

FIGURE 16–16 Switching models of a bipolar transistor. In (b) the transistor is off (in cutoff). In (c) it is on (in saturation). Both states are modeled in (d), where the "imperfect" switch has a saturation voltage drop of 0.2 V.

16–16c). Note that v_{BE} is still 0.7 V, as in the active state. A *saturation voltage* of $v_{CE} = 0.2$ V is not exactly a closed switch, but it is close enough for many applications. A model of a bipolar transistor as an imperfect switch is shown in Figure 16–16d (the details of the base have been suppressed). This combines the cutoff model (switch open) and the saturation model (switch closed).

The behavior of an FET as a switch is similar, as shown by the models in Figure 16–17. The principal difference is that the FET is modeled by a resistance $r_{ds\ (on)}$ in the "on" state (see Figure 16–17c) rather than by a voltage drop. The control is by the gate-source voltage: the FET is "off" for

$$v_{GS} \le V_P, \tag{16.9}$$

and it is "on" for

$$v_{GS} = 0. \tag{16.10}$$

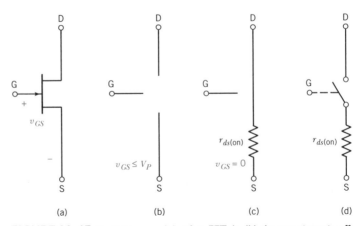

FIGURE 16–17 Switching models of an FET. In (b) the transistor is off, and in (c) it is on. Both states are modeled in (d), where the "imperfect" switch has an "on" resistance $r_{DS\ (on)}$.

The pinch-off voltage V_P typically ranges from -10 V to -1 V, and $r_{ds\ (on)}$ ranges from 30 Ω to 500 Ω, depending on the FET. Some FETs with special geometries—VFETs, DFETs, and HEXFETs—have an $r_{ds\ (on)}$ of only a few ohms.

The remaining sections in this chapter look at some applications of transistors as switches.

16.8   CHOPPER CIRCUITS

A chopper circuit uses a switch to periodically short a signal to ground.* We look at two applications of chopper circuits, one using a bipolar transistor for the switch, and one using an FET for the switch.

Amplitude Modulator

One way to generate an amplitude-modulated (AM) signal is to use a chopper followed by a bandpass filter, as shown in Figure 16–18a. (See Section 6.4 for the details of a bandpass filter.) The chopper uses a bipolar transistor as a switch to periodically short the output v_o to nearly ground, as indicated by the model in Figure 16–18b. When the switch is open, no current flows, there is no voltage drop across the 1-kΩ resistor, and v_o equals the modulating input signal v_m^+. When the

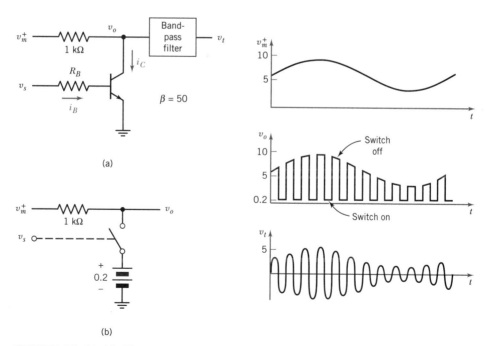

FIGURE 16–18 (a) Chopper circuit using a bipolar transistor as a switch. The output of the bandpass filter is an AM signal. (b) Transistor modeled by a switch with saturation voltage drop.

---

*The chopper probably gets its name from analogy to a shutter that chops a light beam into a sequence of short samples.

switch is closed, $v_o = 0.2$ V. Note that the v_o waveform is modulated only on the top. By removing the low frequencies, the bandpass filter causes the v_t to be symmetric about zero, and by removing the high frequencies, it causes the corners to become rounded. The result is an amplitude-modulated sinusoidal carrier (see Section 11.2).

Now we consider the conditions necessary to turn the transistor off and on. When the control voltage v_s is zero, the transistor is "off" (see the model in Figure 16–16b). Thus $i_b = 0$, there is no voltage drop across R_B, and $v_b = 0$, which satisfies Equation (16.7) for cutoff.

When $v_s = 10$ V, we must choose R_B so $i_b \geq i_C/\beta$, and the transistor will be "on" (see the model in Figure 16–16c). The maximum i_C is for $v_m^+ = 10$ V; then $i_C = (10 - 0.2)/1$ kΩ = 9.8 mA. Also, $i_B = (v_S - 0.7)/R_B = 9.3$ V$/R_B$. For $\beta = 50$, Equation (16.8) requires $i_B \leq i_C/\beta = 9.8$ mA$/50 = 196$ μA. This is satisfied for $R_B \leq 9.3$ V$/196$ μA $= 47.4$ kΩ. Choose $R_B = 47$ kΩ. A smaller R_B would increase i_B, putting the transistor "deeper" into saturation and slowing its switching time.

Chopper-Stabilized Amplifier

Because of the dc offset in amplifiers (see Sections 8.7 and 14.9), it is difficult to accurately amplify small dc voltages such as those generated by thermocouples (see Section 8.1). The offset can be avoided by turning the dc voltage into an ac voltage with a chopper, amplifying the ac voltage with an ac-coupled amplifier, and then recovering the dc voltage with a peak detector. Figure 16–19a shows such a *chopper-*

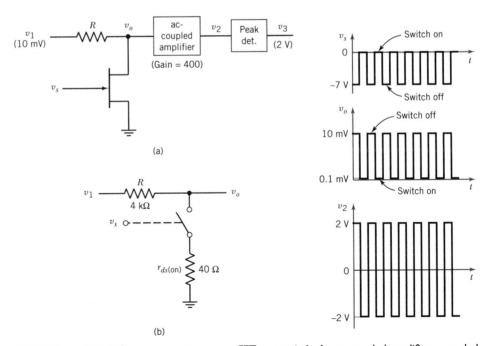

(a)

(b)

FIGURE 16–19 (a) Chopper circuit using an FET as a switch. An ac-coupled amplifier preceded by a chopper and followed by a peak detector is *chopper stabilized*. It can accurately amplify small dc voltages. (b) Chopper with FET modeled by an ideal switch in series with 40 Ω of "on" resistance.

stabilized amplifier with a gain of 200 amplifying an input voltage of 10 mV dc. Because a small input voltage is involved, the 0.2-V saturation voltage of a bipolar transistor used as a switch would be intolerable. Therefore the chopper in Figure 16–19a uses a field-effect transistor (FET), which has no saturation voltage. In Figure 16–19b the FET is modeled by a switch with series resistance $r_{ds\ (on)}$. Since the parameters of FETs vary much more than those of bipolar transistors, it is essential to use a specification sheet when designing with an FET. We will use a 2N4860, which has the following specifications: $V_P = -6$ V and $r_{ds\ (on)} = 40\ \Omega.$*

The FET is turned off by setting v_s equal to or less than V_P, say $v_s = -7$ V. The FET is turned on (made to have minimum resistance $r_{ds\ (on)} = 40\ \Omega$) by setting $v_s = 0$. The FET is modeled by an imperfect switch in Figure 16–19b. With the switch closed, the output voltage is determined by the voltage-divider relationship $v_2 = v_1 r_{ds\ (on)}/(R + r_{ds\ (on)})$. The value of R is chosen to make v_2 some desired fraction of v_1, say 1/100. Solving for R, we have $R = 99 r_{ds\ (on)} = 4\ k\Omega$.

16.9   POWER INVERTER

At a remote site where commercial power is unavailable, 115-V ac power equipment can be powered from a car battery by using a power inverter—a circuit that converts dc power to ac power efficiently. The inverter in Figure 16–20a uses two transistors as switches in converting 12 V dc to 115 V ac. The control voltages† v_{s1} and v_{s2} turn on the transistors alternately to apply an ac voltage to a voltage step-up transformer. The voltage supplied to the load R_L is a square wave rather than the sine wave provided by commercial power. However, much ac equipment will tolerate a square wave.

EXAMPLE 16.4 The inverter circuit in Figure 16–20 is to be designed to provide 100 W of power at 118 V ac.

Given: The minimum β of the transistors is 25. The control voltages v_{s1} and v_{s2} alternate between zero and 10 V at a 60-Hz rate.

Find: the transformer turns ratio N to provide 118 V ac, the necessary current and power ratings of the transistors, the value of R_B to assure saturation of the transistors, and the efficiency of the converter.

Solution: The transformer essentially has two primaries, one with voltage v_1 and one with voltage v_2. The 12-V battery voltage is placed across v_1 when transistor Q_1 is on and across v_2 when Q_2 is on. Because of the transistor's saturation voltage of 0.2 V,‡ v_1 is only $12 - 0.2 = 11.8$ V when Q_1 is on. Then if the corresponding v_L is to be 118 V, the turns ratio must be $N = 118/11.8$

---

*The symbol $V_{GS\ (off)}$ is often used for the pinch-off voltage.
†The control voltages might be provided by an astable multivibrator oscillator (see Section 14.10).
‡For a power transistor, the saturation voltage would perhaps be closer to 0.5 V.

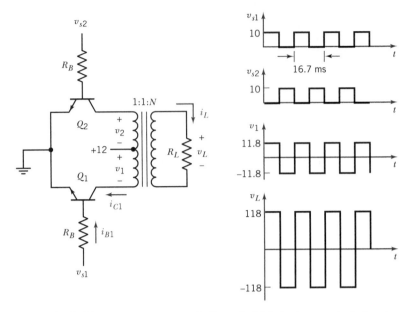

FIGURE 16–20 (a) Power inverter. In Example 16.4 the inverter is designed to convert power at 12 V dc to power at 118 V ac. The output voltage here is a square wave rather than a sine wave.

$= \underline{10}$. When Q_2 is on, $v_2 = -11.8$ V, and $v_L = -118$ V (see the v_L waveform in Figure 16–20).

For a power of $v_L i_L = 100$ W dissipated in the load, the corresponding secondary current must be $i_L = 100$ W/118 V $= 0.85$ A. Then the primary current is $i_{C1} = Ni_L = 10 \times 0.85$ A $= 8.5$ A. Therefore the transistors should be rated for at least $\underline{10\ \text{A}}$. When the transistor is on, it dissipates 0.2 V $\times i_{C1} = 1.7$ W as a result of the saturation voltage. Since the transistor is on half the time, it dissipates an average 0.85 W and should be rated for at least $\underline{1\ \text{W}}$. (The averaging applied here assumes the 60-Hz switching is fast compared with the thermal time constant of the transistor.)

According to Equation (16.8), the transistor is in saturation if $i_{B1} > i_{C1}/\beta = 8.5$ A/25 $= 340$ mA. When the transistor is on, $i_{B1} = (v_{s1} - v_{B1})/R_B = (10-0.7)/R_B$. Therefore $R_B < 9.3$ V/340 mA $= 27.4\ \Omega$. Then let $R_B = \underline{25\ \Omega}$, so $i_{B1} = 372$ mA. When it carries this current, R_B dissipates $i_{B1}^2 R_B = (0.372\ \text{A})^2 \times 25\ \Omega = 3.46$ W. Since it carries current only half the time, it dissipates an average 1.73 W and should be rated for at least 2 W.

The *efficiency* of a power converter such as an inverter circuit is the ratio of the output power to the total input power provided to the circuit. The total power dissipated is 0.85 W in each transistor, 1.73 W in each R_B, and 100 W in the load for a total of 107.8 W. Since all this power must come from the 12-V battery, the input power is 107.8 W. The output power is 100 W. Therefore the efficiency is 100/107.8 $= \underline{92.8\%}$.

16.10   DIODE LOGIC CIRCUITS

Logic circuits are inherently switching circuits since they deal with only two signal voltage levels. In this section diodes and transistors are used as switches to construct some basic logic functions—OR, AND, NOT, and NAND. We have chosen *diode-resistor logic* and *diode-transistor logic* as representative of logic circuit concepts. These are similar to *transistor-transistor logic* (TTL), but they are simpler for instructional purposes. Other logic families such as emitter-coupled logic (ECL) and complementary MOSFET (CMOS) logic also use transistors as switches. CMOS logic is described in the next section, and ECL is illustrated briefly in problem E16.4.

Let an input or an output be defined as "high" (logic 1) when the voltage is between 4 and 5 V. An input or output is defined to be "low" (logic 0) when the voltage is between 0 and 1 V. Consider the OR gate in Figure 16–21. The diodes are oriented so that either A or B can pull the output high by passing current down through R. Because of the 0.7-V diode drop, the output is 4.3 V (high) when either input is 5 V (high).

Consider the AND gate in Figure 16–22. The diodes are oriented so that either input can pull the output low by passing current down through R. Because of the diode drop, the output is 0.7 V (low) when either input is 0 V (low). Then the output is high only when both A and B are high. Because the low voltages are different at the input and the output, this AND gate can't be cascaded—it can't feed another gate like itself. This is true of the OR gate in Figure 16–21 also.

The power dissipated by the OR gate or the AND gate is determined by the value of R. Suppose $R = 1\ k\Omega$. Then when the AND gate has $v_C = 0.7$ V, the current from the 5-V supply is $(5\ V - 0.7\ V)/1\ k\Omega = 4.3$ mA, and the power is 5 V $\times$ 4.3 mA $=$ 21.5 mW. Suppose there is a capacitive load of $C = 15$ pF on the output. In order to go from a low to a high at the output, time is required to charge this C by current through R. The transition time is about $t_T = 2.2RC = 2.2 \times 1\ k\Omega \times 15$ pF $=$ 33 ns. This can be made shorter by decreasing R and increasing the current, but the price is more power. Hence we have a speed-power tradeoff that is typical of most logic families.

A *logic inverter* (logic NOT function) requires input and output to move in opposite directions, something that can't be done with ordinary diodes. Figure

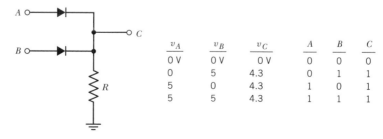

v_A	v_B	v_C	A	B	C
0 V	0 V	0 V	0	0	0
0	5	4.3	0	1	1
5	0	4.3	1	0	1
5	5	4.3	1	1	1

FIGURE 16–21 Logic OR gate. The diodes are oriented so that either A or B can pull the output voltage high. Voltage between 0 V and 1 V represents a logic "0." Voltage between 4 V and 5 V represents a logic "1."

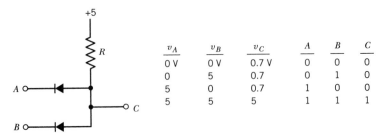

v_A	v_B	v_C	A	B	C
0 V	0 V	0.7 V	0	0	0
0	5	0.7	0	1	0
5	0	0.7	1	0	0
5	5	5	1	1	1

FIGURE 16–22 Logic AND gate. The diodes are oriented so that either A or B can pull the output low. Then C is high only when both A and B are high.

16–23 shows a logic inverter using a bipolar transistor as a switch. R_B is chosen so that the transistor is saturated ($V_C = 0.2$ V) when the input is the minimum high level ($V_A = 4$ V). The reader should confirm that $R_B = 34$ kΩ will assure saturation if the transistor has a β of at least 50. An input level between 0 and 0.3 V assures cutoff, and $v_C = 5$ V.

A NAND gate consists of an AND gate followed by a logic inverter, as shown in Figure 16–24. One reason for combining the two functions is that the transistor serves as a *buffer*; it isolates the output from the input. This allows the input and output to have the same voltage levels—0.2 V for low and 5 V for high—and these levels won't change no matter how many NAND gates are cascaded. Also, it can be shown that any logic function can be performed by a combination of NAND gates.

The extra diode D_1 in the NAND gate is necessary to hold the transistor off when A or B is low. For $v_A = 0.2$ V (low), then $v_1 = 0.9$ V (one diode drop above v_A). But the base voltage must be held below 0.3 V to assure cutoff. The 0.7-V drop provided by D_1 reduces the base voltage to $0.9 - 0.7 = 0.2$ V. As in the design of the logic inverter, R_B in the NAND gate is chosen so that the transistor is saturated when A and B are both high (both diodes are off).

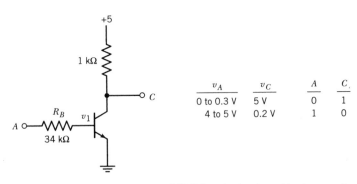

v_A	v_C	A	C
0 to 0.3 V	5 V	0	1
4 to 5 V	0.2 V	1	0

FIGURE 16–23 Logic inverter (NOT function) using a bipolar transistor as a switch. C is high when A is not high.

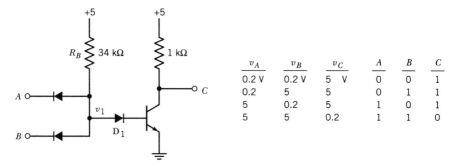

v_A	v_B	v_C	A	B	C
0.2 V	0.2 V	5 V	0	0	1
0.2	5	5	0	1	1
5	0.2	5	1	0	1
5	5	0.2	1	1	0

FIGURE 16–24 Logic NAND gate consisting of an AND gate followed by a logic inverter. Notice the input and output voltage levels are compatible; therefore this gate can be cascaded.

16.11 MOSFET LOGIC

In Section 15.9 we were introduced to field-effect transistors (FETs) that use a diode junction to isolate the gate. These JFETs are used primarily in amplifying applications. FETs used in switching applications usually have a construction that uses silicon oxide (glass) to isolate the gate. The gate is metal, the insulation is oxide, and the transistor body is silicon. Therefore the transistor is called a *metal-oxide-silicon FET*, or *MOSFET*.

In this section we look at logic circuits that use MOSFETs for switches. The simplest (and smallest) use n-channel MOSFETs, or *NMOS FETs*. Low-power logic circuits use both n-channel and p-channel FETs at the expense of more area. This is called *complementary-MOS* or *CMOS logic*. Finally we will see how *programmable logic devices* (*PLDs*) are constructed with MOSFETs.

NMOS Logic

The symbol for an *n*-channel MOSFET is shown in Figure 16–25a. The arrow pointing into the symbol indicates the polarity of the channel semiconductor: *n*-type.

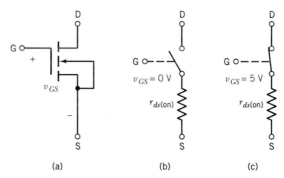

(a)   (b)   (c)

FIGURE 16–25 NMOS enhancement-type FET. (a) Symbol, (b) Model for off state, (c) Model for on state.

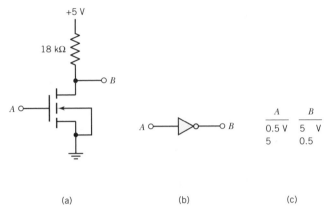

(a)   (b)   (c)

FIGURE 16–26 NMOS logic inverter. (a) Circuit diagram, (b) Logic symbol, (c) Truth table.

The two breaks in the central line of the symbol indicate that it is an *enhancement-type* FET. This means that with $v_{GS} = 0$ the channel is nonconducting, and a positive voltage applied from gate (G) to source (S) will enhance its conduction. For the case shown here, the path from drain (D) to source is an open circuit when $v_{GS} \approx 0$, and the path has a small resistance $r_{ds\ (on)}$ when $v_{GS} = 5$ V (see Figure 16–25b and c). A typical value for $r_{ds\ (on)}$ is 2kΩ, but it can be made as low as 10Ω in applications where the MOSFET must conduct 10 or 20 mA.

As we saw in the previous section, a switching transistor together with a resistor can form a logic inverter. Figure 16–26a shows an inverter using an NMOS FET. When $A = 0.5$ V (close to zero), the FET is off, no current flows, and there is no voltage drop across the resistor. Therefore $B = 5$ V. When $A = 5$ V, the FET acts as a resistor $r_{ds\ (on)}$. For $r_{ds\ (on)} = 2$kΩ, the voltage divider gives $B = 5\ Vr_{ds\ (on)}/(18\ \text{k}\Omega + r_{ds\ (on)}) = 0.5$ V. This inverter action is summarized by the truth table in Figure 16–26c.

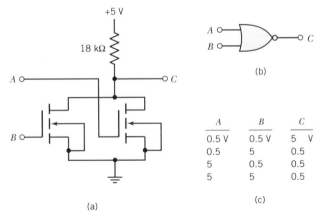

(a)   (c)

FIGURE 16–27 NMOS logic NOR gate. (a) Circuit diagram, (b) Logic symbol, (c) Truth table.

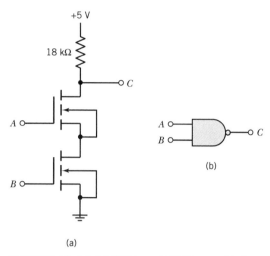

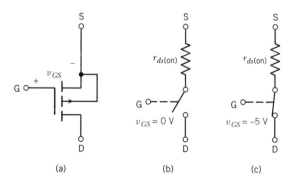

FIGURE 16–28 NMOS logic NAND gate. (a) Circuit diagram, (b) Logic symbol.

A NOR gate can be formed by putting two FETs in parallel, as in Figure 16–27a. Here we are using switches to perform logic as we did in Figure 12–15. A path from C to ground is completed if either FET is on— if A or B is 5 V (see Figure 16–27b). This is a NOR function, as represented in Figure 16–27b.

Figure 16–28a shows the configuration to form a NAND gate. Both A and B must be 5 V to turn both FETs on and pull C low. How much power does the gate dissipate? When both FETs conduct, the current is about 5 V/20 kΩ = 0.25 mA, and the power is 0.25 mA × 5 V = 1.25 mW. If current is conducted half the time on average, then the average dissipation per gate is about 0.62 mW.

CMOS Logic

The power dissipated by a logic gate can be greatly reduced by using both NMOS and PMOS FETs. Figure 16–29a shows the symbol for a p-channel MOSFET (or PMOS FET). Note that the arrow points out of the symbol. The symbol is usually

FIGURE 16–29 PMOS enhancement-type FET. (a) Symbol, (b) Model for off state, (c) Model for on state.

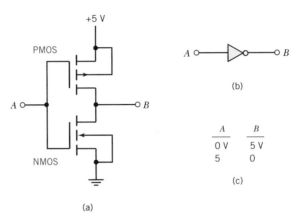

FIGURE 16–30 CMOS logic inverter. (a) Circuit diagram, (b) Logic symbol, (c) Truth table.

oriented with the source (S) at the top so current flows downward. Again, this is an enhancement-type FET, so no current flows when $v_{GS} \approx 0$ (see Figure 16–29b). When $v_{GS} = -5$ V (that is, G is 5 V below S), then the FET is on, and current can flow downward (see Figure 16–29c).

In complementary MOS (or CMOS) logic, both NMOS and PMOS FETs are used, as in the inverter circuit in Figure 16–30a. When $A = 0$ V (5 V below the PMOS FET source), the PMOS FET is on, and the NMOS FET is off. Therefore $B = 5$ V. When $A = 5$ V, the reverse is true, and $B = 0$ V. The important point here is that one or the other of the FETs is off in each case, and no current flows. Virtually no power is dissipated while $A = 0$ V or $A = 5$ V.

There is a brief moment as A passes between the two levels that both FETs conduct at the same time, passing a few μA. This leads to a power dissipation that increases with the switching frequency of the gate. There is also a frequency-dependent power dissipation connected with the charging and discharging of capacitance [See Equations (13.14) and (13.15)]. Therefore CMOS logic provides no power savings if the logic is operated near its frequency limit. But there are great savings if the logic is often idle.

The CMOS logic configuration for a NOR gate is shown in Figure 16–31a. This is the same as the NMOS NOR gate in Figure 16–27a with two PMOS FETs

TABLE 16–1 Truth Table for Circuit in Figure 16–31

A	B	Q_1	Q_2	Q_3	Q_4	C
0V	0V	on	on	off	off	5V
0	5	on	off	off	on	0
5	0	off	on	on	off	0
5	5	off	off	on	on	0

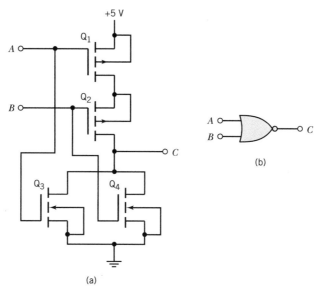

FIGURE 16–31 CMOS logic NOR gate. (a) Circuit diagram, (b) Logic symbol.

replacing the resistor. Table 16–1 shows that, for all four combinations of A and B, the output C is pulled either high or low, but not both. Similarly the CMOS logic configuration in Figure 16–32a realizes a NAND gate. The reader should form a truth table as in Table 16–1 to confirm this.

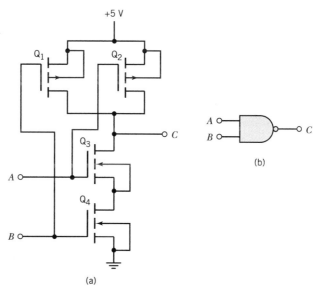

FIGURE 16–32 CMOS logic NAND gate. (a) Circuit diagram, (b) Logic symbol.

Programmable Logic Devices

CMOS logic gates are available commercially as dedicated functions in integrated circuits (ICs). But the trend is increasingly toward user-configurable logic IC. These programmable logic devices (PLDs) were described in Section 13.8. PLDs are usually implemented with NMOS or CMOS logic.

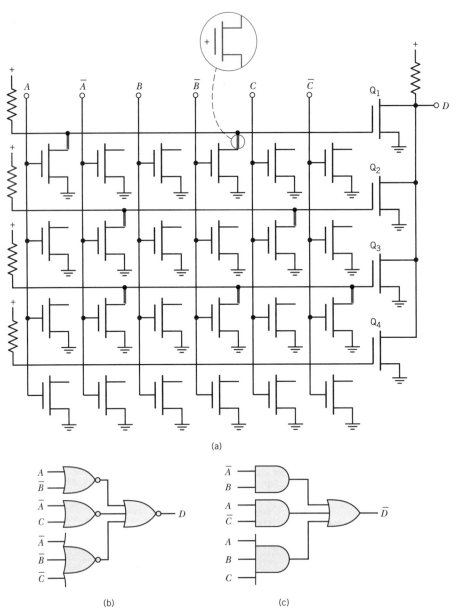

(a)

(b)   (c)

FIGURE 16–33 NMOS programmable logic device. (a) Circuit diagram (note abbreviated MOSFET symbol), (b) Logic symbol, (c) Equivalent logic symbol (using DeMorgan's rule).

The heart of the programmability is the ability to connect different arrangements of signals to the inputs of AND gates, as shown in Figures 13–24 and 13–25. An NMOS implementation is shown in Figure 16–33a. A matrix of FETs are connected horizontally and vertically with metal lines. The user can choose the connections to the horizontal lines, determining the logic function.

In the example here, connections are made so the signal A or the signal $\overline{B}$ can pull the gate of Q_1 down to ground. If the gate of Q_1, Q_2, Q_3, or Q_4 is allowed to go high, then D is pulled low. This logic function is represented by the diagram in Figure 16–33b. By applying DeMorgan's rule [Equations (9.2) and (9.3)], we can convert the logic to that in Figure 16–33c, which has the form of the logic in Figure 13–25.

How are the seven drain connections in Figure 16–33a made? Each of the connections is actually an FET (see inset) that can be made to conduct by applying a positive charge on its gate. This is done electrically with programming lines not shown in Figure 16–33a. Once the charge is deposited, the impedance from the gate to other conductors is high—so high that the charge will remain there for ten years or so. If the circuit is an *erasable PLD* (*EPLD*), the package has a window allowing the user to flood the circuit with ultraviolet light. This causes the insulating semiconductor material to conduct enough that the charge leaks off the gate in about half an hour. The device can then be reprogrammed electrically. The new program is preserved by placing opaque tape over the window.

16.12 SILICON-CONTROLLED RECTIFIERS

Silicon-controlled rectifiers (SCRs) can be used as switches in many situations where a transistor wouldn't be able to handle the voltage or the current. SCRs can be rated as high as 2500 V and 1500 A. Like a transistor, an SCR is turned on by a third terminal called the gate. However, it is not turned off by the gate; like a diode, an SCR turns off when the voltage across it reverses.

Figure 16–34 shows the SCR symbol and a typical SCR characteristic. The three terminals are the anode (A), the cathode (K), and the gate (G). The SCR is

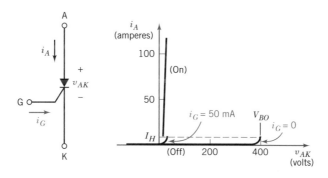

(a)   (b)

FIGURE 16–34 Silicon-controlled rectifier (SCR) symbol and characteristic. The SCR is off until the gate current i_G is pulsed. Then it is on until v_{AK} goes negative.

"off"—conducts no current—as long as the anode-cathode voltage v_{AK} stays less than the *breakover voltage* V_{BO} (see the horizontal part of the characteristic). For the particular characteristic shown, $V_{BO} = 400$ V when the gate current i_G is zero. If v_{AK} does exceed V_{BO}, the SCR conducts, and the voltage drop v_{AK} becomes very small—about one or two volts (see the vertical part of the characteristic). It remains in this "on" state so long as the anode current i_A is greater than the *holding current* I_H. When i_A falls below I_H (typically 100 mA), it returns to the "off" state.

If i_G is increased to the *gate trigger current* I_{GT}—50 mA for this characteristic— V_{BO} is reduced to a few volts, and the SCR goes to the "on" state. Unlike a transistor, the SCR will remain conducting even after i_G returns to zero. This saves significant power in driving the switch (see power dissipated in R_B in Example 16.4). When the anode-cathode voltage is reversed ($v_{AK} \leq 0$), i_A goes to zero, and the SCR is off.

A typical SCR application is in a speed-control circuit for a motor. The speed of a dc motor is roughly proportional to the average armature voltage V_a applied to the motor (see Chapter 19), and an efficient way to reduce V_a is to use a switch to apply voltage for only short intervals. The SCR is used as a switch in series with the motor and an ac source v_1, as shown in Figure 16–35a. The motor is modeled here by an equivalent circuit comprising the armature resistance and inductance R_a and L_a, and a back voltage V_g (see Chapter 19). When the SCR is turned on, $v_a \approx v_1$. When the SCR turns off, the *freewheeling diode* permits i_a to keep flowing, and $v_a \approx 0$.

The waveforms for the speed-control circuit are shown in Figure 16–35b. When v_1 goes positive, the SCR remains off until a *trigger pulse* is applied by v_s (the circuitry to generate the pulses is a study in itself). The pulse develops enough i_G to *fire* the SCR, and it stays on until v_1 goes to zero. The resulting waveform applied to the motor is $v_a = v_1$ when the SCR is on, and $v_a = 0$ elsewhere (this ignores the

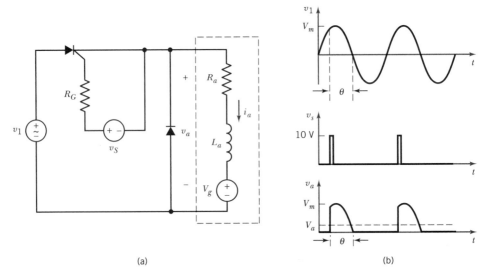

(a)   (b)

FIGURE 16–35 Motor speed control circuit. The average voltage V_a applied to the dc motor's armature is reduced by turning on the SCR for only part of the time v_1 is positive. (See Example 16.5.)

small voltage drops of the SCR and diode). Let V_a be the average of v_a. Then for $v_a = V_m \sin(\omega t)$,

$$V_a = \frac{1}{2\pi} \int_{\pi-\theta}^{\pi} V_m \sin(\omega t) \; d\;(\omega t) = \frac{V_m}{2\pi}(1 - \cos\theta), \qquad (16.11)$$

where V_m is the peak value of v_1, and θ is the *conduction angle*. The largest average voltage is for $\theta = \pi$; then $V_a = V_m/\pi$. For the case shown in Figure 16–35b, $\theta = 2\pi/3$, or 120°, and $V_a = \frac{3}{4} V_m/\pi$. As the trigger pulse is delayed, θ is reduced, V_a is reduced, and the motor speed is reduced.

EXAMPLE 16.5 The speed of a dc motor is to be controlled using the SCR circuit shown in Figure 16–35.

Given: The SCR characteristic is that shown in Figure 16–34. The gate trigger current is $I_{GT} = 50$ mA, and the gate trigger voltage is $V_{GT} = 2$ V. The motor has $R_a = 0.2\ \Omega$, $L_a = 160$ mH, and $V_g = 41$ V (this depends on the motor's load). The ac source voltage is $v_1 = (320\text{ V}) \sin(\omega t)$, where $\omega = 2\pi \times 60$ Hz. The height of the v_s trigger pulses is 10 V, and the conduction angle is $\theta = 90°$.

Find: the value of R_G to fire the SCR (a good rule of thumb is to make $i_G = 3I_{GT}$). Find the average armature voltage V_a and the average armature current I_a.

Solution: When $v_s = 10$ V, then $v_{GK} = V_{GT} = 2$ V, and $i_G = (10\text{ V} - 2\text{ V})/R_G$. For $i_G = 3I_{GT} = 150$ mA, we must choose $R_G = 8\text{ V}/150\text{ mA} = \underline{53\ \Omega}$.

The maximum possible V_a is $V_m/\pi = 320\text{ V}/\pi = 102$ V. For $\theta = 90°$, Equation (16.11) gives $V_a = \underline{51\text{ V}}$.

By assumption $V_g = 41$ V. Then the average voltage across R_a is $V_a - V_g = 10$ V, and the average current is $I_a = 10\text{ V}/R_a = \underline{50\text{ A}}$. The inductor L_a keeps this current going even when the SCR is off. The reader can show the current ripple is only about 10%. ■

16.13   SUMMARY

Both diodes and transistors can be used as switches in electronic circuits. A diode responds to the signal it is switching, becoming a short circuit or open circuit depending on the direction of its current and voltage. A transistor is switched by a third terminal—the base or the gate.

A bipolar transistor is modeled by a switch in series with a 0.2-V saturation voltage. The transistor is on (in saturation) when $i_B > i_C/\beta$, and it is off (in cutoff) when $v_{BE} \leq 0.3$ V. When it is on, a bipolar transistor can conduct current only in one direction—collector to emitter for an *npn* type.

A field-effect transistor (FET) is modeled by a switch in series with a resistance $r_{ds\;(\text{on})}$. It is on when the gate-source voltage v_{GS} is zero. It is off when v_{GS} is below the pinch-off voltage V_P, which ranges from -1 V to -10 V from device to device. When it is on, an FET can conduct current in either direction. Also, an FET serves as a better switch than a bipolar transistor in low-voltage applications.

Diodes and transistors are nearly ideal switches when they are off; "off" current is typically in the nanoamperes. When they are on, these devices are not quite so ideal. A diode has a voltage drop of about 0.7 V, a bipolar transistor has a saturation voltage of about 0.2 V, and an FET has an on resistance between 2 Ω and 500 Ω. Still, when compared with mechanical relays that are nearly perfect shorts when on, semiconductor devices are cheaper, can be switched faster, and are more reliable.

A Zener diode is a diode with a controlled breakdown set typically between 3 V and 50 V. It is used to establish a fixed dc voltage or to limit a varying signal to some peak voltage.

An SCR is useful in switching high power. It becomes conducting when a pulse of current is applied to the gate. It remains on until the voltage from anode to cathode reverses. SCRs with voltage and current ratings in the thousands are available.

Examples of circuits using switching devices are rectifiers, peak detectors, clippers, regulators, dc power supplies, power inverters, chopper-stabilized amplifiers, AM modulators and demodulators, logic circuits, and motor speed controllers.

FOR FURTHER STUDY

A. P. Malvino, *Electronic Principles*, McGraw-Hill, New York, 1984.

P. R. Gray and R. G. Meyer, *Analysis and Design of Analog Integrated Circuits*, Wiley, New York, 1977.

D. A. Hodges and H. G. Jackson, *Analysis and Design of Digital Integrated Circuits*, McGraw-Hill, New York, 1988.

PROBLEMS

Easy Drill Problems (answers at end of chapter)

D16.1. The current waveform $i(t)$ in Figure 16–3c consists of the positive half-cycles of a sinusoid with amplitude I_m. Show that the average value of $i(t)$ is I_m/π, where the average is defined in Equation (16.1).

D16.2. Find the average current I_{dc} provided by the half-wave rectifier in Figure 16–3b for $v_1 = (300 \text{ V}) \sin(\omega t)$ and $R_L = 10 \ \Omega$. What is the peak current the diode must handle?

D16.3. Sketch the waveform $v_2(t)$ for the clipper circuit in Figure D16.3 when $v_1(t) = (5 \text{ V}) \sin(\omega t)$. Assume an ideal-diode characteristic. Repeat the problem for the orientation of the diode reversed.

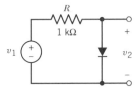

D16.4. Repeat D16.3 using the more complete diode model in Figure 16–15.

D16.5. Sketch the waveform $v_2(t)$ for the clipper circuit in Figure 16–4 when the orientation of the diode is reversed.

D16.6. Choose the largest value of R_d for the voltage regulator in Figure 16–9 for which the load voltage v_L remains 10 V while V_{CC} varies from 11 V to 13 V and the load R_L varies from 10 Ω to 100 Ω. The Zener diode requires $i_Z \geq 50$ mA in order to maintain 10 V. What is the maximum power dissipated by the Zener diode?

D16.7. Suppose the source v_1 in Example 16.2 is 400 Hz rather than 60 Hz. All other specifications are the same. What is the minimum value for C now?

D16.8. How must Equation (16.6) be modified if the 0.7-V diode drop is taken into account?

D16.9. Using diodes and resistors, design a three-input OR logic gate so $D = A + B + C$. (The "$+$" is the Boolean symbol for "OR.")

D16.10. Show that $R_B = 34$ kΩ in Figure 16–24 assures saturation when both A and B are high. The transistor β is 50.

D16.11. Show that in Example 16.5 the ripple is 4 A peak-to-peak. (Use the approximation that V_g and the voltage across R_a remain constant and the voltage across L_a reverses, decreasing i_a, when the SCR is not conducting.)

Application Problems

P16.1. Design a circuit to limit the output voltage v_2 to the range 6 V to 8 V, as shown in the example in Figure P16.1. Use resistors, (ideal) diodes, and voltage sources.

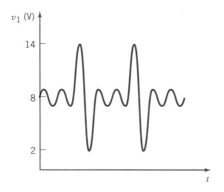

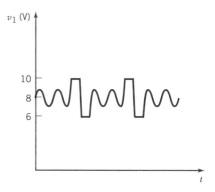

P16.2. Repeat P16.1 using only resistors, Zener diodes, and a $+15$-V supply.

P16.3. The two Zener diodes in Figure P16.3a each have a breakdown voltage of 10 V; the characteristic of I vs V_A is the same as the characteristic in Figure 16–6b. Sketch the characteristic of I vs V_B. Sketch the combined characteristic of I vs V_T. (*Note:* The current is the same, and $V_T = V_A + V_B$.)

Sketch the waveform of $v_2(t)$ in Figure P16.3b for $v_1(t) = (20 \text{ V}) \sin(\omega t)$.

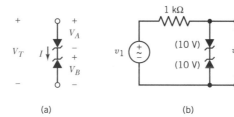

(a) (b)

P16.4. A dc power supply like that in Figure 16–10 is to be designed to provide 5 V at up to 3 A, and the ripple is to be 50 mV peak-to-peak or less. The ac source v_1 is 115 V rms at 60 Hz. Find the turns ratio of the transformer and the minimum value of C.

P16.5. Solve the design problem in Example 16.3 for the case when $v_e = 3\sin(\omega t) + 6$. Assume an ideal diode.

P16.6. For the bipolar transistor chopper circuit in Figure 16–18a, let $v_m^+ = (8\text{ V})\sin(\omega t) + 10$ V, the maximum v_s equal 5 V, and the minimum transistor β equal 100. What is the maximum value of R_B that will assure saturation for $v_s = 5$ V?

P16.7. Suppose that the chopper-stabilized amplifier in Figure 16–19a is to be high precision. The FET has the specifications $V_P = -10$ V and $r_{ds\,(on)} = 30\ \Omega$. Choose a value of R so that the voltage change at v_o is within 0.1% of v_1. Sketch a suitable waveform for v_s.

P16.8. The "integrate-and-dump" circuit in Figure P16.8 acts as an integrator when the FET switch is off (see Section 14.5). When the switch is on, it resets v_2 to zero. Sketch the waveform of v_2 for $v_1 = -5$ V. Given $V_P = -3$ V and $r_{ds\,(on)} = 500\ \Omega$, how completely does the switch discharge the capacitor during the 10 ms?

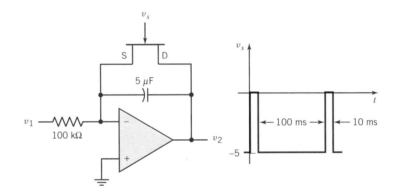

P16.9. A power inverter as in Figure 16–20 is to convert 6 V dc to 115 V ac. The load R_L draws 100 W of power. The switch-control voltages v_{s1} and v_{s2} are 5-V pulses. The transistors have a β of 50 and a saturation voltage of 0.4 V. Find the transformer turns ratio to give the 115 V ac, and find the value of R_B to assure saturation. Find the efficiency of the inverter.

P16.10. Let the AND gate in Figure 16–22 have a capacitance of 15 pF loading its output. Choose R so that the transition time (when both v_A and v_B go to 5 V) is 10 ns. What power is dissipated by the gate when $v_A = 0$ V? Repeat for $t_T = 20$ ns.

P16.11. Let the output of the OR gate in Figure 16–21 be connected to one input of the AND gate in Figure 16–22. When both inputs to the OR gate are at 0 V, what voltage is at the output of the AND gate? What logic level does this correspond to?

P16.12. Let the output of the NAND gate in Figure 16–24 be connected to one input of a second NAND gate. When both inputs to the first NAND gate are 5 V, what voltage is at the output of the second NAND gate? What logic level does this correspond to?

P16.13. Form a truth table for the circuit in Figure 16–32 similar to the one in Table 16–1.

P16.14. Form a truth table for the circuit in Figure P16.14 on p. 559. What is the name of this logic function?

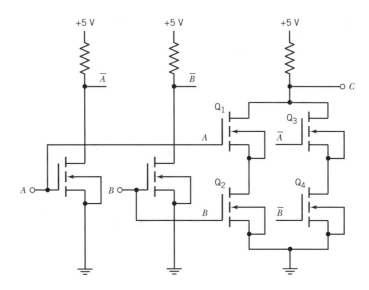

P16.15. Replace the resistors in Figure P16.14 with appropriate arrangements of PMOS FETs so it becomes CMOS logic.

P16.16. On a matrix with columns A, $\overline{A}$, B, $\overline{B}$, C, and $\overline{C}$ and with rows Q_1, Q_2, Q_3, and Q_4, indicate which FETs in Figure 16–33a should be connected to realize the logic function $\overline{D} = \overline{AB}\,\overline{C} + \overline{A}B\overline{C} + \overline{A}\,BC + ABC$.

P16.17. The motor modeled by R_a, L_a, and V_g in Figure 16–35a has a speed of 1000 rpm when the average armature voltage is $V_a = 30$ V. For $v_1 = (230\text{ V})\sin(\omega t)$, find the conduction angle θ that will produce a motor speed of 1000 rpm. What is the minimum (off) breakover voltage for the SCR? For $V_{GT} = 2$ V, $R_G = 53$ Ω, and 10-V, 1.0-ms pulses at v_s, what power is dissipated by R_G?

Extension Problems

E16.1. For the *full-wave rectifier* circuit in Figure E16.1, only two diodes are on at a time. Which two are on when v_2 is positive? What two are on when v_1 is negative? Sketch $v_3(t)$ for $v_2(t) = (14\text{ V})\sin(\omega t)$, where $\omega = 2\pi \times 60$ Hz. Assume ideal diodes. Find the average value of the load current i.

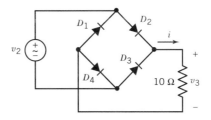

E16.2. The full-wave rectifier in Figure E16.1 can be converted to a dc power supply by putting a capacitor in parallel with the 10-Ω load resistor. Given $v_2(t) = (14\text{ V})$

$\sin(\omega t)$, where $\omega \times 2\pi \times 60$ Hz, choose a capacitor value so that the ripple of the output voltage is 2% or less. Assume ideal diodes. Compare this value with that chosen in Example 16.2.

E16.3. Sometimes the resistor R in the envelope detector in Figure 16–13a is replaced by a current source I. Solve for an upper bound on C in terms of I and v_e if v_3 is to track v_e. Find the maximum C for $I = 0.5$ mA and $v_e = 4\sin(\omega t) + 6$, where $\omega = 2\pi \times 16$ kHz $= 100$ krad/s.

E16.4. The circuit of an emitter-coupled logic (ECL) gate is shown in Figure E16.4. The 10-mA current flows through the transistor(s) with the highest base voltage. Let logic "1" and "0" correspond to -0.8 V and -1.8 V. Find the voltage at C for each of the four combinations of logic levels at A and B. (Neglect base currents.) What is the logic function performed by this gate? Use $V_{BE} = 0.8$ V.

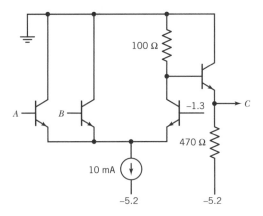

Study Questions

16.1. What is ideal about an ideal diode?

16.2. Name two ways (using different components) to clip a signal.

16.3. What kind of diode usually conducts current in the direction opposed to the arrow?

16.4. The voltage from a dc supply is not absolutely constant. What are the small voltage variations called?

16.5. A peak detector and an envelope detector have the same circuit. What is the distinction between the two?

16.6. Of a bipolar transistor and a field-effect transistor, which is the better switch at high currents? For what current do they perform equally if $r_{ds\ (on)}$ is 100Ω?

16.7. Why is it difficult to amplify small dc voltages?

16.8 Define the efficiency of a dc power supply. Define the efficiency of a power inverter.

16.9. The simplest logic is diode-resistor logic. What is its disadvantage?

16.10. With respect to conducting current, do two switches in series perform an AND function or an OR function?

16.11. NMOS logic uses fewer transistors than CMOS logic. What is the advantage of CMOS logic?

16.12. How is the program stored in a programmable logic device? How is it erased?

ANSWERS TO DRILL PROBLEMS

D16.2. $I_{dc} = 9.55$ A, $I_m = 30$ A

D16.3. v_1 eliminated above zero; v_1 eliminated below zero

D16.4. v_1 eliminated above 0.7 V; v_1 eliminated below -0.7 V

D16.5. v_1 eliminated below 10 V

D16.6. $R_d = 0.952$ Ω, $P = 30.5$ W

D16.7. 12,500 μF

D16.8. $\dfrac{|dv_e/dt|}{v_e - 0.7} < \dfrac{1}{RC}$

D16.9. Add a diode to Figure 16–21.

D16.10. $i_B = 1.1 \times i_C/\beta$

PART THREE

MACHINES
AND
POWER

CHAPTER 17

Plant Power Systems

Engineers of all disciplines are frequently called on to operate and maintain the electric power systems in their plants or at work sites. Not infrequently they help plan or expand them.

For major changes on all but very small systems, it is desirable to engage an electrical engineer consultant who specializes in this kind of work. The local company supplying power will be of help with initial thinking and early planning. At any rate their capabilities and requirements must be considered.

The purpose of this chapter is to assist readers in operating and maintaining their plant power systems, and to enable them to work effectively with the power company and consultants on initial design and future expansion.

The chapter begins with a short consideration of the U.S. power grid and touches briefly on the important issues of energy sources and pollution. It then discusses distribution apparatus, protection, power cost, and power factor correction.

Reader objectives here will vary widely depending on interest, but all engineers may want to become conversant at least with the problems involved in power systems and the production and distribution of electrical energy.

Readers with responsibilities in this area can familiarize themselves in greater detail with

1. plant apparatus
2. system protection

3. interaction with the power company

4. power factor correction

17.1   THE U.S. ELECTRIC POWER SYSTEM

Figure 17–1 shows the general scheme of electric power production and use with a *one-line diagram*. In one-line diagrams a single line represents the three or four conductors of a three-phase system. Note that in the one-line diagram the transformer symbol is usually modified as shown.

Large *central stations* on the left *generate* bulk electric power, which is then *transmitted* to *substations* near points where it will be used. From the substations power is *distributed* to users.

Continuing with Figure 17–1, generation is usually at several thousand volts, often 12 kV or 24 kV. By means of transformers voltage is raised to several hundred thousand for transmission on lines striding across the country. At the ends of the transmission lines, substation transformers step down the voltage for distribution to customers.

In some parts of the country major generation sources are first interconnected to form a *bulk power network* or *vhv* (very high voltage) network at such voltages as 345 kV to 765 kV. These interconnections improve system reliability, allowing quick substitution of one power source for another. Long-distance transmission at these voltages then feeds a *subtransmission* network at perhaps 115 kV or 230 kV which terminates at appropriate points in substations for distributions to customers or customer areas.

Primary distribution voltages (by IEEE standard*) range from 4160 through 34,500 with 12,470 and 13,200 the most common. Such numbers are nominal, actual voltages varying somewhat from these values. Voltage may be reduced further for more detailed distribution, often at 4000 or 2300 V. Large plants may have their own unit substations near loads.

For residential and light commercial use, transformers (often mounted on poles) reduce distribution voltages to the common 220/110 single-phase system. Large commercial users (for example high-rise buildings) and industrial plants are supplied with three-phase power at distribution voltages. Except for very large machinery, most user electrical equipment operates at voltages of 110, 220 or 480 V nominal.

---

*IEEE Standard 141–1986, *Recommended Practice for Electric Power Distribution for Industrial Plants*, commonly known in industry as the "Red Book," provides a thorough and detailed 600-page discussion of its subject matter, with references to other standards and sources as needed. Much of the material presented in this chapter is based on this source.

Standards and industrial practice guides for various aspects of industrial plant power systems are promulgated by the IEEE (Institute of Electrical and Electronics Engineers—the principal professional society of electrical engineers)—and ANSI (American National Standards Institute—a coordinating group for American standards)—and also by UL (Underwriters' Laboratory—a principal electrical testing laboratory), NEMA (National Electrical Manufacturers' Association), the NFPA (National Fire Protection Association), EEI (Edison Electric Institute), and others. See Chapter 21 for a discussion of NEC (the National Electrical Code).

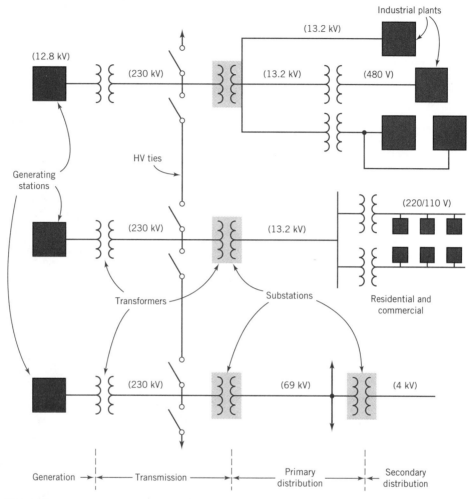

FIGURE 17–1 Most U.S. electric power is generated at a few hundred large central stations, transmitted at very high voltages to substations near users and distributed to users at lower voltages. Voltage numbers in parentheses are typical. The "one-line" diagram (where a single line represents all three phases) is commonly used for power system description.

17.2   ENERGY SOURCES FOR ELECTRIC POWER

Large power station generation is presently cheaper per unit of power and energy delivered than generation by small plants. Exceptions are at remote sites such as solar-powered mountain-top microwave communication repeaters. To such a site, transmission costs would be prohibitive.

If further development reduces the cost of gas fuel cells or direct-conversion solar cells enough, it may become feasible to power houses or perhaps small residential and commercial neighborhoods with their own conversion plants—with consequent competition and flexibility.

A power station may have installed capacity of several hundred to a thousand or more megawatts of power. How much it produces at any instant depends on customer demand—whether users have turned on the switch for their various appliances and loads. Energy sources for such stations are largely hydro power (at large dam sites), fossil fuels (coal and oil), or nuclear fuels. Direct solar conversion, wind, and tidal power are only in their infancy. Wood, waste materials, and volcanism are used in special situations.

There appears to be enough coal in the world for several hundred years of electrical production. Available nuclear resources depend greatly on technological development but *may* be almost inexhaustible. Because of pollution problems—from both nuclear and fossil fuels—and because of their eventual depletion, more attention and effort are being given to so-called alternative or renewable sources. Figure 17–2 provides some statistics of world and U.S. electrical production and resources.

In the past, reflecting many years of cheap electric energy supply, U.S. electrical practices have been somewhat wasteful. Because of the threatened future energy shortage (how far in the future depends on the expert making the analysis and on the outcome of certain international problems with respect to oil) much

1989 Net U.S. electric generation (10^9 kWh)	2784
1979–1989 average U.S. annual generation growth (%)	2.17
1989 installed generating capacity (10^6 kW)	684.6
1979–1989 average annual capacity growth (%)	1.93

U.S. Energy Reserves (quads): Coal 21,400, oil 870, gas 950, uranium 1570 (or 160,000 with breeder technology). (This data is for the period around 1985 and is quite approximate.)

1988 Fuel use for U.S. electric generation (quads): coal 15.85, oil 1.56, gas 2.71, hydro 2.61, nuclear 5.66, other (small), total 28.63.

1988 Net U.S. electricity generation by prime mover (10^9 kWh): coal 1541, oil 144, gas-fired 236, gas turbine 22, nuclear 527, hydro 223, other 12, total 2704.

CONVERSIONS:

Multiply		**by**		**to obtain**	
	Btu		2.9305×10^{-4}		kWh
	Quads		10^{15}		Btu
	bbls oil		5.8×10^{-9}		Quads
	short tons coal		22×10^{-9}		Quads
	cu ft gas		1.031×10^{-12}		Quads
	bbls oil		42		gallons
	miles supply train		10,000		tons coal
	therms		10^5		Btu
	calories		4.186		joules
	pounds		453.6		grams

FIGURE 17–2 Energy statistics for recent years, and conversions.

more emphasis is being placed on conservation. There is, for example, considerable activity in *low-head hydro*, utilizing previously unused low-head sites and, particularly, converting old mill-dam sites to supplementary electrical generation. But for any large-scale generation, hydro sites are essentially used up.

Recovering nuclear energy is theoretically possible either by fission (splitting the nuclei of heavy uranium ore derivatives) or fusion (compacting light nuclei like hydrogen). Fission is a proven and heavily utilized technique. Controlled fusion, researched heavily for many years, has recently (late 1991) been demonstrated by a research team in Britain, although commercial use appears still to be decades away. The availability of practically unlimited fusion fuel from sea water makes the process particularly attractive.

17.3   POLLUTION PROBLEMS OF ENERGY SOURCES

Pollution is simply too much of anything—too much noise, for example, or too much garbage, heat in an estuary, smoke, or visually unattractive structures. Every electric generation, transmission, or distribution technology pollutes. Some are worse than others.

In nuclear plants the amount of fuel and waste handled is small—less than about 50 tons per plant per year. Nevertheless some ash (waste material) from nuclear plants emits dangerous radiation for thousands of years and must therefore be handled and guarded over that period in such a way as to minimize health and other hazards.

While the amount of nuclear waste is small, the consequences of dirtying up the environment with it are dire. Engineers may overcome these limitations and find a treatment or even use for this spent fuel. A further problem is decommissioning hot old plants ("hot" in this sense means that some of the parts now emit radiation in various amounts). It is still not entirely proven that nuclear energy can be used continuously over many years. Nuclear plants appear also to hazard malfunctioning, which might affect thousands of people living nearby.

The well-publicized 1980 operator-error accident at the Three Mile Island plant is a case in point. In spite of grievous mistakes in meeting an equipment fault emergency, no hazardous radiation escaped and apparently no one was harmed. But much of the public in that area of Pennsylvania and in other places throughout the United States is thoroughly frightened; economic losses to the region are in the hundreds of millions of dollars.

The 1986 fission-plant disaster at Chernobyl in the Ukraine was caused again by operator error but was far more serious because of inept safety design. It killed 31 people and injured hundreds more, and spread radioactive pollution over much of northeastern Europe, temporarily poisoning food supplies and some crops. In the Ukraine itself much crop land was contaminated for an unknown period into the future.

Experts appear unanimously confident that there is no possibility of a nuclear explosion is these plants. We appear to have already experienced the worst possible catastrophes (reactor meltdown); one well handled and the other handled very poorly.

At present the fusion process appears to hazard no polluting waste. But it seems likely that the plants themselves will be radioactive when decommissioned.

In consequence of these problems and of citizen protest and resultant government regulation, planning and construction of new nuclear plants in the United States has almost stopped. Nuclear design and construction costs are prohibitive in many cases—to a large extent because of technological difficulties of safety regulation and some regulatory ineptness. Yet other parts of the world continue successful and apparently economical nuclear electrical development and increasing use. In the United States in the early 1990s about 15% of installed electric capacity is already nuclear; and roughly 20% of generation is now nuclear. There are approximately 100 operating nuclear electric plants in the United States.

Each large coal- or oil-fired plant, in contrast with nuclear plants, must handle several million tons of fuel per year, bringing it often from great distances by railroad, pipeline, tanker, and barge. Similar amounts of gaseous emissions and ash require disposal. Stack emissions pollute the air, make breathing harmful, and apparently injure lakes, streams, and vegetation. Ocean problems and the greenhouse effect in the upper atmosphere are not well understood. Most of these things are disputed among experts.

Fossil fuel production and transportation are high-risk areas for industrial deaths, accidents, and health hazards. The risk from nuclear plants is not as well understood, is yet somewhat speculative, and seems to represent a low probability of exceedingly serious events. Fossil fuel procurement, use, and pollution are a present problem involving annually hundreds of fatalities and injury to thousands.

In addition to exercising their individual civic responsibilities, most engineers feel a professional obligation to contribute intelligently to solving the kinds of problems discussed above. Their everyday duties tend to develop an almost unique skill in looking and planning ahead. Engineers also follow an ethical code which requires them to hold the public free from harm due to their work.

17.4   PLANT DISTRIBUTION SYSTEMS

Plant electric loads vary widely but in general include motors—mostly induction—lighting, electric heating, cooling, electroplating or other electrochemical processes, small appliances, and tools. Voltage-reducing transformers are required for part of the load and so may constitute loads in themselves for parts of the system. Some plants use electric welding extensively.

Each load must be connected to the plant power source in some way. This connection network, with its switching, protection, and metering is called a *plant distribution system* and, together with some of the loads, constitutes the plant electric power system.

Figure 17–3 shows a one-line diagram of a typical small plant distribution scheme. Typical voltages are noted in parentheses. Power comes from power company lines on the left, in this case at 13,200 V, is transformed to 480 V, and flows to loads at the right side of the diagram via various circuit breakers (special switches to be described below). Loads are shown sectionalized into four groups, each with its own feeder line and protected by its own breaker.

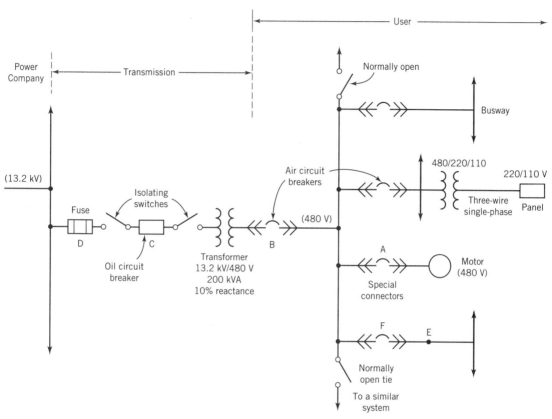

FIGURE 17–3 Small plant distribution system. Load is sectionalized by means of switches so that trouble in one part can be isolated, allowing other parts to be supplied.

In the top load section the breaker feeds a 480-V busway system (in a certain part of the plant) which provides a Tinker-Toy-like covered bus system into which various loads can be plugged or tapped. The term *bus* indicates a heavy conductor (often rectangular in cross section) to which various loads can be connected. The second section includes, among other loads, another transformer to provide 220/110-V single-phase power for lighting and small loads including hand tools and smaller motors. The third section is dedicated solely to a large, 480-V induction motor. Section 4, at the bottom, provides three-phase, 480-V power to various loads in another part of the plant.

Consider the following features of this design moving from left to right. An oil circuit breaker allows the whole system to be completely disconnected from the source in case of emergency, or for required maintenance. The two isolating switches allow the breaker itself to be electrically isolated and safely removed for service or replacement. Similarly an air circuit breaker on the low-voltage side, in connection with the oil circuit breaker, allows the transformer to be isolated. The double arrow connectors on each side of the air circuit breaker symbols indicate that these breakers are built so that they can be withdrawn physically from their mountings, automatically disconnecting and isolating them for service or replacement.

The system shown in Figure 17–3 could be more extensive in an actual situation. In larger plants there may be several 13,200-V transformers, each with its own system of loads for some particular part of the plant, each drawing power from the common primary high-voltage circuit. With these multiple transformers and sectionalized 480-V buses, if one of the main transformers becomes inoperative, the system loads shown can still be supplied by closing a tie circuit breaker (at the top or bottom of the diagram) so that the disabled 480-V bus is fed through another transformer. This feature will also permit scheduled down time for maintenance on the transformers and some of the switching equipment associated with them.

Other configurations are possible for plant distribution systems, some providing for duplication of high-voltage supply lines on the left to further enhance system reliability.

Design goals for a plant electrical system are

1. safety for life and property;
2. reliability and continuity of service;
3. maintainability, simplicity, and ease of operation;
4. good voltage regulation;
5. minimizing cost of installation and operation; and
6. flexibility—particularly with respect to future plant expansion.

In most plants continuous improvement, expansion, and change go on. It is important to see that piecemeal changes do not needlessly destroy excellence designed into the original electrical system, based on the six goals listed above. As production moves into new areas or increases, plants become crowded. Electrical systems become less satisfactory unless provision has been made for orderly expansion.

It should be understood by engineers involved in plant electrical systems that these systems are almost always designed and operated under the requirements of the NEC (National Electrical Code). The NEC provides very detailed requirements for almost all features of plant systems, for example, distribution conductors, switches and protective devices, provisions for appliances of every kind, motors and motor controllers, lighting installations, special installations such as those in hazardous environments, transformer installations, and grounding. Engineers responsible for plant electrical systems will want to familiarize themselves with the general provisions of this code, and possibly with local or other codes that may legally govern their systems. Chapter 21 discusses the NEC further.

17.5   SWITCHES AND CIRCUIT BREAKERS

The application of various types of switches in a power system appears at first glance to be simple and straightforward. But this is not the case. Switches are used for different purposes including: to disconnect equipment (for example, the isolating switches of Figure 17–3), to interrupt load currents (for example, turning off equipment such as motors or lights), for safety purposes (such as interrupting fault—short-circuit—currents), for transferring equipment from one source of power to another, and so on. They have definite current and voltage limits which must be taken into account.

High-voltage or high-current switches are usually called *circuit breakers*—particularly if they open fault currents automatically or can be operated remotely. Circuit breakers may be air-break devices of various kinds or, for higher voltage and current ratings, oil immersed. *Fuses* are an important type of interrupting device in power systems. *Contactors* are heavy-duty switches intended to be used more or less continually, for example, the switches in a motor controller.

To get some idea of the technical problems of switches, consider a simple knife switch as shown in Figure 17–4a. A copper blade is moved up and down by hand to make or break the circuit between two posts to which conductors connect. At the pivot the blade rotates between two tightly spring-compressed copper members. These wiping contacts must be able to carry rated current. At the other end, the blade is forced in between two similar contacts, under squeeze pressure of some sort, or is pulled out of them. Figure 17–4b shows a circuit diagram and switch function.

Suppose the power source to be broken is 480 V and the load draws a normal current of 50 A from it through the switch. All parts of the switch, and the wires connecting load and switch, must be able to withstand a 50-A current and some certain voltage to ground, possibly a voltage of 277 V. When the switch is opened it must break 50 A. As the blade comes away from the contacting surfaces on the

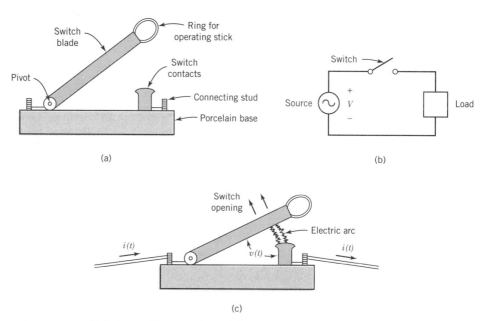

(a)

(b)

(c)

FIGURE 17–4 Switches and breakers have definite limits in the currents they can break, as well as the voltages and currents they can withstand.

right-hand side (Figure 17–4c), an electric arc is drawn in the air between blade and contacts. These arcs must dissipate considerable energy.

The arc is caused principally by two phenomena—first, air has a finite dielectric strength and will break down or ionize and permit current to flow through it if the voltage gradient is more than about 30 kV/cm, which is bound to be the case as the switch opens. Second, in all circuits there is some inertial (inductive) effect which prevents current from being stopped instantly.

This switching energy lost in the arc could be calculated as

$$W = \int_{t=0}^{t=t_0} vi\ dt, \qquad (17.1)$$

where the contact is broken at $t = 0$, the arc goes out at $t = t_0$, and where v and i are the instantaneous voltage across the switch and the current through it.

To keep the energy integral as small as possible, a switch should open very quickly. It must open wide enough to prevent the circuit voltage from maintaining the arc. It must be constructed heavily enough so that the heating and burning of the contacts by the arc integral does not destroy them. Thus switches have current and voltage *ratings*—maximum values up to which the switch is safe to use. Used within these ratings the switch will serve for a useful life period.

If a dc circuit is heavily inductive it will try to maintain its current. Thus it can build up a larger voltage than the supply and burn the switch that is attempting to break it even more badly. Inductive circuits contribute a stored energy of $\frac{1}{2}LI^2$ J to the arc. For this kind of circuit it is customary to specify an especially heavy switch, or take measures to dissipate the inductive energy in some other manner.

Providing a switch able to break the normal load current, however, may not be sufficient. All electrical circuits are subject to *faults*—to troubles of various types. Lightning may strike some part of the circuit, insulation may fail somewhere, wires may inadvertently come together, someone may make a serious mistake in operating the system or run a fork truck into an electrical installation. Well-designed, -constructed, and -maintained systems should have few faults. But even the best system will have trouble occasionally and these faults must be protected against electrically. Most faults are short circuits of various kinds.

Consider, for example, the three-phase 480-V circuit of Figure 17–5a. The three sources in this circuit are shown as the secondaries of three transformers. Figure 17–5b shows the one-line *equivalent-wye* diagram using the usual source symbol. Whether the system being modeled is wye or delta, one-line diagrams are often calculated as equivalent-wye in which the single line carries the voltage of the actual lines against ground and actual balanced line current. Power involved in the single line is then tripled for the entire three-phase system.

The designation ''480-V three-phase'' always refers to line-to-line voltage. Thus the line-to-ground voltage of this delta system will be $480/\sqrt{3}$ or 277 V. Algebra will show that the equivalent-wye impedance Z_e for a delta is the delta $Z_{\text{phase}}/3$. For a wye system, of course, the equivalent Z is the same as the actual phase Z. The ground connection across the bottom is usually considered to have zero impedance, but in unbalanced-fault-current calculations a small impedance here may be important.

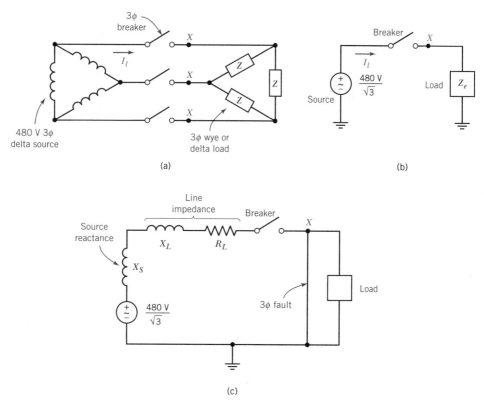

FIGURE 17–5 "Equivalent wye" one-line diagram (b) for three-phase circuit (a). Fault calculation with this diagram (c).

Now suppose the switch is closed and there is a sudden fault at the load with point x accidentally grounded (Figure 17–5c). A large fault current will flow, limited only by the two impedances of the source (transformers) and line. In practice, the reactance in ac machinery is much larger than the resistance, so fault-current calculations are frequently made using only the reactance. However, in low-voltage lines resistance can be important. These calculations are approximate. The two impedances have been represented as $X_S + X_L + R_L$. (See problem P17.6 for a numerical example.)

The point here is that a large fault current will flow, probably somewhere between 10 and 20 times the normal load current. Now if the switch were to be opened to stop the flow of fault current it would need to have a much higher current rating than normal load current. To use a switch to interrupt higher currents than it is rated for will at best shorten its life. In some cases, such as this fault current, attempting to open a woefully inadequate switch may start fires or injure personnel. In extreme cases a switch will be welded across by the arc so that it cannot interrupt the current. Apparently some thought must be given to what currents a switch may have to interrupt. Engineers provide for foreseeable fault currents.

Switches or breakers are, in practice, three-pole devices which open or close all three phase conductors simultaneously.

Simple switches like the knife switch considered up to this point are useful for breaking only small currents at low voltages. But they are used extensively to open circuits in which little or no current is flowing. Such isolating switches are simply not opened under load conditions. In Figure 17–3, for example, the two switches isolating the power circuit breaker need never be opened under load conditions (that is, while current is passing through them). The current, normal or fault, could be interrupted with the breaker first. Then the isolating switches can be opened.

For interrupting large currents and for higher voltages, *circuit breakers*, heavier-duty switches, are used. Their contacts are often made of flat blocks of graphite or of springy copper leaves, which are forced together under spring pressure for closure, and pulled rapidly apart, again by spring action, to break the current. They can be operated by hand, like the knife switch, or opened by automatic means, often remotely, with an electrical control circuit which operates a magnetic trip device in the breaker. Many can also be closed remotely.

Modern switch gear is *dead front*—that is, circuit breakers are totally enclosed and can be operated by hand with little or no danger of the operator touching high-voltage conductors. But they could still be dangerous to operate if applied well beyond their ratings. Many of them can also be tripped by thermal or magnetic devices within the breaker, set so that for sustained currents over a certain value the breaker will open automatically.

The ratings of high-voltage breakers are increased by immersing the contacts in oil. Oil tends to suppress the arc more quickly, absorbs some of the arc energy thus relieving the contacts, and cools the breaker. Oil breakers need regular maintenance of oil and contacts, and of mechanical parts. Many low-voltage air circuit breakers are of an integral type and cannot be disassembled for user maintenance.

The ratings of air breakers can be increased by forcing air between the contacts to blow out the arc. All breakers are operated with stored spring energy to speed up the breaking action and put out the arc quickly.

Fuses are simply another means of circuit interruption. More will be said about them in Section 17.7.

17.6   MOTOR LOADS; SYSTEM VOLTAGE

Motors larger than a few horsepower (and some large groups of smaller motors) present two difficulties to power systems. First, starting currents are several times as large as running currents (factors of 5 or 6 are typical estimates for ac motors under some conditions). Typical motors require very roughly 1 kVA/hp, and so their starting requirement can be about 5 kVA/hp.

Large machines may take ten or more seconds to reach their operating speed, during most of which period larger than normal currents flow. The current surge drops voltage momentarily for other equipment. These voltage dips are especially serious for lights and sensitive equipment such as computers. To minimize starting

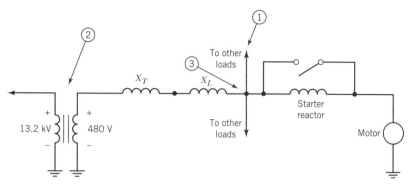

FIGURE 17–6 Voltage regulation is calculated frequently using only reactance values.

current, larger motors are provided with *starters* which reduce starting voltage by one-half or so, often by means of an autotransformer.

Figure 17–6 shows an equivalent single phase of a radial system in which a 13,200-V primary feeds loads at 480 V through a transformer and short distribution line. Large motor starting currents drawn through transformer and line impedances X_T and X_L could drop the voltage at the motor and therefore for other loads fed from that line. A sufficiently low-impedance transformer and line must be provided to keep the load voltages within established tolerances.

EXAMPLE 17.1 In Figure 17–6 assume X_T and X_L are, respectively, 0.15 Ω and 0.01 Ω; no-load voltage is 480 V and full-load current is 150 A at 90% pf.

Find

1. $V_{\text{full load}}$ supplied to motor (starter bypassed).
2. Voltage regulation at motor.
3. V supplied to starter if starting current is $2\frac{1}{2}$ times running current with pf of 50%.
4. Voltage regulation on start.

Solution

1. $X = X_L + X_T = 0.1 + 0.15 = 0.25 \ \Omega$
$V_x = IX = 150 \ \underline{/-26°}$ A $\times 0.25 \ \underline{/90°} \ \Omega$
$= 37.5 \ \underline{/64°}$ V; $V_{\text{full load}} = 480 - 37.5$
$(\cos - j\sin) \ 64° = 480 - 16.4 - j33.7$
$= 464.8 \ \underline{/-4.2°}$ V. ANS

2. Reg $= (V_{n1} - V_{f1})/V_{f1}$
$= 480 - 464.8)/464.8 = 3.3\%$ ANS

3. $I_{\text{start}} = 2.5 \times 150 = 375 \ \underline{/-60°}$ A.
$V_x = IX = 375 \ \underline{/-60°} \times 0.25 \ \underline{/90°}$
$= 93.75 \ \underline{/30°}$ V. $V_{\text{start1}} = 480 - 93.75 \ \underline{/30°}$ V.
$V_{\text{start}} = 480 - (81.2 + j46.9) = 401.6$ V. ANS

4. Reg $= (480 - 401.6)/401.6 = 19.5\%$. ANS

The second difficulty is that in the event of system faults, motors tend for several cycles to act as generators, converting their inertial energy to electrical energy and increasing fault currents. The next section looks briefly at faults.

Unlike resistors, motors will take whatever current is needed to produce the torque required at their shafts. At lower voltages it takes more current to produce a given torque. If the transformer and line are inadequate for the running currents of the motor and other equipment, lowered voltage will cause more current and motors may overheat, reducing life. Further, the maximum torque available from the commonly used induction motor is about proportional to the square of voltage applied. A 10% reduction in voltage reduces available torque 19%.

Thus planning for proper voltages at all loads under normal and starting conditions is essential. Difficulties with electrical systems can often be traced to expansions which have overloaded lines and transformers to the point where voltages are reduced below acceptable limits. Voltmeter measurements—particularly when starting motors—will help find these conditions.

Some loads have greater voltage tolerance than others. Standard motors are usually designed for plus or minus 10% voltage tolerance. Ratings can be deceptive. Many "115-V" motors are designed for nominal 120-V systems and will accept voltages of 104 to 127 V. Incandescent lamps are notoriously sensitive to voltage. Although IEEE Standard 141 recommends that nominal 120-V lamps be kept within 106 to 125 V, at 125 V life is reduced 42%, and at 110 V light is reduced 26%. The current version of IEEE Standard 141 (and other ANSI standards to which it refers) will provide voltage information on many loads. Motor manufacturers, their specifications, and literature should be consulted also in these matters. For any equipment the manufacturer is generally a definitive source of application information.

Voltage regulation problems are solved by reconnecting loads among distribution lines, by increasing line and transformer capacities, by keeping loads with widely different voltage tolerances on separate lines, by time sharing of intermittent loads (especially for motor starting), and by eliminating or converting inessential loads. Heating, for example, can be conveniently converted to gas or steam in some situations.

IEEE Standard 141 provides typical voltage profiles for plants, allotting voltage drops among high-voltage feeder, transformer, and plant wiring.

Large transformers commonly are provided with four additional primary taps permitting 2.5% and 5% voltage variation above and below rated voltage. Some transformers are provided with taps which can be switched under load with automatic equipment.

On occasion voltage troubles are traceable (with a voltmeter) to widely varying power company voltages. Some contracts provide limits within which supply voltage can be expected to range. Cooperation between industrial user and power company can usually eliminate these difficulties.

17.7   SYSTEM PROTECTION

Section 17.5 noted that although well-engineered and -maintained power systems should have few faults, all will have some, and these must be protected against. Most faults are short circuits—line-to-line or line-to-ground—either in lines and ducts or in motors and other equipment. In nearly all cases the worst faults (highest current) are three-phase, *bolted faults*—meaning by that term that all three-phase conductors come together with negligible resistance between them. Such fault currents must be interrupted (stopped) before they can do serious damage. A system is usually considered adequately protected if it can handle and interrupt that kind of fault current without damage to itself or surroundings and without injury to personnel.

To appreciate some of the problems of system protection, consider again Figure 17–3, the simple radial distribution scheme serving four groups of 480-V loads. Protection equipment (breakers and their automatic tripping devices) disconnect the faulty equipment or circuit from the supply when a short-circuit fault occurs.

Suppose the 480-V motor requires a normal running current of 100 A. The air breaker labeled "A" might be set to open with currents of 125 A or higher. It automatically disconnects the motor on substantial and continued overload. But some provision must be made to allow starting current of perhaps 500 A to flow for a reasonable time. Article 430–35 of the NEC permits this motor protection (breaker A) to be shunted during starting, under certain specified conditions. Thus starting will not trip breaker A and, set at 125 to 150% of running requirements, it will give good protection for substantial overloads. Breakers serving this purpose are usually provided with a thermal trip element so that they will not respond to short surges. On a large motor itself there may be a thermal tripping provision for sustained small overloads.

But suppose a bolted fault occurred at the motor. A large current—perhaps ten times motor rated—could flow through the breaker, which would tend to open rapidly, and the sooner the better. For until the fault is cleared the other three low-voltage load groups will be adversely affected. Thus many breakers are provided with both thermal trips to protect against overloads, and with faster-acting overcurrent trips to protect against faults. If that is the case for breaker A, it would have both these capabilities specified in its rating.

Consider breaker B. It will normally be carrying all the current of all four load groups, say 400 A. Thus its action cannot substitute for breaker A to protect against motor overloads. It could, however, protect the branch circuit of breaker A against serious faulting currents. But this would be undesirable since if B opens, all the low-voltage supply for the plant is shut down. A should open first, clearing the fault on the 480-V system so that the rest of the plant can keep operating. However, if a rare short outage or dip could be accepted, it might be preferable for economic reasons, in the simple system shown, that breaker B be depended on for all fault protection, since for the cost of one high-rated breaker all four subcircuits can be provided for.

Breaker C and one of its isolating switches appear redundant and might be eliminated in this distribution scheme. But there may be some justification for redundancy in case of trouble with breaker B.

From this example it can be seen that breakers should be set up to trip selectively, isolating a fault as close to the trouble as possible so as not to shut down other loads or at least to bring them back as quickly as possible. Many breakers have adjustable delay settings so that proper tripping sequence can be made to occur.

If breaker B is to be depended on for the entire low-voltage fault protection duty, trip times for large currents would have to be arranged for B to open first, followed by A opening after the current has been broken, and followed again by B reclosing (or being reclosed by hand if necessary) to quickly restore power to the other load groups. Note that in this case breaker A must be able to withstand the entire fault current until breaker B opens the circuit. Thus breakers (and other equipment such as busways) must also have a *withstand rating* for the maximum current they can withstand for a short time even though not operating. Huge magnetic forces are created by fault currents which can twist and destroy current-carrying members of buses and machines.

A breaker opens (and some reclose) by means of an internal magnetic circuit, which mechanically trips the operating mechanism when enough current is supplied to the coil of the magnet.

Small, simple breakers operate on the current flowing through their contacts. But many large breakers operate on specially supplied tripping power. These breakers are made to operate in the complex ways discussed above, through the use of *relays* to actuate them. A relay is a light, electrically operated switch, which, under the right conditions and at the right time, switches the tripping power onto the breaker to be opened. The desired function (for example, time delay) is built into the relay and on most is adjustable over some specified range.

Relays are designated according to the functions they provide, for example, overcurrent, time-overcurrent, differential, underfrequency, and so on.

Most people are familiar with *fuses* as replaceable elements which open faulted circuits by safely melting. Fuses are initially a good deal cheaper than breakers for the same ratings, and they are cheaper to maintain if fault currents do not occur very often and if the delay in reclosure needed to isolate and replace the fuse can be tolerated. The fuse at D in Figure 17–3 would be a better design for most cases than having a breaker at C. Some fuses not only break fault currents but will limit their peak values. Fuses are often relied on for the most severe fault conditions.

For what actual number of fault amperes should a system be protected? The maximum fault current that could occur is calculated for each part of a distribution system and for each piece of equipment that must withstand it. This is a somewhat complicated calculation which should probably be left to a power consultant. But to show at least the flavor of this thinking a simple fault-current calculation is made in the next section.

17.8   PER-UNIT CALCULATIONS AND RATINGS

Section 5.6 looked briefly at transformers, and they will be covered in more detail in Chapter 18. But for the present note that transformers, like distribution buses (lines and ducts) introduce a series impedance which drops voltage and limits current. For many calculations it is satisfactory to take this as a pure reactance. Hence transformers have a reactance rating, usually expressed in percent rather than ohms, a useful concept for fault calculations. This percent style of impedance description is called *per unit*.

To determine per unit values a kilovolt-ampere base is first established, normally full rating of the device or system being considered. Sometimes 100 MVA is used for power system analysis.

EXAMPLE 17.2 Suppose in the system of Figure 17–3 the 13,200/480-V transformer is rated at 200 kVA and has a 10% reactance.

Find: the actual transformer series reactance in a one-line equivalent wye diagram.

Solution: In a wye-equivalent diagram for a three-phase system the equivalent wye voltage (voltage to ground) is $13.2 \text{ kv}/\sqrt{3} = 7.62$ kV. kVA for the equivalent one-line system is $200/3 = 66.7$. On the primary side, current for full rated kVA would be $66.7 \text{ kVA}/7.62 \text{ kV} = 8.75$ A. Base load impedance (again referred to the primary side) would thus be V/I or $7620/8.75 = \underline{871 \ \Omega}$. Thus the transformer would have a reactance (referred to the primary side) of 10% of 871 or 87 Ω. ◾

If a solid fault were to occur at the air circuit breaker B, fault current on the primary side would be limited by transformer reactance to $V/X = 7620/87 = 88$ A. Fault current on the secondary side would be increased by the turns ratio (13,200/480) to about 2409 A. Actually the supply lines from the power company would provide some additional reactance, as would any equipment between this source and the transformer. This additional reactance is neglected for this simple illustration. Any lines on the secondary side before the fault would also contribute further reactance and resistance to limit fault current.

To make the above calculation over again but on the secondary side at 480 V, while the 10% reactance would convert to a different ohmic reactance, fault-current results for either side would not change. It is instructive to make this calculation.

Suppose next that there is an effective 48-Ω wye-equivalent reactance in the 13,200-V supply line. Using the reasoning above and continuing with a 200-kVA base rating (for all three phases) 48/871 would constitute 0.055 per unit reactance. On the secondary side this reactance would be $0.055 \times 277 \text{ V}/(66.7 \text{ kVA}/277 \text{ V}) = 0.063 \ \Omega$. To make the fault-current calculation now, add the 0.063 Ω to the 0.115 Ω of the transformer (10% referred to the low-voltage side) and divide this number into 277 V to get the secondary fault current of about 1556 A. (These calculations are quite approximate.)

The usefulness of the per unit system is that *if the same base* is used (200 kVA in this case) per unit values can be added directly without any consideration of which side of the transformer the calculation is being made on. Thus fault calculations can be made without reference to transformer ratios or voltages at various points in the system. With base kVA taken as 200, fault limited by 15.5% reactance (10% for the transformer + 5.5% for the 13,200-V supply line) will draw 200/0.155 or 1290 kVA. For example, short-circuit current at a 277-V point on the system will be $(1290/3)/0.277 = 1552$ A. Any kVA base can be used but it must be common throughout the system. The rating of the limiting transformer is often used.

Power engineers would go further with their per unit system and say that with 15.5% fault reactance there would be a per unit current of $1/0.155 = 6.45$. (100%

current on the secondary side is $66.7/0.277 = 241$ A; fault current $= 6.45 \times 241 = 1554$ A.)

Where the reactance of a line or transformer is given for some other base value, it must be converted to the system base selected. For example if a transformer were rated at 250 kVA and 10% reactance, but used on a 200-kVA base system, its reactance would be converted by the ratio of system to device base. Thus its effective reactance rating would be $10\% \times 200/250$ or 8%.

One other point will be of interest to the nonspecialist in connection with fault current—the effect of large rotating machines. As Chapter 20 will show, essentially all motors act as generators if driven mechanically while connected to an electric source. Thus inertia causes them to contribute to initial fault currents. Suppose, for example, that in the distribution system of Figure 17–3 a fault occurred at point E on the low-voltage bus. To calculate the fault current experienced by breaker F, first find the current supplied from the source much as was done above. Because of the longer low-voltage run, perhaps a secondary impedance should be added, reducing current from that calculated in the preceding example. (A low-voltage bus usually has a significant resistive component.) But the motor beyond point A would also act as a 480-V generator providing additional fault current. So there are at least two sources contributing to the fault current at point E. To take this into consideration first assume that all sources (including large motors) are connected in parallel. Then combine the resulting series and parallel per unit reactances by the usual means.

A brief look into the current version of IEEE Standard 141 will suggest how much more complex this subject can be beyond the simple illustration given above. It is important that someone consider these matters carefully when large systems are designed or modified.

17.9   SYSTEM METERING; CURRENT AND POTENTIAL TRANSFORMERS

Meters are essential to determine whether an electrical system is operating satisfactorily and, if not, what needs to be adjusted or where the problem is. Some meters or instruments are permanently built into the system, as at switchboards. Other are used portably for checking where and when needed.

System voltage and various currents are always measured. Power is usually measured and sometimes reactive power. At any rate there should be some easy method of determining power factor. Often elapsed-time meters show how long some major piece of equipment has been run. They are useful to schedule maintenance or make system studies. Temperature is often measured to determine whether some device is being properly cooled or is overloaded.

A large portion of equipment failures occur because insulation breaks down. Meggers or other portable devices for high-voltage testing or leakage measuring of insulation are useful. Their use in maintenance may help predict the end of equipment life. Insulation life is shortened by high temperatures. Ohmmeters of various kinds are useful in testing.

The practice in low-voltage or low-current circuits is to obtain a meter, or select a meter scale, appropriate for the current or voltage to be measured. Where possible the unknown quantity appears in the upper half of the scale.

In high-voltage or high-current power circuits, voltages and currents to be measured are first changed by *instrument transformers* to values more convenient to measure. For this purpose most ammeters are rated at 5 A and voltmeters at 120 V full scale. If a voltage of say 13,800 V is to be measured, a *voltage transformer* is placed across the voltage to reduce it so that it can be measured with a 120-V meter. In permanent installations the meter is usually calibrated on its face in the actual voltages being looked at by the transformer.

Current transformers provide for measuring large currents and are often a single turn in the primary with many turns in the secondary depending on the current reduction ratio needed. Such a device will be recognized as a voltage step-up transformer with a larger ratio. Incorrectly used, it produces dangerous voltages and destroys its own insulation. Current transformers must *never* be open circuited. They are closed on an ammeter or short circuited. If the meter is to be adjusted or replaced, the transformer is first short circuited on the secondary side. Used in this manner current transformers present a very low impedance on the primary side, to the current being measured, as any ammeter should. In addition to transforming currents, these devices conveniently isolate measurement circuits from high-voltage lines and equipment.

17.10   POWER COMPANY REQUIREMENTS AND BILLING

In effect the power company, in its monopoly position, agrees to furnish all electric power needed in its region, at a price approved by the State. The State in return agrees that the price allowed will be enough for a fair return on investment. While this arrangement is sometimes a matter of legal disputes, court action, and political problems, it works surprisingly well.

Thus the company is concerned to plan ahead for new power requirements. Since it may take years to make major changes in available power, industrial users should be in touch with their power supplier as early as possible on their own new or expanded needs.

The first customer requirement is a source of constant voltage with an adequate supply of power behind it. A power company has certain voltage standards it attempts to maintain and customers should know what these are. As more and heavier loads are added to a supply line, its voltage drops and also varies a good deal more than when it is lightly loaded. There should be clear understanding of what the company can expect as a load on its lines and what the customer can expect in the way of voltage maintenance.

Usually a distribution line that supplies one customer must be shared with a number of others. Thus the power company is properly concerned with the nature of any customer's load (kVA requirements, power factor, transients such as large motor starting, etc.) and how it varies during a 24-hour period, since it may affect voltage supply to neighboring customers.

An important consideration on both sides is the fault impedance of power company supply lines and of the customer's distribution systems. A cooperative effort to solve mutual problems normally brings out the best results.

Another touchy point for some kinds of loads is the company's policy of *load shedding* in emergencies. When part of the region's power supply is lost, it is considered better to disconnect some loads than to have the entire system go down so that no one has any electric power. Large customers may be asked to assist in this load shedding by shutting down parts of their own plant. Customers that have serious problems with interrupted power (for example, in some chemical or metallurgical processes) should work these things out ahead of time.

Power companies usually have many different rate schedules, some quite complex. The basic elements are a charge for the fuel used and calculated from kWh usage and a charge for the money invested in the power plant and equipment calculated from the maximum demand. In addition, there are often small minor elements tacked on as requirements of the State, for conservation, cleaner fuel, changes in fuel cost, and so on. For example, a typical Eastern company had more than 10 schedules, among them the following (here simplified a bit and designated "X-2") and calculated on a monthly basis:

Schedule X-2 (effective January 1, 1992)

base monthly charge		$16.02
demand charge	per peak kW	$ 9.49
energy charge	per kWh	$ 0.03307

the total monthly charge will be the sum of these three

The following additional provisions are appended:

1. Rates may be raised if purchased power rates to company are increased.
2. The state DPU (regulatory agency) may authorize increased adjustments for fuel cost changes.
3. Demand is established as the larger of the following: (i) the greatest 15-min energy peak during the month as measured in kilowatts; (ii) 90% of the greatest kVA peak averaged over a 15-min period.
4. Other adjustments may be made for small adders to the energy or power rates for such things as conservation and load management costs, and oil conservation adjustments.

EXAMPLE 17.3 An industrial plant uses 17,460 kWh in a certain month, and during that period the maximum demand averaged over a 15-min period in kW is 97 and in kVA is 124.

Find: the proper power billing for that month based on schedule X-2 above.

Solution: The demand charge will be based on 90% of 124 kVA = 111.6 (since this is greater than 97). 111.6 × 9.49/kW or = kVA = \$1.059. Energy charge = 17,460 × 0.03307/kW = \$577.40. So billing is 16.02 (base charge) + 1,059 + 577.40 = <u>\$1,652.50</u>.   ▨

It is interesting to enquire how the extra expense to the power company of taking care of poor-power-factor loads is provided for in this schedule. Consider par 3 in the above schedule. Other power companies may have explicit percentage penalties called out for low power factor as suggested in the following section.

17.11   POWER FACTOR CORRECTION

Consider the plant load shown in Figure 17–7a. This diagram could be a single-phase system or one line of the wye equivalent for three phase. The voltage is fixed in the example at 440 V (line-to-line) or 254 V line-to-ground. Suppose the plant requires 150 kW and this phase consequently handles 50 kW. With 100% power factor, the load for I_L on the power source is 197 A. But if the power factor is 78%, I_L (or I_S) would have to be 252 A. Thus the poorer the power factor the more current must be drawn from a supply for a given load power. Poor power factors come mainly from the ubiquitous induction motor. These draw lagging currents. The industrial problem is always lagging power factors.

Power companies supply electric power via transformers, transmission lines, and generators. Each device must be able to handle the currents required. The

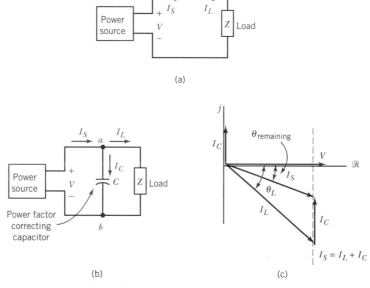

(a)

(b)   (c)

FIGURE 17–7 Power factor is improved (brought closer to 100%) by par-alleling capacitors with the lagging load.

more current, the larger and more expensive the equipment and the greater I^2R losses. So it is advantageous to the power company to have its customers run with a good power factor—nearly 100%. Power rate schedules provide dollar penalties in one way or another for poor power factor. This is implicit in Schedule X-2 par 4. In an efficient economy users pay the cost of their own services.

The plant engineer can correct power factor by placing capacitors across his load. Connected between points a and b of Figure 17–7b, the power-factor-correcting capacitor draws a leading current to offset the lagging portion of load current. According to Kirchhoff's current law, $I_S = I_C + I_L$. The phasor diagram of Figure 17–7c shows how the addition of I_C to I_L reduces (shortens) the source current I_S. The diagram shows that as the capacitor's size and current change, the tip of the I_S phasor will move up and down the dashed vertical line. This is an application of the resonance theory discussed in Section 6.4.

Clearly, minimum source current I_S will occur at 100% power factor. The I_C needed then is simply $I_L \sin \theta$, using the absolute value of θ. From this current calculate the reactance of the capacitor, X_C, as V/I_C—or for the 78% power factor (pf) example, 254 V/158 A = 1.62 Ω. And $C = 1/\omega X = 1650 \mu F$. If the annualized cost of the capacitor is less than the penalty costs in the plant's power bills, the plant engineer might recommend installation of the capacitor. But this is not the most economical value of capacitance to use.

Because the cosine of small angles is so nearly 1.00 there is little benefit to be gained by correcting the power factor to 100%. In going from 90 to 100% in Figure 17–7c, the length of I_S will not be reduced very much. One rule of thumb is to correct it to 90%. Or if sufficient savings are involved, they can be calculated for various amounts of correction and the capacitor selected for maximum dollar savings.

EXAMPLE 17.4 A small industrial plant is supplied by a 440-V power line. In the circuit of Figure 17–7a the load symbol **Z** represents one-third of total plant load, which is 50 kW at 78% pf. Annual power cost without penalty is $18,000. This particular power company assesses low-power-factor penalties for any month in which the power factor falls below a specified amount for more than 15 min in any day as follows:

for power factors below 75%—to be negotiated;

for 0.75 < pf < 0.85   —a penalty of 17%;

for 0.85 < pf < 0.90   —a penalty of 6%.

Assume capacitors for use in this nominal 254-V service are available in 10 kVARC units at $1700 per 10 kVARC installed. Their expected life is 15 years. Minimum acceptable rate of return (MARR) in this situation is 24% before taxes.

Find: Should the power factor be corrected and if so to what level? How much capacitance is needed in kVARC and in microfarads? How much money will be saved?

Solution: *Phasor Diagram Method.* The three possibilities are no correction, correcting to 85%, correcting to 90%. The annual costs of each possibility are as follows:

With no correction: Penalty is $0.17 \times 18{,}000 = \$3060$.

Correcting to 85% (Figure 17–8): I_L, the current drawn by the load, will be $50{,}000/(440 \times \sqrt{3} \times 0.78) = 252.3$ A lagging the voltage by arccos $0.78 = 38.7°$. Its in-phase component (horizontal in Figure 17–8a) is $252.3 \times 0.78 = 196.8$ A. Its out-of-phase component (vertical) will be $196.8 \tan 38.7° = 157.7$ A. At 85% I_S, the supply current, will lag the voltage by $31.8°$ (see method above for 78% situation), and its out-of-phase component will be 122.0 A. By KCL (node above point a in diagram) I_C, the capacitor current, will be $157.7 - 122.0 = 35.7$ A. And by Ohm's law $X_C = V/I = 254/35.7 = 7.12 \ \Omega$. $C = 1/\omega X_C = 1/(377 \times 7.12) = 373 \ \mu$F. This will be the size of the capacitor needed to correct to 85% power factor. Its kVA rating will be $V \times I/1000 = 0.254 \times 35.7 = 9.1$ kVARC. So one \$1700 capacitor is needed. The cost for three (three phases) will be \$5100.

To annualize this cost over 15 years at 24% interest (see any engineering economy text or tables) the factor $A/P = 0.2499$. Hence annualized cost is $0.2499 \times 5100 = \$1275$. But there will still be a penalty cost of $0.06 \times 18{,}000 = \$1080$. So total annual cost of this option is $1080 + 1275 = \$2355$.

Correcting to 90%: By similar analysis $I_S = 218.6$ A at $-25.8°$. $I_C = 52.0$ A and $C = 543 \ \mu$F. kVARC = 13.2, necessitating two units per phase costing \$3400. Annualized cost is \$850, or \$2550 for all phases. There is no penalty for this option.

Clearly, *correcting to 85%* has the lowest cost of these three options and should be recommended. Money saved will be ($1020 - \$785) \times 3 or \$705 per year. It is important to recognize in practical situations that there may be other options to be devised and analyzed. For example, are differently sized capacitors available? Would it be economically advantageous to use 440-V capacitors line-to-line?

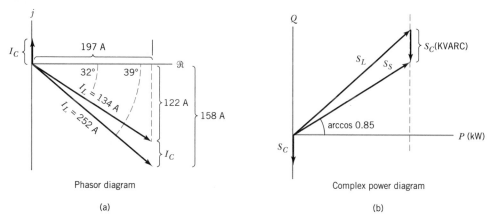

Phasor diagram

(a)

Complex power diagram

(b)

FIGURE 17–8 Power factor correction can be calculated with either phasor or complex power diagrams.

Another way of solving this problem is with a *Complex Power Diagram* (Figure 17–8b).

Correcting to 85%: The phase angle of the load is arccos 0.78 = 38.7°. Reactive power = 50 tan 38.7 = 40.1 kVAR. Similarly at 85% kVAR = 50 tan 31.8° = 31.0, so the capacitor must absorb the difference or must provide 9.1 kVARC. And similarly for the 90% case. This is surely a preferred method once the procedure is understood. But it obscures to some extent what is actually happening with the circuit currents and impedances.   ▓

It is interesting to observe that if only higher-voltage capacitors are available, the kVARC would have to be higher for the same degree of correction (problem D17.9).

17.12   SUMMARY

The U.S. electric power system comprises hundreds of large generating plants (coal, oil, nuclear, or hydro) which are connected into a high-voltage national grid transmitting power to the entire country. These lines terminate at substations where voltages are reduced somewhat and from which radial supply lines distribute electricity to neighborhoods and to large users. Secondary distribution (lower-voltage) lines feed smaller customers and transformers supplying commercial and residential users.

All energy sources for producing electricity are polluting in one way or another. Many have been consistently hazardous to human life over the years; others are potentially so. All of these effects, from both nuclear and fossil fuel burning, are not yet well understood.

Plant electric power distribution systems provide for each of the loads: proper voltage, current capacity, switching, protection against faults, and sometimes metering. Their principal elements include transformers, buses of various kinds, switches and circuit breakers, protective relays, starters and variable frequency drives and meters.

Switches and circuit breakers must be designed and specified to withstand not only rated load currents (for breaking or simply passing) but also possible fault currents. Determining the magnitude of possible fault currents is an important and complex analysis.

High currents and voltages are metered through current and voltage transformers with standard-scale meters. Current transformers must never be open circuited.

Motors constitute a large portion of most industrial loads and pose special problems. They require starting currents which are far larger than those developed for full-load running conditions. Many require some type of soft start at reduced voltages. They supply power to the system to exacerbate fault conditions, while often needing special protection for the motor itself. The increasing use of electronic variable-frequency and dc drives for them complicates the system and pollutes it with harmonics.

The per unit and equivalent-wye methods of power system calculation and depiction greatly improve analysis.

Billing for power use is inherently complex but, thought out, provides opportunities for economizing, including power factor correction through the use of banks of capacitors near poor-power-factor load centers.

FOR FURTHER STUDY

Ollie I. Elgerd, *Basic Electric Power Engineering*, Addison-Wesley, Reading MA, 1977.
William D. Stevenson, Jr., *Elements of Power Systems Analysis*, McGraw-Hill, New York, 1982.
IEEE Standard 141-1986 (or current edition), *Recommended Practice for Electric Power Distribution for Industrial Plants* (the "Red Book"), IEEE, New York.
Energy Information Administration, DOE. *Annual Energy Review* (current edition), Washington, DC 20585.

PROBLEMS

Easy Drill Problems (answers at end of chapter)

D17.1. **(a)** What was the 1989 average efficiency of the nonnuclear generating stations in the United States? (Use material from Figure 17–2. Assume the nuclear fuel use in 1989 was about what it was in 1988.)

    (b) Calculate efficiency from fuel to user by taking into account a 10% loss in transmission and an additional 8% loss in distribution.

D17.2. What was the U.S. *load factor* in 1988? (Electrical energy that was produced divided by amount that could have been produced if the existing plants had been run for 24 hours, 365 days a year.)

D17.3. **(a)** If all U.S. generation were from coal and 28.63 quads of energy continued annually to be used for this purpose, and other uses of coal were forbidden, for how many years would our coal supply last?

    (b) Assume 15% of coal mined is exported. Recalculate your answer to (a).

D17.4 **(a)** What is the heat pollution in btu/day caused by U.S. electric usage?

    (b) What percentage is this of the solar insolescence on our U.S. area? (For this rough calculation assume 1 kW/sq yd striking the United States at an average angle of 45° for an average time of 6 hours per day, on an area of 3×10^6 sq mi.)

D17.5. Assume 350,000 MW, very roughly half of installed U.S. capacity, is converted to solar. How much land area will be required for solar collectors in acres? in square miles? Base your calculations first on an available 1 kW/sq m and unlimited efficiency. (*Note:* 2471 acres = 10,000 sq m; 640 acres = 1 sq mi.)

D17.6. **(a)** Now consider the insolescence figure in D17.5 to be a peak value under cloudless conditions with the sun overhead and that conversion efficiency will further reduce available electrical output. Assume that the average continually available power of a 24-h, 365-day period is 2% of peak insolescence. How many kWh per year will one square mile of solar collecting area supply?

(b) How many square miles will be needed to provide 3 quads of U.S. electrical energy output?

D17.7. In the distribution system of Figure 17–3 the four 480-V load groups on the right-hand side take respectively 34 kVA at 84% pf, 20 at 97%, 60 at 78%, and 38 at 89%.
(a) What is the current and power factor that the 13,200-V supply furnishes?
(b) To what percentage of rated current is the main 200-kVA transformer loaded?

D17.8. An American plant has a three-phase load of 300 kVA at 80% pf lagging.
(a) Sketch the complex power diagram for the plant.
(b) Find the total value of capacitance in kVARC needed to correct the power factor to 100%.
(c) Specify this capacitance in μF per phase if the plant and capacitor nominal voltage is 900 V.
(*Hint:* Find current I_C.)

D17.9. Assume that in the previous problem you will have to settle for a 1000-V capacitor. Will this fact change any answers? (*Hint:* Power-factor-correcting capacitors are rated in kVARC—$V \times I$.)

D17.10. A transformer is rated at 75 kVA 13,200/480 V. It has a reactance, per phase, referred to the primary side, of 230 Ω. What is its per unit reactance?

Application Problems

P17.1. Make a spread sheet chart showing the cost of electricity generation in cents/kWh for plant costs per kW of $400, $600, $800, $1000, $1200 and for fuel costs of $0.50, $1.00, $2.00, $3.00, $4.00, $5.00 per million Btu. Assume a capital cost of 15% per year, an effective life of 30 years, a power plant capacity factor (average plant loading) of 70%, a plant efficiency of 33%. (*Note:* The capital recovery factor A/P for 15% and 30 years is 0.1523.)

P17.2. List the significant factors that you feel are left out of the data provided for P17.1. Assume some reasonable cost figures for these factors and recalculate a few points on your chart. How much difference do they make in percent? (For example, a 1000-MW power plant may employ 100 people at an average total cost of $75,000 per employee.)

P17.3. How sensitive to life (number of years plant will be used) is your analysis in P17.1? Using interest tables check several points in your analysis for other lives (20, 25, 35, and 40 years are suggested).

P17.4. Transmission and distribution costs are sometimes assumed to add about 40% to generation costs. Assume that the average transmission distance is 80 miles. Assume further that for the 40% assumption, this cost is divided about equally among four areas: distribution costs, transmission right-of-way and line construction, transmission line energy losses, substations. Make an improved mathematical model that will predict the percent additional cost to be added to generation as a function of the distance between the distribution load center and the generating facilities.

P17.5. In the switch of Figure 17–4c, assume that a dc current of 100 A is to be broken, the break takes 3 s, and the current decays uniformly. Assume further that the voltage across the switch rises from 0 to 1000 V uniformly during break. The specific heat of copper is 0.0917 (cal/g)/°C.
(a) Find the energy dissipated in the arc.
(b) Assume the metal portions of the switch are copper and weigh 5 lb and that they absorb 50% of this energy. To what temperature will they rise if there is no appreciable radiation or conduction during the 3-s break?

(c) Recalculate (b) assuming that all the energy absorbed goes first for at least 3 s into the 5% of the switch metal nearest the arc.

(d) Repeat (c) assuming that by spring action and forced air blast the arc is suppressed within 0.02 s.

P17.6. In the low-voltage "equivalent-wye" diagram of Figure 17–5c, $X_S = 0.30\ \Omega$, $X_L = 0.02$, $R_L = 0.04$, and $I_{\text{full load}} = 160$ A.

(a) Find the fault current.

(b) What is the ratio of fault current to full load current?

P17.7. Suppose in P17.6 the ratio of X to R in the source (transformer) is 7.4. Find the values of R_S (not shown in sketch) and recalculate the fault current and its ratio to full load current. What percentage error would neglecting either or both R's make in the calculation?

P17.8. In Figure 17–5c remove the short circuit, and using the data of P17.6 and P17.7, calculate the voltage regulation for a no-fault situation in three ways—using only the reactances, including R_L, including both R's.

P17.9. The motor in Figure 17–3 has a full-load requirement of 75 kVA at 80% or 25 kVA per phase. Assume the other three load groups take a three-phase total of 150 kVA at 90%. Find the voltage regulation in volts and percent.

P17.10. On reduced-voltage start, the motor starter takes 500% of full load kVA, at 50% pf. To what voltage will the 480-V supply dip? Assume the current requirements of the motor and other loads are approximately independent of supply voltage. (*Note:* These calculations are approximate and it is customary to use nominal voltages to calculate kVAs and currents.)

P17.11. An industrial single-phase load requires 300 kVA at 70% lagging.

(a) What size capacitor in kVARC is required to correct the power factor to 90%? to 94%?

(b) Assume the voltage is 1300 V. Specify these capacitances in microfarads.

P17.12. Repeat problem P17.11 assuming the load specified there is three-phase balanced delta. Find the values of the three capacitors to be connected line-to-ground.

P17.13. A plant power system calculation is made on a per unit base of 200 kVA. A 100-kVA transformer has an impedance of 10%. What is its per unit impedance in the plant system calculation?

Extension Problems

E17.1. For problem P17.12 assume capacitors are available to you at 750 V in 5-kVARC units at $1050 per unit. Assume further that the penalty for poor power factor is 3% for pf between 85 and 90%, 6% between 80 and 85%, 10% between 75 and 80%, and 15% between 70 and 75%. Assume further that the annual cost of power for your plant is $50,000 before any penalties. You may also assume that the minimum acceptable rate of return in your company is 24% before taxes and that the appropriate 15-year A/P factor for converting capital costs to annual costs is 0.2499. How much power factor correction should be included? What will it cost? What will be the net savings per annum?

Study Questions

1. What are "one-line" and "equivalent-wye" diagrams? Why are they used?

2. Why, do you suppose, is electric power presently generated at large central stations?

3. What are the three major fuel sources for electricity generation? Name also some others.

4. Why are voltages transformed so drastically between generation, transmission, distribution, and use? What is the device on which this transformation depends?

5. Why do engineers go to the expense of interconnecting power systems with each other? What two savings justify this expense?

6. What does the term "low-head hydro" mean?

7. What pollutants are produced by using the three major fuels? How much is produced by a typical plant?

8. Roughly what percentages of the electric power and energy produced in the United States today are from nuclear sources?

9. What is a busway? To what would you compare it in the average home distribution system?

10. What are the purposes of using circuit breakers in distribution systems?

11. List the principal goals of distribution system design. In what relative order of importance would you place them?

12. Why is the NEC important to plant engineers?

13. Describe the voltage and current limitations of switches. What troubles might develop if they are exceeded?

14. How does use of transformers with significant amounts of reactance (leakage reactance) relieve circuit breaker ratings?

15. Give two reasons why motors are especially important in distribution system analysis.

16. Name several changes that might be considered to improve poor voltage regulation in a plant.

17. In the practice of "system protection" what exactly is being protected? What is being protected against?

18. What is a "bolted" fault?

19. What are the advantages and disadvantages of fuses in comparison with resettable circuit breakers?

20. What is the "per unit" system? What advantages does it offer in analyzing distribution systems?

21. Why are current and potential transformers used in metering practice? What is the special safety practice always observed with a current transformer?

22. What is the difference between a "demand charge" and an "energy charge"? Justify the use of each.

23. Why is power factor correction worthwhile to the power company? And often to the industrial user?

ANSWERS TO DRILL PROBLEMS

D17.1. 33.5%, 27.7%

D17.2. 46.0%

D17.3. 747 yr, 650 yr

D17.4. 78.443×10^{12} BTU/day, less than 2 parts in 1000

D17.5. 35.0×10^6 acres, 54.69×10^3 sq mi.

D17.6. 453.8×10^6, 1937

D17.7. 6.58, 85.5, 75.2

D17.8. 180, 196

D17.9. 200 kVARC, μF unchanged

D17.10. 9.9%

CHAPTER 18

Transformers and Magnetics

Chapter 4 introduced the elementary concepts of transformers; these simple relations between primary and secondary voltages or primary and secondary currents on the one hand and turns ratio on the other are entirely adequate for much transformer work. In addition to those ideas it would be useful for the engineer responsible for using and operating transformer equipment to understand problems of heating and losses, voltage regulation, and voltage and current-rating limitation. This chapter provides such further development and also a basic explanation of magnetics and the magnetic circuit. The magnetic-circuit material and many of the rating concepts will be applicable to other kinds of electrical equipment as well as transformers.

The transformer idea—two or more coils or windings linked magnetically, with or without a core to shape and enhance the magnetic flux—is used in many applications to transform voltages, currents, and impedances, and to isolate circuits from each other electrically. Transformers vary in size from fingernail size communication units to multiton giants for power stations.

Transformers rely on the interaction between electrical and magnetic phenomena for their operation. There are no moving parts (except possibly for tap-changing switches, or for cooling fans and pumps in very large power transformers). The next chapter will show that rotating machinery—generators and motors—

depends on interactions among electrical, magnetic, and mechanical phenomena. In motors and generators, of course, there is mechanical movement.

Readers may take as their objectives here to develop a working understanding of:

1. transformers, their losses, tests, voltage, current, models, temperature limitations
2. magnetic circuits, their analogous behavior to electric circuits.

18.1   TRANSFORMER BASICS

It will be helpful to review the elementary transformer discussion of Section 4.11.

Figure 18–1a shows the usual transformer diagram with primary and secondary *windings*, and iron core suggested by the vertical lines. Figure 18–1b illustrates an actual core and windings. This closed-circuit core provides a path for magnetic flux (designated ϕ). Lines of magnetic flux flow around and through the iron core. This flux is considered always to be made up of unbroken lines—every line is closed on itself. Figure 18–6a conveys some feeling for the line concept.

Electric currents flow in the copper wires of the two windings, which are of

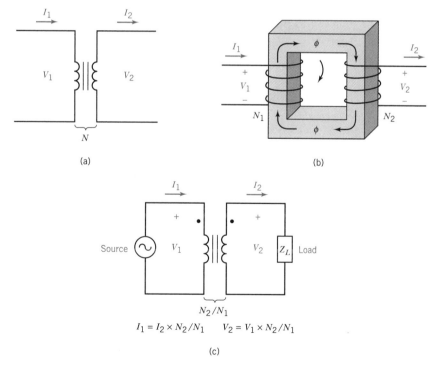

FIGURE 18–1   Basic transformer symbol (a), construction (b), and use (c). Transformers couple two circuits magnetically, transforming voltage, current, and impedance. $V_2 = V_1(N_2/N_1)$.

course insulated from the iron core and from each other. (In a practical transformer the windings are usually on top of each other on one leg of the core.)*

It will be seen that the flux ϕ *links* the turns of each winding. The basic law of transformer action (Faraday's law) is that if a flux ϕ links a winding of N turns, then a voltage is generated in that winding proportional to the rate of change of flux and the number of linking turns, or

$$v = N \frac{d\phi}{dt}. \tag{18.1}$$

Chapter 4 showed that if a voltage V_1 is impressed on the *primary* (the side to which power is supplied) a voltage V_2 will appear on the *secondary* (the side from which power is taken). Applying Equation (18.1) to the primary and secondary windings,

$$v_1 = N_1 \frac{d\phi}{dt},$$

$$v_2 = N_2 \frac{d\phi}{dt}.$$

Then going to phasor notation, and since the same ϕ flows through each coil,

$$V_2 = V_1 N_2/N_1. \tag{18.2}$$

N_2/N_1 is the transformer *turns ratio.* In this book this secondary-to-primary turns ratio will be consistently used.† The same transformer can be connected either *step-up* or *step-down* depending on whether the larger number of turns is used as the secondary or primary.

Because of the relative direction of the two windings (Figure 18–1a,c) there is possibly a 180° phase shift between V_1 and V_2. This ambiguity in the symbol is removed by polarity marks: the two "dots" [sketch (a)] or the 2 + signs [sketch (c)]. Voltages at the dots rise and fall together; and when current is flowing into one dot, it is flowing out of the other. This is usually of no practical importance and can be neglected. But where it does matter (particularly in three-phase transformer connections) it is easily taken into account. For this reason, where it can be done without confusion, the voltage and current relations in this chapter will be described in magnitudes rather than in bold-face phasor notation.

In Figure 18–1c primary voltage V_1 gives rise to secondary voltage V_2 according

---

*The term "coil" is often used for each winding (or sometimes both together), particularly before they are mounted on the core. This use corresponds to the earlier use of "coil" for inductance.

†We adopt this definition of turns ratio for temporary convenience here. The term *turn ratio* or *turns ratio* is variously defined in standard publications, but engineers usually use the ratio of the larger number of turns to the smaller.

to Equation (18.2). This secondary voltage produces current

$$I_2 = V_2/Z_L. \qquad (18.3)$$

This secondary current in turn produces a primary current

$$I_1 = I_2 N_2/N_1. \qquad (18.4)$$

Since, according to the above equations, the transformer transforms both voltage and current, it is not difficult to show that it transforms impedances, so that the impedance looking into the primary is

$$Z_p = Z_L/(N_2/N_1)^2. \qquad (18.5)$$

For instance if N_2/N_1 were equal to 0.5 in Figure 18–1c, and Z_L were 100 Ω, the source on the primary side would see a V/I ratio of 100 $\Omega/(N_2/N_1)^2 = 400$ Ω. Note that if the secondary's voltage is smaller, its current is larger than that in the primary. The larger impedance is on the side with the higher voltage and lower current. Because voltage and current are transformed oppositely, the product of V and I—the kVA of the transformer—will be the same on either side.

For an iron-core transformer, to a first approximation, the impedance of the secondary circuit includes only the load impedance. The same is true of the impedance of the primary circuit—it consists only of the transformed load impedance plus any source impedance. The simple model thus described with no winding resistance or winding reactance, and with perfect current and voltage ratios in their relation to the turns ratio, is called an *ideal transformer*.

Greater accuracy can be achieved by including in each circuit the resistance of the winding plus its leakage reactance, and that will be done in the next section.

18.2   TRANSFORMER MODEL

Figure 18–2 shows the circuit model used in this book for transformers, when something more exact than Equations (18.2)–(18.4) is needed. While no model is perfect, this one will produce excellent analytical results for most situations. Problem solving with this model can develop a strong intuitive understanding of transformer losses, efficiency, and voltage regulation.

The model's principal limitation is lack of information on the effect of higher voltages. Transformers have a very definite voltage limitation. Their voltage ratings should not be exceeded by more than about 10%, and even then with care to see that they do not overheat. For larger transformers temperature should be monitored for even smaller increases. Higher voltages rapidly increase exciting current. There is also the possibility of breaking down insulation. Chapter 4 noted that putting dc voltage on a transformer will very possibly destroy it.

Returning to the model of Figure 18–2, there are five elements, R, X, G, B, and N_2/N_1. N_2/N_1 is an *ideal transformer* (described in 18.1 and Chapter 4) with

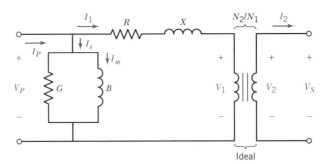

FIGURE 18-2 This simple five-element circuit model allows the engineer to calculate losses, efficiency, and voltage regulation. G represents core (iron) losses, and R copper losses. B provides for magnetizing current and X for leakage reactance.

secondary-to-primary turns ratio N_2/N_1. Ideal means that its windings have no resistance or reactance and no exciting current is required for its operation. The other four quantities take into consideration imperfections in the transformer being modeled.

R and X represent the impedances (resistance and reactance) of the two windings. Impedances of both windings are combined by transforming secondary values with the ratio $1/(N_2/N_1)^2$ and adding them to the primary values. The parallel circuit on the left, composed of resistance and reactance (G and B—not to be confused with the B used as a symbol below for flux density), provides for core losses and magnetizing current. The *exciting current* I_e flows partly through G and partly through B. (The part flowing through the inductor B is called the *magnetizing current*, and produces the core flux.) The resistance marked "G" stands for the core losses (to be described below) and the reactance of "B" is made just large enough to produce the magnetizing current required by the transformer being modeled. (In general $G = 1/R$ and $B = -1/X$—but these are not the same R and X in the series part of the circuit.) Engineers use these reciprocal terms, instead of another R and X, to avoid confusion. Section 3.8 above explained G. B was touched on in Section 4.9.

R and X produce a voltage drop dependent on load current. Power losses in R represent I^2R losses in the resistances or the two windings. X represents a physical quantity called "leakage reactance" caused by leakage flux—a small amount of flux produced by one winding that does not link the other winding. A few of the continuous flux lines complete their path through the air as suggested in Figure 18-1b.

Note that there are two sources of loss in the transformer model—R and G—representing *copper losses* and *core losses*.

EXAMPLE 18.1 A transformer is rated at 2400/220 V and 12 kVA. It can be represented by the standard model (Figure 18-3a) with $R = 9.6 \ \Omega$, $X = 70 \ \Omega$, $G = 20 \ \mu\mho$, $B = -120 \ \mu\mho$, and $N_2/N_1 = 0.09475$. It is operated so as to produce 220 V across a resistive load of $Z_L = 4 \ \Omega$ in the secondary.

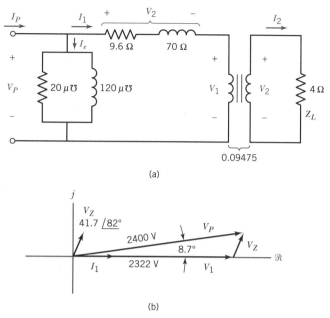

(a)

(b)

FIGURE 18–3 Phasor diagram (b) for V's and I's of model (a). See Example 18.1. V_1 is in phase with V_2, and I_1 with I_2. V_p will lead V_1 slightly.

Find

1. V_1, I_1, I_2, P_{in}, P_{out} neglecting G, B, R, X.
2. These quantities again with all the parameters included. Find $\mathbf{V}_P$ and $\mathbf{I}_P$.
3. Transformer efficiency and voltage regulation.
4. Recalculate part 2 assuming a voltage of 2500 V is applied to the primary (instead of specifying the secondary voltage at 220).

Solution

1. Use the simple model of Chapter 4 (Figure 18–1a): $I_2 = V_2/Z_L = 220/4 = \underline{55 \text{ A}}$. $I_1 = I_P = I_2 \times 0.09475 = \underline{5.21 \text{ A}}$. $V_1 = V_P = 220/0.09475 = \underline{2322 \text{ V}}$. $P_{out} = V \times I$ (for a resistive load) $= 220 \times 55 = \underline{12.1 \text{ kW}}$. Neglecting the imperfect aspects of the transformer, and recognizing the ratio N_2/N_1 as real, $P_1 = V_1 \times I_1 = 2322 \times 5.21 = \underline{12.1 \text{ kW}}$. This result should not surprise us since there are no resistances taken into account in Figure 18–1 to reduce efficiency below 100%.

2. Using the better model of Figures 18–2 and 3, answers are the same from the 4-Ω load back to the left side of N_2/N_1. The voltage $\mathbf{V}_p$ is $\mathbf{V}_1 + \mathbf{I}(R + jX) = 2322 + 5.21 \times (9.6 + j70) = 2400 \underline{/8.7°}$, whence the current through G is $G \times \mathbf{V}_p = 0.048 \underline{/8.7°}$ A. Similarly $\mathbf{I}_B$ is $0.288 \underline{/-81.3°}$ A. Thus $\mathbf{I}_e$ (the exciting current) is the sum of these two or $0.292 \underline{/-71.8°}$ A. And $\mathbf{I}_P$ is the sum if $\mathbf{I}_1$ and $\mathbf{I}_e$ or $5.21 + 0.292 \underline{/-71.8°} = 5.31 \underline{/-3.0°}$ A.

Power out is the same as in part 1 and power in will be considered in part 3. The approximate answers of part 1 were off by only a few percent.

3. Power lost in R is I^2R or $5.21^2 \times 9.6 = 261$ W—the copper losses. Power lost in G—the core losses or iron losses—is $V^2 \times G = 2400^2 \times 20 \times 10^{-6} = 115$ W. Efficiency $= P_{out}/(P_{out} + \text{losses}) = 12.1/(12.1 + 0.261 + 0.115) = \underline{96.8\%}$.

Voltage regulation is a measure of how much the output voltage drops off as load increases from no load to full load, and is defined as $(V_{no\ load} - V_{full\ load})/V_{full\ load}$. Without any load current there would be no drop across R and X. No-load voltage would be $2400 \times 0.09475 = 227.4$ V. At full load the output voltage would be 220 V, a drop of 7.4 V. So the voltage regulation of this transformer is $7.4/220 = \underline{3.4\%}$. All these numbers are realistic for a 12-kVA transformer.

4. Amounts to, "given some applied primary voltage, what voltage and current will exist on the load side?" This is the most common transformer problem. The easiest solution is to assume some reasonable $\mathbf{V}_2$, calculate $\mathbf{V}_P$ as in part 2, and by ratio determine the new $\mathbf{V}_2$. Thus in part 4, using the results of part 2, $V_2 = 220 \times 2500/2400 = 229$ V, and so on for the secondary current.

Figure 18–3b shows a phasor diagram for the above example with $\mathbf{V}_1$ (and $\mathbf{V}_2$) on the real axis. The scale of $\mathbf{V}_Z$ (the voltage drop across R and X) has been exaggerated to make the construction clear. Because of the model's ideal transformer element, $\mathbf{V}_2$ is in phase with $\mathbf{V}_1$, and $\mathbf{I}_2$ in phase with $\mathbf{I}_1$.

In the above example the power factor was 100%—a result of the purely resistive load. From the phasor diagram it can be seen that if load power factor becomes lagging, phasors $\mathbf{I}_1$ and $\mathbf{I}_2$ would be shifted appropriately clockwise and the phasor voltage $\mathbf{V}_Z$ would follow them so that $\mathbf{V}_Z$ would have a much larger effect on the value of $\mathbf{V}_P$. Thus a poor power factor has significant effect on voltage regulation of transformers. (See problem P18.3.)

In operating transformers a voltage is put on the primary, and the usual question is, "what happens at the load?" A direct solution of this problem with the Figure 18–2 model is not simple. But as illustrated in Example 18.1 engineers solve it by first estimating V_2, neglecting R and X. Then for this $\mathbf{V}_2$, $\mathbf{V}_P$ is calculated backward (as in the example above) using the full model. If the newly calculated $\mathbf{V}_P$ is far enough off, $\mathbf{V}_2$ can be adjusted by the ratio new $\mathbf{V}_P$/old $\mathbf{V}_P$ since the model is linear.

18.3   TRANSFORMER RATINGS

A good deal of the thinking in this section can be applied not only to transformers but also to other electrical machines and devices, such as motors.

Transformers are rated in kilovolt-amperes (kVA) at their design frequency and voltage. Voltages cannot be exceeded by much without overheating, but there is no problem about running at lower voltages than rated. Dividing the voltage rating into the kVA gives a current rating—which is what the engineer often needs. Current cannot be increased very much beyond this derived current rating. So at reduced voltages kVA allowed will also be reduced. At lower than rated frequency, voltage ratings must be lowered in proportion to the frequency used. (See Section 18.6.)

The principal limitation of most electrical equipment is temperature. If a transformer or any machine gets too hot, insulation on the windings chars or melts, and a short circuit may destroy the equipment. Sudden short circuits can cause violent explosions from gas pressure built up rapidly by extreme temperatures. But in most cases—in well-designed distribution systems—circuit breakers will open and the failed machine will be automatically disconnected. Some transformers and many motors have temperature sensing elements embedded in them which will open the input circuit until they cool down again. Thus damage of the machine is prevented. But sooner or later most insulation will deteriorate—especially if run at high temperatures.

Temperature rise comes about through introducing heat into a machine. In a transformer heat is developed from losses in the windings and core (in R and G of the transformer model). Heat is dissipated largely through surfaces. One of the reasons the size of a transformer increases with higher kVA ratings is to provide more surface for heat dissipation. Also, if there are higher temperature differences between the surface and ambient air, these surfaces dissipate more heat through a given area. As more and more heat is poured into a machine enclosure through losses, the temperature of the machine rises to whatever temperature is needed to dissipate the losses generated in it. If this temperature is too high they will destruct. Hot summer days or hot locations may take a machine into too high an ambient temperature and require temporary derating, as given in manufacturer's specifications. Temperature limits for transformers are usually expressed as so many degrees rise (above 26°C ambient). 40°C rise is typical, but this figure depends on the kind of insulation used.

Small transformers are air cooled inside and out. The cases of larger ones are often filled with *transformer oil* (an insulating mineral oil) to provide convection of heat from the windings and core to the case for radiation into the surrounding atmosphere. This oil also improves electrical insulation for high-voltage windings. Very large transformers and other machines are equipped with fans and radiators. Rotating machines are usually designed to blow air through themselves.

In transformer maintenance it is important to keep dirt and other obstructions cleaned away to prevent impairment of cooling. Mineral oil loses its insulation strength with time—particularly if water gets into it—and has to be filtered or replaced. Eventually insulations deteriorate and the machine must be rewound. Often lightning hits transformers, or some other disaster occurs, forcing them to be rewound or replaced. It is good practice to check now and then on the temperature of transformers.

From calculations in the above example it will be appreciated that losses in the transformer are caused only by load current (in R) and applied voltage (across G). The phase angle of the load makes no difference to the transformer. So transformers are rated in kilovolt-amperes instead of kilowatts. Allowable current times the allowable voltage yields the allowable kVA. This is somewhat misleading since

current and voltage cannot be traded off over any great range. Excessive voltage
will quickly cause excessive magnetizing currents which can destroy the machine.

EXAMPLE 18.2 A transformer is rated at 62,500/4000 V, 60 Hz, 100 kVA, 10%
reactance.

Find
1. the maximum current that can be taken from its secondary, at 100% pf
   and 71% pf (power factor) lagging;
2. the maximum primary current;
3. the voltage regulation in percent at rated kVA for 100% pf and 71% pf.

Solution
1. Assuming the low-voltage side is the secondary, maximum I_2 = kVA/V_2 =
   100 kVA/4 kV = 25 A. The power factor will make no difference here since
   the limitation is one of current, not power.
2. The corresponding primary current is 25 × 4000/62,500 = 1.6 A. This
   could also be obtained by dividing allowable kVA by 62.5 kV. The exciting
   current must be neglected here, with no data. At full load it will have little
   effect anyway, particularly at high power factors.
3. For 100% pf, I_1 is in phase with V_1, so I_1 = 1.6 $\underline{/0°}$ A. The reactance base
   (see Section 17.8) is V^2/kVA = 62.5^2/100 = 39 kΩ, so X = 10% of this
   or 3.9 kΩ. Voltage drop across X will be 3.9 kΩ × 1.6 A or 6.24 kV. This
   voltage drop figure could have been more directly obtained by simply taking 10% of 62.5 kV. Using a phasor diagram construction similar to Figure
   18–3b, this voltage drop is at 90° lead for 100% pf (neglecting the transformer resistance). At right angles 10% adds (or subtracts) only 0.5% to
   the voltage [$\sqrt{(1^2 + 0.1^2)}$ = 1.005]. Hence voltage regulation at 100% pf
   is close to 0.5%. (Although there are no data given in this problem on
   resistance, it may be $\frac{1}{5}$ or so of X and for this 100% pf case produces an in-
   phase drop. Using some such resistance it is possible to calculate a more
   realistic number for this regulation of 2% or so.)

   For the 71% case a voltage drop of 10% leads the voltage V_1 by only
   45° (see again Figure 18–3b). Hence the voltage will be dropped at full
   load by about 71% of 10% in phase with V_1 and added to it, and a similar
   amount at right angles. Thus the new voltage in terms of V_1 is
   $\sqrt{1.07^2 + .07^2}$, yielding about 7.2% regulation. (The missing resistance
   data will not make as much difference at this load phase angle.)

18.4   TRANSFORMER TESTS

Determining the five elements of the model for a given transformer, Figure 18–2,
can be done easily through two low-power tests—the open-circuit and short-circuit
tests—as shown in Figure 18–4.

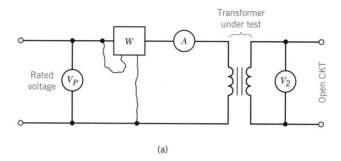

(a)

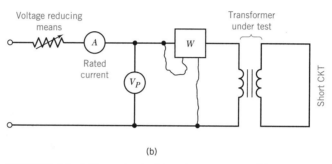

(b)

FIGURE 18–4 Short-circuit (a) and open-circuit (b) tests on a transformer will determine the five model elements of Figure 18–2. The short-circuit test must be performed at reduced voltage.

For each test the primary is equipped with wattmeter, voltmeter, and ammeter. In an *open-circuit test* (performed at rated voltage) with no current in the secondary, there will be no current I_1 and hence no losses in R. All I_P current will be drawn as I_e through the parallel circuit representing the transformer core. From W, V, and A, G and B are determined as in the example below. Also, moving the voltmeter to measure V_2 will establish N_2/N_1, the turns ratio.

A *short-circuit test must be done at reduced voltage.* As shown in Figure 18–4b the secondary is short circuited. Note the interchange of voltmeter and ammeter between the two tests to obtain greater data accuracy. Voltage is raised from some low value until rated current is reached in the ammeter. This happens in most cases at only 10% or so of rated voltage. Under these circumstances the current through G and B will be greatly reduced. Engineers neglect any remaining I_e and assume that all the ammeter current is going through R and X. Again from the readings of W, V, and A the model parameters R and X can be determined.

EXAMPLE 18.3 Tests on a transformer produce the following data. Open circuit: $V_P = 400$ V, $I_P = 2.06$ A, $P = 200$ W, $V_2 = 234$ V: short circuit: $V_P = 23.71$ V, $I_P = 28.75$ A, $P = 165.3$ W.

Find: the five elements of the model in Figure 18–2a.

Solution: From OC test: $N_2/N_1 = V_2/V_P = 234/400 = \underline{0.585}$ (the turns ratio would more commonly be expressed as 1.71), $G = P/V^2 = 200/400^2 = $ <u>1250 u℧</u>, $Y = I/V = 2.06/400 = 5150 \ \mu℧$, $B = \sqrt{(Y^2 - G^2)} = \sqrt{(5150^2 - 1250^2)} = $ <u>4996 $\mu℧$</u>. From SC test: $R = P/I^2 = 165.3/28.75^2 = $ 0.200 Ω, $Z = R + jX = V/I = 23.71/28.75 = 0.8247$. So $X = \sqrt{(Z^2 - R^2)} = \sqrt{(0.8247^2 - 0.200^2)} = $ <u>0.800 Ω</u>. Thus the model is established.   ■

18.5   MAGNETIC CIRCUIT

Figure 18–5a shows an iron magnetic circuit (core) linked with an electrical winding of N turns supplied with current I amperes. It is helpful to visualize a *magnetic flux ϕ* flowing around the core as a result of the electrical current. Experiment shows that for small values of ϕ, doubling the current doubles ϕ. Similarly, doubling the number of turns doubles ϕ. As the product NI increases, ϕ increases proportionately until eventually the iron begins to *saturate*, as shown in Figure 18–5c. After that point, increasing the driving force, the *magnetomotive force* (mmf) or NI, yields very little increase in response ϕ. In electrical circuit analysis, it is helpful to think of a voltage source as a driving force (emf) that pushes current (I) around an electric circuit, with current as the response. Similarly for magnetics, think of the ampere turns, NI, as a magnetic driving force (mmf) pushing flux (ϕ) around the magnetic circuit.

It is interesting to note that sketch 18–5a could be a transformer with an open-circuited secondary, which would act like nothing more than a large iron-core inductor.

Magnetic flux is imperceptible to the five senses and therefore not easy to describe. It is little more than a convenient concept. Flux is known only through its effects [for instance Equation (18.1)]. Flux is measured in units of *webers*. A uniform change of one weber per second in flux linking a one-turn winding will generate one volt across it. Figure 18–5b illustrates the electric circuit analogy—a

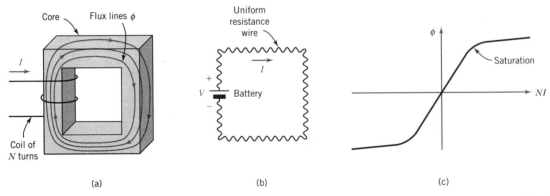

FIGURE 18–5 (a) Magnetic flux ϕ driven around a closed-path iron core by mmf (in units of NI) is analogous to electric current I being driven around a closed electric circuit (b) by emf in units of volts. However, (c) there is a saturation effect in the core, limiting the flux that can be produced.

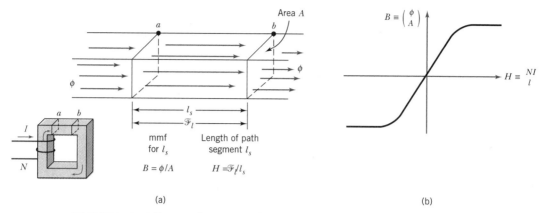

FIGURE 18–6 For a path segment of length l_s and cross-section area A, flux density $B = \phi/A$, and field intensity $H = \mathscr{F}_l/l_s$, where $\mathscr{F}_l$ is the mmf across the segment in ampere-turns. The B-H curve (b) is distinctive for each kind of iron or steel alloy.

battery (with voltage NI) pushing current (flux) through a continuous resistance wire (the core).

This mmf that pushes flux around a closed path is measured in units of ampere-turns—NI. It is also symbolized by NI in much the way voltage is symbolized by V. Figure 18–7 sets out a comparison of magnetic and electric quantities and their units which will be helpful to the beginner in magnetics. It should be studied in connection with Figures 18–5 and 18–6.

Iron and steel are the best "conductors" of ϕ, but magnetic flux will go through anything, including empty space. Very roughly speaking, iron materials are perhaps one or two thousand times better "conductors" than space. Being able to go through space to some extent, flux really needs no core or iron path. Cores are used to increase flux for a given mmf and to guide flux into useful patterns.

Flux density B is calculated as flux per unit cross-sectional area as in Figure 18–6a,

$$B = \phi/A. \tag{18.6}$$

Function	Flow	Force	Flow Per Unit Area	Force Per Unit Length	Resistance to Flow
Electrical quantities, units & symbols	I Current ampere (A)	V EMF volt (V)	i Current density A/m²	$\mathscr{E}$ Electrical field intensity V/m	R Resistance ohm (Ω)
Magnetic quantities, units & symbols	Flux ϕ Weber (Wb)	mmf NI (A-turn)	Flux density T Tesla (Weber/m²)	Magnetic field intensity NI/m (A-turn/m)	$\mathscr{R}$ Reluctance Rel

FIGURE 18–7 Keeping the detailed electrical analogy in mind is helpful in understanding magnetic circuit problems.

Its unit is the tesla (T) or weber per square meter (Wb/m^2). The magnetic field intensity H is calculated, in the same illustration, as mmf per unit length of flux path,

$$H = NI/l. \tag{18.7}$$

Corresponding electrical units, $\imath$—current density in amperes per square meter—and $\mathscr{E}$—electrical field strength in volts per meter—are infrequently used except in specialized work.

The quantity corresponding to R, electrical resistance, is in magnetics $\mathscr{R}$, *reluctance*. $\mathscr{R}$ is defined, correspondingly, as

$$\mathscr{R} = NI/\phi. \tag{18.8}$$

It is not a very useful quantity because the nonlinearity of the ϕ-NI curve (Figure 18–5c) means that it isn't a constant like most resistances. Instead, problems are worked from a curve somewhat like Figure 18–5c. But because such a curve would be different for every core shape, it is first normalized using B and H as in Figure 18–6b. The curve represents B as a function of H for a given magnetic material. A core of any shape made out of it can be analyzed (using the B-H curve) for the relation of ϕ to NI simply by considering its cross-sectional area and path length. Figure 18–8 provides three typical quantitative B-H curves for core materials.

EXAMPLE 18.4 A square, symmetrical, silicon-sheet-steel core, as shown in Figure 18–5a, has outside dimensions $0.2 \times 0.2 \times 0.03$ m; inside dimensions are 0.14×0.14 m. A 500-turn winding is wrapped around one leg with current of 270 mA. The B-H curve for this core is given in Figure 18–8.

Find: B, H, ϕ for this core.

Solution: The cross-sectional area through which flux passes is $A = 0.03 \times 0.03 = 9 \times 10^{-4}$ m^2. Average path length (take mean between outside and inside paths), $1 = 4 \times 0.17 = 0.68$ m [or $(0.8 + 0.56)/2 = 0.68$]. $NI = 500 \times 0.270 = 135$ NI. (This is the total mmf driving flux around the core.) $H = NI/l = 135/0.68 = \underline{199 \text{ NI/m}}$. From the curve for SS steel of Figure 18–8, entering with H of 199 yields a B of about $\underline{1.01 \text{ T}}$ (or 1.01 Wb/m^2). $\phi = B \times A = 1.01 \times 0.0009 = \underline{0.909 \text{ mWb}}$.

In rotating machinery (motors and generators) current-carrying wires move across magnetic fields, as Chapter 20 will detail. Air gaps, through which wires can slip, must therefore be provided in a machine's magnetic circuits. In practical motors there is a small gap (about 0.1 in.) between the rotating rotor and magnetic pole structure mounted on the frame. To simulate this in problem solving, cut a gap across one leg of the core in Figure 18–5a as shown in Figure 18–9. This produces a series magnetic circuit. The air gap portion of the flux path is in series with the iron core portion.

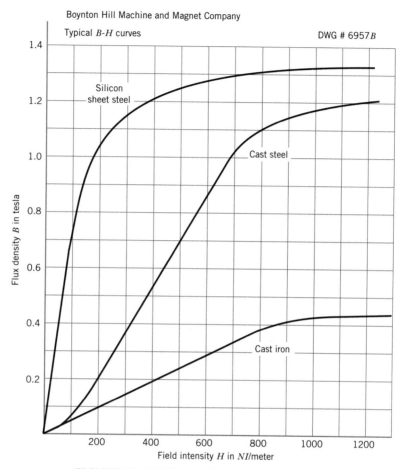

FIGURE 18-8 *B-H* curves for three core materials.

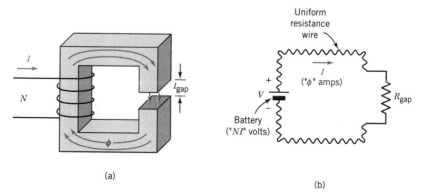

FIGURE 18-9 (a) "Air gaps" in magnetic circuits produce a high reluctance portion effectively in series with the iron portion. *NI* drops across each portion sum to the total magnetic circuit *NI* in a manner analogous to KVL (b).

KCL suggests (because of the continuous lines) that the flux in the gap is the same as the flux in the core. KVL suggests that the mmf supplied by the winding turns and their current is the sum of the mmf drops across the iron path and the air path. Figure 18–9b shows the electric equivalent with the continuous resistance wire (simulating the core) broken and a larger resistor inserted in this break (representing the reluctance of the gap).

Engineers habitually speak of air and air gap, but they should really speak of space since air has a negligible effect on magnetics. "Air" (or really, space) has a linear B-H relationship; no B-H curve is needed. An algebraic relation will suffice,

$$B = \mu_0 \times H. \tag{18.9}$$

The physical constant μ_0, called the *permeability of free space*, has an mks value of $4\pi \times 10^{-7}$ henrys per meter.

To solve a problem in which a given mmf is imposed on a core with air gap is not as simple as it looks. In a similar series electric circuit, for example, Figure 18–9b, resistances would be added. But here in the magnetic circuit, the reluctance of the core is nonlinear and therefore addition is forbidden. A tedious but not difficult solution is to (i) guess the division of mmf between core and gap, (ii) solve for the ϕ of each, and (iii) guess again and repeat until the two ϕ's are essentially equal. Since flux is continuous the two fluxes must be the same. For the present let us take a somewhat quicker problem.

EXAMPLE 18.5 This continues Example 18.4.

In the core arrangement of the previous problem, a gap of length 3 mm is sawed normally across one leg.

Find: the total winding current needed to maintain the same flux density in the core.

Solution: With the same B, all parts of the previous solution for the iron portion of the path will be unchanged (neglecting the tiny shortening of the iron path). B_{air} will equal $B_{iron} = 1.01$ T.*

Using Equation (18.9), $H = B/\mu_0 = 1.01/(4\pi \times 10^{-7}) = 804{,}000$ NI/m. NI needed for air gap is $H \times 1 = 804{,}000 \times 0.003 = 2411NI$. The total NI needed for both iron and air is $135 + 2411 = 2546$. $I = NI/N = 2546/500 = \underline{5.92\ A}$.

18.6   MORE ON HOW TRANSFORMERS WORK

With the skill gained in magnetic circuits from these examples we revisit transformers.

---

*Flux lines near the edge of the gap fringe or bow out, making the average area of the gap slightly larger than the cross-sectional area of the iron. But this effect is small if other gap dimensions are much larger than its length. It will be neglected here.

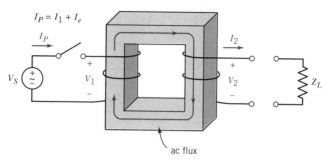

FIGURE 18–10 Primary transformer current I_P is composed of two parts—the load current I_1 and a much smaller exciting current I_e. $I_1 = I_2(N_2/N_1)$.

Consider the transformer of Figure 18–10 to have no R or X and to be un-loaded—no secondary current flows for the moment. The *magnetizing inductance* (L_m) seen when looking into the primary of an unloaded transformer is the B element in the model of Figure 18–2, and $B = -1/\omega L_m$.

Close a voltage source V_S on the primary as shown in the circuit of Figure 18–10. Primary current begins to flow in the winding, producing mmf, and setting up an increasing magnetic flux ϕ in the core. Suppose for a moment that instead of the sinusoidal source shown, the voltage applied is dc (as earlier suggested this is a very bad idea but is used briefly here for explanation). The increase of flux in the core will generate a back voltage $v_1 = N_1 \, d\phi/dt$ [the law expressed in Equation (18.1)]. This voltage V_1 has a polarity opposing V_S. If applied voltage is constant dc, the rate of change of flux will be constant—so the current must increase constantly at a uniform rate.

Now it is clear why transformers are destroyed by dc. Looking at Figure 18–5c, everything is fine until the knee of the saturation curve is reached. At this point V_1, the back voltage, can no longer be maintained without much greater increments of magnetizing current to produce the required $d\phi/dt$. Note that in the primary circuit of Figure 18–10, KVL requires that V_1 equal the applied source voltage V_S. But with the opposing voltage V_1 attempting to drop off, V_S produces ever-increasing current I_P. Current is finally limited only by the small resistance of the primary winding—a resistance neglected so far. These too large magnetizing currents produce large I^2R losses. The transformer burns up.

Now take V_S to be a 60-Hz cosinusoidal ac as shown in Figure 18–10. V_1 must be identical to it. Hence the flux will be sinusoidal so that Equation (18.1) can generate V_1 as a cosinusoidal voltage to match V_S. Current (I_e) will adjust itself for this to happen in a manner similar to the paragraph above, except that with ac it need not continually increase. Note that because of the derivative in Equation (18.1), a sine wave of flux will produce a cosine wave of voltage. (The derivative of sin is cos.) The current wave will be in phase with the flux. Thus magnetizing current lags voltage by 90°. (It will appear below that there must be some small in-phase component in the total exciting current to supply iron losses, but it is ne-glected for the moment.) This explains nicely the operation of coils and their in-ductive nature. While Chapter 6 simply defined an inductance as a two-terminal element that obeyed the law $v = L \, di/dt$, this new understanding provides a rea-sonable physical explanation for the behavior of practical inductors.

We can note also that if V_S is large enough (at its peaks) to take the core into saturation, I_e will be nonsinusoidal.

At this point there is a small exciting current I_e flowing in the primary winding, and magnetic flux flowing in the core. Both are sinusoidal in nature but approximately 90° behind the voltage V_1, which is the same as the applied voltage V_S. The secondary is still open circuited. But the same flux that links the primary links also the secondary. Equation (18.1) applies there also and is the same except for N. Hence $V_2 = V_1 \times N_2/N_1$.

Now connect the load Z_L to the secondary and allow current I_2 to flow. The secondary winding produces an mmf $N_2 \times I_2$ which opposes the sinusoidal flux changes, tending to reduce the flux set up by I_e in the primary winding. But this tends to reduce V_1 [Equation (18.1)], allowing more current to flow in the primary. The increase of primary current beyond the original I_e is called I_1. I_1 will increase until $I_1 \times N_1$ just equals $I_2 \times N_2$, canceling out the secondary's mmf and leaving I_e in total control of the flux. In order for this to happen I_1 will be in phase with I_2. Hence the ratio $I_1/I_2 = N_2/N_1$. Thus the amount of flux in a transformer core is dependent only on voltage. Load has substantially no effect.

18.7   CORE LOSSES

Section 18.2 above discussed practical transformer copper and iron losses in terms of the model. Copper losses are the I^2R losses in the two windings. These are modeled by R. In the core there are two types of iron losses—*hysteresis* and *eddy currents*. These are both modeled by G.

The way in which iron materials augment flux density B for a given H force is to reorient groups (domains) of their molecular-sized particles. These particles and the domains themselves function as small magnets as a result of electron orbit orientation. Figure 18–11a represents the randomly oriented magnetic areas in unmagnetized iron. Figure 18–11b shows a completely saturated bar with all its

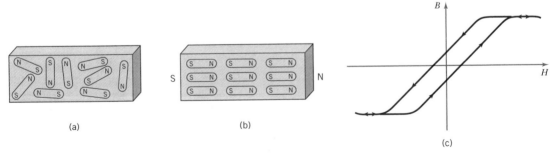

FIGURE 18–11 (a) (b) Hysteresis loss is caused by realignment of minute magnet areas (domains) in the core material as the alternating mmf progresses through its cycle. Examined carefully (c), the B-H curve includes a thin "hysteresis loop." For every ac cycle in which the loop is traversed a certain amount of energy is lost.

magnetic particles or areas lined up. Between these two extremes, as more and more H is applied, B moves up the *B-H* curve (Figure 18–6 or 18–8), more and more of these little magnets within the material being progressively aligned. When there are no more particles left unaligned, saturation is complete.

If ac is applied to a winding, the flux continually reverses itself sinusoidally, and these particles are, in effect, turned back and forth. The particles tend to stick and must be forced around. *Hysteresis* losses are the energy lost in these reversals. The power lost in hysteresis will be proportional to frequency for a given flux density B, since a certain amount of energy is lost for each reversal.

A more complete *B-H* curve is the *hysteresis loop* of Figure 18–11c. Figure 18–6 neglects this detail; the return curve is actually slightly different from the ascending curve. A small area is encircled which can be shown to be proportional to the losses. Core steel manufacturers endeavor to have as thin a hysteresis loop as possible.

On a somewhat more sophisticated basis, we can think in terms of *magnetic domains*, recognizing that within entire small domains the core material is similarly aligned, and that the progressive magnetization and demagnetization of the iron core during an ac cycle is carried out by a corresponding shifting of domain boundaries.

Eddy currents, the second contributor to core loss, are electrical currents flowing in the iron core material. Figure 18–12a illustrates this for a cross-sectional slice taken from any of the cores previously shown. Within the plane of this slice

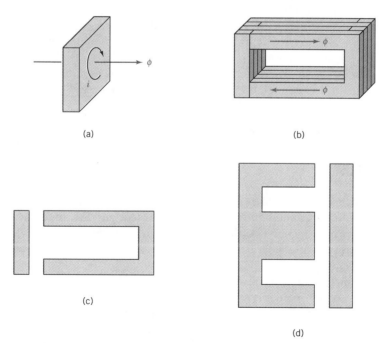

(a)

(b)

(c)

(d)

FIGURE 18–12 Undesirable electrical "eddy currents" (a) flow in the core material itself, causing further core losses. Eddy currents and losses are minimized by assembling cores from thin laminations (b, c, d) or compressing them from powdered materials.

any closed path surrounds flux, and is in effect a coil turn linking the flux. It is conductive since iron conducts. But these paths are linked by changing flux. Equation (18.1) will produce voltages and therefore eddy currents around any such path within the core. The I^2R loss in the iron (not to be confused with the transformer's usual I^2R loss in the windings) is that part of the core loss called "eddy current loss."

One solution to hysteresis is to buy better steel for cores. Reducing flux density also helps. As with most engineering designs there is an economic tradeoff here.

Eddy current losses are minimized by laminating the core (Figure 18–12). Alternating-current cores are built up of thin sheets (*laminations*) electrically insulated from each other by varnish, or sometimes by metallurgical pickling. Thicknesses of 0.020 in. are typical. These *laminations* make eddy current paths thinner, with much smaller losses than a solid core. Thin cross sections contain less flux and therefore lower voltage and current. Laminated construction also has the advantage that transformer cores can be made from stacks of properly shaped punchings. Each coil is wound by machine on a bobbin—usually made of impregnated paper. The core is then assembled around these bobbin coils.

In place of the rectangular core illustrated in this chapter, many transformers have a three-legged core (Figure 18–12d) with both windings on the center leg. These two core types are assembled from C's and I's or from E's and I's as shown in Figure 18–12b and c. The direction of the E or C is alternated from layer to layer.

Another popular core type is the *toroid*, or doughnut-shaped core, which conserves the expensive core material used for higher-frequency applications where hysteresis losses are particularly important. Instead of laminating, toroids are usually made with powdered iron particles sintered together or held together with some nonconducting matrix material. Toroids are wound by automatic machines throwing wire shuttles through the core. Cores of this type may also be tailored for certain magnetic characteristics such as are used in some computer memories.

18.8   SUMMARY

Transformer action is based on Faraday's law. From this it can be deduced that V_2/V_1 and I_1/I_2 are equal to the turns ratio N_2/N_1, and the impedance seen looking into the primary is equal to the load impedance divided by the turns ratio squared. In addition to the load current I_1, the primary includes a small out-of-phase magnetizing current.

Transformers can be simply modeled with the circuit of Figure 18–2, to determine voltage regulation, losses, and efficiency. The elements of this model are quickly determined with open-circuit and short-circuit tests for a given transformer.

Transformers are rated in terms of kVA and voltages. A current rating is derived by dividing these. Voltage ratings cannot be exceeded by more than a few percent without excessive magnetizing current and heating.

Magnetic circuits (in transformers, motors, or other magnetic equipment) are analogous to and analyzed like electric circuits, except for the effect of magnetic saturation limiting the flux (magnetic current) such a circuit can handle. Air gaps require special handling in these circuits. *B-H* curves for particular iron materials relate flux density to magnetic field intensity in the core. Using core dimensions, these numbers can be turned into flux and ampere-turns for the application at hand.

Magnetic core losses include hysteresis and eddy currents.

FOR FURTHER STUDY

Ralph J. Smith and Richard C. Dorf, *Circuits, Devices and Systems*, 5th ed., Chapter 21, John Wiley & Sons Inc., New York, 1992.

PROBLEMS

Easy Drill Problems (answers at end of chapter)

D18.1. A transformer has a turns ratio (secondary to primary) of 0.6667. It is desired to operate a 200-Ω resistive load at 150 V. Find the primary and secondary currents, the source voltage, the power delivered to the primary from the source, and the impedance the source sees looking into the primary winding. (*Note:* With no more data than this it will be necessary to assume a perfect transformer model such as Figure 18–1a.)

D18.2. A transformer has a rating of 500/125 V, 10 kVA, 60 Hz. It is loaded with an impedance of 10 Ω at 75% pf. The source voltage applied to the primary winding is 480 V. Find kVA delivered to load, load voltage, load current, primary current, power delivered to load, power factor of primary, impedance the source sees looking into primary.

D18.3. A winding of 300 turns of wire has a flux passing through it of $\phi(t)$ Wb such that $\phi(t)$ is triangular, varying in 1 min from 0 to 0.0125 Wb, and then back to zero in the next minute, and repeating this pattern indefinitely. Plot quantitatively, as a function of time, two cycles of the voltage that appears at the winding terminals. On the same axes plot $\phi(t)$.

D18.4. Using the same winding as in D18.3, assume that the flux varies sinusoidally between the same limits and with the same period. The sinusoid is centered about 0.0125/2 Wb, with zero slope at the top and bottom of the wave. The wave is always positive. Plot again two cycles of $v(t)$, the voltage induced in the winding. On the same axes plot $\phi(t)$.

D18.5. The resistances of primary and secondary windings of a transformer are respectively 5.12 Ω and 0.0641 Ω. The turns ratio is 0.1023. Find the value of R in the standard model (Figure 18–2) for this transformer.

D18.6 A transformer is rated at 13,800/480 V, 100 kVA, 60 Hz. Find the allowable primary and secondary currents at supply voltage of 13,800 at 100% power factor. Repeat for a power factor of 50%.

D18.7. In the magnetic core of Figure 18–10a a sinusoidal flux flows whose maximum value is 2 mWb. The outside dimensions of the core are 0.4 × 0.4 m. Its inside dimensions are 0.3×0.3 m. Find the maximum value of B in tesla for cores of the following depths: 0.1, 0.05, and 0.035 m. Tabulate your calculations and answers.

D18.8. For the situation of D18.7 find the maximum value of H in $NI/$m for 0.05-m-thick cores made of silicon sheet steel, cast steel, and cast iron. Use curves of Figure 18–8.

D18.9. Assume a winding of 100 turns is wound on the core of problem D18.8. What is the winding's rms voltage for each of the cases of that problem?

Application Problems

P18.1. A transformer is rated 1000/250 V, 6.25 kVA, 60 Hz.
  (a) Find (based on ratings) the load impedance for full load, the primary and secondary currents, and the impedance that source sees looking into primary.
  (b) For this transformer, using the standard model of Figure 18–2: N_2/N_1 is 0.2624, $R = 4\ \Omega$, $X = 40\ \Omega$, $G = 100\ \mu\mho$, $B = 500\ \mu\mho$. Find copper loss, iron loss, efficiency.

P18.2. Continuing with the transformer of P18.1,
  (a) for 100% power factor find the voltage regulation and draw a larger phasor diagram to scale showing V_1, I_1, V drop on Z (X and R), and V_P.
  (b) Repeat for 70% and 50% power factors.
  (c) What is the worst power factor angle for this transformer from the standpoint of regulation?

P18.3. A 4000-V transformer has the following model constants for N_2/N_1, R, X, G, B: 0.0581, 1 and 16 Ω, 20 and 250 $\mu\mho$. A 3800-V source is connected to the primary and an inductive load of 2.4 Ω at 70% pf to the secondary. Find load current, load kVA, load power, drop in secondary voltage when load is connected.

P18.4. Calculate and plot the efficiency of the transformer of problem P18.3 vs percent full load current. Assume source voltage is 4000 and full load rating 100 kVA.

P18.5. For the same transformer and a 4000-V source plot voltage regulation in percent versus pf in percent.

P18.6. A transformer's nameplate and ratings have been lost. It is known that this type of transformer is usually rated for a 40°C rise above 26°C. Explain how a reasonable voltage rating can be determined on open circuit by running up voltage applied to one winding while observing the waveform of I_e. (*Hint:* Consider the effect of curve 18–5c on the waveform of exciting current.) Explain how a reasonable current rating can be determined at rated voltage by observing transformer temperature as load is increased.

P18.7. A transformer is rated 13,800/480 V, 50 kVA, 60 Hz with 40°C rise above 26°C.
  (a) What maximum secondary current would you expect to be able to draw safely?
  (b) The same model transformer is now to be used at rated voltage in a desert situation where temperatures can rise to 105°F. To what maximum value of secondary current would the transformer have to be derated? (Assume that under rated conditions two-thirds of the losses are copper losses. Assume that heat dissipated from surfaces is proportional to temperature rise above ambient.)

P18.8. Using the circuits of Figure 18–4, an engineer performs a short-circuit test at rated current, and an open-circuit test at rated voltage to determine the model parameters of a transformer, and develops the following data: OC: $V_P = 120$, $V_2 = 424$ V, $P = 123$ W, $I = 4.24$ A; SC: $V = 17.4$, $W = 211$, $A = 83$. Calculate N_2/N_1, G, B, R, X.

P18.9. A toroidal (doughnut-shaped) core has outside diameter of 0.4 m and inside diameter of 0.3 m. A 60-Hz sinusoidal flux whose maximum value is 2 mW flows in it, and a winding of 100 turns is wound on it. It has a *B-H* characteristic similar to the silicon-sheet-steel curve of Figure 18–8. Find the maximum values of *B* and *H*. Find the rms values of voltage and current in the winding.

P18.10. An air gap is sawn across one side of the core of problem P18.9, removing 2 mm of material. To what rms value will the current have to be raised to keep the same maximum value of flux?

Extension Problems

E18.1. In problem P18.8 the transformer was tested from the low side. Predict the test results from the high side (that is, test it at say 424 V on the high-voltage winding). What are the parameters now? Tabulate the seven test results and the five model parameters with the corresponding values in P18.8. (*Hint:* Draw the circuit results of P18.8 and put this model circuit into Figure 18–4 to test.)

E18.2. A transformer has three windings of 100, 200, and 300 turns. Place a 100-V source on the 300-turn winding as primary; load the 200-turn winding with a resistor of 100 Ω and the 100-turn winding with an inductor of 100 Ω. Find the primary current in magnitude and phase. Draw a phasor diagram showing all three currents and voltages. (*Hint:* Think through Section 18.6.)

E18.3. An engineer wishes to measure the two components of core loss on a transformer. She runs open-circuit tests under two conditions with results as follows: Test #1: 60 Hz, 240 V, 144 W; Test #2: 50 Hz, 200 V, 108 W. Find and tabulate the hysteresis and eddy current losses under the two conditions. (*Hint:* How much, if any, has the flux density in the core changed between the two tests? Consider the eddy current path as a transformer winding with a fixed resistance load.)

E18.4. A 240-V 5-kVA transformer designed for 60 Hz has copper losses of 143 W, and iron losses of 62 W. You estimate that 40 W of the iron loss are due to hysteresis and the rest to eddy current. It is desired to use this transformer on 50 Hz.

(a) If the transformer were to be derated in voltage to keep the same maximum flux density, what voltage rating should it have?

(b) You decide to operate the transformer at 50 Hz, 220 V, 23 A. Assume that hysteresis loss is proportional to frequency and (within 10% of rated voltage) to the 1.6 power of voltage for a given frequency. Assume that eddy current loss is proportional to the square of voltage. Find the three losses and the new transformer efficiency.

E18.5. A silicon-sheet-steel core similar to that in Figure 18–12b has outside dimensions of 12.5 × 8.5 cm and inside dimensions of 6.5 × 2.5 cm. Its thickness is 5 cm. A winding of 500 turns is wound on one leg with a dc current of 175 mA.

(a) Find the flux density and total flux.

(b) The core is now sawn in two all the way across so that there are two 3-mm air gaps in it. The same winding and current are used. Find the flux density and total flux.

E18.6. A cast-steel two-window core is made from punchings shown in Figure 18–12d. When assembled the outside dimensions are 12 × 8 cm and the thickness is 6 cm. Each window is 2 × 4 cm and so placed that the center leg has twice the width of the other flux-conducting portions. A winding of 300 turns is placed on the center leg. What dc current must flow in it to produce a flux density of 0.5 T in the outside legs? What is the flux density in the center leg?

E18.7. A core identical to that of problem E18.6 is wound with a 300-turn winding on one of the side legs. A dc current of 350 mA flows in the winding. Find the flux density and total flux in each portion of the core.

E18.8. (a) With the same core and winding setup of problem E18.7, a 1-mm gap is sawn across the center leg. Find the flux density and total flux in each portion of the core.

(b) Adjust the width of the outer leg only (not the horizontal portions) so that the flux in the center path and the outer path are equal.

Study Questions

1. Transformers rely on the interaction between what two sets of physical effects?

2. How are the primary and secondary windings linked?

3. Using the phenomenon described by Equation (18.1), show that the primary and secondary voltages are related by the turns ratio.

4. What is the difference between an ideal transformer and a practical transformer?

5. Describe, using Equations (18.2)–(18.4) how a transformer transforms impedances.

6. Assuming that the (model) impedance of a given transformer is purely reactive, demonstrate by a sketch that the phase angle of the load makes a difference in voltage regulation.

7. Why does a given transformer winding (used as the primary) have such a definite voltage limitation? What happens if it is exceeded?

8. Why does a transformer have a current limitation? What happens if it is exceeded?

9. Can the voltage limitation be exceeded if the current is reduced below rating?

10. Can the opposite tradeoff be made?

11. What changes should be made in the transformer design to allow it to dissipate more losses?

12. Why must voltage ratings be reduced if frequency is reduced?

13. What two advantages does surrounding the core and windings with transformer oil provide?

14. Between what two media do the radiators of large transformers transfer heat? Where does this heat come from?

15. Why are transformers rated in kVA instead of kW?

16. What are the two standard transformer tests? Describe them. What instrumentation is used?

17. How can two low-power tests test a transformer to its full high-power ratings?

18. Is there any important aspect of a transformer's operation engineers might like to prove that is not provided by the two standard tests?

19. Carefully trace the cause-and-effect relationships in terms of currents and voltages of a loaded transformer, starting with the power source connected to the primary winding and accounting for all currents and voltages and the magnetic flux.

20. What are the two sources of core loss? Describe each carefully.

21. What steps are taken in transformer design to minimize each of these two losses?

22. Explain in terms of transformer theory why the current through an inductance lags the voltage across it by 90°.

23. Why are different size porcelain bushings often used on the same transformer to provide an insulated path through the steel case?

ANSWERS TO DRILL PROBLEMS

D18.1. 0.50 A, 0.75 A, 225 V, 112.5 W, 450 Ω

D18.2. 144 kVa, 120 V, 12 A, 3 A, 1.08 kW, 75%, 160 Ω

D18.3. $V_{max} = \pm 62.5$ mV

D18.4. $V_{max} = \pm 98.2$ mV

D18.5. 11.25 Ω

D18.6. 7.25 A, 208 A—at any pf

D18.7. 0.4, 0.8, 1.14 Wb/m^2

D18.8. 115, 565 NI/m, impossible for cast iron

D18.9. 53.3 V independent of material

CHAPTER 19

Motors

It has been estimated that two-thirds of the electric energy used in the United States goes into motors of one kind or another. Modern houses generally contain 30 to 60 motors for such purposes as fans, furnace control, door openers, pumps, dishwashers, spits, polishers, refrigerators, ovens, dehumidifiers, hair dryers, and on and on. Industrial motor uses are even wider and run from tiny, finger-sized, subfractional-horsepower motors through giants developing thousands of horsepower.

Motors convert electrical power (voltage and current) into shaft torque and rotation.

Motors are divided by power rating into two principal classes: *integral horsepower* (1, 2, 3, 5, 10, etc.) and *fractional horsepower* (motors physically smaller than a standard 1-hp, 1700–1800 rpm machine). Very small motors are sometimes referred to as *subfractional-horsepower* motors.

The objective of this somewhat qualitative introductory chapter is to develop a broad but clear understanding of electric motors in general, their characteristics, and operation in circuits. Quantitative analysis and many of the physical details are left for Chapter 20—Rotating Machinery Basics. We start with a typical example of an induction motor and load, and from this look into power in and out, efficiency, general construction, torque-speed characteristics, and some basic circuit considerations. The chapter continues with a discussion of other types of motors, modern variable-speed drives, and ends with some remarks on motor selection and operation.

Motors work by the interaction between electrical currents and magnetic fields.

Readers or instructors may prefer to take up the analysis of Chapter 20 before the external engineering considerations of Chapter 19, and there is no reason that should not be done. It is suggested, however, that those who take that order at least scan the figures and their legends in this chapter first. For most it will be easier to appreciate and use the theory of Chapter 20 after being introduced to the external characteristics and common application practices of motors in this chapter.

Objectives in studying this chapter can be to develop an appreciation for motors including:

1. concept of speed-torque relation
2. automatic adjustment of output torque to load requirements
3. starting requirements
4. motor circuit model
5. three principal types
6. variable frequency drives

19.1   A TYPICAL INDUCTION MOTOR AND LOAD EXAMPLE

Figure 19–1 diagrams a polyphase induction motor driving an industrial fan. Three-phase 440-V power supplies the motor. Power flows from left to right. Most of this electrical power is converted to mechanical power in the motor, which then drives

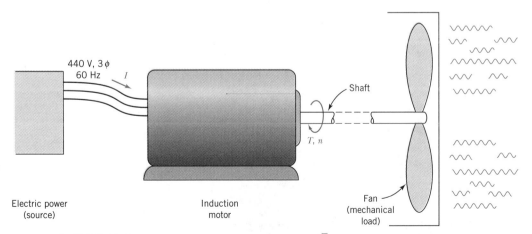

FIGURE 19–1 Motors convert electrical power *in* ($\sqrt{3}\ VI \cos \theta$) into mechanical shaft power *out* (T ω_m). Shaft speed ω_m is in radians per second and should not be confused with electrical radian frequency. T is in units of torque, newton-meters.

the fan through a rotating shaft. Both motor and fan have losses; neither is 100% efficient.

Three-phase electrical power into the motor can be expressed

$$P_{in} = \sqrt{3}\ VI \cos \theta, \tag{19.1}$$

in watts. Mechanical power out is

$$P_{out} = T\omega_m, \tag{19.2}$$

where if torque T is in newton-meters and if mechanical rotational speed ω_m is in radians per second, power is again in watts. In the United States horsepower is still a very common unit of mechanical power. There are 746 W in 1 hp, so that

$$P_{horsepower} = P_{kilowatts}/0.746. \tag{19.3}$$

EXAMPLE 19.1 The fully loaded 440-V, three-phase motor of Figure 19–1 draws 12.4 A at 86% power factor (pf). Its speed is 1730 rpm and its torque output to the mechanical load is 29.1 lb-ft. 29.1 lb-ft × 1.356 = 39.5 Nm.

Find: power in, power out in kilowatts and horsepower, and efficiency.

Solution: $P_{in} = \sqrt{3}\ \text{VI} \cos \theta = \sqrt{3} \times 440 \times 12.4 \times 0.86 = \underline{8.13\ \text{kW}}.$ $\omega_m = (1730/60) \times 2\pi = 181.2$ rad/s. Whence $P_{out} = T\omega_m = 39.5 \times 181.2 = \underline{7.15\text{kW}}.$ $P_{out} = 7.15/0.746 = \underline{9.58\ \text{hp}}.$ And eff $= P_{out}/P_{in} = 7.15/8.13 = \underline{88.0\%}.$ ■

Figure 19–2 shows the two main parts of motors—a fixed hollow cylinder or frame called the *stator,* and a smaller rotating (usually more or less solid) cylinder

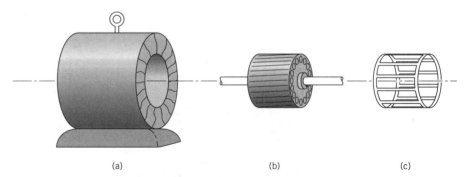

(a)   (b)   (c)

FIGURE 19–2 Motors have an iron cylindrical fixed part (a), the frame or stator, and an iron rotor (b) that turns inside the stator. Longitudinal slots on the inside stator face and the rotor face are wound with electrical conductors. In the squirrel-cage induction motor the electrical rotor portion is a cast aluminum cage (c), which has no external connections.

called the *rotor*. A rotor revolves inside its stator. (In dc motors the frame is referred to as the *field* and the rotating part as the *armature*.)

Both the stator and rotor are made primarily of magnetic iron to provide a path for flux, similar to the magnetic circuits of the last chapter. For ac machines, electrical conductors are wound in longitudinal slots on the inside surface of the stator. Rotors have various kinds of construction, but almost all have provision for electrical currents in them.

The rotor construction of the common ac machine in Figure 19–1 and 2 is called a *squirrel cage*. Figure 19–2c shows the electrical part of the rotor (the squirrel cage) with the iron stripped away. In this case instead of using conductors wound in slots, the rotor's squirrel-cage electrical conductor system is cast of aluminum directly into slots in the iron.

19.2   TORQUE-SPEED CHARACTERISTICS

When a motor is connected to a proper electrical supply, electromagnetic action produces a torque on its shaft, most of which the shaft supplies to the mechanical load. This supplied torque is some function of motor speed. The *torque-speed characteristic* (*T-n*) shown in Figure 19–3a is typical for induction motors. The symbols ω_m (in rad/s) and n (usually in rpm but sometimes in rps) are used to represent shaft speed. Other types of motors have different *T-n* characteristics.

Continuing the motor-fan example, Figure 19–3a shows that when first connected, the motor exerts some starting torque T_s, which begins to turn the fan. As the motor speeds up, greater torque is developed until some maximum torque T_m

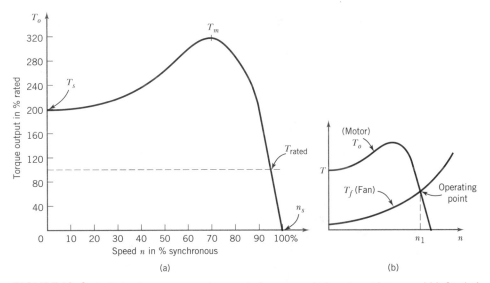

FIGURE 19–3 An induction motor produces a shaft torque, which varies with its speed (a). Similarly the torque required by a mechanical load may vary with its speed (b). If these curves are superposed, the speed at which a given motor will drive a given load can be found at their intersection.

is reached. After this point torque rapidly decreases with speed until at *synchronous speed* n_s (a term explained below) no torque is produced.

Figure 19–3b shows the torque required to turn the fan load at different speeds—a *load T-n* diagram—superposed upon the motor's *T-n* curve. At start (zero speed) very little torque is needed to turn the fan—only enough to overcome bearing friction. As the fan speeds up and starts to move more and more air, the torque required to turn it increases rapidly.

With the two diagrams superposed it will be seen that the extra torque available to accelerate the fan is the difference between motor output torque T_o and the back torque T_f required to move the fan and air. The load will start and begin to speed up because motor starting torque at zero speed is greater than load torque needed. The acceleration could be calculated as $d\omega_m/dt = (T_o - T_f)/J$, where J is the moment of inertia of the entire system. The two curves cross at an *operating point* at speed n_1 as shown. Beyond this point no further acceleration is possible; the motor-load combination will run continuously at that speed. If some transient event should produce a greater speed, the fan would require more torque than the motor is providing and slow the system down. If the speed is below this point, more torque is provided than the fan needs, and the system speeds back up to n_1.

19.3　STARTING AND CONTROLLING MOTORS

Small motors (less than about 1 hp or so) are started by connecting them directly to the line. Motors starting in this way take several times as much current as they do when running at full load. Motors of one common form, if started across the line, draw a starting current of 600% of (full load) running current. Household refrigerator motors are usually about 0.5 hp, but even this small a motor, starting across the line, dims lights in some houses with inadequate wire size.

For integral horsepower motors it is usual to limit current by starting at reduced voltage. The controller (or starter) in Figure 19–4 accomplishes this voltage reduction during start and is often simply an autotransformer. Motors are typically started at about half the rated voltage or a little more.

Torque produced by an induction motor at any given speed is roughly proportional to the square of the voltage applied. Thus the curve of Figure 19–3a applies only at rated voltage. A parametric curve of similar shape but at only one-quarter the torques would pretty well represent a half-voltage situation.

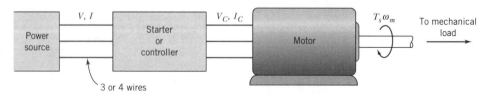

FIGURE 19–4 For integral horsepower motors a "controller" is interposed between motor and line to provide for starting and other control features. In circuit planning, the motor and controller are considered as a unit.

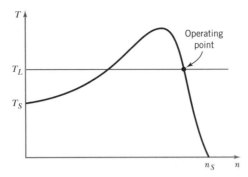

FIGURE 19–5 Horizontal load *T-n* curve marked T_1 is typical of friction loads. Motors with *T-n* curve as shown cannot start this load without special provision such as a clutch.

Starting torques are an important consideration. Figure 19–5 shows the *T-n* curve for a frictional load, which has a nearly constant torque requirement (conveyors are good examples). When such a load is connected to the induction motor being considered, it is apparent from the curves that although the motor is perfectly capable of driving the load once started, it cannot start it from standstill.

A drop of a few percent in voltage supply will prevent the starting of some motor-load combinations. Under these conditions, loads can be coupled to the motor with clutches which are engaged after the motor reaches operating speed. Or, alternatively, other types of motors or controllers with a better starting characteristic than squirrel-cage induction motors (used directly on 60 Hz) could be applied.

The induction-motor *T-n* curve of Figure 19–3 shows that this type, once started, is a nearly constant-speed machine. Chapter 20 shows that the synchronous speed n_s for 60-Hz supply can have only the values 3600 rpm, 1800, 1200, 900, or any submultiple of 3600 rpm. (For other than 60 Hz, these speeds are proportional to the frequency supplied.) The motors run only a few percent below synchronous.

So it is possible to control induction-motor speed and improve starting by controlling the frequency of power applied to the terminals. Section 19.10 below discusses this procedure under the heading Variable-Frequency Drives.

All but the oldest starters are automatic, requiring the operator to do no more than press a button to start or stop a motor.

19.4   MOTOR CIRCUIT MODELS

For amplifiers and other devices, it was found useful in earlier chapters to develop circuit models to understand and predict their reciprocal effects on the circuit to which they are connected. In the same way, Figure 19–6a is a reasonably good, simple motor model which can help in thinking about many problems. Of all ap-

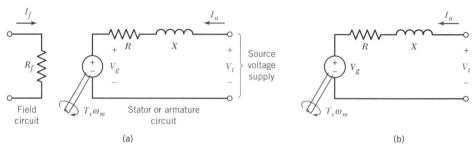

FIGURE 19–6 Most motors can be approximately modeled as in (a). The circuit shown is one phase and ground of an equivalent wye diagram. Squirrel-cage induction motors have no field connection (b). V_g is roughly proportional to speed. $V_t - V_g$ drives current through the armature winding. As a heavier mechanical load slows the machine, more current is taken from the line (source), producing more torque to meet the load requirement.

plications, this model is least satisfactory for the important induction motor, but it can be modified slightly to provide some general usefulness even for that purpose.

Most motors have two more or less separate circuits—field circuit and armature circuit, as shown in Figure 19–6a. In the dc machine the field is on the stator, the frame, and doesn't rotate. The armature, which in any machine is the heavy-current part, is on the rotor and turns—an unfortunate disadvantage since some kind of heavy-current, rotating connections must be made from frame to rotor. Alternating-current machines are configured in the opposite way, with field on the rotor and a stationary armature on the stator. The induction machine, however, has only one circuit to which power-line electrical connections are made, and that is on the stator.

The armature carries nearly all the supply current drawn. Note that the armature of the machine shown in Figure 19–6 is a Thevenin circuit comprising voltage source, V_g, and impedance represented here by X and R. X can be neglected for steady-state analysis of dc machines, and R can be neglected for much ac machine analysis. While essential, the field circuit has only a few percent of the armature's current or power. It will be seen that the model is a hybrid, showing the mechanical shaft output (shaft speed and torque) emerging directly from the voltage source symbol in the armature electrical circuit. This location is intended to show that most of the electrical power going into the "source" V_g is converted to mechanical output power.

The model in Figure 19–6b omits the field circuit but is otherwise similar. It can be used as a crude model for induction motors running near rated speed. This is the motor discussed in the original example. As noted earlier the squirrel-cage rotor has no electrical connection to it, its voltage being induced magnetically by stator currents. Chapter 20 will suggest a more sophisticated circuit model for the induction motor—one based on the transformer. In the meantime Figure 19–6b will be useful to develop a general understanding of induction motors in circuits.

V_t, the line or terminal voltage, and V_g, the back voltage or generated voltage, which opposes the applied voltage V_t, are ac for ac motors and dc for dc motors. Under normal speed and load these two voltages are typically within 10% of each other or closer. For instance, V_t might be 220 V and V_g 200 V. Thus there may be typically only 20 V difference to drive current through the armature circuit impedance. In some ac motors the two voltage magnitudes may be the same, only a phase difference existing to drive current through the stator. Furthermore, V_g is generally proportional to speed, or at least decreases as speed falls off, and for a given voltage, torque is approximately proportional to the product of motor voltage and current.

The important thing that this model tells us is the relation between speed, load, and current drawn. Suppose, for a dc machine such as that shown in Figure 19–6a, rated voltage has been applied to the motor and it is running under approximately full rated load and at a uniform speed. It will be drawing rated current. Now suppose that additional torque is suddenly required by the load. As a simple example, the load might be a rock-crushing machine into which a large boulder is suddenly thrown. According to the right-hand side of Figure 19–3a, the motor is providing the rated torque at rated speed. If more torque is required the system will slow down, motor torque will go up (backing up on the curve), and the system will settle at a lower steady-state speed where the motor torque output meets the load requirement.

When the motor slows down, V_g is reduced and more current will flow into the armature circuit from the supply lines since the difference between V_t and V_g is what makes current flow through R and X. And it is this increased current that increases the torque output.

Similarly, if the torque requirement of the load becomes less (rock has been crushed) the motor will be providing more torque than the load needs and the system will speed up until the two torques are again in balance. This increased speed will increase V_g, thus reducing current to reduce torque.

Thus an electric motor, under normal conditions, tends to be a self-regulating device as far as speed and torque and current drawn are concerned. Of course, any motor operates within reasonable limits. For example, if the induction motor is loaded too heavily, the maximum torque point on Figure 19–3a will be passed and the machine will simply stop. The model also shows the possibility of overloading the machine by too much of a torque demand, drawing too much current, and overheating.

The model also suggests the large starting currents. At start (zero speed) V_g is zero. Hence all of V_t is across the R and X of the machine and produces an inordinate current. Numerical examples and problems illustrating these effects are given in Chapter 20. The loss in R is turned into heat. A motor will overheat if it is started too frequently or if the armature current I_a is too high. There are field and mechanical losses in motors in addition to this armature loss.

Another important but somewhat more subtle point to be appreciated from this model is the dire effect of running motors at reduced voltage for very long. Since torque is proportional to the product of voltage and current, a motor which has too low a voltage will slow down to allow more current. Such a motor can overheat even if the mechanical load is within its rating.

EXAMPLE 19.2 Using Figure 19–6a as a separately excited dc machine model, with I_f of 1.0 A, R (which is R_a) = 0.4 Ω, and with an applied source voltage of 120

V, the machine draws 40 A from the source and runs at 1200 rpm. (We neglect the X in this steady-state dc analysis.)

Find: V_g, the back voltage or generated voltage.

Solution: Using KVL around the armature circuit, $V_g = V_t - V_R = 120$ V $- i_a R_a = 120 - 16 = \underline{104 \text{ V}}$.   ▰

19.5   MAJOR MOTOR TYPES—INDUCTION MOTORS

Induction, synchronous, and *dc* are the principal integral horsepower motor types.

A good deal has already been said about induction motors in the example at the beginning of this chapter where a polyphase induction motor drove a fan. Because of their inexpensive squirrel-cage rotor construction and lack of connections (brushes and slip rings) to the rotating portion of the machine, these motors are the workhorses of industry. They are cheap, rugged, less affected by environment than other types, and easily maintained and rewound. Already the most common motor, they are taking over even more applications with the new variable-frequency drives discussed below in Section 19.10.

Without a variable-frequency supply, the major disadvantage of these motors for some applications is their nearly constant speed. They also have a somewhat limited torque capability on start at reasonable line currents.

Chapter 20 will show in some detail that polyphase currents in the stator windings of most ac motors produce a rotating magnetic field which spins around the inside of the stator at *synchronous speed*. The term n_s—Figure 19–3a—the synchronous speed, is defined as the speed at which the stator's rotating field spins. For 60-Hz power supply, two-pole motor fields rotate once for every cycle and therefore have a synchronous speed of 3600 rpm, four-pole (two cycles per revolution) 1800 rpm, six-pole 1200, and so on. Alternating-current synchronous speeds are all submultiples of 3600 rpm for 60 Hz, or of $60f$ for other frequencies. Where p is the number of poles,

$$n_s = 120f/p. \qquad (19.4)$$

Induction motors run at full-load speeds slightly below synchronous (by up to about 6%) as suggested in the curve of Figure 19–3a. The amount by which they run below these synchronous speeds is called *slip, S,* and is measured in revolutions per minute or, more often, in percent of synchronous speed,

$$S = (n_s - n)/n_s. \qquad (19.5)$$

EXAMPLE 19.3 A four-pole induction motor operates at 4% slip.

Find: the motor speed.

Solution: There are two pole pairs, so $n_s = 3600/2 = 1800$ rpm. Motor speed is 96% of 1800 or $\underline{1728 \text{ rpm}}$.   ▰

▬▬▬▬▬▬ **EXAMPLE 19.4**

Given: A 50-hp, 1150-rpm motor is loaded to 50% of rated torque. (The expression *rated* is often used by engineers for *full load*. The designation of the motor as a "50-hp" machine indicates its rated power output. It is important to recognize that a 50-hp machine may be delivering some other amount of shaft power.)

Find

1. the slip in rpm and in percent;
2. the motor speed in rps and rpm;
3. the largest torque load (in percent of rated) that can be just started if directly coupled to the motor shaft;
4. what is its full-load torque in N-m and in lb-ft?
5. what is the largest torque load it can start in N-m?

Solution

1. n_s is 1200 rpm (nearest synchronous speed just above rated speed). Torque is closely proportional to slip near rated load (see Figure 19–3); for 100% load $S = 50$ rpm, so for 50% torque or load $S = 25$ rpm, or $S = 25/1200 = \underline{2.1\%}$.
2. The speed $n = 97.9\% \times 1200$ rpm $= 19.6$ rps or $\underline{1175 \text{ rpm}}$.
3. According to Figure 19–3a $\underline{200\% \text{ of rated T}}$ can be started.
4. $P_{out} = T\,\omega_m$; so $T = P_{out}/\omega_m = 50 \times 746/(1150 \times 2\pi/60) = \underline{310 \text{ N-m}}$ or $310 \times 39.36/12 \times 0.2247 = \underline{228 \text{ lb-ft}}$.
5. $T_s = 200\% \times 310 = \underline{620 \text{ N-m}}$. ▬

The resistance of the rotor has a significant effect on starting torque and running slip. Increasing the resistance of the rotor in design increases the starting torque, as shown in the *T-n* curves of Figure 19–7a, an advantage for high-torque loads. But this will also increase the losses and slip. Thus for 60-Hz stators with squirrel-cage rotors, motor speed is already determined when they are designed. There is no very practical way to control it in a given machine. These motors are regarded as almost constant-speed devices. But some motor stators have two sets of windings to produce two possible speeds.

With a given motor almost any nearly constant load speed can be attained by belting or gearing the motor to the load. Such a solution will, however, change the torque available, inversely to the speed transformation. Belting and gearing also have additional mechanical losses, and present cost and maintenance problems.

The great majority of induction motors are of the squirrel-cage type, and this is what is almost always meant by the term *induction motor*. However, in contrast to this squirrel-cage machine, another class of induction motor has a *wound rotor*. Rotor conductors are wound of wire in surface slots. These windings are usually connected in wye and the three terminals brought out through *slip rings* and *brushes* as in Figure 19–7b. Stationary carbon brushes make running contact with rotating slip rings as a means of making electrical connections to the rotating part of the machine. A variable wye-connected resistor is attached to the wye rotor leads. By this means it is possible to change the resistance of the rotor with the effect shown in the *T-n* curves of Figure 19–7a. Up to a point, the higher the resistance the greater the starting torque. Once a load is started, an operator (or automatic controller) cuts

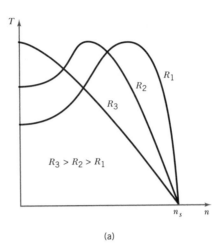

(a)

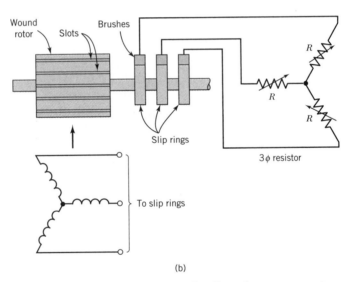

(b)

FIGURE 19–7 Sketch (a) illustrates the effect of more rotor resistance on the induction motor's *T-n* curve—particularly on starting torque. The wound-rotor motor (b) allows rotor circuit resistance to be varied externally to the machine, permitting stronger starts while still allowing for good efficiency at running speed.

out the resistance for better running efficiency and less speed variation with load. Some limited speed control, at the expense of efficiency, is also possible by this means. Section 19.10 will cover the speed control possibilities for squirrel-cage motors through use of variable-frequency power supply.

A last important characteristic of induction motors to note is their lagging power factor. It is the myriad of induction motors in industry that produce the necessity for power-factor correction procedures discussed in Section 17.11. Well-loaded, moderate-sized motors run at power factors between 80% and 90%, and down to 50% or worse at very low loading.

19.6 dc MOTORS

Industry has used dc motors for years because of their outstanding speed control capabilities, down to essentially zero revolutions per minute, and because of their good starting and low-speed torque.

Direct-current motors are usually much more expensive than standard induction motors. In contrast to induction machines, the high-current part of the direct-current machine is on the rotor (called the armature in dc machines). In getting from the supply into the armature these high currents must pass through brushes bearing on a *commutator,* a rotating-switch device. Wear on the commutator and brushes adds substantially to maintenance cost and reduces reliability.

Direct-current motors have the further disadvantage that large dc power supplies are generally not available unless especially provided for. In the last few years cheaper and higher-power electronic control equipment has been developed which converts ac to dc right at the motor. Thus small- and moderate-sized motors can be supplied in effect, right from ac mains, eliminating the need for special power supplies. Such an electronic controller can also readily adjust input voltage to the motor.

Shunt motors have the field shunted across (placed in parallel with) the armature as shown in Figure 19–8c. These dc motors provide a family of T-n characteristics as shown in Figure 19–8a. Each of the curves is quite like the right side of the induction-motor T-n curve of Figure 19-3a. Their parameter is the voltage supplied to the armature, V_t. Varying the field current I_f by means of a rheostat (variable-power resistor) will also have a speed controlling effect over a lesser range of speeds. But field current is small and therefore allows a smooth, cheap, and easy means of control. By varying supply voltage and/or field current the T-n curve can be moved horizontally to any speed desired, as shown. This wide speed-control range is the shunt motor's great advantage.

Direct-current shunt motors have one dangerous proclivity that every user should be aware of. If, when a motor is very lightly loaded, the field circuit is accidentally broken or disconnected, the motor tends to run away, going faster and faster until it flies apart. Large shunt machines are always provided with open-field protection.

In the dc *series* motor a few heavy field turns provide the excitation and are connected in series with the armature so that the same current runs through both as in Figure 19–8d. This scheme provides an unusual and very useful T-n characteristic (Figure 19–8b)—particularly good for loads that need high torque at low speeds. Typical applications are traction motors in electric railroads or vehicles and also motors for electric drills and hand tools. (See also Section 19.8.) Series motors are also capable of dangerous speeds if unloaded. They are usually coupled directly to their loads permanently. In tools like electric drills gearing losses prevent runaway.

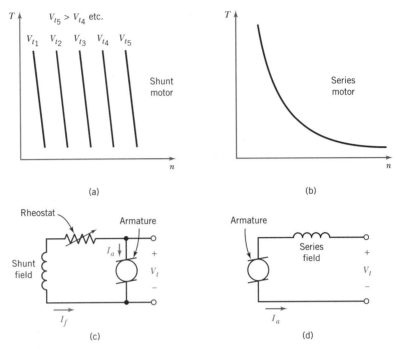

FIGURE 19–8 (a, c) *T-n* curve of dc shunt motor can be moved horizontally by varying the supply voltage V_t, or to a more limited extent by adjusting field current I_f. The dc series motor (b, d) has a particularly good *T-n* characteristic for trains or other electric vehicles.

19.7   SYNCHRONOUS MOTORS

These ac motors are designed to run at exactly synchronous speeds, for the United States 3600 rpm, 1800, 1200, and any submultiple of 3600 rpm [see Equation (19.4)]. But synchronous motor speed can be adjusted by supplying variable frequency. Large synchronous motors have the same rotating-field stator as the induction motor. But they also require a dc field. There are also various forms of small single-phase, fractional-horsepower machines for such applications as electric clocks, timers, and the like. The theory of Chapter 20 will show that in general any motor can also function as a generator. The large synchronous machine, used as a generator at coal, nuclear, and hydro power plants, is the source of electrical power that sustains civilization today.

Nonsynchronous motor types, when load torque is increased, were shown to develop increased torque by slowing up, thus allowing more current to be drawn from the supply, producing more torque. *T-n* curves developed so far illustrate this point. Synchronous motors, on the other hand, run at a constant speed regardless of load—at least within their torque ratings. How do they react to provide more torque when needed? A truly satisfying answer to this question is postponed to Chapter 20. But for the moment it can be simply said that when a heavier load is applied, the rotor falls back in mechanical phase—usually a few degrees—and continues at synchronous speed.

An interesting analogy would be two bar magnets on opposite sides of a piece of window glass. As one magnet is continuously rotated the other follows, but with a slightly lagging angle. If the following magnet is "loaded" mechanically by putting some viscous material on the glass, or in some other way, the fall-back angle will increase.

An engineer would suspect that there is a limit to possible fall-back angles, which is certainly right. If the motor is torqued too far (that is, too much load torque added to the motor shaft) it will fall out of synchronism, slow up, and eventually stop altogether. For large motors this occurrence is somewhat catastrophic, producing huge fault currents and popping circuit breakers. For small clock motors it is hardly noticed except in the error of the clock.

The synchronous motor produces torque *only* at its synchronous speed. Hence these types cannot start themselves. They are usually provided with starting windings similar to an induction motor to make them self-starting. For starting under these conditions the dc field supply for the rotor is turned off. Like large induction motors they are started at reduced voltages. In some applications their loads can be used to start them.

With these unusual starting requirements and their rotating dc fields with slip rings, large synchronous motors are expensive. They have one redeeming feature, however. Their power factor can be easily controlled simply by adjusting the low-current dc field. This includes the ability to draw large leading currents, making them useful for power factor correction in a manner similar to capacitors. They have often been used where there is a large, constant mechanical load. The power-factor-correcting capability can then be used as an auxiliary advantage and further cost justification. There are still some synchronous motor installations in use in industry where the machine, now called a *synchronous capacitor,* was installed solely for power factor correction before power capacitors became less expensive.

Reluctance motors are small, synchronous, single-phase induction motors that when started run up on the torque-speed curve of Figure 19–3a in the fashion previously described. But their rotors are deeply notched so that there is a rotor iron pole structure to match the rotating field poles of the stator winding. As the motor approaches synchronous speed it has a tendency to jump into synchronism in such a way that its iron rotor poles fall into phase just behind the stator rotating electrical poles generated by stator currents. Thus for light loading the reluctance motor runs synchronously. If slightly overloaded it may slip out of synchronism and run induction at a lower speed. Mechanical timing devices are a major application. Designers using these motors are primarily concerned with obtaining a torque characteristic (usually expressed in inch-ounces) that will be adequate for their mechanical load.

19.8   SINGLE-PHASE AND SMALL MOTORS

Fractional-horsepower and smaller motors are almost exclusively single phase. And there are a few larger single-phase applications where it is uneconomical to provide three-phase power, such as to scattered oil wells or farms.

It is inconvenient in houses and most commercial locations to provide three phase for small motors. Fortunately, specially designed induction motors will run almost as well on single-phase power as on polyphase once started, as the theory of Chapter 20 will detail. The problem is that it requires polyphase currents to start an induction motor. Special means must be provided then for single-phase starting.

There are a number of kinds of fractional-horsepower motors (motors smaller than a standard 1700- or 1800-rpm 1-hp motor) that are induction motors, and that differ from each other mostly in how a second (auxiliary) phase is generated and handled. Many have a capacitor (often housed in a small cylindrical case mounted on top) in series with a second winding. Some use a high-resistance auxiliary winding. In some a centrifugal switch disconnects the auxiliary winding when the motor reaches 60% or so of its rated speed. Others leave some or all of a ''second phase'' current connected. By disconnecting the second winding, it can be made of light wire suitable for intermittent duty only. Typically 400% of rated torque can be realized in this way on start. Starting belt-driven compressors and some other machines can require high torques.

Centrifugal switches in this application are the Achilles heel of reliability. Also capacitors, particularly in high-temperature surroundings, tend to fail before other parts of the motor.

Another common single-phase, fractional-horsepower machine is the *universal motor*. It is essentially a dc series motor with the stator laminated to reduce ac eddy current losses. (The rotor of almost any machine is laminated since the flux in a rotor is alternating as it spins.) The term *universal* refers to its ability to run on ac or dc. It has the disadvantage of requiring a commutator and brushes, with relatively short life between maintenance times. Its great advantages are cheapness and lightness for a given power and torque. And it has the excellent low-speed torque characteristic of series motors. So for such applications as electric hand tools it is ideal— many hand tools are used for such short periods of time that a modest life in hours of use can extend over several years. Weight is important as well as cost in such an application. Any dangerous runaway characteristic is prevented by manufacturing them to be integrally connected to their loads.

Small motors to be used as an integral part of other pieces of equipment— household mixers for example—are produced and sold as ''motor parts'' consisting of a rotor and the stator shell with windings but no end bells or bearings. The manufacturer designs his appliance to include the motor pieces as integral parts of whatever machine he is manufacturing.

Shaded-pole (induction) motors start without a second winding by providing a shorting turn (copper strap) around part of each iron field pole (at the inside winding face) in the stator (Figure 19–9). The transformer action of this turn changes the phase of the magnetic flux and produces a rotating-field effect which, while inexpensive to build and light in weight, is inefficient. It is suited mostly to either very intermittent use or quite low-power motors.

Torque, for a given voltage, tends to be roughly proportional to current. The smaller the current, the smaller the wire size and the smaller a motor can be constructed. Since power output is $T\omega_m$, motors of a given power tend to be smaller and therefore cheaper the faster they run. Taking advantage of this fact, *gear motors* are electric motors with built-in gear-reduction trains. They are common in fractional-horsepower sizes and frequently provide for right-angle shafting.

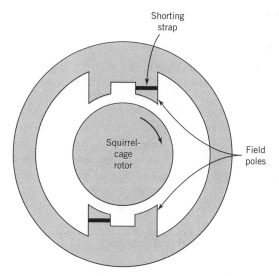

FIGURE 19-9 In a shaded-pole motor, parts of the explicit pole structure are surrounded by a *shorting turn,* a copper strap, which produces an out-of-phase flux component to make the field crudely rotate.

Stepper motors provide successions of very small incremental turns—typically a degree or two. They are used for precise speed control, or position control for devices such as the heads on computer disk drives. Pulses of power provide the stepping action. They couple easily into computer controlled systems.

19.9 MOTOR STANDARDS

Users of electric motors and designers of products which will incorporate them should be aware of the helpful, detailed standards and wide range of motor classes which have been established by various agencies. This has been contributed to most extensively by NEMA, the National Electrical Manufacturers Association. Their work is intended to promote the public interest as well as the manufacturers' by simplifying communication between user and manufacturer, buyer and seller, and by ensuring that motors can be purchased to recognized performance standards. To this end NEMA has categorized motors into many classes, specifying, for example, detailed frame sizes with mounting dimensions and shaft heights, starting currents and torques, standard voltages and their tolerances, speed regulation, voltage limitations, and so on—for various types of motors in a wide range of horsepowers.

Many manufacturers produce motors in these standard classes and sizes. It is a good deal more economical, where possible, to design around a standard motor than to specify a special machine of some kind. Continued procurement is also facilitated by this practice.

Another agency, The Underwriters' Laboratory (UL), concerns itself with non-hazardous design and application of motors, issuing approvals for equipment which passes its tests. These approvals are normally considered to apply to complete devices rather than to the motor alone, although the motor is an important element. See also the material on the National Electrical Code in Chapter 21.

For some applications *explosion-proof* specifications and testing are required. UL is active in this phase of safety also. The U.S. Bureau of Mines has for many years done extensive work in the design, specification, and testing of explosion-proof motors, as part of their work in explosion-proof equipment of all kinds. "Explosion proof" means that an explosion can take place inside a machine without propagating itself to an outside explosive atmosphere. Typical applications are in soft coal mines, flour mills, and environments subject to gasoline or similar fumes and gases.

19.10   VARIABLE-FREQUENCY DRIVES

Of the two-thirds of U.S. electric power that is used for motors, it is estimated that about one-half of this goes into induction motors driving fans and pumps. These motors are all nearly constant speed if supplied by a constant 60-Hz source (see speed-torque curve in Figure 19–3a). Adjustments in flow rate are made either by turning the motor on and off (not very satisfactory) or more commonly by mechanically throttling the fan or pump's output (quite inefficient).

For fans it can be shown that great mechanical energy savings are possible (typically half or even two-thirds) if the fan can be run at slower speeds. (See torque curve of Figure 19–3b and recognize that shaft power in is $T\omega_m$.) Similar but somewhat reduced energy savings are available for many pump installations. It should be clearly recognized that these savings are available only if the fan or pump is to be run for some significant portion of time at less than full rating. But this is usually the case since almost all air-conditioning and many pumping installations must be overdesigned to ensure adequate capacity. Furthermore, ventilation and air-conditioning needs, a major application area, vary greatly with weather and with space use.

To accomplish this variation of fan or pump speed the controller concept of Figure 19–4 can be expanded into a more sophisticated device called a *variable-frequency drive.* This controller accepts 60-Hz power input and provides a power output whose frequency is adjustable over a wide range. It has already been seen that induction-motor speed is nearly constant at a value just below the synchronous speed n_s given by Equation (19.4). To stay within motor ratings, output voltage should be reduced in proportion to frequency, and this is also accomplished by the controller. In some cases it may be desirable to have the frequency-change sequence be automatic or be controlled by some digital control equipment as in textile plants.

---

*The term *drive* to most engineers includes both controller and motor. But in much modern discussion the term is used to suggest only the controller. To be clear this book will use the terms *controller* and *motor* separately where possible.

Alternatively the controller may convert ac input to dc output to allow the use of dc machines. Sophisticated controllers that accomplish this can also provide for variable-voltage dc output to control speed and torque with some precision. One-thousand-to-one smooth speed control is possible.

Whether equipped with ac or dc motors, these sophisticated devices provide a soft start, meaning a gentle start mechanically and one which does not take excessive starting current from the supply. Such a start may be mandated by the process—in bottling for example. It also significantly reduces wear and tear on mechanical and electrical components.

These variable-frequency controllers may cost several times as much as the motor they control. (For example, at this writing a 5-hp class-b induction motor costs about $350. A 5-hp variable-frequency controller costs around $1500.) But the controller cost should really be compared to possible energy savings. And it is generally felt that energy prices will continue to increase as present coal and oil deposits are depleted and world population and industrialization increase. (See problem P19.4.)

Variable-frequency controllers can also be applied to synchronous motors.

An important problem in retrofitting or specifying variable-frequency controllers is cooling. Motors may not have enough cooling capacity if run at slow speeds for extended periods of time. In some fan applications the motor is cooled by the air flow itself.

19.11   SELECTING AND MAINTAINING MOTORS

Two factors make for some bewilderment in specifying motors for specific applications: First, there are many types and sizes of motors available, a few of the most important of these having been discussed above. Second, there are numerous important parts of a motor's environment that may properly affect the kind and rating of the motor chosen—to mention some of the more important: the power supply available, the torque and speed needed for the load, starting and acceleration requirements, duty cycle to be expected (how many starts are required per hour, for example), ambient temperature, dust or explosive vapors present, maintenance capability of work group, types of motors presently used and maintained in plant, and dependability of motor equipment sources.

The motor and its controller should be considered together (and for a large motor also consider at the same time any protection needed). The advantages of variable-frequency drives should not be overlooked. Whether the motor is part of a product to be produced in volume or is for some factory floor application, the engineers will want to talk to one or more motor manufacturers about the possibilities and their recommendations. Electric motor manufacture is a very competitive business in the United States. Handbooks and other literature sources may be helpful, but the state of the art in motors is changing rapidly.

In sizing motors it should be kept in mind that lightly loaded motors suffer in efficiency and power factor. Most motors can be overloaded substantially for short periods of time without harm. "Substantially" is 25% or even 50%. A "short period of time" is one in which the motor does not have time to get too hot—for a 5- or 10-hp motor, usually 5 min or so. Like transformers, motors are primarily limited by temperature. Too high a temperature suffered for too long a period of time causes their insulation to deteriorate. For some applications a need for frequent starting may dictate a larger machine.

Motor maintenance like any other maintenance is principally a matter of systematic, scheduled attention. Most small electric motors today are built with lifetime lubrication devices. Except in poor environments, little or no attention is needed for them until they must be replaced. Larger motors usually have lubrication requirements specified by the maker. All motors benefit in extended life and reliability by being kept reasonably free of dirt that would impede their cooling. Moisture is usually bad for them also.

Explosion-proof design relies on the integrity of motor cases. The parts of these cases (often halves) are designed with very wide flanges, to allow for the fact that an internal explosion will force them apart and allow hot gas to blow through to the outside. The idea is to have the flanges wide enough (several inches) so that escaping burning gases are cooled below a temperature which can ignite combustible gases outside the motor. Close spacing of flange closure bolts limits flange separation. Good maintenance practice requires that when these machines are reassembled the flanges be clean and undamaged. All bolts must be replaced and properly torqued. No holes can be left unplugged in the case.

Special classes of motors are produced for special environments, such as drip proof (all case openings in the motor case are in the bottom half) or splash proof (liquid drops or solid particles projected at not less than 100° from the vertical cannot enter the motor) or totally enclosed.

Many motors are costly enough to make rewinding economically worthwhile when their insulation finally fails. Most cities of any size have motor repair shops which can perform this work, often on a rapid emergency basis. Bearings can be replaced and commutators redressed.

It is common to provide *thermal protection* for many motors by including in the windings a bimetallic thermal switch which will open the motor circuit when the windings get too hot. Most of these automatically reclose when the motor cools.

19.12 SUMMARY

Motors convert electrical power into mechanical shaft power ($P_{\text{shaft}} = \omega_m T$, which gives watts output where speed is in rad/s and torque in N-m). Connected to a power source, except for synchronous machines, torque output is a function of speed, and speed will establish itself at the point where motor torque and the torque the load requires are equal. Each type of motor is characterized by its own torque-speed relation, displayed as a *T-n* diagram.

Induction motors, rugged and inexpensive, are by far the most common type in industry. Motors larger than about 1 hp are almost always polyphase. The induction motor runs at nearly constant speed, a few percent below some synchronous speed n_s. For 60-Hz supply n_s in rpm can be 3600, 1800, 1200, etc. Single-phase motors must be provided with special starting equipment, which may be troublesome at times. Induction motors run with a poor power factor.

Synchronous motors develop torque at synchronous speed by falling back slightly in mechanical (and electrical) phase. They are provided with starting in some other mode, usually as an induction motor.

Direct-current motors require maintained commutators, and a dc supply unless run on an electrical inverter system. The shunt machine's principal advantage is that speed can be controlled over a wide range with good torque. Direct-current series motors are ideal for traction and other loads requiring high torque at low speeds and low torque at high speeds. Properly constructed series machines can be operated on either dc or ac as universal motors and are much used in tools like electric drills.

Motor efficiency is generally in the range of 60% to 90% with smaller and lightly loaded motors in the lower end, or worse, and some very large motors better. 75% is often taken as a crude estimate. Motor loading is limited by temperature rise from losses, making possible for many types considerable overloading for short periods. Motors tend to draw sufficient power from the line to accommodate their loads. Thus if run at less than rated voltage they overheat when fully loaded.

Induction motors and some other types can be advantageously started, controlled, and run with electronic power supplies which convert 60-Hz three-phase ac to variable-frequency ac (or dc). These electronic drives are expensive but often provide great power savings for fan and pump operation. They overcome to some degree the fixed speed of the induction machine and provide softer starts.

FOR FURTHER STUDY

Ollie I. Elgerd, *Basic Electric Power Engineering,* Addison-Wesley, Reading, MA, 1977.

A. E. Fitzgerald et al., *Electrical Machinery,* 4th ed. McGraw-Hill, New York, 1983.

Leander W. Matsch and J. Derald Morgan, *Electromagnetic and Electromechanical Machines,* 3rd ed., Harper & Row, New York, 1986.

Ralph J. Smith and Richard C. Dorf, *Circuits, Devices and Systems,* 5th ed., Chapters 22–24, John Wiley & Sons Inc., New York, 1992.

PROBLEMS

Easy Drill Problems (answers at end of chapter)

D19.1. A three-phase motor is rated at 10 hp, 220 V, 1750 rpm at 85% pf. How many kW of mechanical power does the motor produce at full load? Assuming it is 80% efficient, how many electrical kW does it require?

D19.2. Find the line current of the motor in D19.1.

D19.3. Find the shaft torque delivered to the load by the motor of D19.1.

D19.4. Find the slip in percent for the motor of D19.1.

D19.5. (a) At what speed would a 24-pole synchronous machine operate?
(b) Recalculate for a power supply frequency of 50 Hz.

D19.6. The motor of Example 19.4 is furnished with a starter which reduces applied voltage to 60% of its rated value. What is the largest torque load it can just start?

D19.7. An 80-hp induction motor operates under full load at a slip of 5%. The motor is a four-pole machine. Find the rated speed, half-load speed, no-load speed. (Assume negligible no-load losses.)

D19.8. Find for the machine of D19.7 the same three speeds if it is driven from a 50-Hz supply.

D19.9. Using the dc shunt model of Figure 19–8c, calculate the efficiency of a 220-V machine whose R_a (armature resistance) is 0.3 ohm, I_f (field current) is 1 A, I_a (armature current) is 50 A, and whose mechanical losses are 300 W. (*Note:* $P_{out} = P_{in}$ minus the three losses)

D19.10. Using Figure 19–6a as a separately excited dc machine model, with I_f of 1.5 A, R (which is R_a) = 0.5 ohm, and with an applied source voltage of 120 V, the machine draws 40 A from the source and runs at 1200 rpm. (Neglect the X in this steady-state dc analysis.) What is V_g, the back voltage or generated voltage?

D19.11. Using the dc machine and data of Example 19.2, assume further that there are mechanical losses of 225 watts. Calculate the shaft power out in kW and hp. (*Hint:* Power converted by a motor into mechanical power is the electrical power put into the source symbol in the armature circuit of the model. Power converted supplies the mechanical losses and the shaft power out. There are 746 W in a horsepower.)

Application Problems

P19.1. The motor in D19.1 is supplied with 85% of the voltage for which it is rated. Assume power factor and efficiency remain about constant. For a mechanical load requiring the same full-load torque, find the line current drawn. What percentage of rated line current will this be?

P19.2. At what speed will the motor operating under the conditions of P19.1 run? Make a plot of speed versus voltage supply for voltages from 70% to 100% of rated. (Assume a straight-line *T-n* characteristic in this region.)

P19.3. (a) For this same motor, plot line current versus supply voltage.
(b) Assume that half the rated losses of the motor are caused by a resistance that the line current encounters and that the remainder of losses are independent of line current. Plot losses in watts versus line voltage (70–100% rated) under the assumption of rated full-load torque.

P19.4. An existing 5-hp induction motor runs at about 1750 rpm driving a fan at approximately three-quarters load for 3000 hours a year. It is estimated that 40% of the energy used by the fan can be saved by installing a variable-frequency controller at a cost of $1500.00. Expected life for the controller is 10 years. Electric power costs 10 cents per kWh. Your company's current minimum acceptable rate of return (MARR) is 20% before taxes. Is the new controller installation justified and if so how much will it save per year? Assume that the motor alone has an average efficiency of 75% and that motor and controller together have a nearly constant efficiency of 73%.

P19.5. Resolve P19.4 considering that the new soft start and lower speed capability will extend the life of the recently installed motor and ancillary mechanical equipment (costing $950) from 10 to 20 years.

P19.6. Repeat D19.9 if the same machine is now run with 200-V supply, and consequently, at full load draws an armature current of 55 A.

P19.7. Continuing with the separately excited dc motor of D19.10, assume that for a given excitation (I_f) the shaft torque is proportional to I_a. The load torque is substantially independent of speed. To what value should the source voltage be adjusted to slow the motor and load down to 850 rpm?

P19.8. Use the circuit model 19–6a with data: R_a = 0.4 ohms, I_f = 1.5 A, V_g = 120 V at 1200 rpm. Connect the machine mechanically to a gasoline engine and run it as a separately excited generator at 1000 rpm. What will the no-load terminal voltage be? (*Note:* V_g, the generated voltage, is proportional to speed for a given field strength, whether the machine is being used as a motor or a generator.)

P19.9. Now connect a 3-ohm industrial heater to the machine terminals. Find the voltage and current on the heater.

P19.10. Remove the load and reconnect the field in shunt (R_f = 80 ohms) (called *self-excited* since the generator supplies its own field excitation). What will the open circuit voltage be? (*Note:* dc machines usually have enough "residual magnetism" to enable them to build up voltage when self-excited.)

P19.11. In Example 19.3 change the given motor specification from 1150 rpm to 1140 rpm and repeat the problem.

Extension Problems

E19.1. Assume that the graph of Figure 19–3a applies to a 50-hp, 1170-rpm motor. The motor with its connected fan load has a moment of inertia of 25 kg-m². Assume that the fan torque requirement, Figure 19–3b, can be approximated as $T = 0.05 + 1.053\ n^2$, where T is expressed as a fraction of rated motor torque and n as a fraction of motor synchronous speed. Assume for this problem that the motor is started across the line.
(a) How long will it take the motor to come up to 85% of synchronous speed?
(b) Plot motor speed versus time from start.

E19.2. Suppose that a controller is provided so that the motor in problem E19.1 is started on half voltage to limit the starting current drawn and that the controller automatically switches over to full voltage at 60% of synchronous speed. Repeat problem E19.1 under these circumstances.

E19.3. Using the loss and temperature rise concepts of Section 18.3, assume that machine losses under rated conditions are 25% due to losses proportional to voltage supply level, 25% to fixed mechanical losses, and 50% to losses in winding resistance due to load currents.
(a) Assume a rated rise of 40°C. Make a plot of temperature rise versus supply voltage from 70% to 100% total.
(b) Make a plot of temperature rise versus load from 80% to 120% rated.

Study Questions

1. Motors work by interaction between what two physical phenomena?
2. How many watts are equivalent to 1 hp?
3. Sketch carefully a typical induction-motor torque-speed (T-n) curve. Add several additional curves to the same sketch showing the effect of rotor resistance on starting torque.
4. Sketch on the same axes two T-n curves for the same induction motor, the second with two-thirds the voltage supply of the first.
5. Why is it bad practice to run a motor on voltages lower than its rating for any extended period of time?

6. Why are motors larger than about 1 hp started at reduced voltages?

7. What are the three principal types of large motors?

8. Sketch typical shunt and series dc motor *T-n* curves.

9. What is the shunt dc motor's greatest advantage?

10. Why is the unusual *T-n* curve of the series motor advantageous for traction. (Consider, for example, a train crossing the Rocky Mountains.)

11. List the six highest synchronous motor speeds available in the United States in rpm and rps.

12. How are synchronous motors usually started?

13. Explain how a universal motor's *T-n* characteristic (essentially the same as the series dc motor characteristic) is advantageous in operating a geared electric hand drill.

14. What does the term *explosion proof* mean as applied to motors? By what means is this characteristic obtained in design?

15. Give three reasons why it is disadvantageous to use an ac motor which is much larger than needed for its application. Which of these reasons apply to dc motors?

16. What is a thermal protector and how is it used in motor design?

17. Why does reducing a fan's speed save large quantities of electrical energy?

18. List some advantages and disadvantages of applying variable-frequency controllers.

ANSWERS TO DRILL PROBLEMS

D19.1. 7.46 kW, 9.33 kW

D19.2. 28.8A

D19.3. 40.7 N-m (or 30.0 lb-ft)

D19.4. 2.8%

D19.5. 300 rpm, 250 rpm

D19.6. 223 N-m; or 72%

D19.7. 1710 rpm, 1755 rpm, nearly 1800 rpm

D19.8. 1425 rpm, 1463 rpm, nearly 1500 rpm

D19.9. 88.7%

D19.10. 100 V

D19.11. 104 V, 4.16 kW, 3.86 kW, 5.17 hp

CHAPTER **20**

Rotating Machinery Basics

This chapter presents basic theory to show how the external motor characteristics described in Chapter 19 come about. In addition to being of strong technical interest to the engineer, this theory will help in retention of Chapter 19 material. The purpose of that chapter was to introduce in a somewhat qualitative way the various types of motors, their characteristics, and their applications. Those wishing to start with the present chapter before the last one should at least scan the figures and captions of Chapter 19 first.

The mechanism of losses and efficiency is generally about the same for all types of motors. Yet it is less complex to deal with in the dc machine. The same can be said for speed and speed regulation in many types of motors. Therefore this chapter goes into some detail with dc motor calculations. Quantitative experience developed with this material will carry over by analogy to ac motors.

The chapter also contains a small amount of material on rotating machines as generators.

A centerpiece in understanding rotating machines is seeing that *all* machines, whether operating as motors or generators, are generating voltage and *at the same time* experiencing motor forces. Section 20.1 begins with this point.

The reader's general objective in this chapter is

1. to master details of linear and cylindrical dc machines so as to reinforce the broader machine understanding developed in Chapter 19
2. to satisfy himself or herself about the general principles of the three main ac machine types.

20.1 THE LAWS OF MOTOR AND GENERATOR ACTION

Figure 20–1 shows a conductor in a magnetic field of density B webers per square meter (or B *tesla*). The arrows (flux lines) of Figure 20–1a suggest the field flowing out of a north magnetic pole and into a south pole through the space between

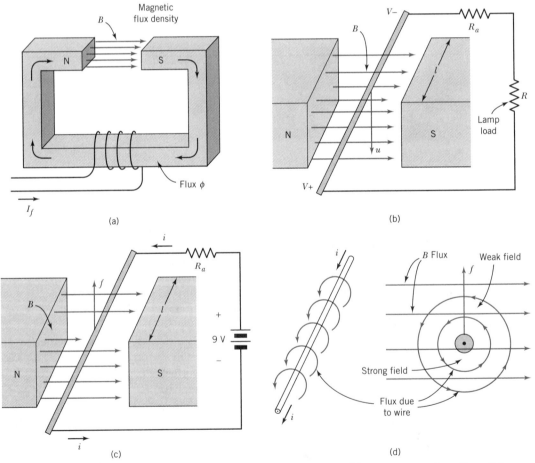

FIGURE 20–1 A conductor in a magnetic field (a) if pulled across it as in (b), generates a voltage $V = ulB$ between its ends. If carrying a current it experiences a force $f = IlB$ N as in (c, d). Conductor tends to move out of strong field. V generated will tend to oppose velocity u.

them, as described in the magnetic circuits of Section 18.5. This field is of course continuous and not actually composed of individual lines. The lines in the figure serve to indicate field direction. Also, by their density, they suggest flux density.

Figure 20–1b illustrates the *law of generator action*: if a conductor is pulled downward, with velocity u meters per second, across flux lines of density B, then a voltage V is generated in the wire according to the expression

$$V = ulB, \tag{20.1}$$

where V is in volts, u in meters per second, B in tesla, and l in meters; l is the length of the conductor within the field.*

This law can be taken as experimental. It assumes that field, wire, and motion are all at right angles to one another. It is possible to express the law more generally in vector notation to allow for angles other than 90°, but practical machines are all constructed so that every factor is at right angles to the others. No other possibility is considered here.† If a load, for example, a proper-sized lamp, were connected across the ends of the wire, current would flow and the lamp would light.

EXAMPLE 20.1 In Figure 20–1b, $l = 0.2$ m, $B = 1.3$ T, $u = 10$ m/s. The flux-cutting conductor has a resistance of 3.5 Ω. The rest of the wiring has a resistance of 0.5 Ω. This total of 4 Ω circuit resistance is represented by the resistance R_a. A 96-Ω load is connected across the ends of the moving wire, outside the magnetic field.

Find
1. voltage generated in the wire,
2. current flowing in the load,
3. power transmitted to the load.

Solution
1. $V = ulB = 10 \times 0.2 \times 1.3 = \underline{2.6 \text{ V}}$.
2. $I = V/R = 2.6/100 = \underline{26 \text{ mA}}$.
3. $P = I^2 R = 0.026^2 \times 96 = \underline{64.9 \text{ mW}}$.

Figure 20–1c illustrates the *law of motor action*†:

$$f = IlB, \tag{20.2}$$

which states that a current-carrying conductor in a magnetic field experiences a force f on it proportional to the current I, flux density B, and conductor length l.

---

*A form of Faraday's law.

†For completeness, the relations in vector form are

$$V = \mathbf{B} \cdot (\mathbf{l} \times \mathbf{u}),$$

where the **l** is directed from voltage minus to plus, and [for Equation 20.2)]

$$\mathbf{f} = I(\mathbf{l} \times \mathbf{B}).$$

‡Ampere's law.

The direction of the force is at right angles to the flux and to the current. Metric units are again used. The force f is in newtons. A newton is about one-quarter pound of force or more exactly 0.2248 lb. This law is also an experimental result. As with the generator law it is assumed here that the current and flux are always at right angles to each other. The previous footnote again gives the more general vector form.

EXAMPLE 20.2 In Figure 20–1c the battery power source provides 9 V, R_a is again 4 Ω, the length of the conductor in the field is 0.2 m. $B = 1.3$ T. The wire is constrained from moving.

Find: the motor force pushing the wire upward.

Solution: $i = V/R = 9/4 = 2.25$ A. $f = IlB = 2.25 \times 0.2 \times 1.3 = 0.59$ N.

Note in the first example with the generator that if current is allowed to flow, the conductor in the field will experience force on it in accordance with Equation (20.2), the motor law. Engineers would suspect that this force is opposing the movement described by velocity u, and experiment shows them right. Thus power provided the 95-Ω resistance is not free, but is provided to the machine mechanically through force times velocity (fu). Some of the power generated also goes to losses such as the i^2R in the conductor itself.

Similarly in the motor example, if the constraint preventing motion is removed, the wire will move, responding to the force generated by Equation (20.2). Voltage will then be generated in it in accordance with the generator law, Equation (20.1). This voltage would oppose the battery voltage, requiring that power be put into the wire segment under the poles. Some of this power is converted into mechanical power when movement occurs.

The experimental facts of what direction an *I-l-B* force takes and what polarity the *u-l-B* voltage has can be usefully embodied in the following rules of thumb (they are both right-hand rules—that is, use the right hand to implement them): In Figure 20–1b place the fingers of the right hand in the position of the flux lines and pointing in the direction of the arrows. Place the hand in such a position that the moving conductor falls into the palm. *The thumb points to the positive end* of the conductor. (These rules would be automatically provided by the vector notation of the previous footnote.)

For motor action (Figure 20–1c), place the fingers of the right hand in the position of the flux lines and pointing in the direction of their arrows. Place the hand in such a position that the current in the conductor enters the palm perpendicularly. *The thumb points in the direction of f.* (What happens to the direction of force if both field and current are reversed at the same time?)

Another useful way of describing this last result is illustrated in Figure 20–1d. A conductor with current produces a field about itself (place right thumb in direction of current; fingers curling around conductor show flux direction). When this conductor is placed in another field B, the two fields add and subtract from each other as shown. The conductor tends to be pushed toward the weaker field.

The rules proposed here are right-hand rules. Some engineers use different versions, but they must all have the same result. Analysts are advised to find methods that appeal to them and forget all others.

These simple examples are somewhat impractical since not much movement is possible within the dimensions of the pole faces given. Real machines must be constructed in such a way as to make movement of conductors in magnetic fields more or less continuous. Cylindrical geometry can accomplish this.

But it is helpful to first consider another particularly good problem to apply the above laws, one which illustrates motor and generator action very clearly.

20.2 THE RAILROAD-TRACK PROBLEM

In Figure 20–2 a smooth bar can slide longitudinally, without appreciable friction, along two long railroad tracks separated by 2 m. By means of the electrical switch on the right-hand end, the tracks can be connected either to a 10-V dc power source or to a 10-Ω load resistor R_L. The bar has a resistance of 2 Ω, R_a in Figure 20–2c. Track resistance is negligible. The vertical component of the earth's magnetic field goes into the ground at this point as indicated by the small crosses (feathered end

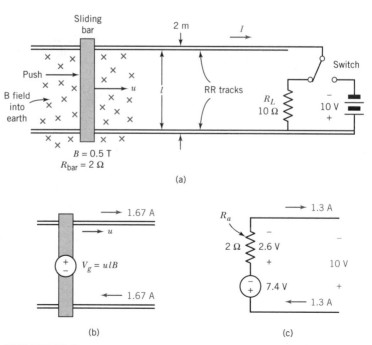

FIGURE 20–2 The classical "Railroad Track" problem. Sliding bar acts as a *linear motor* or *generator*. See example in text. Any electrical power absorbed by the V_g source is converted to mechanical power.

of the flux line arrows) and for problem purposes has a density of 0.5 T.* Thus the bar can be pushed along by a small locomotive or, with the 10-V source connected, can act as a locomotive itself, pulling a train of cars.

EXAMPLE 20.3 The 10-Ω resistor R_L in Figure 20–2a is switched in. A locomotive (with insulated wheels) pushes the bar to the right at velocity $u = 20$ m/s.

Find

1. the voltage generated in the bar, and the current, voltage, and power of the 10-Ω resistor.
2. What is the pushing force and power the locomotive produces to move the bar?
3. What efficiency of power conversion obtains from the locomotive to the resistor?

Solution

1. For the bar, $V_g = ulB = 20 \times 2 \times 0.5 = \underline{20 \text{ V}}$. $I = V/R = 20/(2 + 10) = \underline{1.67 \text{ A}}$. For the resistor $V_R = RI = 10 \times 1.667 = \underline{16.67 \text{ V}}$. For the resistor $P_R = i^2R = 1.667^2 \times 10 = \underline{27.8 \text{ W}}$.
2. $f = IlB = 1.667 \times 2 \times 0.5 = \underline{1.667 \text{ N}}$. For the locomotive $P_{mech} = fu = 1.667 \times 20 = \underline{33.3 \text{ W}}$.
3. eff $= P_{out}/P_{in} = 27.8/33.3 = \underline{83\%}$.

Voltage induced in the moving bar when it cuts the lines of force can be easily visualized as a series voltage source (see Figure 20–2b). Each element of bar length actually produces its own small contribution to *u-l-B* voltage. These elements are distributed along the whole length as little generators. But it is convenient to add them all together in the middle as has been done in the figure. This sum for the whole bar has been labeled V_g, the *generated* or *back voltage*. Similarly the 2-Ω resistance of the bar is distributed over its length, but it is modeled here as a lumped resistor R_a in Figure 20–2c.

It is interesting (and left to the reader) to find how and where the 5.5 W was lost. In an actual machine there are two additional sources of loss besides the one involved here.

EXAMPLE 20.4 The switch in Figure 20–2 is now moved to the right to connect the 10-V source. No locomotive or mechanical load is yet connected to the bar. The mass *m* of the bar is 1 kg. The bar is stationary at the start.

Find

1. *I* and *f* just after the switch is thrown. Find also the acceleration a at that time. In what direction will the bar tend to move?

---

*Apparently these tracks are in the northern hemisphere. The earth's magnetic "north pole," really a south magnetic pole, is so-called because it is near the north geographic pole. The actual magnetic density on the earth's surface is only a few tens of microtesla.

2. To what ultimate speed will the bar accelerate?
3. Suppose the bar now encounters friction of 0.3 N. To what steady speed will the bar settle?
4. The friction continues, and a string of cars is coupled to the bar requiring a drawbar pull of 1 N. To what steady-state speed will the bar settle now? What is the power efficiency from 10-V source to drawbar of the first car?

Solution

1. $I = V/R$ (there is no motion yet and no generated voltage) $= 10/2 = \underline{5 \text{ A}}$. $f = IlB = 5 \text{ A} \times 2 \text{ m} \times 0.5 \text{ T} = \underline{5 \text{ N}}$. To find the acceleration a, $f = ma$, so $a = f/m = 5 \text{ N}/1 \text{ kg} = \underline{5 \text{ m/s}^2}$. With current flowing up, the field caused by the bar will be down on the right, increasing density there, so the bar will tend to *move to the left*.

2. The bar will accelerate as long as there is any force generated (that is, any current to generate it). So it will accelerate until a back voltage V_g of 10 V is generated to stop all current flow. $V = ulB$, so $u = V/lB = 10 \text{ V}/(2\text{m} \times 0.5 \text{ T}) = \underline{10 \text{ m/s}}$.

3. The bar must slow down until it is producing 0.3 N force: $f = IlB$, so $I = f/lB = 0.3 \text{ N}/(2 \text{ m} \times 0.5 \text{ T}) = 0.3 \text{ A}$. But $10 \text{ V} - V_g = IR = 0.3 \text{ A} \times 2 \ \Omega$, so $V_g = 9.4 \text{ V}$. Velocity $u = V/lB = 9.4 \text{ V}/(2 \text{ m} \times 0.5 \text{ T}) = \underline{9.4 \text{ m/s}}$.

4. Proceeding as in part 3, we have $I = f/lB = 1.3 \text{ A}$. $V = 10 - 1.3 \times 2 = 7.4 \text{ V}$ (see Figure 20–2c) and $u = \underline{7.4 \text{ m/s}}$. $P_{in} = VI = 10 \times 1.3 = 13 \text{ W}$. $P_{out} = uf = 7.4 \times 1 = 7.4 \text{ W}$. Eff $= P_{out}/P_{in} = 7.4/13 = \underline{57\%}$. ∎

This last example illustrates mechanical losses in track friction as well as electrical losses in bar resistance. In general, motors have both electrical and mechanical losses. When these are subtracted from the input, the remainder is what this linear motor converts to useful mechanical power to pull the cars.

An interesting and useful way to look at this conversion, for example part 4, is shown in Figure 20–2c. The 7.4-V source represents the generated voltage (ulB). 1.3 A going into the bar gives (loss) power to the 2-Ω bar resistance. But the 7.4-V "source" is similarly absorbing power since current flows into its positive terminal. This absorbed power (1.3 A $\times$ 7.4 V = 9.62 W) is the part of the electrical power supplied that is *converted into mechanical power*. Mechanical power lost to rail friction is $P_{mech} = uf_{friction} = 7.4 \text{ m/s} \times 0.3 \text{ N} = 2.22 \text{ W}$. And if the mechanical losses are subtracted from the converted power, drawbar output $P_{out} = 9.62 - 2.22 = 7.4 \text{ W}$.

Parts 3 and 4 of Example 20.4 show clearly how motor speed is determined. If the drag on the motor is 0.3 N as in part 3, the motor speeds up from the start ($f = ma$) until sufficient V_g is generated [Equation (20.1)] to allow just enough current to produce 0.3 N of force [Equation (20.2)]. If more force is required, as in part 4, the motor does not have enough current and is slowed down until V_g is right to allow the proper current for the new drawbar force. In other words, motors slow down or speed up until their generator action is just right. V_g is the key to motor speed.

The next section turns to developing more practical, cylindrical motors.

20.3   DIRECT-CURRENT MACHINES

Chapter 19 defined the *stator* as the nonmoving part of a motor—its frame. The *rotor* is the rotating portion, inside the stator. In dc machines the rotor is called an *armature*. The stator of a dc machine is usually called its *field*.

Figure 20–3a shows the field circuit and magnetic circuit of a cylindrical machine. These two-dimensional figures actually represent, of course, three-dimensional field and armature components. Field poles correspond closely in depth to the length of the armature cylinder. Field current I_f in field coils on each pole produces magnetic flux which crosses two air gaps between armature and poles and completes its path through the iron of the machine frame. Flux density B in the gaps can be increased or decreased by adjusting field current. As shown in Figure 20–3b, conductors are wound in slots of the rotating armature. There are usually several conductors in each slot but the figure shows only one. Each of these conductors acts, under the field poles, just like the sliding bar of the railroad track

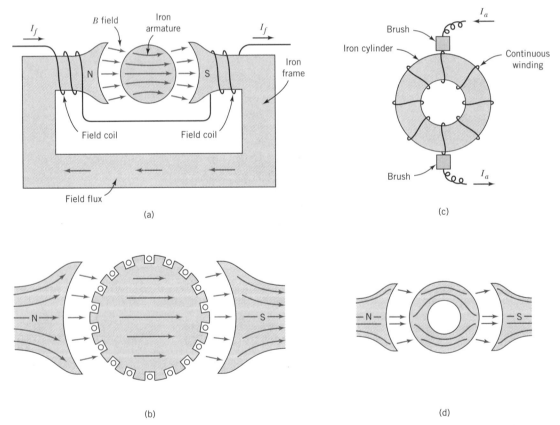

(a)

(c)

(b)

(d)

FIGURE 20–3 Practical machines have cylindrical geometry so conductors continue in magnetic field. The *ring armature* (c, d) makes for easy understanding, but solid armature is more common (b) and works essentially the same way.

problem in Section 20.2. Both motor action and generator action take place for each of them.

For purposes of easy explanation, it is useful to temporarily modify this solid cylindrical armature somewhat, into a *ring armature*, as shown in Figure 20–3c. The ring is a hollow cylinder of magnetic iron. A single continuous winding of copper wire encircles the ring and is closed on itself as shown. Note that, where it can, magnetic flux will go through the iron ring rather than air (Figure 20–3d). Thus on the ring, only outside portions of conductors will be exposed to the magnetic field.

The wire winding is of course insulated from the armature. Remove the outside portion of this insulation so that two *brushes* can bear on the rotating armature conductors at points in between the poles (top and bottom) as in Figure 20–4a. Note that the current I_a entering through the top brush splits in half, $\frac{1}{2}I_a$ moving downward around each side of the iron ring. These two halves reunite and leave the rotating armature at the bottom brush.

Those portions of the winding in the magnetic field under the pole faces are called *active conductors*. If the winding turns are followed carefully from the top brush down on either side, it is seen that on the right side currents in the active conductors are all going into the paper. This is suggested in the notation by small crosses representing the feathered ends of current arrows. Active conductors on the left side have their currents coming out, this time represented by dots which are the points of the arrows as they emerge from the paper sketch. Note further that, using the right-hand rule for force direction, forces on all active conductors (on both sides of the ring) are pushing the armature in a clockwise direction. Thus the total torque on the armature is

$$T = \tfrac{1}{2}N_a \, rlBI_a, \tag{20.3}$$

where N_a is the total number of active conductors and r is the radius of the armature in meters.

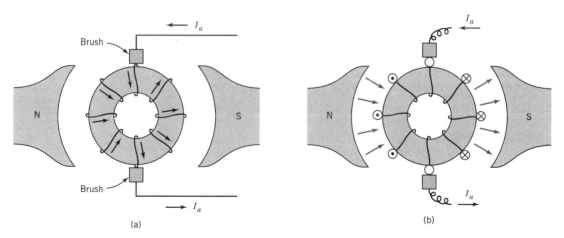

(a)

(b)

FIGURE 20–4 Armature current I_a splits into halves to move through the armature windings (a). Only outer conductors are exposed to magnetic field (b), and polarities are such that all conductors push armature clockwise.

EXAMPLE 20.5 A dc machine has 24 slots with two-thirds of slots active, B in air gaps of 0.9 T, armature radius to conductors of 0.5 m, and armature current of 100 A, half of which passes through each conductor. The armature has an active length of 0.9 m.

Find

1. torque produced by the motor.
2. If the machine is running at $n = 1500$ rpm and mechanical losses are 5% of the power converted, what horsepower is being developed?

Solution

1. There are $N_a = \frac{2}{3} \times 24 = 16$ active conductors. Force per conductor is $f_c = IlB = 50 \text{ A} \times 0.9 \text{ m} \times 0.9 \text{ T} = 40.5 \text{ N}$. Torque for entire motor is $T = N_a f_c r = 16 \times 40.5 \text{ N} \times 0.5 \text{ m} = \underline{324 \text{ N-m}}$ (but some of this goes into friction).
2. Power developed is $P_{\text{out}} = T\omega_m (1\text{-mech losses}) = 324 \text{ N-m} \times 2\pi \times (1500 \text{ rpm}/60 \text{ s/min}) \times 0.95 = 48.41 \text{ kW}$; hp = kW/0.746 = 48.41/0.746 = $\underline{64.81 \text{ hp}}$.

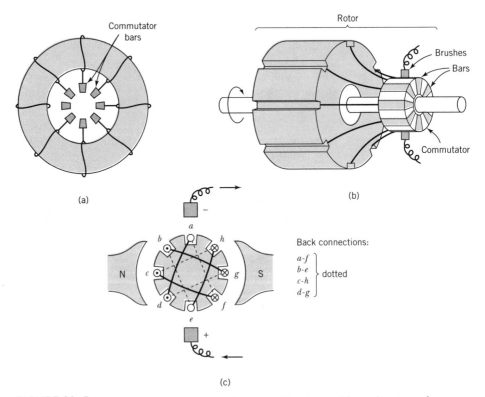

FIGURE 20–5 Rotating commutator bars take wear of brushes and keep direction of current constant under field poles. On drum armature, front and back connections of conductors (c) provide always for proper series connections of active conductors.

Mechanical engineers may object that rubbing carbon brushes on copper conductors will quickly ruin the motor. To get around this problem a *commutator* is used, as shown in Figures 20–5a and b. Copper bars are connected to each turn (or each few turns) as in Figure 20–5a. These bars are laid out on an insulating cylinder on the armature shaft (Figure 20–5b). Bars are insulated from each other and the shaft. Thus graphite brushes bear on the bars instead of on the armature conductors, but are connected to the same points of the winding as before. The commutator takes the wear.

This construction makes dc machines expensive. Brushes wear fairly rapidly on the segmented commutator and require frequent maintenance and adjustment.

To convert the ring armature back to a *drum armature* is simple. There is no hole in the middle, but the outside of the armature is still the same, and active conductors are unchanged (Figure 20–5c). Their ends are simply connected together across the diameter to maintain the proper direction of current under the two poles and keep them in two series strings. Note that the brushes will periodically touch two segments at once, but the conductor between these two segments is not active and so no voltage is short circuited. In the cylindrical armature, series strings of active conductors are made up of conductors from both sides of the machine. The winding is still continuous, and incoming current still splits in half. Practical machines usually have more than one conductor per slot. These conductors are arranged in series by connections across the ends of the rotor. This raises the machine's voltage for a given speed.

More than one pair of poles is often provided. The number of pole pairs will affect brush provisions, coil connections, and current division in the armature.

20.4   CIRCUIT MODEL REVISITED

The previous chapter introduced the circuit model shown in Figure 20–6a. In Figure 20–6b a *field rheostat* (variable resistor) has been added to control field current, and field and armature paralleled on a single voltage supply source V_t. The model's function is not to represent the machine physically but electrically; yet from the dc machine description above it can be seen how closely the model does follow machine construction. The field circuit is the winding on the field poles. In the armature circuit R_a is the series/parallel resistance of that winding between brushes, and V_g is the voltage of those active conductors in series. For ac machines (or dynamic analysis of dc machines) it is important to add field and armature inductive reactances in series with their resistances. For ac machines, reactance X frequently predominates over resistance R to the extent that R can be neglected in many analyses; R must of course be considered in loss calculations.

Direct-current machine losses are grouped into three categories: field loss, armature loss, and mechanical loss. These are respectively the power loss in R_f (including any rheostat); in R_a; and in friction, windage, and magnetic losses. This mechanical loss is sometimes taken as very approximately proportional to speed

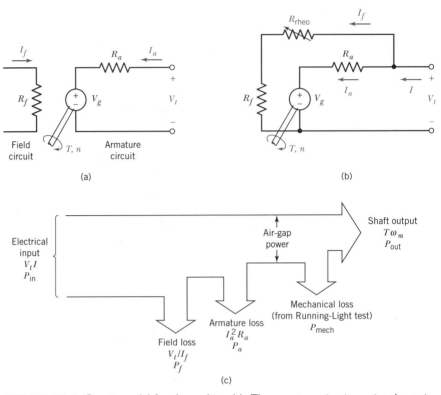

FIGURE 20–6 Circuit model for dc machine (a). The armature circuit carries the main motor current. Parallel field connection in (b) makes a *shunt* machine. In a typical machine V_t might be 220 V and V_g 205 V. (c) diagrams a breakdown of machine power and losses.

over a limited range. As in the linear railroad track problem, the *air-gap power,** is $I_a V_g$, which is converted to mechanical power. Except for the part of this that goes into mechanical losses, it comes out the shaft as output power ($T\omega_m$).

Figure 20–6c is helpful in visualizing these loss and efficiency concepts. Power in this drawing moves from left to right. Electrical power, $V_t I$, is supplied to the motor. Progressive loss elements split off from it at the bottom of the diagram until the remaining output power, on the right, appears as mechanical shaft power.

EXAMPLE 20.6 A 220-V, 50-A dc shunt motor (field paralleled with armature) has $R_a = 0.3\ \Omega$, $R_f + R_{\text{rheostat}} = R = 110\ \Omega$. Its mechanical losses are 876 W. See Figure 20–6b.

Find: losses, input power, output power, and efficiency.

---

*By the rather odd term *air-gap power*, engineers mean the electrical power which is converted into mechanical power. In the model this is the power absorbed by the back voltage source marked V_g. Some of this mechanical power goes into mechanical losses and the rest into shaft output. (In induction motors this term also includes the rotor copper losses.)

Solution: $P_{\text{in}} = V_t I = 220\text{ V} \times 50\text{ A} = \underline{11\text{ kW}}$. $I_f = V/R = 220\text{ V}/110\ \Omega = 2\text{ A}$. $I_a = I_{in} - I_f = 50 - 2 = 48\text{ A}$. $P_a = I_a^2 R = 48^2 \times 0.3\ \Omega = \underline{691\text{ W}}$ armature loss. $P_f = VI = 220\text{ V} \times 2\text{ A} = \underline{440\text{ W}}$ field loss. Total losses = $P_f + P_a + P_{\text{mech}} = 440 + 691 + 876 = \underline{2007\text{ W}}$. $V_g = V_{\text{in}} - I_a R_a = 220\text{ V} - 48\text{ A} \times 0.3\ \Omega = 206\text{ V}$. $P_{\text{out}} = P_{\text{air}}\text{ gap} - P_{\text{mech}} = I_g V_g - P_{\text{mech}} = 48\text{ A} \times 206\text{ V} - 876 = \underline{9.0\text{ kW}}$. Eff $= P_{\text{out}}/P_{\text{in}} = 9.0\text{ kW}/11\text{ kW} = \underline{82.0\%}$.

A *magnetization curve* can be made of V_g vs I_f for any particular speed. This is easily accomplished experimentally by driving the machine as a generator without load. Figure 20–7 is an example taken for a particular machine at 1800 rpm. Since V_g is proportional to speed, Figure 20–7 will provide data for any speed by simply adjusting its ordinates by the ratio (new speed)/1800. The curve bends over for high field currents. This is the magnetic saturation effect discussed in Chapter 18. Motors tend to be operated almost entirely in the linear part of the curve.

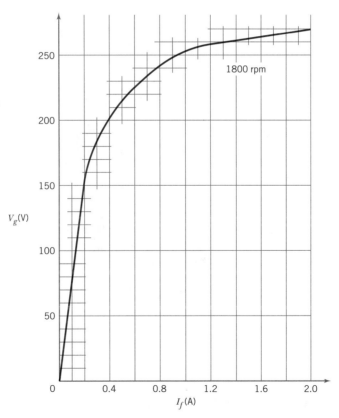

FIGURE 20–7 Magnetization curve for a specific machine taken at 1800 rpm. See Figure 20–6 for meaning of V_g and I_f. V_g is proportional to speed.

EXAMPLE 20.7 The dc machine of the last example has a magnetization curve as in Figure 20–7, and is driven as a generator at 1500 rpm. A load of 40 Ω is connected. 1.5 A field current is separately supplied.

Find: load current and voltage.

Solution: From Figure 20–7, for 1.5 A field current V_g = 264 V at 1800 rpm, or at 1500 rpm, 264 V × 1500/1800 = 220 V. $I_a = V_g/(R_L + R_a)$ = 220 V/40.3 Ω = 5.46 A. Then $V_t = V_g - I_a R_a$ = 220 V − 5.46 A × 0.3 Ω = 218 V.

Motor speed is found from V_g and a little reasoning, as seen in the following example.

EXAMPLE 20.8 The same machine (Examples 20.6 and 20.7) is now run as a motor, Figure 20–6b, supplied with V_T = 220 V, with field adjusted for I_f = 1 A. Without load the motor draws I_{supply} = 5 A. The motor is then loaded until it draws 43 A at full load.

Find
1. the motor's speed under conditions of full load and no load,
2. the speed regulation from no load to full load.

Solution
1. At full load $I_a = I_{supply} - I_f$ = 43 − 1 = 42 A. $V_g = V_t - I_a R_a$ = 220 V − 0.3 Ω × 42 A = 207 V. At 1800 rpm and I_f of 1 A, V_g would be 252 V (from magnetization curve of Figure 20–7). So by proportion n/1800 = 207 V/252 V, whence n = 1479 rpm at full load. Similarly at no load (I_{total} = 5 A), V_g is 219 V and n = 1800 rpm × 219 V/252 V = 1564 rpm.
2. Speed reg = $(n_{nl} - n_{fl})/n_{fl}$ = (1564 rpm − 1479 rpm)/1479 rpm = 6.4%. (The exact answers here are sensitive to curve reading.)

This last problem illustrates the *running light test* engineers use to measure mechanical losses for a given motor. The motor is run with no load (nothing coupled to shaft) and the field adjusted for the speed at which mechanical loss data are desired. All the air-gap power (876 W in the above example, 4 A × 219 V) must be going into mechanical losses since there is no output.

Experiment with variations of the last example will show that in speed control of dc shunt motors, setting the field rheostat for higher I_f lowers the speed.

In another kind of dc motor, the series motor, instead of many field turns and low current, a few turns of heavy wire are used on each pole to create the field flux. Figure 20–6a still holds, but instead of the shunt arrangement of Figure 20–6b, the field is run in series with the armature as in Figure 20–8. The field and armature currents are the same. Details of the effects of this arrangement are left for end of chapter exercises. But it is apparent that this series configuration produces large torques at low speeds (the common armature and field current high) and low torques at high speeds (V_g high and the current low). This motor tends to *run away* at light loads and is therefore never run without being coupled to some load. It

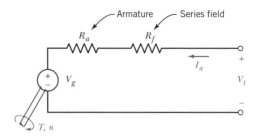

FIGURE 20-8 Series dc motor circuit model. The field is wound with a few turns of wire heavy enough to carry the armature current. Armature and field reactances (X's) are not considered in the static analysis of dc machines.

can also be run on ac since both field and armature will reverse at the same time, continuing torque in the same direction. For ac machines the stator is laminated to reduce eddy current losses. Machines to be run on both ac or dc are called *universal motors.*

Compound motors have both series and shunt fields and partake somewhat of the character of both. The series field can be *aiding* or *opposing* the dominant shunt field depending on the characteristics desired. In compound generators the series field always aids and provides for overcoming voltage drop as load increases.

20.5   THE ALTERNATING-CURRENT ROTATING FIELD

As noted in Chapter 19 several important types of ac motors produce a rotating magnetic field inside their stators (Figure 20-9a). Nothing in the stator rotates mechanically. Alternating current variations in the stator windings produce areas on the inside stator ring where flux is coming out of the iron (north pole) or into it (south pole), and these areas rotate around the inside, as suggested in the figure by the small moving circles marked "N" and "S." Depending on the winding there may be two, four, six, or more *virtual poles* rotating, always in pairs and always alternating between north and south poles. Figures 20-9b–f show how this is accomplished for a single pole pair with two phases of current 90° apart.

In Figure 20-9b a single circular coil with sinusoidal current $i_1(t)$ produces a horizontal flux field B_1, which oscillates back and forth sinusoidally with current. At the peak of the ac cycle, flux is maximum and pointing to the right. As the ac cycle declines flux becomes smaller. Eventually with the ac cycle in its negative half cycle, the flux arrow will point to the left and grow larger in that direction.

Figure 20-9c shows a similarly driven coil, with current i_2, positioned so that its oscillating field B_2 is vertically oriented. The waveforms of these two currents are 90° apart as in Figure 20-9e. If the coils are mounted in stator together (Figure 20-9d), their two fields add vectorially as shown for time t_a in Figure 20-9e. At a later time t_b (as marked on the figure) the resultant vector will have rotated coun-

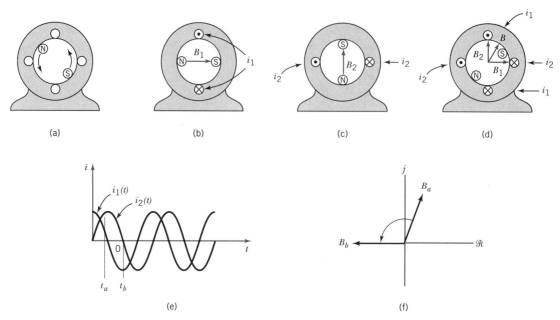

(a)   (b)   (c)   (d)

(e)   (f)

FIGURE 20–9 Alternating current motor windings provide a *rotating field* inside the stator (a). Sketches (b) through (f) show how this can be done with two phases 90° apart in both space and time. Practical motors are wound three phase with 60° space and time separation.

terclockwise to the position shown in Figure 20–9f. Thus the magnetic field inside the stator rotates counterclockwise once for every cycle (1/60 s) or at a speed of 3600 rpm.

Practical motors are three phase instead of the two used for explanation here. But three three-phase coils spaced 60° apart will produce the same result. In a stator wired to produce four poles, each of the six coils (three for each pole pair) connects armature conductors 90° apart instead of 180°, and the resultant magnetic field will rotate once in two cycles or 1800 rpm.* This reasoning can be carried on for any number of pole pairs desired. For 60-Hz power the field will rotate at a speed in rpm of 3600 divided by the number of pole pairs. Or, expressed algebraically,

$$\omega_{\text{m-syn}} = \omega_{\text{supply}}/\tfrac{1}{2}p, \qquad (20.4)$$

where p is the number of poles and $\tfrac{1}{2}p$ the number of pole pairs, synchronous mechanical speed $\omega_{\text{m-syn}}$ is expressed in mechanical radians per second, and ω_{supply}

*In the construction of practical motors, armature windings are fabricated on jigs as coils with the required number of turns. These coils are then mounted into the stator (or armature) slots and held there by wedges. Coil terminals (at the ends of rotor or stator) are then connected together and to the supply lines or commutator bars as appropriate. The coils are often said to span so many mechanical degrees.

is the line frequency of the electrical supply in radians per second. This may also be expressed as

$$n_s = 120f/p, \tag{20.4'}$$

where n_s is the synchronous mechanical speed in revolutions per minute and f the line frequency in hertz.

In a practical machine each coil may also be distributed over more than two slots.

20.6   SYNCHRONOUS MOTORS

To utilize the rotating-field stator described above it is necessary only to mount a dc magnet rotor on a shaft inside to follow the stator field (Figure 20–10a). North and south poles of the rotor are attracted to rotating south and north poles of the stator field. The rotor magnet is excited by dc current supplied via slip rings.

Assume for the moment that the rotor revolves at synchronous speed. If the rotor and the rotating field of the stator are perfectly lined up, as in Figure 20–10a, forces of attraction between north and south poles are entirely radial. Under this condition the motor develops no torque. But as load torque is placed on the shaft, the rotor *falls back* through an angle θ (Figure 20–10b). This falling back in angle establishes a circumferential component to the force between unlike poles, a component that will produce torque. As fallback increases, this useful force component increases.

The T vs θ curve is approximately sinusoidal for most machines (Figure 20–10c). When the rotor has fallen back to be aligned with the next pole (the 180°

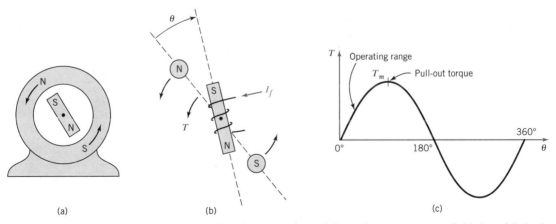

(a)   (b)   (c)

FIGURE 20–10 In a *synchronous* machine (a) a dc rotor magnet follows the rotating stator field. As it falls back with mechanical loading, through angle θ, more torque is developed by a more circumferential direction of interpole forces. This T versus θ curve is nearly sinusoidal (c).

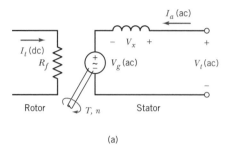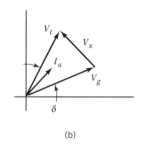

FIGURE 20–11 For a two-pole synchronous machine the fallback angle θ becomes the *power angle* δ between V_t and V_g in the phasor diagram, driving current into the motor armature (stator) with difference voltage V_x.

point on the curve) there will again be zero torque. Any further fallback will tend to push the rotor in the wrong direction. Note also that there is a maximum torque T_m, called the "pull-out" torque.

Thus the synchronous motor develops torque by falling back in angle while still running at synchronous speed. But if too large a load torque is applied to the shaft—more than the pull-out torque marked on Figure 20–10c—the machine falls out of synchronism and stops.

A synchronous motor's circuit model (Figure 20–11a), includes a dc field and ac stator. The term *field* used here refers to the dc rotor magnet and must not be confused with the rotating field produced by stator currents. The stator circuit shows only one of three identical phases. V_g is produced by flux from the rotor field poles cutting stator conductors. Because of the rotor's synchronous speed, V_g has the same frequency as V_t. X is much larger than R in stator windings so R can be neglected for some purposes. (In this hybrid model the rotating shaft comes out of the armature winding on the nonrotating stator.)

In the phasor diagram of Figure 20–11b, the mechanical fallback angle becomes the electrical *power angle,* δ. (In machines whose stator windings produce more than two rotating poles, the mechanical angle is smaller than the electrical angle by the factor $\frac{1}{2}p$, where again p is the number of poles and $\frac{1}{2}p$ the number of pole pairs. Engineers distinguish between *electrical degrees* and *mechanical degrees*.) By KVL V_X, the voltage across the stator impedance is the difference of applied and back voltages, $V_t - V_g$. I_a is V_X/X and will lag V_X by about 90° (or less if the winding's resistance is considered). V_X is normally on the order of 10% of applied voltage. So the phase angle between input voltage and current is small if the two large voltage vectors are equal—that is, if the field current has been adjusted so that $V_g = V_t$ in magnitude.

Adjusting V_g with I_f so that the two voltages mentioned above are not equal brings out one useful peculiarity of these motors: the motor's power factor can be changed. The phasor diagram of Figure 20–11b illustrates this nicely, if the length of V_g is mentally allowed to become longer or shorter than V_t. Increasing field current (and V_g) will cause I_a to lead (see problem E20.4). Thus the synchronous motor can be made to appear capacitive and can, if desired, be supplying mechanical power at the same time.

As noted in Chapter 19, there is a starting problem. Suppose the motor is standing still with rotor field poles excited with dc, and three-phase power closed on the stator winding so that a rotating field suddenly starts. Because of inertia the heavy rotor cannot jump instantly from 0 to 1200 rpm or whatever the synchronous speed is. The torque between the rotating field poles and now-stationary rotor poles will thus alternate in direction 120 times a second. No net torque is developed.

So a synchronous motor cannot start synchronously, and if this is attempted it will simply remain stationary. Some other starting procedure is required. The common solution is to start the rotor as an induction motor. Conductors, much as in an induction motor (described in the next section), are built into the rotor. It is essential that the field be turned off during start. This induction-motor effect brings the motor near enough to synchronous speed that when the field is connected it will jump into synchronism. Like any large induction motor, synchronous machines are started at reduced voltage.

The type of motor described above is used mainly for large constant-speed loads which are seldom shut down. Driving dc generators is a typical use. Their power factor correcting ability is an added bonus for such applications and may help to economically justify their substantially increased cost over induction motors.

Using variable-frequency drive controllers will permit precise speed control ranging from approximately 30% to 120% of the 60-Hz synchronous speed.

Synchronous motors are an exception to the earlier generalization that motors develop torque by running a little slower, as is so clearly exemplified in the dc and induction motors discussed earlier. Keeping a constant (synchronous) speed this motor develops more torque, when needed, by falling back in phase, increasing its power angle but continuing to run at the same speed.

When the torque load exceeds the maximum torque available (Figure 20–10c), the motor will slow down and stop. From the circuit model it can be seen that when slowing down V_g will then rotate clockwise with respect to V_t, producing huge fault currents when these phasors are 180° apart. Protective circuit breakers operate. The machine must be restarted.

Synchronous machines used as generators are called *alternators* and are the source of electric power at thermal stations and dams. It is important to *synchronize* these machines when they are paralleled on a common bus. Suppose in the one-line diagram of Figure 20–12 that the power station is running at a certain level of output with units 1 and 2 connected to the bus and supplying power. (One-line, equivalent-wye diagrams were originally discussed in Chapter 17.) But at this time more power is needed. Another steam turbine and its coupled alternator marked "unit #3" are started up, and the breaker is to be thrown to connect its three leads to the station bus, thus paralleling it with the two units already running and connected. Extreme caution is needed in closing this switch, as described below.

The circuit breaker connecting unit #3 to the bus cannot be closed until the third unit's phase is adjusted so that its voltage maxima coincide in time reasonably well with the bus maxima. Figure 20–12b illustrates this point. The phase of unit #3 with respect to the bus must be allowed to rotate until it nearly coincides with V_{bus}. This is what is meant by "synchronizing two machines" or "synchronizing a

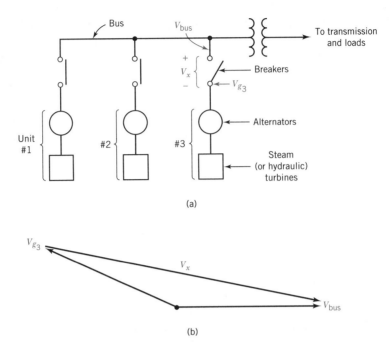

(a)

(b)

FIGURE 20–12 One-line diagram (a) of three synchronous machines used as alternators (ac generators) at a power station. The phase of each generator must be synchronized to the phase of the bus before closing its breaker. (b) illustrates the large voltage difference V_x possible between a nonsynchronized machine and the bus.

machine to the bus." Otherwise there could be a surge of current as large as $2\,V_{\text{bus}}/(X_3 + X_{\text{bus}})$, where X_3 is the third machine's reactance.*

Some large plants generate part of their own power with fuels like waste blast furnace gas. The units so powered must be synchronized before being connected. Often this is done with automatic control equipment, but in some older plants it must still be accomplished by hand switching while watching a synchroscope. Some very large synchronous motors which are started by their loads are also synchronized in this manner before being connected to the ac line.

Other (and smaller) synchronous motors, such as reluctance motors, do not require dc fields.

20.7   INDUCTION MOTORS

Induction motors use the same kind of rotating-field stator described above with synchronous motors but use a quite different rotor, which follows the more common action of slowing down slightly to increase torque.

*In this example X_{bus} would be the paralleled reactances of machines #1 and #2 again parelleled with any bus reactance to the right of the machines. As a worst case, engineers sometimes assume an *infinite bus*, meaning that the bus impedance is negligible and its voltage fixed.

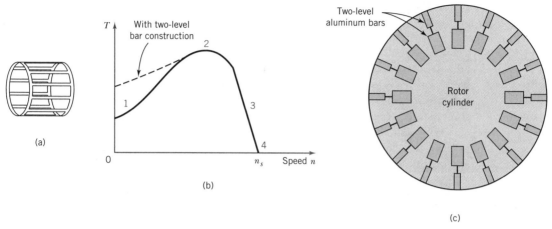

FIGURE 20–13 In an induction motor the cast aluminum "squirrel cage" (a) is not seen because it is imbedded in the rotor iron. Torque characteristic (b) can be improved for start by two-level bar construction.

The rotor conductors or bars of an induction motor are usually cast from aluminum in the squirrel-cage form (Figure 20–13a) directly in the slots of a laminated iron rotor. Integrally cast end rings short-circuit the bars. Being surrounded by a great deal of iron, much of the cage structure cannot be seen. There is no electrical connection to the rotor, no brushes, no slip rings. In smaller motors aluminum fan blades are often cast integrally with the squirrel cage.

Stator rotating field poles passing the rotor bars induce u-l-B voltages in them, where u is the *relative* speed between bars and field. These voltages produce rotor current, which in other motors is supplied externally through slip rings or commutator. Note that before the machine starts to turn, rotor voltages and their resultant currents are at 60 Hz regardless of the number of poles in the stator winding. (If there are twice as many poles—four for example—the rotating speed is half as fast—1800 rpm.) Because the rotor cage is surrounded by magnetic iron, rotor circuits are highly inductive. Bar currents lag voltages heavily until the machine gets to about 70% of synchronous speed.

With these bar currents immersed in the magnetic field of the stator (even though it is rotating) i-l-B forces apply torque to the rotor in such a way that it will follow in the direction of the rotating field. So the motor starts and begins to accelerate. The task now is to explain the T-n curve of Figure 20–13b.

Torque is weakened at low speeds by the reactance of the rotor. This can be explained by noting that the maximum voltage induced in a bar occurs near the moment when a pole, the maximum part of the field, is passing. But because of the large X_{rotor}, bar currents lag bar voltages significantly. Then because of the rotation, bar currents will also lag the maximum parts of the stator field. This reduces the i-l-B product integrated over all bars, and therefore reduces torque.

Note also that as rotor bars begin to turn and speed up, the frequency of voltage induced in them goes down, according to

$$f_{\text{rotor}} = 60S, \tag{20.5}$$

where S is the *slip*, defined as

$$S = (n_s - n)/n_s. \qquad (20.6)$$

(If slip is to be expressed in percent, this result must be multiplied by 100.) As n increases, S decreases, and the reduction of f_{rotor} makes inductance progressively less important. Bar currents come more into phase with the field poles. Net torque is increased.

Some slip is always needed, since at synchronous speed no bars would cut flux, no voltage would be generated in the rotor, and there could be no torque. Therefore torque decreases to zero as synchronous speed is approached.

EXAMPLE 20.9 A two-pole induction motor is turning at first at 100 rpm and again at 3450 rpm.

Find: the slip for both speeds.

Solution: From equation (20.4′), n_s is 3600 rpm. Slip $S =$ (3600 rpm $-$ 100 rpm)/3600 rpm $= 0.972$ or $\underline{97.2\%}$. For 3450 rpm $S =$ (3600 $-$ 3450)/3600 $= \underline{4.2\%}$.

Following the *T-n* curve of Figure 20–13b from left to right: at point #1 the motor is just starting to accelerate from a standstill. As the speed goes up and frequency comes down, two opposing changes take place: first u (the relative speed between bars and rotating field) decreases, reducing the voltage generated in each bar. Second, as frequency goes down rotor X decreases. Thus rotor current remains about constant. But rotor power factor starts to improve and so torque improves. At point #2 rotor reactance has become small; the rotor's resistance has become the major part of its impedance, and decreasing slip and frequency can no longer reduce rotor impedance very much. From this point on voltage generated in the bars and bar currents continue to fall as slip decreases. The straight-line portion of the curve (point #3) is a region where X is negligible compared to rotor resistance R, V is proportional to slip, and therefore rotor I and T are proportional to slip. At point #4, n_s, there is no longer any relative motion between rotor bars and rotating field. No voltage is generated in the rotor. No rotor current flows. Torque is zero.

Because at low speeds the rotor's X is a good deal larger than R, an increase in rotor resistance would improve power factor angle faster than it decreases rotor current. Improving rotor power factor will increase starting torque. Hence a wound-rotor motor with slip rings to allow temporary additional resistance to be inserted into the rotor circuit, as discussed briefly in Chapter 19, can be used to improve starting torque.

A more common means of increasing rotor resistance to improve starting torque is to cast the bars in a cross section as shown in Figure 20–13c. At low speeds (higher rotor frequencies) the low-R innermost bar portions experience higher inductance (more iron surrounds them) so that most bar current stays in the higher-R outside portion of the cross section. Effective R of the rotor is greater for starting but becomes small again (for higher-efficiency, rated-speed operation) as slip decreases.

At no load induction motors run at nearly synchronous speed. Looking at Figure 20–13b: as the shaft is more heavily loaded the machine slows slightly with a rapid increase of torque somewhat reminiscent of the dc shunt motor. Point #3 is the region of normal-load operation. Note that this is still very nearly synchronous speed. Induction motors can be considered almost constant-speed machines.

In order to compare the induction motor with other machines, Figure 19–6b presented a quite approximate circuit model. A much better model is that of Figure 20–14, reminiscent of the earlier transformer circuit model (Figure 18–3). The circuits shown might represent one phase of an equivalent wye. With the rotor blocked a squirrel-cage motor acts as a short-circuited transformer, with currents limited by primary and secondary reactance and resistance. As the motor is allowed to speed up, secondary currents are reduced because of lower relative speed (u) between rotating field and rotor. This can be allowed for in the model by making the voltage V_g, induced into the stator by rotor currents a function of slip S.

In Figure 20–14a and 14b, R_1 and X_1 represent stator resistance and reactance, R_2 and X_2 rotor resistance and reactance referred to the stator (primary) by the usual factor of the turns ratio squared; B_ϕ provides for magnetizing current to produce the stator's rotating field. The current through this parallel element can be considered the no-load current of the motor also. Unlike the closed-core transformer, the induction motor operates with significant air gaps in the magnetic path, and so magnetizing current is much larger, approximately 40% of a machine's rated current. For this reason it is not very accurate to add the R's and X's of the two sides as was done for transformers in Figure 18–3.

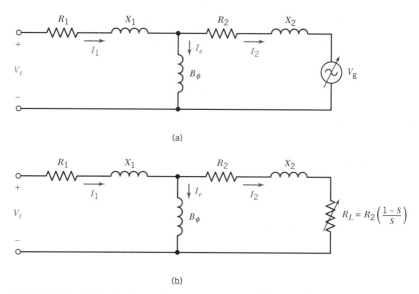

(a)

(b)

FIGURE 20–14 Induction motors can be analyzed with this transformer-like model in which rotor constants R_2 and X_2 are referred to the stator side. The *exciting current* I_e, is the no-load current and is about 40% of the full load current. The voltage V_g generated in the stator by rotor currents is a function of slip S and can be represented by the source symbol or by an equivalent resistance.

The voltage induced in the stator by the rotor currents, V_g, has been replaced in Figure 20–14b by an equivalent resistor $R_L = R_2[(1 - S)/S]$. The power put into either of these elements is the power converted from electrical to mechanical— that is, shaft output plus mechanical losses. To make the form $(1 - S)/S$ reasonable, note that at standstill, the machine is like a short-circuited transformer. At synchronous speed the equivalent transformer is open circuited, and current I_2 zero. It can be observed from the model that rotor losses (in R_2) are proportional to slip, and torque proportional to V_t^2 for a given slip.

20.8   SINGLE-PHASE INDUCTION MOTORS

Squirrel-cage rotors of single-phase machines follow some form of rotating field. But from the analysis of Figure 20–9 it would seem impossible to generate a rotating field with one phase. The following analysis will show that a single-phase, pulsating field (Figure 20–9b) is, in fact, equivalent to two half-strength fields rotating in opposite directions.

Figure 20–15a extends the usual T-n diagram to the left into negative speed regions. This means the rotor is being driven by some external means in the "wrong" direction, against the rotating field. In this situation, slip becomes larger than 100% [Equation (20.6)], rotor frequency goes above 60 Hz, and rotor reactance becomes even worse. Compared to starting conditions, rotor reactance X is twice as great at synchronous speed in the wrong direction. Rotor current is half what it was at standstill. Its phase angle with respect to the rotating field is even worse. So torque at reverse synchronous speed is less than half that at zero speed.

Observe next, as shown in Figure 20–16, that an *oscillating* (nonrotating) vector field, produced by a single coil, can be represented mathematically by two *oppositely rotating* field vectors each half the size of the original nonrotating field. In Figures 20–16a–d, vector 1 rotates clockwise and vector 2 counterclockwise. Suppose each rotating vector has a length of $\frac{1}{2}B_{max}$. Then in Figure 20–16a, the vectors, being 120° apart, will produce a vertical sum of $\frac{1}{2}B_{max}$. There is no other component of their sum than the vertical. Similarly a little later (60 electrical degrees later or $\frac{1}{360}$ s), in Figure 20–16b, both component vectors are vertical and their sum is B_{max}. In Figure 20–16c the vectors oppose each other and have a zero sum. In Figure 20–16d their sum is again $\frac{1}{2}B_{max}$ but this time in a downward direction. Thus the two rotating half-amplitude vectors produce a perfect sinusoidal oscillation when added together. The two rotating half size vectors and the oscillating full size vector are mathematically equivalent. Either can be substituted for the other.

A squirrel-cage rotor can be put into a single-phase stator with its field oscillating as in Figure 20–9b. Both half-amplitude rotating fields are pulling on the rotor to start it in opposite directions and canceling each other exactly. If the rotor is pushed in the clockwise direction, field 1 is favored and accelerates the rotor further, moving to the right up the T-n curve of Figure 20–15a. At the same time the dragging torque effect of field 2 moves to the left on the curve and lessens, so its effect is less than the accelerating torque of 1. As the motor goes faster and faster clockwise its accelerating force increases and the drag force decreases even further.

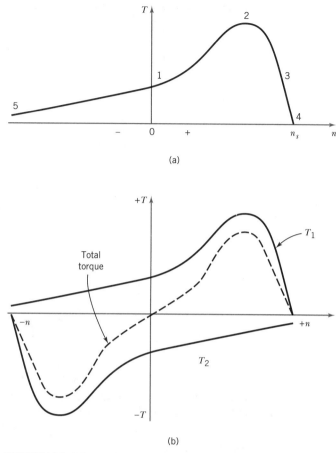

(a)

(b)

FIGURE 20–15 Induction motor *T-n* curve (a) can be extended into negative speed region, where torque continues to decrease with increase in rotor frequency. (b) shows the total combined torques of the two counter-rotating fields of Figure 20–16.

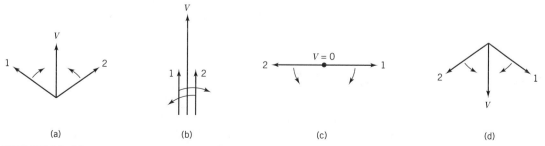

(a)   (b)   (c)   (d)

FIGURE 20–16 A single-phase (pulsating, not rotating) magnetic field can be exactly represented by two counter-rotating half-length vectors. Coupled with *T-n* curves of Figure 20–15, this explains single-phase induction motor torque production.

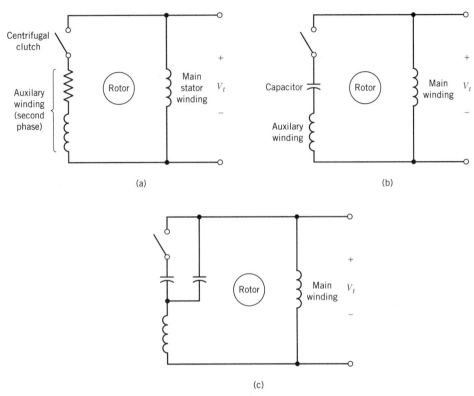

FIGURE 20–17 There are many different types of single-phase induction motors, distinguished by how they provide the auxiliary starting phase and whether it is disconnected on run.

At point #3 of Figure 20–13, the drag of 2 is small. Superposing the forward and reverse torques as in Figure 20–15b generates the T-n curve for a single-phase motor. Thus a single-phase motor will run very well once started.

To start this motor a second phase is generated (usually temporarily) by putting resistance, or more often a capacitor, in series with the line as in Figure 20–17. The resistance is provided by using higher-resistance wire for one winding. In most motors all or most of the second-phase current is removed with a centrifugal switch once the motor is up to 70% or so of its speed. This second-phase winding can then be of light, nonbulky wire since it will operate for only a few seconds per start.

20.9 VARIABLE-FREQUENCY DRIVES

Controllers for the variable-frequency drives discussed in Chapter 19 first convert three-phase (or in some cases single-phase) ac to dc with diode rectifiers (Figure 20–18). In most instances these diode rectifiers are SCRs (silicon control rectifiers discussed briefly in Chapter 16) which permit phase control and hence easy control of the dc voltage output.

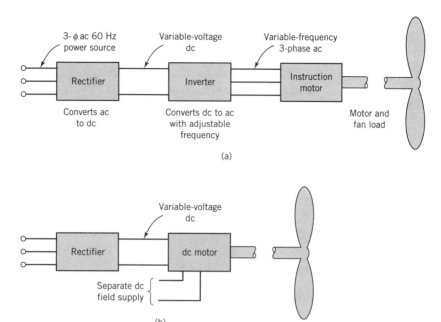

FIGURE 20–18 A variable-frequency drive (a) and a dc drive from ac mains (b). DC output voltage from electronic rectifier circuit (a) can be controlled by the phase of its SCR gates. The electronic inverter circuit reconverts the dc to a variable-frequency three-phase ac to drive an induction motor at variable speeds. Alternatively, as in (b), the dc produced can drive a dc motor directly.

The resultant "dc" voltage has considerable ripple in it. In some cases (Figure 20–18b) this output is fed directly to a dc motor and the voltage used to control its speed. Ripple is usually of little consequence to motors unless it contributes to overheating. High reactance of a motor smoothes current out easily. Simple phase control of the SCRs directly controls dc motor speed by controlling armature voltage. Some other arrangement must be made to supply the field, but for such a low current this is not expensive.

In variable-frequency drives an induction motor (sometimes synchronous motor) is used. Before being supplied to the motor, the dc output of the SCR three-phase rectifier is converted back to ac by an *inverter* circuit, described in Chapter 16. The frequency of the inverter is adjustable over ranges from nearly 0 to 150% or more of synchronous speed. Thus wide-range speed control and soft starts are accomplished for cheap, rugged induction motors. Alternating-current voltage output from inverters is proportional to dc voltage supplied to the inverter. The motor can be brought gently up to some selected speed from standstill by gradually increasing the inverter's frequency. Selected speeds can be adjusted over wide ranges, bounded on top by motor mechanical limitations, and on the low end primarily by motor cooling problems.

In the rectifier-inverter system for driving induction motors, it is essential to reduce the ac voltage applied to the motor in proportion to frequency, since motor reactance will have decreased accordingly.

Efficiency of these systems is typically over 85% at full speed, and it is almost as high at lower speeds. But their power factors on the original ac line continue to be poor at lower speeds.

20.10   SUMMARY

When turning, motors generate a back voltage V_g, which is nearly equal to their applied terminal voltage. Each current-carrying armature conductor also experiences a motor force proportional to the current and the magnetic field surrounding it. Motor torque is the product of this force and the radius summed over all conductors. The relations between V_g and this force are reduced to their simplest, and most clearly seen, in the traditional "railroad track" problem, or linear motor.

The dc machine's rotating armature is connected to the nonrotating world by brushes bearing on a commutator. This rotating switch keeps conductor current under the field poles always in the same direction, for consistent torque.

Air-gap power is converted by the dc armature from electrical to mechanical form to provide mechanical losses and shaft output. Losses can be divided into field (I^2R_f), armature (I^2R_a), and mechanical (which includes magnetic losses). Mechanical losses are determined by a running light test—operating the machine with uncoupled shaft, in which case all air-gap power goes into losses.

A magnetization curve, V_g versus I_f, taken at some specific speed, can be used to predict motor speed under various amounts of armature current loading; from this, speed regulation between no load and full load can be calculated.

The nonrotating armatures (stators) of most ac machines are mounted on the frame. Rotating rotor structures of various kinds generate back voltage in the stator windings. Applied three-phase power produces a rotating magnetic field inside the stator, which interacts with field currents (or magnetic effects) to produce torque. Squirrel-cage induction motor rotors have no external connections but develop voltages (and currents) from transformerlike relation to the stator. The dc fields of synchronous machines must follow the stator rotating field exactly. These motors produce torque by falling back slightly in angle, changing the electrical phase of V_g and thus permitting more armature current.

Single-phase induction motors are analyzed by splitting the pulsating stator field into two counter-rotating parts. A (usually temporary) second phase is generated by capacitors or otherwise to favor one direction at start.

Variable-frequency electronic drives enable dc machines to run from ac mains, or provide controllable frequency for induction motors, easing the starting problem and providing adjustable speed.

FOR FURTHER STUDY

Leander W. Matsch and J. Derald Morgan, *Electromagnetic and Electromechanical Machines,* 3rd ed., Harper & Row, New York, 1986.

Ralph J. Smith and Richard C. Dorf, *Circuits, Devices and Systems,* 5th ed., Chapters 22–24, John Wiley & Sons Inc., New York, 1992.

PROBLEMS

Easy Drill Problems (answers at end of chapter)

(*Note:* SI units are so helpful here that it is suggested that other units be converted into SI for manipulation and then reconverted for solution if that is necessary.)

D20.1. A wire carrying 20 A is positioned normally in a magnetic field of strength 0.5 T. What is the force on each inch of this wire in ounces?

D20.2. The wire of D20.1 is allowed to move, in response to the force, at a velocity of 5 in/s. What voltage is generated in each inch of the wire? Sketch this situation and indicate voltage polarity, current, and motion directions.

D20.3. What power is put into each inch of the wire of D20.2 by the source which is driving the 20-A current through it?

D20.4. Show that the mechanical power each inch of the conductor exerts in D20.3 is equal to the electric power put into it.

D20.5. For an induction motor, make a chart showing for two to ten poles (there must be an even number!)
  (a) the speed of a rotating field in rps,
  (b) the number of poles passing a given bar in poles per second with locked rotor,
  (c) the frequency induced in the rotor bars at standstill,
  (d) the frequency induced in the rotor bars when it has gotten up to one-half synchronous speed.

D20.6. In a classic railroad track situation as in Figure 20–2a, $l = 2$ m, $B = 0.6$ T into the ground. The bar is used as a locomotive to pull seven cars filled with engineering students having a party. Track friction is negligible. Track resistance is negligible but bar $R = 2$ Ω. A 12-V battery is connected to the tracks. Neglecting any friction, find the ultimate speed of the train.

D20.7. Consider now that the bar and cars have a total friction of 2 N. Re-solve D20.6.

D20.8. From D20.7, find the bar's efficiency of power conversion from electrical input to drawbar output. Assume one-third of the 2-N friction is in the bar itself.

D20.9. A small, single-phase ac generator is designed to operate at 10 kW, 240 V, 0.8 pf lagging. It has an R_a of 0.3 Ω and an X_a of 1.2 Ω. Sketch a circuit model of the generator with values and draw phasor diagrams for rated-current operation at 0.8 lagging, unity, and 0.8 leading power factors. Find V_g for each case, with $V_t = 240$ V.

Applications Problems

P20.1. Consider now that the track of D20.7 has a resistance of 1 Ω per mile of roadbed (for both tracks taken together). Neglecting inertia plot the trains' distance from start (in kilometers) vs time.

P20.2. In the railroad track problem of Figure 20–2a suppose that the vertical field component were 10^{-5} T (a more realistic figure for the earth's field). Track resistance and friction are negligible, bar resistance is 2 Ω, the switch of Figure 20–2a is now thrown to the right, connecting the battery, and the whole machine is used as a motor.
  (a) Find initial force on bar and its ultimate speed.
  (b) Would you make any comment on the reasonability of the situation?

P20.3. For Example 20.5 (Section 20.3)

(a) Calculate the terminal voltage applied to this motor.

(b) Assume that the armature winding (from brush to brush) has a resistance of 0.5 Ω. Find the terminal voltage under this assumption.

(c) What is the power supplied by the source to the armature in (a) and (b) above?

P20.4. For the motor of Example 20.6 (Section 20.4) the magnetization curve of Figure 20–7 applies.

(a) Find the motor speed.

(b) Find torque output.

P20.5. Direct-current motor speeds can be controlled over wide ranges by adjusting either applied armature voltage or field current. Assume for the motor of Figure 20–6a that a constant-torque load (assume it includes any motor mechanical torque losses) of 50 Nm is connected to the shaft, and a constant field current of 2 A is provided. R_a is 0.5 Ω. The magnetization curve of Figure 20–7 applies. Make a plot of speed n vs V_t. (See E20.6 for a continuation.)

P20.6. A 220-V series motor (Figure 20–8) has R_a of 0.3 Ω and R_f of 0.2 Ω. The magnetization curve of Figure 20–7 applies if the currents on the horizontal axis are multiplied by 40. Find the speed of the motor for a light load ($I = 5$ A) and for full load ($I = 40$ A).

P20.7. Find the torque output for the two series motor load conditions of P20.6.

(a) First neglect mechanical losses.

(b) Assume that under the 40-A load condition mechanical losses of about 4% occur.

P20.8. Plot a *T-n* curve for the motor of P20.6 neglecting mechanical losses.

P20.9. (See Figure 20–6.) A dc shunt motor has R_a of 0.3 Ω, $R_f = 100$ Ω, $R_{rheo} = 50$ Ω; the motor is powered with 200 V and at full load draws 50.3 A from the line. At no load, with speed adjusted to be the same as at full load, $I_{line} = 6$ A when R_{rheo} is 10 Ω. Calculate the power in, power out, losses, and efficiency for the full load condition.

P20.10. In the manner of Figure 20–9 demonstrate with trigonometry that three-phase windings spaced 60° apart will produce

(a) a constant amplitude,

(b) smoothly rotating field, with

(c) a speed in rps equal to $\omega/2\pi$.

P20.11. The equivalent-wye, per phase ratings of a synchronous motor are 300 V, 15 A, and an X_a of 10 Ω. No-load, 100% pf field current is 2 A. For operation as a *synchronous condenser* (used only to correct power factor, with no mechanical load) estimate the field current required for full-rated KVAR. (*Hint:* Make a phasor diagram similar to Figure 20–11b.)

Extension Problems

E20.1. The train of P20.1 starts out on a clear track at time 0 with friction of 2 N as previously specified. But a heavy snow storm of 3 in./h begins at the same time. Friction is increased by 1 N per inch of snow encountered. No plows are available. Plot a curve of distance *x* in miles traveled from start versus time for the first 3 h of the trip; $B = 0.6$ T still, $l = 2$ m, and $V_{batt} = 12$ V, $R_{track} = 1$ Ω per mile.

E20.2. In Example 20.6 (Section 20.4) R_a is 0.3 Ω. It should be possible to save copper by using smaller size wire in the slots. Suppose the wire cross section were cut in half, thereby doubling resistance between brushes. (Assume that any increase in iron weight due to smaller slots can be offset costwise by the lower reluctance field circuit.) The original armature coils (winding) weighed 435 lb. Assume copper wire costs $0.80/lb and electric power $0.105/kWh. The motor is run 3450 h/year. MARR for your company is 20% before taxes. Motor life is expected to be 12 years.* Is the change in copper content justified? How much does it save? What other matters should be considered in this design alternative for conductor size?

E20.3. Devise a spacing for three-phase conductors in an ac stator that will produce a four-pole machine with rotating field of 1800 rpm. Demonstrate that your solution will actually accomplish this. (See Figure 20–9.)

E20.4. A 1000-hp 4-pole synchronous motor uses the essentially linear part of the $V_g - I_f$ curve such that (equivalent phase to ground) $V_g = 231 I_f$, for peak values. It is driven with an applied voltage (phase rms to ground) of 2309 V. Its power angle is 10°, and reactance per phase is 21.3 Ω.

(a) Make a tabulation of power factor, total power drawn, and total torque—all vs I_f—for field currents up to 12 A.

(b) Assume constant torques of 0 and 4000 N-m. For both these cases make a sketch of power angle and power factor vs excitation current.

(c) What effect would it have on these answers to consider the machine a six-pole design?

E20.5. For the machine of E20.4, assume $I_f = 14$ A. What is the pull-out torque for this machine (the maximum torque it could just produce) for either the six-pole or the four-pole version?

E20.6. Continuing problem P20.5, for a constant V_t of 100 V, plot speed vs field current.

E20.7. A three-phase wye-connected 220-V, 10-hp, 60-Hz, six-pole induction motor has the following constants in ohms per phase referred to the stator (Figure 20–14b); R_1 0.294, R_2 0.144, X_1 0.503, X_2 0.209, X_ϕ 13.25. (X_ϕ is the reciprocal of B_ϕ.) For a slip of 2% find speed, output torque and power, stator current, power factor, and efficiency—all when the motor is operated at rated voltage and frequency. Assume that friction, windage, and core losses are 400 W.[†]

Study Questions

1. State in words and by equation the laws of motor and generator action.

2. Describe two "railroad track problem" situations in which either of these laws operates without the other. Are these practical motor or generator applications?

3. Compare Figures 20–2 and 20–4, and make a chart pairing the corresponding parts of each.

4. Can you conceive of a design in which all the conductors on an armature could be "active" conductors?

5. An automobile is traveling east at 60 mph in Australia. Considering the voltage developed by the front bumper in cutting the earth's magnetic field, which side of the car is positive? What effect will changing direction have?

---

*See Appendix C for simple cost of money calculation methods.

[†]Taken from A. E. Fitzgerald, Charles Kingsley, Jr., and Alexander Kusko, *Electric Machinery*, 3rd ed. McGraw-Hill Book Company, New York, 1971. By permission.

6. Is there a force on the electric wires of your house because of the earth's magnetic field?

7. Why does a dc shunt motor speed up if its field is reduced? If its armature voltage is increased?

8. Sketch a two-layer, 3-D chart comparing the terms armature and field to the terms rotor and stator and the terms ac and dc. Put ×'s in the appropriate boxes.

9. What is the difference between a "slip rings and brushes" system and a "commutator and brushes" system?

10. Describe applications for each machine construction system of the previous question.

11. Why are electric machines designed in a cylindrical configuration—as opposed to the linear machine of, for example, the railroad track problem?

12. With reference to the shunt motor of Figure 20–6b, write and test a computer program that will accept V_t, I, R_{field}, R_a, speed in revolutions per minute, and mechanical losses in watts; and will furnish power out in kilowatts and horsepower, along with the three components of loss, efficiency, and shaft torque.

13. Describe how a combination of two out-of-phase pulsating fields can produce a rotating field inside an ac motor stator.

14. When a mechanical load requires more torque, how does a synchronous motor respond to provide it?

15. Why does a synchronous motor have a maximum (or pull-out) torque beyond which it will simply stop?

16. How are synchronous motors started?

17. Induction motor action can be explained also in a way similar to synchronous motors— as the interaction of two sets of magnetic poles. Rotor currents product a set of poles (rotating at a speed determined by rotor frequency) which are attracted by the rotating stator poles. Develop this explanation.

18. Explain how at rated speed a single-phase source can be made to drive a squirrel-cage rotor.

19. How is the single-phase induction motor started? Describe several means of implementing this idea.

ANSWERS TO DRILL PROBLEMS

D20.1. 0.914 oz/in

D20.2. 1.61 mV

D20.3. 32.3 mW

D20.6. 10 m/s

D20.7. 7.23 m/s

D20.8. 48%

D20.9. V_g = 293 at 8.0°, 264 at 13.8°, 223 at 15.5°

CHAPTER 21

Electrical Safety

Safety in dealing with electrical matters is an important consideration, but not difficult once the problem is understood. However, electricity and much electrical equipment are hazardous to human life and limb, and often a fire hazard as well.

Specific attention must be given to safety. Industrial and other organizations, and managers and leaders at all levels in charge of their operations, are legally liable for their performance with respect to safety.

Extensive, specific, and detailed procedures (codes) have been worked out— and are usually legally mandatory—for constructing and operating safe electrical systems.

This chapter will first examine the nature of the safety problem, then consider some codes, particularly the National Electrical Code, and finally look at a few special accident-exposure situations.

The reader's objective for this chapter is to become so conversant with and convinced about electrical hazards, and the basic preventive ideas presented, that he or she will immediately make a lifelong commitment to assume responsibility for safety in work areas and for employees and the public.

21.1   ELECTRICAL SAFETY PROBLEM

Electrical systems, like almost any other engineering work, present a number of hazards both to those working on them and to the general public. In addition there are hazards to property resulting in fire losses of hundreds of millions of dollars annually.

Statistics published by the federal government show that there are roughly 1000 deaths a year in the United States due to electrical accidents. Electrical wiring and distribution equipment cause about 100,000 fires a year. Motors and appliances add another 60,000 or so. The *Fire Protection Handbook* of the National Fire Protection Association estimates that in a recent five-year period there were about 800,000 building fires caused by electrical malfunctions. Electrical causes produced about $1.5 billion loss in that same period. The handbook holds that one-fifth of all industrial fires in the United States were ignited by electrical arc or overload.

The engineer, who is professionally dedicated to the service of his fellow man, finds major humanitarian considerations facing him or her in connection with these accidents. Also a great deal of national economic waste is involved, again a very unprofessional situation. Because of this property damage and injury to persons, industrial firms lose millions of dollars in additional insurance premiums and litigation costs and awards.

As with most other industrial accidents, electrical accidents are with few exceptions preventable through practical, economic measures. In most instances money spent on prevention will be returned by reduced losses and insurance rates. Many firms harshly judge professional employees and managers who permit accidents to occur, accidents which may have severe economic consequences along with other disadvantages.

21.2   AN ELECTRICAL ACCIDENT

There is probably no such thing as a typical electrical accident to either persons or property. As in other fields these occurrences are insidious and appear to happen most often in some manner which has not been considered or recently considered. Each comes as a surprise. For one reason or another vigilance has been allowed to lapse. The following experience illustrates these points. Except for minor details it is taken from an actual case.

A small one-story plant in the country had planned and was constructing a high-bay addition to one end of its building. Fred Jones, Civil Engineer and a local general contractor, was performing the work. The crane and crane crew to handle heavy loads at the site were provided by a small subcontractor, the Brandon brothers. They had rented an old crawler crane from an equipment company. One of them had a crane operator's license. They had done rigging work in the area for quite a few years.

This plant was supplied with three-phase power at 13,800 V by a quarter-mile overhead line terminating on the outside of the building where a 13,800/220-V transformer sat on a concrete pad. Power line and transformer belonged to the power company.

About halfway through the construction, on a Saturday afternoon when no other work was being done, the Brandons, with two neighbors, Paul and John Mancini, decided to straighten up some of the work area—certainly a generally commendable endeavor. Bundled metal roof trusses had been delivered and dropped on the wrong part of the site. Picking these up with the crane, the crew proceeded around the building with them to the construction area, Matt Brandon in the crane cab and the other three walking with hands on the load to steady it. The boom had been raised high to support such a heavy load.

When the crane passed down the narrow road in front of the building, the plant manager, Harold Smith, who was also putting in some overtime, came out of the front door and reminded the crane men that new shrubbery had been put in and they should watch carefully that the load did not damage it. Proceeding slowly under these conditions Matt ran the crane's cable directly into the power line where it crossed the road. According to court testimony, there were several cannonlike booms. Two of the hook men were thrown several feet; Paul Mancini was killed outright and Harry Brandon severely injured.

What went wrong here? Accidents are almost never caused by a single fault. Had any of several mistakes been avoided, this unfortunate affair would not have happened. Any one of the professionals or foremen connected with the work could have prevented it.

1. Laws and codes were violated in that Matt Brandon, an experienced crane operator, was working in the presence of a high-voltage power line without having designated a man whose primary duty was to watch the crane and line so he could give timely warning.

2. The operator and hookmen allowed their attention to be distracted from safety considerations by the shrubbery problem.

3. None of the men operating the crane and steadying its load was adequately trained. One-time training is not enough. It must be kept fresh and up-to-date.

4. Harold Smith, the plant manager and an engineer, did not keep safety considerations in his mind or use the obvious opportunity in front of the plant to call them to the workmen's attention again.

5. Fred Jones, the general contractor, had no plan and had made no provisions for meeting an obvious safety hazard—the presence of a crawler crane on a job with a high-voltage line. Could this be considered engineering? It is good practice (called out in applicable codes and laws) to mark off appropriate areas on the ground with tape, or by some other means to make hazardous areas apparent to all concerned. There are other common precautions.

6. Neither the crane owner nor the contractors were using up-to-date safety technology. Cranes can be equipped with insulating links in their cables above the load.

7. The power company involved had a similar crane accident a year or so before. Consequently Engineer Jonas Middleton, area coordinator for the power company, urged the owner to let them shut down the power, or underground it, during critical parts of the construction period. But Smith was unwilling to do so for economic reasons. Middleton, perhaps the most aware of the danger of any of the parties, did nothing else.

8. Among other lesser matters, no warning signs were posted in the crane cab. These are legally required.

9. Clearly in this matter there was a breakdown in the chain of command. The general contractor allowed work to take place when his foreman was not present. The owner and purchaser of the construction had apparently dropped his sense of responsibility for the safety on his own property. Matt Brandon wasn't taking care of his employees, who were actually his own brother and friends.

The power company discovered afterward that all three of the phase lines had been run into sequentially, blowing high-voltage fuses several hundred yards away. At the trial resulting from subsequent litigation, Brandon, the crane operator, testified that when the loud booms came to his ears and he saw his hookmen thrown aside, he could only assume that someone was shooting at them. Thus it was apparent that any crane/power-line safety concern had not crossed his mind that afternoon.

The kinds of problems and attitudes illustrated here seem to be present in some degree in almost every electrical accident.

21.3   ELECTRICAL DANGERS TO PERSONS

Engineers must ensure that electrical systems are constructed and operated in such a manner as to protect their employees and to protect the general public. There are two kinds of problems: first to protect those who work on the systems and are *qualified* to do so. This includes training workers at all levels in safe methods and attitudes. Proper work equipment is provided. The working environment must be made and kept a safe one. Training is particularly important for managers and leaders, and also for engineering students.

The second problem is to protect those employees who are not qualified to work on electrical systems, and to protect the general public. The essence of this protection is: *to prevent unqualified persons from having access to live parts of electrical systems,* or to substantially any part of an electrical system running at more than 600 V. The National Electrical Code (discussed below) defines a "qualified person" as "one familiar with the construction and operation of the equipment and the hazards involved." But needless to say, even qualified persons need protection against electrical hazards.

There are three kinds of bodily hazards to be protected against:

1. death or injury by electrocution—that is, by actually coming in contact with voltages above about 20 or 30 V; generally speaking, the higher the voltage the more dangerous it is;

2. being burned by electrical arcs and sparks—these are frequently caused by short circuits when wires are being handled live;

3. injury to eyes either from the radiation of an intense arc near the eye, or from molten metal or such being thrown into the eye from the nearby explosive force of a short circuit or similar trouble.

To these peculiarly electrical hazards might be added two related ones:

4. falling or a similar blow (from a ladder for instance) caused by electric shock or other electrical trouble; this may involve two or more persons working together;
5. injury from electrically driven or controlled machinery—for example, from a motor being inadvertently started.

The human body appears to be itself controlled electrically, through the brain and nerves, by tiny currents and quite small voltages. When much larger currents and voltages are encountered in electrical accidents, normal bodily action can be seriously disrupted. There appear to be wide variations from one individual to another in how much current and voltage can be withstood. Some individuals can sense the presence of as little as 12 V with their moistened fingers. OSHA* requires live parts operating at 50 V or more to be guarded.

The surprising insulation strength of some individuals should not lull the engineer into a relaxed attitude. As much as a megohm has been measured arm-to-arm under some conditions. But this can be as low as a few thousand ohms, particularly when the skin is sweaty or oily. Some authorities have taken 5 mA as a safe limit of allowable current through the human body. Others feel this is too high. There is always the problem of persons in abnormal health being abnormally affected.

The table of Figure 21–1 suggests approximate 60-Hz current levels at which significant physiological effects are seen in persons. Let-go current is that value at which the victim is no longer able to let go of the electric wire he is touching. His muscle control is locked, overridden by the current. Fibrillation is an abnormal, noncoordinated train of repetitive, ineffectual heart contractions—a point at which serious injury or death may result.

It should be noted that even a barely perceptible shock may startle an individual into some such mechanical difficulty as falling off a ladder or scaffold, or even into a stronger electrical contact with some high-voltage part. Thus those employees who must work around energized electrical equipment must be provided sufficient space for safe work. Narrow clearances are dangerous. In addition they must be trained in safe working habits—both initially and periodically as such work continues. In general these employees must come under the definition of "qualified persons" given above.

Other employees and the general public must be denied any access to electrical systems, except under rigidly specified safety conditions, such as are in effect for house wiring and household appliances. A peculiar problem here is the tendency of boys to climb poles and explore the interior of fenced-off areas.

When an electrical accident has occurred, first "survey the scene" to determine if the area is safe. Is there more than one victim? Are there others in the vicinity who can help? Start appropriate first aid and have someone immediately

*The Occupational Safety and Health Administration of the federal government, an agency with widespread and penetrating authority and responsibility for promoting safety and safe conditions in industry.

60-Hz Current (mA)	Effects, Limb-to-limb, l-s
1	Threshold of perception
5	Accepted as maximum harmless current intensity
16	"Let-go" current
50	Pain, possible fainting, exhaustion, mechanical injury, heart and respiratory functions continue
100–2000 or 3000	Fibrillation, respiratory center intact
6000 or more	Intermittent sustained myocardiac contraction; temporary respiratory paralysis; burns if current density is high

FIGURE 21–1 The approximate effects of electric current on the human body. Table compiled by T. A. Fuller from various sources. (Thesis, Worcester Polytechnic Institute, 1974, used by permission.)

call EMS personnel (Emergency Medical Services) dialing 911, or take what steps are prearranged for your plant.

If the victim is still in contact with high voltage, don't allow anyone except a qualified person to touch or attempt to free him or her. Can the high voltage be turned off quickly?

It is important for a number of those who work with high-voltage equipment to be trained in administering appropriate first aid including CPR. This training can be arranged by your local American Red Cross chapter.

21.4   FIRE HAZARDS

Shoddily designed or maintained electrical equipment is prone to overheating. Poor electric heaters may be inadequately protected from contact with flammable material. Poor installation practices may bring hot electrical equipment too close to flammable walls or enclosures.

A *fault* is some malfunction in an electrical system. Aside from the poor practices of the preceding paragraph, most electrical fires are started by fault conditions. Chapter 17—particularly Sections 17.5 and 17.7—touched on the need for and practices of protection for electrical systems. Protection is the provision for safe handling of fault situations. Overcurrent is the most common fault or fault result.

Engineers provide for disconnecting the supply by tripping breakers automatically when too much current is detected.

The most common source of electrical fires in houses is overcurrent heating the wires built into house walls. Many house circuits are rated at 15 A. Too many appliances connected to a single circuit overload the wiring. Or alternatively a faulty plug makes poor contact in an outlet and heats up the outlet itself, destroying insulation and producing a short circuit at the outlet. Another common fire source is a poorly maintained appliance—a waffle iron or television set—burning itself up. The overcurrent device that should protect against most of these problems is a fuse or circuit breaker. Householders are often tempted to disable these devices temporarily, a dangerous practice.

In industry motors are frequent offenders, particularly in dusty locations. Also many industrial power systems have enough capacity behind them to allow for fire-setting faults which are still too small to trip breakers.

Electrical fires are also particularly dangerous because of the hazardous fumes that overheated or ignited electrical insulation gives off. Electrical equipment is often sequestered in vaults or other out-of-the-way places. Once trouble starts they may be difficult to get to. In a fire at a new Boston hotel, which fortunately injured no one, it was discovered that an emergency generating system had been installed in the same room in which the faulty main system was burning.

21.5   THE NATIONAL ELECTRICAL CODE AND ITS USE

The *National Electrical Code* (NEC), a nationally accepted guide to safe installation and operation of electrical conductors and equipment, is the basis for essentially all other codes and electrical safety laws. It was first promulgated by a consortium of fire insurance companies (later the NFPA—National Fire Protection Association) before the year 1900 to improve and standardize electrical practices. The code is presently revised every three or four years and is now a national standard—accepted by the American National Standards Institute—and designated ANSI C1-1978 (for the edition of that year). It sells nearly a million copies for each revision, the most widely adopted code of standard practices in the United States.

Fire insurers themselves have no enforcement authority. But it would be difficult to get fire insurance without complying. However, many groups who do have enforcement authority have adopted the code and require it to be followed.

A volume of several hundred pages, the NEC is a very detailed statement of how electrical systems should be connected, protected, and designed. Licensed electricians are required to be generally familiar with its provisions and familiar in detail with those provisions that cover the area of their own work. Engineers, unless engaged in the design of specific systems and equipment, leave a great deal of this detail to electricians. The wise manager or professional will, however, be generally aware of the codes and laws governing his operation. He or she cannot escape the responsibility for seeing that they are complied with in every detail.

In the space available it is not possible to discuss very much of the NEC, but the following remarks will give its flavor. The code is heavily concerned with fire prevention, as its origin would suggest. So-called *ampacities*, the safe-current carrying capacities of wire classed by type and size, by insulation material, and by environment are specified, chosen to prevent overheating. One of the most common immediate causes of electrical fires is poor splices and connections. Means of splicing and connecting are covered in detail. Since motor overheating is also a common cause of fire, as is the effect of motor current requirements on the rest of the system, the use and specification of motors is detailed.

The code is particularly concerned with grounding requirements, to minimize the voltages to which humans are exposed. Minimum working clearances are specified. Similarly protection of circuits in the sense of automatic interrupting devices (fuses and circuit breakers) is provided for. The code provides for isolating dangerous circuits from the public by, among other means, elevation to minimum clearances.

The NEC has a section (710) addressed to use of circuits "Over 600 Volts, Nominal." But the code is really designed for *users* of electricity and therefore doesn't pretend to "cover" high-voltage systems of *producers*. For example, electrical utility systems or communications utilities are not covered. This point is often misunderstood. It is not that the basic safety concepts provided for in the NEC are unimportant at higher voltages or in complex power or communication systems. It is simply that the NEC is not designed to consider the *additional* safety complexities of such systems.

Since it is concerned with systems and their interconnections for safety of life and property, the code does not address the internal design or construction of devices themselves. It simply specifies that equipment connected should be "suitable for installation and use in conformity with the provisions of this Code. Suitability of equipment may be evidenced by listing or labelling" [Article 110–3(a)(1)]. The article then goes on to discuss general elements of suitability such as heating, rating, abnormal operation, and so on. The next section describes other provisions for equipment or device rating and listing.

21.6   OTHER CODES AND LAWS

While the NEC is the key to electrical safety practices throughout the United States, there are other codes bearing on electrical matters. Many of these can be found in lists of publications by ANSI and IEEE. Most are inspired by, taken from, or bear very close relation to the NEC. Many are quite specialized, as for example the ANSI Crane Code.

The NESC (National Electrical Safety Code—ANSI C2) sets forth comprehensive safety requirements for public utilities and higher-voltage installations. Originated in 1913 by the National Bureau of Standards, it is now sponsored, maintained, and published by the IEEE. This work is much used in state regulatory bodies as a safety standard. Parts of it can be found copied verbatim into state laws or regulations.

The Underwriters Laboratory (UL), founded in 1894, is chartered as an independent not-for-profit organization testing for public safety. It maintains and operates laboratories for the examination and testing of devices, systems, and materials to determine their relation to life, fire, casualty hazards, and crime prevention. Most consumer devices (for example, extension cords and toasters) have the familiar UL marking or sticker on them indicating that they have been tested by the UL and found to comply with UL safety requirements. There are other testing groups. In a sense these equipment tests complement the NEC wiring and user distribution system requirements.

As noted above, the NEC and similar codes and listings have, in themselves, no enforceable legal status. But in most states and some other jurisdictions it has been felt necessary to protect the public from poor and unsafe electrical installations. Consequently throughout most of the United States, jurisdictions such as cities, counties, and states have enacted laws stating that the NEC (or other codes) are binding in their territories. Usually strict compliance with the NEC is adequate. But in some cases jurisdictions have modified the requirements (as for example, "where they do not conflict with other statutes") to make them either stronger or weaker. Occasional situations are found where a governmental agency publishes its own regulations, copied almost word for word from the NEC, the NESC, or other ANSI codes. But now and then some important number is changed, with no particular warning.

The federal government has recently entered this area through OSHA, adopting wordings wholesale from other codes. This procedure is, of course, administratively economical and tends in at least some degree to minimize added details for those who have been using the standard codes for years. But the wording must be carefully analyzed for unexpected changes.

21.7   SOME PARTICULAR HAZARDS

Especially in dwellings and laboratories, water or dampness is dangerous because it provides a good ground for the feet or other part of the body. The hand, or any other body surface, touching a live electrical part completes the circuit for a shock to ground.

Kitchens and bathrooms are especially dangerous. Codes require that metal surfaces of such devices as electric stoves or washing machines be grounded so that if one side of the electrical circuit should short out on the case, the circuit will be opened by its protective devices. But if this ground comes off or is disconnected (the appliance will work satisfactorily without the ground connection) it becomes especially dangerous for a person grounded by dampness. Householders should check periodically to see that these grounds are intact. The same problem can occur with such tools as electric drills which should be grounded through their three-wire power cables. For this reason most modern drills are made with an insulating plastic case. The chuck is still dangerous.

Pole climbing by girls and boys continues to be a surprisingly frequent cause of deaths and injury and one that is difficult to guard against. Children need to be instructed and reinstructed from an early age about the dangers of poles and other electrical structures. Utilities and other industrial companies that allow their outside plant to become overgrown or run-down invite such problems.

Crane/power-line accidents, in the nature of the introductory sketch of this chapter, are sadly common. Hoist cables or booms contacting uninsulated power lines kill or permanently maim approximately 500 workers a year. The problem seems a simple one and easily attacked by properly training crane operators. But in view of the limited attention span of humans involved in other problems, such training is not enough in itself. Every engineer or other professional involved in these operations must take the initiative to see that appropriate precautions are taken in every case.

Construction sites, indoor or outdoor, are especially hazardous. Often it is not economically feasible to make standard safety provisions on a temporary basis. People are involved in construction who are not part of the regular organization. Many ad hoc physical and organizational arrangements have to be made. Responsibility slips between organizational cracks. But effective engineers are men and women who take charge of what is going on around them and see that it goes right—especially where safety is involved.

Laboratory work also tends to be somewhat ad hoc. In school laboratories where students use voltages over 50 V, specific provisions for safety need to be made. Fortunately with engineering students this in itself is a good opportunity for instruction. Students are first taught simple safety measures including respiration and resuscitation. Work with voltages over 50 V or so is seldom permitted by one person alone. A group member is usually designated to be in charge of safety precautions.

21.8   SUMMARY

Electrical safety is everyone's business, but it is a particular responsibility of professionals and managers, who carry great moral and legal responsibility for accidents related to engineering activities and to the results of engineering work.

Because of the thousand or so deaths each year in the United States due to electrical accidents and millions of dollars in property damage, employers do not look lightly on poor accident performance by their professionals.

The human body can safely withstand very little voltage and at most only a few milliamperes of current. Even nondamaging electrical contact can induce secondary accidents such as falling off ladders. Short circuits and arcs must be prevented from injuring persons by molten metal splash or radiation.

Extensive codes of safe practice, particularly the NEC, exist for the guidance of engineers and others in the design and operation of electrical systems and products. Some combination of these codes is legally mandated in every jurisdiction.

Practicing engineers are alert to recognize potentially hazardous electrical situations or activities—for example, dampness, out-of-routine activity, laboratory work, and many others. When something is wrong around him the engineer takes charge and straightens it out before accident or other trouble occurs.

FOR FURTHER STUDY

Wilford I. Summers, Ed., *The National Electrical Code Handbook*, current edition, National Fire Protection Association, Boston. This publication includes the entire NEC plus detailed comments on many parts of it.

National Fire Protection Association, *National Electrical Code*, ANSI C-1, current edition, NFPA, Boston.

Institute of Electrical and Electronic Engineers (IEEE), *National Electrical Safety Code*, ANSI C2, current edition, IEEE, New York.

Study Questions

1. Why is electrical safety important to every engineer? How might it have an adverse effect on his or her career?

2. Very roughly, how many deaths a year in the United States are attributed to electrical accidents?

3. About what is the annual dollar loss in the United States from electrically started fires?

4. List five kinds of dangers to persons (from electricity or electrical machinery) that must be guarded against.

5. Why is contact with electricity dangerous physiologically to the human body?

6. What in your opinion should be the minimum separation of electrical emergency equipment from the regular equipment it is intended to replace?

7. What is the difference in purpose of the NEC and the NESC?

8. What does the blue "UL" marking on appliances mean?

9. By what two mechanisms do the provisions of the non-governmentally produced NEC and NESC become legally binding?

10. Through what agency has the federal government entered the electrical safety field?

11. Why is a pole line especially dangerous to the public?

12. Under what circumstances are moving cranes an electrical hazard?

13. Explain why dampness or water makes electrical systems extra hazardous.

14. Why is it necessary to take special safety precautions at construction sites?

15. What safety steps would you recommend be taken by an electrical engineering teacher responsible for a laboratory? Answer for either
    (a) an electronics lab that never uses any potentials higher than 24 V; or
    (b) a motors lab that uses voltages up to 220 V.

16. Although the matter is in some dispute, 50 V and 5 mA are sometimes accepted as standard tolerances for the human body. What assumption does this make about bodily resistance? How does this assumption fit with measured values?

APPENDIX **A**

Determinants

Circuit analysis (whether dc or ac) not infrequently results in sets of simultaneous equations. Except in the simplest cases, these are best solved by *matrix* (or *determinant*) methods. This appendix is provided to refresh readers on the algebraic procedure.

Matrix methods are advantageous because they are systematic, essentially "no-thought" procedures. (Engineers and other professionals conserve their ideational energy for more important problems.) Furthermore, systems of simultaneous equations become very unwieldy when more than three variables are involved, and nearly impossible for more than four. Even second-order systems are difficult with complex numbers. Yet systems of any size can be solved with computer methods, and these computer procedures are all approached from the standard equation format learned in the matrix method.

A.1   EVALUATING DETERMINANTS

Figure A–1a is a *matrix*, an array of numbers in a specific pattern. Figure A–1b shows a *determinant* made up of the same numerical matrix as Figure A–1a, but this time enclosed in two vertical lines. The lines enclosing the matrix mean: "take the

$$
\begin{array}{ccc} 3 & 2 & 4 \\ 1 & 5 & 7 \\ 9 & 8 & 0 \end{array} \qquad
\begin{vmatrix} 3 & 2 & 4 \\ 1 & 5 & 7 \\ 9 & 8 & 0 \end{vmatrix} \qquad
\begin{vmatrix} r & s \\ t & u \end{vmatrix} \Rightarrow ru - ts \qquad
\begin{vmatrix} 2 & 3 \\ 4 & 5 \end{vmatrix} = -2 \qquad
\begin{vmatrix} 2 & -3 \\ 4 & 5 \end{vmatrix} = 22
$$

$$\quad\;\; (a) \qquad\qquad (b) \qquad\qquad\;\; (c) \qquad\qquad\;\; (d) \qquad\qquad\;\; (e)$$

FIGURE A–1 A matrix and several determinants. The second-order determinant (c) is evaluated as shown. (d, e) carry out this same evaluation numerically.

determinant of the matrix." The word *determinant* refers either to a single number derived from the 3×3 matrix (as described below) or to the combination of matrix and two vertical lines. In this latter sense determinants may be of any size from 2×2 up. Figures A–1c through A–1e show second-order determinants (both numerical and algebraic) that have been evaluated to extract the single number they represent (or single algebraic expression).

A second-order determinant is *evaluated* as the product on the descending diagonal minus the product on the ascending one. A Δ is commonly used to represent a determinant. Thus for Figure A–1c $\Delta = ru - ts$, yielding a single number. Using this scheme, the reader can verify that determinants in Figures A–1d and e would evaluate as -2 and $+22$, respectively.

Figure A–2 illustrates a good method for finding the determinant of a third-order matrix: the matrix in a is resolved into three terms as in b. Each of the terms is derived from an element in the top row of the original matrix with the element's *minor* second-order determinant. Minors are found by striking out the row and column of the element (*a, b,* or *c* in this case) as shown in Figures 2d and e. Thus the third-order determinant is evaluated (in this case) as the sum of the three top row numbers (*with signs alternated*) times their respective minors. Completing this work by evaluating the minors we have the result shown in Figure A–2c:

$$\Delta = aei - ahf - bdi + bgf + cdh - cge. \tag{A.1}$$

$$
\begin{vmatrix} a & b & c \\ d & e & f \\ g & h & i \end{vmatrix} = a \begin{vmatrix} e & f \\ h & i \end{vmatrix} - b \begin{vmatrix} d & f \\ g & i \end{vmatrix} + c \begin{vmatrix} d & e \\ g & h \end{vmatrix} \qquad (b)
$$

$$
= aei - ahf - bdi + bgf + cdh - cge \qquad (c)
$$

$$(a)$$

$$
\begin{array}{ccc} \cancel{a} & \cancel{b} & \cancel{c} \\ d & e & f \\ g & h & i \end{array} \qquad
\begin{array}{ccc} \cancel{a} & b & \cancel{c} \\ d & e & f \\ g & h & i \end{array} \qquad
\begin{vmatrix} + & - & + \\ - & + & - \\ + & - & + \end{vmatrix}
$$

$$\qquad (d) \qquad\qquad (e) \qquad\qquad (f)$$

FIGURE A–2 Third-order (and higher) determinants can be evaluated by "minors" as in (b) with the result (c). (d) and (e) show how the minors of the first two terms in (b) are derived. This method can be carried out on any row or column of the original determinant with the same result. Signs will alternate as shown in (b) and (f).

However, instead of using the top row, *a, b, c,* these third-order arrays may be evaluated on any row or column. In evaluation, the sign of the upper left-hand element is $+$. The sign of any other elements for evaluation may be determined by alternation (as shown in 2f) while moving horizontally or vertically from the upper left-hand corner of the matrix. It is suggested that the reader evaluate this determinant on several other rows or columns to show that the result will be the same as Equation (A.1).

This method of evaluation by minors can be used also with higher-order determinants, although it becomes tedious.

A.2 STANDARD CIRCUIT EQUATIONS

We use the dc example of Figure A–3a for simplicity. A three-mesh circuit yields three simultaneous equations (Figure A–3b). (For methods of generating these equations from the circuit see Chapter 3.)

For generality in this discussion, we use the algebraic form of Figure A–1c, where *X, Y, Z* are the three unknown currents (the *I*'s in equations 3b) and the *a, b, c, d* coefficients are determined from the circuit in accordance with the methods

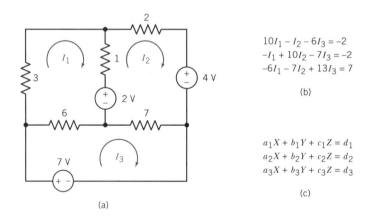

$$10I_1 - I_2 - 6I_3 = -2$$
$$-I_1 + 10I_2 - 7I_3 = -2$$
$$-6I_1 - 7I_2 + 13I_3 = 7$$

(b)

$$a_1 X + b_1 Y + c_1 Z = d_1$$
$$a_2 X + b_2 Y + c_2 Z = d_2$$
$$a_3 X + b_3 Y + c_3 Z = d_3$$

(c)

(a)

$$\Delta = \begin{vmatrix} a_1 & b_1 & c_1 \\ a_2 & b_2 & c_2 \\ a_3 & b_3 & c_3 \end{vmatrix}, \ \Delta_1 = \begin{vmatrix} d_1 & b_1 & c_1 \\ d_2 & b_2 & c_2 \\ d_3 & b_3 & c_3 \end{vmatrix}, \ \Delta_2 = \begin{vmatrix} a_1 & d_1 & c_1 \\ a_2 & d_2 & c_2 \\ a_3 & d_3 & c_3 \end{vmatrix}, \ \Delta_3 = \begin{vmatrix} a_1 & b_1 & d_1 \\ a_2 & b_2 & d_2 \\ a_3 & b_3 & d_3 \end{vmatrix}$$

(d)

$$X + \Delta_1/\Delta; \ Y = \Delta_2/\Delta; \ Z = \Delta_3/\Delta$$

(e)

FIGURE A–3 Equations in standard form (b) are written for each mesh of the three-mesh circuit (a). (c) gives the general form. The coefficients of (c) yield four determinants (d), which can be solved for the three unknowns as in (e).

of Chapters 3 and 4. The unknowns are always written in the same order in all equations. The unknowns (X, Y, Z) and the coefficients (a, b, c, d with their subscripts) may be either real or complex numbers. In general there will be n equations for a circuit with n meshes. It is also possible to have these equations in terms of voltage unknowns, as for example with nodal analysis.

A.3   SOLVING EQUATIONS

The procedure for solving these three equations manually with determinants, to find the three variables, is shown in Figures A–3d and e, and is set out in detail in the four steps below.

1. Write the equations in the systematic form of Figure A–3b—with unknowns in standard order on the left and constant terms on the right. Then, neglecting the unknowns, we have an array made up of the coefficients:

$$a_1 \quad b_1 \quad c_1 = d_1$$
$$a_2 \quad b_2 \quad c_2 = d_2$$
$$a_3 \quad b_3 \quad c_3 = d_3$$

2. As shown in Figure A–3d, form the four *determinants*, Δ, Δ_1, Δ_2, Δ_3 whose elements are assembled from the coefficients. Each of these four matrices is a 3×3 *array* of 9 numbers. Note that if there had been four equations there would then be five matrices each consisting of a 4×4 array of 16 numbers, and so on. It is easy to see the scheme of their generation. The determinant Δ is derived from the array of coefficients on the left-hand side of the equations, when they are written in standard form. The other determinants with subscripts 1, 2, 3 are the same except that one column of coefficients is replaced by the "d" column from the right-hand side. For instance in the matrix that yields Δ_1, the first column is replaced; in the Δ_2 the second is replaced, and so on.

3. Evaluate the determinant of each of these matrices (or as many as are needed to find the circuit information desired). Evaluation is carried out as shown in Section A.1 above. This will yield four specific numbers, Δ, Δ_1, Δ_2, Δ_3.

4. Solve for X, Y, and Z as follows:

$$X = \Delta_1/\Delta,$$
$$Y = \Delta_2/\Delta, \qquad\qquad (A.2)$$
$$Z = \Delta_3/\Delta.$$

These methods work perfectly well also with complex coefficients, yielding complex answers for the variables. The calculations must be made in complex algebra, as discussed in Appendix B, and as used in the ac chapters of the text.

This same method will work for higher-order systems. [Some sophisticated hand calculators have the capability of solving simultaneous equations automatically. For a third-order system, the 12 coefficients (a, b, c, d, with their subscripts) are entered in order. Software is available for many small computers to do this also.]

As a drill it is suggested the reader complete the solution of the circuit in Figure A–3. The intermediate determinants will be found to be $\Delta = 353$, $\Delta_1 = 197$, $\Delta_2 = 234$, $\Delta_3 = 407$, whence $I_1 = 0.558$ A, $I_2 = 0.663$ A, $I_3 = 1.153$ A.

APPENDIX B

Complex Numbers

Steady-state alternating-current circuits are most easily described and calculated with complex numbers. These special numbers represent voltages, currents, impedances, and complex power. The present appendix is provided to assist readers in reviewing this area of mathematics. It will also be a convenient source for formulas and procedures. Most readers will have had some instruction on complex numbers in earlier scholastic algebra courses. But there is enough material here for those new to the subject to learn what they need to know to perform basic ac calculations.

Those who have not already done so should first familiarize themselves with the early part of Chapter 4 (before Section 4.1), where a simple ac circuit is solved by this method. The usefulness of the following material will then be apparent. Helpful problems for drill material will be found at the ends of Chapters 4 and 5.

B.1   COMPLEX NUMBERS

$(3 + j4)$ and $(5 - j5)$ are examples of *complex numbers*. A complex number is simply an ordered pair of real numbers. The symbol j, called the *complex operator*, is placed before the second of the pair to distinguish it from the first. There are certain special algebraic rules about how to add, multiply, subtract, divide, and exponentiate these numbers as set forth below. (The parentheses are used simply for clarity here, and are not part of the numbers.)

Let us call the first complex number **U**, and the second **V**. Then

$$\mathbf{U} = 3 + j4, \qquad \mathbf{V} = 5 - j5.$$

Note that the complex numbers **U** and **V** are set out here in boldface type to distinguish them from simple real variables. We will use boldface notation consistently throughout this text to avoid confusion. But in ordinary engineering problem solving, complex quantities can be easily distinguished by the context in which they are used. Working through the examples and problems of Chapters 4 and 5 the reader will quickly develop the ability to separate complex and simple real quantities with no need for special notation.

U and **V** can also be displayed graphically, on *complex axes*, as shown in Figure B–1. The *real axis* (horizontal) is designated by the script $\mathcal{R}$ symbol; the *j-axis* (vertical) by the small *j*. The numbers **U** and **V** are the solid arrow lines connecting the origin with the cartesian points (3, 4) and (5, −5); they can also be thought of as the points at the ends of the arrows with their two cartesian numbers.*

The designation of **U** and **V** as 3 + *j*4 and 5 − *j*5 is called *rectangular form*. From the sketches of Figure B–1 it is clear that either number could also be designated in *polar form*, by giving the arrow length and the value of the angle *θ*. From the geometry, $U = \sqrt{3^2 + 4^2}$, and so in polar form **U** = 5 at an angle of 53.1°. The two parts of the polar form are often called *magnitude* and *angle*. Similarly, in polar form, **V** = 7.07 at an angle of −45°. Particularly in the electrical use of

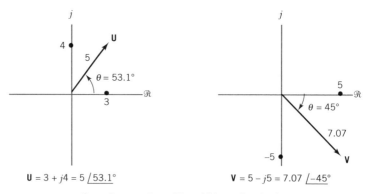

$$\mathbf{U} = 3 + j4 = 5\,\underline{/53.1°} \qquad\qquad \mathbf{V} = 5 - j5 = 7.07\,\underline{/-45°}$$

FIGURE B–1 Complex numbers **U** and **V** can be displayed on complex axes or written in rectangular or polar form.

*Many texts use the letter *i* (instead of *j*) for the complex operator. In this text we prefer to reserve the symbol *i* for instantaneous currents. Scholastic texts usually categorize the first member of a complex pair as the *real number* and the second as the *imaginary*. They also refer to the *imaginary axis*. To avoid confusion we will use the terms *real number*, *j number*, and *j axis*. (As is frequently observed in circuit analysis, the *j* number is just as "real"—in the English language sense—as the real number.)

complex numbers, the notations $5 \underline{/53.1°}$ is often used for the polar form, and will be the usual notation in this text. In spoken form it is called, "5 at an angle of 53.1 degrees amperes" or volts of whatever quantity it represents.* Note also that the *magnitude* of the polar form, also designated U above (in the examples it is 5 or 7.70), is not itself a complex number; nor is the angle (37° or 45°). The *combination* of the two is complex.

Thus a complex number can be displayed in any of three ways: (i) in rectangular form, (ii) in polar form, or (iii) graphically on complex axes.

B.2   COMPLEX ARITHMETIC OPERATIONS

To generalize on all this, let

$$\mathbf{X} = A + jB = X\underline{/\theta_x}$$

and

$$\mathbf{Y} = P + jQ = Y\underline{/\theta_y},$$

$$(B.1)$$

where the left-hand $\mathbf{X}$ and $\mathbf{Y}$ are complex numbers themselves (each has two parts) and the X and Y on the right-hand sides of the equations are the polar magnitudes of complex $\mathbf{X}$ and complex $\mathbf{Y}$. The respective θ's are the angles associated with each complex number. The reader will quickly find this double use of notation natural and useful and not need to use separate designators here as we do with the boldface symbols.

In circuit analysis it is frequently necessary to add, subtract, multiply, and divide complex numbers. This is done as follows:

$$\mathbf{X} + \mathbf{Y} = (A + P) + j(B + Q),$$

$$\mathbf{X} - \mathbf{Y} = (A - P) + j(B - Q),$$

$$\mathbf{X} * \mathbf{Y} = (X * Y)\underline{/\theta_x + \theta_y}$$

$$\mathbf{X}/\mathbf{Y} = (X/Y)\underline{/\theta_x - \theta_y}$$

$$(B.2)$$

For exponentiation, suppose it is desired to raise X to the 1.8 power:

$$\mathbf{X}^{1.8} = (X^{1.8}\underline{/1.8 * \theta_x}).$$

$$(B.3)$$

As suggested in these formulas, addition and subtraction are carried out in the rectangular form; multiplication, division, and exponentiation most easily in the polar form. But it is possible to multiply in the rectangular form by remembering that j^2 is -1.

---

*A more mathematically sophisticated notation, using the concept of a complex exponential, is introduced in Section 6.3 for those readers wishing to pursue this topic further.

The reader can verify that for the complex numbers **U** and **V** (Section B.1) the results of addition and multiplication are $8 - j1$ or $8.06\,\underline{/-7.1°}$, and $35.4\,\underline{/8.1°}$ or $35 + j5.0$.

When the rules for arithmetic are consistently applied to complex numbers, the complex operator j appears to take on a character of its own: $j^2 = -1$, and $1/j = -j$. And as a coefficient, j has the effect of "rotating" the complex quantity it multiplies 90° counterclockwise; j is itself a complex number with polar form $1\,\underline{/90°}$ and rectangular form $0 + j1$. The reader can appreciate these points by experimenting with the three forms of representation.

In circuit work, it is often necessary to convert polar into rectangular numbers and vice versa. Using the complex number **X** from Equation (B.1):

$$A = X \cos \theta_x,$$
$$B = X \sin \theta_x,$$
$$X = \sqrt{A^2 + B^2}, \tag{B.4}$$

and

$$\theta_x = \arctan(B/A),$$

where again X is the polar magnitude of **X**, and X, θ, A, and B are in themselves all real numbers. Most engineering calculators will make these conversions automatically.

DRILL PROBLEM Two complex numbers **R** and **S** are to be added, subtracted, multiplied, divided, and the first exponentiated by 3.5: **R** $= 1 - j2$, **S** $= 3 + j6$.

Find: the five answers in three forms; rectangular, polar, and sketched on complex axes; answers are left to the reader.

Engineers handling problems with complex numbers find the sketch on complex axes particularly useful to protect against mistakes in calculating. It is usually worth sketching numbers out this way so that they can be easily visualized and their approximate components estimated. The experienced engineer may often omit the sketch on paper because he has this graphic representation clearly laid out in his thinking. It is useful when learning this material to make the sketches most of the time.

Examples of complex number calculation appear in Chapters 4 and 5 and throughout the rest of the book.

APPENDIX C

Engineering Economy

C.1 THE NATURE AND IMPORTANCE OF ENGINEERING ECONOMY CALCULATIONS

Suppose an engineer is faced with the problem of selecting and recommending a motor for a certain drive application. His or her solution will depend on the motors available, their characteristics, and the characteristics and requirements of the driven machinery. However, no less important a consideration is the *cost* of these machines both to procure and install and also to operate.

The importance of dollar costs in deciding which motor to use can be looked at from a competitive standpoint. Whether the design objective is some product to be produced and sold, or whether it is a production station in a plant, or a structure, or whatever, if the engineer usually designs or selects in a needlessly expensive way, then competitors will replace him in that product marketplace—and eventually in the marketplace for his own services.

Perhaps more fundamentally, the importance of dollar costs can also be looked at by recognizing that money, while unimportant in itself, represents and commands goods and services of all kinds, and that *minimizing* dollars expended for some purposes is really therefore minimizing the use of U.S. resources to accomplish the task at hand. We all have more if we use resources efficiently.

The engineer's *decision making* about the motor will involve, then (i) determining exactly what is needed technically, (ii) what machines (considering both technical characteristics and costs) are available for consideration, and then (iii) comparing the best possibilities to select a particularly good one.

Thus the expert practicing engineer in just about every phase of the work is consciously developing two or more *alternatives* which he must compare. The dollar costs of the alternatives is the basis for this comparison. Dollar costs are the only simple measures by which all elements of diverse alternatives can be compared.

Engineering economy is the thought procedure by which dollar considerations are factored into engineering decision making. These decisions may involve, for example, how something should be designed, by what processes it should be produced or constructed, whether or when a machine or process should be replaced by a better one, what mixture of ingredients is best suited for some purpose, whether or when some particular work should be undertaken, and so on.

Texts on engineering economy will develop the above ideas far more fully. But before leaving these points we should observe that successful engineering depends primarily on the engineer's skill in *developing the right alternatives* to consider in the first place. He or she is particularly concerned not to miss any unusually good possibilities. In most cases, engineering economy comparison of the alternatives, once these are chosen, is largely mechanical. It is assumed that all engineers can carry out these comparisons as easily as they use the multiplication table.

C.2   PRESENT ECONOMY

In comparing alternatives it is frequently unnecessary to consider costs that are the same for both. Suppose the engineer's choice comes down to two motors in the example above. If motors A and B have the same life expectancy and the same power and maintenance costs, and will accomplish the same task equally well, and either will be paid for in a lump sum when purchased, the decision comes down to simply comparing first costs to procure and install. There is no time element in the decision making.

Engineering economy problems in which the time element need not be considered are called *present economy*. Usually, if all funds involved are to be expended over a period of time shorter than one year, interest and time differences are neglected. The decision is made on a simple present-economy comparison of all costs.

C.3   THE TIME VALUE OF MONEY

In much engineering work, however, there are significant time differences between alternatives. As a very simple example of this suppose that the engineer selecting motors comes up with two alternatives X and Y, either of which will serve equally well. For motor X he pays $1000 now, or for motor Y he pays $600 now and $600 more two years from now. Simply adding the sums expended would say that X costs $1000 and Y $1200. But it would be evidently unfair to directly compare the second $600 to either the first $600 payment or to the $1000 payment. For the $400 saved initially in alternative Y could be earning interest (or doing other work) for the first two years.

To make a proper comparison between X and Y we move the second $600 payment ahead two years to the present time. To do this it must be multiplied by a P/F factor (to be explained in detail below). Let us assume that a proper rate of interest to use in this case is 25%. We will see below that the P/F factor for 2 years and 25% is 0.640. Thus bringing the second $600 payment back to the present adjusts it to $600 \times 0.640 = $384. So the "present worth" of alternative-Y costs is $600 + $384 = $984 and it is less expensive than the $1000 alternative. The reasonableness of the P/F factor can be seen from this: $384 at 25% will be worth 1.25 $\times$ 384 = $480 in one year. Similarly $480 $\times$ 1.25 for a second year will be $600. The interest rate i used here is often referred to as the MARR (minimum acceptable rate of return).

To consider a slightly more interesting example, suppose the engineer we left selecting motors above comes up with two good alternatives, A and B. Each is expected to perform the needed task equally well. Motor A costs $1200 installed, and is expected to use $950 worth of electricity per annum. Its maintenance requirements are negligible. Motor B costs only $650 installed, but will use $1023 for electricity and require $100 maintenance work per year. The expected useful lives for A and B are respectively 12 and 8 years. Assume further that each alternative has a salvage value at life's end of 10% of its installed cost.

Figure C–1 shows the relations between alternatives graphically in two *cash flow diagrams*. The task is to compare these two series of expenditures, these two cash flow charts. But how can the dollar amounts at different points in time be added or compared?

Again, we could simply add the amounts in the A chart (subtracting the $120 salvage, a negative cost), and compare with the same procedure in the B alternative. But this would be incorrect since, for example, $950 to be expended 12 years from now is not worth nearly as much today as $950 required only one year from now.

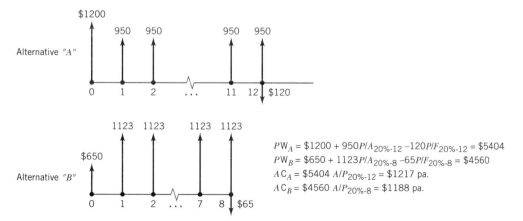

$$PW_A = \$1200 + 950P/A_{20\%\text{-}12} - 120P/F_{20\%\text{-}12} = \$5404$$
$$PW_B = \$650 + 1123P/A_{20\%\text{-}8} - 65P/F_{20\%\text{-}8} = \$4560$$
$$AC_A = \$5404 \ A/P_{20\%\text{-}12} = \$1217 \text{ pa.}$$
$$AC_B = \$4560 \ A/P_{20\%\text{-}8} = \$1188 \text{ pa.}$$

FIGURE C–1 *Cash Flow* diagrams show clearly the "streams of costs" for alternatives A and B. Comparing equivalent "present worths" (PW) is invalid because of their different lives. Instead, PW's are converted into AC's—equivalent "uniform annual costs."

Or looked at another way, if we set aside $950 today for use 12 years from now, we would have 12 years of interest income on it (representing the goods or services it could provide during that time).

To overcome this problem the engineer first decides at what point on the chart (what point in time) he wishes to make the comparison. Let us make another *present worth* (PW) comparison by moving all sums to the present time (marked 0 on the chart for the "end of the zeroth year"). We assume 20% as an appropriate interest rate (these rates are always per annum unless otherwise specified). Sums are moved about in time to the zeroth year with two "factors" (P/A and P/F) as shown in the following equations and explained below:

$$PW_A = 1200 + 950P/A_{(20\% - 12)} - 120P/F_{(20\% - 12)},$$

$$PW_A = 1200 + 950 \times 4.439 - 120 \times 0.1122 = \$5404.$$

Similarly $PW_B = \$4560$.

But it is unfair to compare these two PW numbers since the first provides a motor for 12 years and the second for only 8. Thus we are not comparing equivalent alternatives. There are various ways to overcome this difficulty. We will calculate a *uniform annualized cost* (AC) by converting the two numbers found as follows:

$$AC_A = \$5405 \times A/P_{(20\% - 12)} = 5404 \times 0.2253 = \$1217,$$

$$AC_B = \$4560 \times A/P_{(20\% - 8)} = 4560 \times 0.2606 = \$1188.$$

(This procedure inherently assumes that for the next four years alternative B if chosen can be repeated. We will neglect this point here.) Alternative B saves $29 a year over A. Thus alternative B is the more economic and would presumably be chosen, unless there are other factors to be considered not yet contained in the analysis.

C.4   TIME-VALUE-OF-MONEY FACTORS

To explain the above example and generalize on it we introduce the notation of Figure C–2. P is a present sum, F a future sum, i the appropriate interest rate for the periods considered (usually per annum), and n the number of periods (usually years). The factor $F/P_{(i-n)}$ will convert a present sum P forward over n years to an *equivalent* future sum F at rate of interest i. [Note that the parenthetical designator $(i - n)$ on the factor F/P is not algebraic but is meant only to designate for what interest rate and over how many years the factor is to be calculated.]

Using a 10% interest rate and two years, for example, $100 would be worth $110 in one year and $121 in two years and so on. Thus mathematically $F/P_{(i-n)} = (1 + i)^n$. Inverting, $P/F_{(i-n)} = 1/(1 + i)^n$, and will move sums back in time on the cash flow chart, as illustrated in the previous example.

Continuing the notation in Figure C–2, A is one of a series of end-of-period equal payments. It is customary to collect all payments and receipts in any one year as if they were made at the end of the year. The factor $P/A_{(i-n)}$ will convert n of

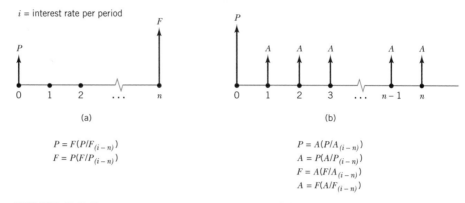

$$P = F(P/F_{(i-n)})$$
$$F = P(F/P_{(i-n)})$$

$$P = A(P/A_{(i-n)})$$
$$A = P(A/P_{(i-n)})$$
$$F = A(F/A_{(i-n)})$$
$$A = F(A/F_{(i-n)})$$

FIGURE C–2 Engineering economy notation is straightforward. *P* is a present amount, *F* a future amount, and *A* one of a series of uniform end-of-period (end-of-year) payments. The interest rate is *i* per period, and the number of periods (usually years) is *n*. If the diagram represents costs, costs are positive (up) and income negative (down).

these uniform payments to a single present worth *P* at the year ending before the first payment is made. It can be shown* that, for a given *i* and *n*,

$$F/P = (1 + i)^n,$$
$$P/A = [(1 + i)^n - 1]/[i(1 + i)^n] \qquad (C.1)$$
$$F/A = [(1 + i)^n - 1]/i.$$

These three formulas can be easily programmed on a hand calculator with memory to yield the six principal time-value-of-money factors (three by inversion).

Until recently it has been customary to look up these factors in tables like Figure C–3. Note that this table has been constructed for 20% interest only and for 1 to 20 years or periods. Complete tables, available in any text, have a page for each value of interest and run up to 50 years or more. Computer programming, once in, is much faster than table use, induces fewer errors, and avoids any need for interpolation between years and interest rates.

C.5   FACTOR USE

Readers wishing to program the three equations (C.1) on their calculators can check the programming work with Figure C–3. On some small computers it is necessary to recognize that $A^B = EXP_{10}(B * LOG(A))$.

The motor selection example in Section C.3 above can be used to check understanding of factor use. Some other examples will be found scattered throughout the text.

---

*An excellent and unusually clear text is Donald G. Newnan, *Engineering Economic Analysis* (current edition), Engineering Press, San Jose. There are many others.

N	F/P	P/F	F/A	A/F	A/P	P/A	N
1	1.200	0.8333	1.000	1.000	1.200	0.8333	1
2	1.440	0.6944	2.200	0.4545	0.6545	1.528	2
3	1.728	0.5787	3.640	0.2747	0.4747	2.106	3
4	2.074	0.4823	5.368	0.1863	0.3863	2.589	4
5	2.488	0.4019	7.442	0.1344	0.3344	2.991	5
6	2.986	0.3349	9.930	0.1007	0.3007	3.326	6
7	3.583	0.2791	12.92	0.0774	0.2774	3.605	7
8	4.300	0.2326	16.50	0.0606	0.2606	3.837	8
9	5.160	0.1938	20.80	0.0481	0.2481	4.031	9
10	6.192	0.1615	25.96	0.0385	0.2385	4.192	10
11	7.430	0.1346	32.15	0.0311	0.2311	4.327	11
12	8.916	0.1122	39.58	0.0253	0.2253	4.439	12
13	10.70	0.0935	48.50	0.0206	0.2206	4.533	13
14	12.84	0.0779	59.20	0.0169	0.2169	4.611	14
15	15.41	0.0649	72.04	0.0139	0.2139	4.675	15
16	18.49	0.0541	87.44	0.0114	0.2114	4.730	16
17	22.19	0.0451	105.9	0.0094	0.2094	4.775	17
18	26.62	0.0376	128.1	0.0078	0.2078	4.812	18
19	31.95	0.0313	154.7	0.0065	0.2065	4.843	19
20	38.34	0.0261	186.7	0.0054	0.2054	4.870	20

FIGURE C–3 Factor table for 20% interest.

There are many ramifications to this subject, but if the costs of alternatives can be displayed as cash flow lines (as in Figure C–1) any problem can be solved. It is important that no costs (or incomes) be omitted unless equal for all alternatives.

Examples of engineering economy throughout this book are all calculated "before taxes" (without regard to the effect of corporate income tax). This is a common procedure in many simple analyses.

C.6   GOING FURTHER

Texts in engineering economy extend the above elementary ideas in several directions; for example, how to select a proper rate of interest (one approach is to use the company's cost of capital), alternatives with continuous variation between them, rate-of-return analysis, payback period, the effect of taxes and depreciation methods, replacement analysis, and benefit-cost ratio analysis—used particularly for public works.

But in general, if all the items of cash income and outgo are faithfully entered on the diagram, the solution of engineering economy problems remains the same—simply compare the two (or more) charts on the basis of AC or PW. In considering taxes the "tax savings" created by expenditures, and "tax penalties" of saving must be considered in their effect on cash flow. Also, the IRS does not allow the cost of machines to be "expensed" (entered as a cost) in the year they are purchased, but appropriate "depreciation" can be charged for each year of their lives.

APPENDIX D

Laplace Transforms

The Laplace transform is a way of representing a nonperiodic time function by the amplitudes of the many frequency components composing it. These are, in general, complex frequencies $s = j\omega + \sigma$ that are continuous in ω. We will start by representing a periodic time function by a series of sinusoids and let the period become infinite (the function becomes nonperiodic).

D.1   FOURIER SERIES

A periodic function $a(t)$ with period T_0 can be expressed as a sum of sinusoids:

$$a(t) = A_0 + 2 \sum_{n=1}^{\infty} A_n \cos(2\pi n f_0 t + \theta_n), \tag{D.1}$$

where $f_0 = 1/T_0$ is the fundamental frequency. This sum of terms is called the *Fourier series* for $a(t)$. The *Fourier coefficients* A_n and the angles θ_n can be found as follows:

$$B_n = f_0 \int_0^{T_0} a(t) \sin(2\pi n f_0 t) \, dt, \qquad C_n = f_0 \int_0^{T_0} a(t) \cos(2\pi n f_0 t) \, dt$$

$$A_n = \sqrt{B_n^2 + C_n^2}, \qquad \theta_n = -\tan^{-1}(B_n / C_n).$$

Consider the periodic function

$$a(t) = e^{-3t}, \qquad 0 < t < T_0$$

with $T_0 = 1$ shown in Figure D.1d. The coefficients A_n and the phases θ_n are given in this case by

$$A_n = \left| \frac{f_0}{3 + j2\pi n f_0} \right| (1 - e^{-3T_0}) = \frac{f_0}{\sqrt{9 + (2\pi n f_0)^2}} (1 - e^{-3T_0}),$$

$$\theta_n = \text{ang} \left[\frac{f_0}{3 + j2\pi n f_0} \right] = -\tan^{-1}(2\pi n f_0 / 3).$$

Figures D.1a through D.1c show how the sum of sinusoids approaches the exponential waveform as the number of terms increases.

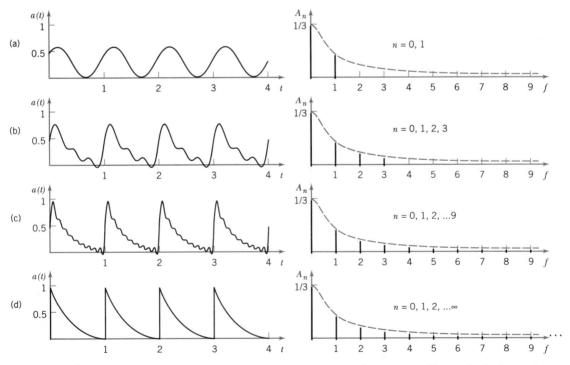

FIGURE D–1 Periodic waveforms $a(t)$ with period T_0 and the corresponding Fourier coefficients A_n. As the number of coefficients increases from (a) to (d), the waveforms approach a periodic exponential e^{-3t}.

D.2 FOURIER TRANSFORM

We would like to be able to express nonperiodic waveforms as the sum of sinusoids. Such waveforms can be thought of as periodic functions with infinite period $T_0 \to \infty$. Then the fundamental goes to an infinitesimal: $f_0 \to df$, $nf_0 \to f$, and

$$A_n \to A(f) = \left| \frac{df}{3 + j2\pi f} \right|.$$

Figure D.2 shows how the amplitudes A_n become a continuous function $A(f)$ as the period increases. The periodic function $a(t)$ approaches an isolated exponential $g(t)$:

$$a(t) \to g(t) = \begin{cases} 0, & t < 0 \\ e^{-3t}, & t \geq 0. \end{cases} \tag{D.2}$$

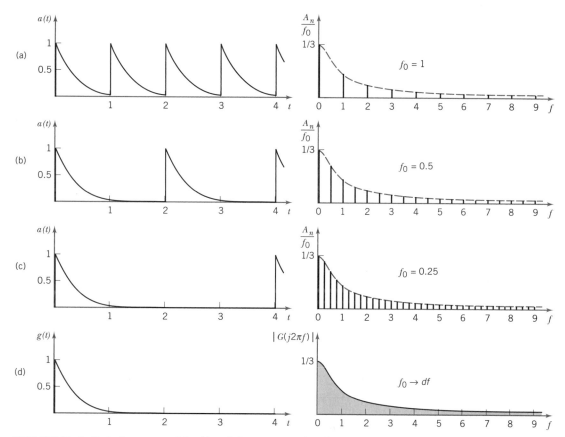

FIGURE D-2 Periodic exponentials $a(t)$ and the corresponding Fourier coefficients A_n. As the period T_0 increases from (a) to (c), the fundamental frequency $f_0 = 1/T_0$ decreases. In (d) the period becomes infinite, the time waveform becomes the nonperiodic $g(t) = e^{-3t}$, and the discrete coefficients become the continuous Fourier transform $G(j\omega)$.

In the limit, the summation in Equation (D.1) becomes an integral:

$$g(t) = 2 \int_0^\infty A(f) \cos(2\pi ft + \theta(f))$$

$$= \int_{-\infty}^\infty \left| \frac{1}{3 + j2\pi f} \right| \cos(2\pi ft + \theta(f)) \, df$$

$$= \int_{-\infty}^\infty \left| \frac{1}{3 + j2\pi f} \right| e^{j2\pi ft + \theta(t)} \, df$$

$$= \int_{-\infty}^\infty \frac{1}{3 + j2\pi f} e^{j2\pi ft} df$$

$$= \frac{1}{2\pi} \int_{-\infty}^\infty \frac{1}{3 + j\omega} e^{j\omega t} \, d\omega$$

$$= \frac{1}{2\pi} \int_{-\infty}^\infty G(j\omega) \, e^{j\omega t} d\omega, \qquad (D.3)$$

where

$$G(j\omega) = \frac{1}{3 + j\omega} \qquad (D.4)$$

is the *Fourier transform* of the time function $g(t)$ given in Equation (D.2). In general, the Fourier transform can be found by the integral

$$G(j\omega) = \int_{-\infty}^\infty g(t) e^{-j\omega t} dt. \qquad (D.5)$$

(See the reference by Lathi at the end of Chapter 7.) The reader should confirm that the $g(t)$ given in Equation (D.2) yields the $G(j\omega)$ given in Equation (D.4). Equation (D.3) is the reverse operation of Equation (D.5) and is called the *inverse Fourier transform*. Engineers seldom have to carry out these integrations because extensive tables of Fourier transforms are available.

D.3   LAPLACE TRANSFORM

One problem with the Fourier transform is that it is nonunique; many time functions have the same Fourier transform. Another problem is that a Fourier transform may become infinite for finite values of ω. The Laplace transform makes minor modifications to the Fourier transform to solve these problems. First, it applies only to time functions that are zero for $t < 0$; this makes the Laplace transform unique. Second, it replaces the imaginary frequency $j\omega$ by the complex frequency $s = j\omega + \sigma$. In effect, this builds time functions from sinusoids of the form $e^{\sigma t} \cos \omega t$ rather than sinusoids cos ωt of constant amplitude. Then σ can be chosen so the transform remains finite for finite values of ω. Substituting s for $j\omega$ in Equation (D.5), we get the Laplace transform:

$$G(s) = \int_{-\infty}^{\infty} g(t) e^{-st} dt, \tag{D.6}$$

and substituting s for $j\omega$ in Equation (D.3), we get the inverse Laplace transform:

$$g(t) = \frac{1}{2\pi j} \int_{-j\infty + \sigma}^{j\infty + \sigma} G(s) e^{st} ds, \tag{D.7}$$

For example, consider the time function in Equation (D.2). Using the unit step function

$$u(t) = \begin{cases} 0, & t < 0 \\ 1, & t > 0, \end{cases}$$

we can express Equation (D.2) as $g(t) = e^{-3t} u(t)$. Substituting s for $j\omega$ in Equation (D.4), we get the Laplace transform $G(s) = 1/(3 + s)$. In general, $1/(s + a)$ is the Laplace transform of $e^{-at} u(t)$. If we let $a = 0$, then we have $1/s$ as the Laplace transform of the unit step $u(t)$. Note that $1/j\omega$ is not the Fourier transform of $u(t)$ because of the problem at $\omega = 0$. This is an example where the Laplace transform is simpler than the Fourier transform. These and other Laplace transforms are listed in Figure D–3. See references at the end of Chapter 7 for more extensive tables.

EXAMPLE D.1

Given: the Laplace transform

$$G(s) = \frac{1}{s^4 + 4s^3 + 6s^2 + 8s}, \tag{D.8}$$

Find: the corresponding time function $g(t)$ [the inverse Lapalce transform of $G(s)$].

Solution: Factor the denominator:

$$G(s) = \frac{8}{s(s + 2)(s^2 + 2s + 4)}.$$

Express the product as a sum of terms. The numerator of each term must be a polynomial in s one order lower than the polynomial in the denominator.

$$G(s) = \frac{A}{s} + \frac{B}{s + 2} + \frac{Cs + D}{s^2 + 2s + 4}, \tag{D.9}$$

Laplace Transform	Time Function
$1/s$	$u(t)$
$1/s^2$	$t\,u(t)$
$1/(s+a)$	$e^{-at}\,u(t)$
$1/(s+a)^2$	$te^{-at}\,u(t)$
$1/(s^2+a^2)$	$(1/a)\sin(at)\,u(t)$
$s/(s^2+a^2)$	$\cos(at)\,u(t)$
$1/(s^2+2\alpha s+\omega_0^2)$	$(1/\omega_N)\,e^{-\alpha t}\sin(\omega_N t)\,u(t),\ \omega_0>\alpha,\ \omega_N=\sqrt{\omega_0^2-\alpha^2}$
$s/(s^2+2\alpha s+\omega_0^2)$	$e^{-\alpha t}[\cos(\omega_N t)-(\alpha/\omega_N)\sin(\omega_N t)]\,u(t),\ \omega_0>\alpha$

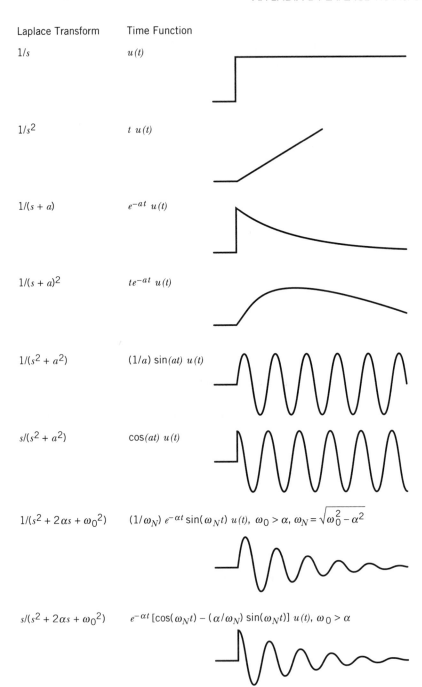

FIGURE D–3 Table of Laplace transforms for a few time functions. Note that the time functions are zero for $t < 0$.

where A, B, C, and D are to be determined. Give the terms a common denominator.

$$G(s) = \frac{As^3 + 4As^2 + 6As + 8A}{s(s + 2)(s^2 + 2s + 4)} + \frac{Bs^3 + 2Bs^2 + 4Bs}{s(s + 2)(s^2 + 2s + 4)} + \frac{Cs^3 + 2Cs^2 + Ds^2 + 2Ds}{s(s + 2)(s^2 + 2s + 4)}$$

$$= \frac{(A + B + C)s^3 + (4A + 2B + 2C + D)s^2 + (6A + 4B + 2D)s + 8A}{s(s + 2)(s^2 + 2s + 4)}. \qquad \text{(D.10)}$$

If Equation (D.10) is to equal Equation (D.8), then

$$A + B + C = 0,$$

$$4A + 2B + 2C + D = 0,$$

$$6A + 4B + 2D = 0,$$

$$8A = 8.$$

From the last equation, $A = 1$. Solving the first three equations simultaneously, we get $B = -0.5$, $C = -0.5$, and $D = -2$. Then from Equation (D.9),

$$G(s) = \frac{1}{s} - \frac{0.5}{s + 2} - \frac{0.5s}{s^2 + 2s + 4} - \frac{2}{s^2 + 2s + 4}.$$

From Figure D.3, the first two terms correspond to the time functions $u(t)$ and $-0.5e^{-2t}u(t)$. The last two terms have a denominator of the form $s^2 + 2as + \omega_0^2$, where $\alpha = 1$ and $\omega_0 = 2$. Then $\omega_N = (4 - 1)^{0.5} = 1.73$, and Figure D.3 gives

$$\begin{aligned}
g(t) &= u(t) - 0.5e^{-2t}u(t) - 0.5e^{-t}[\cos(1.73t) - (1/1.73)\sin(1.73t)]u(t) \\
&\quad - (2/1.73)e^{-\alpha t}\sin(\omega_N t)\,u(t) \\
&= u(t) - 0.5e^{-2t}u(t) - e^{-t}[0.5\cos(1.73t) + 0.867\sin(1.73t)]\,u(t) \\
&= u(t) - 0.5e^{-2t}u(t) + e^{-t}\cos(1.73t + 120°)\,u(t).
\end{aligned}$$

This time function is plotted in Figure D.4.

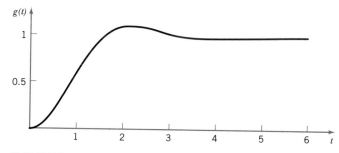

FIGURE D–4 Time waveform corresponding to the Laplace transform $G(s) = 1/(s^4 + 4s^3 + 6s^2 + 8s)$.

Index

SYMBOLS

A	Ampere	gpm	gallons per minute
A	absorbed (power)	h	height
A	area	H	henry
A	complex unloaded gain	H	feedback gain
A_v	unloaded voltage gain	H	magnetic field intensity
A_i	current gain	**H**	complex transfer function
AC	annual cost	hp	horsepower
b	bit	Hz	hertz (cycles per second)
B	transistor base	i	current
B	magnetic flux density	I	dc or rms current
B	susceptance	**I**	complex current
B_m	message bandwidth	**I***	I conjugate
B_n	noise bandwidth	I_a	armature current
B_s	system bandwidth	I_B	input bias current
B_t	transmission bandwidth	I_e	exciting current
B_{3dB}	3dB bandwidth	I_f	field current
BTU	British thermal unit	I_n	Norton current
c	3×10^3 m/s (speed of light)	I_p	primary current
C	Centigrade	I_{ph}	phase current
C	coulomb	I_{DSS}	saturation current (FET)
C	transistor collector	j	(imaginary)
C	capacitance	J	joule
C_{CB}	collector-to-base capacitance	k	10^3 (kilo-)
d	10^{-1} (deci-)	k	coupling coefficient
d	distance	kVA	kilovolt-ampere
D	diode	kVAR	kVA reactive
D	battery size	kVARC	capacitive kVAR
dB	decibel	kWH	kilowatt-hour
e	2.71828 . . .	l	length
E	transistor emitter	L	inductance
eff	efficiency	lb	pound of force
f	frequency (in Hz)	m	meter
f	force	m	10^{-3} (mili-)
F	farad	M	10^6 (mega-)
F	controller gain	M	mutual inductance
F	complex controller gain	n	10^{-9} (nano-)
ft	foot	n	rms noise
f_b	baud (symbol rate)	n	rotational speed (in rpm)
f_c	carrier frequency	n_s	synchronous speed
f_L	lower cutoff frequency	N	newton
f_s	sampling frequency	N_a	number active conductors
f_U	upper cutoff frequency	N_o	noise spectral density
G	generated (power)	n_q	quantization noise
G	conductance	NI	amp-turns, MMF
G	complex loaded gain	p	10^{-12} (pico-)
G_o	mid-band gain	p	number of poles
G_v	loaded voltage gain	p	power
g_m	transconductance	P	average power

P_a	armature power loss	v_t	transmitted signal
p_e	error probability	V_t	terminal voltage
P_f	field power loss	V_{Th}	Thevenin voltage
P_{mech}	mechanical power loss	V_{oc}	open-circuit voltage
pf	power factor	V_P	pinch-off voltage (FET)
PW	present worth	V_p	primary voltage
q	charge	V_p	peak voltage
Q	transistor	V_{ph}	phase voltage
Q	charge	V_q	quantization interval
Q	reactive power	V_{IO}	input offset voltage
R	resistance	VAR	volt-ampere reactive
R_a	armature resistance	W	watt
R_f	field resistance	X	reactance
R_i	input impedance	$\mathbf{Y}$	complex admittance
R_L	load impedance	$\mathbf{Z}$	complex impedance
R_n	Norton impedance	Z_o	characteristic impedance
R_o	output impedance	α	damping coefficient
R_s	source impedance	α	current gain (transistor)
R_{Th}	Thevenin impedance	α	temperature coefficient
r_e	emitter resistance	β	current gain (transistor)
r_s	source resistance (FET)	Γ	reflection coefficient
rad	radian	δ	power angle
rpm	revolutions per minute	Λ	determinate
rps	revolutions per second	Λf	frequency deviation (FM)
s	second	ΔV	symbol difference (PCM)
s	complex frequency	ϵ	dielectric constant
S	siemans	ζ	damping ratio
S	slip	θ	phase angle
$\mathbf{S}$	complex power	λ	wavelength
SNR	signal-to-noise ratio	μ	10^{-6} (micro-)
t	time	μ_0	permeability of free space,
t_h	hold time		$4\pi \times 10^{-7}$ H/m
t_r	transition time	π	$3.14159\ldots$
t_{pd}	propagation delay	ρ	resistivity
t_{su}	setup time	Σ	sum
T	period	τ	exponential time constant
T	temperature	ϕ	phase margin
T	torque	ϕ	magnetic flux
T_s	sampling interval	ω	frequency (in rad/sec)
u	linear velocity	ω_m	mechanical rotational
$u(t)$	unit step function		speed (in rad/sec)
v	voltage	ω_o	resonant frequency
V	volts	ω_o	break point frequency
V	dc or rms voltage	ω_o	carrier frequency
$\mathbf{V}$	complex voltage	ω_m	message frequency
v_c	carrier signal	ω_N	natural frequency
v_m	message signal	Ω	ohm
v_r	received signal	$\mho$	mho